INTRODUCTION TO LINEAR ALGEBRA

Fifth Edition

GILBERT STRANG
Massachusetts Institute of Technology

WELLESLEY - CAMBRIDGE PRESS
Box 812060 Wellesley MA 02482

Introduction to Linear Algebra, 5th Edition

ISBN 978-0-9802327-7-6

LaTeX typesetting by Ashley C. Fernandes (info@problemsolvingpathway.com)
Printed in the United States of America

9 8 7 6 5 4 3

QA184.S78 2016 512'.5 93-14092

Other texts from Wellesley - Cambridge Press

Computational Science and Engineering, Gilbert Strang ISBN 978-0-9614088-1-7

Wavelets and Filter Banks, Gilbert Strang and Truong Nguyen ISBN 978-0-9614088-7-9

Introduction to Applied Mathematics, Gilbert Strang ISBN 978-0-9614088-0-0

Calculus Third Edition (2017), Gilbert Strang ISBN 978-0-9802327-5-2

Algorithms for Global Positioning, Kai Borre & Gilbert Strang ISBN 978-0-9802327-3-8

Essays in Linear Algebra, Gilbert Strang ISBN 978-0-9802327-6-9

Differential Equations and Linear Algebra, Gilbert Strang ISBN 978-0-9802327-9-0

An Analysis of the Finite Element Method, 2008 edition, Gilbert Strang and George Fix ISBN 978-0-9802327-0-7

Wellesley - Cambridge Press
Box 812060
Wellesley MA 02482 USA
www.wellesleycambridge.com

linearalgebrabook@gmail.com
math.mit.edu/∼gs
phone (781) 431-8488
fax (617) 253-4358

The website for this book is **math.mit.edu/linearalgebra**.
The Solution Manual can be printed from that website.

Course material including syllabus and exams and also videotaped lectures are available on the book website and the teaching website: **web.mit.edu/18.06**

Linear Algebra is included in MIT's OpenCourseWare site **ocw.mit.edu**.
This provides video lectures of the full linear algebra course 18.06 and 18.06 SC.

The front cover captures a central idea of linear algebra.
$A\boldsymbol{x} = \boldsymbol{b}$ is solvable when $\boldsymbol{b}$ is in the (red) column space of A.
One particular solution $\boldsymbol{y}$ is in the (yellow) row space: $A\boldsymbol{y} = \boldsymbol{b}$.
Add any vector $\boldsymbol{z}$ from the (green) nullspace of A: $A\boldsymbol{z} = \boldsymbol{0}$.
The complete solution is $\boldsymbol{x} = \boldsymbol{y} + \boldsymbol{z}$. Then $A\boldsymbol{x} = A\boldsymbol{y} + A\boldsymbol{z} = \boldsymbol{b}$.

The cover design was the inspiration of Lois Sellers and Gail Corbett.

Table of Contents

Preface

I am happy for you to see this Fifth Edition of Introduction to Linear Algebra. This is the text for my video lectures on MIT's OpenCourseWare (**ocw.mit.edu** and also **YouTube**). I hope those lectures will be useful to you (maybe even enjoyable !).

Hundreds of colleges and universities have chosen this textbook for their basic linear algebra course. A sabbatical gave me a chance to prepare two new chapters about probability and statistics and understanding data. Thousands of other improvements too—probably only noticed by the author... Here is a new addition for students and all readers:

> Every section opens with a brief summary to explain its contents. When you read a new section, and when you revisit a section to review and organize it in your mind, those lines are a quick guide and an aid to memory.

Another big change comes on this book's website **math.mit.edu/linearalgebra**. That site now contains solutions to the Problem Sets in the book. With unlimited space, this is much more flexible than printing short solutions. There are three key websites :

ocw.mit.edu Messages come from thousands of students and faculty about linear algebra on this OpenCourseWare site. The 18.06 and 18.06 SC courses include video lectures of a complete semester of classes. Those lectures offer an independent review of the whole subject based on this textbook—the professor's time stays free and the student's time can be 2 a.m. (The reader doesn't have to be in a class at all.) Six million viewers around the world have seen these videos (*amazing*). I hope you find them helpful.

web.mit.edu/18.06 This site has homeworks and exams (with solutions) for the current course as it is taught, and as far back as 1996. There are also review questions, Java demos, Teaching Codes, and short essays (*and the video lectures*). My goal is to make this book as useful to you as possible, with all the course material we can provide.

math.mit.edu/linearalgebra This has become an active website. It now has Solutions to Exercises—with space to explain ideas. There are also new exercises from many different sources—practice problems, development of textbook examples, codes in MATLAB and *Julia* and *Python*, plus whole collections of exams (18.06 and others) for review.

Please visit this linear algebra site. *Send suggestions to* **linearalgebrabook@gmail.com**

The Fifth Edition

The cover shows the **Four Fundamental Subspaces**—the row space and nullspace are on the left side, the column space and the nullspace of A^{T} are on the right. It is not usual to put the central ideas of the subject on display like this! When you meet those four spaces in Chapter 3, you will understand why that picture is so central to linear algebra.

Those were named the Four Fundamental Subspaces in my first book, and they start from a matrix A. Each row of A is a vector in n-dimensional space. When the matrix has m rows, each column is a vector in m-dimensional space. The crucial operation in linear algebra is to take ***linear combinations of column vectors***. This is exactly the result of a matrix-vector multiplication. *Ax is a combination of the columns of A.*

When we take *all* combinations Ax of the column vectors, we get the *column space*. If this space includes the vector b, we can solve the equation $Ax = b$.

May I call special attention to Section 1.3, where these ideas come early—with two specific examples. You are not expected to catch every detail of vector spaces in one day! But you will see the first matrices in the book, and a picture of their column spaces. There is even an *inverse matrix* and its connection to calculus. You will be learning the language of linear algebra in the best and most efficient way: by using it.

Every section of the basic course ends with a large collection of review problems. They ask you to use the ideas in that section—-the dimension of the column space, a basis for that space, the rank and inverse and determinant and eigenvalues of A. Many problems look for computations by hand on a small matrix, and they have been highly praised. The *Challenge Problems* go a step further, and sometimes deeper. Let me give four examples:

Section 2.1. Which row exchanges of a Sudoku matrix produce another Sudoku matrix?

Section 2.7: If P is a permutation matrix, why is some power P^k equal to I ?

Section 3.4: If $Ax = b$ and $Cx = b$ have the same solutions for every b, does A equal C ?

Section 4.1: What conditions on the four vectors r, n, c, ℓ allow them to be bases for the row space, the nullspace, the column space, and the left nullspace of a 2 by 2 matrix?

The Start of the Course

The equation $Ax = b$ uses the language of linear combinations right away. The vector Ax is *a combination of the columns of A*. The equation is asking for *a combination that produces b*. The solution vector x comes at three levels and all are important:

1. ***Direct solution*** to find x by forward elimination and back substitution.

2. ***Matrix solution*** using the inverse matrix: $x = A^{-1}b$ (if A has an inverse).

3. ***Particular solution*** (to $Ay = b$) plus ***nullspace solution*** (to $Az = 0$).

That vector space solution $x = y + z$ is shown on the cover of the book.

Direct elimination is the most frequently used algorithm in scientific computing. The matrix A becomes triangular—then solutions come quickly. We also see bases for the four subspaces. But don't spend forever on practicing elimination ... good ideas are coming.

The speed of every new supercomputer is tested on $Ax = b$: pure linear algebra. But even a supercomputer doesn't want the inverse matrix: *too slow*. Inverses give the simplest formula $x = A^{-1}b$ but not the top speed. And everyone must know that determinants are even slower—there is no way a linear algebra course should begin with formulas for the determinant of an n by n matrix. Those formulas have a place, but not first place.

Structure of the Textbook

Already in this preface, you can see the style of the book and its goal. That goal is serious, to explain this beautiful and useful part of mathematics. You will see how the applications of linear algebra reinforce the key ideas. This book moves gradually and steadily from *numbers* to *vectors* to *subspaces*—each level comes naturally and everyone can get it.

Here are 12 points about learning and teaching from this book:

1. Chapter 1 starts with vectors and dot products. If the class has met them before, focus quickly on linear combinations. Section 1.3 provides three independent vectors whose combinations fill all of 3-dimensional space, and three dependent vectors in a plane. ***Those two examples are the beginning of linear algebra.***

2. Chapter 2 shows the row picture and the column picture of $Ax = b$. The heart of linear algebra is in that connection between the rows of A and the columns of A: the same numbers but very different pictures. Then begins the algebra of matrices: an elimination matrix E multiplies A to produce a zero. The goal is to capture the whole process—start with A, multiply by E's, end with U.

   Elimination is seen in the beautiful form $A = LU$. The ***lower triangular*** L holds the forward elimination steps, and U is ***upper triangular*** for back substitution.

3. Chapter 3 is linear algebra at the best level: ***subspaces***. The column space contains all linear combinations of the columns. The crucial question is: ***How many of those columns are needed***? The answer tells us the dimension of the column space, and the key information about A. We reach the Fundamental Theorem of Linear Algebra.

4. With more equations than unknowns, it is almost sure that $Ax = b$ has no solution. We cannot throw out every measurement that is close but not perfectly exact! When we solve by ***least squares***, the key will be the matrix $A^{\mathrm{T}}A$. This wonderful matrix appears everywhere in applied mathematics, when A is rectangular.

5. ***Determinants*** give formulas for all that has come before—Cramer's Rule, inverse matrices, volumes in n dimensions. We don't need those formulas to compute. They slow us down. But $\det A = 0$ tells when a matrix is singular: this is the key to eigenvalues.

6. ***Section* 6.1 *explains eigenvalues for* 2 *by* 2 *matrices*.** Many courses want to see eigenvalues early. It is completely reasonable to come here directly from Chapter 3, because the determinant is easy for a 2 by 2 matrix. *The key equation is* $Ax = \lambda x$.

   Eigenvalues and eigenvectors are an astonishing way to understand a square matrix. They are not for $Ax = b$, they are for dynamic equations like $du/dt = Au$. The idea is always the same: *follow the eigenvectors*. In those special directions, A acts like a single number (the eigenvalue λ) and the problem is one-dimensional.

   An essential highlight of Chapter 6 is ***diagonalizing a symmetric matrix***. When all the eigenvalues are positive, the matrix is "positive definite". This key idea connects the whole course—positive pivots and determinants and eigenvalues and energy. I work hard to reach this point in the book and to explain it by examples.

7. Chapter 7 is new. It introduces ***singular values*** and ***singular vectors***. They separate all martices into simple pieces, ranked in order of their importance. You will see one way to compress an image. Especially you can analyze a matrix full of data.

8. Chapter 8 explains ***linear transformations***. This is geometry without axes, algebra with no coordinates. When we choose a basis, we reach the best possible matrix.

9. Chapter 9 moves from real numbers and vectors to complex vectors and matrices. The Fourier matrix F is the most important complex matrix we will ever see. And the ***Fast Fourier Transform*** (multiplying quickly by F and F^{-1}) is revolutionary.

10. Chapter 10 is full of applications, more than any single course could need:

    10.1 *Graphs and Networks*—leading to the edge-node matrix for Kirchhoff's Laws

    10.2 *Matrices in Engineering*—differential equations parallel to matrix equations

    10.3 *Markov Matrices*—as in Google's *PageRank* algorithm

    10.4 *Linear Programming*—a new requirement $x \geq 0$ and minimization of the cost

    10.5 *Fourier Series*—linear algebra for functions and digital signal processing

    10.6 *Computer Graphics*—matrices move and rotate and compress images

    10.7 *Linear Algebra in Cryptography*—this new section was fun to write. The Hill Cipher is not too secure. It uses modular arithmetic: integers from 0 to $p - 1$. Multiplication gives $4 \times 5 \equiv 1\,(mod\ 19)$. For decoding this gives $4^{-1} \equiv 5$.

11. How should computing be included in a linear algebra course? It can open a new understanding of matrices—every class will find a balance. MATLAB and *Maple* and *Mathematica* are powerful in different ways. *Julia* and *Python* are free and directly accessible on the Web. Those newer languages are powerful too !

    Basic commands begin in Chapter 2. Then Chapter 11 moves toward professional algorithms.You can upload and download codes for this course on the website.

12. Chapter 12 on Probability and Statistics is new, with truly important applications. When random variables are not independent we get covariance matrices. Fortunately they are symmetric positive definite. The linear algebra in Chapter 6 is needed now.

The Variety of Linear Algebra

Calculus is mostly about one special operation (the derivative) and its inverse (the integral). Of course I admit that calculus could be important .... But so many applications of mathematics are discrete rather than continuous, digital rather than analog. The century of data has begun! You will find a light-hearted essay called "Too Much Calculus" on my website. ***The truth is that vectors and matrices have become the language to know.***

Part of that language is the wonderful variety of matrices. Let me give three examples:

Symmetric matrix	***Orthogonal matrix***	***Triangular matrix***
$\begin{bmatrix} 2 & -1 & 0 & 0 \\ -1 & 2 & -1 & 0 \\ 0 & -1 & 2 & -1 \\ 0 & 0 & -1 & 2 \end{bmatrix}$	$\frac{1}{2}\begin{bmatrix} 1 & 1 & 1 & 1 \\ 1 & -1 & 1 & -1 \\ 1 & 1 & -1 & -1 \\ 1 & -1 & -1 & 1 \end{bmatrix}$	$\begin{bmatrix} 1 & 1 & 1 & 1 \\ 0 & 1 & 1 & 1 \\ 0 & 0 & 1 & 1 \\ 0 & 0 & 0 & 1 \end{bmatrix}$

A key goal is learning to "read" a matrix. You need to see the meaning in the numbers. This is really the essence of mathematics—patterns and their meaning.

I have used *italics* and ***boldface*** to pick out the key words on each page. I know there are times when you want to read quickly, looking for the important lines.

May I end with this thought for professors. You might feel that the direction is right, and wonder if your students are ready. ***Just give them a chance***! Literally thousands of students have written to me, frequently with suggestions and surprisingly often with thanks. They know this course has a purpose, because the professor and the book are on their side. Linear algebra is a fantastic subject, enjoy it.

Help With This Book

The greatest encouragement of all is the feeling that you are doing something worthwhile with your life. Hundreds of generous readers have sent ideas and examples and corrections (and favorite matrices) that appear in this book. *Thank you all.*

One person has helped with every word in this book. He is Ashley C. Fernandes, who prepared the LaTeX files. It is now six books that he has allowed me to write and rewrite, aiming for accuracy and also for life. Working with friends is a happy way to live.

Friends inside and outside the MIT math department have been wonderful. Alan Edelman for *Julia* and much more, Alex Townsend for the flag examples in 7.1, and Peter Kempthorne for the finance example in 7.3 : those stand out. Don Spickler's website on cryptography is simply excellent. I thank Jon Bloom, Jack Dongarra, Hilary Finucane, Pavel Grinfeld, Randy LeVeque, David Vogan, Liang Wang, and Karen Willcox. The "eigenfaces" in 7.3 came from Matthew Turk and Jeff Jauregui. And the big step to singular values was accelerated by Raj Rao's great course at Michigan.

This book owes so much to my happy sabbatical in Oxford. Thank you, Nick Trefethen and everyone. Especially you the reader! Best wishes in your work.

Background of the Author

This is my 9th textbook on linear algebra, and I hesitate to write about myself. It is the mathematics that is important, and the reader. The next paragraphs add something brief and personal, as a way to say that textbooks are written by people.

I was born in Chicago and went to school in Washington and Cincinnati and St. Louis. My college was MIT (and my linear algebra course was *extremely abstract*). After that came Oxford and UCLA, then back to MIT for a very long time. I don't know how many thousands of students have taken 18.06 (more than 6 million when you include the videos on ***ocw.mit.edu***). The time for a fresh approach was right, because this fantastic subject was only revealed to math majors—**we needed to open linear algebra to the world.**

I am so grateful for a life of teaching mathematics, more than I could possibly tell you.

Gilbert Strang

PS I hope the next book (2018 ?) will include *Learning from Data*. This subject is growing quickly, especially "deep learning". By knowing a function on a training set of old data, we approximate the function on new data. The approximation only uses one simple nonlinear function $f(x) = \max(0, x)$. It is n matrix multiplications that we optimize to make the learning deep: $\boldsymbol{x}_1 = f(A_1\boldsymbol{x} + \boldsymbol{b}_1), \boldsymbol{x}_2 = f(A_2\boldsymbol{x}_1 + \boldsymbol{b}_2), \ldots, \boldsymbol{x}_n = f(A_n\boldsymbol{x}_{n-1} + \boldsymbol{b}_n)$. Those are $n-1$ hidden layers between the input $\boldsymbol{x}$ and the output $\boldsymbol{x}_n$—which approximates $F(\boldsymbol{x})$ on the training set.

THE MATRIX ALPHABET

A	Any Matrix	P	Permutation Matrix
B	Basis Matrix	P	Projection Matrix
C	Cofactor Matrix	Q	Orthogonal Matrix
D	Diagonal Matrix	R	Upper Triangular Matrix
E	Elimination Matrix	R	Reduced Echelon Matrix
F	Fourier Matrix	S	Symmetric Matrix
H	Hadamard Matrix	T	Linear Transformation
I	Identity Matrix	U	Upper Triangular Matrix
J	Jordan Matrix	U	Left Singular Vectors
K	Stiffness Matrix	V	Right Singular Vectors
L	Lower Triangular Matrix	X	Eigenvector Matrix
M	Markov Matrix	Λ	Eigenvalue Matrix
N	Nullspace Matrix	Σ	Singular Value Matrix

Chapter 1

Introduction to Vectors

The heart of linear algebra is in two operations—both with vectors. We add vectors to get $\boldsymbol{v} + \boldsymbol{w}$. We multiply them by numbers c and d to get $c\boldsymbol{v}$ and $d\boldsymbol{w}$. Combining those two operations (adding $c\boldsymbol{v}$ to $d\boldsymbol{w}$) gives the ***linear combination*** $c\boldsymbol{v} + d\boldsymbol{w}$.

Linear combination $$c\boldsymbol{v} + d\boldsymbol{w} = c\begin{bmatrix} 1 \\ 1 \end{bmatrix} + d\begin{bmatrix} 2 \\ 3 \end{bmatrix} = \begin{bmatrix} c + 2d \\ c + 3d \end{bmatrix}$$

Example $\boldsymbol{v} + \boldsymbol{w} = \begin{bmatrix} 1 \\ 1 \end{bmatrix} + \begin{bmatrix} 2 \\ 3 \end{bmatrix} = \begin{bmatrix} 3 \\ 4 \end{bmatrix}$ is the combination with $c = d = 1$

Linear combinations are all-important in this subject! Sometimes we want one particular combination, the specific choice $c = 2$ and $d = 1$ that produces $c\boldsymbol{v} + d\boldsymbol{w} = (4, 5)$. Other times we want *all the combinations* of $\boldsymbol{v}$ and $\boldsymbol{w}$ (coming from all c and d).

The vectors $c\boldsymbol{v}$ lie along a line. When $\boldsymbol{w}$ is not on that line, **the combinations** $c\boldsymbol{v} + d\boldsymbol{w}$ **fill the whole two-dimensional plane**. Starting from four vectors $\boldsymbol{u}, \boldsymbol{v}, \boldsymbol{w}, \boldsymbol{z}$ in four-dimensional space, their combinations $c\boldsymbol{u} + d\boldsymbol{v} + e\boldsymbol{w} + f\boldsymbol{z}$ are likely to fill the space—but not always. The vectors and their combinations could lie in a plane or on a line.

Chapter 1 explains these central ideas, on which everything builds. We start with two-dimensional vectors and three-dimensional vectors, which are reasonable to draw. Then we move into higher dimensions. The really impressive feature of linear algebra is how smoothly it takes that step into n-dimensional space. Your mental picture stays completely correct, even if drawing a ten-dimensional vector is impossible.

This is where the book is going (into n-dimensional space). The first steps are the operations in Sections 1.1 and 1.2. Then Section 1.3 outlines three fundamental ideas.

1.1 ***Vector addition*** $\boldsymbol{v} + \boldsymbol{w}$ ***and linear combinations*** $c\boldsymbol{v} + d\boldsymbol{w}$.

1.2 ***The dot product*** $\boldsymbol{v} \cdot \boldsymbol{w}$ ***of two vectors and the length*** $\|\boldsymbol{v}\| = \sqrt{\boldsymbol{v} \cdot \boldsymbol{v}}$.

1.3 ***Matrices*** A, ***linear equations*** $A\boldsymbol{x} = \boldsymbol{b}$, ***solutions*** $\boldsymbol{x} = A^{-1}\boldsymbol{b}$.

1.1 Vectors and Linear Combinations

1 $3\boldsymbol{v} + 5\boldsymbol{w}$ is a typical **linear combination** $c\boldsymbol{v} + d\boldsymbol{w}$ of the vectors $\boldsymbol{v}$ and $\boldsymbol{w}$.

2 For $\boldsymbol{v} = \begin{bmatrix} 1 \\ 1 \end{bmatrix}$ and $\boldsymbol{w} = \begin{bmatrix} 2 \\ 3 \end{bmatrix}$ that combination is $3\begin{bmatrix} 1 \\ 1 \end{bmatrix} + 5\begin{bmatrix} 2 \\ 3 \end{bmatrix} = \begin{bmatrix} 3+10 \\ 3+15 \end{bmatrix} = \begin{bmatrix} 13 \\ 18 \end{bmatrix}$.

3 The vector $\begin{bmatrix} 2 \\ 3 \end{bmatrix} = \begin{bmatrix} 2 \\ 0 \end{bmatrix} + \begin{bmatrix} 0 \\ 3 \end{bmatrix}$ goes across to $x = 2$ and up to $y = 3$ in the xy plane.

4 The combinations $c\begin{bmatrix} 1 \\ 1 \end{bmatrix} + d\begin{bmatrix} 2 \\ 3 \end{bmatrix}$ fill the whole xy plane. They produce every $\begin{bmatrix} x \\ y \end{bmatrix}$.

5 The combinations $c\begin{bmatrix} 1 \\ 1 \\ 1 \end{bmatrix} + d\begin{bmatrix} 2 \\ 3 \\ 4 \end{bmatrix}$ fill a **plane** in xyz space. Same plane for $\begin{bmatrix} 1 \\ 1 \\ 1 \end{bmatrix}, \begin{bmatrix} 3 \\ 4 \\ 5 \end{bmatrix}$.

6 But $\begin{matrix} c + 2d = 1 \\ c + 3d = 0 \\ c + 4d = 0 \end{matrix}$ has no solution because its right side $\begin{bmatrix} 1 \\ 0 \\ 0 \end{bmatrix}$ is not on that plane.

"You can't add apples and oranges." In a strange way, this is the reason for vectors. We have two separate numbers v_1 and v_2. That pair produces a ***two-dimensional vector*** $\boldsymbol{v}$:

Column vector $\boldsymbol{v}$ $\quad \boldsymbol{v} = \begin{bmatrix} v_1 \\ v_2 \end{bmatrix} \quad \begin{matrix} v_1 = \text{first component of } \boldsymbol{v} \\ v_2 = \text{second component of } \boldsymbol{v} \end{matrix}$

We write $\boldsymbol{v}$ as a ***column***, not as a row. The main point so far is to have a single letter $\boldsymbol{v}$ (in ***boldface italic***) for this pair of numbers v_1 and v_2 (in *lightface italic*).

Even if we don't add v_1 to v_2, we do ***add vectors***. The first components of $\boldsymbol{v}$ and $\boldsymbol{w}$ stay separate from the second components:

VECTOR ADDITION $\quad \boldsymbol{v} = \begin{bmatrix} v_1 \\ v_2 \end{bmatrix}$ and $\boldsymbol{w} = \begin{bmatrix} w_1 \\ w_2 \end{bmatrix}$ add to $\boldsymbol{v} + \boldsymbol{w} = \begin{bmatrix} v_1 + w_1 \\ v_2 + w_2 \end{bmatrix}$.

Subtraction follows the same idea: *The components of $\boldsymbol{v} - \boldsymbol{w}$ are $v_1 - w_1$ and $v_2 - w_2$.*

The other basic operation is *scalar multiplication*. Vectors can be multiplied by 2 or by -1 or by any number c. To find $2\boldsymbol{v}$, multiply each component of $\boldsymbol{v}$ by 2:

SCALAR MULTIPLICATION $\quad 2\boldsymbol{v} = \begin{bmatrix} 2v_1 \\ 2v_2 \end{bmatrix} = \boldsymbol{v} + \boldsymbol{v} \quad -\boldsymbol{v} = \begin{bmatrix} -v_1 \\ -v_2 \end{bmatrix}$.

The components of $c\boldsymbol{v}$ are cv_1 and cv_2. The number c is called a "scalar".

Notice that the sum of $-\boldsymbol{v}$ and $\boldsymbol{v}$ is the zero vector. This is $\mathbf{0}$, which is not the same as the number zero! The vector $\mathbf{0}$ has components 0 and 0. Forgive me for hammering away at the difference between a vector and its components. Linear algebra is built on these operations $\boldsymbol{v} + \boldsymbol{w}$ and $c\boldsymbol{v}$ and $d\boldsymbol{w}$—***adding vectors and multiplying by scalars***.

Linear Combinations

Now we combine addition with scalar multiplication to produce a "**linear combination**" of $\boldsymbol{v}$ and $\boldsymbol{w}$. Multiply $\boldsymbol{v}$ by c and multiply $\boldsymbol{w}$ by d. Then add $c\boldsymbol{v} + d\boldsymbol{w}$.

> ***The sum of $c\boldsymbol{v}$ and $d\boldsymbol{w}$ is a*** *linear combination $c\boldsymbol{v} + d\boldsymbol{w}$.*

Four special linear combinations are: sum, difference, zero, and a scalar multiple $c\boldsymbol{v}$:

$$\begin{aligned} 1\boldsymbol{v} + 1\boldsymbol{w} &= \text{sum of vectors in Figure 1.1a} \\ 1\boldsymbol{v} - 1\boldsymbol{w} &= \text{difference of vectors in Figure 1.1b} \\ 0\boldsymbol{v} + 0\boldsymbol{w} &= \textbf{\textit{zero vector}} \\ c\boldsymbol{v} + 0\boldsymbol{w} &= \text{vector } c\boldsymbol{v} \text{ in the direction of } \boldsymbol{v} \end{aligned}$$

The zero vector is always a possible combination (its coefficients are zero). Every time we see a "space" of vectors, that zero vector will be included. This big view, taking *all* the combinations of $\boldsymbol{v}$ and $\boldsymbol{w}$, is linear algebra at work.

The figures show how you can visualize vectors. For algebra, we just need the components (like 4 and 2). That vector $\boldsymbol{v}$ is represented by an arrow. The arrow goes $v_1 = 4$ units to the right and $v_2 = 2$ units up. It ends at the point whose x, y coordinates are $4, 2$. This point is another representation of the vector—so we have three ways to describe $\boldsymbol{v}$:

Represent vector $\boldsymbol{v}$	Two numbers	Arrow from $(0, 0)$	Point in the plane

We add using the numbers. We visualize $\boldsymbol{v} + \boldsymbol{w}$ using arrows:

Vector addition (head to tail) ***At the end of*** $\boldsymbol{v}$***, place the start of*** $\boldsymbol{w}$.

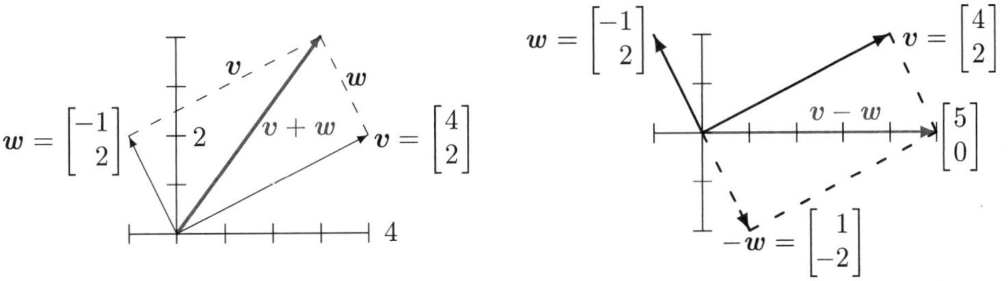

$$\boldsymbol{v} + \boldsymbol{w} = \begin{bmatrix} 4 \\ 2 \end{bmatrix} + \begin{bmatrix} -1 \\ 2 \end{bmatrix} = \begin{bmatrix} 3 \\ 4 \end{bmatrix} \qquad \boldsymbol{v} - \boldsymbol{w} = \begin{bmatrix} 4 \\ 2 \end{bmatrix} - \begin{bmatrix} -1 \\ 2 \end{bmatrix} = \begin{bmatrix} 5 \\ 0 \end{bmatrix}$$

Figure 1.1: Vector addition $\boldsymbol{v} + \boldsymbol{w} = (3,\ 4)$ produces the diagonal of a parallelogram. The reverse of $\boldsymbol{w}$ is $-\boldsymbol{w}$. The linear combination on the right is $\boldsymbol{v} - \boldsymbol{w} = (5,\ 0)$.

We travel along $\boldsymbol{v}$ and then along $\boldsymbol{w}$. Or we take the diagonal shortcut along $\boldsymbol{v} + \boldsymbol{w}$. We could also go along $\boldsymbol{w}$ and then $\boldsymbol{v}$. In other words, $\boldsymbol{w} + \boldsymbol{v}$ **gives the same answer as** $\boldsymbol{v} + \boldsymbol{w}$. These are different ways along the parallelogram (in this example it is a rectangle).

Vectors in Three Dimensions

A vector with two components corresponds to a point in the xy plane. The components of $\boldsymbol{v}$ are the coordinates of the point: $x = v_1$ and $y = v_2$. The arrow ends at this point (v_1, v_2), when it starts from $(0,0)$. Now we allow vectors to have three components (v_1, v_2, v_3).

The xy plane is replaced by three-dimensional xyz space. Here are typical vectors (still column vectors but with three components):

$$\boldsymbol{v} = \begin{bmatrix} 1 \\ 1 \\ -1 \end{bmatrix} \quad \text{and} \quad \boldsymbol{w} = \begin{bmatrix} 2 \\ 3 \\ 4 \end{bmatrix} \quad \text{and} \quad \boldsymbol{v} + \boldsymbol{w} = \begin{bmatrix} 3 \\ 4 \\ 3 \end{bmatrix}.$$

The vector $\boldsymbol{v}$ corresponds to an arrow in 3-space. Usually the arrow starts at the "origin", where the xyz axes meet and the coordinates are $(0,0,0)$. The arrow ends at the point with coordinates v_1, v_2, v_3. There is a perfect match between the ***column vector*** and the ***arrow from the origin*** and the ***point where the arrow ends***.

The vector (x, y) in the plane is different from $(x, y, 0)$ in 3-space !

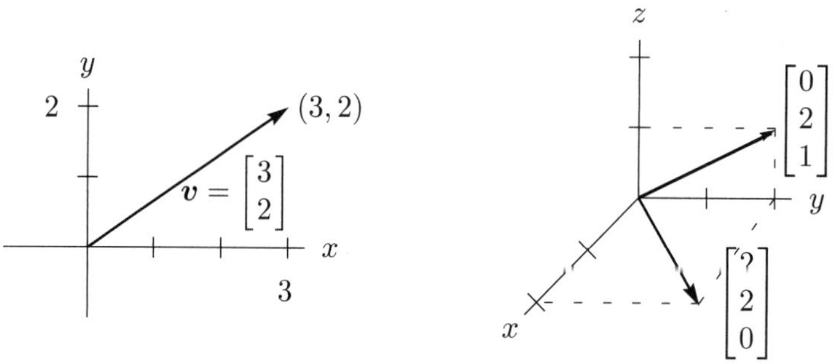

Figure 1.2: Vectors $\begin{bmatrix} x \\ y \end{bmatrix}$ and $\begin{bmatrix} x \\ y \\ z \end{bmatrix}$ correspond to points (x, y) and (x, y, z).

> ***From now on*** $\quad \boldsymbol{v} = \begin{bmatrix} 1 \\ 1 \\ -1 \end{bmatrix} \quad$ ***is also written as*** $\quad \boldsymbol{v} = (1, 1, -1)$.

The reason for the row form (in parentheses) is to save space. But $\boldsymbol{v} = (1, 1, -1)$ is not a row vector! It is in actuality a column vector, just temporarily lying down. The row vector $[\,1 \;\; 1 \;\; -1\,]$ is absolutely different, even though it has the same three components. That 1 by 3 row vector is the "transpose" of the 3 by 1 column vector $\boldsymbol{v}$.

In three dimensions, $\boldsymbol{v} + \boldsymbol{w}$ is still found a component at a time. The sum has components $v_1 + w_1$ and $v_2 + w_2$ and $v_3 + w_3$. You see how to add vectors in 4 or 5 or n dimensions. When $\boldsymbol{w}$ starts at the end of $\boldsymbol{v}$, the third side is $\boldsymbol{v} + \boldsymbol{w}$. The other way around the parallelogram is $\boldsymbol{w} + \boldsymbol{v}$. Question: Do the four sides all lie in the same plane? *Yes*. And the sum $\boldsymbol{v} + \boldsymbol{w} - \boldsymbol{v} - \boldsymbol{w}$ goes completely around to produce the ______ vector.

A typical linear combination of three vectors in three dimensions is $\boldsymbol{u} + 4\boldsymbol{v} - 2\boldsymbol{w}$:

Linear combination
Multiply by $1, 4, -2$
Then add

$$\begin{bmatrix} 1 \\ 0 \\ 3 \end{bmatrix} + 4 \begin{bmatrix} 1 \\ 2 \\ 1 \end{bmatrix} - 2 \begin{bmatrix} 2 \\ 3 \\ -1 \end{bmatrix} = \begin{bmatrix} 1 \\ 2 \\ 9 \end{bmatrix}.$$

The Important Questions

For one vector $\boldsymbol{u}$, the only linear combinations are the multiples $c\boldsymbol{u}$. For two vectors, the combinations are $c\boldsymbol{u} + d\boldsymbol{v}$. For three vectors, the combinations are $c\boldsymbol{u} + d\boldsymbol{v} + e\boldsymbol{w}$. Will you take the big step from *one* combination to **all combinations**? Every c and d and e are allowed. Suppose the vectors $\boldsymbol{u}, \boldsymbol{v}, \boldsymbol{w}$ are in three-dimensional space:

1. What is the picture of *all* combinations $c\boldsymbol{u}$?
2. What is the picture of *all* combinations $c\boldsymbol{u} + d\boldsymbol{v}$?
3. What is the picture of *all* combinations $c\boldsymbol{u} + d\boldsymbol{v} + e\boldsymbol{w}$?

The answers depend on the particular vectors $\boldsymbol{u}$, $\boldsymbol{v}$, and $\boldsymbol{w}$. If they were zero vectors (a very extreme case), then every combination would be zero. If they are typical nonzero vectors (components chosen at random), here are the three answers. This is the key to our subject:

1. The combinations $c\boldsymbol{u}$ fill a ***line through*** $(\mathbf{0}, \mathbf{0}, \mathbf{0})$.
2. The combinations $c\boldsymbol{u} + d\boldsymbol{v}$ fill a ***plane through*** $(\mathbf{0}, \mathbf{0}, \mathbf{0})$.
3. The combinations $c\boldsymbol{u} + d\boldsymbol{v} + e\boldsymbol{w}$ fill ***three-dimensional space***.

The zero vector $(0, 0, 0)$ is on the line because c can be zero. It is on the plane because c and d could both be zero. The line of vectors $c\boldsymbol{u}$ is infinitely long (forward and backward). It is the plane of all $c\boldsymbol{u} + d\boldsymbol{v}$ (combining two vectors in three-dimensional space) that I especially ask you to think about.

Adding all $c\boldsymbol{u}$ on one line to all $d\boldsymbol{v}$ on the other line fills in the plane in Figure 1.3.

When we include a third vector $\boldsymbol{w}$, the multiples $e\boldsymbol{w}$ give a third line. **Suppose that third line is not in the plane of $\boldsymbol{u}$ and $\boldsymbol{v}$.** Then combining all $e\boldsymbol{w}$ with all $c\boldsymbol{u} + d\boldsymbol{v}$ fills up the whole three-dimensional space.

This is the typical situation! **Line**, then **plane**, then **space**. But other possibilities exist. When $\boldsymbol{w}$ happens to be $c\boldsymbol{u} + d\boldsymbol{v}$, that third vector $\boldsymbol{w}$ is in the plane of the first two. The combinations of $\boldsymbol{u}, \boldsymbol{v}, \boldsymbol{w}$ will not go outside that $\boldsymbol{uv}$ plane. We do not get the full three-dimensional space. Please think about the special cases in Problem 1.

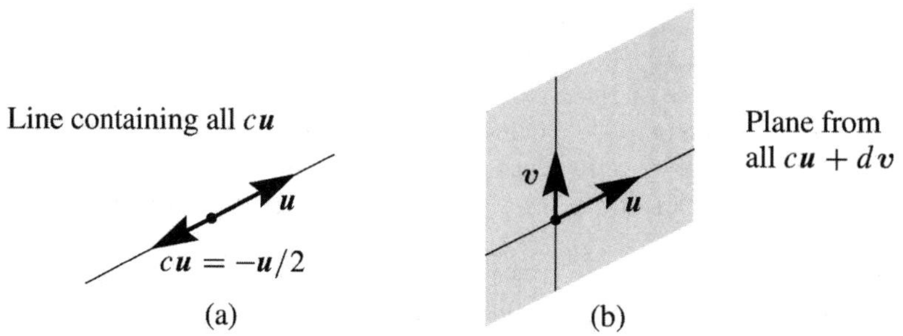

Figure 1.3: (a) Line through $\boldsymbol{u}$. (b) The plane containing the lines through $\boldsymbol{u}$ and $\boldsymbol{v}$.

■ REVIEW OF THE KEY IDEAS ■

1. A vector $\boldsymbol{v}$ in two-dimensional space has two components v_1 and v_2.
2. $\boldsymbol{v} + \boldsymbol{w} = (v_1 + w_1,\ v_2 + w_2)$ and $c\boldsymbol{v} = (cv_1, cv_2)$ are found a component at a time.
3. A linear combination of three vectors $\boldsymbol{u}$ and $\boldsymbol{v}$ and $\boldsymbol{w}$ is $c\boldsymbol{u} + d\boldsymbol{v} + e\boldsymbol{w}$.
4. Take *all* linear combinations of $\boldsymbol{u}$, or $\boldsymbol{u}$ and $\boldsymbol{v}$, or $\boldsymbol{u}, \boldsymbol{v}, \boldsymbol{w}$. In three dimensions, those combinations typically fill a line, then a plane, then the whole space $\mathbf{R}^3$.

■ WORKED EXAMPLES ■

1.1 A The linear combinations of $\boldsymbol{v} = (1, 1, 0)$ and $\boldsymbol{w} = (0, 1, 1)$ fill a plane in $\mathbf{R}^3$. *Describe that plane.* Find a vector that is *not* a combination of $\boldsymbol{v}$ and $\boldsymbol{w}$—not on the plane.

Solution The plane of $\boldsymbol{v}$ and $\boldsymbol{w}$ contains all combinations $c\boldsymbol{v} + d\boldsymbol{w}$. The vectors in that plane allow any c and d. The plane of Figure 1.3 fills in between the two lines.

$$\text{Combinations} \quad c\boldsymbol{v} + d\boldsymbol{w} = c\begin{bmatrix} 1 \\ 1 \\ 0 \end{bmatrix} + d\begin{bmatrix} 0 \\ 1 \\ 1 \end{bmatrix} = \begin{bmatrix} c \\ c+d \\ d \end{bmatrix} \text{ fill a plane.}$$

Four vectors in that plane are $(0, 0, 0)$ and $(2, 3, 1)$ and $(5, 7, 2)$ and $(\pi, 2\pi, \pi)$. The second component $c + d$ is always the sum of the first and third components. Like most vectors, $(1, 2, 3)$ *is **not** in the plane, because* $2 \neq 1 + 3$.

Another description of this plane through $(0, 0, 0)$ is to know that $\boldsymbol{n} = (1, -1, 1)$ is **perpendicular** to the plane. Section 1.2 will confirm that 90° angle by testing dot products: $\boldsymbol{v} \cdot \boldsymbol{n} = 0$ and $\boldsymbol{w} \cdot \boldsymbol{n} = 0$. Perpendicular vectors have zero dot products.

1.1 B For $\boldsymbol{v} = (1, 0)$ and $\boldsymbol{w} = (0, 1)$, describe all points $c\boldsymbol{v}$ with $(\mathbf{1})$ *whole numbers* c $(\mathbf{2})$ *nonnegative* numbers $c \geq 0$. Then add all vectors $d\boldsymbol{w}$ and describe all $c\boldsymbol{v} + d\boldsymbol{w}$.

Solution

$(\mathbf{1})$ The vectors $c\boldsymbol{v} = (c, 0)$ with whole numbers c are **equally spaced points** along the x axis (the direction of $\boldsymbol{v}$). They include $(-2,0), (-1,0), (0,0), (1,0), (2,0)$.

$(\mathbf{2})$ The vectors $c\boldsymbol{v}$ with $c \geq 0$ fill a ***half-line***. It is the positive x axis. This half-line starts at $(0,0)$ where $c = 0$. It includes $(100, 0)$ and $(\pi, 0)$ but not $(-100, 0)$.

$(\mathbf{1}')$ Adding all vectors $d\boldsymbol{w} = (0, d)$ puts a vertical line through those equally spaced $c\boldsymbol{v}$. We have infinitely many ***parallel lines*** from (*whole number c, any number d*).

$(\mathbf{2}')$ Adding all vectors $d\boldsymbol{w}$ puts a vertical line through every $c\boldsymbol{v}$ on the half-line. Now we have a ***half-plane***. The right half of the xy plane has any $x \geq 0$ and any y.

1.1 C Find two equations for c and d so that **the linear combination** $c\boldsymbol{v} + d\boldsymbol{w}$ **equals** $\boldsymbol{b}$:

$$\boldsymbol{v} = \begin{bmatrix} 2 \\ -1 \end{bmatrix} \qquad \boldsymbol{w} = \begin{bmatrix} -1 \\ 2 \end{bmatrix} \qquad \boldsymbol{b} = \begin{bmatrix} 1 \\ 0 \end{bmatrix}.$$

Solution In applying mathematics, many problems have two parts:

1 *Modeling part* Express the problem by a set of equations.

2 *Computational part* Solve those equations by a fast and accurate algorithm.

Here we are only asked for the first part (the equations). Chapter 2 is devoted to the second part (the solution). Our example fits into a fundamental model for linear algebra:

$$\text{Find } n \text{ numbers } \quad c_1, \ldots, c_n \quad \text{so that} \quad c_1\boldsymbol{v}_1 + \cdots + c_n\boldsymbol{v}_n = \boldsymbol{b}.$$

For $n = 2$ we will find a formula for the $\boldsymbol{c}$'s. The "elimination method" in Chapter 2 succeeds far beyond $n = 1000$. For n greater than 1 billion, see Chapter 11. Here $n = 2$:

Vector equation
$c\boldsymbol{v} + d\boldsymbol{w} = \boldsymbol{b}$

$$c\begin{bmatrix} 2 \\ -1 \end{bmatrix} + d\begin{bmatrix} -1 \\ 2 \end{bmatrix} = \begin{bmatrix} 1 \\ 0 \end{bmatrix}$$

The required equations for c and d just come from the two components separately:

Two ordinary equations

$$\begin{aligned} 2c - d &= 1 \\ -c + 2d &= 0 \end{aligned}$$

Each equation produces a line. The two lines cross at the solution $c = \dfrac{2}{3}, d = \dfrac{1}{3}$. Why not see this also as a **matrix equation**, since that is where we are going:

2 by 2 matrix

$$\begin{bmatrix} \mathbf{2} & \mathbf{-1} \\ \mathbf{-1} & \mathbf{2} \end{bmatrix} \begin{bmatrix} c \\ d \end{bmatrix} = \begin{bmatrix} 1 \\ 0 \end{bmatrix}.$$

Problem Set 1.1

Problems 1–9 are about addition of vectors and linear combinations.

1 Describe geometrically (line, plane, or all of $\mathbf{R}^3$) all linear combinations of

$$\text{(a)}\ \begin{bmatrix}1\\2\\3\end{bmatrix} \text{ and } \begin{bmatrix}3\\6\\9\end{bmatrix} \qquad \text{(b)}\ \begin{bmatrix}1\\0\\0\end{bmatrix} \text{ and } \begin{bmatrix}0\\2\\3\end{bmatrix} \qquad \text{(c)}\ \begin{bmatrix}2\\0\\0\end{bmatrix} \text{ and } \begin{bmatrix}0\\2\\2\end{bmatrix} \text{ and } \begin{bmatrix}2\\2\\3\end{bmatrix}$$

2 Draw $\boldsymbol{v} = \begin{bmatrix}4\\1\end{bmatrix}$ and $\boldsymbol{w} = \begin{bmatrix}-2\\2\end{bmatrix}$ and $\boldsymbol{v}+\boldsymbol{w}$ and $\boldsymbol{v}-\boldsymbol{w}$ in a single xy plane.

3 If $\boldsymbol{v} + \boldsymbol{w} = \begin{bmatrix}5\\1\end{bmatrix}$ and $\boldsymbol{v} - \boldsymbol{w} = \begin{bmatrix}1\\5\end{bmatrix}$, compute and draw the vectors $\boldsymbol{v}$ and $\boldsymbol{w}$.

4 From $\boldsymbol{v} = \begin{bmatrix}2\\1\end{bmatrix}$ and $\boldsymbol{w} = \begin{bmatrix}1\\2\end{bmatrix}$, find the components of $3\boldsymbol{v} + \boldsymbol{w}$ and $c\boldsymbol{v} + d\boldsymbol{w}$.

5 Compute $\boldsymbol{u} + \boldsymbol{v} + \boldsymbol{w}$ and $2\boldsymbol{u} + 2\boldsymbol{v} + \boldsymbol{w}$. How do you know $\boldsymbol{u}, \boldsymbol{v}, \boldsymbol{w}$ lie in a plane?

These lie in a plane because $\boldsymbol{w} = c\boldsymbol{u} + d\boldsymbol{v}$. Find c and d

$$\boldsymbol{u} = \begin{bmatrix}1\\2\\3\end{bmatrix}, \quad \boldsymbol{v} = \begin{bmatrix}-3\\1\\-2\end{bmatrix}, \quad \boldsymbol{w} = \begin{bmatrix}2\\-3\\-1\end{bmatrix}.$$

6 Every combination of $\boldsymbol{v} = (1, -2, 1)$ and $\boldsymbol{w} = (0, 1, -1)$ has components that add to ______. Find c and d so that $c\boldsymbol{v} + d\boldsymbol{w} = (3, 3, -6)$. Why is $(3, 3, 6)$ impossible?

7 In the xy plane mark all nine of these linear combinations:

$$c\begin{bmatrix}2\\1\end{bmatrix} + d\begin{bmatrix}0\\1\end{bmatrix} \quad \text{with} \quad c = 0, 1, 2 \quad \text{and} \quad d = 0, 1, 2.$$

8 The parallelogram in Figure 1.1 has diagonal $\boldsymbol{v} + \boldsymbol{w}$. What is its other diagonal? What is the sum of the two diagonals? Draw that vector sum.

9 If three corners of a parallelogram are $(1, 1)$, $(4, 2)$, and $(1, 3)$, what are all three of the possible fourth corners? Draw two of them.

Problems 10–14 are about special vectors on cubes and clocks in Figure 1.4.

10 Which point of the cube is $\boldsymbol{i} + \boldsymbol{j}$? Which point is the vector sum of $\boldsymbol{i} = (1, 0, 0)$ and $\boldsymbol{j} = (0, 1, 0)$ and $\boldsymbol{k} = (0, 0, 1)$? Describe all points (x, y, z) in the cube.

11 Four corners of this unit cube are $(0, 0, 0)$, $(1, 0, 0)$, $(0, 1, 0)$, $(0, 0, 1)$. What are the other four corners? Find the coordinates of the center point of the cube. The center points of the six faces are ______. The cube has how many edges?

12 *Review Question.* In xyz space, where is the plane of all linear combinations of $\boldsymbol{i} = (1, 0, 0)$ and $\boldsymbol{i} + \boldsymbol{j} = (1, 1, 0)$?

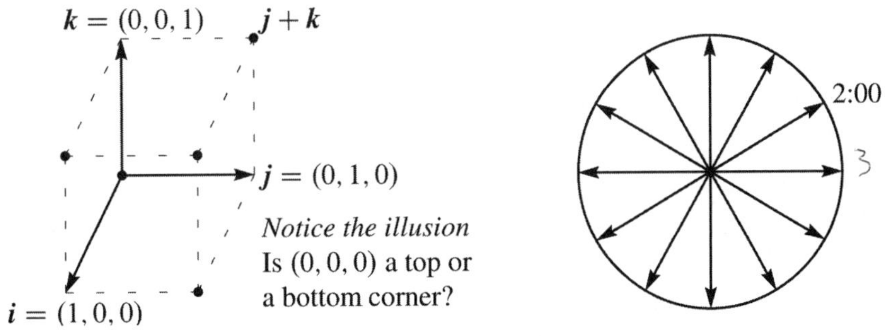

Figure 1.4: Unit cube from $\boldsymbol{i}, \boldsymbol{j}, \boldsymbol{k}$ and twelve clock vectors.

13 (a) What is the sum $\boldsymbol{V}$ of the twelve vectors that go from the center of a clock to the hours 1:00, 2:00, ..., 12:00?

(b) If the 2:00 vector is removed, why do the 11 remaining vectors add to 8:00?

(c) What are the x, y components of that 2:00 vector $\boldsymbol{v} = (\cos\theta, \sin\theta)$?

14 Suppose the twelve vectors start from 6:00 at the bottom instead of $(0, 0)$ at the center. The vector to 12:00 is doubled to $(0, 2)$. The new twelve vectors add to ____.

Problems 15–19 go further with linear combinations of $\boldsymbol{v}$ and $\boldsymbol{w}$ (Figure 1.5a).

15 Figure 1.5a shows $\frac{1}{2}\,\boldsymbol{v} + \frac{1}{2}\,\boldsymbol{w}$. Mark the points $\frac{3}{4}\,\boldsymbol{v} + \frac{1}{4}\,\boldsymbol{w}$ and $\frac{1}{4}\,\boldsymbol{v} + \frac{1}{4}\,\boldsymbol{w}$ and $\boldsymbol{v} + \boldsymbol{w}$.

16 Mark the point $-\boldsymbol{v} + 2\boldsymbol{w}$ and any other combination $c\boldsymbol{v} + d\boldsymbol{w}$ with $c + d = 1$. Draw the line of all combinations that have $c + d = 1$.

17 Locate $\frac{1}{3}\,\boldsymbol{v} + \frac{1}{3}\,\boldsymbol{w}$ and $\frac{2}{3}\,\boldsymbol{v} + \frac{2}{3}\,\boldsymbol{w}$. The combinations $c\boldsymbol{v} + c\boldsymbol{w}$ fill out what line?

18 Restricted by $0 \le c \le 1$ and $0 \le d \le 1$, shade in all combinations $c\boldsymbol{v} + d\boldsymbol{w}$.

19 Restricted only by $c \ge 0$ and $d \ge 0$ draw the "cone" of all combinations $c\boldsymbol{v} + d\boldsymbol{w}$.

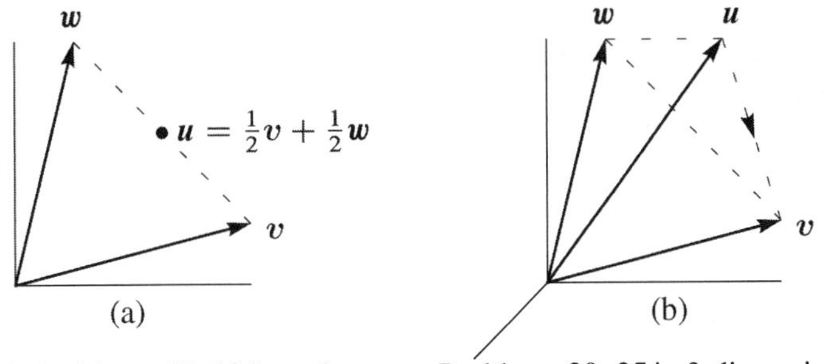

Figure 1.5: Problems **15–19** in a plane   Problems **20–25** in 3-dimensional space

Problems 20–25 deal with u, v, w in three-dimensional space (see Figure 1.5b).

20 Locate $\frac{1}{3}\boldsymbol{u} + \frac{1}{3}\boldsymbol{v} + \frac{1}{3}\boldsymbol{w}$ and $\frac{1}{2}\boldsymbol{u} + \frac{1}{2}\boldsymbol{w}$ in Figure 1.5b. Challenge problem: Under what restrictions on c, d, e, will the combinations $c\boldsymbol{u} + d\boldsymbol{v} + e\boldsymbol{w}$ fill in the dashed triangle? To stay in the triangle, one requirement is $c \geq 0, d \geq 0, e \geq 0$.

21 The three sides of the dashed triangle are $\boldsymbol{v} - \boldsymbol{u}$ and $\boldsymbol{w} - \boldsymbol{v}$ and $\boldsymbol{u} - \boldsymbol{w}$. Their sum is ______. Draw the head-to-tail addition around a plane triangle of $(3, 1)$ plus $(-1, 1)$ plus $(-2, -2)$.

22 Shade in the pyramid of combinations $c\boldsymbol{u} + d\boldsymbol{v} + e\boldsymbol{w}$ with $c \geq 0, d \geq 0, e \geq 0$ and $c + d + e \leq 1$. Mark the vector $\frac{1}{2}(\boldsymbol{u} + \boldsymbol{v} + \boldsymbol{w})$ as inside or outside this pyramid.

23 If you look at *all* combinations of those $\boldsymbol{u}, \boldsymbol{v}$, and $\boldsymbol{w}$, is there any vector that can't be produced from $c\boldsymbol{u} + d\boldsymbol{v} + e\boldsymbol{w}$? Different answer if $\boldsymbol{u}, \boldsymbol{v}, \boldsymbol{w}$ are all in ______.

24 Which vectors are combinations of $\boldsymbol{u}$ and $\boldsymbol{v}$, and *also* combinations of $\boldsymbol{v}$ and $\boldsymbol{w}$?

25 Draw vectors $\boldsymbol{u}, \boldsymbol{v}, \boldsymbol{w}$ so that their combinations $c\boldsymbol{u} + d\boldsymbol{v} + e\boldsymbol{w}$ fill only a line. Find vectors $\boldsymbol{u}, \boldsymbol{v}, \boldsymbol{w}$ so that their combinations $c\boldsymbol{u} + d\boldsymbol{v} + e\boldsymbol{w}$ fill only a plane.

26 What combination $c\begin{bmatrix} \mathbf{1} \\ \mathbf{2} \end{bmatrix} + d\begin{bmatrix} \mathbf{3} \\ \mathbf{1} \end{bmatrix}$ produces $\begin{bmatrix} \mathbf{14} \\ \mathbf{8} \end{bmatrix}$? Express this question as two equations for the coefficients c and d in the linear combination.

Challenge Problems

27 How many corners does a cube have in 4 dimensions? How many 3D faces? How many edges? A typical corner is $(0, 0, 1, 0)$. A typical edge goes to $(0, 1, 0, 0)$.

28 Find vectors $\boldsymbol{v}$ and $\boldsymbol{w}$ so that $\boldsymbol{v} + \boldsymbol{w} = (4, 5, 6)$ and $\boldsymbol{v} - \boldsymbol{w} = (2, 5, 8)$. This is a question with ______ unknown numbers, and an equal number of equations to find those numbers.

29 Find *two different combinations* of the three vectors $\boldsymbol{u} = (1, 3)$ and $\boldsymbol{v} = (2, 7)$ and $\boldsymbol{w} = (1, 5)$ that produce $\boldsymbol{b} = (0, 1)$. Slightly delicate question: If I take any three vectors $\boldsymbol{u}, \boldsymbol{v}, \boldsymbol{w}$ in the plane, will there always be two different combinations that produce $\boldsymbol{b} = (0, 1)$?

30 The linear combinations of $\boldsymbol{v} = (a, b)$ and $\boldsymbol{w} = (c, d)$ fill the plane unless ______. Find four vectors $\boldsymbol{u}, \boldsymbol{v}, \boldsymbol{w}, \boldsymbol{z}$ with four components each so that their combinations $c\boldsymbol{u} + d\boldsymbol{v} + e\boldsymbol{w} + f\boldsymbol{z}$ produce all vectors (b_1, b_2, b_3, b_4) in four-dimensional space.

31 Write down three equations for c, d, e so that $c\boldsymbol{u} + d\boldsymbol{v} + e\boldsymbol{w} = \boldsymbol{b}$. Can you somehow find c, d, e for this $\boldsymbol{b}$?

$$\boldsymbol{u} = \begin{bmatrix} 2 \\ -1 \\ 0 \end{bmatrix} \quad \boldsymbol{v} = \begin{bmatrix} -1 \\ 2 \\ -1 \end{bmatrix} \quad \boldsymbol{w} = \begin{bmatrix} 0 \\ -1 \\ 2 \end{bmatrix} \quad \boldsymbol{b} = \begin{bmatrix} 1 \\ 0 \\ 0 \end{bmatrix}.$$

1.2 Lengths and Dot Products

1 The "dot product" of $\boldsymbol{v} = \begin{bmatrix} 1 \\ 2 \end{bmatrix}$ and $\boldsymbol{w} = \begin{bmatrix} 4 \\ 5 \end{bmatrix}$ is $\boldsymbol{v} \cdot \boldsymbol{w} = (1)(4) + (2)(5) = 4 + 10 = \mathbf{14}.$

2 $\boldsymbol{v} = \begin{bmatrix} 1 \\ 3 \\ 2 \end{bmatrix}$ and $\boldsymbol{w} = \begin{bmatrix} 4 \\ -4 \\ 4 \end{bmatrix}$ are perpendicular because $\boldsymbol{v} \cdot \boldsymbol{w}$ is zero: $(1)(4) + (3)(-4) + (2)(4) = \mathbf{0}.$

3 The length squared of $\boldsymbol{v} = \begin{bmatrix} 1 \\ 3 \\ 2 \end{bmatrix}$ is $\boldsymbol{v} \cdot \boldsymbol{v} = 1 + 9 + 4 = 14$. **The length is** $||\boldsymbol{v}|| = \sqrt{\mathbf{14}}.$

4 Then $\boldsymbol{u} = \dfrac{\boldsymbol{v}}{||\boldsymbol{v}||} = \dfrac{\boldsymbol{v}}{\sqrt{14}} = \dfrac{1}{\sqrt{14}} \begin{bmatrix} 1 \\ 3 \\ 2 \end{bmatrix}$ has length $||\boldsymbol{u}|| = \mathbf{1}$. Check $\dfrac{1}{14} + \dfrac{9}{14} + \dfrac{4}{14} = 1.$

5 The angle θ between $\boldsymbol{v}$ and $\boldsymbol{w}$ has $\cos\theta = \dfrac{\boldsymbol{v} \cdot \boldsymbol{w}}{||\boldsymbol{v}|| \; ||\boldsymbol{w}||}.$

6 The angle between $\begin{bmatrix} 1 \\ 0 \end{bmatrix}$ and $\begin{bmatrix} 1 \\ 1 \end{bmatrix}$ has $\cos\theta = \dfrac{1}{(1)(\sqrt{2})}$. That angle is $\theta = 45^\circ$.

7 All angles have $|\cos\theta| \leq 1$. So all vectors have $\boxed{|\boldsymbol{v} \cdot \boldsymbol{w}| \leq ||\boldsymbol{v}|| \; ||\boldsymbol{w}||}.$

The first section backed off from multiplying vectors. Now we go forward to define the *"dot product"* of $\boldsymbol{v}$ and $\boldsymbol{w}$. This multiplication involves the separate products $v_1 w_1$ and $v_2 w_2$, but it doesn't stop there. Those two numbers are added to produce one number $\boldsymbol{v} \cdot \boldsymbol{w}$.

This is the geometry section (lengths of vectors and cosines of angles between them).

The ***dot product*** or ***inner product*** of $\boldsymbol{v} = (v_1, v_2)$ and $\boldsymbol{w} = (w_1, w_2)$ is the number $\boldsymbol{v} \cdot \boldsymbol{w}$:

$$\boldsymbol{v} \cdot \boldsymbol{w} = v_1 w_1 + v_2 w_2. \tag{1}$$

Example 1 The vectors $\boldsymbol{v} = (4, 2)$ and $\boldsymbol{w} = (-1, 2)$ have a *zero* dot product:

Dot product is zero
Perpendicular vectors

$$\begin{bmatrix} 4 \\ 2 \end{bmatrix} \cdot \begin{bmatrix} -1 \\ 2 \end{bmatrix} = -4 + 4 = 0.$$

In mathematics, zero is always a special number. For dot products, it means that *these two vectors are perpendicular*. The angle between them is 90°. When we drew them in Figure 1.1, we saw a rectangle (not just any parallelogram). The clearest example of perpendicular vectors is $\boldsymbol{i} = (1, 0)$ along the x axis and $\boldsymbol{j} = (0, 1)$ up the y axis. Again the dot product is $\boldsymbol{i} \cdot \boldsymbol{j} = 0 + 0 = 0$. Those vectors $\boldsymbol{i}$ and $\boldsymbol{j}$ form a right angle.

The dot product of $\boldsymbol{v} = (1, 2)$ and $\boldsymbol{w} = (3, 1)$ is 5. Soon $\boldsymbol{v} \cdot \boldsymbol{w}$ will reveal the angle between $\boldsymbol{v}$ and $\boldsymbol{w}$ (not 90°). Please check that $\boldsymbol{w} \cdot \boldsymbol{v}$ is also 5.

The dot product $\boldsymbol{w} \cdot \boldsymbol{v}$ ***equals*** $\boldsymbol{v} \cdot \boldsymbol{w}$. The order of $\boldsymbol{v}$ and $\boldsymbol{w}$ makes no difference.

Example 2 Put a weight of 4 at the point $x = -1$ (left of zero) and a weight of 2 at the point $x = 2$ (right of zero). The x axis will balance on the center point (like a see-saw). The weights balance because the dot product is $(4)(-1) + (2)(2) = 0$.

This example is typical of engineering and science. The vector of weights is $(w_1, w_2) = (4, 2)$. The vector of distances from the center is $(v_1, v_2) = (-1, 2)$. The weights times the distances, w_1v_1 and w_2v_2, give the "moments". The equation for the see-saw to balance is $w_1v_1 + w_2v_2 = 0$.

Example 3 Dot products enter in economics and business. We have three goods to buy and sell. Their prices are (p_1, p_2, p_3) for each unit—this is the "price vector" $\boldsymbol{p}$. The quantities we buy or sell are (q_1, q_2, q_3)—positive when we sell, negative when we buy. *Selling* q_1 *units at the price* p_1 *brings in* q_1p_1. The total income (quantities q times prices p) is ***the dot product*** $\boldsymbol{q} \cdot \boldsymbol{p}$ ***in three dimensions***:

$$\textbf{\textit{Income}} = (q_1, q_2, q_3) \cdot (p_1, p_2, p_3) = q_1p_1 + q_2p_2 + q_3p_3 = \textbf{\textit{dot product}}.$$

A zero dot product means that "the books balance". Total sales equal total purchases if $\boldsymbol{q} \cdot \boldsymbol{p} = 0$. Then $\boldsymbol{p}$ is perpendicular to $\boldsymbol{q}$ (in three-dimensional space). A supermarket with thousands of goods goes quickly into high dimensions.

Small note: Spreadsheets have become essential in management. They compute linear combinations and dot products. What you see on the screen is a matrix.

Main point For $\boldsymbol{v} \cdot \boldsymbol{w}$, multiply each v_i times w_i. Then $\boldsymbol{v} \cdot \boldsymbol{w} = v_1w_1 + \cdots + v_nw_n$.

Lengths and Unit Vectors

An important case is the dot product of a vector *with itself*. In this case $\boldsymbol{v}$ equals $\boldsymbol{w}$. When the vector is $\boldsymbol{v} = (1, 2, 3)$, the dot product with itself is $\boldsymbol{v} \cdot \boldsymbol{v} = \|\boldsymbol{v}\|^2 = 14$:

Dot product $\boldsymbol{v} \cdot \boldsymbol{v}$
Length squared

$$\|\boldsymbol{v}\|^2 = \begin{bmatrix} 1 \\ 2 \\ 3 \end{bmatrix} \cdot \begin{bmatrix} 1 \\ 2 \\ 3 \end{bmatrix} = 1 + 4 + 9 = \mathbf{14}.$$

Instead of a 90° angle between vectors we have 0°. The answer is not zero because $\boldsymbol{v}$ is not perpendicular to itself. The dot product $\boldsymbol{v} \cdot \boldsymbol{v}$ gives the *length of* $\boldsymbol{v}$ *squared*.

DEFINITION The ***length*** $\|\boldsymbol{v}\|$ of a vector $\boldsymbol{v}$ is the square root of $\boldsymbol{v} \cdot \boldsymbol{v}$:

$$\textbf{length} = \|\boldsymbol{v}\| = \sqrt{\boldsymbol{v} \cdot \boldsymbol{v}} = \left(v_1^2 + v_2^2 + \cdots + v_n^2\right)^{1/2}.$$

In two dimensions the length is $\sqrt{v_1^2 + v_2^2}$. In three dimensions it is $\sqrt{v_1^2 + v_2^2 + v_3^2}$. By the calculation above, the length of $\boldsymbol{v} = (1, 2, 3)$ is $\|\boldsymbol{v}\| = \sqrt{14}$.

Here $\|\boldsymbol{v}\| = \sqrt{\boldsymbol{v} \cdot \boldsymbol{v}}$ is just the ordinary length of the arrow that represents the vector. If the components are 1 and 2, the arrow is the third side of a right triangle (Figure 1.6). The Pythagoras formula $a^2 + b^2 = c^2$ connects the three sides: $1^2 + 2^2 = \|\boldsymbol{v}\|^2$.

For the length of $\boldsymbol{v} = (1, 2, 3)$, we used the right triangle formula twice. The vector $(1, 2, 0)$ in the base has length $\sqrt{5}$. This base vector is perpendicular to $(0, 0, 3)$ that goes straight up. So the diagonal of the box has length $\|\boldsymbol{v}\| = \sqrt{5 + 9} = \sqrt{14}$.

The length of a four-dimensional vector would be $\sqrt{v_1^2 + v_2^2 + v_3^2 + v_4^2}$. Thus the vector $(1, 1, 1, 1)$ has length $\sqrt{1^2 + 1^2 + 1^2 + 1^2} = 2$. This is the diagonal through a unit cube in four-dimensional space. That diagonal in n dimensions has length $\sqrt{n}$.

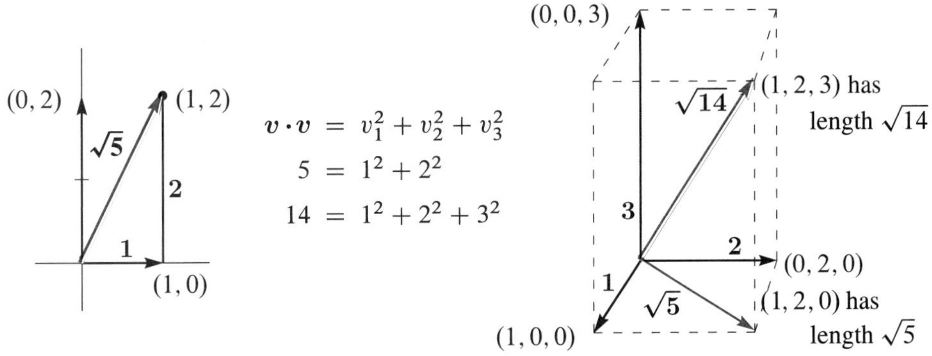

Figure 1.6: The length $\sqrt{\boldsymbol{v} \cdot \boldsymbol{v}}$ of two-dimensional and three-dimensional vectors.

The word "**unit**" is always indicating that some measurement equals "one". The unit price is the price for one item. A unit cube has sides of length one. A unit circle is a circle with radius one. Now we see the meaning of a "unit vector".

DEFINITION ***A unit vector u is a vector whose length equals one***. Then $\boldsymbol{u} \cdot \boldsymbol{u} = \mathbf{1}$.

An example in four dimensions is $\boldsymbol{u} = \left(\frac{1}{2}, \frac{1}{2}, \frac{1}{2}, \frac{1}{2}\right)$. Then $\boldsymbol{u} \cdot \boldsymbol{u}$ is $\frac{1}{4} + \frac{1}{4} + \frac{1}{4} + \frac{1}{4} = 1$. We divided $\boldsymbol{v} = (1, 1, 1, 1)$ by its length $\|\boldsymbol{v}\| = 2$ to get this unit vector.

Example 4 The standard unit vectors along the x and y axes are written $\boldsymbol{i}$ and $\boldsymbol{j}$. In the xy plane, the unit vector that makes an angle "theta" with the x axis is $(\cos\theta, \sin\theta)$:

$$\textbf{Unit vectors} \quad \boldsymbol{i} = \begin{bmatrix} 1 \\ 0 \end{bmatrix} \quad \text{and} \quad \boldsymbol{j} = \begin{bmatrix} 0 \\ 1 \end{bmatrix} \quad \text{and} \quad \boldsymbol{u} = \begin{bmatrix} \cos\theta \\ \sin\theta \end{bmatrix}.$$

When $\theta = 0$, the horizontal vector $\boldsymbol{u}$ is $\boldsymbol{i}$. When $\theta = 90^\circ$ (or $\frac{\pi}{2}$ radians), the vertical vector is $\boldsymbol{j}$. At any angle, the components $\cos\theta$ and $\sin\theta$ produce $\boldsymbol{u} \cdot \boldsymbol{u} = 1$ because

$\cos^2\theta + \sin^2\theta = 1$. These vectors reach out to the unit circle in Figure 1.7. Thus $\cos\theta$ and $\sin\theta$ are simply the coordinates of that point at angle θ on the unit circle.

Since $(2, 2, 1)$ has length 3, the vector $\left(\frac{2}{3}, \frac{2}{3}, \frac{1}{3}\right)$ has length 1. Check that $\boldsymbol{u} \cdot \boldsymbol{u} = \frac{4}{9} + \frac{4}{9} + \frac{1}{9} = 1$. For a unit vector, **divide any nonzero vector $\boldsymbol{v}$ by its length $\|\boldsymbol{v}\|$.**

Unit vector $\boldsymbol{u} = \boldsymbol{v}/\|\boldsymbol{v}\|$ **is a unit vector in the same direction as $\boldsymbol{v}$.**

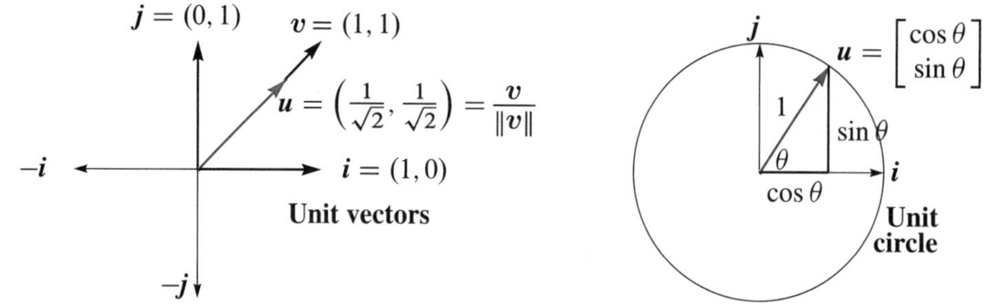

Figure 1.7: The coordinate vectors $\boldsymbol{i}$ and $\boldsymbol{j}$. The unit vector $\boldsymbol{u}$ at angle 45° (left) divides $\boldsymbol{v} = (1, 1)$ by its length $\|\boldsymbol{v}\| = \sqrt{2}$. The unit vector $\boldsymbol{u} = (\cos\theta, \sin\theta)$ is at angle θ.

The Angle Between Two Vectors

We stated that perpendicular vectors have $\boldsymbol{v} \cdot \boldsymbol{w} = 0$. The dot product is zero when the angle is 90°. To explain this, we have to connect angles to dot products. Then we show how $\boldsymbol{v} \cdot \boldsymbol{w}$ finds the angle between any two nonzero vectors $\boldsymbol{v}$ and $\boldsymbol{w}$.

Right angles ***The dot product is $\boldsymbol{v} \cdot \boldsymbol{w} = 0$ when $\boldsymbol{v}$ is perpendicular to $\boldsymbol{w}$.***

Proof When $\boldsymbol{v}$ and $\boldsymbol{w}$ are perpendicular, they form two sides of a right triangle. The third side is $\boldsymbol{v} - \boldsymbol{w}$ (the hypotenuse going across in Figure 1.8). The *Pythagoras Law* for the sides of a right triangle is $a^2 + b^2 = c^2$:

$$\textbf{Perpendicular vectors} \quad \|\boldsymbol{v}\|^2 + \|\boldsymbol{w}\|^2 = \|\boldsymbol{v} - \boldsymbol{w}\|^2 \tag{2}$$

Writing out the formulas for those lengths in two dimensions, this equation is

$$\textbf{Pythagoras} \quad \left(v_1^2 + v_2^2\right) + \left(w_1^2 + w_2^2\right) = (v_1 - w_1)^2 + (v_2 - w_2)^2. \tag{3}$$

The right side begins with $v_1^2 - 2v_1w_1 + w_1^2$. Then v_1^2 and w_1^2 are on both sides of the equation and they cancel, leaving $-2v_1w_1$. Also v_2^2 and w_2^2 cancel, leaving $-2v_2w_2$. (In three dimensions there would be $-2v_3w_3$.) Now divide by -2 to see $\boldsymbol{v} - \boldsymbol{w} = 0$:

$$0 = -2v_1w_1 - 2v_2w_2 \quad \text{which leads to} \quad \boldsymbol{v_1w_1 + v_2w_2 = 0}. \tag{4}$$

Conclusion Right angles produce $\boldsymbol{v} \cdot \boldsymbol{w} = 0$. The dot product is zero when the angle is $\theta = 90°$. Then $\cos\theta = 0$. The zero vector $\boldsymbol{v} = \boldsymbol{0}$ is perpendicular to every vector $\boldsymbol{w}$ because $\boldsymbol{0} \cdot \boldsymbol{w}$ is always zero.

Now suppose $v \cdot w$ is **not zero**. It may be positive, it may be negative. The sign of $v \cdot w$ immediately tells whether we are below or above a right angle. The angle is less than $90°$ when $v \cdot w$ is positive. The angle is above $90°$ when $v \cdot w$ is negative. The right side of Figure 1.8 shows a typical vector $v = (3, 1)$. The angle with $w = (1, 3)$ is less than $90°$ because $v \cdot w = 6$ is positive.

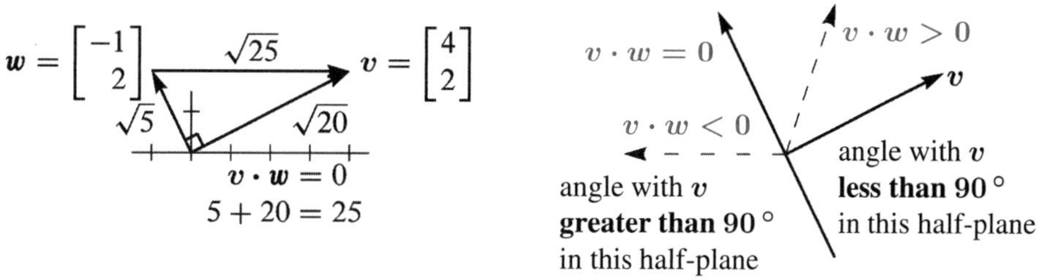

Figure 1.8: Perpendicular vectors have $v \cdot w = 0$. Then $\|v\|^2 + \|w\|^2 = \|v - w\|^2$.

The borderline is where vectors are perpendicular to v. On that dividing line between plus and minus, $(1, -3)$ is perpendicular to $(3, 1)$. The dot product is zero.

The dot product reveals the exact angle θ. For unit vectors u and U, the sign of $u \cdot U$ tells whether $\theta < 90°$ or $\theta > 90°$. More than that, ***the dot product $u \cdot U$ is the cosine of θ.*** This remains true in n dimensions.

Unit vectors u and U at angle θ have $u \cdot U = \cos\theta$. **Certainly $|u \cdot U| \leq 1$.**

Remember that $\cos\theta$ is never greater than 1. It is never less than -1. *The dot product of unit vectors is between -1 and 1.* **The cosine of θ is revealed by $u \cdot U$.**

Figure 1.9 shows this clearly when the vectors are $u = (\cos\theta, \sin\theta)$ and $i = (1, 0)$. The dot product is $u \cdot i = \cos\theta$. That is the cosine of the angle between them.

After rotation through any angle α, these are still unit vectors. The vector $i = (1, 0)$ rotates to $(\cos\alpha, \sin\alpha)$. The vector u rotates to $(\cos\beta, \sin\beta)$ with $\beta = \alpha + \theta$. Their dot product is $\cos\alpha\cos\beta + \sin\alpha\sin\beta$. From trigonometry this is $\cos(\beta - \alpha) = \cos\theta$.

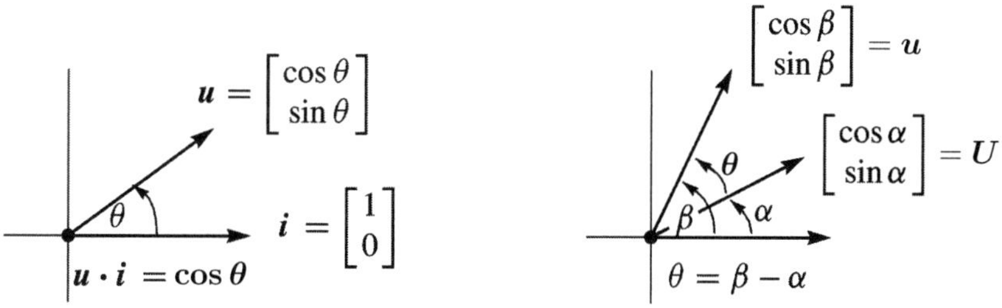

Figure 1.9: Unit vectors : $u \cdot U$ is the cosine of θ (the angle between).

What if $\boldsymbol{v}$ *and* $\boldsymbol{w}$ *are not unit vectors*? Divide by their lengths to get $\boldsymbol{u} = \boldsymbol{v}/\|\boldsymbol{v}\|$ and $\boldsymbol{U} = \boldsymbol{w}/\|\boldsymbol{w}\|$. Then the dot product of those unit vectors $\boldsymbol{u}$ and $\boldsymbol{U}$ gives $\cos\theta$.

COSINE FORMULA If $\boldsymbol{v}$ and $\boldsymbol{w}$ are nonzero vectors then $\dfrac{\boldsymbol{v}\cdot\boldsymbol{w}}{\|\boldsymbol{v}\|\,\|\boldsymbol{w}\|} = \cos\theta.$ (5)

Whatever the angle, this dot product of $\boldsymbol{v}/\|\boldsymbol{v}\|$ with $\boldsymbol{w}/\|\boldsymbol{w}\|$ never exceeds one. That is the ***"Schwarz inequality"*** $|\boldsymbol{v}\cdot\boldsymbol{w}| \le \|\boldsymbol{v}\|\,\|\boldsymbol{w}\|$ for dot products—or more correctly the Cauchy-Schwarz-Buniakowsky inequality. It was found in France and Germany and Russia (and maybe elsewhere—it is the most important inequality in mathematics).

Since $|\cos\theta|$ never exceeds 1, the cosine formula gives two great inequalities:

SCHWARZ INEQUALITY	$\lvert\boldsymbol{v}\cdot\boldsymbol{w}\rvert \le \|\boldsymbol{v}\|\,\|\boldsymbol{w}\|$
TRIANGLE INEQUALITY	$\|\boldsymbol{v}+\boldsymbol{w}\| \le \|\boldsymbol{v}\| + \|\boldsymbol{w}\|$

Example 5 Find $\cos\theta$ for $\boldsymbol{v} = \begin{bmatrix} 2 \\ 1 \end{bmatrix}$ and $\boldsymbol{w} = \begin{bmatrix} 1 \\ 2 \end{bmatrix}$ and check both inequalities.

Solution The dot product is $\boldsymbol{v}\cdot\boldsymbol{w} = 4$. Both $\boldsymbol{v}$ and $\boldsymbol{w}$ have length $\sqrt{5}$. The cosine is $4/5$.

$$\cos\theta = \frac{\boldsymbol{v}\cdot\boldsymbol{w}}{\|\boldsymbol{v}\|\,\|\boldsymbol{w}\|} = \frac{4}{\sqrt{5}\sqrt{5}} = \frac{4}{5}.$$

By the Schwarz inequality, $\boldsymbol{v}\cdot\boldsymbol{w} = 4$ is less than $\|\boldsymbol{v}\|\,\|\boldsymbol{w}\| = 5$. By the triangle inequality, side $3 = \|\boldsymbol{v}+\boldsymbol{w}\|$ is less than side 1 + side 2. For $\boldsymbol{v}+\boldsymbol{w} = (3,3)$ the three sides are $\sqrt{18} < \sqrt{5}+\sqrt{5}$. Square this triangle inequality to get $18 < 20$.

Example 6 The dot product of $\boldsymbol{v} = (a,b)$ and $\boldsymbol{w} = (b,a)$ is $2ab$. Both lengths are $\sqrt{a^2+b^2}$. The Schwarz inequality $\boldsymbol{v}\cdot\boldsymbol{w} \le \|\boldsymbol{v}\|\,\|\boldsymbol{w}\|$ says that $\boldsymbol{2ab \le a^2+b^2}$.

This is more famous if we write $x = a^2$ and $y = b^2$. The "geometric mean" $\sqrt{\boldsymbol{xy}}$ is not larger than the "arithmetic mean" = average $\frac{1}{2}(\boldsymbol{x}+\boldsymbol{y})$.

Geometric mean $\le$ **Arithmetic mean** $\quad ab \le \dfrac{a^2+b^2}{2}$ becomes $\sqrt{xy} \le \dfrac{x+y}{2}$.

Example 5 had $a = 2$ and $b = 1$. So $x = 4$ and $y = 1$. The geometric mean $\sqrt{xy} = 2$ is below the arithmetic mean $\frac{1}{2}(1+4) = 2.5$.

Notes on Computing

MATLAB, Python *and* Julia *work directly with whole vectors, not their components.* When $\boldsymbol{v}$ and $\boldsymbol{w}$ have been defined, $\boldsymbol{v}+\boldsymbol{w}$ is immediately understood. Input $\boldsymbol{v}$ and $\boldsymbol{w}$ as rows—the prime $'$ transposes them to columns. $2\boldsymbol{v}+3\boldsymbol{w}$ becomes $2*\boldsymbol{v}+3*\boldsymbol{w}$. The result will be printed unless the line ends in a semicolon.

MATLAB $v = [2 \;\; 3 \;\; 4]'$; $w = [1 \;\; 1 \;\; 1]'$; $u = 2 * v + 3 * w$

The dot product $v \cdot w$ is ***a row vector times a column vector (use * instead of $\cdot$)***:

Instead of $\begin{bmatrix}1\\2\end{bmatrix} \cdot \begin{bmatrix}3\\4\end{bmatrix}$	we more often see $[1 \;\; 2]\begin{bmatrix}3\\4\end{bmatrix}$	or $v' * w$

The length of v is known to MATLAB as norm (v). This is sqrt $(v' * v)$. Then find the cosine from the dot product $v' * w$ and the angle (in radians) that has that cosine:

Cosine formula	cosine $= v' * w/(\text{norm}\,(v) * \text{norm}\,(w))$
The arc cosine	angle $=$ acos (cosine)

An M-file would create a new function **cosine** (v, w). Python and Julia are open source.

■ REVIEW OF THE KEY IDEAS ■

1. The dot product $v \cdot w$ multiplies each component v_i by w_i and adds all $v_i w_i$.
2. The length $\|v\|$ is the square root of $v \cdot v$. Then $u = v/\|v\|$ is a ***unit vector***: length 1.
3. The dot product is $v \cdot w = 0$ when vectors v and w are perpendicular.
4. The cosine of θ (the angle between any nonzero v and w) never exceeds 1:

$$\textbf{Cosine} \quad \cos\theta = \frac{v \cdot w}{\|v\|\|w\|} \qquad \textit{Schwarz inequality} \quad |v \cdot w| \leq \|v\|\,\|w\|.$$

■ WORKED EXAMPLES ■

1.2 A For the vectors $v = (3, 4)$ and $w = (4, 3)$ test the Schwarz inequality on $v \cdot w$ and the triangle inequality on $\|v + w\|$. Find $\cos\theta$ for the angle between v and w. Which v and w give *equality* $|v \cdot w| = \|v\|\,\|w\|$ and $\|v + w\| = \|v\| + \|w\|$?

Solution The dot product is $v \cdot w = (3)(4) + (4)(3) = 24$. The length of v is $\|v\| = \sqrt{9 + 16} = 5$ and also $\|w\| = 5$. The sum $v + w = (7, 7)$ has length $7\sqrt{2} < 10$.

Schwarz inequality	$\|v \cdot w\| \leq \|v\|\,\|w\|$	is $24 < 25$.
Triangle inequality	$\|v + w\| \leq \|v\| + \|w\|$	is $7\sqrt{2} < 5 + 5$.
Cosine of angle	$\cos\theta = \frac{24}{25}$	Thin angle from $v = (3, 4)$ to $w = (4, 3)$

Equality: One vector is a multiple of the other as in $w = cv$. Then the angle is $0°$ or $180°$. In this case $|\cos\theta| = 1$ and $|v \cdot w|$ *equals* $\|v\|\,\|w\|$. If the angle is $0°$, as in $w = 2v$, then $\|v + w\| = \|v\| + \|w\|$ (both sides give $3\|v\|$). This $v, 2v, 3v$ triangle is flat!

1.2 B Find a unit vector $\boldsymbol{u}$ in the direction of $\boldsymbol{v} = (3,4)$. Find a unit vector $\boldsymbol{U}$ that is perpendicular to $\boldsymbol{u}$. How many possibilities for $\boldsymbol{U}$?

Solution For a unit vector $\boldsymbol{u}$, divide $\boldsymbol{v}$ by its length $\|\boldsymbol{v}\| = 5$. For a perpendicular vector $\boldsymbol{V}$ we can choose $(-4,3)$ since the dot product $\boldsymbol{v}\cdot\boldsymbol{V}$ is $(3)(-4)+(4)(3)=0$. For a *unit* vector perpendicular to $\boldsymbol{u}$, divide $\boldsymbol{V}$ by its length $\|\boldsymbol{V}\|$:

$$\boldsymbol{u} = \frac{\boldsymbol{v}}{\|\boldsymbol{v}\|} = \left(\frac{3}{5},\frac{4}{5}\right) \qquad \boldsymbol{U} = \frac{\boldsymbol{V}}{\|\boldsymbol{V}\|} = \left(-\frac{4}{5},\frac{3}{5}\right) \qquad \boldsymbol{u}\cdot\boldsymbol{U} = 0$$

The only other perpendicular unit vector would be $-\boldsymbol{U} = (\frac{4}{5},-\frac{3}{5})$.

1.2 C Find a vector $\boldsymbol{x} = (c,d)$ that has dot products $\boldsymbol{x}\cdot\boldsymbol{r} = 1$ and $\boldsymbol{x}\cdot\boldsymbol{s} = 0$ with two given vectors $\boldsymbol{r} = (2,-1)$ and $\boldsymbol{s} = (-1,2)$.

Solution Those two dot products give linear equations for c and d. Then $\boldsymbol{x} = (c,d)$.

$$\begin{array}{lll} \boldsymbol{x}\cdot\boldsymbol{r} = 1 & \text{is} & 2c - d = 1 \\ \boldsymbol{x}\cdot\boldsymbol{s} = 0 & \text{is} & -c + 2d = 0 \end{array}$$ **The same equations as in Worked Example 1.1 C**

Comment on n equations for $\boldsymbol{x} = (x_1,\ldots,x_n)$ in n-dimensional space

Section 1.1 would start with columns $\boldsymbol{v}_j$. The goal is to produce $x_1\boldsymbol{v}_1 + \cdots + x_n\boldsymbol{v}_n = \boldsymbol{b}$. This section would start from rows $\boldsymbol{r}_i$. Now the goal is to find $\boldsymbol{x}$ with $\boldsymbol{x}\cdot\boldsymbol{r}_i = b_i$.

Soon the $\boldsymbol{v}$'s will be the columns of a matrix A, and the $\boldsymbol{r}$'s will be the rows of A. Then the (one and only) problem will be to solve $A\boldsymbol{x} = \boldsymbol{b}$.

Problem Set 1.2

1 Calculate the dot products $\boldsymbol{u}\cdot\boldsymbol{v}$ and $\boldsymbol{u}\cdot\boldsymbol{w}$ and $\boldsymbol{u}\cdot(\boldsymbol{v}+\boldsymbol{w})$ and $\boldsymbol{w}\cdot\boldsymbol{v}$:

$$\boldsymbol{u} = \begin{bmatrix} -.6 \\ .8 \end{bmatrix} \qquad \boldsymbol{v} = \begin{bmatrix} 4 \\ 3 \end{bmatrix} \qquad \boldsymbol{w} = \begin{bmatrix} 1 \\ 2 \end{bmatrix}.$$

2 Compute the lengths $\|\boldsymbol{u}\|$ and $\|\boldsymbol{v}\|$ and $\|\boldsymbol{w}\|$ of those vectors. Check the Schwarz inequalities $|\boldsymbol{u}\cdot\boldsymbol{v}| \le \|\boldsymbol{u}\|\,\|\boldsymbol{v}\|$ and $|\boldsymbol{v}\cdot\boldsymbol{w}| \le \|\boldsymbol{v}\|\,\|\boldsymbol{w}\|$.

3 Find unit vectors in the directions of $\boldsymbol{v}$ and $\boldsymbol{w}$ in Problem 1, and the cosine of the angle θ. Choose vectors $\boldsymbol{a},\boldsymbol{b},\boldsymbol{c}$ that make $0°$, $90°$, and $180°$ angles with $\boldsymbol{w}$.

4 For any *unit* vectors $\boldsymbol{v}$ and $\boldsymbol{w}$, find the dot products (actual numbers) of

(a) $\boldsymbol{v}$ and $-\boldsymbol{v}$ (b) $\boldsymbol{v}+\boldsymbol{w}$ and $\boldsymbol{v}-\boldsymbol{w}$ (c) $\boldsymbol{v}-2\boldsymbol{w}$ and $\boldsymbol{v}+2\boldsymbol{w}$

5 Find unit vectors $\boldsymbol{u}_1$ and $\boldsymbol{u}_2$ in the directions of $\boldsymbol{v} = (1,3)$ and $\boldsymbol{w} = (2,1,2)$. Find unit vectors $\boldsymbol{U}_1$ and $\boldsymbol{U}_2$ that are perpendicular to $\boldsymbol{u}_1$ and $\boldsymbol{u}_2$.

6 (a) Describe every vector $\boldsymbol{w} = (w_1, w_2)$ that is perpendicular to $\boldsymbol{v} = (2, -1)$.

(b) All vectors perpendicular to $\boldsymbol{V} = (1, 1, 1)$ lie on a _____ in 3 dimensions.

(c) The vectors perpendicular to both $(1, 1, 1)$ and $(1, 2, 3)$ lie on a _____.

7 Find the angle θ (from its cosine) between these pairs of vectors:

(a) $\boldsymbol{v} = \begin{bmatrix} 1 \\ \sqrt{3} \end{bmatrix}$ and $\boldsymbol{w} = \begin{bmatrix} 1 \\ 0 \end{bmatrix}$ (b) $\boldsymbol{v} = \begin{bmatrix} 2 \\ 2 \\ -1 \end{bmatrix}$ and $\boldsymbol{w} = \begin{bmatrix} 2 \\ -1 \\ 2 \end{bmatrix}$

(c) $\boldsymbol{v} = \begin{bmatrix} 1 \\ \sqrt{3} \end{bmatrix}$ and $\boldsymbol{w} = \begin{bmatrix} -1 \\ \sqrt{3} \end{bmatrix}$ (d) $\boldsymbol{v} = \begin{bmatrix} 3 \\ 1 \end{bmatrix}$ and $\boldsymbol{w} = \begin{bmatrix} -1 \\ -2 \end{bmatrix}$.

8 True or false (give a reason if true or find a counterexample if false):

(a) If $\boldsymbol{u} = (1, 1, 1)$ is perpendicular to $\boldsymbol{v}$ and $\boldsymbol{w}$, then $\boldsymbol{v}$ is parallel to $\boldsymbol{w}$.

(b) If $\boldsymbol{u}$ is perpendicular to $\boldsymbol{v}$ and $\boldsymbol{w}$, then $\boldsymbol{u}$ is perpendicular to $\boldsymbol{v} + 2\boldsymbol{w}$.

(c) If $\boldsymbol{u}$ and $\boldsymbol{v}$ are perpendicular unit vectors then $\|\boldsymbol{u} - \boldsymbol{v}\| = \sqrt{2}$. *Yes*!

9 The slopes of the arrows from $(0, 0)$ to (v_1, v_2) and (w_1, w_2) are v_2/v_1 and w_2/w_1. **Suppose the product** $v_2 w_2 / v_1 w_1$ **of those slopes is** -1. Show that $\boldsymbol{v} \cdot \boldsymbol{w} = 0$ and the vectors are perpendicular. (The line $y = 4x$ is perpendicular to $y = -\frac{1}{4}x$.)

10 Draw arrows from $(0, 0)$ to the points $\boldsymbol{v} = (1, 2)$ and $\boldsymbol{w} = (-2, 1)$. Multiply their slopes. That answer is a signal that $\boldsymbol{v} \cdot \boldsymbol{w} = 0$ and the arrows are _____.

11 If $\boldsymbol{v} \cdot \boldsymbol{w}$ is negative, what does this say about the angle between $\boldsymbol{v}$ and $\boldsymbol{w}$? Draw a 3-dimensional vector $\boldsymbol{v}$ (an arrow), and show where to find all $\boldsymbol{w}$'s with $\boldsymbol{v} \cdot \boldsymbol{w} < 0$.

12 With $\boldsymbol{v} = (1, 1)$ and $\boldsymbol{w} = (1, 5)$ choose a number c so that $\boldsymbol{w} - c\boldsymbol{v}$ is perpendicular to $\boldsymbol{v}$. Then find the formula for c starting from *any* nonzero $\boldsymbol{v}$ and $\boldsymbol{w}$.

13 Find nonzero vectors $\boldsymbol{v}$ and $\boldsymbol{w}$ that are perpendicular to $(1, 0, 1)$ and to each other.

14 Find nonzero vectors $\boldsymbol{u}, \boldsymbol{v}, \boldsymbol{w}$ that are perpendicular to $(1, 1, 1, 1)$ and to each other.

15 The geometric mean of $x = 2$ and $y = 8$ is $\sqrt{xy} = 4$. The arithmetic mean is larger: $\frac{1}{2}(x + y) =$ _____. This would come in Example 6 from the Schwarz inequality for $\boldsymbol{v} = (\sqrt{2}, \sqrt{8})$ and $\boldsymbol{w} = (\sqrt{8}, \sqrt{2})$. Find $\cos\theta$ for this $\boldsymbol{v}$ and $\boldsymbol{w}$.

16 **How long is the vector** $\boldsymbol{v} = (1, 1, \ldots, 1)$ **in 9 dimensions**? Find a unit vector $\boldsymbol{u}$ in the same direction as $\boldsymbol{v}$ and a unit vector $\boldsymbol{w}$ that is perpendicular to $\boldsymbol{v}$.

17 What are the cosines of the angles α, β, θ between the vector $(1, 0, -1)$ and the unit vectors $\boldsymbol{i}, \boldsymbol{j}, \boldsymbol{k}$ along the axes? Check the formula $\cos^2\alpha + \cos^2\beta + \cos^2\theta = 1$.

Problems 18–28 lead to the main facts about lengths and angles in triangles.

18 The parallelogram with sides $\boldsymbol{v} = (4, 2)$ and $\boldsymbol{w} = (-1, 2)$ is a rectangle. Check the Pythagoras formula $a^2 + b^2 = c^2$ which is for ***right triangles only***:

$$(\text{length of } \boldsymbol{v})^2 + (\text{length of } \boldsymbol{w})^2 = (\text{length of } \boldsymbol{v} + \boldsymbol{w})^2.$$

19 (Rules for dot products) These equations are simple but useful:

(1) $\boldsymbol{v} \cdot \boldsymbol{w} = \boldsymbol{w} \cdot \boldsymbol{v}$ **(2)** $\boldsymbol{u} \cdot (\boldsymbol{v} + \boldsymbol{w}) = \boldsymbol{u} \cdot \boldsymbol{v} + \boldsymbol{u} \cdot \boldsymbol{w}$ **(3)** $(c\boldsymbol{v}) \cdot \boldsymbol{w} = c(\boldsymbol{v} \cdot \boldsymbol{w})$

Use **(2)** with $\boldsymbol{u} = \boldsymbol{v} + \boldsymbol{w}$ to prove $\|\boldsymbol{v} + \boldsymbol{w}\|^2 = \boldsymbol{v} \cdot \boldsymbol{v} + 2\boldsymbol{v} \cdot \boldsymbol{w} + \boldsymbol{w} \cdot \boldsymbol{w}$.

20 The "Law of Cosines" comes from $(\boldsymbol{v} - \boldsymbol{w}) \cdot (\boldsymbol{v} - \boldsymbol{w}) = \boldsymbol{v} \cdot \boldsymbol{v} - 2\boldsymbol{v} \cdot \boldsymbol{w} + \boldsymbol{w} \cdot \boldsymbol{w}$:

Cosine Law $$\|\boldsymbol{v} - \boldsymbol{w}\|^2 = \|\boldsymbol{v}\|^2 - 2\|\boldsymbol{v}\|\,\|\boldsymbol{w}\| \cos\theta + \|\boldsymbol{w}\|^2.$$

Draw a triangle with sides $\boldsymbol{v}$ and $\boldsymbol{w}$ and $\boldsymbol{v} - \boldsymbol{w}$. Which of the angles is θ ?

21 The ***triangle inequality*** says: (length of $\boldsymbol{v} + \boldsymbol{w}$) $\leq$ (length of $\boldsymbol{v}$) + (length of $\boldsymbol{w}$).

Problem 19 found $\|\boldsymbol{v} + \boldsymbol{w}\|^2 = \|\boldsymbol{v}\|^2 + 2\boldsymbol{v} \cdot \boldsymbol{w} + \|\boldsymbol{w}\|^2$. Increase that $\boldsymbol{v} \cdot \boldsymbol{w}$ to $\|\boldsymbol{v}\|\,\|\boldsymbol{w}\|$ to show that $\|$**side 3**$\|$ can not exceed $\|$**side 1**$\| + \|$**side 2**$\|$:

Triangle inequality $$\|\boldsymbol{v} + \boldsymbol{w}\|^2 \leq (\|\boldsymbol{v}\| + \|\boldsymbol{w}\|)^2 \quad \textbf{or} \quad \|\boldsymbol{v} + \boldsymbol{w}\| \leq \|\boldsymbol{v}\| + \|\boldsymbol{w}\|.$$

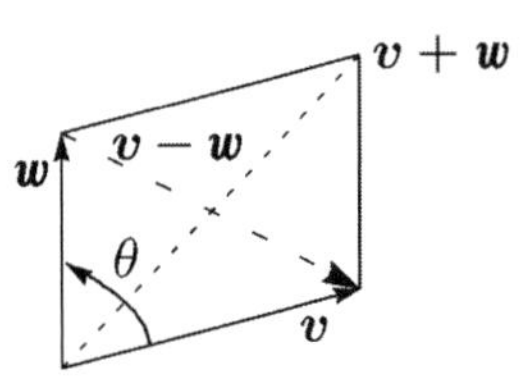

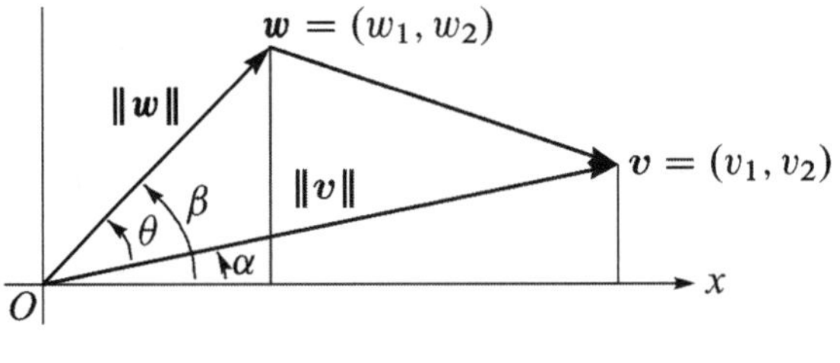

22 The Schwarz inequality $|\boldsymbol{v} \cdot \boldsymbol{w}| \leq \|\boldsymbol{v}\|\,\|\boldsymbol{w}\|$ by algebra instead of trigonometry:

(a) Multiply out both sides of $(v_1 w_1 + v_2 w_2)^2 \leq \left(v_1^2 + v_2^2\right)\left(w_1^2 + w_2^2\right)$.

(b) Show that the difference between those two sides equals $(v_1 w_2 - v_2 w_1)^2$. This cannot be negative since it is a square—so the inequality is true.

23 The figure shows that $\cos\alpha = v_1/\|\boldsymbol{v}\|$ and $\sin\alpha = v_2/\|\boldsymbol{v}\|$. Similarly $\cos\beta$ is _____ and $\sin\beta$ is _____. The angle θ is $\beta - \alpha$. Substitute into the trigonometry formula $\cos\beta\cos\alpha + \sin\beta\sin\alpha$ for $\cos(\beta - \alpha)$ to find $\cos\theta = \boldsymbol{v} \cdot \boldsymbol{w}/\|\boldsymbol{v}\|\,\|\boldsymbol{w}\|$.

24 One-line proof of the inequality $|\boldsymbol{u} \cdot \boldsymbol{U}| \le 1$ for unit vectors (u_1, u_2) and (U_1, U_2):

$$|\boldsymbol{u} \cdot \boldsymbol{U}| \le |u_1|\,|U_1| + |u_2|\,|U_2| \le \frac{u_1^2 + U_1^2}{2} + \frac{u_2^2 + U_2^2}{2} = 1.$$

Put $(u_1, u_2) = (.6, .8)$ and $(U_1, U_2) = (.8, .6)$ in that whole line and find $\cos\theta$.

25 Why is $|\cos\theta|$ never greater than 1 in the first place?

26 (*Recommended*) Draw a parallelogram

27 Parallelogram with two sides $\boldsymbol{v}$ and $\boldsymbol{w}$. Show that the squared diagonal lengths $\|\boldsymbol{v} + \boldsymbol{w}\|^2 + \|\boldsymbol{v} - \boldsymbol{w}\|^2$ add to the sum of four squared side lengths $2\|\boldsymbol{v}\|^2 + 2\|\boldsymbol{w}\|^2$.

28 If $\boldsymbol{v} = (1, 2)$ draw all vectors $\boldsymbol{w} = (x, y)$ in the xy plane with $\boldsymbol{v} \cdot \boldsymbol{w} = x + 2y = 5$. Why do those $\boldsymbol{w}$'s lie along a line? Which is the shortest $\boldsymbol{w}$?

29 (*Recommended*) If $||\boldsymbol{v}|| = 5$ and $||\boldsymbol{w}|| = 3$, what are the smallest and largest possible values of $||\boldsymbol{v} - \boldsymbol{w}||$? What are the smallest and largest possible values of $\boldsymbol{v} \cdot \boldsymbol{w}$?

Challenge Problems

30 Can three vectors in the xy plane have $\boldsymbol{u} \cdot \boldsymbol{v} < 0$ and $\boldsymbol{v} \cdot \boldsymbol{w} < 0$ and $\boldsymbol{u} \cdot \boldsymbol{w} < 0$? I don't know how many vectors in xyz space can have all negative dot products. (Four of those vectors in the plane would certainly be impossible ...).

31 Pick any numbers that add to $x + y + z = 0$. Find the angle between your vector $\boldsymbol{v} = (x, y, z)$ and the vector $\boldsymbol{w} = (z, x, y)$. Challenge question: Explain why $\boldsymbol{v} \cdot \boldsymbol{w}/\|\boldsymbol{v}\|\|\boldsymbol{w}\|$ is always $-\frac{1}{2}$.

32 How could you prove $\sqrt[3]{xyz} \le \frac{1}{3}(x+y+z)$ (geometric mean $\le$ arithmetic mean) ?

33 Find 4 perpendicular unit vectors of the form $(\pm\frac{1}{2}, \pm\frac{1}{2}, \pm\frac{1}{2}, \pm\frac{1}{2})$: Choose $+$ or $-$.

34 Using $\boldsymbol{v} = \mathsf{randn}(3, 1)$ in MATLAB, create a random unit vector $\boldsymbol{u} = \boldsymbol{v}/\|\boldsymbol{v}\|$. Using $V = \mathsf{randn}(3, 30)$ create 30 more random unit vectors $\boldsymbol{U}_j$. What is the average size of the dot products $|\boldsymbol{u} \cdot \boldsymbol{U}_j|$? In calculus, the average is $\int_0^\pi |\cos\theta| d\theta/\pi = 2/\pi$.

1.3 Matrices

1 $A = \begin{bmatrix} 1 & 2 \\ 3 & 4 \\ 5 & 6 \end{bmatrix}$ is a 3 by 2 matrix : $m = 3$ rows and $n = 2$ columns.

2 $A\boldsymbol{x} = \begin{bmatrix} 1 & 2 \\ 3 & 4 \\ 5 & 6 \end{bmatrix} \begin{bmatrix} x_1 \\ x_2 \end{bmatrix}$ is a **combination of the columns** $A\boldsymbol{x} = x_1 \begin{bmatrix} 1 \\ 3 \\ 5 \end{bmatrix} + x_2 \begin{bmatrix} 2 \\ 4 \\ 6 \end{bmatrix}$.

3 The 3 components of $A\boldsymbol{x}$ are dot products of the 3 rows of A with the vector $\boldsymbol{x}$:

Row at a time $\begin{bmatrix} 1 & 2 \\ 3 & 4 \\ 5 & 6 \end{bmatrix} \begin{bmatrix} 7 \\ 8 \end{bmatrix} = \begin{bmatrix} 1 \cdot 7 + 2 \cdot 8 \\ 3 \cdot 7 + 4 \cdot 8 \\ 5 \cdot 7 + 6 \cdot 8 \end{bmatrix} = \begin{bmatrix} 23 \\ 53 \\ 83 \end{bmatrix}$.

4 Equations in matrix form $A\boldsymbol{x} = \boldsymbol{b}$: $\begin{bmatrix} 2 & 5 \\ 3 & 7 \end{bmatrix} \begin{bmatrix} x_1 \\ x_2 \end{bmatrix} = \begin{bmatrix} b_1 \\ b_2 \end{bmatrix}$ replaces $\begin{matrix} 2x_1 + 5x_2 = b_1 \\ 3x_1 + 7x_2 = b_2 \end{matrix}$.

5 The solution to $A\boldsymbol{x} = \boldsymbol{b}$ can be written as $\boldsymbol{x} = A^{-1}\boldsymbol{b}$. But some matrices don't allow A^{-1}.

This section starts with three vectors $\boldsymbol{u}, \boldsymbol{v}, \boldsymbol{w}$. I will combine them using *matrices*.

$$\textbf{Three vectors} \qquad \boldsymbol{u} = \begin{bmatrix} 1 \\ -1 \\ 0 \end{bmatrix} \qquad \boldsymbol{v} = \begin{bmatrix} 0 \\ 1 \\ -1 \end{bmatrix} \qquad \boldsymbol{w} = \begin{bmatrix} 0 \\ 0 \\ 1 \end{bmatrix}.$$

Their linear combinations in three-dimensional space are $x_1\boldsymbol{u} + x_2\boldsymbol{v} + x_3\boldsymbol{w}$:

Combination of the vectors

$$x_1 \begin{bmatrix} 1 \\ -1 \\ 0 \end{bmatrix} + x_2 \begin{bmatrix} 0 \\ 1 \\ -1 \end{bmatrix} + x_3 \begin{bmatrix} 0 \\ 0 \\ 1 \end{bmatrix} = \begin{bmatrix} x_1 \\ x_2 - x_1 \\ x_3 - x_2 \end{bmatrix}. \qquad (1)$$

Now something important: *Rewrite that combination using a matrix.* The vectors $\boldsymbol{u}, \boldsymbol{v}, \boldsymbol{w}$ go into the columns of the matrix A. That matrix "*multiplies*" the vector (x_1, x_2, x_3) :

Matrix times vector
Combination of columns

$$A\boldsymbol{x} = \begin{bmatrix} 1 & 0 & 0 \\ -1 & 1 & 0 \\ 0 & -1 & 1 \end{bmatrix} \begin{bmatrix} x_1 \\ x_2 \\ x_3 \end{bmatrix} = \begin{bmatrix} x_1 \\ x_2 - x_1 \\ x_3 - x_2 \end{bmatrix}. \qquad (2)$$

The numbers x_1, x_2, x_3 are the components of a vector $\boldsymbol{x}$. The matrix A times the vector $\boldsymbol{x}$ is the **same** as the combination $x_1\boldsymbol{u} + x_2\boldsymbol{v} + x_3\boldsymbol{w}$ of the three columns in equation (1).

This is more than a definition of $A\boldsymbol{x}$, because the rewriting brings a crucial change in viewpoint. At first, the numbers x_1, x_2, x_3 were multiplying the vectors. Now the matrix is multiplying those numbers. **The matrix A acts on the vector $\boldsymbol{x}$.** The output $A\boldsymbol{x}$ is a **combination $\boldsymbol{b}$ of the columns of** A.

To see that action, I will write b_1, b_2, b_3 for the components of $A\boldsymbol{x}$:

$$A\boldsymbol{x} = \begin{bmatrix} 1 & 0 & 0 \\ -1 & 1 & 0 \\ 0 & -1 & 1 \end{bmatrix} \begin{bmatrix} x_1 \\ x_2 \\ x_3 \end{bmatrix} = \begin{bmatrix} \boldsymbol{x_1} \\ \boldsymbol{x_2 - x_1} \\ \boldsymbol{x_3 - x_2} \end{bmatrix} = \begin{bmatrix} b_1 \\ b_2 \\ b_3 \end{bmatrix} = \boldsymbol{b}. \qquad (3)$$

The input is $\boldsymbol{x}$ and the output is $\boldsymbol{b} = A\boldsymbol{x}$. This A is a "**difference matrix**" because $\boldsymbol{b}$ contains differences of the input vector $\boldsymbol{x}$. The top difference is $x_1 - x_0 = x_1 - 0$.

Here is an example to show differences of $\boldsymbol{x} = (1, 4, 9)$: squares in $\boldsymbol{x}$, odd numbers in $\boldsymbol{b}$.

$$\boldsymbol{x} = \begin{bmatrix} 1 \\ 4 \\ 9 \end{bmatrix} = \text{squares} \qquad A\boldsymbol{x} = \begin{bmatrix} 1-0 \\ 4-1 \\ 9-4 \end{bmatrix} = \begin{bmatrix} 1 \\ 3 \\ 5 \end{bmatrix} = \boldsymbol{b}. \qquad (4)$$

That pattern would continue for a 4 by 4 difference matrix. The next square would be $x_4 = 16$. The next difference would be $x_4 - x_3 = 16 - 9 = 7$ (the next odd number). The matrix finds all the differences $1, 3, 5, 7$ at once.

Important Note: Multiplication a row at a time. You may already have learned about multiplying $A\boldsymbol{x}$, a matrix times a vector. Probably it was explained differently, using the rows instead of the columns. The usual way takes the dot product of each row with $\boldsymbol{x}$:

$$\begin{array}{l} A\boldsymbol{x} \textbf{ is also} \\ \textbf{dot products} \\ \textbf{with rows} \end{array} \quad A\boldsymbol{x} = \begin{bmatrix} 1 & 0 & 0 \\ -1 & 1 & 0 \\ 0 & -1 & 1 \end{bmatrix} \begin{bmatrix} x_1 \\ x_2 \\ x_3 \end{bmatrix} = \begin{bmatrix} (1,\ 0,\ 0) \cdot (x_1, x_2, x_3) \\ (-1, 1, 0) \cdot (x_1, x_2, x_3) \\ (0, -1, 1) \cdot (x_1, x_2, x_3) \end{bmatrix}. \qquad (5)$$

Those dot products are the same x_1 and $x_2 - x_1$ and $x_3 - x_2$ that we wrote in equation (3). The new way is to work with $A\boldsymbol{x}$ *a column at a time*. Linear combinations are the key to linear algebra, and the output $A\boldsymbol{x}$ is a linear combination of the **columns** of A.

With numbers, you can multiply $A\boldsymbol{x}$ by rows. With letters, columns are the good way. Chapter 2 will repeat these rules of matrix multiplication, and explain the ideas.

Linear Equations

One more change in viewpoint is crucial. Up to now, the numbers x_1, x_2, x_3 were known. The right hand side $\boldsymbol{b}$ was not known. We found that vector of differences by multiplying A times $\boldsymbol{x}$. **Now we think of $\boldsymbol{b}$ as known and we look for $\boldsymbol{x}$.**

Old question: Compute the linear combination $x_1\boldsymbol{u} + x_2\boldsymbol{v} + x_3\boldsymbol{w}$ to find $\boldsymbol{b}$.
New question: Which combination of $\boldsymbol{u}, \boldsymbol{v}, \boldsymbol{w}$ produces a particular vector $\boldsymbol{b}$?

This is the *inverse problem*—to find the input $\boldsymbol{x}$ that gives the desired output $\boldsymbol{b} = A\boldsymbol{x}$. You have seen this before, as a system of linear equations for x_1, x_2, x_3. The right hand sides of the equations are b_1, b_2, b_3. I will now solve that system $A\boldsymbol{x} = \boldsymbol{b}$ to find x_1, x_2, x_3:

Equations $Ax = b$	$\begin{aligned} x_1 \quad\quad\; &= b_1 \\ -x_1 + x_2 \quad &= b_2 \\ -x_2 + x_3 &= b_3 \end{aligned}$	**Solution** $x = A^{-1}b$	$\begin{aligned} x_1 &= b_1 \\ x_2 &= b_1 + b_2 \\ x_3 &= b_1 + b_2 + b_3. \end{aligned}$	(6)

Let me admit right away—most linear systems are not so easy to solve. In this example, the first equation decided $x_1 = b_1$. Then the second equation produced $x_2 = b_1 + b_2$. *The equations can be solved in order* (top to bottom) *because A is a triangular matrix.*

Look at two specific choices 0, 0, 0 and 1, 3, 5 of the right sides b_1, b_2, b_3:

$$\boldsymbol{b} = \begin{bmatrix} 0 \\ 0 \\ 0 \end{bmatrix} \text{ gives } \boldsymbol{x} = \begin{bmatrix} 0 \\ 0 \\ 0 \end{bmatrix} \qquad \boldsymbol{b} = \begin{bmatrix} 1 \\ 3 \\ 5 \end{bmatrix} \text{ gives } \boldsymbol{x} = \begin{bmatrix} 1 \\ 1+3 \\ 1+3+5 \end{bmatrix} = \begin{bmatrix} 1 \\ 4 \\ 9 \end{bmatrix}.$$

The first solution (all zeros) is more important than it looks. In words: *If the output is $\boldsymbol{b} = \mathbf{0}$, then the input must be $\boldsymbol{x} = \mathbf{0}$.* That statement is true for this matrix A. It is not true for all matrices. Our second example will show (for a different matrix C) how we can have $C\boldsymbol{x} = \mathbf{0}$ when $C \neq 0$ and $\boldsymbol{x} \neq \mathbf{0}$.

This matrix A is "**invertible**". From $\boldsymbol{b}$ we can recover $\boldsymbol{x}$. We write $\boldsymbol{x}$ as $A^{-1}\boldsymbol{b}$.

The Inverse Matrix

Let me repeat the solution $\boldsymbol{x}$ in equation (6). A sum matrix will appear!

$$A\boldsymbol{x} = \boldsymbol{b} \text{ is solved by } \begin{bmatrix} x_1 \\ x_2 \\ x_3 \end{bmatrix} = \begin{bmatrix} b_1 \\ b_1 + b_2 \\ b_1 + b_2 + b_3 \end{bmatrix} = \begin{bmatrix} 1 & 0 & 0 \\ 1 & 1 & 0 \\ 1 & 1 & 1 \end{bmatrix} \begin{bmatrix} b_1 \\ b_2 \\ b_3 \end{bmatrix}. \qquad (7)$$

If the differences of the x's are the b's, the sums of the b's are the x's. That was true for the odd numbers $\boldsymbol{b} = (1, 3, 5)$ and the squares $\boldsymbol{x} = (1, 4, 9)$. It is true for all vectors. **The sum matrix in equation (7) is the inverse A^{-1} of the difference matrix A.**

Example: The differences of $\boldsymbol{x} = (1, 2, 3)$ are $\boldsymbol{b} = (1, 1, 1)$. So $\boldsymbol{b} = A\boldsymbol{x}$ and $\boldsymbol{x} = A^{-1}\boldsymbol{b}$:

$$A\boldsymbol{x} = \begin{bmatrix} 1 & 0 & 0 \\ -1 & 1 & 0 \\ 0 & -1 & 1 \end{bmatrix} \begin{bmatrix} \mathbf{1} \\ \mathbf{2} \\ \mathbf{3} \end{bmatrix} = \begin{bmatrix} \mathbf{1} \\ \mathbf{1} \\ \mathbf{1} \end{bmatrix} \qquad A^{-1}\boldsymbol{b} = \begin{bmatrix} 1 & 0 & 0 \\ 1 & 1 & 0 \\ 1 & 1 & 1 \end{bmatrix} \begin{bmatrix} \mathbf{1} \\ \mathbf{1} \\ \mathbf{1} \end{bmatrix} = \begin{bmatrix} \mathbf{1} \\ \mathbf{2} \\ \mathbf{3} \end{bmatrix}$$

Equation (7) for the solution vector $\boldsymbol{x} = (x_1, x_2, x_3)$ tells us two important facts:

1. For every $\boldsymbol{b}$ there is one solution to $A\boldsymbol{x} = \boldsymbol{b}$. **2.** The matrix A^{-1} produces $\boldsymbol{x} = A^{-1}\boldsymbol{b}$.

The next chapters ask about other equations $A\boldsymbol{x} = \boldsymbol{b}$. Is there a solution? How to find it?

Note on calculus. Let me connect these special matrices to calculus. The vector $\boldsymbol{x}$ changes to a function $x(t)$. The differences $A\boldsymbol{x}$ become the *derivative* $dx/dt = b(t)$. In the inverse direction, the sums $A^{-1}\boldsymbol{b}$ become the *integral* of $b(t)$. **Sums of differences are like integrals of derivatives.**

The Fundamental Theorem of Calculus says : ***integration is the inverse of differentiation*** .

$$\boldsymbol{Ax = b \text{ and } x = A^{-1}b} \qquad \frac{dx}{dt} = b \text{ and } x(t) = \int_0^t b\,dt. \tag{8}$$

The differences of squares $0, 1, 4, 9$ are odd numbers $1, 3, 5$. The derivative of $x(t) = t^2$ is $2t$. A perfect analogy would have produced the even numbers $b = 2, 4, 6$ at times $t = 1, 2, 3$. But differences are not the same as derivatives, and our matrix A produces not $2t$ but $2t - 1$:

$$\textbf{Backward} \qquad x(t) - x(t-1) = t^2 - (t-1)^2 = t^2 - (t^2 - 2t + 1) = 2t - 1. \tag{9}$$

The Problem Set will follow up to show that "forward differences" produce $2t + 1$. The best choice (not always seen in calculus courses) is a **centered difference** that uses $x(t+1) - x(t-1)$. Divide that Δx by the distance Δt from $t-1$ to $t+1$, which is 2:

$$\textbf{Centered difference of } \boldsymbol{x(t) = t^2} \qquad \frac{(t+1)^2 - (t-1)^2}{2} = 2t \quad \text{exactly.} \tag{10}$$

Difference matrices are great. Centered is the best. Our second example is *not invertible*.

Cyclic Differences

This example keeps the same columns $\boldsymbol{u}$ and $\boldsymbol{v}$ but changes $\boldsymbol{w}$ to a new vector $\boldsymbol{w}^*$:

$$\textbf{Second example} \qquad \boldsymbol{u} = \begin{bmatrix} 1 \\ -1 \\ 0 \end{bmatrix} \quad \boldsymbol{v} = \begin{bmatrix} 0 \\ 1 \\ -1 \end{bmatrix} \quad \boldsymbol{w}^* = \begin{bmatrix} \mathbf{-1} \\ 0 \\ 1 \end{bmatrix}.$$

Now the linear combinations of $\boldsymbol{u}$, $\boldsymbol{v}$, $\boldsymbol{w}^*$ lead to a **cyclic difference matrix** C:

$$\textbf{Cyclic} \qquad C\boldsymbol{x} = \begin{bmatrix} 1 & 0 & \mathbf{-1} \\ -1 & 1 & 0 \\ 0 & -1 & 1 \end{bmatrix} \begin{bmatrix} x_1 \\ x_2 \\ x_3 \end{bmatrix} = \begin{bmatrix} x_1 - x_3 \\ x_2 - x_1 \\ x_3 - x_2 \end{bmatrix} = \boldsymbol{b}. \tag{11}$$

This matrix C is not triangular. It is not so simple to solve for $\boldsymbol{x}$ when we are given $\boldsymbol{b}$. Actually it is impossible to find *the* solution to $C\boldsymbol{x} = \boldsymbol{b}$, because the three equations either have **infinitely many solutions** (sometimes) or else **no solution** (usually) :

$$\begin{array}{l} C\boldsymbol{x} = \mathbf{0} \\ \textbf{Infinitely} \\ \textbf{many } \boldsymbol{x} \end{array} \qquad \begin{bmatrix} x_1 - x_3 \\ x_2 - x_1 \\ x_3 - x_2 \end{bmatrix} = \begin{bmatrix} 0 \\ 0 \\ 0 \end{bmatrix} \text{ is solved by all vectors } \begin{bmatrix} x_1 \\ x_2 \\ x_3 \end{bmatrix} = \begin{bmatrix} c \\ c \\ c \end{bmatrix}. \tag{12}$$

Every constant vector like $\boldsymbol{x} = (3, 3, 3)$ has zero differences when we go cyclically. The undetermined constant c is exactly like the $+C$ that we add to integrals. The cyclic differences cycle around to $x_1 - x_3$ in the first component, instead of starting from $x_0 = 0$.

The more likely possibility for $Cx = b$ is **no solution** x at all:

$$Cx = b \qquad \begin{bmatrix} x_1 - x_3 \\ x_2 - x_1 \\ x_3 - x_2 \end{bmatrix} = \begin{bmatrix} 1 \\ 3 \\ 5 \end{bmatrix} \qquad \begin{array}{l} \text{Left sides add to } 0 \\ \text{Right sides add to } 9 \\ \textit{No solution } x_1, x_2, x_3 \end{array} \qquad (13)$$

Look at this example geometrically. No combination of u, v, and w^* will produce the vector $b = (1, 3, 5)$. The combinations don't fill the whole three-dimensional space. The right sides must have $b_1 + b_2 + b_3 = 0$ to allow a solution to $Cx = b$, because the left sides $x_1 - x_3$, $x_2 - x_1$, and $x_3 - x_2$ always add to zero. Put that in different words :

All linear combinations $x_1 u + x_2 v + x_3 w^*$ **lie on the plane given by** $b_1 + b_2 + b_3 = 0$.

This subject is suddenly connecting algebra with geometry. Linear combinations can fill all of space, or only a plane. We need a picture to show the crucial difference between u, v, w (the first example) and u, v, w^* (all in the same plane).

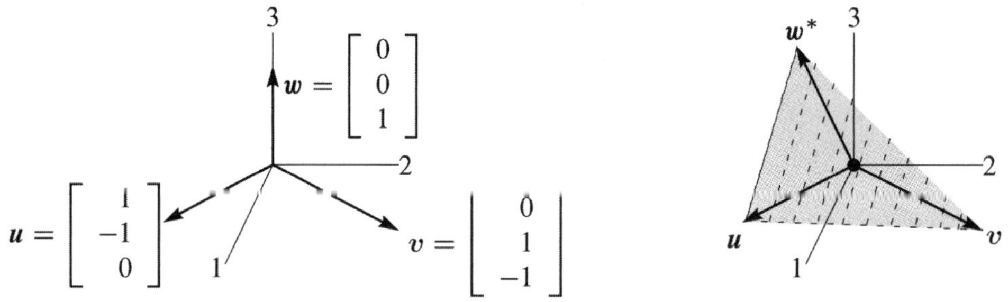

Figure 1.10: Independent vectors u, v, w. Dependent vectors u, v, w^* in a plane.

Independence and Dependence

Figure 1.10 shows those column vectors, first of the matrix A and then of C. The first two columns u and v are the same in both pictures. If we only look at the combinations of those two vectors, we will get a two-dimensional plane. **The key question is whether the third vector is in that plane**:

Independence w is not in the plane of u and v.
Dependence w^* *is* in the plane of u and v.

The important point is that the new vector w^* is a linear combination of u and v:

$$u + v + w^* = 0 \qquad w^* = \begin{bmatrix} -1 \\ 0 \\ 1 \end{bmatrix} = -u - v. \qquad (14)$$

All three vectors $\boldsymbol{u}, \boldsymbol{v}, \boldsymbol{w}^*$ have components adding to zero. Then all their combinations will have $b_1 + b_2 + b_3 = 0$ (as we saw above, by adding the three equations). This is the equation for the plane containing all combinations of $\boldsymbol{u}$ and $\boldsymbol{v}$. By including $\boldsymbol{w}^*$ we get *no new vectors* because $\boldsymbol{w}^*$ is already on that plane.

The original $\boldsymbol{w} = (0, 0, 1)$ is not on the plane: $0 + 0 + 1 \neq 0$. The combinations of $\boldsymbol{u}, \boldsymbol{v}, \boldsymbol{w}$ fill the whole three-dimensional space. We know this already, because the solution $\boldsymbol{x} = A^{-1}\boldsymbol{b}$ in equation (6) gave the right combination to produce any $\boldsymbol{b}$.

The two matrices A and C, with third columns $\boldsymbol{w}$ and $\boldsymbol{w}^*$, allowed me to mention two key words of linear algebra: independence and dependence. The first half of the course will develop these ideas much further—I am happy if you see them early in the two examples:

$\boldsymbol{u}, \boldsymbol{v}, \boldsymbol{w}$ are **independent**. No combination except $0\boldsymbol{u} + 0\boldsymbol{v} + 0\boldsymbol{w} = \boldsymbol{0}$ gives $\boldsymbol{b} = \boldsymbol{0}$.

$\boldsymbol{u}, \boldsymbol{v}, \boldsymbol{w}^*$ are **dependent**. Other combinations like $\boldsymbol{u} + \boldsymbol{v} + \boldsymbol{w}^*$ give $\boldsymbol{b} = \boldsymbol{0}$.

You can picture this in three dimensions. The three vectors lie in a plane or they don't. Chapter 2 has n vectors in n-dimensional space. *Independence or dependence* is the key point. The vectors go into the columns of an n by n matrix:

Independent columns: $A\boldsymbol{x} = \boldsymbol{0}$ has one solution. A is an **invertible matrix**.

Dependent columns: $C\boldsymbol{x} = \boldsymbol{0}$ has many solutions. C is a **singular matrix**.

Eventually we will have n vectors in m-dimensional space. The matrix A with those n columns is now *rectangular* (m by n). Understanding $A\boldsymbol{x} = \boldsymbol{b}$ is the problem of Chapter 3.

■ REVIEW OF THE KEY IDEAS ■

1. **Matrix times vector**: $A\boldsymbol{x} =$ **combination of the columns of** A.

2. The solution to $A\boldsymbol{x} = \boldsymbol{b}$ is $\boldsymbol{x} = A^{-1}\boldsymbol{b}$, when A is an invertible matrix.

3. The cyclic matrix C has no inverse. Its three columns lie in the same plane. Those dependent columns add to the zero vector. $C\boldsymbol{x} = \boldsymbol{0}$ has many solutions.

4. This section is looking ahead to key ideas, not fully explained yet.

■ WORKED EXAMPLES ■

1.3 A Change the southwest entry a_{31} of A (row 3, column 1) to $a_{31} = \mathbf{1}$:

$$A\boldsymbol{x} = \boldsymbol{b} \qquad \begin{bmatrix} 1 & 0 & 0 \\ -1 & 1 & 0 \\ \mathbf{1} & -1 & 1 \end{bmatrix} \begin{bmatrix} x_1 \\ x_2 \\ x_3 \end{bmatrix} = \begin{bmatrix} x_1 \\ -x_1 + x_2 \\ \boldsymbol{x_1} - x_2 + x_3 \end{bmatrix} = \begin{bmatrix} b_1 \\ b_2 \\ b_3 \end{bmatrix}.$$

Find the solution x for any b. From $x = A^{-1}b$ read off the inverse matrix A^{-1}.

Solution Solve the (linear triangular) system $A\boldsymbol{x} = \boldsymbol{b}$ from top to bottom:

$$\begin{array}{ll} \text{first} & x_1 = b_1 \\ \text{then} & x_2 = b_1 + b_2 \\ \text{then} & x_3 = \phantom{b_1 +{}} b_2 + b_3 \end{array} \qquad \text{This says that } \boldsymbol{x} = A^{-1}\boldsymbol{b} = \begin{bmatrix} 1 & 0 & 0 \\ 1 & 1 & 0 \\ 0 & 1 & 1 \end{bmatrix} \begin{bmatrix} b_1 \\ b_2 \\ b_3 \end{bmatrix}$$

This is good practice to see the columns of the inverse matrix multiplying b_1, b_2, and b_3. The first column of A^{-1} is the solution for $\boldsymbol{b} = (1, 0, 0)$. The second column is the solution for $\boldsymbol{b} = (0, 1, 0)$. The third column $\boldsymbol{x}$ of A^{-1} is the solution for $A\boldsymbol{x} = \boldsymbol{b} = (0, 0, 1)$.

The three columns of A are still independent. They don't lie in a plane. The combinations of those three columns, using the right weights x_1, x_2, x_3, can produce any three-dimensional vector $\boldsymbol{b} = (b_1, b_2, b_3)$. Those weights come from $\boldsymbol{x} = A^{-1}\boldsymbol{b}$.

1.3 B This E is an **elimination matrix**. E has a subtraction and E^{-1} has an addition.

$$\boldsymbol{b} = E\boldsymbol{x} \qquad \begin{bmatrix} b_1 \\ b_2 \end{bmatrix} = \begin{bmatrix} x_1 \\ x_2 - \ell\, x_1 \end{bmatrix} = \begin{bmatrix} \mathbf{1} & 0 \\ -\boldsymbol{\ell} & \mathbf{1} \end{bmatrix} \begin{bmatrix} x_1 \\ x_2 \end{bmatrix} \qquad E = \begin{bmatrix} \mathbf{1} & 0 \\ -\boldsymbol{\ell} & \mathbf{1} \end{bmatrix}$$

The first equation is $x_1 = b_1$. The second equation is $x_2 - \ell x_1 = b_2$. The inverse will *add* ℓb_1 to b_2, because the elimination matrix *subtracted* :

$$\boldsymbol{x} = E^{-1}\boldsymbol{b} \qquad \begin{bmatrix} x_1 \\ x_2 \end{bmatrix} = \begin{bmatrix} b_1 \\ \ell b_1 + b_2 \end{bmatrix} = \begin{bmatrix} \mathbf{1} & 0 \\ \boldsymbol{\ell} & \mathbf{1} \end{bmatrix} \begin{bmatrix} b_1 \\ b_2 \end{bmatrix} \qquad E^{-1} = \begin{bmatrix} \mathbf{1} & 0 \\ \boldsymbol{\ell} & \mathbf{1} \end{bmatrix}$$

1.3 C Change C from a cyclic difference to a **centered difference** producing $x_3 - x_1$:

$$C\boldsymbol{x} = \boldsymbol{b} \qquad \begin{bmatrix} 0 & 1 & 0 \\ -1 & 0 & 1 \\ 0 & -1 & 0 \end{bmatrix} \begin{bmatrix} x_1 \\ x_2 \\ x_3 \end{bmatrix} = \begin{bmatrix} x_2 - 0 \\ x_3 - x_1 \\ 0 - x_2 \end{bmatrix} = \begin{bmatrix} b_1 \\ b_2 \\ b_3 \end{bmatrix}. \tag{15}$$

$C\boldsymbol{x} = \boldsymbol{b}$ can only be solved when $b_1 + b_3 = x_2 - x_2 = 0$. That is a plane of vectors $\boldsymbol{b}$ in three-dimensional space. Each column of C is in the plane, the matrix has no inverse. So this plane contains all combinations of those columns (which are all the vectors $C\boldsymbol{x}$).

I included the zeros so you could see that this C produces "centered differences". Row i of $C\boldsymbol{x}$ is x_{i+1} (*right of center*) minus x_{i-1} (*left of center*). Here is 4 by 4:

$$\begin{array}{l} C\boldsymbol{x} = \boldsymbol{b} \\ \textbf{Centered} \\ \textbf{differences} \end{array} \qquad \begin{bmatrix} 0 & 1 & 0 & 0 \\ -1 & 0 & 1 & 0 \\ 0 & -1 & 0 & 1 \\ 0 & 0 & -1 & 0 \end{bmatrix} \begin{bmatrix} x_1 \\ x_2 \\ x_3 \\ x_4 \end{bmatrix} = \begin{bmatrix} x_2 - 0 \\ x_3 - x_1 \\ x_4 - x_2 \\ 0 - x_3 \end{bmatrix} = \begin{bmatrix} b_1 \\ b_2 \\ b_3 \\ b_4 \end{bmatrix} \tag{16}$$

Surprisingly this matrix is now invertible! The first and last rows tell you x_2 and x_3. Then the middle rows give x_1 and x_4. It is possible to write down the inverse matrix C^{-1}. But 5 by 5 will be singular (*not invertible*) again . . .

Problem Set 1.3

1 Find the linear combination $3\boldsymbol{s}_1 + 4\boldsymbol{s}_2 + 5\boldsymbol{s}_3 = \boldsymbol{b}$. Then write $\boldsymbol{b}$ as a matrix-vector multiplication $S\boldsymbol{x}$, with $3, 4, 5$ in $\boldsymbol{x}$. Compute the three dot products (row of S) $\cdot\, \boldsymbol{x}$:

$$\boldsymbol{s}_1 = \begin{bmatrix} 1 \\ 1 \\ 1 \end{bmatrix} \quad \boldsymbol{s}_2 = \begin{bmatrix} 0 \\ 1 \\ 1 \end{bmatrix} \quad \boldsymbol{s}_3 = \begin{bmatrix} 0 \\ 0 \\ 1 \end{bmatrix} \text{ go into the columns of } S.$$

2 Solve these equations $S\boldsymbol{y} = \boldsymbol{b}$ with $\boldsymbol{s}_1, \boldsymbol{s}_2, \boldsymbol{s}_3$ in the columns of S:

$$\begin{bmatrix} 1 & 0 & 0 \\ 1 & 1 & 0 \\ 1 & 1 & 1 \end{bmatrix} \begin{bmatrix} y_1 \\ y_2 \\ y_3 \end{bmatrix} = \begin{bmatrix} 1 \\ 1 \\ 1 \end{bmatrix} \text{ and } \begin{bmatrix} 1 & 0 & 0 \\ 1 & 1 & 0 \\ 1 & 1 & 1 \end{bmatrix} \begin{bmatrix} y_1 \\ y_2 \\ y_3 \end{bmatrix} = \begin{bmatrix} 1 \\ 4 \\ 9 \end{bmatrix}.$$

S is a sum matrix. The sum of the first 5 odd numbers is ____.

3 Solve these three equations for y_1, y_2, y_3 in terms of c_1, c_2, c_3:

$$S\boldsymbol{y} = \boldsymbol{c} \qquad \begin{bmatrix} 1 & 0 & 0 \\ 1 & 1 & 0 \\ 1 & 1 & 1 \end{bmatrix} \begin{bmatrix} y_1 \\ y_2 \\ y_3 \end{bmatrix} = \begin{bmatrix} c_1 \\ c_2 \\ c_3 \end{bmatrix}.$$

Write the solution $\boldsymbol{y}$ as a matrix $A = S^{-1}$ times the vector $\boldsymbol{c}$. Are the columns of S independent or dependent?

4 Find a combination $x_1\boldsymbol{w}_1 + x_2\boldsymbol{w}_2 + x_3\boldsymbol{w}_3$ that gives the zero vector with $x_1 = 1$:

$$\boldsymbol{w}_1 = \begin{bmatrix} 1 \\ 2 \\ 3 \end{bmatrix} \quad \boldsymbol{w}_2 = \begin{bmatrix} 4 \\ 5 \\ 6 \end{bmatrix} \quad \boldsymbol{w}_3 = \begin{bmatrix} 7 \\ 8 \\ 9 \end{bmatrix}.$$

Those vectors are (independent) (dependent). The three vectors lie in a ____. The matrix W with those three columns is *not invertible*.

5 The rows of that matrix W produce three vectors (*I write them as columns*):

$$\boldsymbol{r}_1 = \begin{bmatrix} 1 \\ 4 \\ 7 \end{bmatrix} \quad \boldsymbol{r}_2 = \begin{bmatrix} 2 \\ 5 \\ 8 \end{bmatrix} \quad \boldsymbol{r}_3 = \begin{bmatrix} 3 \\ 6 \\ 9 \end{bmatrix}.$$

Linear algebra says that these vectors must also lie in a plane. There must be many combinations with $y_1\boldsymbol{r}_1 + y_2\boldsymbol{r}_2 + y_3\boldsymbol{r}_3 = \boldsymbol{0}$. Find two sets of y's.

6 Which numbers c give dependent columns so a combination of columns equals zero ?

$$\begin{bmatrix} 1 & 1 & 0 \\ 3 & 2 & 1 \\ 7 & 4 & c \end{bmatrix} \qquad \begin{bmatrix} 1 & 0 & c \\ 1 & 1 & 0 \\ 0 & 1 & 1 \end{bmatrix} \qquad \begin{bmatrix} c & c & c \\ 2 & 1 & 5 \\ 3 & 3 & 6 \end{bmatrix} \quad \begin{matrix} \text{maybe} \\ \text{always} \\ \text{independent for } c \neq 0\,? \end{matrix}$$

7 If the columns combine into $A\boldsymbol{x} = \mathbf{0}$ then each of the rows has $\boldsymbol{r} \cdot \boldsymbol{x} = 0$:

$$\begin{bmatrix} \boldsymbol{a}_1 & \boldsymbol{a}_2 & \boldsymbol{a}_3 \end{bmatrix} \begin{bmatrix} x_1 \\ x_2 \\ x_3 \end{bmatrix} = \begin{bmatrix} 0 \\ 0 \\ 0 \end{bmatrix} \qquad \textbf{By rows} \begin{bmatrix} \boldsymbol{r}_1 \cdot \boldsymbol{x} \\ \boldsymbol{r}_2 \cdot \boldsymbol{x} \\ \boldsymbol{r}_3 \cdot \boldsymbol{x} \end{bmatrix} = \begin{bmatrix} 0 \\ 0 \\ 0 \end{bmatrix}.$$

The three rows also lie in a plane. Why is that plane perpendicular to $\boldsymbol{x}$?

8 Moving to a 4 by 4 difference equation $A\boldsymbol{x} = \boldsymbol{b}$, find the four components x_1, x_2, x_3, x_4. Then write this solution as $\boldsymbol{x} = A^{-1}\boldsymbol{b}$ to find the inverse matrix:

$$A\boldsymbol{x} = \begin{bmatrix} 1 & 0 & 0 & 0 \\ -1 & 1 & 0 & 0 \\ 0 & -1 & 1 & 0 \\ 0 & 0 & -1 & 1 \end{bmatrix} \begin{bmatrix} x_1 \\ x_2 \\ x_3 \\ x_4 \end{bmatrix} = \begin{bmatrix} b_1 \\ b_2 \\ b_3 \\ b_4 \end{bmatrix} = \boldsymbol{b}.$$

9 What is the *cyclic* 4 by 4 difference matrix C? It will have 1 and -1 in each row and each column. Find all solutions $\boldsymbol{x} = (x_1, x_2, x_3, x_4)$ to $C\boldsymbol{x} = \mathbf{0}$. The four columns of C lie in a "three-dimensional hyperplane" inside four-dimensional space.

10 A *forward* difference matrix Δ is *upper* triangular:

$$\Delta \boldsymbol{z} = \begin{bmatrix} -1 & 1 & 0 \\ 0 & -1 & 1 \\ 0 & 0 & -1 \end{bmatrix} \begin{bmatrix} z_1 \\ z_2 \\ z_3 \end{bmatrix} = \begin{bmatrix} z_2 - z_1 \\ z_3 - z_2 \\ 0 - z_3 \end{bmatrix} = \begin{bmatrix} b_1 \\ b_2 \\ b_3 \end{bmatrix} = \boldsymbol{b}.$$

Find z_1, z_2, z_3 from b_1, b_2, b_3. What is the inverse matrix in $\boldsymbol{z} = \Delta^{-1}\boldsymbol{b}$?

11 Show that the forward differences $(t+1)^2 - t^2$ are $2t+1$ = *odd numbers*. As in calculus, the difference $(t+1)^n - t^n$ will begin with the derivative of t^n, which is ______.

12 The last lines of the Worked Example say that the 4 by 4 centered difference matrix in (16) *is* invertible. Solve $C\boldsymbol{x} = (b_1, b_2, b_3, b_4)$ to find its inverse in $\boldsymbol{x} = C^{-1}\boldsymbol{b}$.

Challenge Problems

13 The very last words say that the 5 by 5 centered difference matrix *is not* invertible. Write down the 5 equations $C\boldsymbol{x} = \boldsymbol{b}$. Find a combination of left sides that gives zero. What combination of b_1, b_2, b_3, b_4, b_5 must be zero? (The 5 columns lie on a "4-dimensional hyperplane" in 5-dimensional space. *Hard to visualize.*)

14 If (a, b) is a multiple of (c, d) with $abcd \neq 0$, *show that* (a, c) *is a multiple of* (b, d). This is surprisingly important; two columns are falling on one line. You could use numbers first to see how a, b, c, d are related. The question will lead to:

If $\begin{bmatrix} a & b \\ c & d \end{bmatrix}$ has dependent rows, then it also has dependent columns.

Chapter 2

Solving Linear Equations

2.1 Vectors and Linear Equations

1 The **column picture of** $A\boldsymbol{x} = \boldsymbol{b}$: a combination of n columns of A produces the vector $\boldsymbol{b}$.

2 This is a vector equation $A\boldsymbol{x} = x_1\boldsymbol{a}_1 + \cdots + x_n\boldsymbol{a}_n = \boldsymbol{b}$: the columns of A are $\boldsymbol{a}_1, \boldsymbol{a}_2, \ldots, \boldsymbol{a}_n$.

3 When $\boldsymbol{b} = \boldsymbol{0}$, a combination $A\boldsymbol{x}$ of the columns is *zero*: one possibility is $\boldsymbol{x} = (0, \ldots, 0)$.

4 The **row picture of** $A\boldsymbol{x} = \boldsymbol{b}$: m equations from m rows give m planes meeting at $\boldsymbol{x}$.

5 A dot product gives the equation of each plane: $(\textbf{row 1}) \cdot \boldsymbol{x} = b_1, \ \ldots, (\textbf{row } \boldsymbol{m}) \cdot \boldsymbol{x} = b_m$.

6 When $\boldsymbol{b} = \boldsymbol{0}$, all the planes $(\textbf{row } \boldsymbol{i}) \cdot \boldsymbol{x} = 0$ go through the center point $\boldsymbol{x} = (0, 0, \ \ldots, \ 0)$.

The central problem of linear algebra is to solve a system of equations. Those equations are linear, which means that the unknowns are only multiplied by numbers—we never see x times y. Our first linear system is small. But you will see how far it leads:

Two equations
Two unknowns

$$\begin{aligned} x \ - \ 2y &= 1 \\ 3x \ + \ 2y &= 11 \end{aligned} \qquad (1)$$

We begin *a row at a time.* The first equation $\boldsymbol{x} - \boldsymbol{2y} = \boldsymbol{1}$ produces a straight line in the xy plane. The point $x = 1, y = 0$ is on the line because it solves that equation. The point $x = 3, y = 1$ is also on the line because $3 - 2 = 1$. If we choose $x = 101$ we find $y = 50$.

The slope of this particular line is $\frac{1}{2}$, because y increases by 1 when x changes by 2. But slopes are important in calculus and this is linear algebra!

Figure 2.1 will show that first line $x - 2y = 1$. The second line in this "row picture" comes from the second equation $3x + 2y = 11$. You can't miss the point $x = 3, y = 1$ where the two lines meet. *That point* $(3, 1)$ *lies on both lines and solves both equations.*

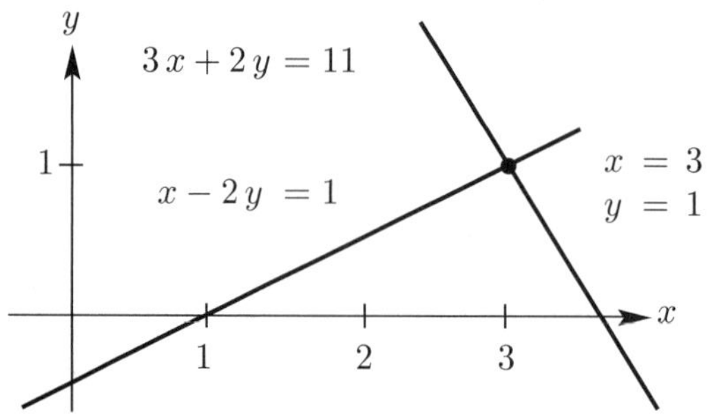

Figure 2.1: *Row picture*: The point $(3, 1)$ where the lines meet solves both equations.

ROWS ***The row picture shows two lines meeting at a single point (the solution).***

Turn now to the column picture. I want to recognize the same linear system as a "vector equation". Instead of numbers we need to see *vectors*. If you separate the original system into its columns instead of its rows, you get a vector equation:

Combination equals b $$x \begin{bmatrix} 1 \\ 3 \end{bmatrix} + y \begin{bmatrix} -2 \\ 2 \end{bmatrix} = \begin{bmatrix} 1 \\ 11 \end{bmatrix} = \boldsymbol{b}. \qquad (2)$$

This has two column vectors on the left side. The problem is *to find the combination of those vectors that equals the vector on the right.* We are multiplying the first column by x and the second column by y, and adding. With the right choices $x = 3$ and $y = 1$ (the same numbers as before), this produces 3 (***column 1***) $+ 1$ (***column 2***) $= \boldsymbol{b}$.

COLUMNS ***The column picture combines the column vectors on the left side to produce the vector b on the right side.***

Figure 2.2 is the "column picture" of two equations in two unknowns. The first part shows the two separate columns, and that first column multiplied by 3. This multiplication by a *scalar* (a number) is one of the two basic operations in linear algebra:

Scalar multiplication $$3 \begin{bmatrix} 1 \\ 3 \end{bmatrix} = \begin{bmatrix} 3 \\ 9 \end{bmatrix}.$$

If the components of a vector $\boldsymbol{v}$ are v_1 and v_2, then $c\boldsymbol{v}$ has components cv_1 and cv_2.

The other basic operation is *vector addition*. We add the first components and the second components separately. The vector sum is $(1, 11)$, the desired vector $\boldsymbol{b}$.

Vector addition $$\begin{bmatrix} 3 \\ 9 \end{bmatrix} + \begin{bmatrix} -2 \\ 2 \end{bmatrix} = \begin{bmatrix} 1 \\ 11 \end{bmatrix}.$$

The right side of Figure 2.2 shows this addition. Two vectors are in black. The sum along the diagonal is the vector $\boldsymbol{b} = (1, 11)$ on the right side of the linear equations.

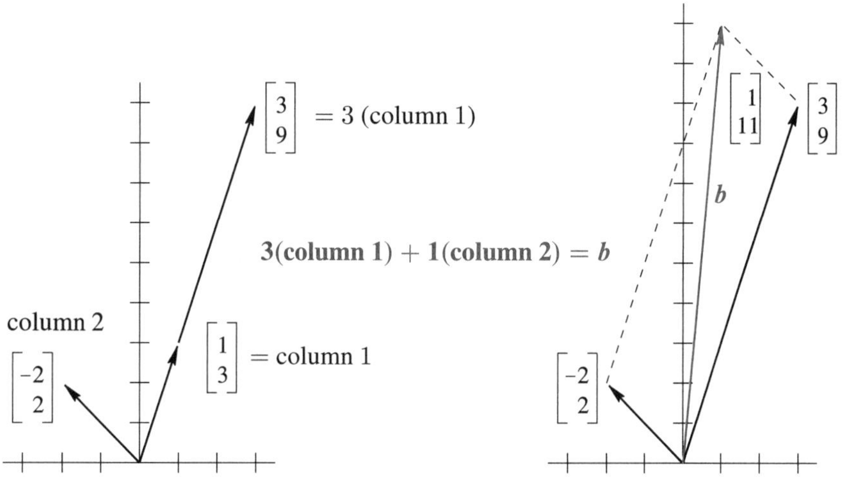

Figure 2.2: *Column picture*: A combination of columns produces the right side (1, 11).

To repeat: The left side of the vector equation is a ***linear combination*** of the columns. The problem is to find the right coefficients $x = 3$ and $y = 1$. We are combining scalar multiplication and vector addition into one step. That step is crucially important, because it contains both of the basic operations: ***Multiply by* 3 *and* 1*, then add.***

$$\textbf{Linear combination} \qquad 3\begin{bmatrix}1\\3\end{bmatrix}+\begin{bmatrix}-2\\2\end{bmatrix}=\begin{bmatrix}1\\11\end{bmatrix}.$$

Of course the solution $x = 3, y = 1$ is the same as in the row picture. I don't know which picture you prefer! I suspect that the two intersecting lines are more familiar at first. You may like the row picture better, but only for one day. My own preference is to combine column vectors. It is a lot easier to see a combination of four vectors in four-dimensional space, than to visualize how four hyperplanes might possibly meet at a point. (*Even one hyperplane is hard enough. . .*)

The ***coefficient matrix*** on the left side of the equations is the 2 by 2 matrix A:

$$\textbf{Coefficient matrix} \qquad A=\begin{bmatrix}1 & -2\\3 & 2\end{bmatrix}.$$

This is very typical of linear algebra, to look at a matrix by rows and by columns. Its rows give the row picture and its columns give the column picture. Same numbers, different pictures, same equations. We combine those equations into a matrix problem $A\boldsymbol{x} = \boldsymbol{b}$:

$$\begin{matrix}\textbf{Matrix equation}\\ A\boldsymbol{x}=\boldsymbol{b}\end{matrix} \qquad \begin{bmatrix}1 & -2\\3 & 2\end{bmatrix}\begin{bmatrix}x\\y\end{bmatrix}=\begin{bmatrix}1\\11\end{bmatrix}.$$

The row picture deals with the two rows of A. The column picture combines the columns. The numbers $x = 3$ and $y = 1$ go into $\boldsymbol{x}$. Here is matrix-vector multiplication:

Dot products with rows
Combination of columns $\quad A\boldsymbol{x} = \boldsymbol{b}$ is $\quad \begin{bmatrix} 1 & -2 \\ 3 & 2 \end{bmatrix} \begin{bmatrix} 3 \\ 1 \end{bmatrix} = \begin{bmatrix} 1 \\ 11 \end{bmatrix}.$

Looking ahead This chapter is going to solve n equations in n unknowns (for any n). I am not going at top speed, because smaller systems allow examples and pictures and a complete understanding. You are free to go faster, as long as **matrix multiplication and inversion** become clear. Those two ideas will be the keys to invertible matrices.

I can list four steps to understanding elimination using matrices.

1. Elimination goes from A to a triangular U by a sequence of matrix steps E_{ij}.

2. The triangular system is solved by ***back substitution***: working bottom to top.

3. In matrix language A is factored into $LU =$ (lower triangular) (upper triangular).

4. Elimination succeeds if A is invertible. (But it may need row exchanges.)

The most-used algorithm in computational science takes those steps (MATLAB calls it **lu**). Its quickest form is *backslash*: $\boldsymbol{x} = A \backslash \boldsymbol{b}$. But linear algebra goes beyond square invertible matrices! For m by n matrices, $A\boldsymbol{x} = \boldsymbol{0}$ may have many solutions. Those solutions will go into a **vector space**. The **rank** of A leads to the **dimension** of that vector space.

All this comes in Chapter 3, and I don't want to hurry. But I must get there.

Three Equations in Three Unknowns

The three unknowns are x, y, z. We have three linear equations:

$$A\boldsymbol{x} = \boldsymbol{b} \qquad \begin{aligned} x + 2y + 3z &= 6 \\ 2x + 5y + 2z &= 4 \\ 6x - 3y + z &= 2 \end{aligned} \tag{3}$$

We look for numbers x, y, z that solve all three equations at once. Those desired numbers might or might not exist. For this system, they do exist. When the number of unknowns matches the number of equations, in this case $3 = 3$, there is *usually* one solution.

Before solving the problem, we visualize it both ways:

ROW ***The row picture shows three planes meeting at a single point.***

COLUMN ***The column picture combines three columns to produce*** $\mathbf{b} = (6, 4, 2)$.

In the row picture, each equation produces a *plane* in three-dimensional space. The first plane in Figure 2.3 comes from the first equation $x + 2y + 3z = 6$. That plane crosses the x and y and z axes at the points $(6, 0, 0)$ and $(0, 3, 0)$ and $(0, 0, 2)$. Those three points solve the equation and they determine the whole plane.

The vector $(x, y, z) = (0, 0, 0)$ does not solve $x + 2y + 3z = 6$. Therefore that plane does not contain the origin. The plane $x + 2y + 3z = 0$ does pass through the origin, and it is parallel to $x + 2y + 3z = 6$. When the right side increases to 6, the parallel plane moves away from the origin.

The second plane is given by the second equation $2x + 5y + 2z = 4$. *It intersects the first plane in a line* $\boldsymbol{L}$. The usual result of two equations in three unknowns is a line $\boldsymbol{L}$ of solutions. (Not if the equations were $x + 2y + 3z = 6$ and $x + 2y + 3z = 0$.)

The third equation gives a third plane. It cuts the line $\boldsymbol{L}$ at a single point. That point lies on all three planes and it solves all three equations. It is harder to draw this triple intersection point than to imagine it. The three planes meet at the solution (which we haven't found yet). **The column form will now show immediately why** $z = 2$.

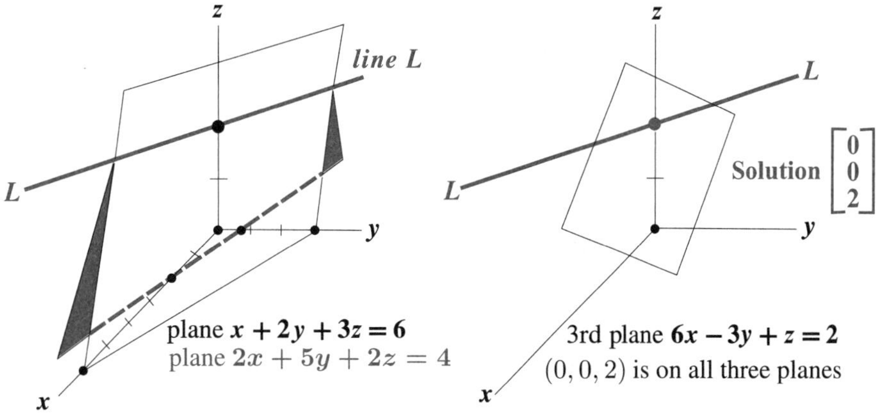

Figure 2.3: *Row picture*: Two planes meet at a line L. Three planes meet at a point.

The column picture starts with the vector form of the equations $A\boldsymbol{x} = \boldsymbol{b}$:

$$\textbf{Combine columns} \qquad x\begin{bmatrix}1\\2\\6\end{bmatrix} + y\begin{bmatrix}2\\5\\-3\end{bmatrix} + z\begin{bmatrix}3\\2\\1\end{bmatrix} = \begin{bmatrix}6\\4\\2\end{bmatrix} = \boldsymbol{b}. \qquad (4)$$

The unknowns are the coefficients x, y, z. We want to multiply the three column vectors by the correct numbers x, y, z to produce $\boldsymbol{b} = (6, 4, 2)$.

Figure 2.4 shows this column picture. Linear combinations of those columns can produce any vector $\boldsymbol{b}$! The combination that produces $\boldsymbol{b} = (6, 4, 2)$ is just 2 times the third column. *The coefficients we need are* $x = 0$, $y = 0$, *and* $z = 2$.

The three planes in the row picture meet at that same solution point $(0, 0, 2)$:

Correct combination
$(x, y, z) = (\mathbf{0}, \mathbf{0}, \mathbf{2})$

$$\mathbf{0}\begin{bmatrix}1\\2\\6\end{bmatrix} + \mathbf{0}\begin{bmatrix}2\\5\\-3\end{bmatrix} + \mathbf{2}\begin{bmatrix}3\\2\\1\end{bmatrix} = \begin{bmatrix}6\\4\\2\end{bmatrix}.$$

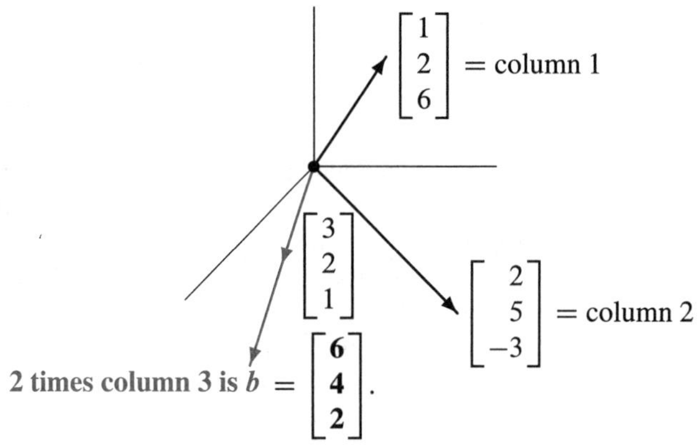

Figure 2.4: *Column picture: Combine the columns with weights* $(x, y, z) = (0, 0, 2)$.

The Matrix Form of the Equations

We have three rows in the row picture and three columns in the column picture (plus the right side). The three rows and three columns contain nine numbers. *These nine numbers fill a* 3 *by* 3 *matrix* A:

$$\textbf{\textit{The "coefficient matrix" in } } A\boldsymbol{x} = \boldsymbol{b} \textbf{\textit{ is}} \qquad A = \begin{bmatrix} 1 & 2 & 3 \\ 2 & 5 & 2 \\ 6 & -3 & 1 \end{bmatrix}.$$

The capital letter A stands for all nine coefficients (in this square array). The letter $\boldsymbol{b}$ denotes the column vector with components $6, 4, 2$. The unknown $\boldsymbol{x}$ is also a column vector, with components x, y, z. (We use boldface because it is a vector, $\boldsymbol{x}$ because it is unknown.) By rows the equations were (3), by columns they were (4), and by matrices they are (5):

$$\textbf{\textit{Matrix equation }} A\boldsymbol{x} = \boldsymbol{b} \qquad \begin{bmatrix} 1 & 2 & 3 \\ 2 & 5 & 2 \\ 6 & -3 & 1 \end{bmatrix} \begin{bmatrix} x \\ y \\ z \end{bmatrix} = \begin{bmatrix} 6 \\ 4 \\ 2 \end{bmatrix}. \tag{5}$$

Basic question: **What does it mean to "multiply A times $\boldsymbol{x}$"?** We can multiply by rows or by columns. Either way, $A\boldsymbol{x} = \boldsymbol{b}$ must be a correct statement of the three equations. You do the same nine multiplications either way.

Multiplication by rows

$A\boldsymbol{x}$ comes from **dot products**, each row times the column $\boldsymbol{x}$:

$$A\boldsymbol{x} = \begin{bmatrix} (\textbf{\textit{row 1}}) \cdot \boldsymbol{x} \\ (\textbf{\textit{row 2}}) \cdot \boldsymbol{x} \\ (\textbf{\textit{row 3}}) \cdot \boldsymbol{x} \end{bmatrix}. \tag{6}$$

Multiplication by columns	Ax is a ***combination of column vectors***: $Ax = x$ ***(column 1)*** $+\ y$ ***(column 2)*** $+\ z$ ***(column 3)***. (7)

When we substitute the solution $\boldsymbol{x} = (0,0,2)$, the multiplication $A\boldsymbol{x}$ produces $\boldsymbol{b}$:

$$\begin{bmatrix} 1 & 2 & 3 \\ 2 & 5 & 2 \\ 6 & -3 & 1 \end{bmatrix} \begin{bmatrix} 0 \\ 0 \\ 2 \end{bmatrix} = \textit{2 times column 3} = \begin{bmatrix} 6 \\ 4 \\ 2 \end{bmatrix}.$$

The dot product from the first row is $(1,2,3) \cdot (0,0,2) = 6$. The other rows give dot products 4 and 2. ***This book sees*** $A\boldsymbol{x}$ ***as a combination of the columns of*** A.

Example 1 Here are 3 by 3 matrices A and I = identity, with three 1's and six 0's:

$$A\boldsymbol{x} = \begin{bmatrix} 1 & 0 & 0 \\ 1 & 0 & 0 \\ 1 & 0 & 0 \end{bmatrix} \begin{bmatrix} 4 \\ 5 \\ 6 \end{bmatrix} = \begin{bmatrix} 4 \\ 4 \\ 4 \end{bmatrix} \qquad I\boldsymbol{x} = \begin{bmatrix} 1 & 0 & 0 \\ 0 & 1 & 0 \\ 0 & 0 & 1 \end{bmatrix} \begin{bmatrix} 4 \\ 5 \\ 6 \end{bmatrix} = \begin{bmatrix} 4 \\ 5 \\ 6 \end{bmatrix}$$

If you are a row person, the dot product of $(1,0,0)$ with $(4,5,6)$ is 4. If you are a column person, the linear combination $A\boldsymbol{x}$ is 4 times the first column $(1,1,1)$. In that matrix A, the second and third columns are zero vectors.

The other matrix I is special. It has ones on the "main diagonal". *Whatever vector this matrix multiplies, that vector is not changed.* This is like multiplication by 1, but for matrices and vectors. The exceptional matrix in this example is the 3 by 3 ***identity matrix***:

$$I = \begin{bmatrix} 1 & 0 & 0 \\ 0 & 1 & 0 \\ 0 & 0 & 1 \end{bmatrix} \quad \text{always yields the multiplication} \quad I\boldsymbol{x} = \boldsymbol{x}\,.$$

Matrix Notation

The first row of a 2 by 2 matrix contains a_{11} and a_{12}. The second row contains a_{21} and a_{22}. The first index gives the row number, so that a_{ij} is an entry in row i. The second index j gives the column number. But those subscripts are not very convenient on a keyboard! Instead of a_{ij} we type $A(i,j)$. ***The entry*** $a_{57} = A(5,7)$ ***would be in row 5, column 7.***

$$A = \begin{bmatrix} a_{11} & a_{12} \\ a_{21} & a_{22} \end{bmatrix} = \begin{bmatrix} A(1,1) & A(1,2) \\ A(2,1) & A(2,2) \end{bmatrix}.$$

For an m by n matrix, the row index i goes from 1 to m. The column index j stops at n. There are mn entries $a_{ij} = A(i,j)$. A square matrix of order n has n^2 entries.

Multiplication in MATLAB

I want to express A and $\boldsymbol{x}$ and their product $A\boldsymbol{x}$ using MATLAB commands. This is a first step in learning that language (and others). I begin by defining A and $\boldsymbol{x}$. A vector $\boldsymbol{x}$ in $\mathbf{R}^n$ is an n by 1 matrix (as in this book). Enter matrices *a row at a time*, and use a semicolon to signal the end of a row. Or enter by columns and transpose by $'$:

$$A = [1 \quad 2 \quad 3; \quad 2 \quad 5 \quad 2; \quad 6 \ -3 \quad 1]$$
$$\boldsymbol{x} = [0 \quad 0 \quad 2]' \quad \textbf{or} \quad \boldsymbol{x} = [0\,;0\,;2]$$

Here are three ways to multiply $A\boldsymbol{x}$ in MATLAB. In reality, $A * \boldsymbol{x}$ is the good way to do it. MATLAB is a high level language, and it works with matrices:

$$\textbf{\textit{Matrix multiplication}} \qquad b = A * x$$

We can also pick out the first row of A (as a smaller matrix !). The notation for that 3 by 3 submatrix is $A(1,:)$. **Here the colon symbol : keeps all columns of row 1**.

$$\textbf{\textit{Row at a time}} \qquad \boldsymbol{b} = [\,A(1,:) * \boldsymbol{x}\,;\ A(2,:) * \boldsymbol{x}\,;\ A(3,:) * \boldsymbol{x}\,]$$

Each entry of $\boldsymbol{b}$ is a dot product, row times column, 1 by 3 matrix times 3 by 1 matrix.

The other way to multiply uses the columns of A. The first column is the 3 by 1 submatrix $A(:,1)$. Now the colon symbol : comes first, *to keep all rows of column* 1. This column multiplies $x(1)$ and the other columns multiply $x(2)$ and $x(3)$:

$$\textbf{\textit{Column at a time}} \qquad \boldsymbol{b} = A(:,1) * x(1) + A(:,2) * x(2) + A(:,3) * x(3)$$

I think that matrices are stored by columns. Then multiplying a column at a time will be a little faster. So $A * \boldsymbol{x}$ is actually executed by columns.

Programming Languages for Mathematics and Statistics

Here are five more important languages and their commands for the multiplication $A\boldsymbol{x}$:

Julia	$A * x$	**julialang.org**
Python	dot(A, x)	**python.org**
R	$A \% * \% x$	**r-project.org**
Mathematica	$A\,.\,x$	**wolfram.com/mathematica**
Maple	$A * x$	**maplesoft.com**

Julia, Python, and **R** are free and open source languages. R is developed particularly for applications in statistics. Other software for statistics (SAS, JMP, and many more) is described on Wikipedia's Comparison of Statistical Packages.

Mathematica and **Maple** allow symbolic entries $a, b, x, \ldots$ and not only real numbers. As in MATLAB's Symbolic Toolbox, they work with symbolic expressions like $x^2 - x$. The power of Mathematica is seen in Wolfram Alpha.

Julia combines the high productivity of SciPy or R for technical computing with performance comparable to C or Fortran. It can call Python and C/Fortran libraries. But it doesn't rely on "vectorized" library functions for speed; Julia is designed to be fast.

I entered **juliabox.org**. I clicked *Sign in via Google* to access my gmail space. Then I clicked *new* at the right and chose a Julia notebook. I chose 0.4.5 and not one under development. The Julia command line came up immediately.

As a novice, I computed $1 + 1$. To see the answer I pressed *Shift+Enter*. I also learned that $1.0 + 1.0$ uses floating point, much faster for a large problem. The website **math.mit.edu/linearalgebra** will show part of the power of Julia and Python and R.

Python is a popular general-purpose programming language. When combined with packages like NumPy and the SciPy library, it provides a full-featured environment for technical computing. NumPy has the basic linear algebra commands. Download the Anaconda Python distribution from **https://www.continuum.io** (a prepackaged collection of Python and most important mathematical libraries, with a graphical installer).

R is free software for statistical computing and graphics. To download and install R, go to **r-project.org** (prefix **https://www.**). Commands are prompted by $>$ and R is a scripted language. It works with lists that can be shaped into vectors and matrices.

It is important to recommend RStudio for editing and graphing (and help resources). When you download from **www.RStudio.com**, a window opens for R commands—plus windows for editing and managing files and plots. Tell R the form of the matrix as well as the list of numerical entries:

$> A =$ matrix $(c\,(1, 2, 3, 2, 5, 2, 6, -3, 1),$ nrow $= 3,$ byrow $=$ TRUE)

$> x =$ matrix $(c\,(0, 0, 2),$ nrow $= 3)$

To see A and x, type their names at the new prompt $>$. To multiply type $b = A\,\%*\%\,x$. Transpose by $t(A)$ and use as.matrix to turn a vector into a matrix.

MATLAB and Julia have a cleaner syntax for matrix computations than R. But R has become very familiar and widely used. The website for this book has space for proper demos (including the *Manipulate* command) of **MATLAB** and **Julia** and **Python** and **R**.

■ REVIEW OF THE KEY IDEAS ■

1. The basic operations on vectors are multiplication $c\boldsymbol{v}$ and vector addition $\boldsymbol{v} + \boldsymbol{w}$.
2. Together those operations give *linear combinations* $c\boldsymbol{v} + d\boldsymbol{w}$.
3. Matrix-vector multiplication $A\boldsymbol{x}$ can be computed by dot products, a row at a time. But $A\boldsymbol{x}$ must be understood as a *combination of the columns of* A.
4. Column picture: $A\boldsymbol{x} = \boldsymbol{b}$ asks for a combination of columns to produce $\boldsymbol{b}$.
5. Row picture: Each equation in $A\boldsymbol{x} = \boldsymbol{b}$ gives a line $(n = 2)$ or a plane $(n = 3)$ or a "hyperplane" $(n > 3)$. They intersect at the solution or solutions, if any.

■ WORKED EXAMPLES ■

2.1 A Describe the column picture of these three equations $A\boldsymbol{x} = \boldsymbol{b}$. Solve by careful inspection of the columns (instead of elimination):

$$\begin{aligned} x + 3y + 2z &= -3 \\ 2x + 2y + 2z &= -2 \\ 3x + 5y + 6z &= -5 \end{aligned} \qquad \text{which is} \qquad \begin{bmatrix} 1 & \mathbf{3} & 2 \\ 2 & \mathbf{2} & 2 \\ 3 & \mathbf{5} & 6 \end{bmatrix} \begin{bmatrix} x \\ y \\ z \end{bmatrix} = \begin{bmatrix} -3 \\ -2 \\ -5 \end{bmatrix}.$$

Solution The column picture asks for a linear combination that produces $\boldsymbol{b}$ from the three columns of A. In this example $\boldsymbol{b}$ is *minus the second column*. So the solution is $x = 0, y = -1, z = 0$. To show that $(0, -1, 0)$ is the *only* solution we have to know that "A is invertible" and "the columns are independent" and "the determinant isn't zero."

Those words are not yet defined but the test comes from elimination: We need (and for this matrix we find) a full set of three nonzero pivots.

Suppose the right side changes to $\boldsymbol{b} = (4, 4, 8) =$ sum of the first two columns. Then the good combination has $x = 1, y = 1, z = 0$. The solution becomes $\boldsymbol{x} = (1, 1, 0)$.

2.1 B This system has *no solution.* The planes in the row picture don't meet at a point. ***No combination of the three columns produces b. How to show this ?***

$$\begin{aligned} x + 3y + 5z &= 4 \\ x + 2y - 3z &= 5 \\ 2x + 5y + 2z &= 8 \end{aligned} \qquad \begin{bmatrix} 1 & 3 & 5 \\ 1 & 2 & -3 \\ 2 & 5 & 2 \end{bmatrix} \begin{bmatrix} x \\ y \\ z \end{bmatrix} = \begin{bmatrix} 4 \\ 5 \\ 8 \end{bmatrix} = \boldsymbol{b}$$

Idea Add (equation 1) + (equation 2) − (equation 3). The result is $\mathbf{0} = \mathbf{1}$. This system cannot have a solution. We could say: The vector $(1, 1, -1)$ is orthogonal to all three columns of A but *not* orthogonal to $\boldsymbol{b}$.

(1) Are any two of the three planes parallel? What are the equations of planes parallel to $x + 3y + 5z = 4$?

(2) Take the dot product of each column of A (and also $\boldsymbol{b}$) with $\boldsymbol{y} = (1, 1, -1)$. How do those dot products show that no combination of columns equals $\boldsymbol{b}$?

(3) Find three different right side vectors $\boldsymbol{b}^*$ and $\boldsymbol{b}^{**}$ and $\boldsymbol{b}^{***}$ that *do* allow solutions.

Solution

(1) The planes don't meet at a point, even though no two planes are parallel. For a plane parallel to $x + 3y + 5z = 4$, change the "4". The parallel plane $x + 3y + 5z = 0$ goes through the origin $(0, 0, 0)$. And the equation multiplied by any nonzero constant still gives the same plane, as in $2x + 6y + 10z = 8$.

(2) The dot product of each column of A with $\boldsymbol{y} = (1, 1, -1)$ is *zero*. On the right side, $\boldsymbol{y} \cdot \boldsymbol{b} = (1, 1, -1) \cdot (4, 5, 8) = \mathbf{1}$ is *not zero*. $A\boldsymbol{x} = \boldsymbol{b}$ led to $0 = 1$: **no solution**.

(3) There is a solution when $\boldsymbol{b}$ is a combination of the columns. These three choices of $\boldsymbol{b}$ have solutions including $\boldsymbol{x}^* = (1, 0, 0)$ and $\boldsymbol{x}^{**} = (1, 1, 1)$ and $\boldsymbol{x}^{***} = (0, 0, 0)$:

$$\boldsymbol{b}^* = \begin{bmatrix} 1 \\ 1 \\ 2 \end{bmatrix} = \text{first column} \qquad \boldsymbol{b}^{**} = \begin{bmatrix} 9 \\ 0 \\ 9 \end{bmatrix} = \text{sum of columns} \qquad \boldsymbol{b}^{***} = \begin{bmatrix} 0 \\ 0 \\ 0 \end{bmatrix}$$

Problem Set 2.1

Problems 1–8 are about the row and column pictures of $Ax = b$.

1 With $A = I$ (the identity matrix) draw the planes in the row picture. Three sides of a box meet at the solution $\boldsymbol{x} = (x, y, z) = (2, 3, 4)$:

$$\begin{aligned} 1x + 0y + 0z &= 2 \\ 0x + 1y + 0z &= 3 \\ 0x + 0y + 1z &= 4 \end{aligned} \qquad \text{or} \qquad \begin{bmatrix} 1 & 0 & 0 \\ 0 & 1 & 0 \\ 0 & 0 & 1 \end{bmatrix} \begin{bmatrix} x \\ y \\ z \end{bmatrix} = \begin{bmatrix} 2 \\ 3 \\ 4 \end{bmatrix}.$$

Draw the vectors in the column picture. Two times column 1 plus three times column 2 plus four times column 3 equals the right side $\boldsymbol{b}$.

2 If the equations in Problem 1 are multiplied by $2, 3, 4$ they become $D\boldsymbol{X} = \boldsymbol{B}$:

$$\begin{aligned} 2x + 0y + 0z &= 4 \\ 0x + 3y + 0z &= 9 \\ 0x + 0y + 4z &= 16 \end{aligned} \qquad \text{or} \qquad D\boldsymbol{X} = \begin{bmatrix} 2 & 0 & 0 \\ 0 & 3 & 0 \\ 0 & 0 & 4 \end{bmatrix} \begin{bmatrix} x \\ y \\ z \end{bmatrix} = \begin{bmatrix} 4 \\ 9 \\ 16 \end{bmatrix} = \boldsymbol{B}$$

Why is the row picture the same? Is the solution $\boldsymbol{X}$ the same as $\boldsymbol{x}$? What is changed in the column picture—the columns or the right combination to give $\boldsymbol{B}$?

3 If equation 1 is added to equation 2, which of these are changed: the planes in the row picture, the vectors in the column picture, the coefficient matrix, the solution? The new equations in Problem 1 would be $x = 2$, $x + y = 5$, $z = 4$.

4 Find a point with $z = 2$ on the intersection line of the planes $x + y + 3z = 6$ and $x - y + z = 4$. Find the point with $z = 0$. Find a third point halfway between.

5 The first of these equations plus the second equals the third:

$$\begin{aligned} x + \ y + \ z &= 2 \\ x + 2y + \ z &= 3 \\ 2x + 3y + 2z &= 5. \end{aligned}$$

The first two planes meet along a line. The third plane contains that line, because if x, y, z satisfy the first two equations then they also _____. The equations have infinitely many solutions (the whole line **L**). Find three solutions on **L**.

6 Move the third plane in Problem 5 to a parallel plane $2x + 3y + 2z = 9$. Now the three equations have no solution—*why not*? The first two planes meet along the line **L**, but the third plane doesn't _____ that line.

7 In Problem 5 the columns are $(1, 1, 2)$ and $(1, 2, 3)$ and $(1, 1, 2)$. This is a "singular case" because the third column is _____. Find two combinations of the columns that give $\boldsymbol{b} = (2, 3, 5)$. This is only possible for $\boldsymbol{b} = (4, 6, c)$ if $c =$ _____.

8 Normally 4 "planes" in 4-dimensional space meet at a _____. Normally 4 column vectors in 4-dimensional space can combine to produce $\boldsymbol{b}$. What combination of $(1,0,0,0), (1,1,0,0), (1,1,1,0), (1,1,1,1)$ produces $\boldsymbol{b} = (3,3,3,2)$? What 4 equations for x, y, z, t are you solving?

Problems 9–14 are about multiplying matrices and vectors.

9 Compute each $A\boldsymbol{x}$ by dot products of the rows with the column vector:

$$\text{(a)} \begin{bmatrix} 1 & 2 & 4 \\ -2 & 3 & 1 \\ -4 & 1 & 2 \end{bmatrix} \begin{bmatrix} 2 \\ 2 \\ 3 \end{bmatrix} \qquad \text{(b)} \begin{bmatrix} 2 & 1 & 0 & 0 \\ 1 & 2 & 1 & 0 \\ 0 & 1 & 2 & 1 \\ 0 & 0 & 1 & 2 \end{bmatrix} \begin{bmatrix} 1 \\ 1 \\ 1 \\ 2 \end{bmatrix}$$

10 Compute each $A\boldsymbol{x}$ in Problem 9 as a combination of the columns:

$$\text{9(a) becomes} \quad A\boldsymbol{x} = 2 \begin{bmatrix} 1 \\ -2 \\ -4 \end{bmatrix} + 2 \begin{bmatrix} 2 \\ 3 \\ 1 \end{bmatrix} + 3 \begin{bmatrix} 4 \\ 1 \\ 2 \end{bmatrix} = \begin{bmatrix} \\ \\ \\ \end{bmatrix}.$$

How many separate multiplications for $A\boldsymbol{x}$, when the matrix is "3 by 3"?

11 Find the two components of $A\boldsymbol{x}$ by rows or by columns:

$$\begin{bmatrix} 2 & 3 \\ 5 & 1 \end{bmatrix} \begin{bmatrix} 4 \\ 2 \end{bmatrix} \quad \text{and} \quad \begin{bmatrix} 3 & 6 \\ 6 & 12 \end{bmatrix} \begin{bmatrix} 2 \\ -1 \end{bmatrix} \quad \text{and} \quad \begin{bmatrix} 1 & 2 & 4 \\ 2 & 0 & 1 \end{bmatrix} \begin{bmatrix} 3 \\ 1 \\ 1 \end{bmatrix}.$$

12 Multiply A times $\boldsymbol{x}$ to find three components of $A\boldsymbol{x}$:

$$\begin{bmatrix} 0 & 0 & 1 \\ 0 & 1 & 0 \\ 1 & 0 & 0 \end{bmatrix} \begin{bmatrix} x \\ y \\ z \end{bmatrix} \quad \text{and} \quad \begin{bmatrix} 2 & 1 & 3 \\ 1 & 2 & 3 \\ 3 & 3 & 6 \end{bmatrix} \begin{bmatrix} 1 \\ 1 \\ -1 \end{bmatrix} \quad \text{and} \quad \begin{bmatrix} 2 & 1 \\ 1 & 2 \\ 3 & 3 \end{bmatrix} \begin{bmatrix} 1 \\ 1 \end{bmatrix}.$$

13 (a) A matrix with m rows and n columns multiplies a vector with _____ components to produce a vector with _____ components.

(b) The planes from the m equations $A\boldsymbol{x} = \boldsymbol{b}$ are in _____-dimensional space. The combination of the columns of A is in _____-dimensional space.

14 Write $2x + 3y + z + 5t = 8$ as a matrix A (how many rows?) multiplying the column vector $\boldsymbol{x} = (x, y, z, t)$ to produce $\boldsymbol{b}$. The solutions $\boldsymbol{x}$ fill a plane or "hyperplane" in 4-dimensional space. *The plane is 3-dimensional with no 4D volume.*

Problems 15–22 ask for matrices that act in special ways on vectors.

15 (a) What is the 2 by 2 identity matrix? I times $\begin{bmatrix} x \\ y \end{bmatrix}$ equals $\begin{bmatrix} x \\ y \end{bmatrix}$.

(b) What is the 2 by 2 exchange matrix? P times $\begin{bmatrix} x \\ y \end{bmatrix}$ equals $\begin{bmatrix} y \\ x \end{bmatrix}$.

16 (a) What 2 by 2 matrix R rotates every vector by $90°$? R times $\begin{bmatrix} x \\ y \end{bmatrix}$ is $\begin{bmatrix} y \\ -x \end{bmatrix}$.

(b) What 2 by 2 matrix R^2 rotates every vector by $180°$?

17 Find the matrix P that multiplies (x, y, z) to give (y, z, x). Find the matrix Q that multiplies (y, z, x) to bring back (x, y, z).

18 What 2 by 2 matrix E subtracts the first component from the second component? What 3 by 3 matrix does the same?

$$E\begin{bmatrix} 3 \\ 5 \end{bmatrix} = \begin{bmatrix} 3 \\ 2 \end{bmatrix} \quad \text{and} \quad E\begin{bmatrix} 3 \\ 5 \\ 7 \end{bmatrix} = \begin{bmatrix} 3 \\ 2 \\ 7 \end{bmatrix}.$$

19 What 3 by 3 matrix E multiplies (x, y, z) to give $(x, y, z + x)$? What matrix E^{-1} multiplies (x, y, z) to give $(x, y, z - x)$? If you multiply $(3, 4, 5)$ by E and then multiply by E^{-1}, the two results are (_____) and (_____).

20 What 2 by 2 matrix P_1 projects the vector (x, y) onto the x axis to produce $(x, 0)$? What matrix P_2 projects onto the y axis to produce $(0, y)$? If you multiply $(5, 7)$ by P_1 and then multiply by P_2, you get (_____) and (_____).

21 What 2 by 2 matrix R rotates every vector through $45°$? The vector $(1, 0)$ goes to $(\sqrt{2}/2, \sqrt{2}/2)$. The vector $(0, 1)$ goes to $(-\sqrt{2}/2, \sqrt{2}/2)$. Those determine the matrix. Draw these particular vectors in the xy plane and find R.

22 Write the dot product of $(1, 4, 5)$ and (x, y, z) as a matrix multiplication $A\boldsymbol{x}$. The matrix A has one row. The solutions to $A\boldsymbol{x} = \mathbf{0}$ lie on a _____ perpendicular to the vector _____. The columns of A are only in _____ -dimensional space.

23 In MATLAB notation, write the commands that define this matrix A and the column vectors $\boldsymbol{x}$ and $\boldsymbol{b}$. What command would test whether or not $A\boldsymbol{x} = \boldsymbol{b}$?

$$A = \begin{bmatrix} 1 & 2 \\ 3 & 4 \end{bmatrix} \qquad \boldsymbol{x} = \begin{bmatrix} 5 \\ -2 \end{bmatrix} \qquad \boldsymbol{b} = \begin{bmatrix} 1 \\ 7 \end{bmatrix}$$

24 The MATLAB commands A = eye(3) and v = [3:5]′ produce the 3 by 3 identity matrix and the column vector $(3, 4, 5)$. What are the outputs from A∗v and v′∗v? (Computer not needed!) If you ask for v∗A, what happens?

25 If you multiply the 4 by 4 all-ones matrix A = ones(4) and the column v = ones(4,1), what is A∗v? (Computer not needed.) If you multiply B = eye(4) + ones(4) times w = zeros(4,1) + 2∗ones(4,1), what is B∗w?

Questions 26–28 review the row and column pictures in 2, 3, and 4 dimensions.

26 Draw the row and column pictures for the equations $x - 2y = 0$, $x + y = 6$.

27 For two linear equations in three unknowns x, y, z, the row picture will show (2 or 3) (lines or planes) in (2 or 3)-dimensional space. The column picture is in (2 or 3)-dimensional space. The solutions normally lie on a _____.

28 For four linear equations in two unknowns x and y, the row picture shows four _____. The column picture is in _____-dimensional space. The equations have no solution unless the vector on the right side is a combination of _____.

29 Start with the vector $\boldsymbol{u}_0 = (1, 0)$. Multiply again and again by the same "Markov matrix" $A = [.8\ .3;\ .2\ .7]$. The next three vectors are $\boldsymbol{u}_1$, $\boldsymbol{u}_2$, $\boldsymbol{u}_3$:

$$\boldsymbol{u}_1 = \begin{bmatrix} .8 & .3 \\ .2 & .7 \end{bmatrix} \begin{bmatrix} 1 \\ 0 \end{bmatrix} = \begin{bmatrix} .8 \\ .2 \end{bmatrix} \qquad \boldsymbol{u}_2 = A\boldsymbol{u}_1 = \underline{\qquad} \qquad \boldsymbol{u}_3 = A\boldsymbol{u}_2 = \underline{\qquad}.$$

What property do you notice for all four vectors $\boldsymbol{u}_0, \boldsymbol{u}_1, \boldsymbol{u}_2, \boldsymbol{u}_3$?

Challenge Problems

30 Continue Problem 29 from $\boldsymbol{u}_0 = (1, 0)$ to $\boldsymbol{u}_7$, and also from $\boldsymbol{v}_0 = (0, 1)$ to $\boldsymbol{v}_7$. What do you notice about $\boldsymbol{u}_7$ and $\boldsymbol{v}_7$? Here are two MATLAB codes, with while and for. They plot $\boldsymbol{u}_0$ to $\boldsymbol{u}_7$ and $\boldsymbol{v}_0$ to $\boldsymbol{v}_7$. You can use other languages:

```
u = [1 ; 0]; A = [.8 .3 ; .2 .7];
x = u; k = [0 : 7];
while size(x,2) <= 7
  u = A∗u; x = [x u];
end
plot(k, x)
```

```
v = [0 ; 1]; A = [.8 .3 ; .2 .7];
x = v; k = [0 : 7];
for j = 1 : 7
  v = A∗v; x = [x v];
end
plot(k, x)
```

The $\boldsymbol{u}$'s and $\boldsymbol{v}$'s are approaching a steady state vector $\boldsymbol{s}$. Guess that vector and check that $A\boldsymbol{s} = \boldsymbol{s}$. If you start with $\boldsymbol{s}$, you stay with $\boldsymbol{s}$.

31 Invent a 3 by 3 **magic matrix** M_3 with entries $1, 2, \ldots, 9$. All rows and columns and diagonals add to 15. The first row could be $8, 3, 4$. What is M_3 times $(1, 1, 1)$? What is M_4 times $(1, 1, 1, 1)$ if a 4 by 4 magic matrix has entries $1, \ldots, 16$?

32 Suppose $\boldsymbol{u}$ and $\boldsymbol{v}$ are the first two columns of a 3 by 3 matrix A. Which third columns $\boldsymbol{w}$ would make this matrix singular? Describe a typical column picture of $A\boldsymbol{x} = \boldsymbol{b}$ in that singular case, and a typical row picture (for a random $\boldsymbol{b}$).

33 **Multiplication by A is a "linear transformation".** Those words mean:

If $\boldsymbol{w}$ is a combination of $\boldsymbol{u}$ and $\boldsymbol{v}$, then $A\boldsymbol{w}$ is the same combination of $A\boldsymbol{u}$ and $A\boldsymbol{v}$.

It is this "***linearity***" $A\boldsymbol{w} = cA\boldsymbol{u} + dA\boldsymbol{v}$ that gives us the name *"linear algebra"*.

Problem: If $\boldsymbol{u} = \begin{bmatrix} 1 \\ 0 \end{bmatrix}$ and $\boldsymbol{v} = \begin{bmatrix} 0 \\ 1 \end{bmatrix}$ then $A\boldsymbol{u}$ and $A\boldsymbol{v}$ are the columns of A.

Combine $\boldsymbol{w} = c\boldsymbol{u} + d\boldsymbol{v}$. **If** $\boldsymbol{w} = \begin{bmatrix} \mathbf{5} \\ \mathbf{7} \end{bmatrix}$ **how is $A\boldsymbol{w}$ connected to $A\boldsymbol{u}$ and $A\boldsymbol{v}$**?

34 Start from the four equations $-x_{i+1} + 2x_i - x_{i-1} = i$ (for $i = 1, 2, 3, 4$ with $x_0 = x_5 = 0$). Write those equations in their matrix form $A\boldsymbol{x} = \boldsymbol{b}$. Can you solve them for x_1, x_2, x_3, x_4?

35 A 9 by 9 ***Sudoku matrix*** S has the numbers $1, \ldots, 9$ in every row and every column, and in every 3 by 3 block. For the all-ones vector $\boldsymbol{x} = (1, \ldots, 1)$, what is $S\boldsymbol{x}$?

A better question is: **Which row exchanges will produce another Sudoku matrix**? Also, which exchanges of block rows give another Sudoku matrix?

Section 2.7 will look at all possible permutations (reorderings) of the rows. I can see 6 orders for the first 3 rows, all giving Sudoku matrices. Also 6 permutations of the next 3 rows, and of the last 3 rows. And 6 block permutations of the block rows?

2.2 The Idea of Elimination

1 For $m = n = 3$, there are three equations $Ax = b$ and three unknowns x_1, x_2, x_3.

2 The first two equations are $a_{11}x_1 + \cdots = b_1$ and $a_{21}x_1 + \cdots = b_2$.

3 Multiply the first equation by a_{21}/a_{11} and subtract from the second : then x_1 **is eliminated.**

4 The corner entry a_{11} is the first "pivot" and the ratio a_{21}/a_{11} is the first "multiplier."

5 Eliminate x_1 from every remaining equation i by subtracting a_{i1}/a_{11} times the first equation.

6 Now the last $n - 1$ equations contain $n - 1$ unknowns $x_2, \ldots, x_n$. *Repeat to eliminate* x_2.

7 Elimination breaks down if zero appears in the pivot. Exchanging two equations may save it.

This chapter explains a systematic way to solve linear equations. The method is called ***"elimination"***, and you can see it immediately in our 2 by 2 example. Before elimination, x and y appear in both equations. After elimination, the first unknown x has disappeared from the second equation $8y = 8$:

$$\textbf{Before} \quad \begin{aligned} x - 2y &= 1 \\ 3x + 2y &= 11 \end{aligned} \qquad \textbf{After} \quad \begin{aligned} x - 2y &= 1 \\ 8y &= 8 \end{aligned} \qquad \begin{aligned} &(\textit{multiply equation } 1 \textit{ by } 3) \\ &(\textit{subtract to eliminate } 3x) \end{aligned}$$

The new equation $8y = 8$ instantly gives $y = 1$. Substituting $y = 1$ back into the first equation leaves $x - 2 = 1$. Therefore $x = 3$ and the solution $(x, y) = (3, 1)$ is complete.

Elimination produces an ***upper triangular system***—this is the goal. The nonzero coefficients $1, -2, 8$ form a triangle. That system is solved from the bottom upwards—first $y = 1$ and then $x = 3$. This quick process is called ***back substitution***. It is used for upper triangular systems of any size, after elimination gives a triangle.

Important point: The original equations have the same solution $x = 3$ and $y = 1$. Figure 2.5 shows each system as a pair of lines, intersecting at the solution point $(3, 1)$. After elimination, the lines still meet at the same point. Every step worked with correct equations.

How did we get from the first pair of lines to the second pair? We subtracted 3 times the first equation from the second equation. The step that eliminates x from equation 2 is the fundamental operation in this chapter. We use it so often that we look at it closely:

To eliminate x : Subtract a multiple of equation 1 from equation 2.

Three times $x - 2y = 1$ gives $3x - 6y = 3$. When this is subtracted from $3x + 2y = 11$, the right side becomes 8. The main point is that $3x$ cancels $3x$. What remains on the left side is $2y - (-6y)$ or $8y$, and x is eliminated. **The system became triangular.**

Ask yourself how that multiplier $\ell = 3$ was found. The first equation contains $\mathbf{1}x$. ***So the first pivot was* 1** (the coefficient of x). The second equation contains $\mathbf{3}x$, **so the multiplier was 3**. Then subtraction $3x - 3x$ produced the zero and the triangle.

You will see the multiplier rule if I change the first equation to $4x - 8y = 4$. (Same straight line but the first pivot becomes 4.) The correct multiplier is now $\ell = \frac{3}{4}$. *To find the multiplier, divide the coefficient "* **3** *" to be eliminated by the pivot "* **4** *"*:

$\mathbf{4}x - 8y = 4$	**Multiply equation 1 by $\frac{3}{4}$**	$4x - 8y = 4$
$\mathbf{3}x + 2y = 11$	**Subtract from equation 2**	$8y = 8.$

The final system is triangular and the last equation still gives $y = 1$. Back substitution produces $4x - 8 = 4$ and $4x = 12$ and $x = 3$. We changed the numbers but not the lines or the solution. ***Divide by the pivot to find that multiplier* $\ell = \frac{3}{4}$**:

Pivot	$=$	***first nonzero in the row that does the elimination***
Multiplier	$=$	***(entry to eliminate) divided by (pivot)*** $= \frac{3}{4}$.

The new second equation starts with the second pivot, which is 8. We would use it to eliminate y from the third equation if there were one. *To solve n equations we want n pivots.* ***The pivots are on the diagonal of the triangle after elimination.***

You could have solved those equations for x and y without reading this book. It is an extremely humble problem, but we stay with it a little longer. Even for a 2 by 2 system, elimination might break down. By understanding the possible breakdown (when we can't find a full set of pivots), you will understand the whole process of elimination.

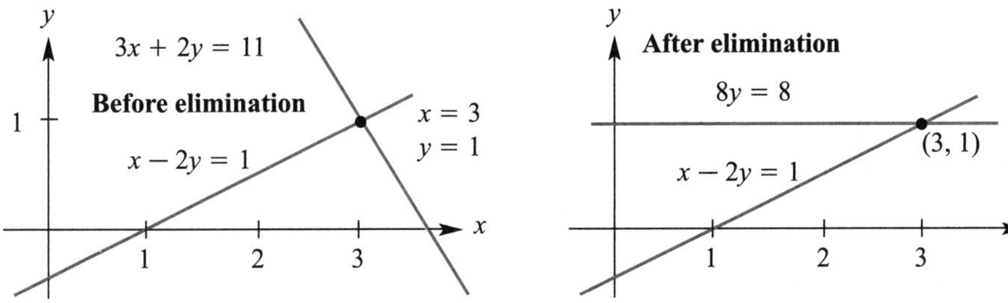

Figure 2.5: Eliminating x makes the second line horizontal. Then $8y = 8$ gives $y = 1$.

Breakdown of Elimination

Normally, elimination produces the pivots that take us to the solution. But failure is possible. At some point, the method might ask us to *divide by zero*. We can't do it. The process has to stop. There might be a way to adjust and continue—or failure may be unavoidable.

Example 1 fails with ***no solution to*** $0y = 8$. Example 2 fails with ***too many solutions to*** $0y = 0$. Example 3 succeeds by exchanging the equations.

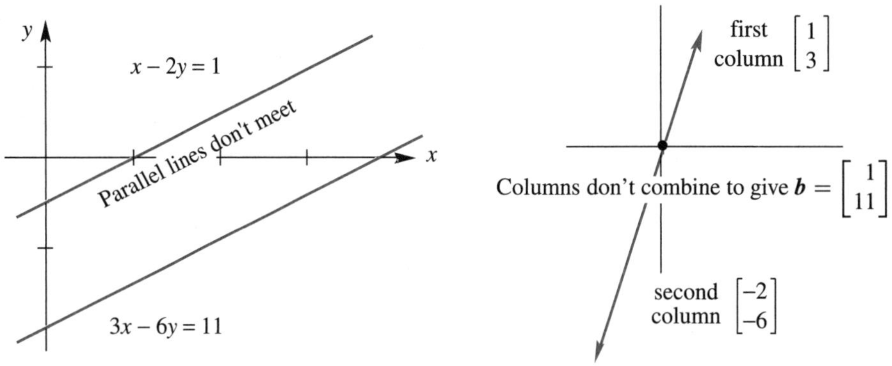

Figure 2.6: Row picture and column picture for Example 1: ***no solution***.

Example 1 ***Permanent failure with no solution***. Elimination makes this clear:

$$\begin{array}{rcl} x - 2y = 1 & \text{Subtract 3 times} & x - 2y = 1 \\ 3x - 6y = 11 & \text{eqn. 1 from eqn. 2} & \mathbf{0}y = 8. \end{array}$$

There is *no* solution to $0y = 8$. Normally we divide the right side 8 by the second pivot, but *this system has no second pivot.* ***(Zero is never allowed as a pivot!)*** The row and column pictures in Figure 2.6 show why failure was unavoidable. If there is no solution, elimination will discover that fact by reaching an equation like $0y = 8$.

The row picture of failure shows parallel lines—which never meet. A solution must lie on both lines. With no meeting point, the equations have no solution.

The column picture shows the two columns $(1, 3)$ and $(-2, -6)$ in the same direction. *All combinations of the columns lie along a line.* But the column from the right side is in a different direction $(1, 11)$. No combination of the columns can produce this right side—therefore no solution.

When we change the right side to $(1, 3)$, failure shows as a whole line of solution points. Instead of no solution, next comes Example 2 with infinitely many.

Example 2 ***Failure with infinitely many solutions. Change*** $\mathbf{b} = (1, 11)$ ***to*** $(1, 3)$.

$$\begin{array}{rccl} x - 2y = 1 & \text{Subtract 3 times} & x - 2y = 1 & \text{Still only} \\ 3x - 6y = 3 & \text{eqn. 1 from eqn. 2} & \mathbf{0}y = \mathbf{0}. & \textbf{one pivot.} \end{array}$$

Every y satisfies $0y = 0$. There is really only one equation $x - 2y = 1$. The unknown y is ***"free"***. After y is freely chosen, x is determined as $x = 1 + 2y$.

In the row picture, the parallel lines have become the same line. Every point on that line satisfies both equations. We have a whole line of solutions in Figure 2.7.

In the column picture, $\boldsymbol{b} = (1, 3)$ is now the same as column 1. So we can choose $x = 1$ and $y = 0$. We can also choose $x = 0$ and $y = -\frac{1}{2}$; column 2 times $-\frac{1}{2}$ equals $\boldsymbol{b}$. Every (x, y) that solves the row problem also solves the column problem.

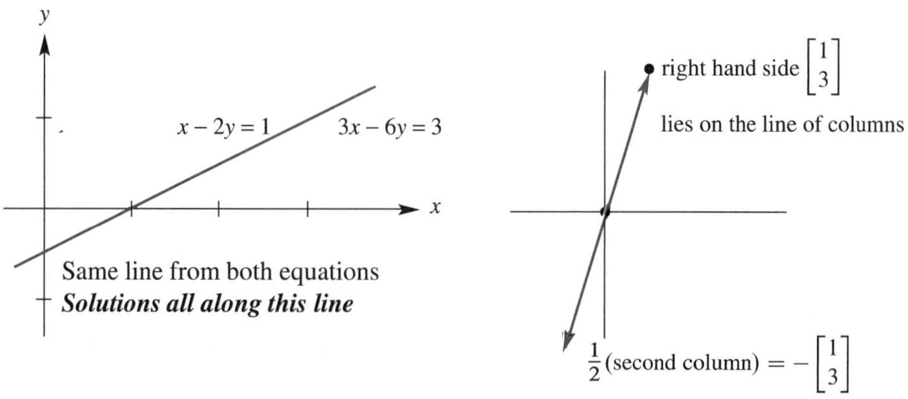

Figure 2.7: Row and column pictures for Example 2: *infinitely many solutions.*

Failure For n equations we do not get n pivots
Elimination leads to an equation $\mathbf{0} \neq \mathbf{0}$ (no solution) or $\mathbf{0} = \mathbf{0}$ (many solutions)

Success comes with n pivots. But we may have to exchange the n equations.

Elimination can go wrong in a third way—but this time it can be fixed. *Suppose the first pivot position contains zero.* We refuse to allow zero as a pivot. When the first equation has no term involving x, we can exchange it with an equation below:

Example 3 ***Temporary failure (zero in pivot). A row exchange produces two pivots***:

Permutation	$\begin{aligned} 0x + 2y &= 4 \\ 3x - 2y &= 5 \end{aligned}$	Exchange the two equations	$\begin{aligned} 3x - 2y &= 5 \\ 2y &= 4. \end{aligned}$

The new system is already triangular. This small example is ready for back substitution. The last equation gives $y = 2$, and then the first equation gives $x = 3$. The row picture is normal (two intersecting lines). The column picture is also normal (column vectors not in the same direction). The pivots 3 and 2 are normal—but a ***row exchange*** was required.

Examples 1 and 2 are ***singular***—there is no second pivot. Example 3 is ***nonsingular***—there is a full set of pivots and exactly one solution. Singular equations have no solution or infinitely many solutions. Pivots must be nonzero because we have to divide by them.

Three Equations in Three Unknowns

To understand Gaussian elimination, you have to go beyond 2 by 2 systems. Three by three is enough to see the pattern. For now the matrices are square—an equal number of rows and columns. Here is a 3 by 3 system, specially constructed so that all elimination steps lead to whole numbers and not fractions:

$$\begin{aligned} \mathbf{2}x + 4y - 2z &= 2 \\ 4x + 9y - 3z &= 8 \\ -2x - 3y + 7z &= 10 \end{aligned} \qquad (1)$$

What are the steps? The first pivot is the boldface **2** (upper left). Below that pivot we want to eliminate the 4. *The first multiplier is the ratio* $4/2 = 2$. Multiply the pivot equation by $\ell_{21} = 2$ and subtract. Subtraction removes the $4x$ from the second equation:

Step 1 Subtract 2 times equation 1 from equation 2. This leaves $y + z = 4$.

We also eliminate $-2x$ from equation 3—still using the first pivot. The quick way is to add equation 1 to equation 3. Then $2x$ cancels $-2x$. We do exactly that, but the rule in this book is to *subtract rather than add.* The systematic pattern has multiplier $\ell_{31} = -2/2 = -1$. Subtracting -1 times an equation is the same as adding:

Step 2 Subtract -1 times equation 1 from equation 3. This leaves $y + 5z = 12$.

The two new equations involve only y and z. The second pivot (in boldface) is 1:

$$\boldsymbol{x} \textbf{ is eliminated} \qquad \begin{aligned} \mathbf{1}y + 1z &= 4 \\ 1y + 5z &= 12 \end{aligned}$$

We have reached a 2 *by* 2 *system.* The final step eliminates y to make it 1 by 1:

Step 3 Subtract equation 2_{new} from 3_{new}. The multiplier is $1/1 = 1$. Then $4z = 8$.

The original $A\boldsymbol{x} = \boldsymbol{b}$ has been converted into an upper triangular $U\boldsymbol{x} = \boldsymbol{c}$:

$$\begin{aligned} 2x + 4y - 2z &= 2 \\ 4x + 9y - 3z &= 8 \\ -2x - 3y + 7z &= 10 \end{aligned} \qquad \begin{matrix} A\boldsymbol{x} = \boldsymbol{b} \\ \text{has become} \\ U\boldsymbol{x} = \boldsymbol{c} \end{matrix} \qquad \begin{aligned} \mathbf{2}x + 4y - 2z &= 2 \\ \mathbf{1}y + 1z &= 4 \\ \mathbf{4}z &= 8. \end{aligned} \qquad (2)$$

The goal is achieved—forward elimination is complete from A to U. ***Notice the pivots* $\mathbf{2, 1, 4}$ *along the diagonal of* U.** The pivots 1 and 4 were hidden in the original system. Elimination brought them out. $U\boldsymbol{x} = \boldsymbol{c}$ is ready for ***back substitution***, which is quick:

$$(4z = 8 \quad \text{gives} \quad z = 2) \qquad (y + z = 4 \quad \text{gives} \quad y = 2) \qquad (\text{equation 1 gives} \quad x = -1)$$

The solution is $(x, y, z) = (-1, 2, 2)$. The row picture has three planes from three equations. All the planes go through this solution. The original planes are sloping, but the last plane $4z = 8$ after elimination is horizontal.

The column picture shows a combination $A\boldsymbol{x}$ of column vectors producing the right side $\boldsymbol{b}$. The coefficients in that combination are $-1, 2, 2$ (the solution):

$$A\boldsymbol{x} = (\mathbf{-1}) \begin{bmatrix} 2 \\ 4 \\ -2 \end{bmatrix} + \mathbf{2} \begin{bmatrix} 4 \\ 9 \\ -3 \end{bmatrix} + \mathbf{2} \begin{bmatrix} -2 \\ -3 \\ 7 \end{bmatrix} \text{ equals } \begin{bmatrix} 2 \\ 8 \\ 10 \end{bmatrix} = \boldsymbol{b}. \qquad (3)$$

The numbers x, y, z multiply columns 1, 2, 3 in $A\boldsymbol{x} = \boldsymbol{b}$ and also in the triangular $U\boldsymbol{x} = \boldsymbol{c}$.

Elimination from A to U

For a 4 by 4 problem, or an n by n problem, elimination proceeds in the same way. Here is the whole idea, column by column from A to U, when Gaussian elimination succeeds.

Column 1. ***Use the first equation to create zeros below the first pivot.***

Column 2. ***Use the new equation 2 to create zeros below the second pivot.***

Columns 3 to n. ***Keep going to find all n pivots and the upper triangular U.***

$$\text{After column 2 we have} \quad \begin{bmatrix} \boldsymbol{x} & x & x & x \\ 0 & \boldsymbol{x} & x & x \\ 0 & 0 & x & x \\ 0 & 0 & x & x \end{bmatrix}. \quad \text{We want} \quad \begin{bmatrix} \boldsymbol{x} & x & x & x \\ & \boldsymbol{x} & x & x \\ & & \boldsymbol{x} & x \\ & & & \boldsymbol{x} \end{bmatrix}. \tag{4}$$

The result of forward elimination is an upper triangular system. It is nonsingular if there is a full set of n pivots (never zero!). *Question*: Which x on the left won't be changed in elimination because the pivot is known? Here is a final example to show the original $A\boldsymbol{x} = \boldsymbol{b}$, the triangular system $U\boldsymbol{x} = \boldsymbol{c}$, and the solution (x, y, z) from back substitution:

$$\begin{aligned} x + y + z &= 6 \\ x + 2y + 2z &= 9 \quad \textbf{Forward} \\ x + 2y + 3z &= 10 \quad \textbf{Forward} \end{aligned} \qquad \begin{aligned} x + y + z &= 6 \\ y + z &= 3 \\ z &= 1 \end{aligned} \qquad \begin{bmatrix} x \\ y \\ z \end{bmatrix} = \begin{bmatrix} 3 \\ 2 \\ 1 \end{bmatrix} \quad \begin{matrix} \textbf{Back} \\ \textbf{Back} \\ \end{matrix}$$

All multipliers are 1. All pivots are 1. All planes meet at the solution $(3, 2, 1)$. The columns of A combine with $3, 2, 1$ to give $\boldsymbol{b} = (6, 9, 10)$. The triangle shows $U\boldsymbol{x} = \boldsymbol{c} = (6, 3, 1)$.

■ REVIEW OF THE KEY IDEAS ■

1. A linear system $(A\boldsymbol{x} = \boldsymbol{b})$ becomes **upper triangular** $(U\boldsymbol{x} = \boldsymbol{c})$ after elimination.
2. We **subtract** ℓ_{ij} times equation j from equation i, to make the (i, j) entry zero.
3. The **multiplier** is $\ell_{ij} = \dfrac{\text{entry to eliminate in row } i}{\text{pivot in row } j}$. **Pivots** can not be zero!
4. When zero is in the pivot position, **exchange rows** if there is a nonzero below it.
5. The upper triangular $U\boldsymbol{x} = \boldsymbol{c}$ is solved by **back substitution** (starting at the bottom).
6. When **breakdown** is permanent, $A\boldsymbol{x} = \boldsymbol{b}$ has no solution or infinitely many.

■ WORKED EXAMPLES ■

2.2 A When elimination is applied to this matrix A, what are the first and second pivots? What is the multiplier ℓ_{21} in the first step (ℓ_{21} times row 1 is *subtracted* from row 2)?

$$A = \begin{bmatrix} 1 & 1 & 0 \\ 1 & 2 & 1 \\ 0 & 1 & 2 \end{bmatrix} \longrightarrow \begin{bmatrix} 1 & 1 & 0 \\ 0 & 1 & 1 \\ 0 & 1 & 2 \end{bmatrix} \longrightarrow \begin{bmatrix} 1 & 1 & 0 \\ 0 & 1 & 1 \\ 0 & 0 & 1 \end{bmatrix} = U.$$

What entry in the $2, 2$ position (instead of 2) would force an exchange of rows 2 and 3? Why is the lower left multiplier $\ell_{31} = 0$, subtracting zero times row 1 from row 3?
If you change the corner entry from $a_{33} = 2$ to $a_{33} = 1$, why does elimination fail?

Solution The first pivot is 1. The multiplier ℓ_{21} is $1, 1$. When 1 times row 1 is subtracted from row 2, the second pivot is revealed as another 1. If the original middle entry had been 1 instead of 2, that would have forced a row exchange.

The multiplier ℓ_{31} is zero because $a_{31} = 0$. A zero at the start of a row needs no elimination. This A is a "*band matrix*". Everything stays zero outside the band.

The last pivot is also 1. So if the original corner entry $a_{33} = 2$ reduced by 1, elimination would produce 0. **No third pivot, elimination fails.**

2.2 B Suppose A is already a ***triangular matrix*** (upper triangular or lower triangular). *Where do you see its pivots*? When does $A\boldsymbol{x} = \boldsymbol{b}$ have exactly one solution for every $\boldsymbol{b}$?

Solution The pivots of a triangular matrix are already set along the main diagonal. *Elimination succeeds when all those numbers are nonzero.* Use ***back*** substitution when A is upper triangular, go ***forward*** when A is lower triangular.

2.2 C Use elimination to reach upper triangular matrices U. Solve by back substitution or explain why this is impossible. What are the pivots (never zero)? Exchange equations when necessary. The only difference is the $-x$ in the last equation.

$$\textbf{Success} \quad \begin{aligned} x + y + z &= 7 \\ x + y - z &= 5 \\ x - y + z &= 3 \end{aligned} \qquad\qquad \textbf{Failure} \quad \begin{aligned} x + y + z &= 7 \\ x + y - z &= 5 \\ -x - y + z &= 3 \end{aligned}$$

Solution For the first system, subtract equation 1 from equations 2 and 3 (the multipliers are $\ell_{21} = 1$ and $\ell_{31} = 1$). The $2, 2$ entry becomes zero, so exchange equations 2 and 3:

$$\textbf{Success} \quad \begin{aligned} x + y + z &= 7 \\ \mathbf{0}y - 2z &= -2 \\ -2y + 0z &= -4 \end{aligned} \quad \text{exchanges into} \quad \begin{aligned} x + y + z &= 7 \\ -\mathbf{2}y + 0z &= -4 \\ -2z &= -2 \end{aligned}$$

Then back substitution gives $z = 1$ and $y = 2$ and $x = 4$. The pivots are $1, -2, -2$.

For the second system, subtract equation 1 from equation 2 as before. Add equation 1 to equation 3. This leaves zero in the $2, 2$ entry *and also below*:

	$x + y + z = \;\; 7$	There is **no pivot in column 2** (it was – column 1)
Failure	$0y - 2z = -2$	A further elimination step gives $\mathbf{0z = 8}$
	$0y + 2z = \;\; 10$	The three planes **don't meet**

Plane 1 meets plane 2 in a line. Plane 1 meets plane 3 in a parallel line. *No solution.*

If we change the "3" in the original third equation to "-5" then elimination would lead to $0 = 0$. There are infinitely many solutions! *The three planes now meet along a whole line.*

Changing 3 to -5 moved the third plane to meet the other two. The second equation gives $z = 1$. Then the first equation leaves $x + y = 6$. **No pivot in column 2 makes y free** (free variables can have any value). Then $x = 6 - y$.

Problem Set 2.2

Problems 1–10 are about elimination on 2 by 2 systems.

1 What multiple ℓ_{21} of equation 1 should be subtracted from equation 2?

$$\begin{aligned} 2x + 3y &= 1 \\ 10x + 9y &= 11. \end{aligned}$$

After elimination, write down the upper triangular system and circle the two pivots. The numbers 1 and 11 don't affect the pivots—use them now in back substitution.

2 Solve the triangular system of Problem 1 by back substitution, y before x. Verify that x times $(2, 10)$ plus y times $(3, 9)$ equals $(1, 11)$. If the right side changes to $(4, 44)$, what is the new solution?

3 What multiple of equation 1 should be *subtracted* from equation 2?

$$\begin{aligned} 2x - 4y &= 6 \\ -x + 5y &= 0. \end{aligned}$$

After this elimination step, solve the triangular system. If the right side changes to $(-6, 0)$, what is the new solution?

4 What multiple ℓ of equation 1 should be subtracted from equation 2 to remove c?

$$\begin{aligned} ax + by &= f \\ cx + dy &= g. \end{aligned}$$

The first pivot is a (assumed nonzero). Elimination produces what formula for the second pivot? What is y? The second pivot is missing when $ad = bc$: singular.

5 Choose a right side which gives no solution and another right side which gives infinitely many solutions. What are two of those solutions?

Singular system
$$\begin{aligned} 3x + 2y &= 10 \\ 6x + 4y &= \end{aligned}$$

6 Choose a coefficient b that makes this system singular. Then choose a right side g that makes it solvable. Find two solutions in that singular case.

$$\begin{aligned} 2x + by &= 16 \\ 4x + 8y &= g. \end{aligned}$$

7 For which numbers a does elimination break down (1) permanently (2) temporarily?

$$\begin{aligned} ax + 3y &= -3 \\ 4x + 6y &= \quad 6. \end{aligned}$$

Solve for x and y after fixing the temporary breakdown by a row exchange.

8 For which three numbers k does elimination break down? Which is fixed by a row exchange? In each case, is the number of solutions 0 or 1 or ∞?

$$\begin{aligned} kx + 3y &= \quad 6 \\ 3x + ky &= -6. \end{aligned}$$

9 What test on b_1 and b_2 decides whether these two equations allow a solution? How many solutions will they have? Draw the column picture for $\boldsymbol{b} = (1, 2)$ and $(1, 0)$.

$$\begin{aligned} 3x - 2y &= b_1 \\ 6x - 4y &= b_2. \end{aligned}$$

10 In the xy plane, draw the lines $x+y = 5$ and $x+2y = 6$ and the equation $y =$ _____ that comes from elimination. The line $5x - 4y = c$ will go through the solution of these equations if $c =$ _____.

Problems 11–20 study elimination on 3 by 3 systems (and possible failure).

11 (Recommended) A system of linear equations can't have exactly two solutions. *Why*?

(a) If (x, y, z) and (X, Y, Z) are two solutions, what is another solution?

(b) If 25 planes meet at two points, where else do they meet?

12 Reduce this system to upper triangular form by two row operations:

$$\begin{aligned} 2x + 3y + z &= 8 \\ 4x + 7y + 5z &= 20 \\ -2y + 2z &= 0. \end{aligned}$$

Circle the pivots. Solve by back substitution for z, y, x.

13 Apply elimination (circle the pivots) and back substitution to solve

$$\begin{aligned} 2x - 3y \phantom{{}+ z} &= 3 \\ 4x - 5y + z &= 7 \\ 2x - y - 3z &= 5. \end{aligned}$$

List the three row operations: Subtract _____ times row _____ from row _____.

14 Which number d forces a row exchange, and what is the triangular system (not singular) for that d? Which d makes this system singular (no third pivot)?

$$\begin{aligned} 2x + 5y + z &= 0 \\ 4x + dy + z &= 2 \\ y - z &= 3. \end{aligned}$$

15 Which number b leads later to a row exchange? Which b leads to a missing pivot? In that singular case find a nonzero solution x, y, z.

$$\begin{aligned} x + by \phantom{{}- z} &= 0 \\ x - 2y - z &= 0 \\ y + z &= 0. \end{aligned}$$

16 (a) Construct a 3 by 3 system that needs two row exchanges to reach a triangular form and a solution.

(b) Construct a 3 by 3 system that needs a row exchange to keep going, but breaks down later.

17 If rows 1 and 2 are the same, how far can you get with elimination (allowing row exchange)? If columns 1 and 2 are the same, which pivot is missing?

Equal rows

$$\begin{aligned} 2x - y + z &= 0 \\ 2x - y + z &= 0 \\ 4x + y + z &= 2 \end{aligned}$$

Equal columns

$$\begin{aligned} 2x + 2y + z &= 0 \\ 4x + 4y + z &= 0 \\ 6x + 6y + z &= 2. \end{aligned}$$

18 Construct a 3 by 3 example that has 9 different coefficients on the left side, but rows 2 and 3 become zero in elimination. How many solutions to your system with $\boldsymbol{b} = (1, 10, 100)$ and how many with $\boldsymbol{b} = (0, 0, 0)$?

19 Which number q makes this system singular and which right side t gives it infinitely many solutions? Find the solution that has $z = 1$.

$$\begin{aligned} x + 4y - 2z &= 1 \\ x + 7y - 6z &= 6 \\ 3y + qz &= t. \end{aligned}$$

20 Three planes can fail to have an intersection point, even if no planes are parallel. The system is singular if row 3 of A is a _____ of the first two rows. Find a third equation that can't be solved together with $x + y + z = 0$ and $x - 2y - z = 1$.

21 Find the pivots and the solution for both systems ($A\boldsymbol{x} = \boldsymbol{b}$ and $K\boldsymbol{x} = \boldsymbol{b}$):

$$\begin{aligned} 2x + y \qquad\qquad &= 0 \\ x + 2y + z \qquad &= 0 \\ y + 2z + t &= 0 \\ z + 2t &= 5 \end{aligned} \qquad\qquad \begin{aligned} 2x - y \qquad\qquad &= 0 \\ -x + 2y - z \qquad &= 0 \\ -y + 2z - t &= 0 \\ -z + 2t &= 5. \end{aligned}$$

22 If you extend Problem 21 following the $1, 2, 1$ pattern or the $-1, 2, -1$ pattern, what is the fifth pivot? What is the nth pivot? K is my favorite matrix.

23 If elimination leads to $x + y = 1$ and $2y = 3$, find three possible original problems.

24 For which two numbers a will elimination fail on $A = \begin{bmatrix} a & 2 \\ a & a \end{bmatrix}$?

25 For which three numbers a will elimination fail to give three pivots?

$$A = \begin{bmatrix} a & 2 & 3 \\ a & a & 4 \\ a & a & a \end{bmatrix} \text{ is singular for three values of } a.$$

26 Look for a matrix that has row sums 4 and 8, and column sums 2 and s:

$$\text{Matrix} = \begin{bmatrix} a & b \\ c & d \end{bmatrix} \qquad \begin{matrix} a + b = 4 & a + c = 2 \\ c + d = 8 & b + d = s \end{matrix}$$

The four equations are solvable only if $s =$ _____. Then find two different matrices that have the correct row and column sums. *Extra credit*: Write down the 4 by 4 system $A\boldsymbol{x} = \boldsymbol{b}$ with $\boldsymbol{x} = (a, b, c, d)$ and make A triangular by elimination.

27 Elimination in the usual order gives what matrix U and what solution to this "lower triangular" system? We are really solving by *forward substitution*:

$$\begin{aligned} 3x \qquad\quad &= 3 \\ 6x + 2y \quad &= 8 \\ 9x - 2y + z &= 9. \end{aligned}$$

28 Create a MATLAB command A(2, :) = ... for the new row 2, to subtract 3 times row 1 from the existing row 2 if the matrix A is already known.

Challenge Problems

29 Find experimentally the average 1st and 2nd and 3rd pivot sizes from MATLAB 's $[L, U] = \mathbf{lu}\,(\mathbf{rand}\,(3))$. The average size $\mathbf{abs}\,(U(1,1))$ is above $\frac{1}{2}$ because **lu** picks the largest available pivot in column 1. Here $A = \mathbf{rand}\,(3)$ has random entries between 0 and 1.

30 If the last corner entry is $A(5,5) = 11$ and the last pivot of A is $U(5,5) = 4$, what different entry $A(5,5)$ would have made A singular?

31 Suppose elimination takes A to U without row exchanges. Then row j of U is a combination of which rows of A? If $A\boldsymbol{x} = \mathbf{0}$, is $U\boldsymbol{x} = \mathbf{0}$? If $A\boldsymbol{x} = \boldsymbol{b}$, is $U\boldsymbol{x} = \boldsymbol{b}$? If A starts out lower triangular, what is the upper triangular U?

32 Start with 100 equations $A\boldsymbol{x} = \mathbf{0}$ for 100 unknowns $\boldsymbol{x} = (x_1, \ldots, x_{100})$. Suppose elimination reduces the 100th equation to $0 = 0$, so the system is "singular".

(a) Elimination takes linear combinations of the rows. So this singular system has the singular property: Some linear combination of the 100 **rows** is _____.

(b) Singular systems $A\boldsymbol{x} = \mathbf{0}$ have infinitely many solutions. This means that some linear combination of the 100 ***columns*** is _____.

(c) Invent a 100 by 100 singular matrix with no zero entries.

(d) For your matrix, describe in words the row picture and the column picture of $A\boldsymbol{x} = \mathbf{0}$. Not necessary to draw 100-dimensional space.

2.3 Elimination Using Matrices

1 The first step multiplies the equations $\boldsymbol{Ax} = \boldsymbol{b}$ by a matrix E_{21} to produce $E_{21}\boldsymbol{Ax} = E_{21}\boldsymbol{b}$.

2 That matrix $E_{21}A$ has a zero in row 2, column 1 because x_1 is eliminated from equation 2.

3 E_{21} is the **identity matrix** (diagonal of 1's) minus the multiplier a_{21}/a_{11} in row 2, column 1.

4 Matrix-matrix multiplication is n matrix-vector multiplications: $\boldsymbol{EA} = [\boldsymbol{Ea_1} \ \dots \ \boldsymbol{Ea_n}]$.

5 We must also multiply $E\boldsymbol{b}$! So E is multiplying the **augmented matrix** $[A\,\boldsymbol{b}] = [\boldsymbol{a}_1 \ \dots \ \boldsymbol{a}_n\,\boldsymbol{b}]$.

6 Elimination multiplies $A\boldsymbol{x} = \boldsymbol{b}$ by $E_{21}, E_{31}, \dots, E_{n1}$, then $E_{32}, E_{42}, \dots, E_{n2}$, and onward.

7 The **row exchange matrix** is not E_{ij} but P_{ij}. To find P_{ij}, exchange rows i and j of I.

This section gives our first examples of **matrix multiplication**. Naturally we start with matrices that contain many zeros. Our goal is to see that matrices *do something*. E acts on a vector $\boldsymbol{b}$ or a matrix A to produce a new vector $E\boldsymbol{b}$ or a new matrix EA.

Our first examples will be "**elimination matrices**." They execute the elimination steps. Multiply the jth equation by ℓ_{ij} and subtract from the ith equation. (This eliminates x_j from equation i.) We need a lot of these simple matrices E_{ij}, one for every nonzero to be eliminated below the main diagonal.

Fortunately we won't see all these matrices E_{ij} in later chapters. They are good examples to start with, but there are too many. They can combine into one overall matrix E that takes all steps at once. The neatest way is to combine all their inverses $(E_{ij})^{-1}$ into one overall matrix $L = E^{-1}$. Here is the purpose of the next pages.

1. To see how each step is a matrix multiplication.

2. To assemble all those steps E_{ij} into one elimination matrix E.

3. To see how each E_{ij} is inverted by its inverse matrix E_{ij}^{-1}.

4. To assemble all those inverses E_{ij}^{-1} (in the right order) into L.

The special property of L is that all the multipliers ℓ_{ij} fall into place. Those numbers are mixed up in E (forward elimination from A to U). They are perfect in L (undoing elimination, returning from U to A). Inverting puts the steps and their matrices E_{ij}^{-1} in the opposite order and that prevents the mixup.

This section finds the matrices E_{ij}. Section 2.4 presents four ways to multiply matrices. Section 2.5 inverts every step. (For elimination matrices we can already see E_{ij}^{-1} here.) Then those inverses go into L.

Matrices times Vectors and $Ax = b$

The 3 by 3 example in the previous section has the short form $A\boldsymbol{x} = \boldsymbol{b}$:

$$\begin{aligned} 2x_1 + 4x_2 - 2x_3 &= 2 \\ 4x_1 + 9x_2 - 3x_3 &= 8 \\ -2x_1 - 3x_2 + 7x_3 &= 10 \end{aligned} \quad \text{is the same as} \quad \begin{bmatrix} 2 & 4 & -2 \\ 4 & 9 & -3 \\ -2 & -3 & 7 \end{bmatrix} \begin{bmatrix} x_1 \\ x_2 \\ x_3 \end{bmatrix} = \begin{bmatrix} 2 \\ 8 \\ 10 \end{bmatrix}. \qquad (1)$$

The nine numbers on the left go into the matrix A. That matrix not only sits beside $\boldsymbol{x}$. *A multiplies $\boldsymbol{x}$*. The rule for "A times $\boldsymbol{x}$" is exactly chosen to yield the three equations.

***Review of** A **times** $\boldsymbol{x}$.* A matrix times a vector gives a vector. The matrix is square when the number of equations (three) matches the number of unknowns (three). Our matrix is 3 by 3. A general square matrix is n by n. Then the vector $\boldsymbol{x}$ is in n-dimensional space.

$$\textbf{\textit{The unknown is}} \quad \boldsymbol{x} = \begin{bmatrix} x_1 \\ x_2 \\ x_3 \end{bmatrix} \quad \textbf{\textit{and the solution is}} \quad \boldsymbol{x} = \begin{bmatrix} -1 \\ 2 \\ 2 \end{bmatrix}.$$

Key point: $A\boldsymbol{x} = \boldsymbol{b}$ represents the row form and also the column form of the equations.

$$\textbf{Column form} \qquad A\boldsymbol{x} = (-1)\begin{bmatrix} 2 \\ 4 \\ -2 \end{bmatrix} + 2\begin{bmatrix} 4 \\ 9 \\ -3 \end{bmatrix} + 2\begin{bmatrix} -2 \\ -3 \\ 7 \end{bmatrix} = \begin{bmatrix} 2 \\ 8 \\ 10 \end{bmatrix} = \boldsymbol{b}. \qquad (2)$$

$A\boldsymbol{x}$ is a combination of the columns of A. To compute each component of $A\boldsymbol{x}$, we use the **row form** of matrix multiplication. ***Components of $A\boldsymbol{x}$ are dot products with rows of A.*** The short formula for that dot product with $\boldsymbol{x}$ uses "sigma notation".

The first component of $A\boldsymbol{x}$ above is $(-1)(2) + (2)(4) + (2)(-2)$.

The ith component of $A\boldsymbol{x}$ is $(\text{row } i) \cdot \boldsymbol{x} = a_{i1}x_1 + a_{i2}x_2 + \cdots + a_{in}x_n$.

This is sometimes written with the sigma symbol as $\sum_{j=1}^{n} a_{ij}x_j$.

$\sum$ is an instruction to add. Start with $j = 1$ and stop with $j = n$. The sum begins with $a_{i1}x_1$ and ends with $a_{in}x_n$. That produces the dot product (row i) $\cdot$ $\boldsymbol{x}$.

One point to repeat about matrix notation: The entry in row 1, column 1 (the top left corner) is a_{11}. The entry in row 1, column 3 is a_{13}. The entry in row 3, column 1 is a_{31}. (Row number comes before column number.) The word "entry" for a matrix corresponds to "component" for a vector. General rule: ***$a_{ij} = A(i, j)$ is in row i, column j.***

Example 1 This matrix has $a_{ij} = 2i + j$. Then $a_{11} = 3$. Also $a_{12} = 4$ and $a_{21} = 5$. Here is $A\boldsymbol{x}$ by rows with numbers and letters:

$$\begin{bmatrix} 3 & 4 \\ 5 & 6 \end{bmatrix} \begin{bmatrix} 2 \\ 1 \end{bmatrix} = \begin{bmatrix} 3 \cdot 2 + 4 \cdot 1 \\ 5 \cdot 2 + 6 \cdot 1 \end{bmatrix} \qquad \begin{bmatrix} a_{11} & a_{12} \\ a_{21} & a_{22} \end{bmatrix} \begin{bmatrix} x_1 \\ x_2 \end{bmatrix} = \begin{bmatrix} a_{11}x_1 + a_{12}x_2 \\ a_{21}x_1 + a_{22}x_2 \end{bmatrix}.$$

A row times a column gives a dot product.

[1]Einstein shortened this even more by omitting the $\sum$. The repeated j in $a_{ij}x_j$ automatically meant addition. He also wrote the sum as $a_i^j x_j$. Not being Einstein, we include the $\sum$.

The Matrix Form of One Elimination Step

$A\boldsymbol{x} = \boldsymbol{b}$ is a convenient form for the original equation. What about the elimination steps? In this example, 2 times the first equation is subtracted from the second equation. On the right side, 2 times the first component of $\boldsymbol{b}$ is subtracted from the second component.

$$\textbf{First step} \qquad \boldsymbol{b} = \begin{bmatrix} 2 \\ \mathbf{8} \\ 10 \end{bmatrix} \quad \text{changes to} \quad \boldsymbol{b}_{\text{new}} = \begin{bmatrix} 2 \\ \mathbf{4} \\ 10 \end{bmatrix}.$$

We want to do that subtraction with a matrix! The same result $\boldsymbol{b}_{\text{new}} = E\boldsymbol{b}$ is achieved when we multiply an "elimination matrix" E times $\boldsymbol{b}$. It subtracts $2b_1$ from b_2:

$$\textbf{\textit{The elimination matrix is}} \qquad E = \begin{bmatrix} 1 & 0 & 0 \\ -2 & 1 & 0 \\ 0 & 0 & 1 \end{bmatrix}.$$

Multiplication by E subtracts 2 times row 1 from row 2. Rows 1 and 3 stay the same:

$$\begin{bmatrix} 1 & 0 & 0 \\ \mathbf{-2} & \mathbf{1} & 0 \\ 0 & 0 & 1 \end{bmatrix} \begin{bmatrix} \mathbf{2} \\ \mathbf{8} \\ 10 \end{bmatrix} = \begin{bmatrix} 2 \\ \mathbf{4} \\ 10 \end{bmatrix} \qquad \begin{bmatrix} 1 & 0 & 0 \\ \mathbf{-2} & \mathbf{1} & 0 \\ 0 & 0 & 1 \end{bmatrix} \begin{bmatrix} \boldsymbol{b_1} \\ \boldsymbol{b_2} \\ b_3 \end{bmatrix} = \begin{bmatrix} b_1 \\ \boldsymbol{b_2 - 2b_1} \\ b_3 \end{bmatrix}$$

The first and third rows of E come from the identity matrix I. They don't change the first and third numbers (2 and 10). The new second component is the number 4 that appeared after the elimination step. This is $b_2 - 2b_1$.

It is easy to describe the "elementary matrices" or "elimination matrices" like this E. Start with the identity matrix I. *Change one of its zeros to the multiplier* $-\ell$:

The ***identity matrix*** has 1's on the diagonal and otherwise 0's. Then $I\boldsymbol{b} = \boldsymbol{b}$ for all $\boldsymbol{b}$. The ***elementary matrix or elimination matrix*** E_{ij} has the extra nonzero entry $-\ell$ in the i, j position. Then E_{ij} subtracts a multiple ℓ of row j from row i.

Example 2 The matrix E_{31} has $-\ell$ in the $3, 1$ position:

$$\textbf{Identity} \quad I = \begin{bmatrix} 1 & 0 & 0 \\ 0 & 1 & 0 \\ 0 & 0 & 1 \end{bmatrix} \qquad \textbf{Elimination} \quad E_{31} = \begin{bmatrix} 1 & 0 & 0 \\ 0 & 1 & 0 \\ -\ell & 0 & 1 \end{bmatrix}.$$

When you multiply I times $\boldsymbol{b}$, you get $\boldsymbol{b}$. But E_{31} subtracts ℓ times the first component from the third component. With $\ell = 4$ this example gives $9 - 4 = 5$:

$$I\boldsymbol{b} = \begin{bmatrix} 1 & 0 & 0 \\ 0 & 1 & 0 \\ 0 & 0 & 1 \end{bmatrix} \begin{bmatrix} 1 \\ 3 \\ 9 \end{bmatrix} = \begin{bmatrix} 1 \\ 3 \\ 9 \end{bmatrix} \quad \text{and} \quad E\boldsymbol{b} = \begin{bmatrix} 1 & 0 & 0 \\ 0 & 1 & 0 \\ \mathbf{-4} & 0 & 1 \end{bmatrix} \begin{bmatrix} \mathbf{1} \\ 3 \\ \mathbf{9} \end{bmatrix} = \begin{bmatrix} 1 \\ 3 \\ \mathbf{5} \end{bmatrix}.$$

What about the left side of $A\boldsymbol{x} = \boldsymbol{b}$? Both sides will be multiplied by this E_{31}. ***The purpose of $\mathbf{E_{31}}$ is to produce a zero in the $(3,1)$ position of the matrix.***

The notation fits this purpose. Start with A. Apply E's to produce zeros below the pivots (the first E is E_{21}). End with a triangular U. We now look in detail at those steps.

First a small point. The vector $\boldsymbol{x}$ stays the same. The solution $\boldsymbol{x}$ is not changed by elimination. (That may be more than a small point.) It is the coefficient matrix that is changed. When we start with $A\boldsymbol{x} = \boldsymbol{b}$ and multiply by E, the result is $EA\boldsymbol{x} = E\boldsymbol{b}$. The new matrix EA is the result of *multiplying E times A.*

Confession The *elimination matrices* E_{ij} are great examples, but you won't see them later. They show how a matrix acts on rows. By taking several elimination steps, we will see how to *multiply matrices* (and the order of the E's becomes important). ***Products and inverses*** are especially clear for E's. It is those two ideas that the book will use.

Matrix Multiplication

The big question is: ***How do we multiply two matrices?*** When the first matrix is E, we know what to expect for EA. This particular E subtracts 2 times row 1 from row 2. The multiplier is $\ell = 2$:

$$EA = \begin{bmatrix} 1 & 0 & 0 \\ \mathbf{-2} & \mathbf{1} & 0 \\ 0 & 0 & 1 \end{bmatrix} \begin{bmatrix} \mathbf{2} & \mathbf{4} & \mathbf{-2} \\ \mathbf{4} & \mathbf{9} & \mathbf{-3} \\ -2 & -3 & 7 \end{bmatrix} = \begin{bmatrix} 2 & 4 & -2 \\ \mathbf{0} & \mathbf{1} & \mathbf{1} \\ -2 & -3 & 7 \end{bmatrix} \quad \text{(with the zero).} \qquad (3)$$

This step does not change rows 1 and 3 of A. Those rows are unchanged in EA—only row 2 is different. *Twice the first row has been subtracted from the second row.* Matrix multiplication agrees with elimination—and the new system of equations is $EA\boldsymbol{x} = E\boldsymbol{b}$.

$EA\boldsymbol{x}$ is simple but it involves a subtle idea. Start with $A\boldsymbol{x} = \boldsymbol{b}$. Multiplying both sides by E gives $E(A\boldsymbol{x}) = E\boldsymbol{b}$. With matrix multiplication, this is also $(EA)\boldsymbol{x} = E\boldsymbol{b}$.

The first was E times Ax, the second is EA times x. They are the same.

Parentheses are not needed. We just write $EA\boldsymbol{x}$.

That rule extends to a matrix C with several column vectors. When multiplying EAC, you can do AC first or EA first. This is the point of an "associative law" like $3 \times (4 \times 5) = (3 \times 4) \times 5$. Multiply 3 times 20, or multiply 12 times 5. Both answers are 60. That law seems so clear that it is hard to imagine it could be false.

The "commutative law" $3 \times 4 = 4 \times 3$ looks even more obvious. But EA is usually different from AE. When E multiplies on the right, it acts on the ***columns*** of A—not the rows. AE actually subtracts 2 times column 2 from column 1. So $EA \neq AE$.

Associative law is true	$A(BC) = (AB)C$
Commutative law is false	**Often** $AB \neq BA$

There is another requirement on matrix multiplication. Suppose B has only one column (this column is $\boldsymbol{b}$). The matrix-matrix law for EB should agree with the matrix-vector law for $E\boldsymbol{b}$. Even more, we should be able to *multiply matrices EB a column at a time*:

If B has several columns b_1, b_2, b_3, then the columns of EB are Eb_1, Eb_2, Eb_3.

Matrix multiplication $\quad AB = A\,[\boldsymbol{b}_1\ \boldsymbol{b}_2\ \boldsymbol{b}_3] = [A\boldsymbol{b}_1\ A\boldsymbol{b}_2\ A\boldsymbol{b}_3].$ (4)

This holds true for the matrix multiplication in (3). If you multiply column 3 of A by E, you correctly get column 3 of EA:

$$\begin{bmatrix} 1 & 0 & 0 \\ -2 & 1 & 0 \\ 0 & 0 & 1 \end{bmatrix} \begin{bmatrix} -2 \\ -3 \\ 7 \end{bmatrix} = \begin{bmatrix} -2 \\ 1 \\ 7 \end{bmatrix} \qquad E(\text{column } j \text{ of } A) = \text{column } j \text{ of } EA.$$

This requirement deals with columns, while elimination is applied to rows. **The next section describes each entry of every product** AB. The beauty of matrix multiplication is that all three approaches (*rows*, *columns*, *whole matrices*) come out right.

The Matrix P_{ij} for a Row Exchange

To subtract row j from row i we use E_{ij}. To exchange or "permute" those rows we use another matrix P_{ij} (a **permutation matrix**). A row exchange is needed when zero is in the pivot position. Lower down, that pivot column may contain a nonzero. By exchanging the two rows, we have a pivot and elimination goes forward.

What matrix P_{23} exchanges row 2 with row 3? We can find it by exchanging rows of the identity matrix I:

$$\textbf{Permutation matrix} \qquad P_{23} = \begin{bmatrix} 1 & 0 & 0 \\ 0 & 0 & 1 \\ 0 & 1 & 0 \end{bmatrix}.$$

This is a ***row exchange matrix***. Multiplying by P_{23} exchanges components 2 and 3 of any column vector. Therefore it also exchanges rows 2 and 3 of any matrix:

$$\begin{bmatrix} 1 & 0 & 0 \\ 0 & 0 & 1 \\ 0 & 1 & 0 \end{bmatrix} \begin{bmatrix} 1 \\ \mathbf{3} \\ \mathbf{5} \end{bmatrix} = \begin{bmatrix} 1 \\ \mathbf{5} \\ \mathbf{3} \end{bmatrix} \quad \text{and} \quad \begin{bmatrix} 1 & 0 & 0 \\ 0 & 0 & 1 \\ 0 & 1 & 0 \end{bmatrix} \begin{bmatrix} 2 & 4 & 1 \\ \mathbf{0} & \mathbf{0} & \mathbf{3} \\ 0 & 6 & 5 \end{bmatrix} = \begin{bmatrix} 2 & 4 & 1 \\ 0 & 6 & 5 \\ \mathbf{0} & \mathbf{0} & \mathbf{3} \end{bmatrix}.$$

On the right, P_{23} is doing what it was created for. With zero in the second pivot position and "6" below it, the exchange puts 6 into the pivot.

Matrices *act*. They don't just sit there. We will soon meet other permutation matrices, which can change the order of several rows. Rows 1, 2, 3 can be moved to 3, 1, 2. Our P_{23} is one particular permutation matrix—it exchanges rows 2 and 3.

Row Exchange Matrix P_{ij} is the identity matrix with rows i and j reversed. When this "**permutation matrix**" P_{ij} multiplies a matrix, it exchanges rows i and j.

$$\textit{\textbf{To exchange equations 1 and 3 multiply by}} \quad P_{13} = \begin{bmatrix} 0 & 0 & 1 \\ 0 & 1 & 0 \\ 1 & 0 & 0 \end{bmatrix}.$$

Usually row exchanges are not required. The odds are good that elimination uses only the E_{ij}. But the P_{ij} are ready if needed, to move a pivot up to the diagonal.

The Augmented Matrix

This book eventually goes far beyond elimination. Matrices have all kinds of practical applications, in which they are multiplied. Our best starting point was a square E times a square A, because we met this in elimination—and we know what answer to expect for EA. The next step is to allow a *rectangular matrix*. It still comes from our original equations, but now it includes the right side $\boldsymbol{b}$.

Key idea: Elimination does the same row operations to A and to $\boldsymbol{b}$. ***We can include b as an extra column and follow it through elimination***. The matrix A is enlarged or "augmented" by the extra column $\boldsymbol{b}$:

$$\textit{\textbf{Augmented matrix}} \qquad \begin{bmatrix} A & \boldsymbol{b} \end{bmatrix} = \begin{bmatrix} 2 & 4 & -2 & \mathbf{2} \\ 4 & 9 & -3 & \mathbf{8} \\ -2 & -3 & 7 & \mathbf{10} \end{bmatrix}.$$

Elimination acts on whole rows of this matrix. The left side and right side are both multiplied by E, to subtract 2 times equation 1 from equation 2. With $\begin{bmatrix} A & \boldsymbol{b} \end{bmatrix}$ those steps happen together:

$$\begin{bmatrix} 1 & 0 & 0 \\ -2 & 1 & 0 \\ 0 & 0 & 1 \end{bmatrix} \begin{bmatrix} 2 & 4 & -2 & \mathbf{2} \\ 4 & 9 & -3 & \mathbf{8} \\ -2 & -3 & 7 & \mathbf{10} \end{bmatrix} = \begin{bmatrix} 2 & 4 & -2 & \mathbf{2} \\ 0 & 1 & 1 & \mathbf{4} \\ -2 & -3 & 7 & \mathbf{10} \end{bmatrix}.$$

The new second row contains $0, 1, 1, 4$. The new second equation is $x_2 + x_3 = 4$. Matrix multiplication works by rows and at the same time by columns:

ROWS Each row of E acts on $\begin{bmatrix} A & \boldsymbol{b} \end{bmatrix}$ to give a row of $\begin{bmatrix} EA & E\boldsymbol{b} \end{bmatrix}$.

COLUMNS E acts on each column of $\begin{bmatrix} A & \boldsymbol{b} \end{bmatrix}$ to give a column of $\begin{bmatrix} EA & E\boldsymbol{b} \end{bmatrix}$.

Notice again that word "acts." This is essential. Matrices do something! The matrix A acts on $\boldsymbol{x}$ to produce $\boldsymbol{b}$. The matrix E operates on A to give EA. The whole process of elimination is a sequence of row operations, alias matrix multiplications. A goes to $E_{21}A$ which goes to $E_{31}E_{21}A$. Finally $E_{32}E_{31}E_{21}A$ is a triangular matrix.

The right side is included in the augmented matrix. The end result is a triangular system of equations. We stop for exercises on multiplication by E, before writing down the rules for all matrix multiplications (including block multiplication).

■ REVIEW OF THE KEY IDEAS ■

1. $A\boldsymbol{x} = x_1$ times column $1 + \cdots + x_n$ times column n. And $(A\boldsymbol{x})_i = \sum_{j=1}^{n} a_{ij}x_j$.
2. Identity matrix $= I$, elimination matrix $= E_{ij}$ using ℓ_{ij}, exchange matrix $= P_{ij}$.
3. Multiplying $A\boldsymbol{x} = \boldsymbol{b}$ by E_{21} subtracts a multiple ℓ_{21} of equation 1 from equation 2. The number $-\ell_{21}$ is the $(2,1)$ entry of the elimination matrix E_{21}.
4. For the augmented matrix $\begin{bmatrix} A & \boldsymbol{b} \end{bmatrix}$, that elimination step gives $\begin{bmatrix} E_{21}A & E_{21}\boldsymbol{b} \end{bmatrix}$.
5. When A multiplies any matrix B, it multiplies each column of B separately.

■ WORKED EXAMPLES ■

2.3 A What 3 by 3 matrix E_{21} subtracts 4 times row 1 from row 2? What matrix P_{32} exchanges row 2 and row 3? If you multiply A on the *right* instead of the left, describe the results AE_{21} and AP_{32}.

Solution By doing those operations on the identity matrix I, we find

$$E_{21} = \begin{bmatrix} 1 & 0 & 0 \\ -4 & 1 & 0 \\ 0 & 0 & 1 \end{bmatrix} \quad \text{and} \quad P_{32} = \begin{bmatrix} 1 & 0 & 0 \\ 0 & 0 & 1 \\ 0 & 1 & 0 \end{bmatrix}.$$

Multiplying by E_{21} on the right side will subtract 4 times **column 2** from **column 1**. Multiplying by P_{32} on the right will exchange **columns 2** and **3**.

2.3 B Write down the augmented matrix $[A \quad b]$ with an extra column:

$$\begin{aligned} x + 2y + 2z &= 1 \\ 4x + 8y + 9z &= 3 \\ 3y + 2z &= 1 \end{aligned}$$

Apply E_{21} and then P_{32} to reach a triangular system. Solve by back substitution. What combined matrix $P_{32}\,E_{21}$ will do both steps at once?

Solution E_{21} removes the 4 in column 1. But zero also appears in column 2:

$$[A \ \ \boldsymbol{b}] = \begin{bmatrix} 1 & 2 & 2 & 1 \\ \mathbf{4} & 8 & 9 & 3 \\ 0 & 3 & 2 & 1 \end{bmatrix} \quad \text{and} \quad E_{21}[A \ \ \boldsymbol{b}] = \begin{bmatrix} 1 & 2 & 2 & 1 \\ \mathbf{0} & \mathbf{0} & 1 & -1 \\ 0 & 3 & 2 & 1 \end{bmatrix}$$

Now P_{32} exchanges rows 2 and 3. Back substitution produces z then y and x.

$$P_{32}\,E_{21}[A \ \ \boldsymbol{b}] = \begin{bmatrix} 1 & 2 & 2 & 1 \\ 0 & 3 & 2 & 1 \\ 0 & 0 & 1 & -1 \end{bmatrix} \quad \text{and} \quad \begin{bmatrix} x \\ y \\ z \end{bmatrix} = \begin{bmatrix} 1 \\ 1 \\ -1 \end{bmatrix}$$

For the matrix $P_{32}\,E_{21}$ that does both steps at once, *apply* P_{32} *to* E_{21}.

One matrix
Both steps
$$P_{32}\,E_{21} = \text{exchange the rows of } E_{21} = \begin{bmatrix} 1 & 0 & 0 \\ 0 & 0 & 1 \\ -4 & 1 & 0 \end{bmatrix}.$$

2.3 C Multiply these matrices in two ways. First, rows of A times columns of B. Second, ***columns of A times rows of B***. That unusual way produces two matrices that add to AB. How many separate ordinary multiplications are needed?

Both ways
$$AB = \begin{bmatrix} 3 & 4 \\ 1 & 5 \\ 2 & 0 \end{bmatrix} \begin{bmatrix} 2 & 4 \\ 1 & 1 \end{bmatrix} = \begin{bmatrix} \mathbf{10} & \mathbf{16} \\ \mathbf{7} & \mathbf{9} \\ \mathbf{4} & \mathbf{8} \end{bmatrix}$$

Solution Rows of A times columns of B are dot products of vectors:

$$(\text{row } 1) \cdot (\text{column } 1) = \begin{bmatrix} 3 & 4 \end{bmatrix} \begin{bmatrix} 2 \\ 1 \end{bmatrix} = \mathbf{10} \quad \text{is the } (1,1) \text{ entry of } AB$$

$$(\text{row } 2) \cdot (\text{column } 1) = \begin{bmatrix} 1 & 5 \end{bmatrix} \begin{bmatrix} 2 \\ 1 \end{bmatrix} = \mathbf{7} \quad \text{is the } (2,1) \text{ entry of } AB$$

We need 6 dot products, 2 multiplications each, 12 in all $(3 \cdot 2 \cdot 2)$. The same AB comes from *columns of A times rows of B*. A column times a row is a matrix.

$$AB = \begin{bmatrix} 3 \\ 1 \\ 2 \end{bmatrix} \begin{bmatrix} 2 & 4 \end{bmatrix} + \begin{bmatrix} 4 \\ 5 \\ 0 \end{bmatrix} \begin{bmatrix} 1 & 1 \end{bmatrix} = \begin{bmatrix} \mathbf{6} & \mathbf{12} \\ \mathbf{2} & \mathbf{4} \\ \mathbf{4} & \mathbf{8} \end{bmatrix} + \begin{bmatrix} \mathbf{4} & \mathbf{4} \\ \mathbf{5} & \mathbf{5} \\ \mathbf{0} & \mathbf{0} \end{bmatrix}$$

Problem Set 2.3

Problems 1–15 are about elimination matrices.

1 Write down the 3 by 3 matrices that produce these elimination steps:

(a) E_{21} subtracts 5 times row 1 from row 2.

(b) E_{32} subtracts -7 times row 2 from row 3.

(c) P exchanges rows 1 and 2, then rows 2 and 3.

2 In Problem 1, applying E_{21} and then E_{32} to $\boldsymbol{b} = (1, 0, 0)$ gives $E_{32}E_{21}\boldsymbol{b} =$ ______. Applying E_{32} before E_{21} gives $E_{21}E_{32}\boldsymbol{b} =$ ______. When E_{32} comes first, row ______ feels no effect from row ______.

3 Which three matrices E_{21}, E_{31}, E_{32} put A into triangular form U?

$$A = \begin{bmatrix} 1 & 1 & 0 \\ 4 & 6 & 1 \\ -2 & 2 & 0 \end{bmatrix} \quad \text{and} \quad E_{32}E_{31}E_{21}A = U.$$

Multiply those E's to get one matrix M that does elimination: $MA = U$.

4 Include $\boldsymbol{b} = (1, 0, 0)$ as a fourth column in Problem 3 to produce $[\,A \;\; \boldsymbol{b}\,]$. Carry out the elimination steps on this augmented matrix to solve $A\boldsymbol{x} = \boldsymbol{b}$.

5 Suppose $a_{33} = 7$ and the third pivot is 5. If you change a_{33} to 11, the third pivot is ______. If you change a_{33} to ______, there is no third pivot.

6 If every column of A is a multiple of $(1, 1, 1)$, then $A\boldsymbol{x}$ is always a multiple of $(1, 1, 1)$. Do a 3 by 3 example. How many pivots are produced by elimination?

7 Suppose E subtracts 7 times row 1 from row 3.

(a) To *invert* that step you should ______ 7 times row ______ to row ______.

(b) What "inverse matrix" E^{-1} takes that reverse step (so $E^{-1}E = I$)?

(c) If the reverse step is applied first (and then E) show that $EE^{-1} = I$.

8 The ***determinant*** of $M = \left[\begin{smallmatrix} a & b \\ c & d \end{smallmatrix}\right]$ is det $M = ad - bc$. Subtract ℓ times row 1 from row 2 to produce a new M^*. Show that det $M^* =$ det M for every ℓ. When $\ell = c/a$, *the product of pivots equals the determinant*: $(a)(d - \ell b)$ equals $ad - bc$.

9 (a) E_{21} subtracts row 1 from row 2 and then P_{23} exchanges rows 2 and 3. What matrix $M = P_{23}E_{21}$ does both steps at once?

(b) P_{23} exchanges rows 2 and 3 and then E_{31} subtracts row 1 from row 3. What matrix $M = E_{31}P_{23}$ does both steps at once? Explain why the M's are the same but the E's are different.

10 (a) What 3 by 3 matrix E_{13} will add row 3 to row 1?

(b) What matrix adds row 1 to row 3 and *at the same time* row 3 to row 1?

(c) What matrix adds row 1 to row 3 and *then* adds row 3 to row 1? 24

11 Create a matrix that has $a_{11} = a_{22} = a_{33} = 1$ but elimination produces two negative pivots without row exchanges. (The first pivot is 1.) 25

12 Multiply these matrices:

$$\begin{bmatrix} 0 & 0 & 1 \\ 0 & 1 & 0 \\ 1 & 0 & 0 \end{bmatrix} \begin{bmatrix} 1 & 2 & 3 \\ 4 & 5 & 6 \\ 7 & 8 & 9 \end{bmatrix} \begin{bmatrix} 0 & 0 & 1 \\ 0 & 1 & 0 \\ 1 & 0 & 0 \end{bmatrix} \qquad \begin{bmatrix} 1 & 0 & 0 \\ -1 & 1 & 0 \\ -1 & 0 & 1 \end{bmatrix} \begin{bmatrix} 1 & 2 & 3 \\ 1 & 3 & 1 \\ 1 & 4 & 0 \end{bmatrix}.$$

13 Explain these facts. If the third column of B is all zero, the third column of EB is all zero (for any E). If the third *row* of B is all zero, the third row of EB might *not* be zero.

14 This 4 by 4 matrix will need elimination matrices E_{21} and E_{32} and E_{43}. What are those matrices?

$$A = \begin{bmatrix} 2 & -1 & 0 & 0 \\ -1 & 2 & -1 & 0 \\ 0 & -1 & 2 & -1 \\ 0 & 0 & -1 & 2 \end{bmatrix}.$$

15 Write down the 3 by 3 matrix that has $a_{ij} = 2i - 3j$. This matrix has $a_{32} = 0$, but elimination still needs E_{32} to produce a zero in the $3, 2$ position. Which previous step destroys the original zero and what is E_{32}?

Problems 16–23 are about creating and multiplying matrices.

16 Write these ancient problems in a 2 by 2 matrix form $A\boldsymbol{x} = \boldsymbol{b}$ and solve them:

(a) X is twice as old as Y and their ages add to 33.

(b) $(x, y) = (2, 5)$ and $(3, 7)$ lie on the line $y = mx + c$. Find m and c.

17 The parabola $y = a + bx + cx^2$ goes through the points $(x, y) = (1, 4)$ and $(2, 8)$ and $(3, 14)$. Find and solve a matrix equation for the unknowns (a, b, c).

18 Multiply these matrices in the orders EF and FE:

$$E = \begin{bmatrix} 1 & 0 & 0 \\ a & 1 & 0 \\ b & 0 & 1 \end{bmatrix} \qquad F = \begin{bmatrix} 1 & 0 & 0 \\ 0 & 1 & 0 \\ 0 & c & 1 \end{bmatrix}.$$

Also compute $E^2 = EE$ and $F^3 = FFF$. You can guess F^{100}.

19 Multiply these row exchange matrices in the orders PQ and QP and P^2:

$$P = \begin{bmatrix} 0 & 1 & 0 \\ 1 & 0 & 0 \\ 0 & 0 & 1 \end{bmatrix} \quad \text{and} \quad Q = \begin{bmatrix} 0 & 0 & 1 \\ 0 & 1 & 0 \\ 1 & 0 & 0 \end{bmatrix}.$$

Find another non-diagonal matrix whose square is $M^2 = I$.

20 (a) Suppose all columns of B are the same. Then all columns of EB are the same, because each one is E times ______ .

(b) Suppose all rows of B are $[\,1 \;\; 2 \;\; 4\,]$. Show by example that all rows of EB are *not* $[\,1 \;\; 2 \;\; 4\,]$. It is true that those rows are ______ .

21 If E adds row 1 to row 2 and F adds row 2 to row 1, does EF equal FE?

22 The entries of A and $\boldsymbol{x}$ are a_{ij} and x_j. So the first component of $A\boldsymbol{x}$ is $\sum a_{1j}x_j = a_{11}x_1 + \cdots + a_{1n}x_n$. If E_{21} subtracts row 1 from row 2, write a formula for

(a) the third component of $A\boldsymbol{x}$

(b) the $(2,1)$ entry of $E_{21}A$

(c) the $(2,1)$ entry of $E_{21}(E_{21}A)$

(d) the first component of $E_{21}A\boldsymbol{x}$.

23 The elimination matrix $E = \begin{bmatrix} 1 & 0 \\ -2 & 1 \end{bmatrix}$ subtracts 2 times row 1 of A from row 2 of A. The result is EA. What is the effect of $E(EA)$? In the opposite order AE, we are subtracting 2 times ______ of A from ______. (Do examples.)

Problems 24–27 include the column b in the augmented matrix $[\,A \;\; b\,]$.

24 Apply elimination to the 2 by 3 augmented matrix $[\,A \;\; \boldsymbol{b}\,]$. What is the triangular system $U\boldsymbol{x} = \boldsymbol{c}$? What is the solution $\boldsymbol{x}$?

$$A\boldsymbol{x} = \begin{bmatrix} 2 & 3 \\ 4 & 1 \end{bmatrix} \begin{bmatrix} x_1 \\ x_2 \end{bmatrix} = \begin{bmatrix} 1 \\ 17 \end{bmatrix}.$$

25 Apply elimination to the 3 by 4 augmented matrix $[\,A \;\; \boldsymbol{b}\,]$. How do you know this system has no solution? Change the last number 6 so there *is* a solution.

$$A\boldsymbol{x} = \begin{bmatrix} 1 & 2 & 3 \\ 2 & 3 & 4 \\ 3 & 5 & 7 \end{bmatrix} \begin{bmatrix} x \\ y \\ z \end{bmatrix} = \begin{bmatrix} 1 \\ 2 \\ 6 \end{bmatrix}.$$

26 The equations $A\boldsymbol{x} = \boldsymbol{b}$ and $A\boldsymbol{x}^* = \boldsymbol{b}^*$ have the same matrix A. What double augmented matrix should you use in elimination to solve both equations at once?

Solve both of these equations by working on a 2 by 4 matrix:

$$\begin{bmatrix} 1 & 4 \\ 2 & 7 \end{bmatrix} \begin{bmatrix} x \\ y \end{bmatrix} = \begin{bmatrix} 1 \\ 0 \end{bmatrix} \quad \text{and} \quad \begin{bmatrix} 1 & 4 \\ 2 & 7 \end{bmatrix} \begin{bmatrix} u \\ v \end{bmatrix} = \begin{bmatrix} 0 \\ 1 \end{bmatrix}.$$

27 Choose the numbers a, b, c, d in this augmented matrix so that there is (a) no solution (b) infinitely many solutions.

$$\begin{bmatrix} A & \boldsymbol{b} \end{bmatrix} = \begin{bmatrix} 1 & 2 & 3 & a \\ 0 & 4 & 5 & b \\ 0 & 0 & d & c \end{bmatrix}$$

Which of the numbers a, b, c, or d have no effect on the solvability?

28 If $AB = I$ and $BC = I$ use the associative law to prove $A = C$.

Challenge Problems

29 Find the triangular matrix E that reduces "*Pascal's matrix*" to a smaller Pascal:

Elimination on column 1 $$E \begin{bmatrix} 1 & 0 & 0 & 0 \\ 1 & 1 & 0 & 0 \\ 1 & 2 & 1 & 0 \\ 1 & 3 & 3 & 1 \end{bmatrix} = \begin{bmatrix} 1 & 0 & 0 & 0 \\ 0 & 1 & 0 & 0 \\ 0 & 1 & 1 & 0 \\ 0 & 1 & 2 & 1 \end{bmatrix}.$$

Which matrix M (multiplying several E's) reduces Pascal all the way to I? Pascal's triangular matrix is exceptional, all of its multipliers are $\ell_{ij} = 1$.

30 Write $M = \left[\begin{smallmatrix} 3 & 4 \\ 5 & 7 \end{smallmatrix}\right]$ as a product of many factors $A = \left[\begin{smallmatrix} 1 & 0 \\ 1 & 1 \end{smallmatrix}\right]$ and $B = \left[\begin{smallmatrix} 1 & 1 \\ 0 & 1 \end{smallmatrix}\right]$.

(a) What matrix E subtracts row 1 from row 2 to make row 2 of EM smaller?

(b) What matrix F subtracts row 2 of EM from row 1 to reduce row 1 of FEM?

(c) Continue E's and F's until (many E's and F's) times (M) is $(A$ or $B)$.

(d) E and F are the inverses of A and B ! Moving all E's and F's to the right side will give you the desired result $\boldsymbol{M}$ **= *product of A's and B's***.
This is possible for integer matrices $M = \left[\begin{smallmatrix} a & b \\ c & d \end{smallmatrix}\right] > 0$ that have $ad - bc = 1$.

31 Find elimination matrices E_{21} then E_{32} then E_{43} to change K into U:

$$E_{43}\, E_{32}\, E_{21} \begin{bmatrix} 1 & 0 & 0 & 0 \\ -a & 1 & 0 & 0 \\ 0 & -b & 1 & 0 \\ 0 & 0 & -c & 1 \end{bmatrix} = I.$$

Apply those three steps to the identity matrix *I*, to multiply $\boldsymbol{E_{43}E_{32}E_{21}}$.

2.4 Rules for Matrix Operations

1 Matrices A with n columns multiply matrices B with n rows: $\boxed{\boldsymbol{A_{m\times n}\,B_{n\times p} = C_{m\times p}}.}$

2 Each entry in $AB = C$ is a dot product: $C_{ij} =$ (row i of A) $\cdot$ (column j of B).

3 This rule is chosen so that **AB times C equals A times BC**. And $(AB)\,\boldsymbol{x} = A(B\,\boldsymbol{x})$.

4 More ways to compute AB: (A times columns of B) (rows of A times B) (*columns times rows*).

5 It is not usually true that $AB = BA$. In most cases A *doesn't commute with* B.

6 Matrices can be multiplied by *blocks*: $A = [A_1\ A_2]$ times $B = \begin{bmatrix} B_1 \\ B_2 \end{bmatrix}$ is $A_1B_1 + A_2B_2$.

I will start with basic facts. A matrix is a rectangular array of numbers or "entries". When A has m rows and n columns, it is an "m by n" matrix. Matrices can be added if their shapes are the same. They can be multiplied by any constant c. Here are examples of $A+B$ and $2A$, for 3 by 2 matrices:

$$\begin{bmatrix} 1 & 2 \\ 3 & 4 \\ 0 & 0 \end{bmatrix} + \begin{bmatrix} 2 & 2 \\ 4 & 4 \\ 9 & 9 \end{bmatrix} = \begin{bmatrix} 3 & 4 \\ 7 & 8 \\ 9 & 9 \end{bmatrix} \quad \text{and} \quad 2\begin{bmatrix} 1 & 2 \\ 3 & 4 \\ 0 & 0 \end{bmatrix} = \begin{bmatrix} 2 & 4 \\ 6 & 8 \\ 0 & 0 \end{bmatrix}.$$

Matrices are added exactly as vectors are—one entry at a time. We could even regard a column vector as a matrix with only one column (so $n = 1$). The matrix $-A$ comes from multiplication by $c = -1$ (reversing all the signs). Adding A to $-A$ leaves the *zero matrix*, with all entries zero. All this is only common sense.

The entry in row i and column j is called a_{ij} or $A(i,j)$. The n entries along the first row are $a_{11}, a_{12}, \ldots, a_{1n}$. The lower left entry in the matrix is a_{m1} and the lower right is a_{mn}. The row number i goes from 1 to m. The column number j goes from 1 to n.

Matrix addition is easy. The serious question is ***matrix multiplication.*** When can we multiply A times B, and what is the product AB? *This section gives* 4 *ways to find* AB. But we cannot multiply when A and B are 3 by 2. They don't pass the following test:

To multiply AB: If A has n columns, B must have n rows.

When A is 3 by 2, the matrix B can be 2 by 1 (a vector) or 2 by 2 (square) or 2 by 20. ***Every column of B is multiplied by A.*** I will begin matrix multiplication the *dot product way*, and return to this *column way*: A times columns of B. Both ways follow this rule:

Fundamental Law of Matrix Multiplication ***AB times C equals A times BC*** (1)

The parentheses can move safely in $(AB)C = A(BC)$. Linear algebra depends on this law.

Suppose A is m by n and B is n by p. We can multiply. The product AB is m by p.

$$(m \times n)(n \times p) = (m \times p) \qquad \begin{bmatrix} m \textbf{ rows} \\ n \textit{ columns} \end{bmatrix} \begin{bmatrix} n \textit{ rows} \\ p \textbf{ columns} \end{bmatrix} = \begin{bmatrix} m \textbf{ rows} \\ p \textbf{ columns} \end{bmatrix}.$$

A row times a column is an extreme case. Then 1 by n multiplies n by 1. The result will be 1 by 1. That single number is the "dot product".

In every case AB is filled with dot products. For the top corner, the $(1,1)$ entry of AB is (row 1 of A) $\cdot$ (column 1 of B). This is the first way, and the usual way, to multiply matrices. ***Take the dot product of each row of A with each column of B.***

1. ***The entry in row i and column j of AB is*** (row i of A) $\cdot$ (column j of B) .

Figure 2.8 picks out the second row ($i = 2$) of a 4 by 5 matrix A. It picks out the third column ($j = 3$) of a 5 by 6 matrix B. Their dot product goes into row 2 and column 3 of AB. The matrix AB has *as many rows as* A (4 rows), and *as many columns as* B.

$$\begin{bmatrix} * & & & \\ a_{i1} & a_{i2} & \cdots & a_{i5} \\ * & & & \\ * & & & \end{bmatrix} \begin{bmatrix} * & * & b_{1j} & * & * & * \\ & & b_{2j} & & & \\ & & \vdots & & & \\ & & b_{5j} & & & \end{bmatrix} = \begin{bmatrix} & & * & & & \\ * & * & (AB)_{ij} & * & * & * \\ & & * & & & \\ & & * & & & \end{bmatrix}$$

A is 4 by 5 $\qquad$ B is 5 by 6 $\qquad$ AB is $(4 \times 5)(5 \times 6) = 4$ by 6

Figure 2.8: Here $i = 2$ and $j = 3$. Then $(AB)_{23}$ is **(row 2)** $\cdot$ **(column 3)** $=$ sum of $a_{2k}b_{k3}$.

Example 1 Square matrices can be multiplied if and only if they have the same size:

$$\begin{bmatrix} 1 & 1 \\ 2 & -1 \end{bmatrix} \begin{bmatrix} 2 & 2 \\ 3 & 4 \end{bmatrix} = \begin{bmatrix} 5 & 6 \\ 1 & 0 \end{bmatrix}.$$

The first dot product is $1 \cdot 2 + 1 \cdot 3 = 5$. Three more dot products give $6, 1$, and 0. Each dot product requires two multiplications—thus eight in all.

If A and B are n by n, so is AB. It contains n^2 dot products, row of A times column of B. Each dot product needs n multiplications, so ***the computation of*** AB ***uses*** $\boldsymbol{n^3}$ ***separate multiplications***. For $n = 100$ we multiply a million times. For $n = 2$ we have $n^3 = 8$.

Mathematicians thought until recently that AB absolutely needed $2^3 = 8$ multiplications. Then somebody found a way to do it with 7 (and extra additions). By breaking n by n matrices into 2 by 2 blocks, this idea also reduced the count to multiply large matrices. Instead of n^3 multiplications the count has now dropped to $n^{2.376}$. Maybe n^2 is possible? But the algorithms are so awkward that scientific computing is done the regular n^3 way.

Example 2 Suppose A is a row vector (1 by 3) and B is a column vector (3 by 1). Then AB is 1 by 1 (only one entry, the dot product). On the other hand B times A (***a column times a row***) is a full 3 by 3 matrix. This multiplication is allowed!

Column times row
$(\boldsymbol{n}\times \mathbf{1})(\mathbf{1}\times \boldsymbol{n}) = (\boldsymbol{n}\times \boldsymbol{n})$

$$\begin{bmatrix}0\\1\\2\end{bmatrix}\begin{bmatrix}1 & 2 & 3\end{bmatrix} = \begin{bmatrix}0 & 0 & 0\\1 & 2 & 3\\2 & 4 & 6\end{bmatrix}.$$

A row times a column is an *"inner"* product—that is another name for dot product. A column times a row is an *"outer"* product. These are extreme cases of matrix multiplication.

The Second and Third Ways: Rows and Columns

In the big picture, A multiplies each column of B. The result is a column of AB. In that column, we are combining the columns of A. ***Each column of AB is a combination of the columns of A.*** That is the column picture of matrix multiplication:

2. Matrix A times every column of B $\qquad A\begin{bmatrix}\boldsymbol{b}_1 \cdots \boldsymbol{b}_p\end{bmatrix} = \begin{bmatrix}A\boldsymbol{b}_1 \cdots A\boldsymbol{b}_p\end{bmatrix}.$

The row picture is reversed. Each row of A multiplies the whole matrix B. The result is a row of AB. **Every row of AB is a combination of the rows of B**:

3. Every row of A times matrix B $\qquad \begin{bmatrix}\textbf{row } i \textbf{ of } A\end{bmatrix}\begin{bmatrix}1 & 2 & 3\\4 & 5 & 6\\7 & 8 & 9\end{bmatrix} = \begin{bmatrix}\textbf{row } i \textbf{ of } AB\end{bmatrix}.$

We see row operations in elimination (E times A). Soon we see columns in $AA^{-1} = I$. The "row-column picture" has the dot products of rows with columns. Dot products are the usual way to multiply matrices by hand: mnp separate steps of multiply/add.

$$\boldsymbol{AB} = (\boldsymbol{m}\times \boldsymbol{n})(\boldsymbol{n}\times \boldsymbol{p}) = (\boldsymbol{m}\times \boldsymbol{p}) \quad \boldsymbol{mp} \textbf{ dot products with } \boldsymbol{n} \textbf{ steps each} \qquad (2)$$

The Fourth Way: Columns Multiply Rows

There is a fourth way to multiply matrices. Not many people realize how important this is. I feel like a magician explaining a trick. Magicians won't do it but mathematicians try. The fourth way was in previous editions of this book, but I didn't emphasize it enough.

4. Multiply columns 1 to n of A times rows 1 to n of B. Add those matrices.

Column 1 of A multiplies row 1 of B. Columns 2 and 3 multiply rows 2 and 3. Then add:

$$\begin{bmatrix}\textbf{col 1} & \textbf{col 2} & \textbf{col 3}\\ \cdot & \cdot & \cdot\\ \cdot & \cdot & \cdot\end{bmatrix}\begin{bmatrix}\textbf{row 1} & \cdot\cdot\cdot\\ \textbf{row 2} & \cdot\cdot\cdot\\ \textbf{row 3} & \cdot\cdot\cdot\end{bmatrix} = (\textbf{col 1})\,(\textbf{row 1}) + (\textbf{col 2})\,(\textbf{row 2}) + (\textbf{col 3})\,(\textbf{row 3}).$$

If I multiply 2 by 2 matrices this column–row way, you will see that AB is correct.

$$AB = \begin{bmatrix} \boldsymbol{a} & b \\ \boldsymbol{c} & d \end{bmatrix} \begin{bmatrix} \boldsymbol{E} & \boldsymbol{F} \\ G & H \end{bmatrix} = \begin{bmatrix} \boldsymbol{aE} + bG & \boldsymbol{aF} + bH \\ \boldsymbol{cE} + dG & \boldsymbol{cF} + dH \end{bmatrix}$$

Add columns of A times rows of B

$$\boldsymbol{AB} = \begin{bmatrix} \boldsymbol{a} \\ \boldsymbol{c} \end{bmatrix} \begin{bmatrix} \boldsymbol{E} & \boldsymbol{F} \end{bmatrix} + \begin{bmatrix} b \\ d \end{bmatrix} \begin{bmatrix} G & H \end{bmatrix} \qquad (3)$$

Column k of A multiplies row k of B. That gives a matrix (not just a number). Then you add those matrices for $k = 1, 2, \ldots, n$ to produce AB.

If AB is (m by n) (n by p) then n matrices will be (*column*) (*row*). They are all m by p. This uses the same mnp steps as in the dot products—but in a new order.

The Laws for Matrix Operations

May I put on record six laws that matrices do obey, while emphasizing a rule they *don't* obey? The matrices can be square or rectangular, and the laws involving $A + B$ are all simple and all obeyed. Here are three addition laws:

$$\begin{aligned} A + B &= B + A && \text{(commutative law)} \\ c(A + B) &= cA + cB && \text{(distributive law)} \\ A + (B + C) &= (A + B) + C && \text{(associative law).} \end{aligned}$$

Three more laws hold for multiplication, but $AB = BA$ is not one of them:

$$\begin{aligned} AB &\neq BA && \text{(the commutative "law" is \textit{usually broken})} \\ A(B + C) &= AB + AC && \text{(distributive law from the left)} \\ (A + B)C &= AC + BC && \text{(distributive law from the right)} \\ A(BC) &= (AB)C && \text{(associative law for } ABC\text{) (\textbf{\textit{parentheses not needed}}).} \end{aligned}$$

When A and B are not square, AB is a different size from BA. These matrices can't be equal—even if both multiplications are allowed. For square matrices, almost any example shows that AB is different from BA:

$$AB = \begin{bmatrix} 0 & 0 \\ 1 & 0 \end{bmatrix} \begin{bmatrix} 0 & 1 \\ 0 & 0 \end{bmatrix} = \begin{bmatrix} 0 & 0 \\ 0 & \mathbf{1} \end{bmatrix} \quad \text{but} \quad BA = \begin{bmatrix} 0 & 1 \\ 0 & 0 \end{bmatrix} \begin{bmatrix} 0 & 0 \\ 1 & 0 \end{bmatrix} = \begin{bmatrix} \mathbf{1} & 0 \\ 0 & 0 \end{bmatrix}.$$

It is true that $AI = IA$. All square matrices commute with I and also with cI. Only these matrices cI commute with all other matrices.

The law $A(B + C) = AB + AC$ is proved a column at a time. Start with $A(\boldsymbol{b} + \boldsymbol{c}) = A\boldsymbol{b} + A\boldsymbol{c}$ for the first column. That is the key to everything—***linearity***. Say no more.

The law $A(BC) = (AB)C$ means that you can multiply BC first or else AB first. The direct proof is sort of awkward (Problem 37) but this law is extremely useful. We highlighted it above; it is the key to the way we multiply matrices.

Look at the special case when $A = B = C =$ square matrix. Then (A *times* A^2) is equal to (A^2 *times* A). The product in either order is A^3. The matrix powers A^p follow the same rules as numbers:

$$\boldsymbol{A^p = AAA \cdots A \;(p \text{ factors})} \qquad (A^p)(A^q) = A^{p+q} \qquad (A^p)^q = A^{pq}.$$

Those are the ordinary laws for exponents. A^3 times A^4 is A^7 (seven factors). But the fourth power of A^3 is A^{12} (twelve A's). When p and q are zero or negative these rules still hold, provided A has a "-1 power"—which is the *inverse matrix* A^{-1}. Then $A^0 = I$ is the identity matrix in analogy with $2^0 = 1$.

For a number, a^{-1} is $1/a$. For a matrix, the inverse is written A^{-1}. (It is *not* I/A, except in MATLAB.) Every number has an inverse except $a = 0$. To decide when A has an inverse is a central problem in linear algebra. Section 2.5 will start on the answer. This section is a Bill of Rights for matrices, to say when A and B can be multiplied and how.

Block Matrices and Block Multiplication

We have to say one more thing about matrices. They can be cut into ***blocks*** (which are smaller matrices). This often happens naturally. Here is a 4 by 6 matrix broken into blocks of size 2 by 2—in this example each block is just I:

4 by 6 matrix
2 by 2 blocks give
2 by 3 block matrix

$$A = \left[\begin{array}{cc|cc|cc} 1 & 0 & 1 & 0 & 1 & 0 \\ 0 & 1 & 0 & 1 & 0 & 1 \\ \hline 1 & 0 & 1 & 0 & 1 & 0 \\ 0 & 1 & 0 & 1 & 0 & 1 \end{array}\right] = \begin{bmatrix} I & I & I \\ I & I & I \end{bmatrix}.$$

If B is also 4 by 6 and the block sizes match, you can add $A + B$ *a block at a time*.

You have seen block matrices before. The right side vector $\boldsymbol{b}$ was placed next to A in the "augmented matrix". Then $[\,A \;\; \boldsymbol{b}\,]$ has two blocks of different sizes. Multiplying by an elimination matrix gave $[\,EA \;\; E\boldsymbol{b}\,]$. No problem to multiply blocks times blocks, when their shapes permit.

Block multiplication If blocks of A can multiply blocks of B, then block multiplication of AB is allowed. Cuts between columns of A match cuts between rows of B.

$$\begin{bmatrix} A_{11} & A_{12} \\ A_{21} & A_{22} \end{bmatrix} \begin{bmatrix} B_{11} \\ B_{21} \end{bmatrix} = \begin{bmatrix} A_{11}B_{11} + A_{12}B_{21} \\ A_{21}B_{11} + A_{22}B_{21} \end{bmatrix}. \tag{4}$$

This equation is the same as if the blocks were numbers (which are 1 by 1 blocks). We are careful to keep A's in front of B's, because BA can be different.

Main point When matrices split into blocks, it is often simpler to see how they act. The block matrix of I's above is much clearer than the original 4 by 6 matrix A.

Example 3 **(Important special case)** Let the blocks of A be its n columns. Let the blocks of B be its n rows. Then block multiplication AB adds up ***columns times rows***:

Columns times rows
$$\begin{bmatrix} | & & | \\ \boldsymbol{a}_1 & \cdots & \boldsymbol{a}_n \\ | & & | \end{bmatrix} \begin{bmatrix} - & \boldsymbol{b}_1 & - \\ & \vdots & \\ - & \boldsymbol{b}_n & - \end{bmatrix} = \begin{bmatrix} \boldsymbol{a}_1\boldsymbol{b}_1 + \cdots + \boldsymbol{a}_n\boldsymbol{b}_n \end{bmatrix}. \qquad (5)$$

This is Rule 4 to multiply matrices. Here is a numerical example:

$$\begin{bmatrix} 1 & 4 \\ 1 & 5 \end{bmatrix} \begin{bmatrix} 3 & 2 \\ 1 & 0 \end{bmatrix} = \begin{bmatrix} 1 \\ 1 \end{bmatrix} \begin{bmatrix} 3 & 2 \end{bmatrix} + \begin{bmatrix} 4 \\ 5 \end{bmatrix} \begin{bmatrix} 1 & 0 \end{bmatrix} = \begin{bmatrix} 3 & 2 \\ 3 & 2 \end{bmatrix} + \begin{bmatrix} 4 & 0 \\ 5 & 0 \end{bmatrix} = \begin{bmatrix} 7 & 2 \\ 8 & 2 \end{bmatrix}.$$

Summary The usual way, rows times columns, gives four dot products (8 multiplications). The new way, columns times rows, gives two full matrices (the same 8 multiplications).

Example 4 (**Elimination by blocks**) Suppose the first column of A contains $1, 3, 4$. To change 3 and 4 to 0 and 0, multiply the pivot row by 3 and 4 and subtract. Those row operations are really multiplications by elimination matrices E_{21} and E_{31} :

One at a time
$$E_{21} = \begin{bmatrix} 1 & 0 & 0 \\ -3 & 1 & 0 \\ 0 & 0 & 1 \end{bmatrix} \quad \text{and} \quad E_{31} = \begin{bmatrix} 1 & 0 & 0 \\ 0 & 1 & 0 \\ -4 & 0 & 1 \end{bmatrix}.$$

The "block idea" is to do both eliminations with one matrix E. That matrix clears out the whole first column of A below the pivot $a = 1$:

$$E = \begin{bmatrix} 1 & 0 & 0 \\ -3 & 1 & 0 \\ -4 & 0 & 1 \end{bmatrix} \quad \text{multiplies} \quad \begin{bmatrix} \mathbf{1} & x & x \\ \mathbf{3} & x & x \\ \mathbf{4} & x & x \end{bmatrix} \quad \text{to give} \quad EA = \begin{bmatrix} \mathbf{1} & x & x \\ \mathbf{0} & y & y \\ \mathbf{0} & z & z \end{bmatrix}.$$

Using inverse matrices, a block matrix E can do elimination on a whole (block) column. Suppose a matrix has four blocks A, B, C, D. Watch how E eliminates C by blocks:

Block elimination
$$\left[\begin{array}{c|c} I & \mathbf{0} \\ \hline -CA^{-1} & I \end{array}\right] \left[\begin{array}{c|c} A & B \\ \hline C & D \end{array}\right] = \left[\begin{array}{c|c} A & B \\ \hline \mathbf{0} & \boldsymbol{D - CA^{-1}B} \end{array}\right]. \qquad (6)$$

Elimination multiplies the first row $[\,A \quad B\,]$ by CA^{-1} (previously c/a). It subtracts from C to get a zero block in the first column. It subtracts from D to get $S = D - CA^{-1}B$.

This is ordinary elimination, a column at a time—using blocks. The pivot block is A. That final block is $D - CA^{-1}B$, just like $d - cb/a$. This is called the ***Schur complement***.

■ REVIEW OF THE KEY IDEAS ■

1. The (i, j) entry of AB is (row i of A) · (column j of B).
2. An m by n matrix times an n by p matrix uses mnp separate multiplications.
3. A times BC equals AB times C (surprisingly important).
4. AB is also the sum of these n matrices : (column j of A) times (row j of B).
5. Block multiplication is allowed when the block shapes match correctly.
6. Block elimination produces the *Schur complement* $D - CA^{-1}B$.

■ WORKED EXAMPLES ■

2.4 A A graph or a network has n nodes. Its **adjacency matrix** S is n by n. This is a 0–1 matrix with $s_{ij} = 1$ when nodes i and j are connected by an edge.

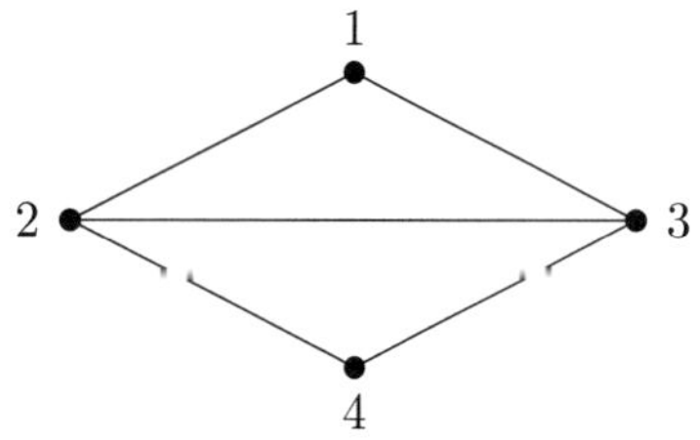

Adjacency matrix
Square and symmetric
for undirected graphs
Edges go both ways

$$S = \begin{bmatrix} 0 & 1 & 1 & 0 \\ 1 & 0 & 1 & 1 \\ 1 & 1 & 0 & 1 \\ 0 & 1 & 1 & 0 \end{bmatrix}$$

The matrix S^2 has a useful interpretation. $(\boldsymbol{S^2})_{ij}$ **counts the walks of length 2** between node i and node j. Between nodes 2 and 3 the graph has two walks: go via 1 or go via 4. From node 1 to node 1, there are also two walks: 1–2–1 and 1–3–1.

$$S^2 = \begin{bmatrix} \mathbf{2} & 1 & 1 & 2 \\ 1 & 3 & \mathbf{2} & 1 \\ 1 & 2 & 3 & 1 \\ 2 & 1 & 1 & 2 \end{bmatrix} \qquad S^3 = \begin{bmatrix} 2 & \mathbf{5} & 5 & 2 \\ 5 & 4 & 5 & 5 \\ 5 & 5 & 4 & 5 \\ 2 & 5 & 5 & 2 \end{bmatrix}$$

Can you find 5 walks of length 3 between nodes 1 and 2 ?

The real question is why S^N counts all the N-step paths between pairs of nodes. Start with S^2 and look at matrix multiplication by dot products:

$$(S^2)_{ij} = (\text{row } i \text{ of } S) \cdot (\text{column } j \text{ of } S) = s_{i1}s_{1j} + s_{i2}s_{2j} + s_{i3}s_{3j} + s_{i4}s_{4j}. \quad (7)$$

If there is a 2-step path $i \to 1 \to j$, the first multiplication gives $s_{i1}s_{1j} = (1)(1) = 1$. If $i \to 1 \to j$ is *not* a path, then either $i \to 1$ is missing or $1 \to j$ is missing. So the multiplication gives $s_{i1}s_{1j} = 0$ in that case.

$(S^2)_{ij}$ is adding up 1's for all the 2-step paths $i \to k \to j$. So it counts those paths. In the same way $S^{N-1}S$ will count N-step paths, because those are $(N-1)$-step paths from i to k followed by one step from k to j. Matrix multiplication is exactly suited to counting paths on a graph—channels of communication between employees in a company.

2.4 B For these matrices, when does $AB = BA$? When does $BC = CB$? When does A times BC equal AB times C? Give the conditions on their entries p, q, r, z:

$$A = \begin{bmatrix} p & 0 \\ q & r \end{bmatrix} \qquad B = \begin{bmatrix} 1 & 1 \\ 0 & 1 \end{bmatrix} \qquad C = \begin{bmatrix} 0 & z \\ 0 & 0 \end{bmatrix}$$

If $p, q, r, 1, z$ are 4 by 4 blocks instead of numbers, do the answers change?

Solution First of all, A times BC *always* equals AB times C. Parentheses are not needed in $A(BC) = (AB)C = ABC$. But we must keep the matrices in this order:

$$\textbf{Usually } \boldsymbol{AB \neq BA} \qquad AB = \begin{bmatrix} p & p \\ q & q+r \end{bmatrix} \qquad BA = \begin{bmatrix} p+q & r \\ q & r \end{bmatrix}.$$

$$\textbf{By chance } \boldsymbol{BC = CB} \qquad BC = \begin{bmatrix} 0 & z \\ 0 & 0 \end{bmatrix} \qquad CB = \begin{bmatrix} 0 & z \\ 0 & 0 \end{bmatrix}.$$

B and C happen to commute. Part of the explanation is that the diagonal of B is I, which commutes with all 2 by 2 matrices. When p, q, r, z are 4 by 4 blocks and 1 changes to I, all these products remain correct. So the answers are the same.

Problem Set 2.4

Problems 1–16 are about the laws of matrix multiplication.

1 A is 3 by 5, B is 5 by 3, C is 5 by 1, and D is 3 by 1. *All entries are* 1. Which of these matrix operations are allowed, and what are the results?

$$BA \qquad AB \qquad ABD \qquad DC \qquad A(B+C).$$

2 What rows or columns or matrices do you multiply to find

(a) the second column of AB?

(b) the first row of AB?

(c) the entry in row 3, column 5 of AB?

(d) the entry in row 1, column 1 of CDE?

3 Add AB to AC and compare with $A(B+C)$:

$$A = \begin{bmatrix} 1 & 5 \\ 2 & 3 \end{bmatrix} \quad \text{and} \quad B = \begin{bmatrix} 0 & 2 \\ 0 & 1 \end{bmatrix} \quad \text{and} \quad C = \begin{bmatrix} 3 & 1 \\ 0 & 0 \end{bmatrix}.$$

4 In Problem 3, multiply A times BC. Then multiply AB times C.

5 Compute A^2 and A^3. Make a prediction for A^5 and A^n:

$$A = \begin{bmatrix} 1 & b \\ 0 & 1 \end{bmatrix} \quad \text{and} \quad A = \begin{bmatrix} 2 & 2 \\ 0 & 0 \end{bmatrix}.$$

6 Show that $(A + B)^2$ is different from $A^2 + 2AB + B^2$, when

$$A = \begin{bmatrix} 1 & 2 \\ 0 & 0 \end{bmatrix} \quad \text{and} \quad B = \begin{bmatrix} 1 & 0 \\ 3 & 0 \end{bmatrix}.$$

Write down the correct rule for $(A + B)(A + B) = A^2 + \_\_\_\_\_ + B^2$.

7 True or false. Give a specific example when false:

(a) If columns 1 and 3 of B are the same, so are columns 1 and 3 of AB.

(b) If rows 1 and 3 of B are the same, so are rows 1 and 3 of AB.

(c) If rows 1 and 3 of A are the same, so are rows 1 and 3 of ABC.

(d) $(AB)^2 = A^2B^2$.

8 How is each row of DA and EA related to the rows of A, when

$$D = \begin{bmatrix} 3 & 0 \\ 0 & 5 \end{bmatrix} \quad \text{and} \quad E = \begin{bmatrix} 0 & 1 \\ 0 & 1 \end{bmatrix} \quad \text{and} \quad A = \begin{bmatrix} a & b \\ c & d \end{bmatrix}?$$

How is each column of AD and AE related to the columns of A?

9 Row 1 of A is added to row 2. This gives EA below. Then column 1 of EA is added to column 2 to produce $(EA)F$:

$$EA = \begin{bmatrix} 1 & 0 \\ 1 & 1 \end{bmatrix} \begin{bmatrix} a & b \\ c & d \end{bmatrix} = \begin{bmatrix} a & b \\ a+c & b+d \end{bmatrix}$$

$$\text{and} \quad (EA)F = (EA) \begin{bmatrix} 1 & 1 \\ 0 & 1 \end{bmatrix} = \begin{bmatrix} a & a+b \\ a+c & a+c+b+d \end{bmatrix}.$$

(a) Do those steps in the opposite order. First add column 1 of A to column 2 by AF, then add row 1 of AF to row 2 by $E(AF)$.

(b) Compare with $(EA)F$. What law is obeyed by matrix multiplication?

10 Row 1 of A is again added to row 2 to produce EA. Then F adds row 2 of EA to row 1. The result is $F(EA)$:

$$F(EA) = \begin{bmatrix} 1 & 1 \\ 0 & 1 \end{bmatrix} \begin{bmatrix} a & b \\ a+c & b+d \end{bmatrix} = \begin{bmatrix} 2a+c & 2b+d \\ a+c & b+d \end{bmatrix}.$$

(a) Do those steps in the opposite order: first add row 2 to row 1 by FA, then add row 1 of FA to row 2.

(b) What law is or is not obeyed by matrix multiplication?

11 This fact still amazes me. If you do a row operation on A and then a column operation, the result is the same as if you did the column operation first. (Try it.) Why is this true?

12 (3 by 3 matrices) Choose the only B so that for every matrix A

(a) $BA = 4A$

(b) $BA = 4B$

(c) BA has rows 1 and 3 of A reversed and row 2 unchanged

(d) All rows of BA are the same as row 1 of A.

13 Suppose $AB = BA$ and $AC = CA$ for these two particular matrices B and C:

$$A = \begin{bmatrix} a & b \\ c & d \end{bmatrix} \quad \text{commutes with} \quad B = \begin{bmatrix} 1 & 0 \\ 0 & 0 \end{bmatrix} \quad \text{and} \quad C = \begin{bmatrix} 0 & 1 \\ 0 & 0 \end{bmatrix}.$$

Prove that $a = d$ and $b = c = 0$. Then A is a multiple of I. The only matrices that commute with B and C and all other 2 by 2 matrices are A = multiple of I.

14 Which of the following matrices are guaranteed to equal $(A - B)^2$: $A^2 - B^2$, $(B - A)^2$, $A^2 - 2AB + B^2$, $A(A - B) - B(A - B)$, $A^2 - AB - BA + B^2$?

15 True or false:

(a) If A^2 is defined then A is necessarily square.

(b) If AB and BA are defined then A and B are square.

(c) If AB and BA are defined then AB and BA are square.

(d) If $AB = B$ then $A = I$.

16 If A is m by n, how many separate multiplications are involved when

(a) A multiplies a vector $\boldsymbol{x}$ with n components?

(b) A multiplies an n by p matrix B?

(c) A multiplies itself to produce A^2 ? Here $m = n$.

17 For $A = \begin{bmatrix} 2 & -1 \\ 3 & -2 \end{bmatrix}$ and $B = \begin{bmatrix} 1 & 0 & 4 \\ 1 & 0 & 6 \end{bmatrix}$, compute these answers *and nothing more*:

(a) column 2 of AB

(b) row 2 of AB

(c) row 2 of $AA = A^2$

(d) row 2 of $AAA = A^3$.

Problems 18–20 use a_{ij} for the entry in row i, column j of A.

18 Write down the 3 by 3 matrix A whose entries are

(a) $a_{ij} = $ minimum of i and j

(b) $a_{ij} = (-1)^{i+j}$

(c) $a_{ij} = i/j$.

19 What words would you use to describe each of these classes of matrices? Give a 3 by 3 example in each class. Which matrix belongs to all four classes?

(a) $a_{ij} = 0$ if $i \neq j$

(b) $a_{ij} = 0$ if $i < j$

(c) $a_{ij} = a_{ji}$

(d) $a_{ij} = a_{1j}$.

20 The entries of A are a_{ij}. Assuming that zeros don't appear, what is

(a) the first pivot?

(b) the multiplier ℓ_{31} of row 1 to be subtracted from row 3?

(c) the new entry that replaces a_{32} after that subtraction?

(d) the second pivot?

Problems 21–24 involve powers of A.

21 Compute A^2, A^3, A^4 and also $A\boldsymbol{v}$, $A^2\boldsymbol{v}$, $A^3\boldsymbol{v}$, $A^4\boldsymbol{v}$ for

$$A = \begin{bmatrix} 0 & 2 & 0 & 0 \\ 0 & 0 & 2 & 0 \\ 0 & 0 & 0 & 2 \\ 0 & 0 & 0 & 0 \end{bmatrix} \quad \text{and} \quad \boldsymbol{v} = \begin{bmatrix} x \\ y \\ z \\ t \end{bmatrix}.$$

22 By trial and error find real nonzero 2 by 2 matrices such that

$$A^2 = -I \qquad BC = 0 \qquad DE = -ED \text{ (not allowing } DE = 0).$$

23 (a) Find a nonzero matrix A for which $A^2 = 0$.

(b) Find a matrix that has $A^2 \neq 0$ but $A^3 = 0$.

24 By experiment with $n = 2$ and $n = 3$ predict A^n for these matrices:

$$A_1 = \begin{bmatrix} 2 & 1 \\ 0 & 1 \end{bmatrix} \quad \text{and} \quad A_2 = \begin{bmatrix} 1 & 1 \\ 1 & 1 \end{bmatrix} \quad \text{and} \quad A_3 = \begin{bmatrix} a & b \\ 0 & 0 \end{bmatrix}.$$

Problems 25–31 use column-row multiplication and block multiplication.

25 Multiply A times I using columns of A (3 by 3) times rows of I.

26 Multiply AB using columns times rows:

$$AB = \begin{bmatrix} 1 & 0 \\ 2 & 4 \\ 2 & 1 \end{bmatrix} \begin{bmatrix} 3 & 3 & 0 \\ 1 & 2 & 1 \end{bmatrix} = \begin{bmatrix} 1 \\ 2 \\ 2 \end{bmatrix} \begin{bmatrix} 3 & 3 & 0 \end{bmatrix} + \underline{\qquad} = \underline{\qquad}.$$

27 Show that the product of upper triangular matrices is always upper triangular:

$$AB = \begin{bmatrix} x & x & x \\ 0 & x & x \\ 0 & 0 & x \end{bmatrix} \begin{bmatrix} x & x & x \\ 0 & x & x \\ 0 & 0 & x \end{bmatrix} = \begin{bmatrix} & & \\ 0 & & \\ 0 & 0 & \end{bmatrix}.$$

Proof using dot products (*Row times column*) (Row 2 of A) $\cdot$ (column 1 of B)$= 0$. Which other dot products give zeros ?

Proof using full matrices (*Column times row*) Draw x's and 0's in (column 2 of A) times (row 2 of B). Also show (column 3 of A) times (row 3 of B).

28 Draw the cuts in A (2 by 3) and B (3 by 4) and AB to show how each of the four multiplication rules is really a block multiplication :

(1) Matrix A times columns of B. **Columns of AB**

(2) Rows of A times the matrix B. **Rows of AB**

(3) Rows of A times columns of B. **Inner products** (numbers in AB)

(4) Columns of A times rows of B. **Outer products** (matrices add to AB)

29 Which matrices E_{21} and E_{31} produce zeros in the $(2, 1)$ and $(3, 1)$ positions of $E_{21}A$ and $E_{31}A$?

$$A = \begin{bmatrix} 2 & 1 & 0 \\ -2 & 0 & 1 \\ 8 & 5 & 3 \end{bmatrix}.$$

Find the single matrix $E = E_{31}E_{21}$ that produces both zeros at once. Multiply EA.

30 Block multiplication says that column 1 is eliminated by

$$EA = \begin{bmatrix} 1 & \mathbf{0} \\ -\boldsymbol{c}/a & I \end{bmatrix} \begin{bmatrix} a & \boldsymbol{b} \\ \boldsymbol{c} & D \end{bmatrix} = \begin{bmatrix} a & \boldsymbol{b} \\ \mathbf{0} & D - \boldsymbol{cb}/a \end{bmatrix}.$$

In Problem 29, what numbers go into $\boldsymbol{c}$ and D and what is $D - \boldsymbol{cb}/a$?

31 With $i^2 = -1$, the product of $(A + iB)$ and $(\boldsymbol{x} + i\boldsymbol{y})$ is $A\boldsymbol{x} + iB\boldsymbol{x} + iA\boldsymbol{y} - B\boldsymbol{y}$. Use blocks to separate the real part without i from the imaginary part that multiplies i:

$$\begin{bmatrix} A & -B \\ ? & ? \end{bmatrix} \begin{bmatrix} \boldsymbol{x} \\ \boldsymbol{y} \end{bmatrix} = \begin{bmatrix} A\boldsymbol{x} - B\boldsymbol{y} \\ ? \end{bmatrix} \begin{matrix} \text{real part} \\ \text{imaginary part} \end{matrix}$$

32 (*Very important*) Suppose you solve $A\boldsymbol{x} = \boldsymbol{b}$ for three special right sides $\boldsymbol{b}$:

$$A\boldsymbol{x}_1 = \begin{bmatrix}1\\0\\0\end{bmatrix} \quad \text{and} \quad A\boldsymbol{x}_2 = \begin{bmatrix}0\\1\\0\end{bmatrix} \quad \text{and} \quad A\boldsymbol{x}_3 = \begin{bmatrix}0\\0\\1\end{bmatrix}.$$

If the three solutions $\boldsymbol{x}_1, \boldsymbol{x}_2, \boldsymbol{x}_3$ are the columns of a matrix X, what is A times X?

33 If the three solutions in Question 32 are $\boldsymbol{x}_1 = (1,1,1)$ and $\boldsymbol{x}_2 = (0,1,1)$ and $\boldsymbol{x}_3 = (0,0,1)$, solve $A\boldsymbol{x} = \boldsymbol{b}$ when $\boldsymbol{b} = (3,5,8)$. Challenge problem: What is A?

34 Find all matrices $A = \left[\begin{smallmatrix}a & b\\ c & d\end{smallmatrix}\right]$ that satisfy $A\left[\begin{smallmatrix}1 & 1\\ 1 & 1\end{smallmatrix}\right] = \left[\begin{smallmatrix}1 & 1\\ 1 & 1\end{smallmatrix}\right]A$.

35 Suppose a "circle graph" has 4 nodes connected (in both directions) by edges around a circle. What is its adjacency matrix S from Worked Example **2.4 A**? What is S^2? Find all the 2-step paths predicted by S^2.

Challenge Problems

36 **Practical question** Suppose A is m by n, B is n by p, and C is p by q. Then the multiplication count is $\boldsymbol{mnp}$ for $AB + \boldsymbol{mpq}$ for $(AB)C$. The same matrix comes from A times BC with $\boldsymbol{mnq} + \boldsymbol{npq}$ separate multiplications. Notice npq for BC.

(a) If A is 2 by 4, B is 4 by 7, and C is 7 by 10, do you prefer $(AB)C$ or $A(BC)$?

(b) With N-component vectors, would you choose $(\boldsymbol{u}^{\mathrm{T}}\boldsymbol{v})\,\boldsymbol{w}^{\mathrm{T}}$ or $\boldsymbol{u}^{\mathrm{T}}(\boldsymbol{v}\boldsymbol{w}^{\mathrm{T}})$?

(c) Divide by $mnpq$ to show that $(AB)C$ is faster when $n^{-1} + q^{-1} < m^{-1} + p^{-1}$.

37 To prove that $\boldsymbol{(AB)\,C = A\,(BC)}$, use the column vectors $\boldsymbol{b}_1, \ldots, \boldsymbol{b}_n$ of B. First suppose that C has only one column $\boldsymbol{c}$ with entries $c_1, \ldots, c_n$:

AB has columns $A\boldsymbol{b}_1, \ldots, A\boldsymbol{b}_n$ and then $(AB)\,\boldsymbol{c}$ equals $c_1 A\boldsymbol{b}_1 + \cdots + c_n A\boldsymbol{b}_n$.

$B\boldsymbol{c}$ has one column $c_1\boldsymbol{b}_1 + \cdots + c_n\boldsymbol{b}_n$ and then $A\,(B\boldsymbol{c})$ equals $A\,(c_1\boldsymbol{b}_1 + \cdots + c_n\boldsymbol{b}_n)$.

Linearity gives equality of those two sums. This proves $(AB)\boldsymbol{c} = A(B\boldsymbol{c})$. The same is true for all other ______ of C. Therefore $(AB)C = A(BC)$. Apply to inverses:

If $BA = I$ and $AC = I$, prove that the left-inverse B equals the right-inverse C.

38 (a) Suppose A has rows $\boldsymbol{a}_1^{\mathrm{T}}, \ldots, \boldsymbol{a}_m^{\mathrm{T}}$. Why does $A^{\mathrm{T}}A$ equal $\boldsymbol{a}_1\boldsymbol{a}_1^{\mathrm{T}} + \cdots + \boldsymbol{a}_m\boldsymbol{a}_m^{\mathrm{T}}$?

(b) If C is a diagonal matrix with $c_1, \ldots, c_m$ on its diagonal, find a similar sum of columns times rows for $A^{\mathrm{T}}CA$. First do an example with $m = n = 2$.

2.5 Inverse Matrices

1 If the square matrix A has an inverse, then both $A^{-1}A = I$ and $AA^{-1} = I$.

2 The *algorithm* to test invertibility is elimination: A must have n (nonzero) pivots.

3 The *algebra* test for invertibility is the determinant of A: $\det A$ must not be zero.

4 The *equation* that tests for invertibility is $A\boldsymbol{x} = \mathbf{0}$: $\boldsymbol{x} = \mathbf{0}$ **must be the only solution.**

5 If A and B (same size) are invertible then so is AB: $\boxed{(AB)^{-1} = B^{-1}A^{-1}.}$

6 $AA^{-1} = I$ is n equations for n columns of A^{-1}. Gauss-Jordan eliminates $[A\ I]$ to $[I\ A^{-1}]$.

7 The last page of the book gives 14 equivalent conditions for a square A to be invertible.

Suppose A is a square matrix. We look for an "***inverse matrix***" A^{-1} of the same size, such that A^{-1} *times* A *equals* I. Whatever A does, A^{-1} undoes. Their product is the identity matrix—which does nothing to a vector, so $A^{-1}A\boldsymbol{x} = \boldsymbol{x}$. *But* A^{-1} *might not exist.*

What a matrix mostly does is to multiply a vector $\boldsymbol{x}$. Multiplying $A\boldsymbol{x} = \boldsymbol{b}$ by A^{-1} gives $A^{-1}A\boldsymbol{x} = A^{-1}\boldsymbol{b}$. ***This is*** $\boldsymbol{x} = A^{-1}\boldsymbol{b}$. The product $A^{-1}A$ is like multiplying by a number and then dividing by that number. A number has an inverse if it is not zero—matrices are more complicated and more interesting. The matrix A^{-1} is called "A inverse."

DEFINITION The matrix A is ***invertible*** if there exists a matrix A^{-1} that "inverts" A:

$$\textbf{Two-sided inverse} \qquad A^{-1}A = I \quad \text{and} \quad AA^{-1} = I. \tag{1}$$

Not all matrices have inverses. This is the first question we ask about a square matrix: Is A invertible? We don't mean that we immediately calculate A^{-1}. In most problems we never compute it! Here are six "notes" about A^{-1}.

Note 1 ***The inverse exists if and only if elimination produces*** n ***pivots*** (row exchanges are allowed). Elimination solves $A\boldsymbol{x} = \boldsymbol{b}$ without explicitly using the matrix A^{-1}.

Note 2 The matrix A cannot have two different inverses. Suppose $BA = I$ and also $AC = I$. Then $B = C$, according to this "proof by parentheses":

$$B(AC) = (BA)C \quad \text{gives} \quad BI = IC \quad \text{or} \quad B = C. \tag{2}$$

This shows that a *left-inverse* B (multiplying from the left) and a *right-inverse* C (multiplying A from the right to give $AC = I$) must be the *same matrix*.

Note 3 If A is invertible, the one and only solution to $A\boldsymbol{x} = \boldsymbol{b}$ is $\boldsymbol{x} = A^{-1}\boldsymbol{b}$:

$$\textbf{\textit{Multiply}} \quad A\boldsymbol{x} = \boldsymbol{b} \quad \textbf{\textit{by}} \quad A^{-1}. \quad \textbf{\textit{Then}} \quad \boldsymbol{x} = A^{-1}A\boldsymbol{x} = A^{-1}\boldsymbol{b}.$$

Note 4 (Important) ***Suppose there is a nonzero vector x such that $Ax = \mathbf{0}$. Then A cannot have an inverse.*** No matrix can bring $\mathbf{0}$ back to $\boldsymbol{x}$.

If A is invertible, then $Ax = \mathbf{0}$ can only have the zero solution $x = A^{-1}\mathbf{0} = \mathbf{0}$.

Note 5 A 2 by 2 matrix is invertible if and only if $ad - bc$ is not zero:

$$\textbf{2 by 2 Inverse:} \quad \begin{bmatrix} a & b \\ c & d \end{bmatrix}^{-1} = \frac{1}{ad - bc} \begin{bmatrix} d & -b \\ -c & a \end{bmatrix}. \qquad (3)$$

This number $ad - bc$ is the *determinant* of A. A matrix is invertible if its determinant is not zero (Chapter 5). The test for n pivots is usually decided before the determinant appears.

Note 6 A diagonal matrix has an inverse provided no diagonal entries are zero:

$$\text{If} \quad A = \begin{bmatrix} d_1 & & \\ & \ddots & \\ & & d_n \end{bmatrix} \quad \text{then} \quad A^{-1} = \begin{bmatrix} 1/d_1 & & \\ & \ddots & \\ & & 1/d_n \end{bmatrix}.$$

Example 1 The 2 by 2 matrix $A = \begin{bmatrix} 1 & 2 \\ 1 & 2 \end{bmatrix}$ is not invertible. It fails the test in Note 5, because $ad - bc$ equals $2 - 2 = 0$. It fails the test in Note 3, because $A\boldsymbol{x} = \mathbf{0}$ when $\boldsymbol{x} = (2, -1)$. It fails to have two pivots as required by Note 1.

Elimination turns the second row of this matrix A into a zero row.

The Inverse of a Product AB

For two nonzero numbers a and b, the sum $a + b$ might or might not be invertible. The numbers $a = 3$ and $b = -3$ have inverses $\frac{1}{3}$ and $-\frac{1}{3}$. Their sum $a + b = 0$ has no inverse. But the product $ab = -9$ does have an inverse, which is $\frac{1}{3}$ times $-\frac{1}{3}$.

For two matrices A and B, the situation is similar. It is hard to say much about the invertibility of $A + B$. But the *product* AB has an inverse, if and only if the two factors A and B are separately invertible (and the same size). The important point is that A^{-1} and B^{-1} come in ***reverse order***:

If A and B are invertible then so is AB. The inverse of a product AB is

$$(AB)^{-1} = B^{-1}A^{-1}. \qquad (4)$$

To see why the order is reversed, multiply AB times $B^{-1}A^{-1}$. Inside that is $BB^{-1} = I$:

$$\textbf{Inverse of } \boldsymbol{AB} \qquad (AB)(B^{-1}A^{-1}) = AIA^{-1} = AA^{-1} = I.$$

We moved parentheses to multiply BB^{-1} first. Similarly $B^{-1}A^{-1}$ times AB equals I.

$B^{-1}A^{-1}$ illustrates a basic rule of mathematics: Inverses come in reverse order. It is also common sense: If you put on socks and then shoes, the first to be taken off are the _____. The same reverse order applies to three or more matrices:

Reverse order $$(ABC)^{-1} = C^{-1}B^{-1}A^{-1}. \tag{5}$$

Example 2 ***Inverse of an elimination matrix.*** If E subtracts 5 times row 1 from row 2, then E^{-1} *adds* 5 times row 1 to row 2:

E subtracts
E^{-1} adds

$$E = \begin{bmatrix} 1 & 0 & 0 \\ \mathbf{-5} & 1 & 0 \\ 0 & 0 & 1 \end{bmatrix} \quad \text{and} \quad E^{-1} = \begin{bmatrix} 1 & 0 & 0 \\ \mathbf{5} & 1 & 0 \\ 0 & 0 & 1 \end{bmatrix}.$$

Multiply EE^{-1} to get the identity matrix I. Also multiply $E^{-1}E$ to get I. We are adding and subtracting the same 5 times row 1. If $AC = I$ then automatically $CA = I$.

For square matrices, an inverse on one side is automatically an inverse on the other side.

Example 3 Suppose F subtracts 4 times row 2 from row 3, and F^{-1} adds it back:

$$F = \begin{bmatrix} 1 & 0 & 0 \\ 0 & 1 & 0 \\ 0 & -4 & 1 \end{bmatrix} \quad \text{and} \quad F^{-1} = \begin{bmatrix} 1 & 0 & 0 \\ 0 & 1 & 0 \\ 0 & 4 & 1 \end{bmatrix}.$$

Now multiply F by the matrix E in Example 2 to find FE. Also multiply E^{-1} times F^{-1} to find $(FE)^{-1}$. Notice the orders FE and $E^{-1}F^{-1}$!

$$FE = \begin{bmatrix} 1 & 0 & 0 \\ -5 & 1 & 0 \\ \mathbf{20} & -4 & 1 \end{bmatrix} \quad \text{is inverted by} \quad E^{-1}F^{-1} = \begin{bmatrix} 1 & 0 & 0 \\ \mathbf{5} & 1 & 0 \\ \mathbf{0} & 4 & 1 \end{bmatrix}. \tag{6}$$

The result is beautiful and correct. The product FE contains "20" but its inverse doesn't. E subtracts 5 times row 1 from row 2. Then F subtracts 4 times the *new* row 2 (changed by row 1) from row 3. ***In this order FE, row 3 feels an effect from row* 1.**

In the order $E^{-1}F^{-1}$, that effect does not happen. First F^{-1} adds 4 times row 2 to row 3. After that, E^{-1} adds 5 times row 1 to row 2. There is no 20, because row 3 doesn't change again. ***In this order $E^{-1}F^{-1}$, row 3 feels no effect from row* 1.**

This is why the next section chooses $A = LU$, to go back from the triangular U to A. The multipliers fall into place perfectly in the lower triangular L.

In elimination order F follows E. In reverse order E^{-1} follows F^{-1}.
$E^{-1}F^{-1}$ is quick. The multipliers* 5, 4 *fall into place below the diagonal of* 1*'s.

Calculating A^{-1} by Gauss-Jordan Elimination

I hinted that A^{-1} might not be explicitly needed. The equation $A\boldsymbol{x} = \boldsymbol{b}$ is solved by $\boldsymbol{x} = A^{-1}\boldsymbol{b}$. But it is not necessary or efficient to compute A^{-1} and multiply it times $\boldsymbol{b}$. *Elimination goes directly to* $\boldsymbol{x}$. And elimination is also the way to calculate A^{-1}, as we now show. The Gauss-Jordan idea is to solve $AA^{-1} = I$, *finding each column of* A^{-1}.

A multiplies the first column of A^{-1} (call that $\boldsymbol{x}_1$) to give the first column of I (call that $\boldsymbol{e}_1$). This is our equation $A\boldsymbol{x}_1 = \boldsymbol{e}_1 = (1, 0, 0)$. There will be two more equations. ***Each of the columns*** $\boldsymbol{x}_1$, $\boldsymbol{x}_2$, $\boldsymbol{x}_3$ ***of*** A^{-1} ***is multiplied by A to produce a column of I:***

3 columns of A^{-1}
$$AA^{-1} = A\begin{bmatrix}\boldsymbol{x}_1 & \boldsymbol{x}_2 & \boldsymbol{x}_3\end{bmatrix} = \begin{bmatrix}\boldsymbol{e}_1 & \boldsymbol{e}_2 & \boldsymbol{e}_3\end{bmatrix} = I. \qquad (7)$$

To invert a 3 by 3 matrix A, we have to solve three systems of equations: $A\boldsymbol{x}_1 = \boldsymbol{e}_1$ and $A\boldsymbol{x}_2 = \boldsymbol{e}_2 = (0, 1, 0)$ and $A\boldsymbol{x}_3 = \boldsymbol{e}_3 = (0, 0, 1)$. Gauss-Jordan finds A^{-1} this way.

The Gauss-Jordan method computes A^{-1} by solving all n equations together. Usually the "augmented matrix" $[A \;\; \boldsymbol{b}]$ has one extra column $\boldsymbol{b}$. Now we have three right sides $\boldsymbol{e}_1, \boldsymbol{e}_2, \boldsymbol{e}_3$ (when A is 3 by 3). They are the columns of I, so the augmented matrix is really the block matrix $[\,A \;\; I\,]$. I take this chance to invert my favorite matrix K, with 2's on the main diagonal and -1's next to the 2's:

$$\begin{bmatrix}K & \boldsymbol{e}_1 & \boldsymbol{e}_2 & \boldsymbol{e}_3\end{bmatrix} = \begin{bmatrix} \mathbf{2} & \mathbf{-1} & \mathbf{0} & 1 & 0 & 0 \\ \mathbf{-1} & \mathbf{2} & \mathbf{-1} & 0 & 1 & 0 \\ \mathbf{0} & \mathbf{-1} & \mathbf{2} & 0 & 0 & 1 \end{bmatrix} \quad \textbf{Start Gauss-Jordan on } K$$

$$\rightarrow \begin{bmatrix} 2 & -1 & 0 & 1 & 0 & 0 \\ \mathbf{0} & \mathbf{\tfrac{3}{2}} & \mathbf{-1} & \mathbf{\tfrac{1}{2}} & \mathbf{1} & \mathbf{0} \\ 0 & -1 & 2 & 0 & 0 & 1 \end{bmatrix} \quad (\tfrac{1}{2}\; \textit{row}\; 1 + \textit{row}\; 2)$$

$$\rightarrow \begin{bmatrix} 2 & -1 & 0 & 1 & 0 & 0 \\ 0 & \tfrac{3}{2} & -1 & \tfrac{1}{2} & 1 & 0 \\ \mathbf{0} & \mathbf{0} & \mathbf{\tfrac{4}{3}} & \mathbf{\tfrac{1}{3}} & \mathbf{\tfrac{2}{3}} & \mathbf{1} \end{bmatrix} \quad (\tfrac{2}{3}\; \textit{row}\; 2 + \textit{row}\; 3)$$

We are halfway to K^{-1}. The matrix in the first three columns is U (upper triangular). The pivots $2, \frac{3}{2}, \frac{4}{3}$ are on its diagonal. Gauss would finish by back substitution. The contribution of Jordan is *to continue with elimination*! He goes all the way to the ***reduced echelon form*** $\boldsymbol{R} = \boldsymbol{I}$. Rows are added to rows above them, to produce ***zeros above the pivots*** :

$$\begin{pmatrix}\text{Zero above}\\ \text{third pivot}\end{pmatrix} \rightarrow \begin{bmatrix} 2 & -1 & 0 & 1 & 0 & 0 \\ \mathbf{0} & \mathbf{\tfrac{3}{2}} & \mathbf{0} & \mathbf{\tfrac{3}{4}} & \mathbf{\tfrac{3}{2}} & \mathbf{\tfrac{3}{4}} \\ 0 & 0 & \tfrac{4}{3} & \tfrac{1}{3} & \tfrac{2}{3} & 1 \end{bmatrix} \quad (\tfrac{3}{4}\; \textit{row}\; 3 + \textit{row}\; 2)$$

$$\begin{pmatrix}\text{Zero above}\\ \text{second pivot}\end{pmatrix} \rightarrow \begin{bmatrix} \mathbf{2} & \mathbf{0} & \mathbf{0} & \mathbf{\tfrac{3}{2}} & \mathbf{1} & \mathbf{\tfrac{1}{2}} \\ 0 & \tfrac{3}{2} & 0 & \tfrac{3}{4} & \tfrac{3}{2} & \tfrac{3}{4} \\ 0 & 0 & \tfrac{4}{3} & \tfrac{1}{3} & \tfrac{2}{3} & 1 \end{bmatrix} \quad (\tfrac{2}{3}\; \textit{row}\; 2 + \textit{row}\; 1)$$

The final Gauss-Jordan step is to divide each row by its pivot. The new pivots are all 1.

We have reached I in the first half of the matrix, because K is invertible. ***The three columns of K^{-1} are in the second half of $[\,I\ K^{-1}\,]$:***

$$\begin{matrix}\text{(divide by 2)}\\ \text{(divide by } \frac{3}{2})\\ \text{(divide by } \frac{4}{3})\end{matrix}\quad \begin{bmatrix} \mathbf{1} & 0 & 0 & \frac{3}{4} & \frac{1}{2} & \frac{1}{4} \\ 0 & \mathbf{1} & 0 & \frac{1}{2} & 1 & \frac{1}{2} \\ 0 & 0 & \mathbf{1} & \frac{1}{4} & \frac{1}{2} & \frac{3}{4} \end{bmatrix} = \begin{bmatrix} I & \boldsymbol{x}_1 & \boldsymbol{x}_2 & \boldsymbol{x}_3 \end{bmatrix} = \begin{bmatrix} I & \boldsymbol{K^{-1}} \end{bmatrix}.$$

Starting from the 3 by 6 matrix $[\,K\ \ I\,]$, we ended with $[\,I\ \ K^{-1}\,]$. Here is the whole Gauss-Jordan process on one line for any invertible matrix A:

Gauss-Jordan ***Multiply $[\,A\ \ I\,]$ by A^{-1} to get $[\,I\ \ A^{-1}]$.***

The elimination steps create the inverse matrix while changing A to I. For large matrices, we probably don't want A^{-1} at all. But for small matrices, it can be very worthwhile to know the inverse. We add three observations about K^{-1}: an important example.

1. K is ***symmetric*** across its main diagonal. Then K^{-1} is also symmetric.

2. K is ***tridiagonal*** (only three nonzero diagonals). But K^{-1} is a dense matrix with no zeros. That is another reason we don't often compute inverse matrices. The inverse of a band matrix is generally a dense matrix.

3. The *product of pivots* is $2\left(\frac{3}{2}\right)\left(\frac{4}{3}\right) = 4$. This number 4 is the ***determinant*** of K.

K^{-1} involves division by the determinant of K
$$K^{-1} = \frac{1}{4}\begin{bmatrix} 3 & 2 & 1 \\ 2 & 4 & 2 \\ 1 & 2 & 3 \end{bmatrix}. \tag{8}$$

This is why an invertible matrix cannot have a zero determinant: we need to divide.

Example 4 Find A^{-1} by Gauss-Jordan elimination starting from $A = \left[\begin{smallmatrix} 2 & 3 \\ 4 & 7 \end{smallmatrix}\right]$.

$$\begin{bmatrix} \boldsymbol{A} & I \end{bmatrix} = \begin{bmatrix} \mathbf{2} & \mathbf{3} & 1 & 0 \\ \mathbf{4} & \mathbf{7} & 0 & 1 \end{bmatrix} \to \begin{bmatrix} 2 & 3 & 1 & 0 \\ \mathbf{0} & \mathbf{1} & \mathbf{-2} & \mathbf{1} \end{bmatrix} \quad (\text{this is } [\,U\ \ L^{-1}\,])$$

$$\to \begin{bmatrix} \mathbf{2} & \mathbf{0} & \mathbf{7} & \mathbf{-3} \\ 0 & 1 & -2 & 1 \end{bmatrix} \to \begin{bmatrix} 1 & 0 & \frac{7}{2} & -\frac{3}{2} \\ 0 & 1 & -\mathbf{2} & \mathbf{1} \end{bmatrix} \quad (\text{this is } [\,I\ \ \boldsymbol{A^{-1}}\,]).$$

Example 5 ***If A is invertible and upper triangular, so is A^{-1}.*** Start with $\boldsymbol{AA^{-1}} = \boldsymbol{I}$.

1 A times *column j of A^{-1}* equals *column j of I*, ending with $n - j$ zeros.

2 Back substitution keeps those $n - j$ zeros at the end of column j of A^{-1}.

3 Put those columns $[* \ldots * \, 0 \ldots 0]^{\mathrm{T}}$ into A^{-1} and that matrix is upper triangular!

$$A^{-1} = \begin{bmatrix} 1 & -1 & 0 \\ \mathbf{0} & 1 & -1 \\ \mathbf{0} & \mathbf{0} & 1 \end{bmatrix}^{-1} = \begin{bmatrix} 1 & 1 & 1 \\ \mathbf{0} & 1 & 1 \\ \mathbf{0} & \mathbf{0} & 1 \end{bmatrix}$$ Columns $j = 1$ and 2 end with $3 - j = 2$ and 1 zeros.

The code for $X =$ **inv**(A) can use **rref**, the reduced row echelon form from Chapter 3:

```
I = eye (n);                 % Define the n by n identity matrix
R = rref ([A  I]);           % Eliminate on the augmented matrix [A  I]
X = R( : , n + 1 : n + n)    % Pick X = A^-1 from the last n columns of R
```

A must be invertible, or elimination cannot reduce it to I (in the left half of R).

Gauss-Jordan shows why A^{-1} is expensive. We solve n equations for its n columns. But all those equations involve the same matrix A on the left side (where most of the work is done). The total cost for A^{-1} is n^3 multiplications and subtractions. To solve a single $A\boldsymbol{x} = \boldsymbol{b}$ that cost (see the next section) is $n^3/3$.

To solve $Ax = b$ without A^{-1}, we deal with *one* column b to find one column x.

Singular versus Invertible

We come back to the central question. Which matrices have inverses? The start of this section proposed the pivot test: ***A^{-1} exists exactly when A has a full set of n pivots.*** (Row exchanges are allowed.) Now we can prove that by Gauss-Jordan elimination:

1. With n pivots, elimination solves all the equations $A\boldsymbol{x}_i = \boldsymbol{e}_i$. The columns $\boldsymbol{x}_i$ go into A^{-1}. Then $AA^{-1} = I$ and A^{-1} is at least a ***right-inverse***.

2. Elimination is really a sequence of multiplications by E's and P's and D^{-1}:

Left-inverse C $$CA = (D^{-1} \cdots E \cdots P \cdots E)A = I. \qquad (9)$$

D^{-1} divides by the pivots. The matrices E produce zeros below and above the pivots. P will exchange rows if needed (see Section 2.7). The product matrix in equation (9) is evidently a ***left-inverse of A***. With n pivots we have reached $A^{-1}A = I$.

The right-inverse equals the left-inverse. That was Note 2 at the start of in this section. So a square matrix with a full set of pivots will always have a two-sided inverse.

Reasoning in reverse will now show that A ***must have n pivots if*** $AC = I$.

1. If A doesn't have n pivots, elimination will lead to a *zero row*.
2. Those elimination steps are taken by an invertible M. *So a row of MA is zero.*
3. If $AC = I$ had been possible, then $MAC = M$. The zero row of MA, times C, gives a zero row of M itself.
4. An invertible matrix M can't have a zero row! A *must* have n pivots if $AC = I$.

That argument took four steps, but the outcome is short and important. C is A^{-1}.

Elimination gives a complete test for invertibility of a square matrix. ***A^{-1} exists (and Gauss-Jordan finds it) exactly when A has n pivots.*** The argument above shows more:

$$\text{If } AC = I \text{ then } CA = I \text{ and } C = A^{-1} \tag{10}$$

Example 6 If L is lower triangular with 1's on the diagonal, so is L^{-1}.

A triangular matrix is invertible if and only if no diagonal entries are zero.

Here L has 1's so L^{-1} also has 1's. Use the Gauss-Jordan method to construct L^{-1} from E_{32}, E_{31}, E_{21}. Notice how L^{-1} contains the strange entry 11, from 3 times 5 minus 4.

Gauss-Jordan on triangular L

$$\begin{bmatrix} \mathbf{1} & \mathbf{0} & \mathbf{0} & 1 & 0 & 0 \\ \mathbf{3} & \mathbf{1} & \mathbf{0} & 0 & 1 & 0 \\ \mathbf{4} & \mathbf{5} & \mathbf{1} & 0 & 0 & 1 \end{bmatrix} = \begin{bmatrix} L & I \end{bmatrix}$$

$$\begin{matrix} \\ \rightarrow \\ \rightarrow \end{matrix} \begin{bmatrix} 1 & 0 & 0 & 1 & 0 & 0 \\ 0 & 1 & 0 & -3 & 1 & 0 \\ 0 & 5 & 1 & -4 & 0 & 1 \end{bmatrix} \quad \begin{matrix} \text{(3 times row 1 from row 2)} \\ \text{(4 times row 1 from row 3)} \\ \text{(then 5 times row 2 from row 3)} \end{matrix}$$

The inverse is still triangular

$$\begin{matrix} \\ \\ \rightarrow \end{matrix} \begin{bmatrix} 1 & 0 & 0 & \mathbf{1} & \mathbf{0} & \mathbf{0} \\ 0 & 1 & 0 & \mathbf{-3} & \mathbf{1} & \mathbf{0} \\ 0 & 0 & 1 & \mathbf{11} & \mathbf{-5} & \mathbf{1} \end{bmatrix} = \begin{bmatrix} I & \boldsymbol{L^{-1}} \end{bmatrix}.$$

Recognizing an Invertible Matrix

Normally, it takes work to decide if a matrix is invertible. The usual way is to find a full set of nonzero pivots in elimination. (Then the nonzero determinant comes from multiplying those pivots.) But for some matrices you can see quickly that they are invertible because every number a_{ii} on their main diagonal dominates the off-diagonal part of that row i.

Diagonally dominant matrices are invertible. Each a_{ii} on the diagonal is larger than the total sum along the rest of row i. On every row,

$$|\boldsymbol{a_{ii}}| > \sum_{j \neq i} |\boldsymbol{a_{ij}}| \quad \text{means that} \quad |a_{ii}| > |a_{i1}| + \cdots (\text{skip } |a_{ii}|) \cdots + |a_{in}|. \tag{11}$$

Examples. A is diagonally dominant $(3 > 2)$. B is not (but still invertible). C is singular.

$$A = \begin{bmatrix} 3 & 1 & 1 \\ 1 & 3 & 1 \\ 1 & 1 & 3 \end{bmatrix} \qquad B = \begin{bmatrix} 2 & 1 & 1 \\ 1 & 2 & 1 \\ 1 & 1 & 3 \end{bmatrix} \qquad C = \begin{bmatrix} 1 & 1 & 1 \\ 1 & 1 & 1 \\ 1 & 1 & 3 \end{bmatrix}$$

Reasoning. Take any nonzero vector $\boldsymbol{x}$. *Suppose its largest component is* $|x_i|$. Then $A\boldsymbol{x} = \mathbf{0}$ is impossible, because row i of $A\boldsymbol{x} = \mathbf{0}$ would need

$$a_{i1}x_1 + \cdots + a_{ii}x_i + \cdots + a_{in}x_n = 0.$$

Those can't add to zero when A is diagonally dominant! The size of $a_{ii}x_i$ (that one particular term) is greater than all the other terms combined:

$$\textbf{All } |x_j| \le |x_i| \qquad \sum_{j \ne i} |a_{ij}x_j| \le \sum_{j \ne i} |a_{ij}|\,|x_i| < |a_{ii}|\,|x_i| \quad \textbf{because } a_{ii} \textbf{ dominates}$$

This shows that $A\boldsymbol{x} = \mathbf{0}$ is only possible when $\boldsymbol{x} = \mathbf{0}$. *So A is invertible*. The example B was also invertible but not quite diagonally dominant: 2 is not larger than $1 + 1$.

■ REVIEW OF THE KEY IDEAS ■

1. The inverse matrix gives $AA^{-1} = I$ and $A^{-1}A = I$.
2. A is invertible if and only if it has n pivots (row exchanges allowed).
3. *Important*. If $A\boldsymbol{x} = \mathbf{0}$ for a nonzero vector $\boldsymbol{x}$, then A has no inverse.
4. The inverse of AB is the reverse product $B^{-1}A^{-1}$. And $(ABC)^{-1} = C^{-1}B^{-1}A^{-1}$.
5. The Gauss-Jordan method solves $AA^{-1} = I$ to find the n columns of A^{-1}. The augmented matrix $\begin{bmatrix} A & I \end{bmatrix}$ is row-reduced to $\begin{bmatrix} I & A^{-1} \end{bmatrix}$.
6. Diagonally dominant matrices are invertible. Each $|a_{ii}|$ dominates its row.

■ WORKED EXAMPLES ■

2.5 A The inverse of a triangular **difference matrix** A is a triangular **sum matrix** S:

$$\begin{bmatrix} A & I \end{bmatrix} = \left[\begin{array}{rrr|rrr} 1 & 0 & 0 & 1 & 0 & 0 \\ -1 & 1 & 0 & 0 & 1 & 0 \\ 0 & -1 & 1 & 0 & 0 & 1 \end{array}\right] \to \left[\begin{array}{rrr|rrr} 1 & 0 & 0 & 1 & 0 & 0 \\ 0 & 1 & 0 & 1 & 1 & 0 \\ 0 & -1 & 1 & 0 & 0 & 1 \end{array}\right]$$

$$\to \left[\begin{array}{rrr|rrr} 1 & 0 & 0 & 1 & 0 & 0 \\ 0 & 1 & 0 & 1 & 1 & 0 \\ 0 & 0 & 1 & 1 & 1 & 1 \end{array}\right] = \begin{bmatrix} I & A^{-1} \end{bmatrix} = \begin{bmatrix} I & \textit{sum matrix} \end{bmatrix}.$$

If I change a_{13} to -1, then all rows of A add to zero. The equation $A\boldsymbol{x} = \mathbf{0}$ will now have the nonzero solution $\boldsymbol{x} = (1, 1, 1)$. A clear signal: ***This new A can't be inverted.***

2.5 B Three of these matrices are invertible, and three are singular. Find the inverse when it exists. Give reasons for noninvertibility (zero determinant, too few pivots, nonzero solution to $A\boldsymbol{x} = \mathbf{0}$) for the other three. The matrices are in the order A, B, C, D, S, E:

$$\begin{bmatrix} 4 & 3 \\ 8 & 6 \end{bmatrix} \quad \begin{bmatrix} 4 & 3 \\ 8 & 7 \end{bmatrix} \quad \begin{bmatrix} 6 & 6 \\ 6 & 0 \end{bmatrix} \quad \begin{bmatrix} 6 & 6 \\ 6 & 6 \end{bmatrix} \quad \begin{bmatrix} 1 & 0 & 0 \\ 1 & 1 & 0 \\ 1 & 1 & 1 \end{bmatrix} \quad \begin{bmatrix} 1 & 1 & 1 \\ 1 & 1 & 0 \\ 1 & 1 & 1 \end{bmatrix}$$

Solution

$$B^{-1} = \frac{1}{4}\begin{bmatrix} 7 & -3 \\ -8 & 4 \end{bmatrix} \quad C^{-1} = \frac{1}{36}\begin{bmatrix} 0 & 6 \\ 6 & -6 \end{bmatrix} \quad S^{-1} = \begin{bmatrix} 1 & 0 & 0 \\ -1 & 1 & 0 \\ 0 & -1 & 1 \end{bmatrix}$$

A is not invertible because its determinant is $4 \cdot 6 - 3 \cdot 8 = 24 - 24 = 0$. D is not invertible because there is only one pivot; the second row becomes zero when the first row is subtracted. E has two equal rows (and the second column minus the first column is zero). In other words $E\boldsymbol{x} = \mathbf{0}$ has the solution $\boldsymbol{x} = (-1, 1, 0)$.

Of course all three reasons for noninvertibility would apply to each of A, D, E.

2.5 C Apply the Gauss-Jordan method to invert this triangular "Pascal matrix" L. You see **Pascal's triangle**—adding each entry to the entry on its left gives the entry below. The entries of L are "binomial coefficients". The next row would be $1, 4, 6, 4, 1$.

$$\textbf{Triangular Pascal matrix} \quad L = \begin{bmatrix} \mathbf{1} & 0 & 0 & 0 \\ \mathbf{1} & \mathbf{1} & 0 & 0 \\ \mathbf{1} & \mathbf{2} & \mathbf{1} & 0 \\ \mathbf{1} & \mathbf{3} & \mathbf{3} & \mathbf{1} \end{bmatrix} = \text{abs(pascal (4,1))}$$

Solution Gauss-Jordan starts with $[\,L \;\; I\,]$ and produces zeros by subtracting row 1:

$$[\,\boldsymbol{L}\;\;\boldsymbol{I}\,] = \left[\begin{array}{cccc|cccc} \mathbf{1} & 0 & 0 & 0 & 1 & 0 & 0 & 0 \\ \mathbf{1} & \mathbf{1} & 0 & 0 & 0 & 1 & 0 & 0 \\ \mathbf{1} & \mathbf{2} & \mathbf{1} & 0 & 0 & 0 & 1 & 0 \\ \mathbf{1} & \mathbf{3} & \mathbf{3} & \mathbf{1} & 0 & 0 & 0 & 1 \end{array}\right] \to \left[\begin{array}{cccc|cccc} 1 & 0 & 0 & 0 & 1 & 0 & 0 & 0 \\ \mathbf{0} & 1 & 0 & 0 & \mathbf{-1} & 1 & 0 & 0 \\ \mathbf{0} & 2 & 1 & 0 & \mathbf{-1} & 0 & 1 & 0 \\ \mathbf{0} & 3 & 3 & 1 & \mathbf{-1} & 0 & 0 & 1 \end{array}\right].$$

The next stage creates zeros below the second pivot, using multipliers 2 and 3. Then the last stage subtracts 3 times the new row 3 from the new row 4:

$$\to \left[\begin{array}{cccc|cccc} 1 & 0 & 0 & 0 & 1 & 0 & 0 & 0 \\ 0 & 1 & 0 & 0 & -1 & 1 & 0 & 0 \\ 0 & \mathbf{0} & 1 & 0 & \mathbf{1} & \mathbf{-2} & 1 & 0 \\ 0 & \mathbf{0} & 3 & 1 & \mathbf{2} & \mathbf{-3} & 0 & 1 \end{array}\right] \to \left[\begin{array}{cccc|cccc} 1 & 0 & 0 & 0 & \mathbf{1} & 0 & 0 & 0 \\ 0 & 1 & 0 & 0 & \mathbf{-1} & \mathbf{1} & 0 & 0 \\ 0 & 0 & 1 & 0 & \mathbf{1} & \mathbf{-2} & \mathbf{1} & 0 \\ 0 & 0 & 0 & 1 & \mathbf{-1} & \mathbf{3} & \mathbf{-3} & \mathbf{1} \end{array}\right] = [\,\boldsymbol{I}\;\;\boldsymbol{L}^{-1}\,].$$

All the pivots were 1! So we didn't need to divide rows by pivots to get I. The inverse matrix L^{-1} looks like L itself, except odd-numbered diagonals have minus signs.

The same pattern continues to n by n Pascal matrices. L^{-1} has "alternating diagonals".

Problem Set 2.5

1 Find the inverses (directly or from the 2 by 2 formula) of A, B, C:

$$A = \begin{bmatrix} 0 & 3 \\ 4 & 0 \end{bmatrix} \quad \text{and} \quad B = \begin{bmatrix} 2 & 0 \\ 4 & 2 \end{bmatrix} \quad \text{and} \quad C = \begin{bmatrix} 3 & 4 \\ 5 & 7 \end{bmatrix}.$$

2 For these "permutation matrices" find P^{-1} by trial and error (with 1's and 0's):

$$P = \begin{bmatrix} 0 & 0 & 1 \\ 0 & 1 & 0 \\ 1 & 0 & 0 \end{bmatrix} \quad \text{and} \quad P = \begin{bmatrix} 0 & 1 & 0 \\ 0 & 0 & 1 \\ 1 & 0 & 0 \end{bmatrix}.$$

3 Solve for the first column (x, y) and second column (t, z) of A^{-1}:

$$\begin{bmatrix} 10 & 20 \\ 20 & 50 \end{bmatrix} \begin{bmatrix} x \\ y \end{bmatrix} = \begin{bmatrix} 1 \\ 0 \end{bmatrix} \quad \text{and} \quad \begin{bmatrix} 10 & 20 \\ 20 & 50 \end{bmatrix} \begin{bmatrix} t \\ z \end{bmatrix} = \begin{bmatrix} 0 \\ 1 \end{bmatrix}.$$

4 Show that $\left[\begin{smallmatrix} 1 & 2 \\ 3 & 6 \end{smallmatrix}\right]$ is not invertible by trying to solve $AA^{-1} = I$ for column 1 of A^{-1}:

$$\begin{bmatrix} 1 & 2 \\ 3 & 6 \end{bmatrix} \begin{bmatrix} x \\ y \end{bmatrix} = \begin{bmatrix} 1 \\ 0 \end{bmatrix} \qquad \left(\begin{array}{l} \textit{For a different } A\text{, could column 1 of } A^{-1} \\ \text{be possible to find but not column 2?} \end{array} \right)$$

5 Find an upper triangular U (not diagonal) with $U^2 = I$ which gives $U = U^{-1}$.

6 (a) If A is invertible and $AB = AC$, prove quickly that $B = C$.

(b) If $A = \left[\begin{smallmatrix} 1 & 1 \\ 1 & 1 \end{smallmatrix}\right]$, find two different matrices such that $AB = AC$.

7 (Important) If A has row 1 + row 2 = row 3, show that A is not invertible:

(a) Explain why $A\boldsymbol{x} = (0, 0, 1)$ cannot have a solution. Add eqn 1 + eqn 2.

(b) Which right sides (b_1, b_2, b_3) might allow a solution to $A\boldsymbol{x} = \boldsymbol{b}$?

(c) In elimination, what happens to equation 3?

8 If A has column 1 + column 2 = column 3, show that A is not invertible:

(a) Find a nonzero solution $\boldsymbol{x}$ to $A\boldsymbol{x} = \boldsymbol{0}$. The matrix is 3 by 3.

(b) Elimination keeps column 1 + column 2 = column 3. Explain why there is no third pivot.

9 Suppose A is invertible and you exchange its first two rows to reach B. Is the new matrix B invertible? How would you find B^{-1} from A^{-1}?

10 Find the inverses (in any legal way) of

$$A = \begin{bmatrix} 0 & 0 & 0 & 2 \\ 0 & 0 & 3 & 0 \\ 0 & 4 & 0 & 0 \\ 5 & 0 & 0 & 0 \end{bmatrix} \quad \text{and} \quad B = \begin{bmatrix} 3 & 2 & 0 & 0 \\ 4 & 3 & 0 & 0 \\ 0 & 0 & 6 & 5 \\ 0 & 0 & 7 & 6 \end{bmatrix}.$$

11 (a) Find invertible matrices A and B such that $A + B$ is not invertible.

(b) Find singular matrices A and B such that $A + B$ is invertible.

12 If the product $C = AB$ is invertible (A and B are square), then A itself is invertible. Find a formula for A^{-1} that involves C^{-1} and B.

13 If the product $M = ABC$ of three square matrices is invertible, then B is invertible. (So are A and C.) Find a formula for B^{-1} that involves M^{-1} and A and C.

14 If you add row 1 of A to row 2 to get B, how do you find B^{-1} from A^{-1}?

Notice the order. The inverse of $B = \begin{bmatrix} 1 & 0 \\ 1 & 1 \end{bmatrix} \begin{bmatrix} & A & \end{bmatrix}$ is ______.

15 Prove that a matrix with a column of zeros cannot have an inverse.

16 Multiply $\left[\begin{smallmatrix} a & b \\ c & d \end{smallmatrix}\right]$ times $\left[\begin{smallmatrix} d & -b \\ -c & a \end{smallmatrix}\right]$. What is the inverse of each matrix if $ad \neq bc$?

17 (a) What 3 by 3 matrix E has the same effect as these three steps? Subtract row 1 from row 2, subtract row 1 from row 3, then subtract row 2 from row 3.

(b) What single matrix L has the same effect as these three reverse steps? Add row 2 to row 3, add row 1 to row 3, then add row 1 to row 2.

18 If B is the inverse of A^2, show that AB is the inverse of A.

19 Find the numbers a and b that give the inverse of $5 * \text{eye}(4) - \text{ones}(4,4)$:

$$\begin{bmatrix} 4 & -1 & -1 & -1 \\ -1 & 4 & -1 & -1 \\ -1 & -1 & 4 & -1 \\ -1 & -1 & -1 & 4 \end{bmatrix}^{-1} = \begin{bmatrix} a & b & b & b \\ b & a & b & b \\ b & b & a & b \\ b & b & b & a \end{bmatrix}.$$

What are a and b in the inverse of $6 * \text{eye}(5) - \text{ones}(5,5)$?

20 Show that $\text{A} = 4 * \text{eye}(4) - \text{ones}(4,4)$ is *not* invertible: Multiply $\text{A} * \text{ones}(4,1)$.

21 There are sixteen 2 by 2 matrices whose entries are 1's and 0's. How many of them are invertible?

Questions 22–28 are about the Gauss-Jordan method for calculating A^{-1}.

22 Change I into A^{-1} as you reduce A to I (by row operations):

$$\begin{bmatrix} A & I \end{bmatrix} = \begin{bmatrix} 1 & 3 & 1 & 0 \\ 2 & 7 & 0 & 1 \end{bmatrix} \quad \text{and} \quad \begin{bmatrix} A & I \end{bmatrix} = \begin{bmatrix} 1 & 4 & 1 & 0 \\ 3 & 9 & 0 & 1 \end{bmatrix}$$

23 Follow the 3 by 3 text example but with plus signs in A. Eliminate above and below the pivots to reduce $[\,A \;\; I\,]$ to $[\,I \;\; A^{-1}\,]$:

$$\begin{bmatrix} A & I \end{bmatrix} = \begin{bmatrix} 2 & 1 & 0 & 1 & 0 & 0 \\ 1 & 2 & 1 & 0 & 1 & 0 \\ 0 & 1 & 2 & 0 & 0 & 1 \end{bmatrix}.$$

24 Use Gauss-Jordan elimination on $[\,U \;\; I\,]$ to find the upper triangular U^{-1}:

$$\boldsymbol{UU^{-1} = I} \qquad \begin{bmatrix} 1 & a & b \\ 0 & 1 & c \\ 0 & 0 & 1 \end{bmatrix} \begin{bmatrix} \boldsymbol{x}_1 & \boldsymbol{x}_2 & \boldsymbol{x}_3 \end{bmatrix} = \begin{bmatrix} 1 & 0 & 0 \\ 0 & 1 & 0 \\ 0 & 0 & 1 \end{bmatrix}.$$

25 Find A^{-1} and B^{-1} (*if they exist*) by elimination on $[\,A \;\; I\,]$ and $[\,B \;\; I\,]$:

$$A = \begin{bmatrix} 2 & 1 & 1 \\ 1 & 2 & 1 \\ 1 & 1 & 2 \end{bmatrix} \quad \text{and} \quad B = \begin{bmatrix} 2 & -1 & -1 \\ -1 & 2 & -1 \\ -1 & -1 & 2 \end{bmatrix}.$$

26 What three matrices E_{21} and E_{12} and D^{-1} reduce $A = \left[\begin{smallmatrix} 1 & 2 \\ 2 & 6 \end{smallmatrix}\right]$ to the identity matrix? Multiply $D^{-1}E_{12}\,E_{21}$ to find A^{-1}.

27 Invert these matrices A by the Gauss-Jordan method starting with $[\,A \;\; I\,]$:

$$A = \begin{bmatrix} 1 & 0 & 0 \\ 2 & 1 & 3 \\ 0 & 0 & 1 \end{bmatrix} \quad \text{and} \quad A = \begin{bmatrix} 1 & 1 & 1 \\ 1 & 2 & 2 \\ 1 & 2 & 3 \end{bmatrix}.$$

28 Exchange rows and continue with Gauss-Jordan to find A^{-1}:

$$\begin{bmatrix} A & I \end{bmatrix} = \begin{bmatrix} 0 & 2 & 1 & 0 \\ 2 & 2 & 0 & 1 \end{bmatrix}.$$

29 True or false (with a counterexample if false and a reason if true):

(a) A 4 by 4 matrix with a row of zeros is not invertible.

(b) Every matrix with 1's down the main diagonal is invertible.

(c) If A is invertible then A^{-1} and A^2 are invertible.

30 (Recommended) Prove that A is invertible if $a \neq 0$ and $a \neq b$ (find the pivots or A^{-1}). Then find three numbers c so that C is not invertible:

$$A = \begin{bmatrix} a & b & b \\ a & a & b \\ a & a & a \end{bmatrix} \qquad C = \begin{bmatrix} 2 & c & c \\ c & c & c \\ 8 & 7 & c \end{bmatrix}.$$

31 This matrix has a remarkable inverse. Find A^{-1} by elimination on $[\,A \;\; I\,]$. Extend to a 5 by 5 "alternating matrix" and guess its inverse; then multiply to confirm.

$$\text{Invert } A = \begin{bmatrix} 1 & -1 & 1 & -1 \\ 0 & 1 & -1 & 1 \\ 0 & 0 & 1 & -1 \\ 0 & 0 & 0 & 1 \end{bmatrix} \text{ and solve } A\boldsymbol{x} = (1,1,1,1).$$

32 Suppose the matrices P and Q have the same rows as I but in any order. They are "permutation matrices". Show that $P - Q$ is singular by solving $(P - Q)\,\boldsymbol{x} = \mathbf{0}$.

33 Find and check the inverses (assuming they exist) of these block matrices:

$$\begin{bmatrix} I & 0 \\ C & I \end{bmatrix} \quad \begin{bmatrix} A & 0 \\ C & D \end{bmatrix} \quad \begin{bmatrix} 0 & I \\ I & D \end{bmatrix}.$$

34 Could a 4 by 4 matrix A be invertible if every row contains the numbers $0, 1, 2, 3$ in some order? What if every row of B contains $0, 1, 2, -3$ in some order?

35 In the Worked Example **2.5 C**, the triangular Pascal matrix L has $L^{-1} = DLD$, where the diagonal matrix D has alternating entries $1, -1, 1, -1$. Then $LDLD = I$, so what is the inverse of $LD =$ pascal $(4, 1)$?

36 The Hilbert matrices have $H_{ij} = 1/(i + j - 1)$. Ask MATLAB for the exact 6 by 6 inverse invhilb (6). Then ask it to compute inv (hilb (6)). How can these be different, when the computer never makes mistakes?

37 (a) Use inv(P) to invert MATLAB's 4 by 4 symmetric matrix $P =$ pascal (4).

(b) Create Pascal's lower triangular $L =$ abs (pascal (4, 1)) and test $P = LL^{\mathrm{T}}$.

38 If $A =$ ones (4) and $\boldsymbol{b} =$ rand (4, 1), how does MATLAB tell you that $A\boldsymbol{x} = \boldsymbol{b}$ has no solution? For the special $\boldsymbol{b} =$ ones (4, 1), which solution to $A\boldsymbol{x} = \boldsymbol{b}$ is found by $A\backslash\boldsymbol{b}$?

Challenge Problems

39 (Recommended) A is a 4 by 4 matrix with 1's on the diagonal and $-a$, $-b$, $-c$ on the diagonal above. Find A^{-1} for this bidiagonal matrix.

40 Suppose E_1, E_2, E_3 are 4 by 4 identity matrices, except E_1 has a, b, c in column 1 and E_2 has d, e in column 2 and E_3 has f in column 3 (below the 1's). Multiply $L = E_1 E_2 E_3$ to show that all these nonzeros are copied into L.

$E_1 E_2 E_3$ is in the *opposite* order from elimination (because E_3 is acting first). But $E_1 E_2 E_3 = L$ is in the *correct* order to invert elimination and recover A.

41 Second difference matrices have beautiful inverses if they start with $T_{11} = 1$ (instead of $K_{11} = 2$). Here is the 3 by 3 tridiagonal matrix T and its inverse:

$$T = \begin{bmatrix} \mathbf{1} & -1 & 0 \\ -1 & 2 & -1 \\ 0 & -1 & 2 \end{bmatrix} \qquad T^{-1} = \begin{bmatrix} 3 & 2 & 1 \\ 2 & 2 & 1 \\ 1 & 1 & 1 \end{bmatrix}$$

One approach is Gauss-Jordan elimination on $[\,T \quad I\,]$. I would rather write T as the product of first differences L times U. The inverses of L and U in Worked Example **2.5 A** are **sum matrices**, so here are $T = LU$ and $T^{-1} = U^{-1}L^{-1}$:

$$T = \underset{\textbf{difference}}{\begin{bmatrix} 1 & & \\ -1 & 1 & \\ 0 & -1 & 1 \end{bmatrix}} \underset{\textbf{difference}}{\begin{bmatrix} 1 & -1 & 0 \\ & 1 & -1 \\ & & 1 \end{bmatrix}} \qquad T^{-1} = \underset{\textbf{sum}}{\begin{bmatrix} 1 & 1 & 1 \\ & 1 & 1 \\ & & 1 \end{bmatrix}} \underset{\textbf{sum}}{\begin{bmatrix} 1 & & \\ 1 & 1 & \\ 1 & 1 & 1 \end{bmatrix}}$$

Question. (**4 by 4**) What are the pivots of T? What is its 4 by 4 inverse? The reverse order UL gives what matrix T^*? What is the inverse of T^*?

42 Here are two more difference matrices, both important. ***But are they invertible***?

$$\textbf{Cyclic } C = \begin{bmatrix} 2 & -1 & 0 & -1 \\ -1 & 2 & -1 & 0 \\ 0 & -1 & 2 & -1 \\ -1 & 0 & -1 & 2 \end{bmatrix} \qquad \textbf{Free ends } F = \begin{bmatrix} 1 & -1 & 0 & 0 \\ -1 & 2 & -1 & 0 \\ 0 & -1 & 2 & -1 \\ 0 & 0 & -1 & 1 \end{bmatrix}.$$

43 *Elimination for a block matrix*: When you multiply the first block row $[A \ B]$ by CA^{-1} and subtract from the second row $[C \ D]$, the "*Schur complement*" S appears:

$$\begin{bmatrix} I & 0 \\ -CA^{-1} & I \end{bmatrix} \begin{bmatrix} A & B \\ C & D \end{bmatrix} = \begin{bmatrix} A & B \\ 0 & S \end{bmatrix} \qquad \begin{matrix} A \text{ and } D \text{ are square} \\ S = D - CA^{-1}B. \end{matrix}$$

Multiply on the right to subtract $A^{-1}B$ times block column 1 from block column 2.

$$\begin{bmatrix} A & B \\ 0 & S \end{bmatrix} \begin{bmatrix} I & -A^{-1}B \\ 0 & I \end{bmatrix} = ? \quad \text{Find } S \text{ for} \quad \begin{bmatrix} A & B \\ C & I \end{bmatrix} = \begin{bmatrix} 2 & 3 & 3 \\ 4 & 1 & 0 \\ 4 & 0 & 1 \end{bmatrix}.$$

The block pivots are** A **and** S. **If they are invertible, so is $[\,A \quad B\,; \ C \quad D\,]$.

44 How does the identity $A(I + BA) = (I + AB)A$ connect the inverses of $I + BA$ and $I + AB$? Those are both invertible or both singular: not obvious.

2.6 Elimination = Factorization: $A = LU$

1 Each elimination step E_{ij} is inverted by L_{ij}. Off the main diagonal change $-\ell_{ij}$ to $+\ell_{ij}$.

2 The whole forward elimination process (with no row exchanges) is inverted by $\boldsymbol{L}$:
$$\boldsymbol{L} = (L_{21}L_{31}\ldots L_{n1})(L_{32}\ldots L_{n2})(L_{43}\ldots L_{n3})\ldots(L_{n\,n-1}).$$

3 That product matrix $\boldsymbol{L}$ is still lower triangular. **Every multiplier $\boldsymbol{\ell_{ij}}$ is in row $\boldsymbol{i}$, column $\boldsymbol{j}$.**

4 The original A is recovered from U by $\boldsymbol{A} = \boldsymbol{LU} =$ (lower triangular)(upper triangular).

5 Elimination on $A\boldsymbol{x} = \boldsymbol{b}$ reaches $U\boldsymbol{x} = \boldsymbol{c}$. Then back-substitution solves $U\boldsymbol{x} = \boldsymbol{c}$.

6 Solving a triangular system takes $n^2/2$ multiply-subtracts. Elimination to find U takes $n^3/3$.

Students often say that mathematics courses are too theoretical. Well, not this section. It is almost purely practical. The goal is to describe Gaussian elimination in the most useful way. Many key ideas of linear algebra, when you look at them closely, are really *factorizations* of a matrix. The original matrix A becomes the product of two or three special matrices. The first factorization—also the most important in practice—comes now from elimination. ***The factors L and U are triangular matrices. The factorization that comes from elimination is A = LU.***

We already know U, the upper triangular matrix with the pivots on its diagonal. The elimination steps take A to U. We will show how reversing those steps (taking U back to A) is achieved by a lower triangular L. ***The entries of*** L ***are exactly the multipliers*** ℓ_{ij}—which multiplied the pivot row j when it was subtracted from row i.

Start with a 2 by 2 example. The matrix A contains $2, 1, 6, 8$. The number to eliminate is 6. ***Subtract 3 times row*** 1 ***from row*** 2. That step is E_{21} in the forward direction with multiplier $\ell_{21} = 3$. The return step from U to A is $L = E_{21}^{-1}$ (an addition using $+3$):

$$\textbf{\textit{Forward from } } A \textbf{\textit{ to }} U: \quad E_{21}A = \begin{bmatrix} 1 & 0 \\ -3 & 1 \end{bmatrix}\begin{bmatrix} 2 & 1 \\ 6 & 8 \end{bmatrix} = \begin{bmatrix} 2 & 1 \\ 0 & 5 \end{bmatrix} = U$$

$$\textbf{\textit{Back from }} U \textbf{\textit{ to }} A: \quad E_{21}^{-1}U = \begin{bmatrix} 1 & 0 \\ 3 & 1 \end{bmatrix}\begin{bmatrix} 2 & 1 \\ 0 & 5 \end{bmatrix} = \begin{bmatrix} 2 & 1 \\ 6 & 8 \end{bmatrix} = A.$$

The second line is our factorization $LU = A$. Instead of E_{21}^{-1} we write L. Move now to larger matrices with many E's. ***Then L will include all their inverses.***

Each step from A to U multiplies by a matrix E_{ij} to produce zero in the (i, j) position. To keep this clear, we stay with the most frequent case—***when no row exchanges are involved.*** If A is 3 by 3, we multiply by E_{21} and E_{31} and E_{32}. The multipliers ℓ_{ij} produce zeros in the $(2, 1)$ and $(3, 1)$ and $(3, 2)$ positions—all below the diagonal. Elimination ends with the upper triangular U.

Now move those E's onto the other side, *where their inverses multiply* U:

$$(E_{32}E_{31}E_{21})A = U \quad \textbf{becomes} \quad \boxed{A = (E_{21}^{-1}E_{31}^{-1}E_{32}^{-1})\,U} \quad \textbf{which is} \quad \boxed{A = LU.}$$

The inverses go in opposite order, as they must. That product of three inverses is L. ***We have reached*** $\boldsymbol{A = LU}$. Now we stop to understand it.

Explanation and Examples

First point: Every inverse matrix E^{-1} is *lower triangular*. Its off-diagonal entry is ℓ_{ij}, to undo the subtraction produced by $-\ell_{ij}$. The main diagonals of E and E^{-1} contain 1's. Our example above had $\ell_{21} = 3$ and $E = \begin{bmatrix} 1 & 0 \\ -3 & 1 \end{bmatrix}$ and $L = E^{-1} = \begin{bmatrix} 1 & 0 \\ 3 & 1 \end{bmatrix}$.

Second point: Equation (2) shows a lower triangular matrix (the product of the E_{ij}) multiplying A. It also shows all the E_{ij}^{-1} multiplying U to bring back A. ***This lower triangular product of inverses is L.***

One reason for working with the inverses is that we want to factor A, not U. The "inverse form" gives $A = LU$. Another reason is that we get something extra, almost more than we deserve. This is the third point, showing that L is exactly right.

Third point: Each multiplier ℓ_{ij} goes directly into its i, j position—*unchanged*—in the product of inverses which is L. Usually matrix multiplication will mix up all the numbers. Here that doesn't happen. The order is right for the inverse matrices, to keep the ℓ's unchanged. The reason is given below in equation (2).

Since each E^{-1} has 1's down its diagonal, the final good point is that L does too.

$\boldsymbol{A = LU}$	***This is elimination without row exchanges***. The upper triangular U has the pivots on its diagonal. The lower triangular L has all 1's on its diagonal. ***The multipliers*** $\boldsymbol{\ell_{ij}}$ ***are below the diagonal of*** $\boldsymbol{L}$.

Example 1 Elimination subtracts $\frac{1}{2}$ times row 1 from row 2. The last step subtracts $\frac{2}{3}$ times row 2 from row 3. The lower triangular L has $\ell_{21} = \frac{1}{2}$ and $\ell_{32} = \frac{2}{3}$. Multiplying LU produces A:

$$A = \begin{bmatrix} 2 & 1 & 0 \\ 1 & 2 & 1 \\ 0 & 1 & 2 \end{bmatrix} = \begin{bmatrix} 1 & 0 & 0 \\ \frac{1}{2} & 1 & 0 \\ 0 & \frac{2}{3} & 1 \end{bmatrix} \begin{bmatrix} 2 & 1 & 0 \\ 0 & \frac{3}{2} & 1 \\ 0 & 0 & \frac{4}{3} \end{bmatrix} = LU.$$

The $(3, 1)$ multiplier is zero because the $(3, 1)$ entry in A is zero. No operation needed.

Example 2 Change the top left entry from 2 in A to 1 in B. The pivots all become 1. The multipliers are all 1. That pattern continues when B is 4 by 4:

$$\textbf{Special pattern} \quad B = \begin{bmatrix} \mathbf{1} & 1 & 0 & 0 \\ 1 & 2 & 1 & 0 \\ 0 & 1 & 2 & 1 \\ 0 & 0 & 1 & 2 \end{bmatrix} = \begin{bmatrix} 1 & & & \\ 1 & 1 & & \\ 0 & 1 & 1 & \\ 0 & 0 & 1 & 1 \end{bmatrix} \begin{bmatrix} 1 & 1 & 0 & 0 \\ & 1 & 1 & 0 \\ & & 1 & 1 \\ & & & 1 \end{bmatrix}.$$

These LU examples are showing something extra, which is very important in practice. Assume no row exchanges. When can we predict *zeros* in L and U?

When a row of A starts with zeros, so does that row of L.

When a column of A starts with zeros, so does that column of U.

If a row starts with zero, we don't need an elimination step. L has a zero, which saves computer time. Similarly, zeros at the *start* of a column survive into U. But please realize: Zeros in the *middle* of a matrix are likely to be filled in, while elimination sweeps forward. We now explain why L has the multipliers ℓ_{ij} in position, with no mix-up.

The key reason why A equals LU: Ask yourself about the pivot rows that are subtracted from lower rows. Are they the original rows of A? *No*, elimination probably changed them. Are they rows of U? *Yes*, the pivot rows never change again. When computing the third row of U, we subtract multiples of earlier rows of U (*not rows of* A!):

$$\text{Row 3 of } U = (\text{Row 3 of } A) - \ell_{31}(\text{Row 1 of } U) - \ell_{32}(\text{Row 2 of } U). \tag{1}$$

Rewrite this equation to see that the row $[\,\ell_{31} \quad \ell_{32} \quad 1\,]$ is multiplying the matrix U:

$$(\text{Row 3 of } A) = \ell_{31}(\text{Row 1 of } U) + \ell_{32}(\text{Row 2 of } U) + 1(\text{Row 3 of } U). \tag{2}$$

This is exactly row** 3 **of $A = LU$. That row of L holds ℓ_{31}, ℓ_{32}, 1. All rows look like this, whatever the size of A. With no row exchanges, we have $A = LU$.

Better balance from LDU $A = LU$ is "unsymmetric" because U has the pivots on its diagonal where L has 1's. This is easy to change. ***Divide U by a diagonal matrix D that contains the pivots.*** That leaves a new triangular matrix with 1's on the diagonal:

$$\text{Split } U \text{ into } \begin{bmatrix} d_1 & & & \\ & d_2 & & \\ & & \ddots & \\ & & & d_n \end{bmatrix} \begin{bmatrix} 1 & u_{12}/d_1 & u_{13}/d_1 & \cdot \\ & 1 & u_{23}/d_2 & \cdot \\ & & \ddots & \vdots \\ & & & 1 \end{bmatrix}.$$

It is convenient (but a little confusing) to keep the same letter U for this new triangular matrix. It has 1's on the diagonal (like L). Instead of the normal LU, the new form has D in the middle: ***Lower triangular L times diagonal D times upper triangular U.***

The triangular factorization can be written $\quad A = LU \quad$ or $\quad A = LDU.$

Whenever you see LDU, it is understood that U has 1's on the diagonal. *Each row is divided by its first nonzero entry—the pivot.* Then L and U are treated evenly in LDU:

$$\begin{bmatrix} 1 & 0 \\ 3 & 1 \end{bmatrix} \begin{bmatrix} 2 & 8 \\ 0 & 5 \end{bmatrix} \quad \text{splits further into} \quad \begin{bmatrix} 1 & 0 \\ 3 & 1 \end{bmatrix} \begin{bmatrix} 2 & \\ & 5 \end{bmatrix} \begin{bmatrix} 1 & 4 \\ 0 & 1 \end{bmatrix}. \tag{3}$$

The pivots 2 and 5 went into D. Dividing the rows by 2 and 5 left the rows $[\,1 \quad 4\,]$ and $[\,0 \quad 1\,]$ in the new U with diagonal ones. The multiplier 3 is still in L.

My own lectures sometimes stop at this point. I go forward to 2.7. The next paragraphs show how elimination codes are organized, and how long they take. If MATLAB (or any software) is available, you can measure the computing time by just counting the seconds.

One Square System = Two Triangular Systems

The matrix L contains our memory of Gaussian elimination. It holds the numbers that multiplied the pivot rows, before subtracting them from lower rows. When do we need this record and how do we use it in solving $A\boldsymbol{x} = \boldsymbol{b}$?

We need L as soon as there is a *right side* $\boldsymbol{b}$. The factors L and U were completely decided by the left side (the matrix A). On the right side of $A\boldsymbol{x} = \boldsymbol{b}$, we use L^{-1} and then U^{-1}. That ***Solve*** step deals with two triangular matrices.

1 ***Factor*** (into L and U, by elimination on the left side matrix A).

2 ***Solve*** (forward elimination on $\boldsymbol{b}$ using L, then back substitution for $\boldsymbol{x}$ using U).

Earlier, we worked on A and $\boldsymbol{b}$ at the same time. No problem with that—just augment to $[\,A \;\; \boldsymbol{b}\,]$. But most computer codes keep the two sides separate. The memory of elimination is held in L and U, to process $\boldsymbol{b}$ whenever we want to. The User's Guide to LAPACK remarks that "This situation is so common and the savings are so important that no provision has been made for solving a single system with just one subroutine."

How does ***Solve*** work on $\boldsymbol{b}$? First, apply forward elimination to the right side (the multipliers are stored in L, use them now). This changes $\boldsymbol{b}$ to a new right side $\boldsymbol{c}$. *We are really solving* $L\boldsymbol{c} = \boldsymbol{b}$. Then back substitution solves $U\boldsymbol{x} = \boldsymbol{c}$ as always. The original system $A\boldsymbol{x} = \boldsymbol{b}$ is factored into ***two triangular systems***:

Forward and backward ***Solve*** $\quad L\boldsymbol{c} = \boldsymbol{b} \quad$ ***and then solve*** $\quad U\boldsymbol{x} = \boldsymbol{c}\,.$ (4)

To see that $\boldsymbol{x}$ is correct, multiply $U\boldsymbol{x} = \boldsymbol{c}$ by L. Then $LU\boldsymbol{x} = L\boldsymbol{c}$ is just $A\boldsymbol{x} = \boldsymbol{b}$.

To emphasize: There is *nothing new* about those steps. This is exactly what we have done all along. We were really solving the triangular system $L\boldsymbol{c} = \boldsymbol{b}$ as elimination went forward. Then back substitution produced $\boldsymbol{x}$. An example shows what we actually did.

Example 3 Forward elimination (downward) on $A\boldsymbol{x} = \boldsymbol{b}$ ends at $U\boldsymbol{x} = \boldsymbol{c}$:

$$A\boldsymbol{x} = \boldsymbol{b} \qquad \begin{aligned} u + 2v &= 5 \\ 4u + 9v &= 21 \end{aligned} \qquad \text{becomes} \qquad \begin{aligned} u + 2v &= 5 \\ v &= 1 \end{aligned} \qquad U\boldsymbol{x} = \boldsymbol{c}$$

The multiplier was 4, which is saved in L. The right side used that 4 to change 21 to 1:

$Lc = b$ The lower triangular system $\begin{bmatrix} 1 & 0 \\ 4 & 1 \end{bmatrix} \begin{bmatrix} c \end{bmatrix} = \begin{bmatrix} 5 \\ 21 \end{bmatrix}$ gave $c = \begin{bmatrix} 5 \\ 1 \end{bmatrix}$.

$Ux = c$ The upper triangular system $\begin{bmatrix} 1 & 2 \\ 0 & 1 \end{bmatrix} \begin{bmatrix} x \end{bmatrix} = \begin{bmatrix} 5 \\ 1 \end{bmatrix}$ gives $x = \begin{bmatrix} 3 \\ 1 \end{bmatrix}$.

L and U can go into the n^2 storage locations that originally held A (now forgettable).

The Cost of Elimination

A very practical question is cost—or computing time. We can solve 1000 equations on a PC. What if $n = 100,000$? (*Is A dense or sparse*?) Large systems come up all the time in scientific computing, where a three-dimensional problem can easily lead to a million unknowns. We can let the calculation run overnight, but we can't leave it for 100 years.

The first stage of elimination produces zeros below the first pivot in column 1. To find each new entry below the pivot row requires one multiplication and one subtraction. *We will count this first stage as n^2 multiplications and n^2 subtractions.* It is actually less, $n^2 - n$, because row 1 does not change.

The next stage clears out the second column below the second pivot. The working matrix is now of size $n-1$. Estimate this stage by $(n-1)^2$ multiplications and subtractions. The matrices are getting smaller as elimination goes forward. The rough count to reach U is the sum of squares $n^2 + (n-1)^2 + \cdots + 2^2 + 1^2$.

There is an exact formula $\frac{1}{3}n\left(n + \frac{1}{2}\right)(n+1)$ for this sum of squares. When n is large, the $\frac{1}{2}$ and the 1 are not important. *The number that matters is* $\frac{1}{3}n^3$. The sum of squares is like the integral of x^2 ! The integral from 0 to n is $\frac{1}{3}n^3$:

Elimination on A requires about $\frac{1}{3}n^3$ multiplications and $\frac{1}{3}n^3$ subtractions.

What about the right side b? Going forward, we subtract multiples of b_1 from the lower components $b_2, \ldots, b_n$. This is $n-1$ steps. The second stage takes only $n-2$ steps, because b_1 is not involved. The last stage of forward elimination takes one step.

Now start back substitution. Computing x_n uses one step (divide by the last pivot). The next unknown uses two steps. When we reach x_1 it will require n steps ($n-1$ substitutions of the other unknowns, then division by the first pivot). The total count on the right side, from b to c to x—*forward to the bottom and back to the top*—is exactly n^2:

$$[(n-1) + (n-2) + \cdots + 1] \ + \ [1 + 2 + \cdots + (n-1) + n] = n^2. \tag{5}$$

To see that sum, pair off $(n-1)$ with 1 and $(n-2)$ with 2. The pairings leave n terms, each equal to n. That makes n^2. The right side costs a lot less than the left side!

Solve ***Each right side needs n^2 multiplications and n^2 subtractions.***

A **band matrix** B has only w nonzero diagonals below and above its main diagonal. The zero entries outside the band stay zero in elimination (they are zero in L and U).

Clearing out the first column needs w^2 multiplications and subtractions (w zeros to be produced below the pivot, each one using a pivot row of length w). Then clearing out all n columns, to reach U, needs no more than nw^2. This saves a lot of time:

Band matrix	***A* to *U*** $\frac{1}{3}n^3$ reduces to nw^2	**Solve** n^2 reduces to $2\,nw$

A tridiagonal matrix (bandwidth $w = 1$) allows very fast computation. Don't store zeros!

The book's website has Teaching Codes to factor A into LU and to solve $A\boldsymbol{x} = \boldsymbol{b}$. Professional codes will look down each column for the *largest available pivot*, to exchange rows and reduce roundoff error.

MATLAB's backslash command $\boldsymbol{x} = A\backslash\boldsymbol{b}$ combines **Factor** and **Solve** to reach $\boldsymbol{x}$.

How long does it take to solve $Ax = b$***?*** For a random matrix of order $n = 1000$, a typical time on a PC is 1 second. The time is multiplied by about 8 when n is multiplied by 2. For professional codes go to **netlib.org**.

According to this n^3 rule, matrices that are 10 times as large (order 10,000) will take a thousand seconds. Matrices of order 100,000 will take a million seconds. This is too expensive without a supercomputer, but remember that these matrices are full. Most matrices in practice are sparse (many zero entries). In that case $A = LU$ is much faster.

■ REVIEW OF THE KEY IDEAS ■

1. Gaussian elimination (with no row exchanges) factors A into L times U.

2. The lower triangular L contains the numbers ℓ_{ij} that multiply pivot rows, going from A to U. The product LU adds those rows back to recover A.

3. On the right side we solve $L\boldsymbol{c} = \boldsymbol{b}$ (forward) and $U\boldsymbol{x} = \boldsymbol{c}$ (backward).

4. ***Factor***: There are $\frac{1}{3}(n^3 - n)$ multiplications and subtractions on the left side.

5. ***Solve***: There are n^2 multiplications and subtractions on the right side.

6. For a band matrix, change $\frac{1}{3}n^3$ to nw^2 and change n^2 to $2wn$.

■ WORKED EXAMPLES ■

2.6 A The lower triangular Pascal matrix L contains the famous "*Pascal triangle*". Gauss-Jordan inverted L in the worked example **2.5 C**. Here we factor Pascal.

The symmetric Pascal matrix P is a product of triangular Pascal matrices L and U. The symmetric P has Pascal's triangle tilted, so each entry is the sum of the entry above and the entry to the left. The n by n symmetric P is pascal (n) in MATLAB.

Problem: *Establish the amazing lower-upper factorization $P = LU$.*

$$\text{pascal}(4) = \begin{bmatrix} \mathbf{1} & \mathbf{1} & \mathbf{1} & \mathbf{1} \\ \mathbf{1} & \mathbf{2} & \mathbf{3} & 4 \\ \mathbf{1} & \mathbf{3} & 6 & 10 \\ \mathbf{1} & 4 & 10 & 20 \end{bmatrix} = \begin{bmatrix} \mathbf{1} & 0 & 0 & 0 \\ \mathbf{1} & \mathbf{1} & 0 & 0 \\ \mathbf{1} & \mathbf{2} & \mathbf{1} & 0 \\ \mathbf{1} & \mathbf{3} & \mathbf{3} & \mathbf{1} \end{bmatrix} \begin{bmatrix} \mathbf{1} & \mathbf{1} & \mathbf{1} & \mathbf{1} \\ 0 & \mathbf{1} & \mathbf{2} & \mathbf{3} \\ 0 & 0 & \mathbf{1} & \mathbf{3} \\ 0 & 0 & 0 & \mathbf{1} \end{bmatrix} = LU.$$

Then predict and check the next row and column for 5 by 5 Pascal matrices.

Solution You could multiply LU to get P. Better to start with the symmetric P and reach the upper triangular U by elimination:

$$P = \begin{bmatrix} 1 & 1 & 1 & 1 \\ 1 & 2 & 3 & 4 \\ 1 & 3 & 6 & 10 \\ 1 & 4 & 10 & 20 \end{bmatrix} \to \begin{bmatrix} 1 & 1 & 1 & 1 \\ 0 & 1 & 2 & 3 \\ 0 & 2 & 5 & 9 \\ 0 & 3 & 9 & 19 \end{bmatrix} \to \begin{bmatrix} 1 & 1 & 1 & 1 \\ 0 & 1 & 2 & 3 \\ 0 & 0 & 1 & 3 \\ 0 & 0 & 3 & 10 \end{bmatrix} \to \begin{bmatrix} 1 & 1 & 1 & 1 \\ 0 & 1 & 2 & 3 \\ 0 & 0 & 1 & 3 \\ 0 & 0 & 0 & 1 \end{bmatrix} = U.$$

The multipliers ℓ_{ij} that entered these steps go perfectly into L. Then $P = LU$ is a particularly neat example. *Notice that every pivot is* 1 on the diagonal of U.

The next section will show how symmetry produces a special relationship between the triangular L and U. For Pascal, U is the "**transpose**" of L.

You might expect the MATLAB command lu (pascal (4)) to produce these L and U. That doesn't happen because the **lu** subroutine chooses the largest available pivot in each column. The second pivot will change from 1 to 3. But a "Cholesky factorization" does no row exchanges: $U =$ chol (pascal (4))

The full proof of $P = LU$ for all Pascal sizes is quite fascinating. The paper "*Pascal Matrices*" is on the course web page **web.mit.edu/18.06** which is also available through MIT's *OpenCourseWare* at **ocw.mit.edu**. These Pascal matrices have so many remarkable properties—we will see them again.

2.6 B The problem is: *Solve* $P\boldsymbol{x} = \boldsymbol{b} = (1, 0, 0, 0)$. This right side = column of I means that $\boldsymbol{x}$ will be the first column of P^{-1}. That is Gauss-Jordan, matching the columns of $PP^{-1} = I$. We already know the Pascal matrices L and U as factors of P:

Two triangular systems $\quad L\boldsymbol{c} = \boldsymbol{b}$ (forward) $\quad U\boldsymbol{x} = \boldsymbol{c}$ (back).

Solution The lower triangular system $L\boldsymbol{c} = \boldsymbol{b}$ is solved *top to bottom*:

$$\begin{aligned} c_1 &= 1 \\ c_1 + c_2 &= 0 \\ c_1 + 2c_2 + c_3 &= 0 \\ c_1 + 3c_2 + 3c_3 + c_4 &= 0 \end{aligned} \quad \text{gives} \quad \begin{aligned} c_1 &= +1 \\ c_2 &= -1 \\ c_3 &= +1 \\ c_4 &= -1 \end{aligned}$$

Forward elimination is multiplication by L^{-1}. It produces the upper triangular system $U\boldsymbol{x} = \boldsymbol{c}$. The solution $\boldsymbol{x}$ comes as always by back substitution, *bottom to top*:

$$\begin{aligned} x_1 + x_2 + \ x_3 + \ x_4 &= \ \ 1 \\ x_2 + 2x_3 + 3x_4 &= -1 \\ x_3 + 3x_4 &= \ \ 1 \\ x_4 &= -1 \end{aligned} \quad \text{gives} \quad \begin{aligned} x_1 &= \mathbf{+4} \\ x_2 &= \mathbf{-6} \\ x_3 &= \mathbf{+4} \\ x_4 &= \mathbf{-1} \end{aligned}$$

I see a pattern in that $\boldsymbol{x}$, but I don't know where it comes from. Try **inv** (**pascal** (4)).

Problem Set 2.6

Problems 1–14 compute the factorization $A = LU$ (and also $A = LDU$).

1 (Important) Forward elimination changes $\left[\begin{smallmatrix}1&1\\1&2\end{smallmatrix}\right]\boldsymbol{x} = \boldsymbol{b}$ to a triangular $\left[\begin{smallmatrix}1&1\\0&1\end{smallmatrix}\right]\boldsymbol{x} = \boldsymbol{c}$:

$$\begin{aligned} x + \ y &= \mathbf{5} \\ x + 2y &= \mathbf{7} \end{aligned} \quad \longrightarrow \quad \begin{aligned} x + \ y &= \mathbf{5} \\ y &= \mathbf{2} \end{aligned} \qquad \begin{bmatrix} 1 & 1 & \mathbf{5} \\ 1 & 2 & \mathbf{7} \end{bmatrix} \longrightarrow \begin{bmatrix} 1 & 1 & \mathbf{5} \\ 0 & 1 & \mathbf{2} \end{bmatrix}$$

That step subtracted $\ell_{21} =$ _____ times row 1 from row 2. The reverse step *adds* ℓ_{21} times row 1 to row 2. The matrix for that reverse step is $L =$ _____. Multiply this L times the triangular system $\left[\begin{smallmatrix}1&1\\0&1\end{smallmatrix}\right]\boldsymbol{x}_1 = \left[\begin{smallmatrix}5\\2\end{smallmatrix}\right]$ to get _____ $=$ _____. In letters, L multiplies $U\boldsymbol{x} = \boldsymbol{c}$ to give _____.

2 Write down the 2 by 2 triangular systems $L\boldsymbol{c} = \boldsymbol{b}$ and $U\boldsymbol{x} = \boldsymbol{c}$ from Problem 1. Check that $\boldsymbol{c} = (5, 2)$ solves the first one. Find $\boldsymbol{x}$ that solves the second one.

3 (Move to 3 by 3) Forward elimination changes $A\boldsymbol{x} = \boldsymbol{b}$ to a triangular $U\boldsymbol{x} = \boldsymbol{c}$:

$$\begin{aligned} x + \ y + \ z &= 5 \\ x + 2y + 3z &= 7 \\ x + 3y + 6z &= 11 \end{aligned} \qquad \begin{aligned} x + \ y + \ z &= 5 \\ y + 2z &= 2 \\ 2y + 5z &= 6 \end{aligned} \qquad \begin{aligned} x + \ y + \ z &= 5 \\ y + 2z &= 2 \\ z &= 2 \end{aligned}$$

The equation $z = 2$ in $U\boldsymbol{x} = \boldsymbol{c}$ comes from the original $x + 3y + 6z = 11$ in $A\boldsymbol{x} = \boldsymbol{b}$ by subtracting $\ell_{31} =$ _____ times equation 1 and $\ell_{32} =$ _____ times the *final* equation 2. Reverse that to recover $[1 \ \ 3 \ \ 6 \ \ 11]$ in the last row of A and $\boldsymbol{b}$ from the final $[1 \ \ 1 \ \ 1 \ \ 5]$ and $[0 \ \ 1 \ \ 2 \ \ 2]$ and $[0 \ \ 0 \ \ 1 \ \ 2]$ in U and $\boldsymbol{c}$:

$$\text{Row 3 of } \begin{bmatrix} A & \boldsymbol{b} \end{bmatrix} = (\ell_{31} \text{ Row } 1 + \ell_{32} \text{ Row } 2 + 1 \text{ Row } 3) \text{ of } \begin{bmatrix} U & \boldsymbol{c} \end{bmatrix}.$$

In matrix notation this is multiplication by L. So $A = LU$ and $\boldsymbol{b} = L\boldsymbol{c}$.

4 What are the 3 by 3 triangular systems $L\boldsymbol{c} = \boldsymbol{b}$ and $U\boldsymbol{x} = \boldsymbol{c}$ from Problem 3? Check that $\boldsymbol{c} = (5, 2, 2)$ solves the first one. Which $\boldsymbol{x}$ solves the second one?

5 What matrix E puts A into triangular form $EA = U$? Multiply by $E^{-1} = L$ to factor A into LU:

$$A = \begin{bmatrix} 2 & 1 & 0 \\ 0 & 4 & 2 \\ 6 & 3 & 5 \end{bmatrix}.$$

6 What two elimination matrices E_{21} and E_{32} put A into upper triangular form $E_{32}E_{21}A = U$? Multiply by E_{32}^{-1} and E_{21}^{-1} to factor A into $LU = E_{21}^{-1}E_{32}^{-1}U$:

$$A = \begin{bmatrix} 1 & 1 & 1 \\ 2 & 4 & 5 \\ 0 & 4 & 0 \end{bmatrix}.$$

7 What three elimination matrices E_{21}, E_{31}, E_{32} put A into its upper triangular form $E_{32}E_{31}E_{21}A = U$? Multiply by E_{32}^{-1}, E_{31}^{-1} and E_{21}^{-1} to factor A into L times U:

$$A = \begin{bmatrix} 1 & 0 & 1 \\ 2 & 2 & 2 \\ 3 & 4 & 5 \end{bmatrix} \qquad L = E_{21}^{-1}E_{31}^{-1}E_{32}^{-1}.$$

8 **This is the problem that shows how the inverses E_{ij}^{-1} multiply to give L.** You see this best when A is already lower triangular with 1's on the diagonal. **Then $U = I$** !

$$A = L = \begin{bmatrix} 1 & 0 & 0 \\ a & 1 & 0 \\ b & c & 1 \end{bmatrix}.$$

The elimination matrices E_{21}, E_{31}, E_{32} contain $-a$ then $-b$ then $-c$.

(a) Multiply $E_{32}E_{31}E_{21}$ to find the single matrix E that produces $EA = I$.

(b) Multiply $E_{21}^{-1}E_{31}^{-1}E_{32}^{-1}$ to bring back L.
The multipliers a, b, c are mixed up in E but perfect in L.

9 *When zero appears in a pivot position, $A = LU$ is not possible*! (We are requiring nonzero pivots in U.) Show directly why these equations are both impossible:

$$\begin{bmatrix} 0 & 1 \\ 2 & 3 \end{bmatrix} = \begin{bmatrix} 1 & 0 \\ \ell & 1 \end{bmatrix} \begin{bmatrix} d & e \\ 0 & f \end{bmatrix} \qquad \begin{bmatrix} 1 & 1 & 0 \\ 1 & 1 & 2 \\ 1 & 2 & 1 \end{bmatrix} = \begin{bmatrix} 1 & & \\ \ell & 1 & \\ m & n & 1 \end{bmatrix} \begin{bmatrix} d & e & g \\ & f & h \\ & & i \end{bmatrix}.$$

These matrices need a row exchange. That uses a "*permutation matrix*" P.

10 Which number c leads to zero in the second pivot position? A row exchange is needed and $A = LU$ will not be possible. Which c produces zero in the third pivot position? Then a row exchange can't help and elimination fails:

$$A = \begin{bmatrix} 1 & c & 0 \\ 2 & 4 & 1 \\ 3 & 5 & 1 \end{bmatrix}.$$

11 What are L and D (the diagonal ***pivot matrix***) for this matrix A? What is U in $A = LU$ and what is the new U in $A = LDU$?

$$\textbf{Already triangular} \qquad A = \begin{bmatrix} 2 & 4 & 8 \\ 0 & 3 & 9 \\ 0 & 0 & 7 \end{bmatrix}.$$

12 A and B are symmetric across the diagonal (because $4 = 4$). Find their triple factorizations LDU and say how U is related to L for these symmetric matrices:

$$\textbf{Symmetric} \qquad A = \begin{bmatrix} 2 & 4 \\ 4 & 11 \end{bmatrix} \quad \text{and} \quad B = \begin{bmatrix} 1 & 4 & 0 \\ 4 & 12 & 4 \\ 0 & 4 & 0 \end{bmatrix}.$$

13 (*Recommended*) Compute L and U for the symmetric matrix A:

$$A = \begin{bmatrix} a & a & a & a \\ a & b & b & b \\ a & b & c & c \\ a & b & c & d \end{bmatrix}.$$

Find four conditions on a, b, c, d to get $A = LU$ with four pivots.

14 This nonsymmetric matrix will have the same L as in Problem **13**:

$$\textbf{Find } L \textbf{ and } U \textbf{ for} \qquad A = \begin{bmatrix} a & r & r & r \\ a & b & s & s \\ a & b & c & t \\ a & b & c & d \end{bmatrix}.$$

Find the four conditions on a, b, c, d, r, s, t to get $A = LU$ with four pivots.

Problems 15–16 use L and U (without needing A) to solve $A\boldsymbol{x} = \boldsymbol{b}$.

15 Solve the triangular system $L\boldsymbol{c} = \boldsymbol{b}$ to find $\boldsymbol{c}$. Then solve $U\boldsymbol{x} = \boldsymbol{c}$ to find $\boldsymbol{x}$:

$$L = \begin{bmatrix} 1 & 0 \\ 4 & 1 \end{bmatrix} \quad \text{and} \quad U = \begin{bmatrix} 2 & 4 \\ 0 & 1 \end{bmatrix} \quad \text{and} \quad \boldsymbol{b} = \begin{bmatrix} 2 \\ 11 \end{bmatrix}.$$

For safety multiply LU and solve $A\boldsymbol{x} = \boldsymbol{b}$ as usual. Circle $\boldsymbol{c}$ when you see it.

16 Solve $L\boldsymbol{c} = \boldsymbol{b}$ to find $\boldsymbol{c}$. Then solve $U\boldsymbol{x} = \boldsymbol{c}$ to find $\boldsymbol{x}$. **What was A?**

$$L = \begin{bmatrix} 1 & 0 & 0 \\ 1 & 1 & 0 \\ 1 & 1 & 1 \end{bmatrix} \quad \text{and} \quad U = \begin{bmatrix} 1 & 1 & 1 \\ 0 & 1 & 1 \\ 0 & 0 & 1 \end{bmatrix} \quad \text{and} \quad \boldsymbol{b} = \begin{bmatrix} 4 \\ 5 \\ 6 \end{bmatrix}.$$

17 (a) When you apply the usual elimination steps to L, what matrix do you reach?

$$L = \begin{bmatrix} 1 & 0 & 0 \\ \ell_{21} & 1 & 0 \\ \ell_{31} & \ell_{32} & 1 \end{bmatrix}.$$

(b) When you apply the same steps to I, what matrix do you get?

(c) When you apply the same steps to LU, what matrix do you get?

18 If $A = LDU$ and also $A = L_1 D_1 U_1$ with all factors invertible, then $L = L_1$ and $D = D_1$ and $U = U_1$. *"The three factors are unique."*

Derive the equation $L_1^{-1} LD = D_1 U_1 U^{-1}$. Are the two sides triangular or diagonal? Deduce $L = L_1$ and $U = U_1$ (they all have diagonal 1's). Then $D = D_1$.

19 *Tridiagonal matrices* have zero entries except on the main diagonal and the two adjacent diagonals. Factor these into $A = LU$ and $A = LDL^{\mathrm{T}}$:

$$A = \begin{bmatrix} 1 & 1 & 0 \\ 1 & 2 & 1 \\ 0 & 1 & 2 \end{bmatrix} \quad \text{and} \quad A = \begin{bmatrix} a & a & 0 \\ a & a+b & b \\ 0 & b & b+c \end{bmatrix}.$$

20 When T is tridiagonal, its L and U factors have only two nonzero diagonals. How would you take advantage of knowing the zeros in T, in a code for Gaussian elimination? Find L and U.

$$\textbf{Tridiagonal} \qquad T = \begin{bmatrix} 1 & 2 & 0 & 0 \\ 2 & 3 & 1 & 0 \\ 0 & 1 & 2 & 3 \\ 0 & 0 & 3 & 4 \end{bmatrix}.$$

21 If A and B have nonzeros in the positions marked by x, which zeros (marked by 0) *stay zero* in their factors L and U?

$$A = \begin{bmatrix} x & x & x & x \\ x & x & x & 0 \\ 0 & x & x & x \\ 0 & 0 & x & x \end{bmatrix} \qquad B = \begin{bmatrix} x & x & x & 0 \\ x & x & 0 & x \\ x & 0 & x & x \\ 0 & x & x & x \end{bmatrix}.$$

22 Suppose you eliminate upwards (almost unheard of). Use the last row to produce zeros in the last column (the pivot is 1). Then use the second row to produce zero above the second pivot. Find the factors in the unusual order $A = UL$.

$$\textbf{Upper times lower} \qquad A = \begin{bmatrix} 5 & 3 & 1 \\ 3 & 3 & 1 \\ 1 & 1 & 1 \end{bmatrix}.$$

23 *Easy but important.* If A has pivots $5, 9, 3$ with no row exchanges, what are the pivots for the upper left 2 by 2 submatrix A_2 (without row 3 and column 3)?

Challenge Problems

24 Which invertible matrices allow $A = LU$ (elimination without row exchanges)? *Good question*! Look at each of the square upper left submatrices A_k of A.

All upper left k by k submatrices A_k must be invertible (**sizes** $k = 1, \ldots, n$).

Explain that answer: A_k factors into ____ because $LU = \begin{bmatrix} L_k & 0 \\ * & * \end{bmatrix} \begin{bmatrix} U_k & * \\ 0 & * \end{bmatrix}$.

25 For the 6 by 6 second difference constant-diagonal matrix K, put the pivots and multipliers into $K = LU$. (L and U will have only two nonzero diagonals, because K has three.) Find a formula for the i, j entry of L^{-1}, by software like MATLAB using inv (L) or by looking for a nice pattern.

$$-1, 2, -1 \textbf{ matrix} \quad K = \begin{bmatrix} 2 & -1 & & & & \\ -1 & \cdot & \cdot & & & \\ & \cdot & \cdot & \cdot & & \\ & & \cdot & \cdot & \cdot & \\ & & & \cdot & \cdot & -1 \\ & & & & -1 & 2 \end{bmatrix} = \text{toeplitz}\,([2 \;\; -1 \;\; 0 \;\; 0 \;\; 0 \;\; 0])$$

26 If you print K^{-1}, it doesn't look so good (6 by 6). But if you print $7K^{-1}$, that matrix looks wonderful. Write down $7K^{-1}$ by hand, following this pattern:

1 Row 1 and column 1 are $(6, 5, 4, 3, 2, 1)$.

2 On and above the main diagonal, row i is i times row 1.

3 On and below the main diagonal, column j is j times column 1.

Multiply K times that $7K^{-1}$ to produce $7I$. Here is $4K^{-1}$ for $n = 3$:

3 by 3 case
The determinant
of this K is 4

$$(K)(4K^{-1}) = \begin{bmatrix} 2 & -1 & 0 \\ -1 & 2 & -1 \\ 0 & -1 & 2 \end{bmatrix} \begin{bmatrix} \mathbf{3} & \mathbf{2} & \mathbf{1} \\ \mathbf{2} & \mathbf{4} & \mathbf{2} \\ \mathbf{1} & \mathbf{2} & \mathbf{3} \end{bmatrix} = \begin{bmatrix} 4 & & \\ & 4 & \\ & & 4 \end{bmatrix}.$$

2.7 Transposes and Permutations

1 The transposes of $A\boldsymbol{x}$ and AB and A^{-1} are $\boldsymbol{x}^{\mathrm{T}}A^{\mathrm{T}}$ and $B^{\mathrm{T}}A^{\mathrm{T}}$ and $(A^{\mathrm{T}})^{-1}$.

2 The dot product (inner product) is $\boldsymbol{x} \cdot \boldsymbol{y} = \boldsymbol{x}^{\mathrm{T}}\boldsymbol{y}$. This is $(1 \times n)(n \times 1) = (1 \times 1)$.
The outer product is $\boldsymbol{x}\boldsymbol{y}^{\mathbf{T}}$ = column times row = $(n \times 1)(1 \times n) = n \times n$ matrix.

3 The idea behind A^{T} is that $A\boldsymbol{x} \cdot \boldsymbol{y}$ equals $\boldsymbol{x} \cdot A^{\mathrm{T}}\boldsymbol{y}$ because $(A\boldsymbol{x})^{\mathrm{T}}\boldsymbol{y} = \boldsymbol{x}^{\mathrm{T}}A^{\mathrm{T}}\boldsymbol{y} = \boldsymbol{x}^{\mathrm{T}}(A^{\mathrm{T}}\boldsymbol{y})$.

4 A **symmetric matrix** has $\boldsymbol{S^{\mathrm{T}} = S}$ (and the product $A^{\mathrm{T}}A$ is always symmetric).

5 An **orthogonal matrix** has $\boldsymbol{Q^{\mathrm{T}} = Q^{-1}}$. The columns of Q are orthogonal unit vectors.

6 A **permutation matrix** $\boldsymbol{P}$ has the same rows as I (in any order). There are $\boldsymbol{n}\,!$ different orders.

7 Then $P\boldsymbol{x}$ puts the components $x_1, x_2, \ldots, x_n$ in that new order. And P^{T} equals P^{-1}.

We need one more matrix, and fortunately it is much simpler than the inverse. It is the ***"transpose"*** of A, which is denoted by A^{T}. *The columns of* A^{T} *are the rows of* A.

When A is an m by n matrix, the transpose is n by m:

Transpose $\quad$ If $A = \begin{bmatrix} 1 & 2 & 3 \\ 0 & 0 & 4 \end{bmatrix}$ then $A^{\mathrm{T}} = \begin{bmatrix} 1 & 0 \\ 2 & 0 \\ 3 & 4 \end{bmatrix}$.

You can write the rows of A into the columns of A^{T}. Or you can write the columns of A into the rows of A^{T}. The matrix "flips over" its main diagonal. The entry in row i, column j of A^{T} comes from row j, column i of the original A:

Exchange rows and columns $\qquad \boldsymbol{(A^{\mathrm{T}})_{ij} = A_{ji}}.$

The transpose of a lower triangular matrix is upper triangular. (But the inverse is still lower triangular.) The transpose of A^{T} is A.

Note MATLAB's symbol for the transpose of A is A'. Typing $[1\ \ 2\ \ 3]$ gives a row vector and the column vector is $\boldsymbol{v} = [1\ \ 2\ \ 3]'$. To enter a matrix M with second column $\boldsymbol{w} = [\ 4\ 5\ 6\ ]'$ you could define $M = [\ \boldsymbol{v}\ \boldsymbol{w}\ ]$. Quicker to enter by rows and then transpose the whole matrix: $M = [1\ \ 2\ \ 3;\ \ 4\ \ 5\ \ 6]'$.

The rules for transposes are very direct. We can transpose $A + B$ to get $(A + B)^{\mathrm{T}}$. Or we can transpose A and B separately, and then add $A^{\mathrm{T}} + B^{\mathrm{T}}$—with the same result.

The serious questions are about the transpose of a product AB and an inverse A^{-1}:

Sum	The transpose of $A+B$ is $A^{\mathrm{T}}+B^{\mathrm{T}}$.	(1)
Product	The transpose of AB is $(AB)^{\mathrm{T}}=B^{\mathrm{T}}A^{\mathrm{T}}$.	(2)
Inverse	The transpose of A^{-1} is $(A^{-1})^{\mathrm{T}}=(A^{\mathrm{T}})^{-1}$.	(3)

Notice especially how $B^{\mathrm{T}}A^{\mathrm{T}}$ comes in reverse order. For inverses, this reverse order was quick to check: $B^{-1}A^{-1}$ times AB produces I. To understand $(AB)^{\mathrm{T}}=B^{\mathrm{T}}A^{\mathrm{T}}$, start with $(A\boldsymbol{x})^{\mathrm{T}}=\boldsymbol{x}^{\mathrm{T}}A^{\mathrm{T}}$ when B is just a vector:

Ax ***combines the columns of*** A ***while*** $x^{\mathrm{T}}A^{\mathrm{T}}$ ***combines the rows of*** A^{T}.

It is the same combination of the same vectors! In A they are columns, in A^{T} they are rows. So the transpose of the column $A\boldsymbol{x}$ is the row $\boldsymbol{x}^{\mathrm{T}}A^{\mathrm{T}}$. That fits our formula $(A\boldsymbol{x})^{\mathrm{T}}=\boldsymbol{x}^{\mathrm{T}}A^{\mathrm{T}}$. Now we can prove the formula $(AB)^{\mathrm{T}}=B^{\mathrm{T}}A^{\mathrm{T}}$, when B has several columns.

If $B=[\boldsymbol{x}_1\ \ \boldsymbol{x}_2]$ has two columns, apply the same idea to each column. The columns of AB are $A\boldsymbol{x}_1$ and $A\boldsymbol{x}_2$. Their transposes appear correctly in the rows of $B^{\mathrm{T}}A^{\mathrm{T}}$:

$$\text{Transposing } AB=\begin{bmatrix} \\ A\boldsymbol{x}_1 & A\boldsymbol{x}_2 & \cdots \\ \\ \end{bmatrix} \text{ gives } \begin{bmatrix} \boldsymbol{x}_1^{\mathrm{T}}A^{\mathrm{T}} \\ \boldsymbol{x}_2^{\mathrm{T}}A^{\mathrm{T}} \\ \vdots \end{bmatrix} \text{ which is } B^{\mathrm{T}}A^{\mathrm{T}}. \tag{4}$$

The right answer $B^{\mathrm{T}}A^{\mathrm{T}}$ comes out a row at a time. Here are numbers in $(AB)^{\mathrm{T}}=B^{\mathrm{T}}A^{\mathrm{T}}$:

$$AB=\begin{bmatrix}1&0\\1&1\end{bmatrix}\begin{bmatrix}5&0\\4&1\end{bmatrix}=\begin{bmatrix}\mathbf{5}&\mathbf{0}\\\mathbf{9}&\mathbf{1}\end{bmatrix} \quad\text{and}\quad B^{\mathrm{T}}A^{\mathrm{T}}=\begin{bmatrix}5&4\\0&1\end{bmatrix}\begin{bmatrix}1&1\\0&1\end{bmatrix}=\begin{bmatrix}\mathbf{5}&\mathbf{9}\\\mathbf{0}&\mathbf{1}\end{bmatrix}.$$

The reverse order rule extends to three or more factors: $(ABC)^{\mathrm{T}}$ equals $C^{\mathrm{T}}B^{\mathrm{T}}A^{\mathrm{T}}$.

If $\boldsymbol{A=LDU}$ ***then*** $\boldsymbol{A^{\mathrm{T}}=U^{\mathrm{T}}D^{\mathrm{T}}L^{\mathrm{T}}}$. ***The pivot matrix has*** $\boldsymbol{D=D^{\mathrm{T}}}$.

Now apply this product rule by transposing both sides of $A^{-1}A=I$. On one side, I^{T} is I. We confirm the rule that $(A^{-1})^{\mathrm{T}}$ *is the inverse of* A^{T}. Their product is I:

$$\textbf{Transpose of inverse}\qquad A^{-1}A=I \quad\text{is transposed to}\quad A^{\mathrm{T}}(A^{-1})^{\mathrm{T}}=I. \tag{5}$$

Similarly $AA^{-1}=I$ leads to $(A^{-1})^{\mathrm{T}}A^{\mathrm{T}}=I$. We can invert the transpose or we can transpose the inverse. Notice especially: A^{T} ***is invertible exactly when*** A ***is invertible.***

Example 1 The inverse of $A=\begin{bmatrix}1&0\\6&1\end{bmatrix}$ is $A^{-1}=\begin{bmatrix}1&0\\-6&1\end{bmatrix}$. The transpose is $A^{\mathrm{T}}=\begin{bmatrix}1&6\\0&1\end{bmatrix}$.

$$(A^{-1})^{\mathrm{T}} \ \textit{and}\ (A^{\mathrm{T}})^{-1} \ \textit{are both equal to}\ \begin{bmatrix}1&-6\\0&1\end{bmatrix}.$$

The Meaning of Inner Products

We know the dot product (inner product) of $\boldsymbol{x}$ and $\boldsymbol{y}$. It is the sum of numbers $x_i y_i$. Now we have a better way to write $\boldsymbol{x} \cdot \boldsymbol{y}$, without using that unprofessional dot. Use matrix notation instead:

^{T} **is inside**	*The dot product or inner product is* $x^{\mathrm{T}}y$	$(1 \times n)(n \times 1)$
^{T} **is outside**	*The rank one product or outer product is* xy^{T}	$(n \times 1)(1 \times n)$

$\boldsymbol{x}^{\mathrm{T}}\boldsymbol{y}$ is a number, $\boldsymbol{x}\boldsymbol{y}^{\mathrm{T}}$ is a matrix. Quantum mechanics would write those as $< \boldsymbol{x}|\boldsymbol{y} >$ (inner) and $|\boldsymbol{x} >< \boldsymbol{y}|$ (outer). Maybe the universe is governed by linear algebra. Here are three more examples where the inner product has meaning:

From mechanics	Work = (Movements) (Forces) = $\boldsymbol{x}^{\mathrm{T}}\boldsymbol{f}$
From circuits	Heat loss = (Voltage drops) (Currents) = $\boldsymbol{e}^{\mathrm{T}}\boldsymbol{y}$
From economics	Income = (Quantities) (Prices) = $\boldsymbol{q}^{\mathrm{T}}\boldsymbol{p}$

We are really close to the heart of applied mathematics, and there is one more point to emphasize. It is the deeper connection between inner products and the transpose of A.

We defined A^{T} by flipping the matrix across its main diagonal. That's not mathematics. There is a better way to approach the transpose. *A^{T} is the matrix that makes these two inner products equal for every $\boldsymbol{x}$ and $\boldsymbol{y}$*:

$$(Ax)^{\mathrm{T}}y = x^{\mathrm{T}}(A^{\mathrm{T}}y) \quad \textbf{Inner product of } Ax \textbf{ with } y = \textbf{Inner product of } x \textbf{ with } A^{\mathrm{T}}y$$

$$\text{Start with } A = \begin{bmatrix} -1 & 1 & 0 \\ 0 & -1 & 1 \end{bmatrix} \qquad \boldsymbol{x} = \begin{bmatrix} x_1 \\ x_2 \\ x_3 \end{bmatrix} \qquad \boldsymbol{y} = \begin{bmatrix} y_1 \\ y_2 \end{bmatrix}$$

On one side we have $A\boldsymbol{x}$ multiplying $\boldsymbol{y}$: $(x_2 - x_1)\,y_1 + (x_3 - x_2)\,y_2$
That is the same as $x_1\,(-y_1) + x_2\,(y_1 - y_2) + x_3\,(y_2)$. Now $\boldsymbol{x}$ is multiplying $A^{\mathrm{T}}\boldsymbol{y}$.

$$A^{\mathrm{T}}\boldsymbol{y} \text{ must be } \begin{bmatrix} -y_1 \\ y_1 - y_2 \\ y_2 \end{bmatrix} \text{ which produces } \quad A^{\mathrm{T}} = \begin{bmatrix} -1 & 0 \\ 1 & -1 \\ 0 & 1 \end{bmatrix} \text{ as expected.}$$

Symmetric Matrices

For a *symmetric matrix*, transposing A to A^{T} produces no change. Then A^{T} equals A. Its (j, i) entry across the main diagonal equals its (i, j) entry. In my opinion, these are the most important matrices of all. We give symmetric matrices the special letter S.

DEFINITION A ***symmetric matrix*** has $S^{\mathrm{T}} = S$. This means that $s_{ji} = s_{ij}$.

$$\textbf{Symmetric matrices} \qquad S = \begin{bmatrix} 1 & 2 \\ 2 & 5 \end{bmatrix} = S^{\mathrm{T}} \quad \text{and} \quad D = \begin{bmatrix} 1 & 0 \\ 0 & 10 \end{bmatrix} = D^{\mathrm{T}}.$$

The inverse of a symmetric matrix is also symmetric. The transpose of S^{-1} is $(S^{-1})^{\mathrm{T}} = (S^{\mathrm{T}})^{-1} = S^{-1}$. That says S^{-1} is symmetric (when S is invertible):

$$\textbf{Symmetric inverses} \qquad S^{-1} = \begin{bmatrix} 5 & -2 \\ -2 & 1 \end{bmatrix} \quad \text{and} \quad D^{-1} = \begin{bmatrix} 1 & 0 \\ 0 & 0.1 \end{bmatrix}.$$

Now we produce a symmetric matrix S by ***multiplying any matrix A by*** $\boldsymbol{A}^{\mathbf{T}}$.

Symmetric Products $A^{\mathrm{T}}A$ and AA^{T} and LDL^{T}

Choose any matrix A, probably rectangular. Multiply A^{T} times A. Then the product $S = A^{\mathrm{T}}A$ is automatically a square symmetric matrix:

$$\textit{The transpose of} \quad A^{\mathrm{T}}A \quad \textit{is} \quad A^{\mathrm{T}}(A^{\mathrm{T}})^{\mathrm{T}} \quad \textit{which is} \quad A^{\mathrm{T}}A \ \textit{again.} \tag{6}$$

That is a quick proof of symmetry for $A^{\mathrm{T}}A$. We could look at the (i,j) entry of $A^{\mathrm{T}}A$. It is the dot product of row i of A^{T} (column i of A) with column j of A. The (j,i) entry is the same dot product, column j with column i. So $A^{\mathrm{T}}A$ is symmetric.

The matrix AA^{T} is also symmetric. (The shapes of A and A^{T} allow multiplication.) But AA^{T} is a different matrix from $A^{\mathrm{T}}A$. In our experience, most scientific problems that start with a rectangular matrix A end up with $A^{\mathrm{T}}A$ or AA^{T} or both. As in least squares.

Example 2 Multiply $A = \begin{bmatrix} -1 & 1 & 0 \\ 0 & -1 & 1 \end{bmatrix}$ and $A^{\mathrm{T}} = \begin{bmatrix} 1 & 0 \\ 1 & -1 \\ 0 & 1 \end{bmatrix}$ in both orders.

$$AA^{\mathrm{T}} = \begin{bmatrix} 2 & -1 \\ -1 & 2 \end{bmatrix} \quad \text{and} \quad A^{\mathrm{T}}A = \begin{bmatrix} 1 & -1 & 0 \\ -1 & 2 & -1 \\ 0 & -1 & 1 \end{bmatrix} \text{ are both symmetric matrices.}$$

The product $A^{\mathrm{T}}A$ is n by n. In the opposite order, AA^{T} is m by m. Both are symmetric, with positive diagonal (*why*?). But even if $m = n$, it is very likely that $A^{\mathrm{T}}A \neq AA^{\mathrm{T}}$. Equality can happen, but it is abnormal.

Symmetric matrices in elimination $\quad S^{\mathrm{T}} = S$ makes elimination faster, because we can work with half the matrix (plus the diagonal). It is true that the upper triangular U is probably not symmetric. ***The symmetry is in the triple product*** $S = LDU$. Remember how the diagonal matrix D of pivots can be divided out, to leave 1's on the diagonal of both L and U:

$$\begin{bmatrix} 1 & 2 \\ 2 & 7 \end{bmatrix} = \begin{bmatrix} 1 & 0 \\ 2 & 1 \end{bmatrix} \begin{bmatrix} 1 & 2 \\ 0 & 3 \end{bmatrix} \qquad LU \text{ misses the symmetry of } S$$

$$\begin{bmatrix} 1 & 2 \\ 2 & 7 \end{bmatrix} = \begin{bmatrix} \mathbf{1} & 0 \\ \mathbf{2} & \mathbf{1} \end{bmatrix} \begin{bmatrix} 1 & 0 \\ 0 & 3 \end{bmatrix} \begin{bmatrix} \mathbf{1} & \mathbf{2} \\ 0 & \mathbf{1} \end{bmatrix} \qquad \begin{array}{l} LDL^{\mathrm{T}} \text{ captures the symmetry} \\ \textit{Now } U \textit{ is the transpose of } L. \end{array}$$

When S is symmetric, the usual form $A = LDU$ becomes $\boldsymbol{S = LDL^{\mathrm{T}}}$. The final U (with 1's on the diagonal) is the transpose of L (also with 1's on the diagonal). The diagonal matrix D containing the pivots is symmetric by itself.

If $S = S^{\mathrm{T}}$ is factored into LDU with no row exchanges, then U is exactly L^{T}.

The symmetric factorization of a symmetric matrix is $S = LDL^{\mathrm{T}}$.

Notice that the transpose of LDL^{T} is automatically $(L^{\mathrm{T}})^{\mathrm{T}}D^{\mathrm{T}}L^{\mathrm{T}}$ which is LDL^{T} again. The work of elimination is cut in half, from $n^3/3$ multiplications to $n^3/6$. The storage is also cut essentially in half. We only keep L and D, not U which is just L^{T}.

Permutation Matrices

The transpose plays a special role for a *permutation matrix*. This matrix P has a single "1" in every row and every column. Then P^{T} is also a permutation matrix—maybe the same as P or maybe different. Any product P_1P_2 is again a permutation matrix.

We now create every P from the identity matrix, by reordering the rows of I.

The simplest permutation matrix is $P = I$ (*no exchanges*). The next simplest are the row exchanges P_{ij}. Those are constructed by exchanging two rows i and j of I. Other permutations reorder more rows. By doing all possible row exchanges to I, we get all possible permutation matrices:

DEFINITION ***A permutation matrix P has the rows of the identity I in any order.***

Example 3 There are six 3 by 3 permutation matrices. Here they are without the zeros:

$$I = \begin{bmatrix} 1 & & \\ & 1 & \\ & & 1 \end{bmatrix} \quad P_{21} = \begin{bmatrix} & 1 & \\ 1 & & \\ & & 1 \end{bmatrix} \quad P_{32}P_{21} = \begin{bmatrix} & 1 & \\ & & 1 \\ 1 & & \end{bmatrix}$$

$$P_{31} = \begin{bmatrix} & & 1 \\ & 1 & \\ 1 & & \end{bmatrix} \quad P_{32} = \begin{bmatrix} 1 & & \\ & & 1 \\ & 1 & \end{bmatrix} \quad P_{21}P_{32} = \begin{bmatrix} & & 1 \\ 1 & & \\ & 1 & \end{bmatrix}.$$

There are $n!$ permutation matrices of order n. The symbol $n!$ means "n factorial," the product of the numbers $(1)(2)\cdots(n)$. Thus $3! = (1)(2)(3)$ which is 6. There will be 24 permutation matrices of order $n = 4$. And 120 permutations of order 5.

There are only two permutation matrices of order 2, namely $\left[\begin{smallmatrix} 1 & 0 \\ 0 & 1 \end{smallmatrix}\right]$ and $\left[\begin{smallmatrix} 0 & 1 \\ 1 & 0 \end{smallmatrix}\right]$.

Important: P^{-1} is also a permutation matrix. Among the six 3 by 3 P's displayed above, the four matrices on the left are their own inverses. The two matrices on the right are inverses of each other. In all cases, a single row exchange is its own inverse. If we repeat the exchange we are back to I. But for $P_{32}P_{21}$, the inverses go in opposite order as always. The inverse is $P_{21}P_{32}$.

More important: $\boldsymbol{P^{-1}}$ ***is always the same as*** $\boldsymbol{P^{\mathrm{T}}}$**.** The two matrices on the right are transposes—and inverses—of each other. When we multiply PP^{T}, the "1" in the first row of P hits the "1" in the first column of P^{T} (since the first row of P is the first column of P^{T}). It misses the ones in all the other columns. So $PP^{\mathrm{T}} = I$.

Another proof of $P^{\mathrm{T}} = P^{-1}$ looks at P as a product of row exchanges. Every row exchange is its own transpose and its own inverse. P^{T} and P^{-1} both come from the product of row exchanges *in reverse order*. So P^{T} and P^{-1} are the same.

Permutations (row exchanges before elimination) lead to $PA = LU$***.***

The $PA = LU$ Factorization with Row Exchanges

We sure hope you remember $A = LU$. It started with $A = (E_{21}^{-1} \cdots E_{ij}^{-1} \cdots)U$. Every elimination step was carried out by an E_{ij} and it was inverted by E_{ij}^{-1}. Those inverses were compressed into one matrix L. The lower triangular L has 1's on the diagonal, and the result is $A = LU$.

This is a great factorization, but it doesn't always work. Sometimes row exchanges are needed to produce pivots. Then $A = (E^{-1} \cdots P^{-1} \cdots E^{-1} \cdots P^{-1} \cdots)U$. Every row exchange is carried out by a P_{ij} and inverted by that P_{ij}. We now compress those row exchanges into a *single permutation matrix* P. This gives a factorization for every invertible matrix A—which we naturally want.

The main question is where to collect the P_{ij}'s. There are two good possibilities—do all the exchanges before elimination, or do them after the E_{ij}'s. The first way gives $PA = LU$. The second way has a permutation matrix P_1 in the middle.

1. The row exchanges can be done *in advance*. Their product P puts the rows of A in the right order, so that no exchanges are needed for PA. ***Then*** $\boldsymbol{PA = LU}$.

2. If we hold row exchanges until *after elimination*, the pivot rows are in a strange order. P_1 puts them in the correct triangular order in U_1. ***Then*** $\boldsymbol{A = L_1P_1U_1}$.

$PA = LU$ is constantly used in all computing. ***We will concentrate on this form.***

The factorization $A = L_1P_1U_1$ might be more elegant. If we mention both, it is because the difference is not well known. Probably you will not spend a long time on either one. Please don't. The most important case has $P = I$, when A equals LU with no exchanges.

This matrix A starts with $a_{11} = 0$. Exchange rows 1 and 2 to bring the first pivot into its usual place. Then go through elimination on PA:

$$\underset{A}{\begin{bmatrix} 0 & 1 & 1 \\ 1 & 2 & 1 \\ 2 & 7 & 9 \end{bmatrix}} \rightarrow \underset{PA}{\begin{bmatrix} 1 & 2 & 1 \\ 0 & 1 & 1 \\ 2 & 7 & 9 \end{bmatrix}} \rightarrow \underset{\ell_{31} = 2}{\begin{bmatrix} 1 & 2 & 1 \\ 0 & 1 & 1 \\ 0 & 3 & 7 \end{bmatrix}} \rightarrow \underset{\ell_{32} = 3}{\begin{bmatrix} 1 & 2 & 1 \\ 0 & 1 & 1 \\ 0 & 0 & 4 \end{bmatrix}}.$$

The matrix PA has its rows in good order, and it factors as usual into LU:

$$\boldsymbol{P} = \begin{bmatrix} 0 & \mathbf{1} & 0 \\ \mathbf{1} & 0 & 0 \\ 0 & 0 & \mathbf{1} \end{bmatrix} \qquad \boldsymbol{PA} = \begin{bmatrix} 1 & 0 & 0 \\ 0 & 1 & 0 \\ 2 & 3 & 1 \end{bmatrix} \begin{bmatrix} 1 & 2 & 1 \\ 0 & 1 & 1 \\ 0 & 0 & 4 \end{bmatrix} = \boldsymbol{LU}. \tag{7}$$

We started with A and ended with U. *The only requirement is invertibility of* A.

If A is invertible, a permutation P will put its rows in the right order to factor $\boldsymbol{PA = LU}$
There must be a full set of pivots after row exchanges for A to be invertible.

In MATLAB, $A([r\ k],:) = A([k\ r],:)$ exchanges row k with row r below it (where the kth pivot has been found). Then the **lu** code updates L and P and the sign of P:

	$A([r\ k],:) = A([k\ r],:);$
This is part of	$L([r\ k], 1:k-1) = L([k\ r], 1:k-1);$
$[L, U, P] =$ **lu** (A)	$P([r\ k],:) = P([k\ r],:);$
	sign $= -$sign

The "**sign**" of P tells whether the number of row exchanges is even (sign $= +1$). An odd number of row exchanges will produce sign $= -1$. At the start, P is I and sign $= +1$. When there is a row exchange, the sign is reversed. The final value of sign is the **determinant of P** and it does not depend on the order of the row exchanges.

For PA we get back to the familiar LU. In reality, a code like **lu** (A) often does not use the first available pivot. Mathematically we can accept a small pivot— anything but zero. **All good codes look down the column for the largest pivot**.

Section 11.1 explains why this "*partial pivoting*" reduces the roundoff error. Then P may contain row exchanges that are not algebraically necessary. Still $PA = LU$.

Our advice is to understand permutations but let the computer do the work. Calculations of $A = LU$ are enough to do by hand, without P. The Teaching Code splu (A) factors $PA = LU$ and splv (A, b) solves $A\boldsymbol{x} = \boldsymbol{b}$ for any invertible A. The program splu on the website stops if no pivot can be found in column k. Then A is not invertible.

■ REVIEW OF THE KEY IDEAS ■

1. The transpose puts the rows of A into the columns of A^{T}. Then $(A^{\mathrm{T}})_{ij} = A_{ji}$.

2. The transpose of AB is $B^{\mathrm{T}}A^{\mathrm{T}}$. The transpose of A^{-1} is the inverse of A^{T}.

3. The dot product is $\boldsymbol{x} \cdot \boldsymbol{y} = \boldsymbol{x}^{\mathrm{T}}\boldsymbol{y}$. Then $(A\boldsymbol{x})^{\mathrm{T}}\boldsymbol{y}$ equals the dot product $\boldsymbol{x}^{\mathrm{T}}(A^{\mathrm{T}}\boldsymbol{y})$.

4. When S is symmetric ($S^{\mathrm{T}} = S$), its LDU factorization is symmetric: $S = LDL^{\mathrm{T}}$.

5. A permutation matrix P has a 1 in each row and column, and $\boldsymbol{P}^{\mathrm{T}} = \boldsymbol{P}^{-1}$.

6. There are $n!$ permutation matrices of size n. *Half even, half odd.*

7. If A is invertible then a permutation P will reorder its rows for $PA = LU$.

■ WORKED EXAMPLES ■

2.7 A Applying the permutation P to the rows of S destroys its symmetry:

$$P = \begin{bmatrix} 0 & 1 & 0 \\ 0 & 0 & 1 \\ 1 & 0 & 0 \end{bmatrix} \qquad S = \begin{bmatrix} 1 & 4 & 5 \\ 4 & \mathbf{2} & 6 \\ 5 & 6 & \mathbf{3} \end{bmatrix} \qquad PS = \begin{bmatrix} 4 & \mathbf{2} & 6 \\ 5 & 6 & \mathbf{3} \\ 1 & 4 & 5 \end{bmatrix}$$

What permutation Q applied to the *columns* of PS will recover symmetry in PSQ? The numbers $1, 2, 3$ must come back to the main diagonal (not necessarily in order). Show that Q is P^{T}, so that **symmetry is saved by** PSP^{T}.

Solution To recover symmetry and put "2" back on the diagonal, column 2 of PS must move to column 1. Column 3 of PS (containing "3") must move to column 2. Then the "1" moves to the $3, 3$ position. The matrix that permutes columns is Q:

$$PS = \begin{bmatrix} 4 & \mathbf{2} & 6 \\ 5 & 6 & \mathbf{3} \\ 1 & 4 & 5 \end{bmatrix} \qquad Q = \begin{bmatrix} 0 & 0 & 1 \\ 1 & 0 & 0 \\ 0 & 1 & 0 \end{bmatrix} \qquad PSQ = \begin{bmatrix} \mathbf{2} & 6 & 4 \\ 6 & \mathbf{3} & 5 \\ 4 & 5 & 1 \end{bmatrix} \text{ is symmetric.}$$

The matrix Q is P^{T}. This choice always recovers symmetry, because PSP^{T} is guaranteed to be symmetric. (Its transpose is again PSP^{T}.) *The matrix Q is also P^{-1}, because the inverse of every permutation matrix is its transpose.*

If D is a diagonal matrix, we are finding that PDP^{T} is also diagonal. When P moves row 1 down to row 3, P^{T} on the right will move column 1 to column 3. The $(1, 1)$ entry moves down to $(3, 1)$ and over to $(3, 3)$.

2.7 B Find the symmetric factorization $S = LDL^{\mathrm{T}}$ for the matrix S above.

Solution To factor S into LDL^{T} we eliminate as usual to reach U:

$$S = \begin{bmatrix} 1 & 4 & 5 \\ 4 & 2 & 6 \\ 5 & 6 & 3 \end{bmatrix} \longrightarrow \begin{bmatrix} 1 & 4 & 5 \\ 0 & -14 & -14 \\ 0 & -14 & -22 \end{bmatrix} \longrightarrow \begin{bmatrix} 1 & 4 & 5 \\ 0 & -14 & -14 \\ 0 & 0 & -8 \end{bmatrix} = U.$$

The multipliers were $\ell_{21} = 4$ and $\ell_{31} = 5$ and $\ell_{32} = 1$. **The pivots $\mathbf{1, -14, -8}$ go into $\boldsymbol{D}$.** When we divide the rows of U by those pivots, L^{T} should appear:

Symmetric factorization when $\boldsymbol{S = S^{\mathrm{T}}}$

$$\boldsymbol{S = LDL^{\mathrm{T}}} = \begin{bmatrix} 1 & 0 & 0 \\ 4 & 1 & 0 \\ 5 & 1 & 1 \end{bmatrix} \begin{bmatrix} 1 & & \\ & -14 & \\ & & -8 \end{bmatrix} \begin{bmatrix} 1 & 4 & 5 \\ 0 & 1 & 1 \\ 0 & 0 & 1 \end{bmatrix}.$$

This matrix S is invertible because *it has three pivots*. Its inverse is $(L^{\mathrm{T}})^{-1}D^{-1}L^{-1}$ and S^{-1} is also symmetric. The numbers 14 and 8 will turn up in the denominators of S^{-1}. The "determinant" of S is the product of the pivots $(1)(-14)(-8) = 112$.

2.7 C For a rectangular A, this ***saddle-point matrix S*** is symmetric and important:

Block matrix from least squares $$\boldsymbol{S}=\begin{bmatrix}\boldsymbol{I} & \boldsymbol{A}\\ \boldsymbol{A}^{\mathrm{T}} & \boldsymbol{0}\end{bmatrix}=\boldsymbol{S}^{\mathrm{T}}\text{ has size } m+n.$$

Apply block elimination to find a **block factorization** $S=LDL^{\mathrm{T}}$. Then test invertibility:

$$\textit{S is invertible} \iff A^{\mathrm{T}}A \textit{ is invertible} \iff Ax\neq 0 \textit{ whenever } x\neq 0$$

Solution The first block pivot is I. Subtract A^{T} times row 1 from row 2:

Block elimination $$S=\begin{bmatrix}I & A\\ A^{\mathrm{T}} & 0\end{bmatrix}\text{ goes to }\begin{bmatrix}I & A\\ 0 & -\boldsymbol{A}^{\mathrm{T}}\boldsymbol{A}\end{bmatrix}.\text{ This is } U.$$

The block pivot matrix D contains I and $-A^{\mathrm{T}}A$. Then L and L^{T} contain A^{T} and A:

Block factorization $$S=LDL^{\mathrm{T}}=\begin{bmatrix}I & 0\\ A^{\mathrm{T}} & I\end{bmatrix}\begin{bmatrix}I & 0\\ 0 & -A^{\mathrm{T}}A\end{bmatrix}\begin{bmatrix}I & A\\ 0 & I\end{bmatrix}.$$

L is certainly invertible, with diagonal 1's. The inverse of the middle matrix involves $(A^{\mathrm{T}}A)^{-1}$. Section 4.2 answers a key question about the matrix $A^{\mathrm{T}}A$:

When is $A^{\mathrm{T}}A$ invertible? *Answer: A must have independent columns.*
Then $Ax=0$ only if $x=0$. Otherwise $Ax=0$ will lead to $A^{\mathrm{T}}Ax=0$.

Problem Set 2.7

Questions 1–7 are about the rules for transpose matrices.

1 Find A^{T} and A^{-1} and $(A^{-1})^{\mathrm{T}}$ and $(A^{\mathrm{T}})^{-1}$ for

$$A=\begin{bmatrix}1 & 0\\ 9 & 3\end{bmatrix}\quad\text{and also}\quad A=\begin{bmatrix}1 & c\\ c & 0\end{bmatrix}.$$

2 Verify that $(AB)^{\mathrm{T}}$ equals $B^{\mathrm{T}}A^{\mathrm{T}}$ but those are different from $A^{\mathrm{T}}B^{\mathrm{T}}$:

$$A=\begin{bmatrix}1 & 0\\ 2 & 1\end{bmatrix}\quad B=\begin{bmatrix}1 & 3\\ 0 & 1\end{bmatrix}\quad AB=\begin{bmatrix}1 & 3\\ 2 & 7\end{bmatrix}.$$

Show also that AA^{T} is different from $A^{\mathrm{T}}A$. But both of those matrices are _____.

3 (a) The matrix $((AB)^{-1})^{\mathrm{T}}$ comes from $(A^{-1})^{\mathrm{T}}$ and $(B^{-1})^{\mathrm{T}}$. *In what order?*

(b) If U is upper triangular then $(U^{-1})^{\mathrm{T}}$ is _____ triangular.

4 Show that $A^2=0$ is possible but $A^{\mathrm{T}}A=0$ is not possible (unless $A=$ zero matrix).

5 (a) The row vector $\boldsymbol{x}^{\mathrm{T}}$ times A times the column $\boldsymbol{y}$ produces what number?

$$\boldsymbol{x}^{\mathrm{T}}A\boldsymbol{y} = \begin{bmatrix} 0 & 1 \end{bmatrix} \begin{bmatrix} 1 & 2 & 3 \\ 4 & 5 & 6 \end{bmatrix} \begin{bmatrix} 0 \\ 1 \\ 0 \end{bmatrix} = \underline{\qquad}.$$

(b) This is the row $\boldsymbol{x}^{\mathrm{T}}A =$ _____ times the column $\boldsymbol{y} = (0, 1, 0)$.

(c) This is the row $\boldsymbol{x}^{\mathrm{T}} = [0 \;\; 1]$ times the column $A\boldsymbol{y} =$ _____.

6 The transpose of a block matrix $M = \begin{bmatrix} \mathbf{A} & \mathbf{B} \\ \mathbf{C} & \mathbf{D} \end{bmatrix}$ is $M^{\mathrm{T}} =$ _____. Test an example. Under what conditions on A, B, C, D is the block matrix symmetric?

7 True or false:

(a) The block matrix $\begin{bmatrix} \mathbf{0} & \mathbf{A} \\ \mathbf{A} & \mathbf{0} \end{bmatrix}$ is automatically symmetric.

(b) If A and B are symmetric then their product AB is symmetric.

(c) If A is not symmetric then A^{-1} is not symmetric.

(d) When A, B, C are symmetric, the transpose of ABC is CBA.

Questions 8–15 are about permutation matrices.

8 Why are there $n!$ permutation matrices of order n?

9 If P_1 and P_2 are permutation matrices, so is P_1P_2. This still has the rows of I in some order. Give examples with $P_1P_2 \neq P_2P_1$ and $P_3P_1 = P_1P_3$.

10 There are 12 "*even*" permutations of $(1, 2, 3, 4)$, with an *even number of exchanges*. Two of them are $(1, 2, 3, 4)$ with no exchanges and $(4, 3, 2, 1)$ with two exchanges. List the other ten. Instead of writing each 4 by 4 matrix, just order the numbers.

11 Which permutation makes PA upper triangular? Which permutations make P_1AP_2 lower triangular? ***Multiplying*** A ***on the*** *right* ***by*** P_2 ***exchanges the*** _____ ***of*** A.

$$A = \begin{bmatrix} 0 & 0 & 6 \\ 1 & 2 & 3 \\ 0 & 4 & 5 \end{bmatrix}.$$

12 Explain why the dot product of $\boldsymbol{x}$ and $\boldsymbol{y}$ equals the dot product of $P\boldsymbol{x}$ and $P\boldsymbol{y}$. Then $(P\boldsymbol{x})^{\mathrm{T}}(P\boldsymbol{y}) = \boldsymbol{x}^{\mathrm{T}}\boldsymbol{y}$ tells us that $P^{\mathrm{T}}P = I$ for any permutation. With $\boldsymbol{x} = (1, 2, 3)$ and $\boldsymbol{y} = (1, 4, 2)$ choose P to show that $P\boldsymbol{x} \cdot \boldsymbol{y}$ is not always $\boldsymbol{x} \cdot P\boldsymbol{y}$.

13 (a) Find a 3 by 3 permutation matrix with $P^3 = I$ (but not $P = I$).

(b) Find a 4 by 4 permutation $\widehat{P}$ with $\widehat{P}^4 \neq I$.

14 If P has 1's on the antidiagonal from $(1, n)$ to $(n, 1)$, describe PAP. Note $P = P^{\mathrm{T}}$.

15 All row exchange matrices are symmetric: $P^{\mathrm{T}} = P$. Then $P^{\mathrm{T}}P = I$ becomes $P^2 = I$. Other permutation matrices may or may not be symmetric.

(a) If P sends row 1 to row 4, then P^{T} sends row _____ to row _____.
When $P^{\mathrm{T}} = P$ the row exchanges come in pairs with no overlap.

(b) Find a 4 by 4 example with $P^{\mathrm{T}} = P$ that moves all four rows.

Questions 16–21 are about symmetric matrices and their factorizations.

16 If $A = A^{\mathrm{T}}$ and $B = B^{\mathrm{T}}$, which of these matrices are certainly symmetric?

(a) $A^2 - B^2$ (b) $(A+B)(A-B)$ (c) ABA (d) $ABAB$.

17 Find 2 by 2 symmetric matrices $S = S^{\mathrm{T}}$ with these properties:

(a) S is not invertible.

(b) S is invertible but cannot be factored into LU (row exchanges needed).

(c) S can be factored into LDL^{T} but not into LL^{T} (because of negative D).

18 (a) How many entries of S can be chosen independently, if $S = S^{\mathrm{T}}$ is 5 by 5?

(b) How do L and D (still 5 by 5) give the same number of choices in LDL^{T}?

(c) How many entries can be chosen if A is *skew-symmetric*? ($A^{\mathrm{T}} = -A$).

19 Suppose A is rectangular (m by n) and S is symmetric (m by m).

(a) Transpose $A^{\mathrm{T}}SA$ to show its symmetry. What shape is this matrix?

(b) Show why $A^{\mathrm{T}}A$ has no negative numbers on its diagonal.

20 Factor these symmetric matrices into $S = LDL^{\mathrm{T}}$. The pivot matrix D is diagonal:

$$S = \begin{bmatrix} 1 & 3 \\ 3 & 2 \end{bmatrix} \quad \text{and} \quad S = \begin{bmatrix} 1 & b \\ b & c \end{bmatrix} \quad \text{and} \quad S = \begin{bmatrix} 2 & -1 & 0 \\ -1 & 2 & -1 \\ 0 & -1 & 2 \end{bmatrix}.$$

21 After elimination clears out column 1 below the first pivot, find the symmetric 2 by 2 matrix that appears in the lower right corner:

$$\text{Start from } S = \begin{bmatrix} 2 & 4 & 8 \\ 4 & 3 & 9 \\ 8 & 9 & 0 \end{bmatrix} \quad \text{and} \quad S = \begin{bmatrix} 1 & b & c \\ b & d & e \\ c & e & f \end{bmatrix}.$$

Questions 22–24 are about the factorizations $PA = LU$ and $A = L_1P_1U_1$.

22 Find the $PA = LU$ factorizations (and check them) for

$$A = \begin{bmatrix} 0 & 1 & 1 \\ 1 & 0 & 1 \\ 2 & 3 & 4 \end{bmatrix} \quad \text{and} \quad A = \begin{bmatrix} 1 & 2 & 0 \\ 2 & 4 & 1 \\ 1 & 1 & 1 \end{bmatrix}.$$

23 Find a 4 by 4 permutation matrix (call it A) that needs 3 row exchanges to reach the end of elimination. For this matrix, what are its factors P, L, and U?

24 Factor the following matrix into $PA = LU$. Factor it also into $A = L_1P_1U_1$ (hold the exchange of row 3 until 3 times row 1 is subtracted from row 2):

$$A = \begin{bmatrix} 0 & 1 & 2 \\ 0 & 3 & 8 \\ 2 & 1 & 1 \end{bmatrix}.$$

25 Prove that the identity matrix cannot be the product of three row exchanges (or five). It can be the product of two exchanges (or four).

26 (a) Choose E_{21} to remove the 3 below the first pivot. Then multiply $E_{21}SE_{21}^{\mathrm{T}}$ to remove both 3's:

$$S = \begin{bmatrix} 1 & 3 & 0 \\ 3 & 11 & 4 \\ 0 & 4 & 9 \end{bmatrix} \quad \text{is going toward} \quad D = \begin{bmatrix} 1 & 0 & 0 \\ 0 & 2 & 0 \\ 0 & 0 & 1 \end{bmatrix}.$$

(b) Choose E_{32} to remove the 4 below the second pivot. Then S is reduced to D by $E_{32}E_{21}SE_{21}^{\mathrm{T}}E_{32}^{\mathrm{T}} = D$. Invert the E's to find L in $S = LDL^{\mathrm{T}}$.

27 If every row of a 4 by 4 matrix contains the numbers $0, 1, 2, 3$ in some order, can the matrix be symmetric?

28 Prove that no reordering of rows and reordering of columns can transpose a typical matrix. (Watch the diagonal entries.)

The next three questions are about applications of the identity $(A\boldsymbol{x})^{\mathrm{T}}\boldsymbol{y} = \boldsymbol{x}^{\mathrm{T}}(A^{\mathrm{T}}\boldsymbol{y})$.

29 Wires go between Boston, Chicago, and Seattle. Those cities are at voltages x_B, x_C, x_S. With unit resistances between cities, the currents between cities are in $\boldsymbol{y}$:

$$\boldsymbol{y} = A\boldsymbol{x} \quad \text{is} \quad \begin{bmatrix} y_{BC} \\ y_{CS} \\ y_{BS} \end{bmatrix} = \begin{bmatrix} 1 & -1 & 0 \\ 0 & 1 & -1 \\ 1 & 0 & -1 \end{bmatrix} \begin{bmatrix} x_B \\ x_C \\ x_S \end{bmatrix}.$$

(a) Find the total currents $A^{\mathrm{T}}\boldsymbol{y}$ out of the three cities.

(b) Verify that $(A\boldsymbol{x})^{\mathrm{T}}\boldsymbol{y}$ agrees with $\boldsymbol{x}^{\mathrm{T}}(A^{\mathrm{T}}\boldsymbol{y})$—six terms in both.

30 Producing x_1 trucks and x_2 planes needs $x_1 + 50x_2$ tons of steel, $40x_1 + 1000x_2$ pounds of rubber, and $2x_1 + 50x_2$ months of labor. If the unit costs y_1, y_2, y_3 are \$700 per ton, \$3 per pound, and \$3000 per month, what are the values of one truck and one plane? Those are the components of $A^{\mathrm{T}}\boldsymbol{y}$.

31 $A\boldsymbol{x}$ gives the amounts of steel, rubber, and labor to produce $\boldsymbol{x}$ in Problem 31. Find A. Then $A\boldsymbol{x} \cdot \boldsymbol{y}$ is the ____ of inputs while $\boldsymbol{x} \cdot A^{\mathrm{T}}\boldsymbol{y}$ is the value of ____.

32 The matrix P that multiplies (x, y, z) to give (z, x, y) is also a rotation matrix. Find P and P^3. The rotation axis $\boldsymbol{a} = (1, 1, 1)$ doesn't move, it equals $P\boldsymbol{a}$. What is the angle of rotation from $\boldsymbol{v} = (2, 3, -5)$ to $P\boldsymbol{v} = (-5, 2, 3)$?

33 Write $A = \begin{bmatrix} 1 & 2 \\ 4 & 9 \end{bmatrix}$ as the product ES of an elementary row operation matrix E and a symmetric matrix S.

34 Here is a new factorization of A into LS: *triangular* (with 1's) *times symmetric*:

$$\text{Start from } A = LDU. \text{ Then } A \text{ equals } L\,(U^{\mathrm{T}})^{-1} \text{ times } S = U^{\mathrm{T}}DU.$$

Why is $L\,(U^{\mathrm{T}})^{-1}$ triangular? Its diagonal is all 1's. Why is $U^{\mathrm{T}}DU$ symmetric?

35 A *group* of matrices includes AB and A^{-1} if it includes A and B. "Products and inverses stay in the group." Which of these sets are groups?
Lower triangular matrices L with 1's on the diagonal, symmetric matrices S, positive matrices M, diagonal invertible matrices D, permutation matrices P, matrices with $Q^{\mathrm{T}} = Q^{-1}$. ***Invent two more matrix groups.***

Challenge Problems

36 A square ***northwest matrix*** B is zero in the southeast corner, below the antidiagonal that connects $(1, n)$ to $(n, 1)$. Will B^{T} and B^2 be northwest matrices? Will B^{-1} be northwest or southeast? What is the shape of BC = ***northwest times southeast***?

37 If you take powers of a permutation matrix, why is some P^k eventually equal to I?

Find a 5 by 5 permutation P so that the smallest power to equal I is P^6.

38 (a) Write down any 3 by 3 matrix M. Split M into $S + A$ where $S = S^{\mathrm{T}}$ is symmetric and $A = -A^{\mathrm{T}}$ is anti-symmetric.

(b) Find formulas for S and A involving M and M^{T}. We want $M = S + A$.

39 Suppose Q^{T} *equals* Q^{-1} (transpose equals inverse, so $Q^{\mathrm{T}}Q = I$).

(a) Show that the columns $q_1, \ldots, q_n$ are unit vectors: $\|\boldsymbol{q}_i\|^2 = 1$.

(b) Show that every two columns of Q are perpendicular: $\boldsymbol{q}_1^{\mathrm{T}}\boldsymbol{q}_2 = 0$.

(c) Find a 2 by 2 example with first entry $q_{11} = \cos\theta$.

The Transpose of a Derivative

Will you allow me a little calculus? It is extremely important or I wouldn't leave linear algebra. (This is really linear algebra for functions $x(t)$.) **The matrix changes to a derivative so $\boldsymbol{A = d/dt}$.** To find the transpose of this unusual A we need to define the inner product between two functions $x(t)$ and $y(t)$.

The inner product changes from the sum of $x_k\, y_k$ to the *integral* of $x(t)\, y(t)$.

Inner product of functions

$$\boldsymbol{x^{\mathrm{T}}y = (x, y) = \int_{-\infty}^{\infty} x(t)\, y(t)\, dt}$$

From this inner product we know the requirement on A^{T}. The word "adjoint" is more correct than "transpose" when we are working with derivatives.

The transpose of a matrix has $(A\boldsymbol{x})^{\mathrm{T}}\boldsymbol{y} = \boldsymbol{x}^{\mathrm{T}}(A^{\mathrm{T}}\boldsymbol{y})$. The adjoint of $A = \dfrac{d}{dt}$ has

$$(Ax, y) = \int_{-\infty}^{\infty} \frac{dx}{dt}\, y(t)\, dt = \int_{-\infty}^{\infty} x(t) \left(-\frac{dy}{dt}\right) dt = (x, A^{\mathrm{T}}y)$$

I hope you recognize integration by parts. The derivative moves from the first function $x(t)$ to the second function $y(t)$. During that move, a minus sign appears. This tells us that ***the transpose of the derivative is minus the derivative***.

The derivative is *antisymmetric*: $\boldsymbol{A = d/dt}$ ***and*** $\boldsymbol{A^{\mathrm{T}} = -d/dt}$. Symmetric matrices have $S^{\mathrm{T}} = S$, antisymmetric matrices have $A^{\mathrm{T}} = -A$. Linear algebra includes derivatives and integrals in Chapter 8, *because those are both linear.*

This antisymmetry of the derivative applies also to centered difference matrices.

$$A = \begin{bmatrix} 0 & 1 & 0 & 0 \\ -1 & 0 & 1 & 0 \\ 0 & -1 & 0 & 1 \\ 0 & 0 & -1 & 0 \end{bmatrix} \quad \text{transposes to} \quad A^{\mathrm{T}} = \begin{bmatrix} 0 & -1 & 0 & 0 \\ 1 & 0 & -1 & 0 \\ 0 & 1 & 0 & -1 \\ 0 & 0 & 1 & 0 \end{bmatrix} = -A.$$

And a forward difference matrix transposes to a backward difference matrix, *multiplied by* -1. In differential equations, the second derivative (acceleration) is symmetric. The first derivative (damping proportional to velocity) is *anti*symmetric.

Chapter 3

Vector Spaces and Subspaces

3.1 Spaces of Vectors

1 The standard n-dimensional space $\mathbf{R}^n$ contains all real column vectors with n components.

2 If $\boldsymbol{v}$ and $\boldsymbol{w}$ are in a **vector space** $\boldsymbol{S}$, every combination $c\boldsymbol{v} + d\boldsymbol{w}$ must be in $\boldsymbol{S}$.

3 The "vectors" in $\boldsymbol{S}$ can be matrices or functions of x. The 1-point space $\boldsymbol{Z}$ consists of $\boldsymbol{x} = \mathbf{0}$.

4 A **subspace** of $\mathbf{R}^n$ is a vector space inside $\mathbf{R}^n$. *Example*: The line $y = 3x$ inside $\mathbf{R}^2$.

5 The **column space** of A contains all combinations of the columns of A: a subspace of $\mathbf{R}^m$.

6 The column space contains all the vectors $A\boldsymbol{x}$. So $A\boldsymbol{x} = \boldsymbol{b}$ is solvable when $\boldsymbol{b}$ is in $\boldsymbol{C}(A)$.

To a newcomer, matrix calculations involve a lot of numbers. To you, they involve vectors. The columns of $A\boldsymbol{x}$ and AB are linear combinations of n vectors—the columns of A. This chapter moves from numbers and vectors to a third level of understanding (the highest level). Instead of individual columns, we look at "spaces" of vectors. Without seeing ***vector spaces*** and especially their ***subspaces***, you haven't understood everything about $A\boldsymbol{x} = \boldsymbol{b}$.

Since this chapter goes a little deeper, it may seem a little harder. That is natural. We are looking inside the calculations, to find the mathematics. The author's job is to make it clear. The chapter ends with the "*Fundamental Theorem of Linear Algebra*".

We begin with the most important vector spaces. They are denoted by $\mathbf{R}^1$, $\mathbf{R}^2$, $\mathbf{R}^3$, $\mathbf{R}^4$, ... Each space $\mathbf{R}^n$ consists of a whole collection of vectors. $\mathbf{R}^5$ contains all column vectors with five components. This is called "5-dimensional space".

DEFINITION ***The space $\mathbf{R}^n$ consists of all column vectors v with n components.***

The components of $\boldsymbol{v}$ are real numbers, which is the reason for the letter $\mathbf{R}$. A vector whose n components are complex numbers lies in the space $\mathbf{C}^n$.

The vector space $\mathbf{R}^2$ is represented by the usual xy plane. Each vector $\boldsymbol{v}$ in $\mathbf{R}^2$ has two components. The word "*space*" asks us to think of all those vectors—the whole plane. Each vector gives the x and y coordinates of a point in the plane: $\boldsymbol{v} = (x, y)$.

Similarly the vectors in $\mathbf{R}^3$ correspond to points (x, y, z) in three-dimensional space. The one-dimensional space $\mathbf{R}^1$ is a line (like the x axis). As before, we print vectors as a column between brackets, or along a line using commas and parentheses:

$$\begin{bmatrix} 4 \\ \pi \end{bmatrix} \text{ is in } \mathbf{R}^2, \quad (1,1,0,1,1) \text{ is in } \mathbf{R}^5, \quad \begin{bmatrix} 1+i \\ 1-i \end{bmatrix} \text{ is in } \mathbf{C}^2.$$

The great thing about linear algebra is that it deals easily with five-dimensional space. We don't draw the vectors, we just need the five numbers (or n numbers).

To multiply $\boldsymbol{v}$ by 7, multiply every component by 7. Here 7 is a "scalar". To add vectors in $\mathbf{R}^5$, add them a component at a time. The two essential vector operations go on *inside the vector space*, and they produce ***linear combinations***:

We can add any vectors in $\mathbf{R}^n$, *and we can multiply any vector* v *by any scalar* c.

"Inside the vector space" means that ***the result stays in the space.*** If $\boldsymbol{v}$ is the vector in $\mathbf{R}^4$ with components $1, 0, 0, 1$, then $2\boldsymbol{v}$ is the vector in $\mathbf{R}^4$ with components $2, 0, 0, 2$. (In this case 2 is the scalar.) A whole series of properties can be verified in $\mathbf{R}^n$. The commutative law is $\boldsymbol{v} + \boldsymbol{w} = \boldsymbol{w} + \boldsymbol{v}$; the distributive law is $c(\boldsymbol{v} + \boldsymbol{w}) = c\boldsymbol{v} + c\boldsymbol{w}$. There is a unique "zero vector" satisfying $\mathbf{0} + \boldsymbol{v} = \boldsymbol{v}$. Those are three of the eight conditions listed at the start of the problem set.

These eight conditions are required of every vector space. There are vectors other than column vectors, and there are vector spaces other than $\mathbf{R}^n$, and all vector spaces have to obey the eight reasonable rules.

A real vector space is a set of "vectors" *together with rules for vector addition and for multiplication by real numbers.* The addition and the multiplication must produce vectors that are in the space. And the eight conditions must be satisfied (which is usually no problem). Here are three vector spaces other than $\mathbf{R}^n$:

M The vector space of ***all real*** 2 ***by*** 2 ***matrices.***
F The vector space of ***all real functions*** $f(x)$.
Z The vector space that consists only of a ***zero vector***.

In $\mathbf{M}$ the "vectors" are really matrices. In $\mathbf{F}$ the vectors are functions. In $\mathbf{Z}$ the only addition is $\mathbf{0} + \mathbf{0} = \mathbf{0}$. In each case we can add: matrices to matrices, functions to functions, zero vector to zero vector. We can multiply a matrix by 4 or a function by 4 or the zero vector by 4. The result is still in $\mathbf{M}$ or $\mathbf{F}$ or $\mathbf{Z}$. The eight conditions are all easily checked.

The function space $\mathbf{F}$ is infinite-dimensional. A smaller function space is $\mathbf{P}$, or $\mathbf{P}_n$, containing all polynomials $a_0 + a_1x + \cdots + a_nx^n$ of degree n.

The space $\mathbf{Z}$ is zero-dimensional (by any reasonable definition of dimension). $\mathbf{Z}$ is the smallest possible vector space. We hesitate to call it $\mathbf{R}^0$, which means no components—you might think there was no vector. The vector space $\mathbf{Z}$ contains exactly *one vector* (zero). No space can do without that zero vector. Each space has its own zero vector—the zero matrix, the zero function, the vector $(0, 0, 0)$ in $\mathbf{R}^3$.

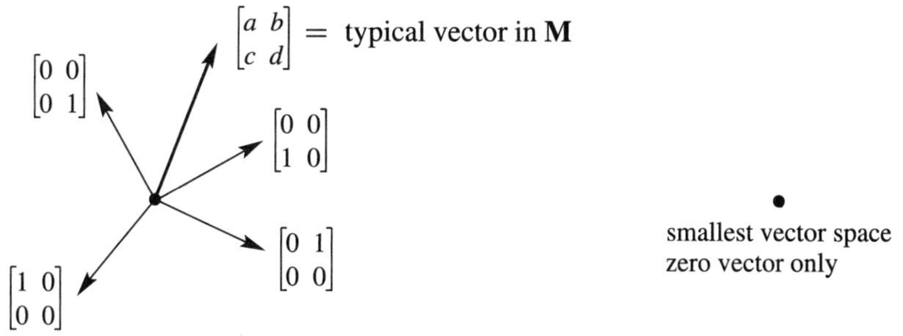

Figure 3.1: "Four-dimensional" matrix space $\mathbf{M}$. The "zero-dimensional" space $\mathbf{Z}$.

Subspaces

At different times, we will ask you to think of matrices and functions as vectors. But at all times, the vectors that we need most are ordinary column vectors. They are vectors with n components—but *maybe not all* of the vectors with n components. There are important vector spaces ***inside*** $\mathbf{R}^n$. Those are ***subspaces*** of $\mathbf{R}^n$.

Start with the usual three-dimensional space $\mathbf{R}^3$. Choose a plane through the origin $(0, 0, 0)$. ***That plane is a vector space in its own right.*** If we add two vectors in the plane, their sum is in the plane. If we multiply an in-plane vector by 2 or -5, it is still in the plane. A plane in three-dimensional space is not $\mathbf{R}^2$ (even if it looks like $\mathbf{R}^2$). The vectors have three components and they belong to $\mathbf{R}^3$. The plane is a vector space ***inside*** $\mathbf{R}^3$.

This illustrates one of the most fundamental ideas in linear algebra. The plane going through $(0, 0, 0)$ is a ***subspace*** of the full vector space $\mathbf{R}^3$.

DEFINITION A ***subspace*** of a vector space is a set of vectors (including $\mathbf{0}$) that satisfies two requirements: ***If*** $\boldsymbol{v}$ ***and*** $\boldsymbol{w}$ ***are vectors in the subspace and*** c ***is any scalar, then***

(i) $\boldsymbol{v} + \boldsymbol{w}$ is in the subspace **(ii)** $c\boldsymbol{v}$ is in the subspace.

In other words, the set of vectors is "closed" under addition $\boldsymbol{v} + \boldsymbol{w}$ and multiplication $c\boldsymbol{v}$ (and $d\boldsymbol{w}$). Those operations leave us in the subspace. We can also subtract, because $-\boldsymbol{w}$ is in the subspace and its sum with $\boldsymbol{v}$ is $\boldsymbol{v} - \boldsymbol{w}$. In short, ***all linear combinations stay in the subspace***.

All these operations follow the rules of the host space, so the eight required conditions are automatic. We just have to check the linear combinations requirement for a subspace.

First fact: ***Every subspace contains the zero vector***. The plane in $\mathbf{R}^3$ has to go through $(0, 0, 0)$. We mention this separately, for extra emphasis, but it follows directly from rule **(ii)**. Choose $c = 0$, and the rule requires $0\boldsymbol{v}$ to be in the subspace.

Planes that don't contain the origin fail those tests. Those planes are not subspaces.

Lines through the origin are also subspaces. When we multiply by 5, or add two vectors on the line, we stay on the line. But the line must go through $(0, 0, 0)$.

Another subspace is all of $\mathbf{R}^3$. The whole space is a subspace (*of itself*). Here is a list of all the possible subspaces of $\mathbf{R}^3$:

(**L**) Any line through $(0, 0, 0)$	($\mathbf{R}^3$) The whole space
(**P**) Any plane through $(0, 0, 0)$	(**Z**) The single vector $(0, 0, 0)$

If we try to keep only *part* of a plane or line, the requirements for a subspace don't hold. Look at these examples in $\mathbf{R}^2$—they are not subspaces.

Example 1 Keep only the vectors (x, y) whose components are positive or zero (this is a quarter-plane). The vector $(2, 3)$ is included but $(-2, -3)$ is not. So rule **(ii)** is violated when we try to multiply by $c = -1$. ***The quarter-plane is not a subspace***.

Example 2 Include also the vectors whose components are both negative. Now we have two quarter-planes. Requirement **(ii)** is satisfied; we can multiply by any c. But rule **(i)** now fails. The sum of $\boldsymbol{v} = (2, 3)$ and $\boldsymbol{w} = (-3, -2)$ is $(-1, 1)$, which is outside the quarter-planes. ***Two quarter-planes don't make a subspace***.

Rules **(i)** and **(ii)** involve vector addition $\boldsymbol{v} + \boldsymbol{w}$ and multiplication by scalars c and d. The rules can be combined into a single requirement—*the rule for subspaces*:

A subspace containing v and w must contain all linear combinations $cv + dw$.

Example 3 Inside the vector space $\mathbf{M}$ of all 2 by 2 matrices, here are two subspaces:

$$(\mathbf{U}) \text{ All upper triangular matrices } \begin{bmatrix} a & b \\ 0 & d \end{bmatrix} \quad (\mathbf{D}) \text{ All diagonal matrices } \begin{bmatrix} a & 0 \\ 0 & d \end{bmatrix}.$$

Add any two matrices in $\mathbf{U}$, and the sum is in $\mathbf{U}$. Add diagonal matrices, and the sum is diagonal. In this case $\mathbf{D}$ is also a subspace of $\mathbf{U}$! Of course the zero matrix is in these subspaces, when a, b, and d all equal zero. $\mathbf{Z}$ is always a subspace.

Multiples of the identity matrix also form a subspace. $2I + 3I$ is in this subspace, and so is 3 times $4I$. The matrices cI form a "line of matrices" inside $\mathbf{M}$ and $\mathbf{U}$ and $\mathbf{D}$.

Is the matrix I a subspace by itself? Certainly not. Only the zero matrix is. Your mind will invent more subspaces of 2 by 2 matrices—write them down for Problem 5.

The Column Space of A

The most important subspaces are tied directly to a matrix A. We are trying to solve $A\boldsymbol{x} = \boldsymbol{b}$. If A is not invertible, the system is solvable for some $\boldsymbol{b}$ and not solvable for other $\boldsymbol{b}$. We want to describe the good right sides $\boldsymbol{b}$—the vectors that *can* be written as A times some vector $\boldsymbol{x}$. Those $\boldsymbol{b}'s$ form the "*column space*" of A.

Remember that $A\boldsymbol{x}$ is a combination of the columns of A. To get every possible $\boldsymbol{b}$, we use every possible $\boldsymbol{x}$. Start with the columns of A and ***take all their linear combinations. This produces the column space of A.*** **It is a vector space made up of column vectors.**

$\boldsymbol{C}(A)$ contains not just the n columns of A, but all their combinations $A\boldsymbol{x}$.

DEFINITION The ***column space*** consists of ***all linear combinations of the columns***. The combinations are all possible vectors $A\boldsymbol{x}$. They fill the column space $\boldsymbol{C}(A)$.

This column space is crucial to the whole book, and here is why. ***To solve $A\boldsymbol{x} = \boldsymbol{b}$ is to express b as a combination of the columns***. The right side $\boldsymbol{b}$ has to be *in the column space* produced by A on the left side, or no solution!

The system $A\boldsymbol{x} = \boldsymbol{b}$ is solvable if and only if b is in the column space of A.

When $\boldsymbol{b}$ is in the column space, it is a combination of the columns. The coefficients in that combination give us a solution $\boldsymbol{x}$ to the system $A\boldsymbol{x} = \boldsymbol{b}$.

Suppose A is an m by n matrix. Its columns have m components (not n). So the columns belong to $\mathbf{R}^m$. ***The column space of A is a subspace of $\mathbf{R}^m$ (not $\mathbf{R}^n$).*** The set of all column combinations $A\boldsymbol{x}$ satisfies rules (**i**) and (**ii**) for a subspace: When we add linear combinations or multiply by scalars, we still produce combinations of the columns. The word "subspace" is justified *by taking all linear combinations*.

Here is a 3 by 2 matrix A, whose column space is a subspace of $\mathbf{R}^3$. The column space of A is a plane in Figure 3.2. With only 2 columns, $\boldsymbol{C}(A)$ can't be all of $\mathbf{R}^3$.

Example 4

$$A\boldsymbol{x} \quad \text{is} \quad \begin{bmatrix} 1 & 0 \\ 4 & 3 \\ 2 & 3 \end{bmatrix} \begin{bmatrix} x_1 \\ x_2 \end{bmatrix} \quad \text{which is} \quad x_1 \begin{bmatrix} 1 \\ 4 \\ 2 \end{bmatrix} + x_2 \begin{bmatrix} 0 \\ 3 \\ 3 \end{bmatrix}.$$

The column space of all combinations of the two columns *fills up a plane in* $\mathbf{R}^3$. We drew one particular $\boldsymbol{b}$ (a combination of the columns). This $\boldsymbol{b} = A\boldsymbol{x}$ lies on the plane. The plane has zero thickness, so most right sides $\boldsymbol{b}$ in $\mathbf{R}^3$ are *not* in the column space. For most $\boldsymbol{b}$ there is no solution to our 3 equations in 2 unknowns.

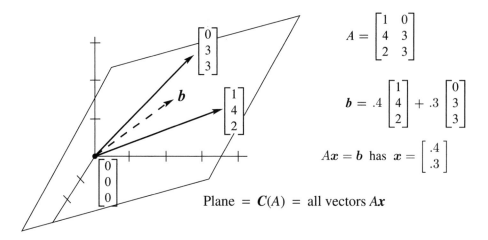

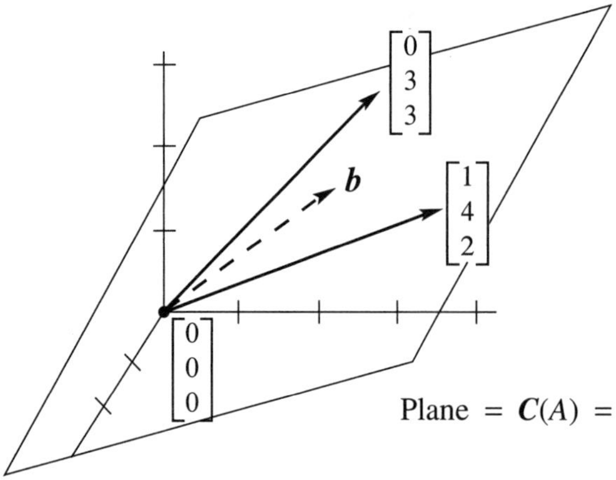

$$A = \begin{bmatrix}1 & 0\\4 & 3\\2 & 3\end{bmatrix}$$

$$b = .4\begin{bmatrix}1\\4\\2\end{bmatrix} + .3\begin{bmatrix}0\\3\\3\end{bmatrix}$$

$$Ax = b \text{ has } x = \begin{bmatrix}.4\\.3\end{bmatrix}$$

Plane $= C(A) =$ all vectors Ax

Figure 3.2: The column space $C(A)$ is a plane containing the two columns. $Ax = b$ is solvable when b is on that plane. Then b is a combination of the columns.

Of course $(0,0,0)$ is in the column space. The plane passes through the origin. There is certainly a solution to $Ax = \mathbf{0}$. That solution, always available, is $x =$ _____.

To repeat, the attainable right sides b are exactly the vectors in the column space. One possibility is the first column itself—take $x_1 = 1$ and $x_2 = 0$. Another combination is the second column—take $x_1 = 0$ and $x_2 = 1$. The new level of understanding is to see *all* combinations—the whole subspace is generated by those two columns.

Notation The column space of A is denoted by $C(A)$. Start with the columns and take all their linear combinations. We might get the whole $\mathbf{R}^m$ or only a subspace.

Important Instead of columns in $\mathbf{R}^m$, we could start with any set $\mathbf{S}$ of vectors in a vector space $\mathbf{V}$. To get a *subspace* $\mathbf{SS}$ of $\mathbf{V}$, we take *all combinations* of the vectors in that set:

$$\begin{aligned}\mathbf{S} &= \text{set of vectors in } \mathbf{V} \text{ (probably } \textit{not} \text{ a subspace)}\\ \mathbf{SS} &= \text{all combinations of vectors in } \mathbf{S} \text{ (definitely a subspace)}\end{aligned}$$

$$\mathbf{SS} = \text{ all } c_1 v_1 + \cdots + c_N v_N = \textbf{the subspace of V} \text{ "spanned" by } \mathbf{S}$$

When $\mathbf{S}$ is the set of columns, $\mathbf{SS}$ is the column space. When there is only one nonzero vector v in $\mathbf{S}$, the subspace $\mathbf{SS}$ is the line through v. *Always* $\mathbf{SS}$ *is the smallest subspace containing* $\mathbf{S}$. This is a fundamental way to create subspaces and we will come back to it.

To repeat: The columns "span" the column space.

The subspace SS is the "span" of S, containing all combinations of vectors in S.

Example 5 Describe the column spaces (they are subspaces of $\mathbf{R}^2$) for

$$I = \begin{bmatrix} 1 & 0 \\ 0 & 1 \end{bmatrix} \quad \text{and} \quad A = \begin{bmatrix} 1 & 2 \\ 2 & 4 \end{bmatrix} \quad \text{and} \quad B = \begin{bmatrix} 1 & 2 & 3 \\ 0 & 0 & 4 \end{bmatrix}.$$

Solution The column space of I is the *whole space* $\mathbf{R}^2$. Every vector is a combination of the columns of I. In vector space language, $\boldsymbol{C}(I)$ is $\mathbf{R}^2$.

The column space of A is only a line. The second column $(2, 4)$ is a multiple of the first column $(1, 2)$. Those vectors are different, but our eye is on vector ***spaces***. The column space contains $(1, 2)$ and $(2, 4)$ and all other vectors $(c, 2c)$ along that line. The equation $A\boldsymbol{x} = \boldsymbol{b}$ is only solvable when $\boldsymbol{b}$ is on the line.

For the third matrix (with three columns) the column space $\boldsymbol{C}(B)$ is all of $\mathbf{R}^2$. Every $\boldsymbol{b}$ is attainable. The vector $\boldsymbol{b} = (5, 4)$ is column 2 plus column 3, so $\boldsymbol{x}$ can be $(0, 1, 1)$. The same vector $(5, 4)$ is also 2(column 1) + column 3, so another possible $\boldsymbol{x}$ is $(2, 0, 1)$. This matrix has the same column space as I—any $\boldsymbol{b}$ is allowed. But now $\boldsymbol{x}$ has extra components and there are more solutions—more combinations that give $\boldsymbol{b}$.

The next section creates a vector space $\boldsymbol{N}(A)$, to describe all the solutions of $A\boldsymbol{x} = \mathbf{0}$. This section created the column space $\boldsymbol{C}(A)$, to describe all the attainable right sides $\boldsymbol{b}$.

■ REVIEW OF THE KEY IDEAS ■

1. $\mathbf{R}^n$ contains all column vectors with n real components.

2. $\mathbf{M}$ (2 by 2 matrices) and $\mathbf{F}$ (functions) and $\mathbf{Z}$ (zero vector alone) are vector spaces.

3. A subspace containing $\boldsymbol{v}$ and $\boldsymbol{w}$ must contain all their combinations $c\boldsymbol{v} + d\boldsymbol{w}$.

4. The combinations of the columns of A form the ***column space*** $\boldsymbol{C}(A)$. Then the column space is "spanned" by the columns.

5. $A\boldsymbol{x} = \boldsymbol{b}$ has a solution exactly when $\boldsymbol{b}$ is in the column space of A.

$$C(A) = \textbf{all combinations of the columns} = \textbf{all vectors } A\boldsymbol{x}.$$

■ WORKED EXAMPLES ■

3.1 A We are given three different vectors $\boldsymbol{b}_1, \boldsymbol{b}_2, \boldsymbol{b}_3$. Construct a matrix so that the equations $A\boldsymbol{x} = \boldsymbol{b}_1$ and $A\boldsymbol{x} = \boldsymbol{b}_2$ are solvable, but $A\boldsymbol{x} = \boldsymbol{b}_3$ is not solvable. How can you decide if this is possible? How could you construct A?

Solution We want to have $\boldsymbol{b}_1$ and $\boldsymbol{b}_2$ in the column space of A. Then $A\boldsymbol{x} = \boldsymbol{b}_1$ and $A\boldsymbol{x} = \boldsymbol{b}_2$ will be solvable. *The quickest way is to make $\boldsymbol{b}_1$ and $\boldsymbol{b}_2$ the two columns of A.* Then the solutions are $\boldsymbol{x} = (1,0)$ and $\boldsymbol{x} = (0,1)$.

Also, we don't want $A\boldsymbol{x} = \boldsymbol{b}_3$ to be solvable. So don't make the column space any larger! Keeping only the columns $\boldsymbol{b}_1$ and $\boldsymbol{b}_2$, the question is:

$$\text{Is } A\boldsymbol{x} = \begin{bmatrix} \boldsymbol{b}_1 & \boldsymbol{b}_2 \end{bmatrix} \begin{bmatrix} x_1 \\ x_2 \end{bmatrix} = \boldsymbol{b}_3 \text{ solvable?} \qquad \text{Is } \boldsymbol{b}_3 \text{ a combination of } \boldsymbol{b}_1 \text{ and } \boldsymbol{b}_2\text{?}$$

If the answer is *no*, we have the desired matrix A. If the answer is *yes*, then it is *not possible* to construct A. When the column space contains $\boldsymbol{b}_1$ and $\boldsymbol{b}_2$, it will have to contain all their linear combinations. So $\boldsymbol{b}_3$ would necessarily be in that column space and $A\boldsymbol{x} = \boldsymbol{b}_3$ would necessarily be solvable.

3.1 B Describe a subspace $\mathbf{S}$ of each vector space $\mathbf{V}$, and then a subspace $\mathbf{SS}$ of $\mathbf{S}$.

$$\begin{aligned}
\mathbf{V}_1 &= \text{all combinations of } (1,1,0,0) \text{ and } (1,1,1,0) \text{ and } (1,1,1,1) \\
\mathbf{V}_2 &= \text{all vectors perpendicular to } \boldsymbol{u} = (1,2,1), \text{ so } \boldsymbol{u} \cdot \boldsymbol{v} = 0 \\
\mathbf{V}_3 &= \text{all symmetric 2 by 2 matrices (a subspace of } \mathbf{M}) \\
\mathbf{V}_4 &= \text{all solutions to the equation } d^4y/dx^4 = 0 \text{ (a subspace of } \mathbf{F})
\end{aligned}$$

Describe each $\mathbf{V}$ two ways: "*All combinations of* ..." "*All solutions of the equations...*"

Solution $\mathbf{V}_1$ starts with three vectors. A subspace $\mathbf{S}$ comes from all combinations of the first two vectors $(1,1,0,0)$ and $(1,1,1,0)$. A subspace $\mathbf{SS}$ of $\mathbf{S}$ comes from all multiples $(c,c,0,0)$ of the first vector. So many possibilities.

A subspace $\mathbf{S}$ of $\mathbf{V}_2$ is the line through $(1,-1,1)$. This line is perpendicular to $\boldsymbol{u}$. The vector $\boldsymbol{x} = (0,0,0)$ is in $\mathbf{S}$ and all its multiples $c\boldsymbol{x}$ give the smallest subspace $\mathbf{SS} = \mathbf{Z}$.

The diagonal matrices are a subspace $\mathbf{S}$ of the symmetric matrices. The multiples cI are a subspace $\mathbf{SS}$ of the diagonal matrices.

$\mathbf{V}_4$ contains all cubic polynomials $y = a + bx + cx^2 + dx^3$, with $d^4y/dx^4 = 0$. The quadratic polynomials give a subspace $\mathbf{S}$. The linear polynomials are one choice of $\mathbf{SS}$. The constants could be $\mathbf{SSS}$.

In all four parts we could take $\mathbf{S} = \mathbf{V}$ itself, and $\mathbf{SS} =$ the zero subspace $\mathbf{Z}$.

Each $\mathbf{V}$ can be described as *all combinations of* .... and as *all solutions of* ....:

$$\begin{aligned}
&\mathbf{V}_1 = \text{all combinations of the 3 vectors} && \mathbf{V}_1 = \text{all solutions of } v_1 - v_2 = 0 \\
&\mathbf{V}_2 = \text{all combinations of } (1,0,-1) \text{ and } (1,-1,1) && \mathbf{V}_2 = \text{all solutions of } \boldsymbol{u} \cdot \boldsymbol{v} = 0. \\
&\mathbf{V}_3 = \text{all combinations of } \left[\begin{smallmatrix} 1 & 0 \\ 0 & 0 \end{smallmatrix}\right], \left[\begin{smallmatrix} 0 & 1 \\ 1 & 0 \end{smallmatrix}\right], \left[\begin{smallmatrix} 0 & 0 \\ 0 & 1 \end{smallmatrix}\right]. && \mathbf{V}_3 = \text{ all solutions } \left[\begin{smallmatrix} a & b \\ c & d \end{smallmatrix}\right] \text{ of } b = c \\
&\mathbf{V}_4 = \text{all combinations of } 1, x, x^2, x^3 && \mathbf{V}_4 = \text{all solutions to } d^4y/dx^4 = 0.
\end{aligned}$$

Problem Set 3.1

The first problems 1–8 are about vector spaces in general. The vectors in those spaces are not necessarily column vectors. In the definition of a *vector space*, vector addition $x + y$ and scalar multiplication cx must obey the following eight rules:

(1) $\boldsymbol{x} + \boldsymbol{y} = \boldsymbol{y} + \boldsymbol{x}$

(2) $\boldsymbol{x} + (\boldsymbol{y} + \boldsymbol{z}) = (\boldsymbol{x} + \boldsymbol{y}) + \boldsymbol{z}$

(3) There is a unique "zero vector" such that $\boldsymbol{x} + \mathbf{0} = \boldsymbol{x}$ for all $\boldsymbol{x}$

(4) For each $\boldsymbol{x}$ there is a unique vector $-\boldsymbol{x}$ such that $\boldsymbol{x} + (-\boldsymbol{x}) = \mathbf{0}$

(5) 1 times $\boldsymbol{x}$ equals $\boldsymbol{x}$

(6) $(c_1 c_2)\boldsymbol{x} = c_1(c_2\boldsymbol{x})$ — (1) to (4) about $\boldsymbol{x} + \boldsymbol{y}$

(7) $c(\boldsymbol{x} + \boldsymbol{y}) = c\boldsymbol{x} + c\boldsymbol{y}$ — (5) to (6) about $c\boldsymbol{x}$

(8) $(c_1 + c_2)\boldsymbol{x} = c_1\boldsymbol{x} + c_2\boldsymbol{x}$. — (7) to (8) connects them

1 Suppose $(x_1, x_2) + (y_1, y_2)$ is defined to be $(x_1 + y_2, x_2 + y_1)$. With the usual multiplication $c\boldsymbol{x} = (cx_1, cx_2)$, which of the eight conditions are not satisfied?

2 Suppose the multiplication $c\boldsymbol{x}$ is defined to produce $(cx_1, 0)$ instead of (cx_1, cx_2). With the usual addition in $\mathbf{R}^2$, are the eight conditions satisfied?

3 (a) Which rules are broken if we keep only the positive numbers $x > 0$ in $\mathbf{R}^1$? Every c must be allowed. The half-line is not a subspace.

(b) The positive numbers with $\boldsymbol{x} + \boldsymbol{y}$ and $c\boldsymbol{x}$ redefined to equal the usual xy and x^c *do* satisfy the eight rules. Test rule 7 when $c = 3, x = 2, y = 1$. (Then $\boldsymbol{x} + \boldsymbol{y} = 2$ and $c\boldsymbol{x} = 8$.) Which number acts as the "zero vector"?

4 The matrix $A = \begin{bmatrix} 2 & -2 \\ 2 & -2 \end{bmatrix}$ is a "vector" in the space $\mathbf{M}$ of all 2 by 2 matrices. Write down the zero vector in this space, the vector $\frac{1}{2}A$, and the vector $-A$. What matrices are in the smallest subspace containing A?

5 (a) Describe a subspace of $\mathbf{M}$ that contains $A = \begin{bmatrix} 1 & 0 \\ 0 & 0 \end{bmatrix}$ but not $B = \begin{bmatrix} 0 & 0 \\ 0 & -1 \end{bmatrix}$.

(b) If a subspace of $\mathbf{M}$ does contain A and B, must it contain I?

(c) Describe a subspace of $\mathbf{M}$ that contains no nonzero diagonal matrices.

6 The functions $\boldsymbol{f}(x) = x^2$ and $\boldsymbol{g}(x) = 5x$ are "vectors" in $\mathbf{F}$. This is the vector space of all real functions. (The functions are defined for $-\infty < x < \infty$.) The combination $3\boldsymbol{f}(x) - 4\boldsymbol{g}(x)$ is the function $\boldsymbol{h}(x) =$ _____.

7 Which rule is broken if multiplying $\boldsymbol{f}(x)$ by c gives the function $f(cx)$? Keep the usual addition $\boldsymbol{f}(x) + \boldsymbol{g}(x)$.

8 If the sum of the "vectors" $\boldsymbol{f}(x)$ and $\boldsymbol{g}(x)$ is defined to be the function $f(g(x))$, then the "zero vector" is $\boldsymbol{g}(x) = x$. Keep the usual scalar multiplication $c\boldsymbol{f}(x)$ and find two rules that are broken.

Questions 9–18 are about the "subspace requirements": $x + y$ and cx (and then all linear combinations $cx + dy$) stay in the subspace.

9 One requirement can be met while the other fails. Show this by finding

(a) A set of vectors in $\mathbf{R}^2$ for which $\boldsymbol{x} + \boldsymbol{y}$ stays in the set but $\frac{1}{2}\boldsymbol{x}$ may be outside.

(b) A set of vectors in $\mathbf{R}^2$ (other than two quarter-planes) for which every $c\boldsymbol{x}$ stays in the set but $\boldsymbol{x} + \boldsymbol{y}$ may be outside.

10 Which of the following subsets of $\mathbf{R}^3$ are actually subspaces?

(a) The plane of vectors (b_1, b_2, b_3) with $b_1 = b_2$.

(b) The plane of vectors with $b_1 = 1$.

(c) The vectors with $b_1 b_2 b_3 = 0$.

(d) All linear combinations of $\boldsymbol{v} = (1, 4, 0)$ and $\boldsymbol{w} = (2, 2, 2)$.

(e) All vectors that satisfy $b_1 + b_2 + b_3 = 0$.

(f) All vectors with $b_1 \leq b_2 \leq b_3$.

11 Describe the smallest subspace of the matrix space $\mathbf{M}$ that contains

(a) $\begin{bmatrix} 1 & 0 \\ 0 & 0 \end{bmatrix}$ and $\begin{bmatrix} 0 & 1 \\ 0 & 0 \end{bmatrix}$ (b) $\begin{bmatrix} 1 & 1 \\ 0 & 0 \end{bmatrix}$ (c) $\begin{bmatrix} 1 & 0 \\ 0 & 0 \end{bmatrix}$ and $\begin{bmatrix} 1 & 0 \\ 0 & 1 \end{bmatrix}$.

12 Let P be the plane in $\mathbf{R}^3$ with equation $x + y - 2z = 4$. The origin $(0, 0, 0)$ is not in P! Find two vectors in P and check that their sum is not in P.

13 Let $\mathbf{P}_0$ be the plane through $(0, 0, 0)$ parallel to the previous plane P. What is the equation for $\mathbf{P}_0$? Find two vectors in $\mathbf{P}_0$ and check that their sum is in $\mathbf{P}_0$.

14 The subspaces of $\mathbf{R}^3$ are planes, lines, $\mathbf{R}^3$ itself, or $\mathbf{Z}$ containing only $(0, 0, 0)$.

(a) Describe the three types of subspaces of $\mathbf{R}^2$.

(b) Describe all subspaces of $\mathbf{D}$, the space of 2 by 2 diagonal matrices.

15 (a) The intersection of two planes through $(0,0,0)$ is probably a _____ in $\mathbf{R}^3$ but it could be a _____. It can't be $\mathbf{Z}$!

(b) The intersection of a plane through $(0,0,0)$ with a line through $(0,0,0)$ is probably a _____ but it could be a _____.

(c) If $\mathbf{S}$ and $\mathbf{T}$ are subspaces of $\mathbf{R}^5$, prove that their intersection $\mathbf{S} \cap \mathbf{T}$ is a subspace of $\mathbf{R}^5$. Here $\mathbf{S} \cap \mathbf{T}$ consists of the vectors that lie in both subspaces. *Check that $\boldsymbol{x}+\boldsymbol{y}$ and $c\boldsymbol{x}$ are in $\mathbf{S} \cap \mathbf{T}$ if $\boldsymbol{x}$ and $\boldsymbol{y}$ are in both spaces.*

16 Suppose $\mathbf{P}$ is a plane through $(0,0,0)$ and $\mathbf{L}$ is a line through $(0,0,0)$. The smallest vector space containing both $\mathbf{P}$ and $\mathbf{L}$ is either _____ or _____.

17 (a) Show that the set of *invertible* matrices in $\mathbf{M}$ is not a subspace.

(b) Show that the set of *singular* matrices in $\mathbf{M}$ is not a subspace.

18 True or false (check addition in each case by an example):

(a) The symmetric matrices in $\mathbf{M}$ (with $A^{\mathrm{T}} = A$) form a subspace.

(b) The skew-symmetric matrices in $\mathbf{M}$ (with $A^{\mathrm{T}} = -A$) form a subspace.

(c) The unsymmetric matrices in $\mathbf{M}$ (with $A^{\mathrm{T}} \neq A$) form a subspace.

Questions 19–27 are about column spaces $C(A)$ and the equation $A\boldsymbol{x} = \boldsymbol{b}$.

19 Describe the column spaces (lines or planes) of these particular matrices:

$$A = \begin{bmatrix} 1 & 2 \\ 0 & 0 \\ 0 & 0 \end{bmatrix} \quad \text{and} \quad B = \begin{bmatrix} 1 & 0 \\ 0 & 2 \\ 0 & 0 \end{bmatrix} \quad \text{and} \quad C = \begin{bmatrix} 1 & 0 \\ 2 & 0 \\ 0 & 0 \end{bmatrix}.$$

20 For which right sides (find a condition on b_1, b_2, b_3) are these systems solvable?

$$\text{(a)} \quad \begin{bmatrix} 1 & 4 & 2 \\ 2 & 8 & 4 \\ -1 & -4 & -2 \end{bmatrix} \begin{bmatrix} x_1 \\ x_2 \\ x_3 \end{bmatrix} = \begin{bmatrix} b_1 \\ b_2 \\ b_3 \end{bmatrix} \qquad \text{(b)} \quad \begin{bmatrix} 1 & 4 \\ 2 & 9 \\ -1 & -4 \end{bmatrix} \begin{bmatrix} x_1 \\ x_2 \end{bmatrix} = \begin{bmatrix} b_1 \\ b_2 \\ b_3 \end{bmatrix}.$$

21 Adding row 1 of A to row 2 produces B. Adding column 1 to column 2 produces C. A combination of the columns of (B or C ?) is also a combination of the columns of A. Which two matrices have the same column _____ ?

$$A = \begin{bmatrix} 1 & 2 \\ 2 & 4 \end{bmatrix} \quad \text{and} \quad B = \begin{bmatrix} 1 & 2 \\ 3 & 6 \end{bmatrix} \quad \text{and} \quad C = \begin{bmatrix} 1 & 3 \\ 2 & 6 \end{bmatrix}.$$

22 For which vectors (b_1, b_2, b_3) do these systems have a solution?

$$\begin{bmatrix} 1 & 1 & 1 \\ 0 & 1 & 1 \\ 0 & 0 & 1 \end{bmatrix} \begin{bmatrix} x_1 \\ x_2 \\ x_3 \end{bmatrix} = \begin{bmatrix} b_1 \\ b_2 \\ b_3 \end{bmatrix} \quad \text{and} \quad \begin{bmatrix} 1 & 1 & 1 \\ 0 & 1 & 1 \\ 0 & 0 & 0 \end{bmatrix} \begin{bmatrix} x_1 \\ x_2 \\ x_3 \end{bmatrix} = \begin{bmatrix} b_1 \\ b_2 \\ b_3 \end{bmatrix}$$

$$\text{and} \quad \begin{bmatrix} 1 & 1 & 1 \\ 0 & 0 & 1 \\ 0 & 0 & 1 \end{bmatrix} \begin{bmatrix} x_1 \\ x_2 \\ x_3 \end{bmatrix} = \begin{bmatrix} b_1 \\ b_2 \\ b_3 \end{bmatrix}.$$

23 (Recommended) If we add an extra column $\boldsymbol{b}$ to a matrix A, then the column space gets larger unless _____. Give an example where the column space gets larger and an example where it doesn't. Why is $A\boldsymbol{x} = \boldsymbol{b}$ solvable exactly when the column space *doesn't* get larger—it is the same for A and $\begin{bmatrix} A & \boldsymbol{b} \end{bmatrix}$?

24 The columns of AB are combinations of the columns of A. This means: *The column space of AB is contained in* (possibly equal to) *the column space of A*. Give an example where the column spaces of A and AB are not equal.

25 Suppose $A\boldsymbol{x} = \boldsymbol{b}$ and $A\boldsymbol{y} = \boldsymbol{b}^*$ are both solvable. Then $A\boldsymbol{z} = \boldsymbol{b} + \boldsymbol{b}^*$ is solvable. What is $\boldsymbol{z}$? This translates into: If $\boldsymbol{b}$ and $\boldsymbol{b}^*$ are in the column space $\boldsymbol{C}(A)$, *then* $\boldsymbol{b} + \boldsymbol{b}^*$ is in $\boldsymbol{C}(A)$.

26 If A is any 5 by 5 invertible matrix, then its column space is _____. Why?

27 True or false (with a counterexample if false):

(a) The vectors $\boldsymbol{b}$ that are not in the column space $\boldsymbol{C}(A)$ form a subspace.

(b) If $\boldsymbol{C}(A)$ contains only the zero vector, then A is the zero matrix.

(c) The column space of $2A$ equals the column space of A.

(d) The column space of $A - I$ equals the column space of A (test this).

28 Construct a 3 by 3 matrix whose column space contains $(1,1,0)$ and $(1,0,1)$ but not $(1,1,1)$. Construct a 3 by 3 matrix whose column space is only a line.

29 If the 9 by 12 system $A\boldsymbol{x} = \boldsymbol{b}$ is solvable for every $\boldsymbol{b}$, then $\boldsymbol{C}(A) =$ _____.

Challenge Problems

30 Suppose $\mathbf{S}$ and $\mathbf{T}$ are two subspaces of a vector space $\mathbf{V}$.

(a) **Definition**: The **sum** $\mathbf{S} + \mathbf{T}$ contains all sums $\boldsymbol{s} + \boldsymbol{t}$ of a vector $\boldsymbol{s}$ in $\mathbf{S}$ and a vector $\boldsymbol{t}$ in $\mathbf{T}$. Show that $\mathbf{S} + \mathbf{T}$ satisfies the requirements (addition and scalar multiplication) for a vector space.

(b) If $\mathbf{S}$ and $\mathbf{T}$ are lines in $\mathbf{R}^m$, what is the difference between $\mathbf{S} + \mathbf{T}$ and $\mathbf{S} \cup \mathbf{T}$? That union contains all vectors from $\mathbf{S}$ or $\mathbf{T}$ or both. Explain this statement: *The span of* $\mathbf{S} \cup \mathbf{T}$ *is* $\mathbf{S} + \mathbf{T}$. (Section 3.5 returns to this word "span".)

31 If $\mathbf{S}$ is the column space of A and $\mathbf{T}$ is $\boldsymbol{C}(B)$, then $\mathbf{S} + \mathbf{T}$ is the column space of what matrix M? The columns of A and B and M are all in $\mathbf{R}^m$. (I don't think $A + B$ is always a correct M.)

32 Show that the matrices A and $\begin{bmatrix} A & AB \end{bmatrix}$ (with extra columns) have the same column space. But find a square matrix with $\boldsymbol{C}(A^2)$ smaller than $\boldsymbol{C}(A)$. Important point:

An n by n matrix has $\boldsymbol{C}(A) = \mathbf{R}^n$ exactly when A is an _____ matrix.

3.2 The Nullspace of A: Solving $Ax = 0$ and $Rx = 0$

1 The **nullspace** $\boldsymbol{N}(A)$ in $\mathbf{R}^n$ contains all solutions $\boldsymbol{x}$ to $A\boldsymbol{x} = \mathbf{0}$. This includes $\boldsymbol{x} = \mathbf{0}$.

2 Elimination (from A to U to R) does not change the nullspace: $\boldsymbol{N}(A) = \boldsymbol{N}(U) = \boldsymbol{N}(R)$.

3 The **reduced row echelon form $\boldsymbol{R}$ = rref($\boldsymbol{A}$)** has all pivots $= 1$, with zeros above and below.

4 If column j of R is free (no pivot), there is a "*special solution*" to $A\boldsymbol{x} = \mathbf{0}$ with $x_j = 1$.

5 Number of pivots $=$ number of nonzero rows in $R =$ **rank $\boldsymbol{r}$**. There are $n - r$ free columns.

6 Every matrix with $m < n$ has nonzero solutions to $A\boldsymbol{x} = \mathbf{0}$ in its nullspace.

This section is about the subspace containing all solutions to $A\boldsymbol{x} = \mathbf{0}$. The m by n matrix A can be square or rectangular. The right hand side is $\boldsymbol{b} = \mathbf{0}$. *One immediate solution is* $\boldsymbol{x} = \mathbf{0}$. For invertible matrices this is the only solution. For other matrices, not invertible, there are nonzero solutions to $A\boldsymbol{x} = \mathbf{0}$. *Each solution $\boldsymbol{x}$ belongs to the nullspace of A.*

Elimination will find all solutions and identify this very important subspace.

The nullspace $\boldsymbol{N}(A)$ consists of all solutions to $Ax = 0$. These vectors x are in $\mathbf{R}^n$.

Check that the solution vectors form a subspace. Suppose $\boldsymbol{x}$ and $\boldsymbol{y}$ are in the nullspace (this means $A\boldsymbol{x} = \mathbf{0}$ and $A\boldsymbol{y} = \mathbf{0}$). The rules of matrix multiplication give $A(\boldsymbol{x} + \boldsymbol{y}) = \mathbf{0} + \mathbf{0}$. The rules also give $A(c\boldsymbol{x}) = c\mathbf{0}$. The right sides are still zero. Therefore $\boldsymbol{x} + \boldsymbol{y}$ and $c\boldsymbol{x}$ are also in the nullspace $\boldsymbol{N}(A)$. Since we can add and multiply without leaving the nullspace, it is a subspace.

To repeat: The solution vectors $\boldsymbol{x}$ have n components. They are vectors in $\mathbf{R}^n$, so *the nullspace is a subspace of* $\mathbf{R}^n$. The column space $\boldsymbol{C}(A)$ is a subspace of $\mathbf{R}^m$.

Example 1 Describe the nullspace of $A = \begin{bmatrix} 1 & 2 \\ 3 & 6 \end{bmatrix}$. This matrix is singular !

Solution Apply elimination to the linear equations $A\boldsymbol{x} = \mathbf{0}$:

$$\begin{aligned} x_1 + 2x_2 &= 0 \\ 3x_1 + 6x_2 &= 0 \end{aligned} \quad \to \quad \begin{aligned} x_1 + 2x_2 &= 0 \\ \mathbf{0} &= \mathbf{0} \end{aligned}$$

There is really only one equation. The second equation is the first equation multiplied by 3. In the row picture, the line $x_1 + 2x_2 = 0$ is the same as the line $3x_1 + 6x_2 = 0$. That line is the nullspace $\boldsymbol{N}(A)$. It contains all solutions (x_1, x_2).

To describe the solutions to $A\boldsymbol{x} = \mathbf{0}$, here is an efficient way. Choose one point on the line (one "***special solution***"). Then all points on the line are multiples of this one. We choose the second component to be $x_2 = 1$ (a special choice). From the equation $x_1 + 2x_2 = 0$, the first component must be $x_1 = -2$. **The special solution is $\boldsymbol{s} = (-2, 1)$.**

Special solution $\quad A\boldsymbol{s} = \mathbf{0}$ The nullspace of $A = \begin{bmatrix} 1 & 2 \\ 3 & 6 \end{bmatrix}$ contains all multiples of $\boldsymbol{s} = \begin{bmatrix} -2 \\ \mathbf{1} \end{bmatrix}$.

This is the best way to describe the nullspace, by computing special solutions to $A\boldsymbol{x} = \mathbf{0}$. **The solution is special because we set the free variable to $x_2 = 1$.**

The nullspace of A consists of all combinations of the special solutions to $Ax = 0$.

Example 2 $x + 2y + 3z = 0$ comes from the 1 by 3 matrix $A = \begin{bmatrix} 1 & 2 & 3 \end{bmatrix}$. Then $A\boldsymbol{x} = \mathbf{0}$ produces a plane. All vectors on the plane are perpendicular to $(1, 2, 3)$. *The plane is the nullspace of A.* There are two free variables y and z: Set to 0 and 1.

$$\begin{bmatrix} 1 & 2 & 3 \end{bmatrix} \begin{bmatrix} x \\ y \\ z \end{bmatrix} = 0 \text{ has two special solutions } \boldsymbol{s}_1 = \begin{bmatrix} -2 \\ \mathbf{1} \\ \mathbf{0} \end{bmatrix} \text{ and } \boldsymbol{s}_2 = \begin{bmatrix} -3 \\ \mathbf{0} \\ \mathbf{1} \end{bmatrix}.$$

Those vectors $\boldsymbol{s}_1$ and $\boldsymbol{s}_2$ lie on the plane $x + 2y + 3z = 0$. All vectors on the plane are combinations of $\boldsymbol{s}_1$ and $\boldsymbol{s}_2$.

Notice what is special about $\boldsymbol{s}_1$ and $\boldsymbol{s}_2$. *The last two components are "free" and we choose them specially as* $1, 0$ *and* $0, 1$. Then the first components -2 and -3 are determined by the equation $A\boldsymbol{x} = \mathbf{0}$.

The solutions to $x + 2y + 3z = \mathbf{6}$ also lie on a plane, but that plane is not a subspace. The vector $\boldsymbol{x} = \mathbf{0}$ is only a solution if $\boldsymbol{b} = \mathbf{0}$. Section 3.3 will show how the solutions to $A\boldsymbol{x} = \boldsymbol{b}$ (if there are any solutions) are shifted away from zero by one particular solution.

The two key steps of this section are (**1**) reducing A to its **row echelon form $\boldsymbol{R}$**
(**2**) finding the **special solutions to $A\boldsymbol{x} = \mathbf{0}$**

The display on page 138 shows 4 by 5 matrices A and R, with 3 pivots.

The equations $A\boldsymbol{x} = \mathbf{0}$ and also $R\boldsymbol{x} = \mathbf{0}$ have $5 - 3 = 2$ special solutions $\boldsymbol{s}_1$ and $\boldsymbol{s}_2$.

Pivot Columns and Free Columns

The first column of $A = \begin{bmatrix} 1 & 2 & 3 \end{bmatrix}$ contains the only pivot, so the first component of $\boldsymbol{x}$ is *not free*. **The free components correspond to columns with no pivots**. The special choice (one or zero) is only for the free variables in the special solutions.

Example 3 Find the nullspaces of A, B, C and the two special solutions to $C\boldsymbol{x} = \mathbf{0}$.

$$\boldsymbol{A} = \begin{bmatrix} 1 & 2 \\ 3 & 8 \end{bmatrix} \quad \boldsymbol{B} = \begin{bmatrix} A \\ 2A \end{bmatrix} = \begin{bmatrix} 1 & 2 \\ 3 & 8 \\ 2 & 4 \\ 6 & 16 \end{bmatrix} \quad \boldsymbol{C} = \begin{bmatrix} A & 2A \end{bmatrix} = \begin{bmatrix} 1 & 2 & 2 & 4 \\ 3 & 8 & 6 & 16 \end{bmatrix}.$$

Solution The equation $A\boldsymbol{x} = \mathbf{0}$ has only the zero solution $\boldsymbol{x} = \mathbf{0}$. *The nullspace is* $\mathbf{Z}$. It contains only the single point $\boldsymbol{x} = \mathbf{0}$ in $\mathbf{R}^2$. This fact comes from elimination:

$$A\boldsymbol{x} = \begin{bmatrix} 1 & 2 \\ 3 & 8 \end{bmatrix} \begin{bmatrix} x_1 \\ x_2 \end{bmatrix} = \begin{bmatrix} 0 \\ 0 \end{bmatrix} \text{ yields } \begin{bmatrix} 1 & 2 \\ 0 & 2 \end{bmatrix} \begin{bmatrix} x_1 \\ x_2 \end{bmatrix} = \begin{bmatrix} 0 \\ 0 \end{bmatrix} \text{ and } \begin{bmatrix} \boldsymbol{x_1} = \mathbf{0} \\ \boldsymbol{x_2} = \mathbf{0} \end{bmatrix}.$$

A is invertible. There are no special solutions. Both columns of this matrix have pivots.

The rectangular matrix B has the same nullspace $\mathbf{Z}$. The first two equations in $B\boldsymbol{x} = \mathbf{0}$ again require $\boldsymbol{x} = \mathbf{0}$. The last two equations would also force $\boldsymbol{x} = \mathbf{0}$. When we add extra equations (giving extra rows), the nullspace certainly cannot become larger. The extra rows impose more conditions on the vectors $\boldsymbol{x}$ in the nullspace.

The rectangular matrix C is different. It has extra columns instead of extra rows. The solution vector $\boldsymbol{x}$ has *four* components. Elimination will produce pivots in the first two columns of C, but **the last two columns of C and U are "free". They don't have pivots:**

Subtract 3 (row 1) from row 2 of C

$$C = \begin{bmatrix} 1 & 2 & 2 & 4 \\ 3 & 8 & 6 & 16 \end{bmatrix} \text{ becomes } U = \begin{bmatrix} \mathbf{1} & 2 & 2 & 4 \\ 0 & \mathbf{2} & 0 & 4 \end{bmatrix}$$

$$\begin{matrix} \uparrow & \uparrow & \uparrow & \uparrow \end{matrix}$$

pivot columns **free columns**

For the free variables x_3 and x_4, we make special choices of ones and zeros. First $x_3 = 1$, $x_4 = 0$ and second $x_3 = 0$, $x_4 = 1$. The pivot variables x_1 and x_2 are determined by the equation $U\boldsymbol{x} = \mathbf{0}$ (or $C\boldsymbol{x} = \mathbf{0}$ or eventually $R\boldsymbol{x} = \mathbf{0}$). We get two special solutions in the nullspace of C. This is also the nullspace of U: elimination doesn't change solutions.

Special solutions $\boldsymbol{Cs = 0}$ $\boldsymbol{Us = 0}$

$$\boldsymbol{s}_1 = \begin{bmatrix} -2 \\ 0 \\ 1 \\ 0 \end{bmatrix} \text{ and } \boldsymbol{s}_2 = \begin{bmatrix} 0 \\ -2 \\ 0 \\ 1 \end{bmatrix} \begin{matrix} \leftarrow \textbf{pivot} \\ \leftarrow \textbf{variables} \\ \leftarrow \textbf{free} \\ \leftarrow \textbf{variables} \end{matrix}$$

The Reduced Row Echelon Form R

When A is rectangular, elimination will not stop at the upper triangular U. We can continue to make this matrix simpler, in two ways. These steps bring us to the best matrix R:

1. *Produce zeros above the pivots*. **Use pivot rows to eliminate upward in R.**
2. *Produce ones in the pivots.* **Divide the whole pivot row by its pivot.**

Those steps don't change the zero vector on the right side of the equation. The nullspace stays the same: $\boldsymbol{N}(A) = \boldsymbol{N}(U) = \boldsymbol{N}(R)$. This nullspace becomes easiest to see when we reach the ***reduced row echelon form*** $\boldsymbol{R} = \textbf{rref}\,(\boldsymbol{A})$. ***The pivot columns of R contain I.***

Reduced form R $\quad U = \begin{bmatrix} 1 & 2 & 2 & 4 \\ 0 & 2 & 0 & 4 \end{bmatrix}$ becomes $R = \begin{bmatrix} \mathbf{1} & \mathbf{0} & 2 & 0 \\ \mathbf{0} & \mathbf{1} & 0 & 2 \end{bmatrix}$.
$\uparrow \; \uparrow$

I subtracted row 2 of U from row 1. Then I multiplied row 2 by $\frac{1}{2}$ to get pivot $= 1$.

Now (**free column 3**) $=$ **2** (**pivot column 1**), so -2 appears in $\boldsymbol{s}_1 = (-2,0,1,0)$. The special solutions are much easier to find from the reduced system $R\boldsymbol{x} = \mathbf{0}$. In each free column of R, I change all the signs to find $\boldsymbol{s}$. Second special solution $\boldsymbol{s}_2 = (0,-2,0,1)$.

Before moving to m by n matrices A and their nullspaces $\boldsymbol{N}(A)$ and special solutions, allow me to repeat one comment. For many matrices, the only solution to $A\boldsymbol{x} = \mathbf{0}$ is $\boldsymbol{x} = \mathbf{0}$. Their nullspaces $\boldsymbol{N}(A) = \mathbf{Z}$ contain only that zero vector: *no* special solutions. The only combination of the columns that produces $\boldsymbol{b} = \mathbf{0}$ is then the "zero combination". The solution to $A\boldsymbol{x} = \mathbf{0}$ is trivial (just $\boldsymbol{x} = \mathbf{0}$) but the idea is not trivial.

This case of a zero nullspace $\mathbf{Z}$ is of the greatest importance. It says that the columns of A are **independent**. No combination of columns gives the zero vector (except the zero combination). All columns have pivots, and no columns are free. You will see this idea of independence again...

Pivot Variables and Free Variables in the Echelon Matrix R

$$\boldsymbol{A} = \begin{bmatrix} p & p & f & p & f \\ | & | & | & | & | \\ | & | & | & | & | \end{bmatrix} \qquad \boldsymbol{R} = \begin{bmatrix} \mathbf{1} & 0 & a & 0 & c \\ 0 & \mathbf{1} & b & 0 & d \\ 0 & 0 & 0 & \mathbf{1} & e \\ 0 & 0 & 0 & 0 & 0 \end{bmatrix} \qquad \boldsymbol{s}_1 = \begin{bmatrix} -a \\ -b \\ \mathbf{1} \\ 0 \\ \mathbf{0} \end{bmatrix} \qquad \boldsymbol{s}_2 = \begin{bmatrix} -c \\ -d \\ \mathbf{0} \\ -e \\ \mathbf{1} \end{bmatrix}$$

3 pivot columns p	I in pivot columns	special $R\boldsymbol{s}_1 = \mathbf{0}$ and $R\boldsymbol{s}_2 = \mathbf{0}$
2 free columns f	F in free columns	take $-a$ to $-e$ from R
to be revealed by R	3 pivots: rank $r = 3$	$R\boldsymbol{s} = \mathbf{0}$ means $A\boldsymbol{s} = \mathbf{0}$

R shows clearly: *column* $3 = a\,$(*column* $1) + b\,$(*column* 2). The same must be true for A. The special solution $\boldsymbol{s}_1$ repeats that combination so $(-\boldsymbol{a}, -\boldsymbol{b}, \mathbf{1}, 0, 0)$ has $R\boldsymbol{s}_1 = \mathbf{0}$. Nullspace of $A =$ Nullspace of $R =$ all combinations of $\boldsymbol{s}_1$ and $\boldsymbol{s}_2$.

Here are those steps for a 4 by 7 ***reduced row echelon matrix R*** with three pivots:

$$\boldsymbol{R} = \begin{bmatrix} \mathbf{1} & \mathbf{0} & x & x & x & \mathbf{0} & x \\ \mathbf{0} & \mathbf{1} & x & x & x & \mathbf{0} & x \\ \mathbf{0} & \mathbf{0} & 0 & 0 & 0 & \mathbf{1} & x \\ 0 & 0 & 0 & 0 & 0 & 0 & 0 \end{bmatrix}$$

Three pivot variables x_1, x_2, x_6
Four free variables x_3, x_4, x_5, x_7
Four special solutions s in $\boldsymbol{N}(\boldsymbol{R})$
The pivot rows and columns contain $\boldsymbol{I}$

Question What are the column space and the nullspace for this matrix R?

Answer The columns of R have four components so they lie in $\mathbf{R}^4$. (Not in $\mathbf{R}^3$!) The fourth component of every column is zero. Every combination of the columns—every vector in the column space—has fourth component zero. *The column space* $\boldsymbol{C}(R)$ *consists of all vectors of the form* $(b_1, b_2, b_3, 0)$. For those vectors we can solve $R\boldsymbol{x} = \boldsymbol{b}$.

The nullspace $\boldsymbol{N}(R)$ is a subspace of $\mathbf{R}^7$. The solutions to $R\boldsymbol{x} = \mathbf{0}$ are all the combinations of the four special solutions—*one for each free variable*:

1. Columns $3, 4, 5, 7$ have no pivots. So the four free variables are x_3, x_4, x_5, x_7.
2. Set one free variable to 1 and set the other three free variables to zero.
3. To find $\boldsymbol{s}$, solve $R\boldsymbol{x} = \mathbf{0}$ for the pivot variables x_1, x_2, x_6.

Counting the pivots leads to an extremely important theorem. Suppose A has more columns than rows. ***With*** $\boldsymbol{n > m}$ ***there is at least one free variable.*** The system $A\boldsymbol{x} = \mathbf{0}$ has at least one special solution. This solution is *not zero*!

> Suppose $A\boldsymbol{x} = \mathbf{0}$ has more unknowns than equations ($\boldsymbol{n > m}$, more columns than rows). There must be at least one free column. **Then** $\boldsymbol{Ax} = \mathbf{0}$ **has nonzero solutions.**

A short wide matrix $(n > m)$ *always has nonzero vectors in its nullspace.* There must be at least $n - m$ free variables, since the number of pivots cannot exceed m. (The matrix only has m rows, and a row never has two pivots.) Of course a row might have *no* pivot—which means an extra free variable. But here is the point: When there is a free variable, it can be set to 1. Then the equation $A\boldsymbol{x} = \mathbf{0}$ has at least a line of nonzero solutions.

The nullspace is a subspace. Its "dimension" is the number of free variables. This central idea—the ***dimension*** of a subspace—is defined and explained in this chapter.

The Rank of a Matrix

The numbers m and n give the size of a matrix—but not necessarily the *true size* of a linear system. An equation like $0 = 0$ should not count. If there are two identical rows in A, the second one disappears in elimination. Also if row 3 is a combination of rows 1 and 2, then row 3 will become all zeros in the triangular U and the reduced echelon form R. We don't want to count rows of zeros. ***The true size of*** A ***is given by its rank.***

> **DEFINITION OF RANK** ***The rank of A is the number of pivots. This number is r.***

That definition is computational, and I would like to say more about the rank r. The final matrix R will have r nonzero rows. Start with a 3 by 4 example of rank $r = 2$:

Four columns
Two pivots

$$A = \begin{bmatrix} 1 & 1 & 2 & 4 \\ 1 & 2 & 2 & 5 \\ 1 & 3 & 2 & 6 \end{bmatrix} \qquad R = \begin{bmatrix} \mathbf{1} & \mathbf{0} & 2 & 3 \\ \mathbf{0} & \mathbf{1} & 0 & 1 \\ 0 & 0 & 0 & 0 \end{bmatrix}.$$

The first two columns of A are $(1, 1, 1)$ and $(1, 2, 3)$, going in different directions. Those will be pivot columns (revealed by R). The third column $(2, 2, 2)$ is a multiple of the first. We won't see a pivot in that third column. The fourth column $(4, 5, 6)$ is the sum of the first three. That fourth column will also have no pivot. The rank of A and R is **2**.

Every "free column" is a combination of earlier pivot columns. It is the special solutions s that tell us those combinations:

$$\begin{array}{ll} \text{Column } 3 = \mathbf{2}\,(\text{column } 1) + \mathbf{0}\,(\text{column } 2) & \boldsymbol{s}_1 = (-\mathbf{2}, -\mathbf{0}, 1, 0) \\ \text{Column } 4 = \mathbf{3}\,(\text{column } 1) + \mathbf{1}\,(\text{column } 2) & \boldsymbol{s}_2 = (-\mathbf{3}, -\mathbf{1}, 0, 1) \end{array}$$

The numbers $2, 0$ in column 3 of R show up in $\boldsymbol{s}_1$ (with signs reversed). And the numbers $3, 1$ in column 4 of R show up in $\boldsymbol{s}_2$ (with signs reversed to $-3, -1$).

Rank One

Matrices of ***rank one*** have only *one pivot*. When elimination produces zero in the first column, it produces zero in all the columns. *Every row is a multiple of the pivot row.* At the same time, every column is a multiple of the pivot column!

$$\text{Rank one matrix} \qquad A = \begin{bmatrix} \mathbf{1} & 3 & 10 \\ \mathbf{2} & 6 & 20 \\ \mathbf{3} & 9 & 30 \end{bmatrix} \longrightarrow R = \begin{bmatrix} \mathbf{1} & 3 & 10 \\ 0 & 0 & 0 \\ 0 & 0 & 0 \end{bmatrix}.$$

The column space of a rank one matrix is "one-dimensional". Here all columns are on the line through $\boldsymbol{u} = (1, 2, 3)$. The columns of A are $\boldsymbol{u}$ and $3\boldsymbol{u}$ and $10\boldsymbol{u}$. Put those numbers into the row $\boldsymbol{v}^{\mathrm{T}} = [\ 1 \quad 3 \quad 10\ ]$ and you have the special rank one form $A = \boldsymbol{u}\boldsymbol{v}^{\mathrm{T}}$:

$$\boldsymbol{A} = \textbf{column times row} = \boldsymbol{u}\boldsymbol{v}^{\mathrm{T}} \qquad \begin{bmatrix} 1 & 3 & 10 \\ 2 & 6 & 20 \\ 3 & 9 & 30 \end{bmatrix} = \begin{bmatrix} \mathbf{1} \\ \mathbf{2} \\ \mathbf{3} \end{bmatrix} \begin{bmatrix} \mathbf{1} & \mathbf{3} & \mathbf{10} \end{bmatrix}$$

With rank one, $A\boldsymbol{x} = \mathbf{0}$ is easy to understand. That equation $\boldsymbol{u}(\boldsymbol{v}^{\mathrm{T}}\boldsymbol{x}) = \mathbf{0}$ leads us to $\boldsymbol{v}^{\mathrm{T}}\boldsymbol{x} = 0$. All vectors $\boldsymbol{x}$ in the nullspace must be orthogonal to $\boldsymbol{v}$ in the row space. This is the geometry when $r = 1$: *row space = line, nullspace = perpendicular plane.*

Example 4 When all rows are multiples of one pivot row, the rank is $r = 1$:

$$\begin{bmatrix} 1 & 3 & 4 \\ 2 & 6 & 8 \end{bmatrix} \text{ and } \begin{bmatrix} 0 & 3 \\ 0 & \mathbf{5} \end{bmatrix} \text{ and } \begin{bmatrix} 5 \\ 2 \end{bmatrix} \text{ and } \begin{bmatrix} 6 \end{bmatrix} \text{ all have rank } 1.$$

For those matrices, the reduced row echelon $R =$ **rref** (A) can be checked by eye:

$$R = \begin{bmatrix} 1 & 3 & 4 \\ 0 & 0 & 0 \end{bmatrix} \text{ and } \begin{bmatrix} 0 & 1 \\ 0 & 0 \end{bmatrix} \text{ and } \begin{bmatrix} 1 \\ 0 \end{bmatrix} \text{ and } \begin{bmatrix} 1 \end{bmatrix} \text{ have only one pivot.}$$

Our second definition of rank will be at a higher level. It deals with entire rows and entire columns—vectors and not just numbers. All three matrices A and U and R have r **independent rows**.

A and U and R also have $\boldsymbol{r}$ **independent columns** (the pivot columns). Section 3.4 says what it means for rows or columns to be independent.

A third definition of rank, at the top level of linear algebra, will deal with *spaces* of vectors. ***The rank r is the "dimension" of the column space. It is also the dimension of the row space***. The great thing is that $\boldsymbol{n - r}$ **is the dimension of the nullspace.**

■ REVIEW OF THE KEY IDEAS ■

1. The nullspace $\boldsymbol{N}(A)$ is a subspace of $\mathbf{R}^n$. It contains all solutions to $A\boldsymbol{x} = \mathbf{0}$.
2. Elimination on A produces a row reduced R with pivot columns and free columns.
3. Every free column leads to a special solution. That free variable is 1, the others are 0.
4. The *rank* r of A is the number of pivots. All pivots are 1's in $R = \mathrm{rref(A)}$.
5. The complete solution to $A\boldsymbol{x} = \mathbf{0}$ is a combination of the $n - r$ special solutions.
6. A always has a free column if $n > m$, giving a *nonzero solution* to $A\boldsymbol{x} = \mathbf{0}$.

■ WORKED EXAMPLES ■

3.2 A Why do A and R have the same nullspace if $EA = R$ and E is invertible?

Solution If $A\boldsymbol{x} = \mathbf{0}$ then $R\boldsymbol{x} = EA\boldsymbol{x} = E\mathbf{0} = \mathbf{0}$
If $R\boldsymbol{x} = \mathbf{0}$ then $A\boldsymbol{x} = E^{-1}R\boldsymbol{x} = E^{-1}\mathbf{0} = \mathbf{0}$
A and R also have the same row space and the same rank.

3.2 B Create a 3 by 4 matrix R whose special solutions to $R\boldsymbol{x} = \mathbf{0}$ are $\boldsymbol{s}_1$ and $\boldsymbol{s}_2$:

$$\boldsymbol{s}_1 = \begin{bmatrix} -3 \\ 1 \\ 0 \\ 0 \end{bmatrix} \quad \text{and} \quad \boldsymbol{s}_2 = \begin{bmatrix} -2 \\ 0 \\ -6 \\ 1 \end{bmatrix} \qquad \begin{array}{l} \text{pivot columns 1 and 3} \\ \text{free variables } x_2 \text{ and } x_4 \end{array}$$

Describe all possible matrices A with this nullspace $\boldsymbol{N}(A) =$ all combinations of $\boldsymbol{s}_1$ and $\boldsymbol{s}_2$.

Solution The reduced matrix R has pivots $= 1$ in columns 1 and 3. There is no third pivot, so row 3 of R is all zeros. The free columns 2 and 4 will be combinations of the pivot columns: $3, 0, 2, 6$ in R come from $-3, -0, -2, -6$ in $\boldsymbol{s}_1$ and $\boldsymbol{s}_2$. **Every $\boldsymbol{A = ER}$.**

Every 3 by 4 matrix has at least one special solution. *These matrices have two.*

$$R = \begin{bmatrix} \mathbf{1} & \mathbf{3} & 0 & \mathbf{2} \\ 0 & \mathbf{0} & \mathbf{1} & \mathbf{6} \\ 0 & 0 & 0 & 0 \end{bmatrix} \quad \text{has} \quad R\boldsymbol{s}_1 = \mathbf{0} \quad \text{and} \quad R\boldsymbol{s}_2 = \mathbf{0}.$$

3.2 C Find the row reduced form R and the rank r of A and B (*those depend on* c). Which are the pivot columns of A? What are the special solutions?

Find special solutions $$A = \begin{bmatrix} 1 & 2 & 1 \\ 3 & 6 & 3 \\ 4 & 8 & c \end{bmatrix} \quad \text{and} \quad B = \begin{bmatrix} c & c \\ c & c \end{bmatrix}.$$

Solution The matrix A has row $2 = 3$ (row 1). The rank of A is $r = 2$ ***except if*** $\boldsymbol{c = 4}$. Row $4 - 4$(row 1) ends in $c - 4$. The pivots are in columns 1 and 3. The second variable x_2 is free. Notice the form of R: Row 3 has moved up into row 2.

$$\boldsymbol{c \neq 4} \quad R = \begin{bmatrix} \mathbf{1} & 2 & 0 \\ 0 & 0 & \mathbf{1} \\ 0 & 0 & 0 \end{bmatrix} \qquad \boldsymbol{c = 4} \quad R = \begin{bmatrix} \mathbf{1} & 2 & 1 \\ 0 & 0 & 0 \\ 0 & 0 & 0 \end{bmatrix}.$$

Two pivots leave one free variable x_2. But when $c = 4$, the only pivot is in column 1 (rank one). The second and third variables are free, producing two special solutions:

$$\boldsymbol{c \neq 4} \quad \text{Special solution } (-2, 1, 0) \qquad \boldsymbol{c = 4} \quad \text{Another special solution } (-1, 0, 1).$$

The 2 by 2 matrix $B = \left[\begin{smallmatrix} c & c \\ c & c \end{smallmatrix}\right]$ has rank $r = 1$ ***except if*** $\boldsymbol{c = 0}$, when the rank is zero!

$$\boldsymbol{c \neq 0} \quad R = \begin{bmatrix} 1 & 1 \\ 0 & 0 \end{bmatrix} \qquad \boldsymbol{c = 0} \quad R = \begin{bmatrix} 0 & 0 \\ 0 & 0 \end{bmatrix} \quad \text{and nullspace} = \mathbf{R}^2.$$

Problem Set 3.2

1 Reduce A and B to their triangular echelon forms U. Which variables are free?

$$\text{(a)}\ A = \begin{bmatrix} 1 & 2 & 2 & 4 & 6 \\ 1 & 2 & 3 & 6 & 9 \\ 0 & 0 & 1 & 2 & 3 \end{bmatrix} \qquad \text{(b)}\ B = \begin{bmatrix} 2 & 4 & 2 \\ 0 & 4 & 4 \\ 0 & 8 & 8 \end{bmatrix}.$$

2 For the matrices in Problem 1, find a special solution for each free variable. (Set the free variable to 1. Set the other free variables to zero.)

3 By further row operations on each U in Problem 1, find the reduced echelon form R. *True or false with a reason*: The nullspace of R equals the nullspace of U.

4 For the same A and B, find the special solutions to $A\boldsymbol{x} = \mathbf{0}$ and $B\boldsymbol{x} = \mathbf{0}$. For an m by n matrix, the number of pivot variables plus the number of free variables is _____. This is the **Counting Theorem**: $\boldsymbol{r + (n - r) = n}$.

$$\text{(a)}\ A = \begin{bmatrix} -1 & 3 & 5 \\ -2 & 6 & 10 \end{bmatrix} \qquad \text{(b)}\ B = \begin{bmatrix} -1 & 3 & 5 \\ -2 & 6 & 7 \end{bmatrix}.$$

Questions 5–14 are about free variables and pivot variables.

5 True or false (with reason if true or example to show it is false):

(a) A square matrix has no free variables.

(b) An invertible matrix has no free variables.

(c) An m by n matrix has no more than n pivot variables.

(d) An m by n matrix has no more than m pivot variables.

6 Put as many 1's as possible in a 4 by 7 echelon matrix U whose pivot columns are

(a) 2, 4, 5 (b) 1, 3, 6, 7 (c) 4 and 6.

7 Put as many 1's as possible in a 4 by 8 *reduced* echelon matrix R so that the free columns are

(a) 2, 4, 5, 6 (b) 1, 3, 6, 7, 8.

8 Suppose column 4 of a 3 by 5 matrix is all zero. Then x_4 is certainly a _____ variable. The special solution for this variable is the vector $\boldsymbol{x} =$ _____ .

9 Suppose the first and last columns of a 3 by 5 matrix are the same (not zero). Then _____ is a free variable. Find the special solution for this variable.

10 Suppose an m by n matrix has r pivots. The number of special solutions is _____ . The nullspace contains only $\boldsymbol{x} = \mathbf{0}$ when $r =$ _____ . The column space is all of $\mathbf{R}^m$ when $r =$ _____ .

11 The nullspace of a 5 by 5 matrix contains only $\boldsymbol{x} = \mathbf{0}$ when the matrix has _____ pivots. The column space is $\mathbf{R}^5$ when there are _____ pivots. Explain why.

12 The equation $x - 3y - z = 0$ determines a plane in $\mathbf{R}^3$. What is the matrix A in this equation? Which variables are free? The special solutions are _____ and _____ .

13 (Recommended) The plane $x - 3y - z = 12$ is parallel to $x - 3y - z = 0$. One particular point on this plane is $(12, 0, 0)$. All points on the plane have the form

$$\begin{bmatrix} x \\ y \\ z \end{bmatrix} = \begin{bmatrix} \ \\ 0 \\ 0 \end{bmatrix} + y \begin{bmatrix} \ \\ 1 \\ 0 \end{bmatrix} + z \begin{bmatrix} \ \\ 0 \\ 1 \end{bmatrix}.$$

14 Suppose column 1 + column 3 + column 5 = $\mathbf{0}$ in a 4 by 5 matrix with four pivots. Which column has no pivot? What is the special solution? Describe $\boldsymbol{N}(A)$.

Questions 15–22 ask for matrices (if possible) with specific properties.

15 Construct a matrix for which $\boldsymbol{N}(A) =$ all combinations of $(2, 2, 1, 0)$ and $(3, 1, 0, 1)$.

16 Construct A so that $\boldsymbol{N}(A) =$ all multiples of $(4, 3, 2, 1)$. Its rank is _____ .

17 Construct a matrix whose column space contains $(1,1,5)$ and $(0,3,1)$ and whose nullspace contains $(1,1,2)$.

18 Construct a matrix whose column space contains $(1,1,0)$ and $(0,1,1)$ and whose nullspace contains $(1,0,1)$ and $(0,0,1)$.

19 Construct a matrix whose column space contains $(1,1,1)$ and whose nullspace is the line of multiples of $(1,1,1,1)$.

20 Construct a 2 by 2 matrix whose nullspace equals its column space. This is possible.

21 Why does no 3 by 3 matrix have a nullspace that equals its column space?

22 If $AB = 0$ then the column space of B is contained in the _____ of A. Why?

23 The reduced form R of a 3 by 3 matrix with randomly chosen entries is almost sure to be _____. What R is virtually certain if the random A is 4 by 3?

24 Show by example that these three statements are generally *false*:

(a) A and A^{T} have the same nullspace.

(b) A and A^{T} have the same free variables.

(c) If R is the reduced form $\textbf{rref}(A)$ then R^{T} is $\textbf{rref}(A^{\mathrm{T}})$.

25 If $\boldsymbol{N}(A) =$ all multiples of $\boldsymbol{x} = (2,1,0,1)$, what is R and what is its rank?

26 If the special solutions to $R\boldsymbol{x} = \mathbf{0}$ are in the columns of these nullspace matrices N, go backward to find the nonzero rows of the reduced matrices R:

$$N = \begin{bmatrix} 2 & 3 \\ 1 & 0 \\ 0 & 1 \end{bmatrix} \quad \text{and} \quad N = \begin{bmatrix} 0 \\ 0 \\ 1 \end{bmatrix} \quad \text{and} \quad N = \begin{bmatrix} \; \end{bmatrix} \quad \text{(empty 3 by 1).}$$

27 (a) What are the five 2 by 2 reduced matrices R whose entries are all 0's and 1's?

(b) What are the eight 1 by 3 matrices containing only 0's and 1's? Are all eight of them reduced echelon matrices R?

28 Explain why A and $-A$ always have the same reduced echelon form R.

29 If A is 4 by 4 and invertible, describe the nullspace of the 4 by 8 matrix $B = [A \;\; A]$.

30 How is the nullspace $\boldsymbol{N}(C)$ related to the spaces $\boldsymbol{N}(A)$ and $\boldsymbol{N}(B)$, if $C = \begin{bmatrix} A \\ B \end{bmatrix}$?

31 Find the reduced row echelon forms R and the rank of these matrices:

(a) The 3 by 4 matrix with all entries equal to 4.

(b) The 3 by 4 matrix with $a_{ij} = i + j - 1$.

(c) The 3 by 4 matrix with $a_{ij} = (-1)^j$.

32 Kirchhoff's Current Law $A^{\mathrm{T}}\boldsymbol{y} = \mathbf{0}$ says that *current in* $=$ *current out* at every node. At node 1 this is $y_3 = y_1 + y_4$. Write the four equations for Kirchhoff's Law at the four nodes (arrows show the positive direction of each y). Reduce A^{T} to R and find three special solutions in the nullspace of A^{T} (4 by 6 matrix).

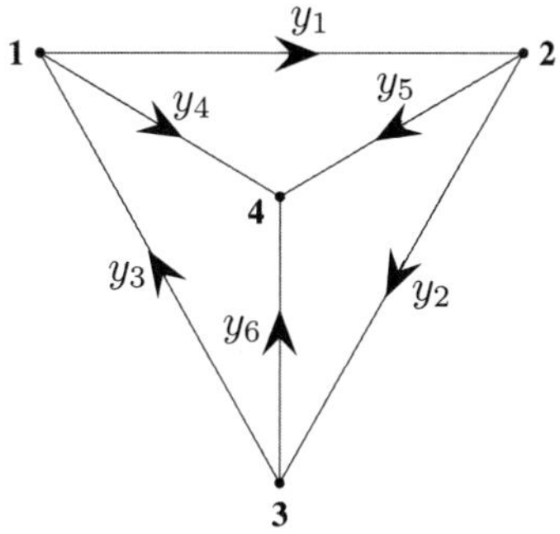

$$A = \begin{bmatrix} -1 & 1 & 0 & 0 \\ 0 & -1 & 1 & 0 \\ 1 & 0 & -1 & 0 \\ -1 & 0 & 0 & 1 \\ 0 & -1 & 0 & 1 \\ 0 & 0 & -1 & 1 \end{bmatrix}$$

33 Which of these rules gives a correct definition of the *rank* of A?

(a) The number of nonzero rows in R.

(b) The number of columns minus the total number of rows.

(c) The number of columns minus the number of free columns.

(d) The number of 1's in the matrix R.

34 Find the reduced R for each of these (block) matrices:

$$A = \begin{bmatrix} 0 & 0 & 0 \\ 0 & 0 & 3 \\ 2 & 4 & 6 \end{bmatrix} \qquad B = \begin{bmatrix} A & A \end{bmatrix} \qquad C = \begin{bmatrix} A & A \\ A & 0 \end{bmatrix}$$

35 Suppose all the pivot variables come *last* instead of first. Describe all four blocks in the reduced echelon form (the block B should be r by r):

$$R = \begin{bmatrix} A & B \\ C & D \end{bmatrix}.$$

What is the nullspace matrix N containing the special solutions?

36 (Silly problem) Describe all 2 by 3 matrices A_1 and A_2, with row echelon forms R_1 and R_2, such that $R_1 + R_2$ is the row echelon form of $A_1 + A_2$. Is is true that $R_1 = A_1$ and $R_2 = A_2$ in this case? Does $R_1 - R_2$ equal $\mathbf{rref}(A_1 - A_2)$?

37 If A has r pivot columns, how do you know that A^{T} has r pivot columns? Give a 3 by 3 example with different column numbers in *pivcol* for A and A^{T}.

38 What are the special solutions to $R\boldsymbol{x} = \mathbf{0}$ and $\boldsymbol{y}^{\mathrm{T}}R = \mathbf{0}$ for these R?

$$R = \begin{bmatrix} 1 & 0 & 2 & 3 \\ 0 & 1 & 4 & 5 \\ 0 & 0 & 0 & 0 \end{bmatrix} \qquad R = \begin{bmatrix} 0 & 1 & 2 \\ 0 & 0 & 0 \\ 0 & 0 & 0 \end{bmatrix}$$

39 Fill out these matrices so that they have rank 1:

$$A = \begin{bmatrix} 1 & 2 & 4 \\ 2 & & \\ 4 & & \end{bmatrix} \quad \text{and} \quad B = \begin{bmatrix} & 9 & \\ 1 & & \\ 2 & 6 & -3 \end{bmatrix} \quad \text{and} \quad M = \begin{bmatrix} a & b \\ c & \end{bmatrix}.$$

40 If A is an m by n matrix with $r = 1$, its columns are multiples of one column and its rows are multiples of one row. The column space is a _____ in $\mathbf{R}^m$. The nullspace is a _____ in $\mathbf{R}^n$. The nullspace matrix N has shape _____.

41 Choose vectors $\boldsymbol{u}$ and $\boldsymbol{v}$ so that $A = \boldsymbol{u}\boldsymbol{v}^{\mathrm{T}}$ = column times row:

$$A = \begin{bmatrix} 3 & 6 & 6 \\ 1 & 2 & 2 \\ 4 & 8 & 8 \end{bmatrix} \quad \text{and} \quad A = \begin{bmatrix} 2 & 2 & 6 & 4 \\ -1 & -1 & -3 & -2 \end{bmatrix}.$$

$A = \boldsymbol{u}\boldsymbol{v}^{\mathrm{T}}$ *is the natural form for every matrix that has rank* $r = 1$.

42 If A is a rank one matrix, the second row of R is _____. Do an example.

Problems 43–45 are about r by r invertible matrices inside A.

43 *If A has rank r, then it has an r by r submatrix S that is invertible.* Remove $m - r$ rows and $n - r$ columns to find an invertible submatrix S inside A, B, and C. You could keep the pivot rows and pivot columns:

$$A = \begin{bmatrix} 1 & 2 & 3 \\ 1 & 2 & 4 \end{bmatrix} \qquad B = \begin{bmatrix} 1 & 2 & 3 \\ 2 & 4 & 6 \end{bmatrix} \qquad C = \begin{bmatrix} 0 & 1 & 0 \\ 0 & 0 & 0 \\ 0 & 0 & 1 \end{bmatrix}.$$

44 Suppose P contains only the r pivot columns of an m by n matrix. Explain why this m by r submatrix P has rank r.

45 Transpose P in Problem 44. Find the r pivot columns of P^{T} (which is r by m). Transposing back, **this produces an r by r invertible submatrix S inside P and A:**

$$\text{For } A = \begin{bmatrix} 1 & 2 & 3 \\ 2 & 4 & 6 \\ 2 & 4 & 7 \end{bmatrix} \text{ find } P \text{ (3 by 2) and then the invertible } S \text{ (2 by 2).}$$

Problems 46–51 show that $\operatorname{rank}(AB)$ is not greater than $\operatorname{rank}(A)$ or $\operatorname{rank}(B)$.

46 Find the ranks of AB and AC (rank one matrix times rank one matrix):

$$A = \begin{bmatrix} 1 & 2 \\ 2 & 4 \end{bmatrix} \quad \text{and} \quad B = \begin{bmatrix} 2 & 1 & 4 \\ 3 & 1.5 & 6 \end{bmatrix} \quad \text{and} \quad C = \begin{bmatrix} 1 & b \\ c & bc \end{bmatrix}.$$

47 The rank one matrix $\boldsymbol{u}\boldsymbol{v}^{\mathrm{T}}$ times the rank one matrix $\boldsymbol{w}\boldsymbol{z}^{\mathrm{T}}$ is $\boldsymbol{u}\boldsymbol{z}^{\mathrm{T}}$ times the number _____. This product $\boldsymbol{u}\boldsymbol{v}^{\mathrm{T}}\boldsymbol{w}\boldsymbol{z}^{\mathrm{T}}$ also has rank one unless _____ $= 0$.

48 (a) Suppose column j of B is a combination of previous columns of B. Show that column j of AB is the same combination of previous columns of AB. Then AB cannot have new pivot columns, so $\mathbf{rank}(\boldsymbol{AB}) \leq \mathbf{rank}(\boldsymbol{B})$.

(b) Find A_1 and A_2 so that $\text{rank}(A_1B) = 1$ and $\text{rank}(A_2B) = 0$ for $B = \left[\begin{smallmatrix}1&1\\1&1\end{smallmatrix}\right]$.

49 Problem 48 proved that $\text{rank}(AB) \leq \text{rank}(B)$. Then the same reasoning gives $\mathbf{rank}(\boldsymbol{B}^{\mathbf{T}}\boldsymbol{A}^{\mathbf{T}}) \leq \text{rank}(A^{\mathrm{T}})$. How do you deduce that $\mathbf{rank}(\boldsymbol{AB}) \leq \mathbf{rank}\ \boldsymbol{A}$?

50 *(Important)* Suppose A and B are n by n matrices, and $AB = I$. Prove from $\text{rank}(AB) \leq \text{rank}(A)$ that the rank of A is n. So A is invertible and B must be its two-sided inverse (Section 2.5). Therefore $BA = I$ *(which is not so obvious!)*.

51 If A is 2 by 3 and B is 3 by 2 and $AB = I$, show from its rank that $BA \neq I$. Give an example of A and B with $AB = I$. For $m < n$, a right inverse is not a left inverse.

52 Suppose A and B have the *same* reduced row echelon form R.

(a) Show that A and B have the same nullspace and the same row space.

(b) We know $E_1A = R$ and $E_2B = R$. So A equals an _____ matrix times B.

53 Express A and then B as the sum of two rank one matrices:

$$\textbf{rank} = \mathbf{2} \qquad A = \begin{bmatrix} 1 & 1 & 0 \\ 1 & 1 & 4 \\ 1 & 1 & 8 \end{bmatrix} \qquad B = \begin{bmatrix} 2 & 2 \\ 2 & 3 \end{bmatrix}.$$

54 Answer the same questions as in Worked Example **3.2 C** for

$$A = \begin{bmatrix} 1 & 1 & 2 & 2 \\ 2 & 2 & 4 & 4 \\ 1 & c & 2 & 2 \end{bmatrix} \quad \text{and} \quad B = \begin{bmatrix} 1-c & 2 \\ 0 & 2-c \end{bmatrix}.$$

55 What is the nullspace matrix N (containing the special solutions) for A, B, C?

$$\text{Block matrices} \quad A = \begin{bmatrix} I & I \end{bmatrix} \quad \text{and} \quad B = \begin{bmatrix} I & I \\ 0 & 0 \end{bmatrix} \quad \text{and} \quad C = \begin{bmatrix} I & I & I \end{bmatrix}.$$

56 *Neat fact* ***Every m by n matrix of rank r reduces to*** $(m$ by $r)$ ***times*** $(r$ by $n)$:

$$A = (\text{pivot columns of } A)\,(\text{first } r \text{ rows of } R) = (\mathbf{COL})(\mathbf{ROW}).$$

Write the 3 by 4 matrix A of all ones as the product of the 3 by 1 matrix from the pivot columns and the 1 by 4 matrix from R.

Challenge Problems

57 Suppose A is an m by n matrix of rank r. Its reduced echelon form is R. Describe exactly the matrix Z (its shape and all its entries) that comes from *transposing the reduced row echelon form of* R^{T}:

$$R = \text{rref}(A) \quad \text{and} \quad Z = (\text{rref}(A^{\mathrm{T}}))^{\mathrm{T}}.$$

58 (Recommended) Suppose R is m by n of rank r, with pivot columns first:

$$R = \begin{bmatrix} I & F \\ 0 & 0 \end{bmatrix}.$$

(a) What are the shapes of those four blocks?

(b) Find a *right-inverse* B with $RB = I$ if $r = m$. The zero blocks are gone.

(c) Find a *left-inverse* C with $CR = I$ if $r = n$. The F and 0 column is gone.

(d) What is the reduced row echelon form of R^{T} (with shapes)?

(e) What is the reduced row echelon form of $R^{\mathrm{T}}R$ (with shapes)?

59 I think that the reduced echelon form of $R^{\mathrm{T}}R$ is always R (except for extra zero rows). Can you do an example when R is 2 by 3? Later we show that $A^{\mathrm{T}}A$ always has the same nullspace as A (a valuable fact).

60 Suppose you allow elementary *column* operations on A as well as elementary row operations (which get to R). What is the "row-and-column reduced form" for an m by n matrix of rank r?

Elimination: The Big Picture

This page explains elimination at the vector level and subspace level, when A is reduced to R. You know the steps and I won't repeat them. Elimination starts with the first pivot. It moves a column at a time (left to right) and a row at a time (top to bottom). As it moves, elimination answers two questions:

Question 1 Is this column a combination of previous columns?

If the column contains a pivot, the answer is no. Pivot columns are "independent" of previous columns. If column 4 has no pivot, it is a combination of columns $1, 2, 3$.

Question 2 Is this row a combination of previous rows?

If the row contains a pivot, the answer is no. Pivot rows are "independent" of previous rows. If row 3 ends up with no pivot, it is a zero row and it is moved to the bottom of R.

It is amazing to me that one pass through the matrix answers both questions. Actually that pass reaches the triangular echelon matrix U, not the reduced echelon matrix R. Then the reduction from U to R goes bottom to top. U tells which columns are combinations of earlier columns (pivots are missing). Then *R tells us what those combinations are.*

In other words, **$\boldsymbol{R}$ tells us the special solutions to $\boldsymbol{Ax} = \mathbf{0}$**. We could reach R from A by different row exchanges and elimination steps, but it will always be the same R (because the special solutions are decided by A). In the language coming soon, R reveals a "basis" for three fundamental subspaces:

The **column space** of A—choose the pivot columns of A as a basis.

The **row space** of A—choose the nonzero rows of R as a basis.

The **nullspace** of A—choose the special solutions to $R\boldsymbol{x} = \mathbf{0}$ (and $A\boldsymbol{x} = \mathbf{0}$).

We learn from elimination the single most important number—**the rank $\boldsymbol{r}$**. That number counts the pivot columns and the pivot rows. Then $n - r$ counts the free columns and the special solutions.

I mention that reducing $[A \;\; I]$ to $[R \;\; E]$ will tell you even more about A—in fact virtually everything (including $EA = R$). The matrix E keeps a record, otherwise lost, of the elimination from A to R. When A is square and invertible, R is I and E is A^{-1}.

3.3 The Complete Solution to $Ax = b$

1 Complete solution to $A\boldsymbol{x} = \boldsymbol{b}$: $\boldsymbol{x} =$ (one particular solution $\boldsymbol{x}_p$) + (any $\boldsymbol{x}_n$ in the nullspace).

2 Elimination on $\begin{bmatrix} A & \boldsymbol{b} \end{bmatrix}$ leads to $\begin{bmatrix} R & \boldsymbol{d} \end{bmatrix}$. Then $A\boldsymbol{x} = \boldsymbol{b}$ is equivalent to $R\boldsymbol{x} = \boldsymbol{d}$.

3 $A\boldsymbol{x} = \boldsymbol{b}$ and $R\boldsymbol{x} = \boldsymbol{d}$ are solvable only when all zero rows of R have zeros in $\boldsymbol{d}$.

4 When $R\boldsymbol{x} = \boldsymbol{d}$ is solvable, one very particular solution $\boldsymbol{x}_p$ has all free variables equal to zero.

5 A has **full column rank** $\boldsymbol{r} = \boldsymbol{n}$ when its nullspace $\boldsymbol{N}(A) =$ zero vector: *no free variables*.

6 A has **full row rank** $\boldsymbol{r} = \boldsymbol{m}$ when its column space $\boldsymbol{C}(A)$ is $\mathbf{R}^m$: $A\boldsymbol{x} = \boldsymbol{b}$ is always solvable.

7 The four cases are $r = m = n$ (A is invertible) and $r = m < n$ (every $A\boldsymbol{x} = \boldsymbol{b}$ is solvable) and $r = n < m$ ($A\boldsymbol{x} = \boldsymbol{b}$ has 1 or 0 solutions) and $r < m, r < n$ (0 or ∞ solutions).

The last section totally solved $A\boldsymbol{x} = \mathbf{0}$. Elimination converted the problem to $R\boldsymbol{x} = \mathbf{0}$. The free variables were given special values (one and zero). Then the pivot variables were found by back substitution. We paid no attention to the right side $\boldsymbol{b}$ because it stayed at zero. The solution $\boldsymbol{x}$ was in the nullspace of A.

Now $\boldsymbol{b}$ is not zero. Row operations on the left side must act also on the right side. $A\boldsymbol{x} = \boldsymbol{b}$ is reduced to a simpler system $R\boldsymbol{x} = \boldsymbol{d}$ with the same solutions. One way to organize that is to ***add b as an extra column of the matrix***. I will *"augment"* A with the right side $(b_1, b_2, b_3) = (1, 6, 7)$ to produce the **augmented matrix** $\begin{bmatrix} A & \boldsymbol{b} \end{bmatrix}$:

$$\begin{bmatrix} 1 & 3 & 0 & 2 \\ 0 & 0 & 1 & 4 \\ 1 & 3 & 1 & 6 \end{bmatrix} \begin{bmatrix} x_1 \\ x_2 \\ x_3 \\ x_4 \end{bmatrix} = \begin{bmatrix} \mathbf{1} \\ \mathbf{6} \\ \mathbf{7} \end{bmatrix} \quad \begin{matrix} \text{has the} \\ \text{augmented} \\ \text{matrix} \end{matrix} \quad \begin{bmatrix} 1 & 3 & 0 & 2 & \mathbf{1} \\ 0 & 0 & 1 & 4 & \mathbf{6} \\ 1 & 3 & 1 & 6 & \mathbf{7} \end{bmatrix} = \begin{bmatrix} A & \boldsymbol{b} \end{bmatrix}.$$

When we apply the usual elimination steps to A, reaching R, we also apply them to $\boldsymbol{b}$.

In this example we subtract row 1 from row 3. Then we subtract row 2 from row 3. This produces a *row of zeros in* R, and it changes $\boldsymbol{b}$ to a new right side $\boldsymbol{d} = (1, 6, 0)$:

$$\begin{bmatrix} 1 & 3 & 0 & 2 \\ 0 & 0 & 1 & 4 \\ 0 & 0 & 0 & 0 \end{bmatrix} \begin{bmatrix} x_1 \\ x_2 \\ x_3 \\ x_4 \end{bmatrix} = \begin{bmatrix} \mathbf{1} \\ \mathbf{6} \\ \mathbf{0} \end{bmatrix} \quad \begin{matrix} \text{has the} \\ \text{augmented} \\ \text{matrix} \end{matrix} \quad \begin{bmatrix} 1 & 3 & 0 & 2 & \mathbf{1} \\ 0 & 0 & 1 & 4 & \mathbf{6} \\ 0 & 0 & 0 & 0 & \mathbf{0} \end{bmatrix} = \begin{bmatrix} R & \boldsymbol{d} \end{bmatrix}.$$

That very last zero is crucial. The third equation has become $0 = 0$. So the equations can be solved. In the original matrix A, the first row plus the second row equals the third row. If the equations are consistent, this must be true on the right side of the equations also! The all-important property of the right side $\boldsymbol{b}$ was $\mathbf{1} + \mathbf{6} = \mathbf{7}$. That led to $0 = 0$.

Here are the same augmented matrices for a general $\boldsymbol{b} = (b_1, b_2, b_3)$:

$$\begin{bmatrix} A & \boldsymbol{b} \end{bmatrix} = \begin{bmatrix} 1 & 3 & 0 & 2 & \boldsymbol{b}_1 \\ 0 & 0 & 1 & 4 & \boldsymbol{b}_2 \\ 1 & 3 & 1 & 6 & \boldsymbol{b}_3 \end{bmatrix} \longrightarrow \begin{bmatrix} 1 & 3 & 0 & 2 & \boldsymbol{b}_1 \\ 0 & 0 & 1 & 4 & \boldsymbol{b}_2 \\ 0 & 0 & 0 & 0 & \boldsymbol{b}_3 - \boldsymbol{b}_1 - \boldsymbol{b}_2 \end{bmatrix} = \begin{bmatrix} R & \boldsymbol{d} \end{bmatrix}$$

Now we get $0 = 0$ in the third equation only if $b_3 - b_1 - b_2 = 0$. This is $b_1 + b_2 = b_3$.

One Particular Solution $A\boldsymbol{x}_p = \boldsymbol{b}$

For an easy solution $\boldsymbol{x}_p$, *choose the free variables to be zero:* $x_2 = x_4 = 0$. Then the two nonzero equations give the two pivot variables $x_1 = 1$ and $x_3 = 6$. Our particular solution to $A\boldsymbol{x} = \boldsymbol{b}$ (and also $R\boldsymbol{x} = \boldsymbol{d}$) is $\boldsymbol{x}_p = (1, 0, 6, 0)$. This particular solution is my favorite: ***free variables* = *zero*, *pivot variables from* d.** The method always works.

For a solution to exist, zero rows in R must also be zero in d. Since I is in the pivot rows and pivot columns of R, the pivot variables in $x_{\text{particular}}$ come from d:

$$R\boldsymbol{x}_p = \begin{bmatrix} \mathbf{1} & 3 & \mathbf{0} & 2 \\ \mathbf{0} & 0 & \mathbf{1} & 4 \\ 0 & 0 & 0 & 0 \end{bmatrix} \begin{bmatrix} \mathbf{1} \\ 0 \\ \mathbf{6} \\ 0 \end{bmatrix} = \begin{bmatrix} \mathbf{1} \\ \mathbf{6} \\ 0 \end{bmatrix}$$

Pivot variables $1, 6$
Free variables $0, 0$
Solution $\boldsymbol{x}_p = (\mathbf{1}, \mathbf{0}, \mathbf{6}, \mathbf{0})$.

Notice how we *choose* the free variables (as zero) and *solve* for the pivot variables. After the row reduction to R, those steps are quick. When the free variables are zero, the pivot variables for $\boldsymbol{x}_p$ are already seen in the right side vector $\boldsymbol{d}$.

$x_{\text{particular}}$	***The particular solution solves***	$A\boldsymbol{x}_p = \boldsymbol{b}$
$x_{\text{nullspace}}$	***The $n - r$ special solutions solve***	$A\boldsymbol{x}_n = \mathbf{0}$.

That particular solution is $(1, 0, 6, 0)$. The two special (nullspace) solutions to $R\boldsymbol{x} = \mathbf{0}$ come from the two free columns of R, by reversing signs of $3, 2$, and 4. ***Please notice how I write the complete solution $x_p + x_n$ to $Ax = b$:***

Complete solution
one $\boldsymbol{x}_p$
many $\boldsymbol{x}_n$

$$\boldsymbol{x} = \boldsymbol{x}_p + \boldsymbol{x}_n = \begin{bmatrix} 1 \\ 0 \\ 6 \\ 0 \end{bmatrix} + x_2 \begin{bmatrix} -3 \\ 1 \\ 0 \\ 0 \end{bmatrix} + x_4 \begin{bmatrix} -2 \\ 0 \\ -4 \\ 1 \end{bmatrix}.$$

Question Suppose A is a square invertible matrix, $m = n = r$. What are $\boldsymbol{x}_p$ and $\boldsymbol{x}_n$?
Answer The particular solution is the one and *only* solution $\boldsymbol{x}_p = A^{-1}\boldsymbol{b}$. There are no special solutions or free variables. $R = I$ has no zero rows. The only vector in the nullspace is $\boldsymbol{x}_n = \mathbf{0}$. The complete solution is $\boldsymbol{x} = \boldsymbol{x}_p + \boldsymbol{x}_n = A^{-1}\boldsymbol{b} + \mathbf{0}$.

We didn't mention the nullspace in Chapter 2, because A was invertible and $\boldsymbol{N}(A)$ contained only the zero vector. Reduction went from $\begin{bmatrix} A & \boldsymbol{b} \end{bmatrix}$ to $\begin{bmatrix} I & A^{-1}\boldsymbol{b} \end{bmatrix}$. The matrix A was reduced all the way to I. Then $A\boldsymbol{x} = \boldsymbol{b}$ became $\boldsymbol{x} = A^{-1}\boldsymbol{b}$ which is $\boldsymbol{d}$. This is a special case here, but square invertible matrices are the ones we see most often in practice. So they got their own chapter at the start of the book.

For small examples we can reduce $\begin{bmatrix} A & \boldsymbol{b} \end{bmatrix}$ to $\begin{bmatrix} R & \boldsymbol{d} \end{bmatrix}$. For a large matrix, MATLAB does it better. One particular solution (not necessarily ours) is $\boldsymbol{x} = A\backslash\boldsymbol{b}$ from backslash. Here is an example with *full column rank*. Both columns have pivots.

Example 1 Find the condition on (b_1, b_2, b_3) for $A\boldsymbol{x} = \boldsymbol{b}$ to be solvable, if

$$A = \begin{bmatrix} 1 & 1 \\ 1 & 2 \\ -2 & -3 \end{bmatrix} \quad \text{and} \quad \boldsymbol{b} = \begin{bmatrix} b_1 \\ b_2 \\ b_3 \end{bmatrix}.$$

This condition puts $\boldsymbol{b}$ in the column space of A. Find the complete $\boldsymbol{x} = \boldsymbol{x}_p + \boldsymbol{x}_n$.

Solution Use the augmented matrix, with its extra column $\boldsymbol{b}$. Subtract row 1 of $\begin{bmatrix} A & \boldsymbol{b} \end{bmatrix}$ from row 2. Then add 2 times row 1 to row 3 to reach $\begin{bmatrix} R & \boldsymbol{d} \end{bmatrix}$:

$$\begin{bmatrix} 1 & 1 & b_1 \\ 1 & 2 & b_2 \\ -2 & -3 & b_3 \end{bmatrix} \to \begin{bmatrix} 1 & 1 & b_1 \\ 0 & 1 & b_2 - b_1 \\ 0 & -1 & b_3 + 2b_1 \end{bmatrix} \to \begin{bmatrix} 1 & 0 & 2b_1 - b_2 \\ 0 & 1 & b_2 - b_1 \\ 0 & 0 & \boldsymbol{b_3 + b_1 + b_2} \end{bmatrix}.$$

The last equation is $0 = 0$ provided $b_3 + b_1 + b_2 = 0$. This is the condition to put $\boldsymbol{b}$ in the column space. Then $A\boldsymbol{x} = \boldsymbol{b}$ will be solvable. The rows of A add to the zero row. So for consistency (these are equations!) the entries of $\boldsymbol{b}$ must also add to zero.

This example has no free variables since $n - r = 2 - 2$. Therefore no special solutions. The nullspace solution is $\boldsymbol{x}_n = \boldsymbol{0}$. The particular solution to $A\boldsymbol{x} = \boldsymbol{b}$ and $R\boldsymbol{x} = \boldsymbol{d}$ is at the top of the final column $\boldsymbol{d}$:

$$\textbf{Only solution to } A\boldsymbol{x} = \boldsymbol{b} \qquad \boldsymbol{x} = \boldsymbol{x}_p + \boldsymbol{x}_n = \begin{bmatrix} 2b_1 - b_2 \\ b_2 - b_1 \end{bmatrix} + \begin{bmatrix} 0 \\ 0 \end{bmatrix}.$$

If $b_3 + b_1 + b_2$ is not zero, there is no solution to $A\boldsymbol{x} = \boldsymbol{b}$ ($\boldsymbol{x}_p$ and $\boldsymbol{x}$ don't exist).

This example is typical of an extremely important case: A has *full column rank*. Every column has a pivot. *The rank is* $r = n$. The matrix is tall and thin ($m \geq n$). Row reduction puts I at the top, when A is reduced to R with rank n:

$$\textbf{Full column rank} \quad R = \begin{bmatrix} I \\ 0 \end{bmatrix} = \begin{bmatrix} n \text{ by } n \text{ identity matrix} \\ m - n \text{ rows of zeros} \end{bmatrix} \tag{1}$$

There are no free columns or free variables. The nullspace is $\mathbf{Z} = \{\text{zero vector}\}$.

We will collect together the different ways of recognizing this type of matrix.

Every matrix A with **full column rank** $(r = n)$ has all these properties:

1. All columns of A are pivot columns.
2. There are no free variables or special solutions.
3. The nullspace $N(A)$ contains only the zero vector $x = 0$.
4. If $Ax = b$ has a solution (it might not) then it has only *one solution*.

In the essential language of the next section, **this A has *independent columns*.** $Ax = 0$ only happens when $x = 0$. In Chapter 4 we will add one more fact to the list: *The square matrix $A^T A$ is invertible when the rank is n.*

In this case the nullspace of A (and R) has shrunk to the zero vector. The solution to $Ax = b$ is *unique* (if it exists). There will be $m - n$ zero rows in R. So there are $m - n$ conditions on b in order to have $0 = 0$ in those rows, and b in the column space. With full column rank, $Ax = b$ has *one* solution or *no* solution ($m > n$ is overdetermined).

The Complete Solution

The other extreme case is full row rank. Now $Ax = b$ has *one or infinitely many* solutions. In this case A must be *short and wide* ($m \le n$). ***A matrix has full row rank if $r = m$.*** "*The rows are independent.*" Every row has a pivot, and here is an example.

Example 2 This system $Ax = b$ has $n = 3$ unknowns but only $m = 2$ equations:

Full row rank
$$\begin{aligned} x + y + z &= 3 \\ x + 2y - z &= 4 \end{aligned} \qquad (\text{rank } r = m = 2)$$

These are two planes in xyz space. The planes are not parallel so they intersect in a line. This line of solutions is exactly what elimination will find. ***The particular solution will be one point on the line. Adding the nullspace vectors x_n will move us along the line in Figure 3.3***. Then $x = x_p + x_n$ gives the whole line of solutions.

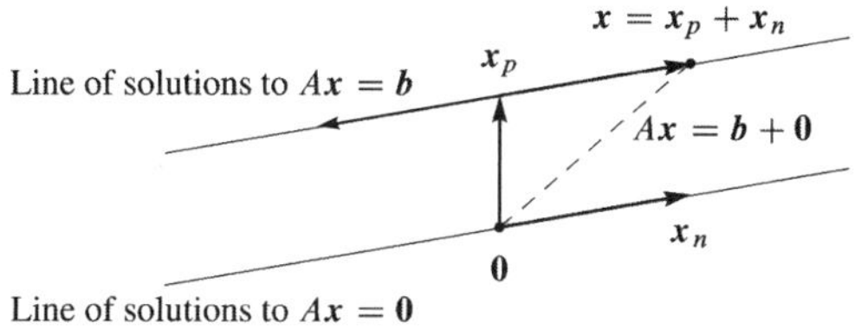

Figure 3.3: Complete solution = *one* particular solution + *all* nullspace solutions.

We find $\boldsymbol{x}_p$ and $\boldsymbol{x}_n$ by elimination on $\begin{bmatrix} A & \boldsymbol{b} \end{bmatrix}$. Subtract row 1 from row 2 and then subtract row 2 from row 1:

$$\begin{bmatrix} 1 & 1 & 1 & \mathbf{3} \\ 1 & 2 & -1 & \mathbf{4} \end{bmatrix} \to \begin{bmatrix} 1 & 1 & 1 & \mathbf{3} \\ 0 & 1 & -2 & \mathbf{1} \end{bmatrix} \to \begin{bmatrix} 1 & 0 & 3 & \mathbf{2} \\ 0 & 1 & -2 & \mathbf{1} \end{bmatrix} = \begin{bmatrix} R & \boldsymbol{d} \end{bmatrix}.$$

The particular solution has free variable $x_3 = 0$. The special solution has $x_3 = 1$:

$\boldsymbol{x}_{\text{particular}}$ comes directly from $\boldsymbol{d}$ on the right side: $\boldsymbol{x}_p = (2, 1, 0)$
$\boldsymbol{x}_{\text{special}}$ comes from the third column (free column) of R: $\boldsymbol{s} = (-3, 2, 1)$

It is wise to check that $\boldsymbol{x}_p$ and $\boldsymbol{s}$ satisfy the original equations $A\boldsymbol{x}_p = \boldsymbol{b}$ and $A\boldsymbol{s} = \mathbf{0}$:

$$\begin{aligned} 2+1 &= 3 \qquad & -3+2+1 &= 0 \\ 2+2 &= 4 \qquad & -3+4-1 &= 0 \end{aligned}$$

The nullspace solution $\boldsymbol{x}_n$ is any multiple of $\boldsymbol{s}$. It moves along the line of solutions, starting at $\boldsymbol{x}_{\text{particular}}$. ***Please notice again how to write the answer***:

Complete solution $$\boldsymbol{x} = \boldsymbol{x}_p + \boldsymbol{x}_n = \begin{bmatrix} 2 \\ 1 \\ 0 \end{bmatrix} + x_3 \begin{bmatrix} -3 \\ 2 \\ 1 \end{bmatrix}.$$

This line of solutions is drawn in Figure 3.3. Any point on the line *could* have been chosen as the particular solution. We chose the point with $x_3 = 0$.

The particular solution is *not* multiplied by an arbitrary constant! The special solution needs that constant, and you understand why—to produce all $\boldsymbol{x}_n$ in the nullspace.

Now we summarize this short wide case of *full row rank*. If $m < n$ the equation $A\boldsymbol{x} = \boldsymbol{b}$ is **underdetermined** (many solutions).

Every matrix A with ***full row rank*** $(\boldsymbol{r} = \boldsymbol{m})$ has all these properties:

1. All rows have pivots, and $\boldsymbol{R}$ **has no zero rows.**
2. $A\boldsymbol{x} = \boldsymbol{b}$ has a **solution for every right side** $\boldsymbol{b}$.
3. The column space is the whole space $\mathbf{R}^m$.
4. There are $n - r = n - m$ special solutions in the nullspace of A.

In this case with m pivots, the rows are "***linearly independent***". So the columns of A^{T} are linearly independent. The nullspace of A^{T} is the zero vector.

We are ready for the definition of linear independence, as soon as we summarize the four possibilities—which depend on the rank. Notice how r, m, n are the critical numbers.

The four possibilities for linear equations depend on the rank r

$\boldsymbol{r = m}$	and	$\boldsymbol{r = n}$	*Square and invertible*	$Ax = b$	has 1 solution
$\boldsymbol{r = m}$	and	$r < n$	*Short and wide*	$Ax = b$	has ∞ solutions
$r < m$	and	$\boldsymbol{r = n}$	*Tall and thin*	$Ax = b$	has 0 or 1 solution
$r < m$	and	$r < n$	*Not full rank*	$Ax = b$	has 0 or ∞ solutions

The reduced R will fall in the same category as the matrix A. In case the pivot columns happen to come first, we can display these four possibilities for R. For $Rx = d$ (and the original $Ax = b$) to be solvable, d must end in $m - r$ zeros. F is the free part of R.

Four types for R	$\begin{bmatrix} I \end{bmatrix}$	$\begin{bmatrix} I & F \end{bmatrix}$	$\begin{bmatrix} I \\ 0 \end{bmatrix}$	$\begin{bmatrix} I & F \\ 0 & 0 \end{bmatrix}$
Their ranks	$\boldsymbol{r = m = n}$	$\boldsymbol{r = m < n}$	$\boldsymbol{r = n < m}$	$\boldsymbol{r < m, r < n}$

Cases 1 and 2 have full row rank $r = m$. Cases 1 and 3 have full column rank $r = n$. Case 4 is the most general in theory and it is the least common in practice.

■ REVIEW OF THE KEY IDEAS ■

1. The rank r is the number of pivots. The matrix R has $m - r$ zero rows.

2. $Ax = b$ is solvable if and only if the last $m - r$ equations reduce to $0 = 0$.

3. One particular solution x_p has all free variables equal to zero.

4. The pivot variables are determined after the free variables are chosen.

5. Full column rank $r = n$ means no free variables: one solution or none.

6. Full row rank $r = m$ means one solution if $m = n$ or infinitely many if $m < n$.

■ WORKED EXAMPLES ■

3.3 A This question connects elimination (**pivot columns and back substitution**) to **column space-nullspace-rank-solvability** (the higher level picture). A has rank 2:

$$Ax = b \quad \text{is} \quad \begin{aligned} x_1 + 2x_2 + 3x_3 + 5x_4 &= b_1 \\ 2x_1 + 4x_2 + 8x_3 + 12x_4 &= b_2 \\ 3x_1 + 6x_2 + 7x_3 + 13x_4 &= b_3 \end{aligned}$$

1. Reduce $[\,A \quad \boldsymbol{b}\,]$ to $[\,U \quad \boldsymbol{c}\,]$, so that $A\boldsymbol{x} = \boldsymbol{b}$ becomes a triangular system $U\boldsymbol{x} = \boldsymbol{c}$.
2. Find the condition on b_1, b_2, b_3 for $A\boldsymbol{x} = \boldsymbol{b}$ to have a solution.
3. Describe the column space of A. Which plane in $\mathbf{R}^3$?
4. Describe the nullspace of A. Which special solutions in $\mathbf{R}^4$?
5. Reduce $[\,U \quad c\,]$ to $[\,R \quad \boldsymbol{d}\,]$: Special solutions from R, particular solution from $\boldsymbol{d}$.
6. Find a particular solution to $A\boldsymbol{x} = (0, 6, -6)$ and then the complete solution.

Solution

1. The multipliers in elimination are 2 and 3 and -1. They take $[\,A \quad \boldsymbol{b}\,]$ into $[\,U \quad \boldsymbol{c}\,]$.

$$\begin{bmatrix} 1 & 2 & 3 & 5 & \mathbf{b_1} \\ 2 & 4 & 8 & 12 & \mathbf{b_2} \\ 3 & 6 & 7 & 13 & \mathbf{b_3} \end{bmatrix} \rightarrow \left[\begin{array}{cccc|c} 1 & 2 & 3 & 5 & \mathbf{b_1} \\ 0 & 0 & 2 & 2 & \mathbf{b_2 - 2b_1} \\ 0 & 0 & -2 & -2 & \mathbf{b_3 - 3b_1} \end{array}\right] \rightarrow \left[\begin{array}{cccc|c} 1 & 2 & 3 & 5 & \mathbf{b_1} \\ 0 & 0 & 2 & 2 & \mathbf{b_2 - 2b_1} \\ 0 & 0 & 0 & 0 & \mathbf{b_3 + b_2 - 5b_1} \end{array}\right]$$

2. The last equation shows the solvability condition $b_3 + b_2 - 5b_1 = 0$. Then $0 = 0$.
3. **First description**: The column space is the plane containing all combinations of the pivot columns $(1, 2, 3)$ and $(3, 8, 7)$. The pivots are in columns 1 and 3. **Second description**: The column space contains all vectors with $b_3 + b_2 - 5b_1 = 0$. That makes $A\boldsymbol{x} = \boldsymbol{b}$ solvable, so $\boldsymbol{b}$ is in the column space. *All columns of A pass this test $b_3 + b_2 - 5b_1 = 0$. This is the equation for the plane in the first description*!
4. The special solutions have free variables $x_2 = 1, x_4 = 0$ and then $x_2 = 0, x_4 = 1$:

Special solutions to $A\boldsymbol{x} = \mathbf{0}$
Back substitution in $U\boldsymbol{x} = \mathbf{0}$
or change signs of 2, 2, 1 in R

$$\boldsymbol{s}_1 = \begin{bmatrix} -2 \\ 1 \\ 0 \\ 0 \end{bmatrix} \qquad \boldsymbol{s}_2 = \begin{bmatrix} -2 \\ 0 \\ -1 \\ 1 \end{bmatrix}$$

The nullspace $\boldsymbol{N}(A)$ in $\mathbf{R}^4$ contains all $\boldsymbol{x}_n = c_1\boldsymbol{s}_1 + c_2\boldsymbol{s}_2$.

5. In the reduced form R, the third column changes from $(3, 2, 0)$ in U to $(0, 1, 0)$. The right side $\boldsymbol{c} = (0, 6, 0)$ becomes $\boldsymbol{d} = (-9, 3, 0)$ showing -9 and 3 in $\boldsymbol{x}_p$:

$$[\,U \quad \boldsymbol{c}\,] = \begin{bmatrix} 1 & 2 & 3 & 5 & \mathbf{0} \\ 0 & 0 & 2 & 2 & \mathbf{6} \\ 0 & 0 & 0 & 0 & \mathbf{0} \end{bmatrix} \longrightarrow [\,R \quad \boldsymbol{d}\,] = \begin{bmatrix} 1 & 2 & 0 & 2 & \mathbf{-9} \\ 0 & 0 & 1 & 1 & \mathbf{3} \\ 0 & 0 & 0 & 0 & \mathbf{0} \end{bmatrix}$$

6. One particular solution $\boldsymbol{x}_p$ has free variables = zero. Back substitute in $U\boldsymbol{x} = \boldsymbol{c}$:

Particular solution to $A\boldsymbol{x}_p = \boldsymbol{b}$
Bring -9 and 3 from the vector $\boldsymbol{d}$
Free variables x_2 and x_4 are zero

$$\boldsymbol{x}_p = \begin{bmatrix} -9 \\ 0 \\ 3 \\ 0 \end{bmatrix}$$

The complete solution to $A\boldsymbol{x} = (0, 6, -6)$ is $\boldsymbol{x} = \boldsymbol{x}_p + \boldsymbol{x}_n = \boldsymbol{x}_p + c_1\boldsymbol{s}_1 + c_2\boldsymbol{s}_2$.

3.3 B Suppose you have this information about the solutions to $A\boldsymbol{x} = \boldsymbol{b}$ for a specific $\boldsymbol{b}$. What does that tell you about m and n and r (and A itself)? And possibly about $\boldsymbol{b}$.

1. There is exactly one solution.
2. All solutions to $A\boldsymbol{x} = \boldsymbol{b}$ have the form $\boldsymbol{x} = \begin{bmatrix} 2 \\ 1 \end{bmatrix} + c\begin{bmatrix} 1 \\ 1 \end{bmatrix}$.
3. There are no solutions.
4. All solutions to $A\boldsymbol{x} = \boldsymbol{b}$ have the form $\boldsymbol{x} = \begin{bmatrix} 1 \\ 1 \\ 0 \end{bmatrix} + c\begin{bmatrix} 1 \\ 0 \\ 1 \end{bmatrix}$
5. There are infinitely many solutions.

Solution In case $\mathbf{1}$, with exactly one solution, A must have full column rank $r = n$. The nullspace of A contains only the zero vector. Necessarily $m \geq n$.

In case $\mathbf{2}$, A must have $n = 2$ columns (and m is arbitrary). With $\begin{bmatrix} 1 \\ 1 \end{bmatrix}$ in the nullspace of A, column 2 is the *negative* of column 1. Also $A \neq 0$: the rank is 1. With $\boldsymbol{x} = \begin{bmatrix} 2 \\ 1 \end{bmatrix}$ as a solution, $\boldsymbol{b} = 2(\text{column } 1) + (\text{column } 2)$. My choice for $\boldsymbol{x}_p$ would be $(1, 0)$.

In case $\mathbf{3}$ we only know that $\boldsymbol{b}$ is not in the column space of A. The rank of A must be less than m. I guess we know $\boldsymbol{b} \neq \mathbf{0}$, otherwise $\boldsymbol{x} = \mathbf{0}$ would be a solution.

In case $\mathbf{4}$, A must have $n = 3$ columns. With $(1, 0, 1)$ in the nullspace of A, column 3 is the negative of column 1. Column 2 must *not* be a multiple of column 1, or the nullspace would contain another special solution. So the rank of A is $3 - 1 = 2$. Necessarily A has $m \geq 2$ rows. The right side $\boldsymbol{b}$ is column $1 +$ column 2.

In case $\mathbf{5}$ with infinitely many solutions, the nullspace must contain nonzero vectors. The rank r must be less than n (not full column rank), and $\boldsymbol{b}$ must be in the column space of A. We don't know if *every* $\boldsymbol{b}$ is in the column space, so we don't know if $r = m$.

3.3 C Find the complete solution $\boldsymbol{x} = \boldsymbol{x_p} + \boldsymbol{x_n}$ by forward elimination on $[A \;\; \boldsymbol{b}]$:

$$\begin{bmatrix} 1 & 2 & 1 & 0 \\ 2 & 4 & 4 & 8 \\ 4 & 8 & 6 & 8 \end{bmatrix} \begin{bmatrix} x_1 \\ x_2 \\ x_3 \\ x_4 \end{bmatrix} = \begin{bmatrix} 4 \\ 2 \\ 10 \end{bmatrix}.$$

Find numbers y_1, y_2, y_3 so that y_1 (row 1) $+ \, y_2$ (row 2) $+ \, y_3$ (row 3) $=$ ***zero row***. Check that $\boldsymbol{b} = (4, 2, 10)$ satisfies the condition $y_1 b_1 + y_2 b_2 + y_3 b_3 = 0$. Why is this the condition for the equations to be solvable and $\boldsymbol{b}$ to be in the column space?

Solution Forward elimination on $[A \;\; \boldsymbol{b}]$ produces a zero row in $[U \;\; \boldsymbol{c}]$. The third equation becomes $0 = 0$ and the equations are consistent (and solvable):

$$\begin{bmatrix} 1 & 2 & 1 & 0 & \mathbf{4} \\ 2 & 4 & 4 & 8 & \mathbf{2} \\ 4 & 8 & 6 & 8 & \mathbf{10} \end{bmatrix} \longrightarrow \begin{bmatrix} 1 & 2 & 1 & 0 & \mathbf{4} \\ 0 & 0 & 2 & 8 & \mathbf{-6} \\ 0 & 0 & 2 & 8 & \mathbf{-6} \end{bmatrix} \longrightarrow \begin{bmatrix} 1 & 2 & 1 & 0 & \mathbf{4} \\ 0 & 0 & 2 & 8 & \mathbf{-6} \\ 0 & 0 & 0 & 0 & \mathbf{0} \end{bmatrix}.$$

Columns 1 and 3 contain pivots. The variables x_2 and x_4 are free. If we set those to zero we can solve (back substitution) for the particular solution or we continue to R.

$R\boldsymbol{x} = \boldsymbol{d}$ shows that the particular solution with free variables $= 0$ is $\boldsymbol{x_p} = (7, 0, -3, 0)$.

$$\begin{bmatrix} 1 & 2 & 1 & 0 & \mathbf{4} \\ 0 & 0 & 2 & 8 & \mathbf{-6} \\ 0 & 0 & 0 & 0 & \mathbf{0} \end{bmatrix} \longrightarrow \begin{bmatrix} 1 & 2 & 1 & 0 & \mathbf{4} \\ 0 & 0 & 1 & 4 & \mathbf{-3} \\ 0 & 0 & 0 & 0 & \mathbf{0} \end{bmatrix} \longrightarrow \begin{bmatrix} 1 & 2 & 0 & -4 & \mathbf{7} \\ 0 & 0 & 1 & 4 & \mathbf{-3} \\ 0 & 0 & 0 & 0 & \mathbf{0} \end{bmatrix}.$$

For the nullspace part $\boldsymbol{x_n}$ with $\boldsymbol{b} = \mathbf{0}$, set the free variables x_2, x_4 to $1, 0$ and also $0, 1$:

Special solutions $\quad s_1 = (-2, 1, 0, 0)$ **and** $s_2 = (4, 0, -4, 1)$

Then the complete solution to $A\boldsymbol{x} = \boldsymbol{b}$ (and $R\boldsymbol{x} = \boldsymbol{d}$) is $\boldsymbol{x}_{\text{complete}} = \boldsymbol{x_p} + c_1 \boldsymbol{s}_1 + c_2 \boldsymbol{s}_2$.

The rows of A produced the zero row from $2(\text{row } 1) + (\text{row } 2) - (\text{row } 3) = (0, 0, 0, 0)$. Thus $\boldsymbol{y} = (2, 1, -1)$. The same combination for $\boldsymbol{b} = (4, 2, 10)$ gives $2(4) + (2) - (10) = 0$.

If a combination of the rows (on the left side) gives the zero row, then the same combination must give zero on the right side. Of course! *Otherwise no solution.*

Later we will say this again in different words: If every column of A is perpendicular to $\boldsymbol{y} = (2, 1, -1)$, then any combination $\boldsymbol{b}$ of those columns must also be perpendicular to $\boldsymbol{y}$. Otherwise $\boldsymbol{b}$ is not in the column space and $A\boldsymbol{x} = \boldsymbol{b}$ is not solvable.

And again: If $\boldsymbol{y}$ is in the nullspace of A^{T} then $\boldsymbol{y}$ must be perpendicular to every $\boldsymbol{b}$ in the column space of A. Just looking ahead...

Problem Set 3.3

1 (Recommended) Execute the six steps of Worked Example **3.3 A** to describe the column space and nullspace of A and the complete solution to $A\boldsymbol{x} = \boldsymbol{b}$:

$$A = \begin{bmatrix} 2 & 4 & 6 & 4 \\ 2 & 5 & 7 & 6 \\ 2 & 3 & 5 & 2 \end{bmatrix} \qquad \boldsymbol{b} = \begin{bmatrix} b_1 \\ b_2 \\ b_3 \end{bmatrix} = \begin{bmatrix} 4 \\ 3 \\ 5 \end{bmatrix}$$

2 Carry out the same six steps for this matrix A with rank one. You will find *two* conditions on b_1, b_2, b_3 for $A\boldsymbol{x} = \boldsymbol{b}$ to be solvable. Together these two conditions put $\boldsymbol{b}$ into the _____ space (two planes give a line):

$$A = \begin{bmatrix} 1 \\ 3 \\ 2 \end{bmatrix} \begin{bmatrix} 2 & 1 & 3 \end{bmatrix} = \begin{bmatrix} 2 & 1 & 3 \\ 6 & 3 & 9 \\ 4 & 2 & 6 \end{bmatrix} \qquad \boldsymbol{b} = \begin{bmatrix} b_1 \\ b_2 \\ b_3 \end{bmatrix} = \begin{bmatrix} 10 \\ 30 \\ 20 \end{bmatrix}$$

Questions 3–15 are about the solution of $A\boldsymbol{x} = \boldsymbol{b}$. Follow the steps in the text to $\boldsymbol{x_p}$ and $\boldsymbol{x_n}$. Start from the augmented matrix with last column $\boldsymbol{b}$.

3 Write the complete solution as $\boldsymbol{x_p}$ plus any multiple of $\boldsymbol{s}$ in the nullspace:

$$\begin{aligned} x + 3y + 3z &= 1 \\ 2x + 6y + 9z &= 5 \\ -x - 3y + 3z &= 5. \end{aligned}$$

4 Find the complete solution (also called the *general solution*) to

$$\begin{bmatrix} 1 & 3 & 1 & 2 \\ 2 & 6 & 4 & 8 \\ 0 & 0 & 2 & 4 \end{bmatrix} \begin{bmatrix} x \\ y \\ z \\ t \end{bmatrix} = \begin{bmatrix} 1 \\ 3 \\ 1 \end{bmatrix}.$$

5 Under what condition on b_1, b_2, b_3 is this system solvable? Include $\boldsymbol{b}$ as a fourth column in elimination. Find all solutions when that condition holds:

$$\begin{aligned} x + 2y - 2z &= b_1 \\ 2x + 5y - 4z &= b_2 \\ 4x + 9y - 8z &= b_3. \end{aligned}$$

6 What conditions on b_1, b_2, b_3, b_4 make each system solvable? Find $\boldsymbol{x}$ in that case:

$$\begin{bmatrix} 1 & 2 \\ 2 & 4 \\ 2 & 5 \\ 3 & 9 \end{bmatrix} \begin{bmatrix} x_1 \\ x_2 \end{bmatrix} = \begin{bmatrix} b_1 \\ b_2 \\ b_3 \\ b_4 \end{bmatrix} \qquad \begin{bmatrix} 1 & 2 & 3 \\ 2 & 4 & 6 \\ 2 & 5 & 7 \\ 3 & 9 & 12 \end{bmatrix} \begin{bmatrix} x_1 \\ x_2 \\ x_3 \end{bmatrix} = \begin{bmatrix} b_1 \\ b_2 \\ b_3 \\ b_4 \end{bmatrix}.$$

7 Show by elimination that (b_1, b_2, b_3) is in the column space if $b_3 - 2b_2 + 4b_1 = 0$.

$$A = \begin{bmatrix} 1 & 3 & 1 \\ 3 & 8 & 2 \\ 2 & 4 & 0 \end{bmatrix}.$$

What combination of the rows of A gives the zero row?

8 Which vectors (b_1, b_2, b_3) are in the column space of A? Which combinations of the rows of A give zero?

$$\text{(a)}\ A = \begin{bmatrix} 1 & 2 & 1 \\ 2 & 6 & 3 \\ 0 & 2 & 5 \end{bmatrix} \qquad \text{(b)}\quad A = \begin{bmatrix} 1 & 1 & 1 \\ 1 & 2 & 4 \\ 2 & 4 & 8 \end{bmatrix}.$$

9 (a) The Worked Example **3.3 A** reached $[\,U \ \ \boldsymbol{c}\,]$ from $[\,A \ \ \boldsymbol{b}\,]$. Put the multipliers into L and verify that LU equals A and $L\boldsymbol{c}$ equals $\boldsymbol{b}$.

(b) Combine the pivot columns of A with the numbers -9 and 3 in the particular solution $\boldsymbol{x}_p$. What is that linear combination and why?

10 Construct a 2 by 3 system $A\boldsymbol{x} = \boldsymbol{b}$ with particular solution $\boldsymbol{x}_p = (2, 4, 0)$ and homogeneous solution $\boldsymbol{x}_n =$ any multiple of $(1, 1, 1)$.

11 Why can't a 1 by 3 system have $\boldsymbol{x}_p = (2, 4, 0)$ and $\boldsymbol{x}_n =$ any multiple of $(1, 1, 1)$?

12 (a) If $A\boldsymbol{x} = \boldsymbol{b}$ has two solutions $\boldsymbol{x}_1$ and $\boldsymbol{x}_2$, find two solutions to $A\boldsymbol{x} = \mathbf{0}$.

(b) Then find another solution to $A\boldsymbol{x} = \mathbf{0}$ and another solution to $A\boldsymbol{x} = \boldsymbol{b}$.

13 Explain why these are all false:

(a) The complete solution is any linear combination of $\boldsymbol{x}_p$ and $\boldsymbol{x}_n$.

(b) A system $A\boldsymbol{x} = \boldsymbol{b}$ has at most one particular solution.

(c) The solution $\boldsymbol{x}_p$ with all free variables zero is the shortest solution (minimum length $\|\boldsymbol{x}\|$). Find a 2 by 2 counterexample.

(d) If A is invertible there is no solution $\boldsymbol{x}_n$ in the nullspace.

14 Suppose column 5 of U has no pivot. Then x_5 is a _____ variable. The zero vector (is) (is not) the only solution to $A\boldsymbol{x} = \mathbf{0}$. If $A\boldsymbol{x} = \boldsymbol{b}$ has a solution, then it has _____ solutions.

15 Suppose row 3 of U has no pivot. Then that row is _____. The equation $U\boldsymbol{x} = \boldsymbol{c}$ is only solvable provided _____. The equation $A\boldsymbol{x} = \boldsymbol{b}$ (*is*) (*is not*) (*might not be*) solvable.

Questions 16–20 are about matrices of "full rank" $r = m$ or $r = n$.

16 The largest possible rank of a 3 by 5 matrix is _____. Then there is a pivot in every _____ of U and R. The solution to $A\boldsymbol{x} = \boldsymbol{b}$ (*always exists*) (*is unique*). The column space of A is _____. An example is $A =$ _____.

17 The largest possible rank of a 6 by 4 matrix is _____. Then there is a pivot in every _____ of U and R. The solution to $A\boldsymbol{x} = \boldsymbol{b}$ (*always exists*) (*is unique*). The nullspace of A is _____. An example is $A =$ _____.

18 Find by elimination the rank of A and also the rank of A^{T}:

$$A = \begin{bmatrix} 1 & 4 & 0 \\ 2 & 11 & 5 \\ -1 & 2 & 10 \end{bmatrix} \quad \text{and} \quad A = \begin{bmatrix} 1 & 0 & 1 \\ 1 & 1 & 2 \\ 1 & 1 & q \end{bmatrix} \text{ (rank depends on } q\text{).}$$

19 Find the rank of A and also of $A^{\mathrm{T}}A$ and also of AA^{T}:

$$A = \begin{bmatrix} 1 & 1 & 5 \\ 1 & 0 & 1 \end{bmatrix} \quad \text{and} \quad A = \begin{bmatrix} 2 & 0 \\ 1 & 1 \\ 1 & 2 \end{bmatrix}.$$

20 Reduce A to its echelon form U. Then find a triangular L so that $A = LU$.

$$A = \begin{bmatrix} 3 & 4 & 1 & 0 \\ 6 & 5 & 2 & 1 \end{bmatrix} \quad \text{and} \quad A = \begin{bmatrix} 1 & 0 & 1 & 0 \\ 2 & 2 & 0 & 3 \\ 0 & 6 & 5 & 4 \end{bmatrix}.$$

21 Find the complete solution in the form $\boldsymbol{x}_p + \boldsymbol{x}_n$ to these full rank systems:

(a) $x + y + z = 4$ (b) $\begin{aligned} x + y + z &= 4 \\ x - y + z &= 4. \end{aligned}$

22 If $A\boldsymbol{x} = \boldsymbol{b}$ has infinitely many solutions, why is it impossible for $A\boldsymbol{x} = \boldsymbol{B}$ (new right side) to have only one solution? Could $A\boldsymbol{x} = \boldsymbol{B}$ have no solution?

23 Choose the number q so that (if possible) the ranks are (a) 1, (b) 2, (c) 3:

$$A = \begin{bmatrix} 6 & 4 & 2 \\ -3 & -2 & -1 \\ 9 & 6 & q \end{bmatrix} \quad \text{and} \quad B = \begin{bmatrix} 3 & 1 & 3 \\ q & 2 & q \end{bmatrix}.$$

24 Give examples of matrices A for which the number of solutions to $A\boldsymbol{x} = \boldsymbol{b}$ is

(a) 0 or 1, depending on $\boldsymbol{b}$

(b) ∞, regardless of $\boldsymbol{b}$

(c) 0 or ∞, depending on $\boldsymbol{b}$

(d) 1, regardless of $\boldsymbol{b}$.

25 Write down all known relations between r and m and n if $A\boldsymbol{x} = \boldsymbol{b}$ has

(a) no solution for some $\boldsymbol{b}$

(b) infinitely many solutions for every $\boldsymbol{b}$

(c) exactly one solution for some $\boldsymbol{b}$, no solution for other $\boldsymbol{b}$

(d) exactly one solution for every $\boldsymbol{b}$.

Questions 26–33 are about Gauss-Jordan elimination (upwards as well as downwards) and the reduced echelon matrix R.

26 Continue elimination from U to R. Divide rows by pivots so the new pivots are all 1. Then produce zeros *above* those pivots to reach R:

$$U = \begin{bmatrix} 2 & 4 & 4 \\ 0 & 3 & 6 \\ 0 & 0 & 0 \end{bmatrix} \quad \text{and} \quad U = \begin{bmatrix} 2 & 4 & 4 \\ 0 & 3 & 6 \\ 0 & 0 & 5 \end{bmatrix}.$$

27 If A is a triangular matrix, when is $R = \text{rref}(A)$ equal to I?

28 Apply Gauss-Jordan elimination to $U\boldsymbol{x} = \boldsymbol{0}$ and $U\boldsymbol{x} = \boldsymbol{c}$. Reach $R\boldsymbol{x} = \boldsymbol{0}$ and $R\boldsymbol{x} = \boldsymbol{d}$:

$$\begin{bmatrix} U & \boldsymbol{0} \end{bmatrix} = \begin{bmatrix} 1 & 2 & 3 & \mathbf{0} \\ 0 & 0 & 4 & \mathbf{0} \end{bmatrix} \quad \text{and} \quad \begin{bmatrix} U & \boldsymbol{c} \end{bmatrix} = \begin{bmatrix} 1 & 2 & 3 & \mathbf{5} \\ 0 & 0 & 4 & \mathbf{8} \end{bmatrix}.$$

Solve $R\boldsymbol{x} = \boldsymbol{0}$ to find $\boldsymbol{x}_n$ (its free variable is $x_2 = 1$). Solve $R\boldsymbol{x} = \boldsymbol{d}$ to find $\boldsymbol{x}_p$ (its free variable is $x_2 = 0$).

29 Apply Gauss-Jordan elimination to reduce to $R\boldsymbol{x} = \mathbf{0}$ and $R\boldsymbol{x} = \boldsymbol{d}$:

$$\begin{bmatrix} U & 0 \end{bmatrix} = \begin{bmatrix} 3 & 0 & 6 & \mathbf{0} \\ 0 & 0 & 2 & \mathbf{0} \\ 0 & 0 & 0 & \mathbf{0} \end{bmatrix} \quad \text{and} \quad \begin{bmatrix} U & \boldsymbol{c} \end{bmatrix} = \begin{bmatrix} 3 & 0 & 6 & \mathbf{9} \\ 0 & 0 & 2 & \mathbf{4} \\ 0 & 0 & 0 & \mathbf{5} \end{bmatrix}.$$

Solve $U\boldsymbol{x} = \mathbf{0}$ or $R\boldsymbol{x} = \mathbf{0}$ to find $\boldsymbol{x}_n$ (free variable $= 1$). What are the solutions to $R\boldsymbol{x} = \boldsymbol{d}$?

30 Reduce to $U\boldsymbol{x} = \boldsymbol{c}$ (Gaussian elimination) and then $R\boldsymbol{x} = \boldsymbol{d}$ (Gauss-Jordan):

$$A\boldsymbol{x} = \begin{bmatrix} 1 & 0 & 2 & 3 \\ 1 & 3 & 2 & 0 \\ 2 & 0 & 4 & 9 \end{bmatrix} \begin{bmatrix} x_1 \\ x_2 \\ x_3 \\ x_4 \end{bmatrix} = \begin{bmatrix} 2 \\ 5 \\ 10 \end{bmatrix} = \boldsymbol{b}.$$

Find a particular solution $\boldsymbol{x}_p$ and all homogeneous solutions $\boldsymbol{x}_n$.

31 Find matrices A and B with the given property or explain why you can't:

(a) The only solution of $A\boldsymbol{x} = \begin{bmatrix} \mathbf{1} \\ \mathbf{2} \\ \mathbf{3} \end{bmatrix}$ is $\boldsymbol{x} = \begin{bmatrix} \mathbf{0} \\ \mathbf{1} \end{bmatrix}$.

(b) The only solution of $B\boldsymbol{x} = \begin{bmatrix} \mathbf{0} \\ \mathbf{1} \end{bmatrix}$ is $\boldsymbol{x} = \begin{bmatrix} \mathbf{1} \\ \mathbf{2} \\ \mathbf{3} \end{bmatrix}$.

32 Find the LU factorization of A and the complete solution to $A\boldsymbol{x} = \boldsymbol{b}$:

$$A = \begin{bmatrix} 1 & 3 & 1 \\ 1 & 2 & 3 \\ 2 & 4 & 6 \\ 1 & 1 & 5 \end{bmatrix} \quad \text{and} \quad \boldsymbol{b} = \begin{bmatrix} 1 \\ 3 \\ 6 \\ 5 \end{bmatrix} \quad \text{and then} \quad \boldsymbol{b} = \begin{bmatrix} 1 \\ 0 \\ 0 \\ 0 \end{bmatrix}.$$

33 The complete solution to $A\boldsymbol{x} = \begin{bmatrix} \mathbf{1} \\ \mathbf{3} \end{bmatrix}$ is $\boldsymbol{x} = \begin{bmatrix} \mathbf{1} \\ \mathbf{0} \end{bmatrix} + c \begin{bmatrix} \mathbf{0} \\ \mathbf{1} \end{bmatrix}$. Find A.

Challenge Problems

34 (Recommended!) Suppose you know that the 3 by 4 matrix A has the vector $\boldsymbol{s} = (2, 3, 1, 0)$ as the only special solution to $A\boldsymbol{x} = \boldsymbol{0}$.

(a) What is the *rank* of A and the complete solution to $A\boldsymbol{x} = \boldsymbol{0}$?

(b) What is the exact row reduced echelon form R of A?

(c) How do you know that $A\boldsymbol{x} = \boldsymbol{b}$ can be solved for all $\boldsymbol{b}$?

35 Suppose K is the 9 by 9 second difference matrix (2's on the diagonal, -1's on the diagonal above and also below). Solve the equation $K\boldsymbol{x} = \boldsymbol{b} = (10, \ldots, 10)$. If you graph $x_1, \ldots, x_9$ above the points $1, \ldots, 9$ on the x axis, I think the nine points fall on a parabola.

36 Suppose $A\boldsymbol{x} = \boldsymbol{b}$ and $C\boldsymbol{x} = \boldsymbol{b}$ have the same (complete) solutions for every $\boldsymbol{b}$. Is it true that A equals C?

37 Describe the column space of a reduced row echelon matrix R.

3.4 Independence, Basis and Dimension

1 **Independent columns** of A: The only solution to $A\boldsymbol{x} = \mathbf{0}$ is $\boldsymbol{x} = \mathbf{0}$. The nullspace is $\boldsymbol{Z}$.

2 Independent vectors: The only zero combination $c_1\boldsymbol{v}_1 + \cdots + c_k\boldsymbol{v}_k = 0$ has all c's $= 0$.

3 A matrix with $m < n$ has **dependent columns**: At least $n-m$ free variables / special solutions.

4 The vectors $\boldsymbol{v}_1, \ldots, \boldsymbol{v}_k$ **span the space** $\boldsymbol{S}$ if $\boldsymbol{S} =$ all combinations of the $\boldsymbol{v}$'s.

5 The vectors $\boldsymbol{v}_1, \ldots, \boldsymbol{v}_k$ are a **basis for** $\boldsymbol{S}$ if they are independent and they span $\boldsymbol{S}$.

6 The **dimension of a space** $\boldsymbol{S}$ is the number of vectors in every basis for $\boldsymbol{S}$.

7 If A is 4 by 4 and invertible, its columns are a basis for $\mathbf{R}^4$. The dimension of $\mathbf{R}^4$ is 4.

This important section is about the true size of a subspace. There are n columns in an m by n matrix. But the true "dimension" of the column space is not necessarily n. The dimension is measured by counting ***independent columns***—and we have to say what that means. We will see that *the true dimension of the column space is the rank* r.

The idea of independence applies to any vectors $\boldsymbol{v}_1, \ldots, \boldsymbol{v}_n$ in any vector space. Most of this section concentrates on the subspaces that we know and use—especially the column space and the nullspace of A. In the last part we also study "vectors" that are not column vectors. They can be matrices and functions; they can be linearly independent (or dependent). First come the key examples using column vectors.

The goal is to understand a ***basis***: **independent vectors that "span the space".**

Every vector in the space is a unique combination of the basis vectors.

We are at the heart of our subject, and we cannot go on without a basis. The four essential ideas in this section (with first hints at their meaning) are:

1. Independent vectors	(*no extra vectors*)
2. Spanning a space	(*enough vectors to produce the rest*)
3. Basis for a space	(*not too many or too few*)
4. Dimension of a space	(*the number of vectors in a basis*)

Linear Independence

Our first definition of independence is not so conventional, but you are ready for it.

DEFINITION The columns of A are *linearly independent* when the only solution to $Ax = 0$ is $x = 0$. ***No other combination*** Ax ***of the columns gives the zero vector.***

The columns are independent when the nullspace $\boldsymbol{N}(A)$ contains only the zero vector. Let me illustrate linear independence (and dependence) with three vectors in $\mathbf{R}^3$:

1. If three vectors are *not* in the same plane, they are independent. No combination of $\boldsymbol{v}_1, \boldsymbol{v}_2, \boldsymbol{v}_3$ in Figure 3.4 gives zero except $0\boldsymbol{v}_1 + 0\boldsymbol{v}_2 + 0\boldsymbol{v}_3$.

2. If three vectors $\boldsymbol{w}_1, \boldsymbol{w}_2, \boldsymbol{w}_3$ are *in the same plane*, they are dependent.

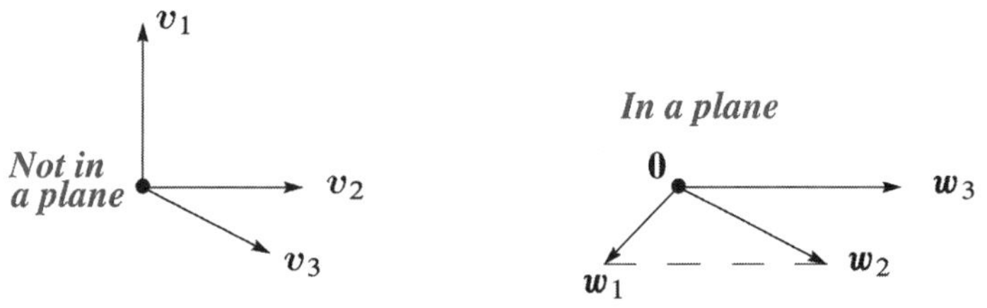

Figure 3.4: Independent vectors $\boldsymbol{v}_1,\ \boldsymbol{v}_2,\ \boldsymbol{v}_3$. Only $0\boldsymbol{v}_1 + 0\boldsymbol{v}_2 + 0\boldsymbol{v}_3$ gives the vector $\mathbf{0}$. Dependent vectors $\boldsymbol{w}_1, \boldsymbol{w}_2, \boldsymbol{w}_3$. The combination $\boldsymbol{w}_1 - \boldsymbol{w}_2 + \boldsymbol{w}_3$ is $(0,0,0)$.

This idea of independence applies to 7 vectors in 12-dimensional space. If they are the columns of A, and independent, the nullspace only contains $\boldsymbol{x} = \mathbf{0}$. None of the vectors is a combination of the other six vectors.

Now we choose different words to express the same idea. The following definition of independence will apply to any sequence of vectors in any vector space. When the vectors are the columns of A, the two definitions say exactly the same thing.

DEFINITION The sequence of vectors $\boldsymbol{v}_1, \ldots, \boldsymbol{v}_n$ is ***linearly independent*** if the only combination that gives the zero vector is $0\boldsymbol{v}_1 + 0\boldsymbol{v}_2 + \cdots + 0\boldsymbol{v}_n$.

Linear independence

$$x_1\boldsymbol{v}_1 + x_2\boldsymbol{v}_2 + \cdots + x_n\boldsymbol{v}_n = \mathbf{0} \quad \text{only happens when all } x\text{'s are zero.} \tag{1}$$

If a combination gives $\mathbf{0}$, when the x's are not all zero, the vectors are *dependent.*

Correct language: "The sequence of vectors is linearly independent." *Acceptable shortcut*: "The vectors are independent." *Unacceptable*: "The matrix is independent."

A sequence of vectors is either dependent or independent. They can be combined to give the zero vector (with nonzero x's) or they can't. So the key question is: Which combinations of the vectors give zero? We begin with some small examples in $\mathbf{R}^2$:

(a) The vectors $(1,0)$ and $(0,1)$ are independent.

(b) The vectors $(1,0)$ and $(1,0.00001)$ are independent.

(c) The vectors $(1,1)$ and $(-1,-1)$ are *dependent.*

(d) The vectors $(1,1)$ and $(0,0)$ are *dependent* because of the zero vector.

(e) In $\mathbf{R}^2$, any three vectors (a,b) and (c,d) and (e,f) are *dependent.*

Geometrically, $(1,1)$ and $(-1,-1)$ are on a line through the origin. They are dependent. To use the definition, find numbers x_1 and x_2 so that $x_1(1,1) + x_2(-1,-1) = (0,0)$. This is the same as solving $A\boldsymbol{x} = \mathbf{0}$:

$$\begin{bmatrix} 1 & -1 \\ 1 & -1 \end{bmatrix} \begin{bmatrix} x_1 \\ x_2 \end{bmatrix} = \begin{bmatrix} 0 \\ 0 \end{bmatrix} \quad \text{for } x_1 = 1 \text{ and } x_2 = 1.$$

The columns are dependent exactly when *there is a nonzero vector in the nullspace*.

If one of the $\boldsymbol{v}$'s is the zero vector, independence has no chance. Why not?

Three vectors in $\mathbf{R}^2$ cannot be independent! One way to see this: the matrix A with those three columns must have a free variable and then a special solution to $A\boldsymbol{x} = \mathbf{0}$. Another way: If the first two vectors are independent, some combination will produce the third vector. See the second highlight below.

Now move to three vectors in $\mathbf{R}^3$. If one of them is a multiple of another one, these vectors are dependent. But the complete test involves all three vectors at once. We put them in a matrix and try to solve $A\boldsymbol{x} = \mathbf{0}$.

Example 1 The columns of this A are dependent. $A\boldsymbol{x} = \mathbf{0}$ has a nonzero solution:

$$A\boldsymbol{x} = \begin{bmatrix} 1 & 0 & 3 \\ 2 & 1 & 5 \\ 1 & 0 & 3 \end{bmatrix} \begin{bmatrix} -3 \\ 1 \\ 1 \end{bmatrix} \quad \text{is} \quad -3 \begin{bmatrix} 1 \\ 2 \\ 1 \end{bmatrix} + 1 \begin{bmatrix} 0 \\ 1 \\ 0 \end{bmatrix} + 1 \begin{bmatrix} 3 \\ 5 \\ 3 \end{bmatrix} = \begin{bmatrix} 0 \\ 0 \\ 0 \end{bmatrix}.$$

The rank is only $r = 2$. *Independent columns produce full column rank $r = n = 3$.*

In that matrix the rows are also dependent. Row 1 minus row 3 is the zero row. For a *square matrix*, we will show that dependent columns imply dependent rows.

Question How to find that solution to $A\boldsymbol{x} = \mathbf{0}$? The systematic way is elimination.

$$A = \begin{bmatrix} 1 & 0 & 3 \\ 2 & 1 & 5 \\ 1 & 0 & 3 \end{bmatrix} \text{ reduces to } R = \begin{bmatrix} 1 & 0 & 3 \\ 0 & 1 & -1 \\ 0 & 0 & 0 \end{bmatrix}.$$

The solution $\boldsymbol{x} = (-3, 1, 1)$ was exactly the special solution. It shows how the free column (column 3) is a combination of the pivot columns. That kills independence!

Full column rank The columns of A are independent exactly when the rank is $r = n$. There are n pivots and no free variables. Only $\boldsymbol{x} = \mathbf{0}$ is in the nullspace.

One case is of special importance because it is clear from the start. Suppose seven columns have five components each ($m = 5$ is less than $n = 7$). Then the columns *must be dependent*. Any seven vectors from $\mathbf{R}^5$ are dependent. The rank of A cannot be larger than 5. There cannot be more than five pivots in five rows. $A\boldsymbol{x} = \mathbf{0}$ has at least $7 - 5 = 2$ free variables, so it has nonzero solutions—which means that the columns are dependent.

Any set of n vectors in $\mathbf{R}^m$ must be linearly dependent if $n > m$.

This type of matrix has more columns than rows—it is short and wide. The columns are certainly dependent if $n > m$, because $A\boldsymbol{x} = \mathbf{0}$ has a nonzero solution.

The columns might be dependent or might be independent if $n \le m$. Elimination will reveal the r pivot columns. *It is those r pivot columns that are independent.*

Note Another way to describe linear dependence is this: "*One vector is a combination of the other vectors.*" That sounds clear. Why don't we say this from the start? Our definition was longer: "*Some combination gives the zero vector, other than the trivial combination with every $x = 0$.*" We must rule out the easy way to get the zero vector. That trivial combination of zeros gives every author a headache. If one vector is a combination of the others, that vector has coefficient $x = 1$.

The point is, our definition doesn't pick out one particular vector as guilty. All columns of A are treated the same. We look at $A\boldsymbol{x} = \mathbf{0}$, and it has a nonzero solution or it hasn't. In the end that is better than asking if the last column (or the first, or a column in the middle) is a combination of the others.

Vectors that Span a Subspace

The first subspace in this book was the column space. Starting with columns $\boldsymbol{v}_1, \ldots, \boldsymbol{v}_n$, the subspace was filled out by including all combinations $x_1\boldsymbol{v}_1 + \cdots + x_n\boldsymbol{v}_n$. *The column space consists of all combinations $A\boldsymbol{x}$ of the columns.* We now introduce the single word "span" to describe this: The column space is ***spanned*** by the columns.

DEFINITION A set of vectors ***spans*** a space if their linear combinations fill the space.

The columns of a matrix span its column space. They might be dependent.

Example 2 $\boldsymbol{v}_1 = \begin{bmatrix} 1 \\ 0 \end{bmatrix}$ and $\boldsymbol{v}_2 = \begin{bmatrix} 0 \\ 1 \end{bmatrix}$ span the full two-dimensional space $\mathbf{R}^2$.

Example 3 $\boldsymbol{v}_1 = \begin{bmatrix} 1 \\ 0 \end{bmatrix}, \boldsymbol{v}_2 = \begin{bmatrix} 0 \\ 1 \end{bmatrix}, \boldsymbol{v}_3 = \begin{bmatrix} 4 \\ 7 \end{bmatrix}$ also span the full space $\mathbf{R}^2$.

Example 4 $\boldsymbol{w}_1 = \begin{bmatrix} 1 \\ 1 \end{bmatrix}$ and $\boldsymbol{w}_2 = \begin{bmatrix} -1 \\ -1 \end{bmatrix}$ only span a line in $\mathbf{R}^2$. So does $\boldsymbol{w}_1$ by itself.

Think of two vectors coming out from $(0,0,0)$ in 3-dimensional space. Generally they span a plane. Your mind fills in that plane by taking linear combinations. Mathematically you know other possibilities: two vectors could span a line, three vectors could span all of $\mathbf{R}^3$, or only a plane. It is even possible that three vectors span only a line, or ten vectors span only a plane. They are certainly not independent!

The columns span the column space. Here is a new subspace—*which is spanned by the rows*. ***The combinations of the rows produce the "row space".***

DEFINITION The ***row space*** of a matrix is the subspace of $\mathbf{R}^n$ spanned by the rows.

The row space of A ***is*** $\boldsymbol{C}(A^{\mathrm{T}})$***. It is the column space of*** A^{T}***.***

The rows of an m by n matrix have n components. They are vectors in $\mathbf{R}^n$—or they would be if they were written as column vectors. There is a quick way to fix that: *Transpose the matrix.* Instead of the rows of A, look at the columns of A^{T}. Same numbers, but now in the column space $\boldsymbol{C}(A^{\mathrm{T}})$. This row space of A is a subspace of $\mathbf{R}^n$.

Example 5 Describe the column space and the row space of A.

$$A = \begin{bmatrix} 1 & 4 \\ 2 & 7 \\ 3 & 5 \end{bmatrix} \text{ and } A^{\mathrm{T}} = \begin{bmatrix} 1 & 2 & 3 \\ 4 & 7 & 5 \end{bmatrix}. \text{ Here } m = 3 \text{ and } n = 2.$$

The column space of A is the plane in $\mathbf{R}^3$ spanned by the two columns of A. *The row space of A is spanned by the three rows of A* (which are columns of A^{T}). This row space is all of $\mathbf{R}^2$. Remember: The rows are in $\mathbf{R}^n$ spanning the row space. The columns are in $\mathbf{R}^m$ spanning the column space. Same numbers, different vectors, different spaces.

A Basis for a Vector Space

Two vectors can't span all of $\mathbf{R}^3$, even if they are independent. Four vectors can't be independent, even if they span $\mathbf{R}^3$. We want ***enough independent vectors to span the space*** (and not more). A "***basis***" is just right.

DEFINITION A ***basis*** for a vector space is a sequence of vectors with two properties:

The basis vectors are linearly independent and they span the space.

This combination of properties is fundamental to linear algebra. Every vector $\boldsymbol{v}$ in the space is a combination of the basis vectors, because they span the space. More than that, the combination that produces $\boldsymbol{v}$ is *unique*, because the basis vectors $\boldsymbol{v}_1, \ldots, \boldsymbol{v}_n$ are independent:

There is one and only one way to write v as a combination of the basis vectors.

Reason: Suppose $\boldsymbol{v} = a_1\boldsymbol{v}_1 + \cdots + a_n\boldsymbol{v}_n$ and also $\boldsymbol{v} = b_1\boldsymbol{v}_1 + \cdots + b_n\boldsymbol{v}_n$. By subtraction $(a_1 - b_1)\boldsymbol{v}_1 + \cdots + (a_n - b_n)\boldsymbol{v}_n$ is the zero vector. From the independence of the $\boldsymbol{v}$'s, each $a_i - b_i = 0$. Hence $a_i = b_i$, and there are not two ways to produce $\boldsymbol{v}$.

Example 6 The columns of $I = \begin{bmatrix} 1 & 0 \\ 0 & 1 \end{bmatrix}$ produce the "standard basis" for $\mathbf{R}^2$.

$$\text{The basis vectors} \quad \boldsymbol{i} = \begin{bmatrix} 1 \\ 0 \end{bmatrix} \quad \text{and} \quad \boldsymbol{j} = \begin{bmatrix} 0 \\ 1 \end{bmatrix} \quad \text{are independent. They span } \mathbf{R}^2.$$

Everybody thinks of this basis first. The vector $\boldsymbol{i}$ goes across and $\boldsymbol{j}$ goes straight up. The columns of the 3 by 3 identity matrix are the standard basis $\boldsymbol{i}$, $\boldsymbol{j}$, $\boldsymbol{k}$. The columns of the n by n identity matrix give the "**standard basis**" for $\mathbf{R}^n$.

Now we find many other bases (infinitely many). The basis is not unique!

Example 7 (Important) The columns of *every invertible n by n matrix* give a basis for $\mathbf{R}^n$:

Invertible matrix Independent columns Column space is $\mathbf{R}^3$	$A = \begin{bmatrix} 1 & 0 & 0 \\ 1 & 1 & 0 \\ 1 & 1 & 1 \end{bmatrix}$	**Singular matrix** Dependent columns Column space $\neq \mathbf{R}^3$	$B = \begin{bmatrix} 1 & 0 & 1 \\ 1 & 1 & 2 \\ 1 & 1 & 2 \end{bmatrix}.$

The only solution to $A\boldsymbol{x} = \mathbf{0}$ is $\boldsymbol{x} = A^{-1}\mathbf{0} = \mathbf{0}$. The columns are independent. They span the whole space $\mathbf{R}^n$—because every vector $\boldsymbol{b}$ is a combination of the columns. $A\boldsymbol{x} = \boldsymbol{b}$ can always be solved by $\boldsymbol{x} = A^{-1}\boldsymbol{b}$. Do you see how everything comes together for invertible matrices? Here it is in one sentence:

The vectors $\boldsymbol{v}_1, \ldots, \boldsymbol{v}_n$ are a ***basis for*** $\mathbf{R}^n$ exactly when they are ***the columns of an n by n invertible matrix***. Thus $\mathbf{R}^n$ has infinitely many different bases.

When the columns are dependent, we keep only the *pivot columns*—the first two columns of B above, with its two pivots. They are independent and they span the column space.

The pivot columns of A are a basis for its column space. The pivot rows of A are a basis for its row space. So are the pivot rows of its echelon form R.

Example 8 This matrix is not invertible. Its columns are not a basis for anything!

One pivot column **One pivot row** $(r = 1)$	$A = \begin{bmatrix} 2 & 4 \\ 3 & 6 \end{bmatrix}$	reduces to $R = \begin{bmatrix} 1 & 2 \\ 0 & 0 \end{bmatrix}.$

Column 1 of A is the pivot column. That column alone is a basis for its column space. The second column of A would be a different basis. So would any nonzero multiple of that column. There is no shortage of bases. One definite choice is the pivot columns.

Notice that the pivot column $(1, 0)$ of this R ends in zero. That column is a basis for the column space of R, but it doesn't belong to the column space of A. The column spaces of A and R are different. Their bases are different. (Their dimensions are the same.)

The row space of A is the *same* as the row space of R. It contains $(2, 4)$ and $(1, 2)$ and all other multiples of those vectors. As always, there are infinitely many bases to choose from. One natural choice is to pick the nonzero rows of R (rows with a pivot). So this matrix A with rank one has only one vector in the basis:

$$\text{Basis for the column space: } \begin{bmatrix} 2 \\ 3 \end{bmatrix}. \quad \text{Basis for the row space: } \begin{bmatrix} 1 \\ 2 \end{bmatrix}.$$

The next chapter will come back to these bases for the column space and row space. We are happy first with examples where the situation is clear (and the idea of a basis is still new). The next example is larger but still clear.

Example 9 Find bases for the column and row spaces of this rank two matrix:

$$R = \begin{bmatrix} 1 & 2 & 0 & 3 \\ 0 & 0 & 1 & 4 \\ 0 & 0 & 0 & 0 \end{bmatrix}.$$

Columns 1 and 3 are the pivot columns. They are a basis for the column space (of R!). The vectors in that column space all have the form $\boldsymbol{b} = (x, y, 0)$. The column space of R is the "xy plane" inside the full 3-dimensional xyz space. That plane is not $\mathbf{R}^2$, it is a subspace of $\mathbf{R}^3$. Columns 2 and 3 are also a basis for the same column space. Which pairs of columns of R are *not* a basis for its column space?

The row space of R is a subspace of $\mathbf{R}^4$. The simplest basis for that row space is the two nonzero rows of R. The third row (the zero vector) is in the row space too. But it is not in a *basis* for the row space. The basis vectors must be independent.

Question Given five vectors in $\mathbf{R}^7$, ***how do you find a basis for the space they span?***

First answer Make them the rows of A, and eliminate to find the nonzero rows of R.

Second answer Put the five vectors into the columns of A. Eliminate to find the pivot columns (of A not R). Those pivot columns are a basis for the column space.

Could another basis have more vectors, or fewer? This is a crucial question with a good answer: *No.* ***All bases for a vector space contain the same*** *number* ***of vectors***.

The number of vectors, in any and every basis, is the "dimension" of the space.

Dimension of a Vector Space

We have to prove what was just stated. There are many choices for the basis vectors, but the *number* of basis vectors doesn't change.

If $\boldsymbol{v}_1, \ldots, \boldsymbol{v}_m$ and $\boldsymbol{w}_1, \ldots, \boldsymbol{w}_n$ are both bases for the same vector space, then $m = n$.

Proof Suppose that there are more $\boldsymbol{w}$'s than $\boldsymbol{v}$'s. From $n > m$ we want to reach a contradiction. The $\boldsymbol{v}$'s are a basis, so $\boldsymbol{w}_1$ must be a combination of the $\boldsymbol{v}$'s. If $\boldsymbol{w}_1$ equals

$a_{11}\boldsymbol{v}_1 + \cdots + a_{m1}\boldsymbol{v}_m$, this is the first column of a matrix multiplication VA:

Each w is a combination of the v's

$$W = \begin{bmatrix} \boldsymbol{w}_1 & \boldsymbol{w}_2 & \dots & \boldsymbol{w}_n \end{bmatrix} = \begin{bmatrix} \boldsymbol{v}_1 & \dots & \boldsymbol{v}_m \end{bmatrix} \begin{bmatrix} a_{11} & & a_{1n} \\ \vdots & & \vdots \\ a_{m1} & & a_{mn} \end{bmatrix} = VA.$$

We don't know each a_{ij}, but we know the shape of A (it is m by n). The second vector $\boldsymbol{w}_2$ is also a combination of the $\boldsymbol{v}$'s. The coefficients in that combination fill the second column of A. The key is that A has a row for every $\boldsymbol{v}$ and a column for every $\boldsymbol{w}$. A is a short wide matrix, since we assumed $n > m$. So $A\boldsymbol{x} = \boldsymbol{0}$ ***has a nonzero solution***.

$A\boldsymbol{x} = 0$ gives $VA\boldsymbol{x} = \boldsymbol{0}$ which is $W\boldsymbol{x} = \boldsymbol{0}$. *A combination of the $\boldsymbol{w}$'s gives zero*! Then the $\boldsymbol{w}$'s could not be a basis—our assumption $n > m$ is **not possible** for two bases.

If $m > n$ we exchange the $\boldsymbol{v}$'s and $\boldsymbol{w}$'s and repeat the same steps. The only way to avoid a contradiction is to have $m = n$. This completes the proof that $m = n$.

The number of basis vectors depends on the space—not on a particular basis. The number is the same for every basis, and it counts the "degrees of freedom" in the space. The dimension of the space $\mathbf{R}^n$ is n. We now introduce the important word ***dimension*** for other vector spaces too.

DEFINITION The ***dimension of a space*** is the ***number of vectors*** in every basis.

This matches our intuition. The line through $\boldsymbol{v} = (1, 5, 2)$ has dimension one. It is a subspace with this one vector $\boldsymbol{v}$ in its basis. Perpendicular to that line is the plane $x + 5y + 2z = 0$. This plane has dimension 2. To prove it, we find a basis $(-5, 1, 0)$ and $(-2, 0, 1)$. The dimension is 2 because the basis contains two vectors.

The plane is the nullspace of the matrix $A = \begin{bmatrix} 1 & 5 & 2 \end{bmatrix}$, which has two free variables. Our basis vectors $(-5, 1, 0)$ and $(-2, 0, 1)$ are the "special solutions" to $A\boldsymbol{x} = \boldsymbol{0}$. The next section shows that the $n - r$ special solutions always give *a basis for the nullspace*. $\boldsymbol{C}(A)$ has dimension r and the nullspace $\boldsymbol{N}(A)$ has dimension $n - r$.

Note about the language of linear algebra We never say "the rank of a space" or "the dimension of a basis" or "the basis of a matrix". Those terms have no meaning. It is the ***dimension of the column space*** that equals the ***rank of the matrix.***

Bases for Matrix Spaces and Function Spaces

The words "independence" and "basis" and "dimension" are not at all restricted to column vectors. We can ask whether three matrices A_1, A_2, A_3 are independent. When they are in the space of all 3 by 4 matrices, some combination might give the zero matrix. We can also ask the dimension of the full 3 by 4 matrix space. (It is 12.)

In differential equations, $d^2y/dx^2 = y$ has a space of solutions. One basis is $y = e^x$ and $y = e^{-x}$. Counting the basis functions gives the dimension 2 for the space of all solutions. (The dimension is 2 because of the second derivative.)

Matrix spaces and function spaces may look a little strange after $\mathbf{R}^n$. But in some way, you haven't got the ideas of basis and dimension straight until you can apply them to "vectors" other than column vectors.

Matrix spaces The vector space $\mathbf{M}$ contains all 2 by 2 matrices. Its dimension is 4.

$$\textbf{One basis is} \quad A_1, A_2, A_3, A_4 = \begin{bmatrix} 1 & 0 \\ 0 & 0 \end{bmatrix}, \begin{bmatrix} 0 & 1 \\ 0 & 0 \end{bmatrix}, \begin{bmatrix} 0 & 0 \\ 1 & 0 \end{bmatrix}, \begin{bmatrix} 0 & 0 \\ 0 & 1 \end{bmatrix}.$$

Those matrices are linearly independent. We are not looking at their columns, but at the whole matrix. Combinations of those four matrices can produce any matrix in $\mathbf{M}$, so they span the space:

$$\begin{array}{l}\textbf{Every } A \textbf{ combines} \\ \textbf{the basis matrices}\end{array} \qquad c_1A_1 + c_2A_2 + c_3A_3 + c_4A_4 = \begin{bmatrix} c_1 & c_2 \\ c_3 & c_4 \end{bmatrix} = A.$$

A is zero only if the c's are all zero—this proves independence of A_1, A_2, A_3, A_4.

The three matrices A_1, A_2, A_4 are a basis for a subspace—the upper triangular matrices. Its dimension is 3. A_1 and A_4 are a basis for the diagonal matrices. What is a basis for the symmetric matrices? Keep A_1 and A_4, and throw in $A_2 + A_3$.

To push this further, think about the space of all n by n matrices. One possible basis uses matrices that have only a single nonzero entry (that entry is 1). There are n^2 positions for that 1, so there are n^2 basis matrices:

The dimension of the whole n by n matrix space is n^2.

The dimension of the subspace of *upper triangular* matrices is $\frac{1}{2}n^2 + \frac{1}{2}n$.

The dimension of the subspace of *diagonal* matrices is n.

The dimension of the subspace of *symmetric* matrices is $\frac{1}{2}n^2 + \frac{1}{2}n$ (why?).

Function spaces The equations $d^2y/dx^2 = 0$ and $d^2y/dx^2 = -y$ and $d^2y/dx^2 = y$ involve the second derivative. In calculus we solve to find the functions $y(x)$:

$$\begin{array}{ll} y'' = 0 & \text{is solved by any linear function } y = cx + d \\ y'' = -y & \text{is solved by any combination } y = c\sin x + d\cos x \\ y'' = y & \text{is solved by any combination } y = ce^x + de^{-x}. \end{array}$$

That solution space for $y'' = -y$ has two basis functions: $\sin x$ and $\cos x$. The space for $y'' = 0$ has x and 1. It is the "nullspace" of the second derivative! The dimension is 2 in each case (these are second-order equations).

The solutions of $y'' = 2$ don't form a subspace—the right side $b = 2$ is not zero. A particular solution is $y(x) = x^2$. The complete solution is $y(x) = x^2 + cx + d$. All those functions satisfy $y'' = 2$. Notice the particular solution plus any function $cx + d$ in the nullspace. A linear differential equation is like a linear matrix equation $A\boldsymbol{x} = \boldsymbol{b}$. But we solve it by calculus instead of linear algebra.

We end here with the space $\mathbf{Z}$ that contains only the zero vector. The dimension of this space is *zero*. ***The empty set*** (containing no vectors) ***is a basis for*** $\mathbf{Z}$. We can never allow the zero vector into a basis, because then linear independence is lost.

■ REVIEW OF THE KEY IDEAS ■

1. The columns of A are ***independent*** if $\boldsymbol{x} = \mathbf{0}$ is the only solution to $A\boldsymbol{x} = \mathbf{0}$.
2. The vectors $\boldsymbol{v}_1, \ldots, \boldsymbol{v}_r$ ***span*** a space if their combinations fill that space.
3. ***A basis consists of linearly independent vectors that span the space.*** Every vector in the space is a *unique* combination of the basis vectors.
4. All bases for a space have the same number of vectors. This number of vectors in a basis is the ***dimension*** of the space.
5. The pivot columns are one basis for the column space. The dimension is r.

■ WORKED EXAMPLES ■

3.4 A Start with the vectors $\boldsymbol{v}_1 = (1, 2, 0)$ and $\boldsymbol{v}_2 = (2, 3, 0)$. **(a)** Are they linearly independent? **(b)** Are they a basis for any space? **(c)** What space $\mathbf{V}$ do they span? **(d)** What is the dimension of $\mathbf{V}$? **(e)** Which matrices A have $\mathbf{V}$ as their column space? **(f)** Which matrices have $\mathbf{V}$ as their nullspace? **(g)** Describe all vectors $\boldsymbol{v}_3$ that complete a basis $\boldsymbol{v}_1, \boldsymbol{v}_2, \boldsymbol{v}_3$ for $\mathbf{R}^3$.

Solution

(a) $\boldsymbol{v}_1$ and $\boldsymbol{v}_2$ are independent—the only combination to give $\mathbf{0}$ is $0\boldsymbol{v}_1 + 0\boldsymbol{v}_2$.

(b) Yes, they are a basis for the space they span.

(c) That space $\mathbf{V}$ contains all vectors $(x, y, 0)$. It is the xy plane in $\mathbf{R}^3$.

(d) The dimension of $\mathbf{V}$ is 2 since the basis contains two vectors.

(e) This $\mathbf{V}$ is the column space of any 3 by n matrix A of rank 2, if every column is a combination of $\boldsymbol{v}_1$ and $\boldsymbol{v}_2$. In particular A could just have columns $\boldsymbol{v}_1$ and $\boldsymbol{v}_2$.

(f) This $\mathbf{V}$ is the nullspace of any m by 3 matrix B of rank 1, if every row is a multiple of $(0, 0, 1)$. In particular take $B = [0 \;\; 0 \;\; 1]$. Then $B\boldsymbol{v}_1 = \mathbf{0}$ and $B\boldsymbol{v}_2 = \mathbf{0}$.

(g) Any third vector $\boldsymbol{v}_3 = (a, b, c)$ will complete a basis for $\mathbf{R}^3$ provided $c \neq 0$.

3.4 B Start with three independent vectors $\boldsymbol{w}_1, \boldsymbol{w}_2, \boldsymbol{w}_3$. Take combinations of those vectors to produce $\boldsymbol{v}_1, \boldsymbol{v}_2, \boldsymbol{v}_3$. Write the combinations in matrix form as $V = WB$:

$$\begin{aligned} \boldsymbol{v}_1 &= \boldsymbol{w}_1 + \boldsymbol{w}_2 \\ \boldsymbol{v}_2 &= \boldsymbol{w}_1 + 2\boldsymbol{w}_2 + \boldsymbol{w}_3 \\ \boldsymbol{v}_3 &= \quad\quad \boldsymbol{w}_2 + c\boldsymbol{w}_3 \end{aligned} \quad \text{which is} \quad \begin{bmatrix} & & \\ \boldsymbol{v}_1 & \boldsymbol{v}_2 & \boldsymbol{v}_3 \\ & & \end{bmatrix} = \begin{bmatrix} & & \\ \boldsymbol{w}_1 & \boldsymbol{w}_2 & \boldsymbol{w}_3 \\ & & \end{bmatrix} \begin{bmatrix} 1 & 1 & 0 \\ 1 & 2 & 1 \\ 0 & 1 & c \end{bmatrix}$$

What is the test on B to see if $V = WB$ has independent columns? If $c \neq 1$ show that $\boldsymbol{v}_1, \boldsymbol{v}_2, \boldsymbol{v}_3$ are linearly independent. If $c = 1$ show that the $\boldsymbol{v}$'s are linearly *dependent.*

Solution The test on V for independence of its columns was in our first definition: *The nullspace of V must contain only the zero vector.* Then $\boldsymbol{x} = (0, 0, 0)$ is the only combination of the columns that gives $V\boldsymbol{x} =$ zero vector.

If $c = 1$ in our problem, we can see *dependence* in two ways. First, $\boldsymbol{v}_1 + \boldsymbol{v}_3$ will be the same as $\boldsymbol{v}_2$. (If you add $\boldsymbol{w}_1 + \boldsymbol{w}_2$ to $\boldsymbol{w}_2 + \boldsymbol{w}_3$ you get $\boldsymbol{w}_1 + 2\boldsymbol{w}_2 + \boldsymbol{w}_3$ which is $\boldsymbol{v}_2$.) In other words $\boldsymbol{v}_1 - \boldsymbol{v}_2 + \boldsymbol{v}_3 = \boldsymbol{0}$—which says that the $\boldsymbol{v}$'s are not independent.

The other way is to look at the nullspace of B. If $c = 1$, the vector $\boldsymbol{x} = (1, -1, 1)$ is in that nullspace, and $B\boldsymbol{x} = \boldsymbol{0}$. Then certainly $WB\boldsymbol{x} = \boldsymbol{0}$ which is the same as $V\boldsymbol{x} = 0$. So the $\boldsymbol{v}$'s are dependent. This specific $\boldsymbol{x} = (1, -1, 1)$ from the nullspace tells us again that $\boldsymbol{v}_1 - \boldsymbol{v}_2 + \boldsymbol{v}_3 = \boldsymbol{0}$.

Now suppose $c \neq 1$. Then the matrix B is invertible. So if $\boldsymbol{x}$ is *any nonzero vector* we know that $B\boldsymbol{x}$ is nonzero. Since the $\boldsymbol{w}$'s are given as independent, we further know that $WB\boldsymbol{x}$ is nonzero. Since $V = WB$, this says that $\boldsymbol{x}$ is *not* in the nullspace of V. In other words $\boldsymbol{v}_1, \boldsymbol{v}_2, \boldsymbol{v}_3$ are independent.

The general rule is "independent $\boldsymbol{v}$'s from independent $\boldsymbol{w}$'s when B is invertible". And if these vectors are in $\mathbf{R}^3$, they are not only independent—they are a basis for $\mathbf{R}^3$. ***"Basis of $\boldsymbol{v}$'s from basis of $\boldsymbol{w}$'s when the change of basis matrix B is invertible."***

3.4 C (***Important example***) Suppose $\boldsymbol{v}_1, \ldots, \boldsymbol{v}_n$ is a basis for $\mathbf{R}^n$ and the n by n matrix A is invertible. Show that $A\boldsymbol{v}_1, \ldots, A\boldsymbol{v}_n$ is also a basis for $\mathbf{R}^n$.

Solution In *matrix language*: Put the basis vectors $\boldsymbol{v}_1, \ldots, \boldsymbol{v}_n$ in the columns of an invertible(!) matrix V. Then $A\boldsymbol{v}_1, \ldots, A\boldsymbol{v}_n$ are the columns of AV. Since A is invertible, so is AV and its columns give a basis.

In *vector language*: Suppose $c_1 A\boldsymbol{v}_1 + \cdots + c_n A\boldsymbol{v}_n = \boldsymbol{0}$. This is $A\boldsymbol{v} = \boldsymbol{0}$ with $\boldsymbol{v} = c_1\boldsymbol{v}_1 + \cdots + c_n\boldsymbol{v}_n$. Multiply by A^{-1} to reach $\boldsymbol{v} = \boldsymbol{0}$. By linear independence of the $\boldsymbol{v}$'s, all $c_i = 0$. This shows that the $A\boldsymbol{v}$'s are independent.

To show that the $A\boldsymbol{v}$'s span $\mathbf{R}^n$, solve $c_1 A\boldsymbol{v}_1 + \cdots + c_n A\boldsymbol{v}_n = \boldsymbol{b}$ which is the same as $c_1\boldsymbol{v}_1 + \cdots + c_n\boldsymbol{v}_n = A^{-1}\boldsymbol{b}$. Since the $\boldsymbol{v}$'s are a basis, this must be solvable.

Problem Set 3.4

Questions 1–10 are about linear independence and linear dependence.

1 Show that $\boldsymbol{v}_1, \boldsymbol{v}_2, \boldsymbol{v}_3$ are independent but $\boldsymbol{v}_1, \boldsymbol{v}_2, \boldsymbol{v}_3, \boldsymbol{v}_4$ are dependent:

$$\boldsymbol{v}_1 = \begin{bmatrix} 1 \\ 0 \\ 0 \end{bmatrix} \quad \boldsymbol{v}_2 = \begin{bmatrix} 1 \\ 1 \\ 0 \end{bmatrix} \quad \boldsymbol{v}_3 = \begin{bmatrix} 1 \\ 1 \\ 1 \end{bmatrix} \quad \boldsymbol{v}_4 = \begin{bmatrix} 2 \\ 3 \\ 4 \end{bmatrix}.$$

Solve $c_1\boldsymbol{v}_1 + c_2\boldsymbol{v}_2 + c_3\boldsymbol{v}_3 + c_4\boldsymbol{v}_4 = \mathbf{0}$ or $A\boldsymbol{x} = \mathbf{0}$. The $\boldsymbol{v}$'s go in the columns of A.

2 (Recommended) Find the largest possible number of independent vectors among

$$\boldsymbol{v}_1 = \begin{bmatrix} 1 \\ -1 \\ 0 \\ 0 \end{bmatrix} \quad \boldsymbol{v}_2 = \begin{bmatrix} 1 \\ 0 \\ -1 \\ 0 \end{bmatrix} \quad \boldsymbol{v}_3 = \begin{bmatrix} 1 \\ 0 \\ 0 \\ -1 \end{bmatrix} \quad \boldsymbol{v}_4 = \begin{bmatrix} 0 \\ 1 \\ -1 \\ 0 \end{bmatrix} \quad \boldsymbol{v}_5 = \begin{bmatrix} 0 \\ 1 \\ 0 \\ -1 \end{bmatrix} \quad \boldsymbol{v}_6 = \begin{bmatrix} 0 \\ 0 \\ 1 \\ -1 \end{bmatrix}$$

3 Prove that if $a = 0$ or $d = 0$ or $f = 0$ (3 cases), the columns of U are dependent:

$$U = \begin{bmatrix} a & b & c \\ 0 & d & e \\ 0 & 0 & f \end{bmatrix}.$$

4 If a, d, f in Question 3 are all nonzero, show that the only solution to $U\boldsymbol{x} = \mathbf{0}$ is $\boldsymbol{x} = \mathbf{0}$. Then the upper triangular U has independent columns.

5 Decide the dependence or independence of

(a) the vectors $(1, 3, 2)$ and $(2, 1, 3)$ and $(3, 2, 1)$

(b) the vectors $(1, -3, 2)$ and $(2, 1, -3)$ and $(-3, 2, 1)$.

6 Choose three independent columns of U. Then make two other choices. Do the same for A.

$$U = \begin{bmatrix} 2 & 3 & 4 & 1 \\ 0 & 6 & 7 & 0 \\ 0 & 0 & 0 & 9 \\ 0 & 0 & 0 & 0 \end{bmatrix} \quad \text{and} \quad A = \begin{bmatrix} 2 & 3 & 4 & 1 \\ 0 & 6 & 7 & 0 \\ 0 & 0 & 0 & 9 \\ 4 & 6 & 8 & 2 \end{bmatrix}.$$

7 If $\boldsymbol{w}_1, \boldsymbol{w}_2, \boldsymbol{w}_3$ are independent vectors, show that the differences $\boldsymbol{v}_1 = \boldsymbol{w}_2 - \boldsymbol{w}_3$ and $\boldsymbol{v}_2 = \boldsymbol{w}_1 - \boldsymbol{w}_3$ and $\boldsymbol{v}_3 = \boldsymbol{w}_1 - \boldsymbol{w}_2$ are *dependent*. Find a combination of the $\boldsymbol{v}$'s that gives zero. Which matrix A in $[\,\boldsymbol{v}_1 \ \boldsymbol{v}_2 \ \boldsymbol{v}_3\,] = [\,\boldsymbol{w}_1 \ \boldsymbol{w}_2 \ \boldsymbol{w}_3\,]\,A$ is singular?

8 If $\boldsymbol{w}_1, \boldsymbol{w}_2, \boldsymbol{w}_3$ are independent vectors, show that the sums $\boldsymbol{v}_1 = \boldsymbol{w}_2 + \boldsymbol{w}_3$ and $\boldsymbol{v}_2 = \boldsymbol{w}_1 + \boldsymbol{w}_3$ and $\boldsymbol{v}_3 = \boldsymbol{w}_1 + \boldsymbol{w}_2$ are *independent*. (Write $c_1\boldsymbol{v}_1 + c_2\boldsymbol{v}_2 + c_3\boldsymbol{v}_3 = \mathbf{0}$ in terms of the $\boldsymbol{w}$'s. Find and solve equations for the c's, to show they are zero.)

9 Suppose $\boldsymbol{v}_1, \boldsymbol{v}_2, \boldsymbol{v}_3, \boldsymbol{v}_4$ are vectors in $\mathbf{R}^3$.

(a) These four vectors are dependent because ______.

(b) The two vectors $\boldsymbol{v}_1$ and $\boldsymbol{v}_2$ will be dependent if ______.

(c) The vectors $\boldsymbol{v}_1$ and $(0,0,0)$ are dependent because ______.

10 Find two independent vectors on the plane $x+2y-3z-t=0$ in $\mathbf{R}^4$. Then find three independent vectors. Why not four? This plane is the nullspace of what matrix?

Questions 11–14 are about the space *spanned* by a set of vectors. Take all linear combinations of the vectors.

11 Describe the subspace of $\mathbf{R}^3$ (is it a line or plane or $\mathbf{R}^3$?) spanned by

(a) the two vectors $(1,1,-1)$ and $(-1,-1,1)$

(b) the three vectors $(0,1,1)$ and $(1,1,0)$ and $(0,0,0)$

(c) all vectors in $\mathbf{R}^3$ with whole number components

(d) all vectors with positive components.

12 The vector $\boldsymbol{b}$ is in the subspace spanned by the columns of A when ______ has a solution. The vector $\boldsymbol{c}$ is in the row space of A when ______ has a solution.

True or false: If the zero vector is in the row space, the rows are dependent.

13 Find the dimensions of these 4 spaces. Which two of the spaces are the same? (a) column space of A, (b) column space of U, (c) row space of A, (d) row space of U:

$$A = \begin{bmatrix} 1 & 1 & 0 \\ 1 & 3 & 1 \\ 3 & 1 & -1 \end{bmatrix} \quad \text{and} \quad U = \begin{bmatrix} 1 & 1 & 0 \\ 0 & 2 & 1 \\ 0 & 0 & 0 \end{bmatrix}.$$

14 $\boldsymbol{v}+\boldsymbol{w}$ and $\boldsymbol{v}-\boldsymbol{w}$ are combinations of $\boldsymbol{v}$ and $\boldsymbol{w}$. Write $\boldsymbol{v}$ and $\boldsymbol{w}$ as combinations of $\boldsymbol{v}+\boldsymbol{w}$ and $\boldsymbol{v}-\boldsymbol{w}$. The two pairs of vectors ______ the same space. When are they a basis for the same space?

Questions 15–25 are about the requirements for a basis.

15 If $\boldsymbol{v}_1, \ldots, \boldsymbol{v}_n$ are linearly independent, the space they span has dimension ______. These vectors are a ______ for that space. If the vectors are the columns of an m by n matrix, then m is ______ than n. If $m = n$, that matrix is ______.

16 Find a basis for each of these subspaces of $\mathbf{R}^4$:

(a) All vectors whose components are equal.

(b) All vectors whose components add to zero.

(c) All vectors that are perpendicular to $(1,1,0,0)$ and $(1,0,1,1)$.

(d) The column space and the nullspace of I (4 by 4).

17 Find three different bases for the column space of $U = \begin{bmatrix} 1 & 0 & 1 & 0 & 1 \\ 0 & 1 & 0 & 1 & 0 \end{bmatrix}$. Then find two different bases for the row space of U.

18 Suppose $\boldsymbol{v}_1, \boldsymbol{v}_2, \ldots, \boldsymbol{v}_6$ are six vectors in $\mathbf{R}^4$.

(a) Those vectors (do)(do not)(might not) span $\mathbf{R}^4$.

(b) Those vectors (are)(are not)(might be) linearly independent.

(c) Any four of those vectors (are)(are not)(might be) a basis for $\mathbf{R}^4$.

19 The columns of A are n vectors from $\mathbf{R}^m$. If they are linearly independent, what is the rank of A? If they span $\mathbf{R}^m$, what is the rank? If they are a basis for $\mathbf{R}^m$, what then? *Looking ahead*: The rank r counts the number of _____ columns.

20 Find a basis for the plane $x - 2y + 3z = 0$ in $\mathbf{R}^3$. Then find a basis for the intersection of that plane with the xy plane. Then find a basis for all vectors perpendicular to the plane.

21 Suppose the columns of a 5 by 5 matrix A are a basis for $\mathbf{R}^5$.

(a) The equation $A\boldsymbol{x} = \mathbf{0}$ has only the solution $\boldsymbol{x} = \mathbf{0}$ because _____ .

(b) If $\boldsymbol{b}$ is in $\mathbf{R}^5$ then $A\boldsymbol{x} = \boldsymbol{b}$ is solvable because the basis vectors _____ $\mathbf{R}^5$.

Conclusion: A is invertible. Its rank is 5. Its rows are also a basis for $\mathbf{R}^5$.

22 Suppose $\mathbf{S}$ is a 5-dimensional subspace of $\mathbf{R}^6$. True or false (example if false):

(a) Every basis for $\mathbf{S}$ can be extended to a basis for $\mathbf{R}^6$ by adding one more vector.

(b) Every basis for $\mathbf{R}^6$ can be reduced to a basis for $\mathbf{S}$ by removing one vector.

23 U comes from A by subtracting row 1 from row 3:

$$A = \begin{bmatrix} 1 & 3 & 2 \\ 0 & 1 & 1 \\ 1 & 3 & 2 \end{bmatrix} \quad \text{and} \quad U = \begin{bmatrix} 1 & 3 & 2 \\ 0 & 1 & 1 \\ 0 & 0 & 0 \end{bmatrix}.$$

Find bases for the two column spaces. Find bases for the two row spaces. Find bases for the two nullspaces. Which spaces stay fixed in elimination?

24 True or false (give a good reason):

(a) If the columns of a matrix are dependent, so are the rows.

(b) The column space of a 2 by 2 matrix is the same as its row space.

(c) The column space of a 2 by 2 matrix has the same dimension as its row space.

(d) The columns of a matrix are a basis for the column space.

25 For which numbers c and d do these matrices have rank 2?

$$A = \begin{bmatrix} 1 & 2 & 5 & 0 & 5 \\ 0 & 0 & c & 2 & 2 \\ 0 & 0 & 0 & d & 2 \end{bmatrix} \quad \text{and} \quad B = \begin{bmatrix} c & d \\ d & c \end{bmatrix}.$$

Questions 26–30 are about spaces where the "vectors" are matrices.

26 Find a basis (and the dimension) for each of these subspaces of 3 by 3 matrices:

(a) All diagonal matrices.
(b) All symmetric matrices ($A^{\mathrm{T}} = A$).
(c) All skew-symmetric matrices ($A^{\mathrm{T}} = -A$).

27 Construct six linearly independent 3 by 3 echelon matrices $U_1, \ldots, U_6$.

28 Find a basis for the space of all 2 by 3 matrices whose columns add to zero. Find a basis for the subspace whose rows also add to zero.

29 What subspace of 3 by 3 matrices is spanned (take all combinations) by

(a) the invertible matrices?
(b) the rank one matrices?
(c) the identity matrix?

30 Find a basis for the space of 2 by 3 matrices whose nullspace contains $(2, 1, 1)$.

Questions 31–35 are about spaces where the "vectors" are functions.

31 (a) Find all functions that satisfy $\frac{dy}{dx} = 0$.
(b) Choose a particular function that satisfies $\frac{dy}{dx} = 3$.
(c) Find all functions that satisfy $\frac{dy}{dx} = 3$.

32 The cosine space $\mathbf{F}_3$ contains all combinations $y(x) = A\cos x + B\cos 2x + C\cos 3x$. Find a basis for the subspace with $y(0) = 0$.

33 Find a basis for the space of functions that satisfy

(a) $\frac{dy}{dx} - 2y = 0$
(b) $\frac{dy}{dx} - \frac{y}{x} = 0$.

34 Suppose $y_1(x), y_2(x), y_3(x)$ are three different functions of x. The vector space they span could have dimension 1, 2, or 3. Give an example of y_1, y_2, y_3 to show each possibility.

35 Find a basis for the space of polynomials $p(x)$ of degree ≤ 3. Find a basis for the subspace with $p(1) = 0$.

36 Find a basis for the space $\mathbf{S}$ of vectors (a, b, c, d) with $a + c + d = 0$ and also for the space $\mathbf{T}$ with $a + b = 0$ and $c = 2d$. What is the dimension of the intersection $\mathbf{S} \cap \mathbf{T}$?

37 If $AS = SA$ for the *shift matrix* S, show that A must have this special form:

$$\text{If } \begin{bmatrix} a & b & c \\ d & e & f \\ g & h & i \end{bmatrix} \begin{bmatrix} 0 & 1 & 0 \\ 0 & 0 & 1 \\ 0 & 0 & 0 \end{bmatrix} = \begin{bmatrix} 0 & 1 & 0 \\ 0 & 0 & 1 \\ 0 & 0 & 0 \end{bmatrix} \begin{bmatrix} a & b & c \\ d & e & f \\ g & h & i \end{bmatrix} \text{ then } A = \begin{bmatrix} a & b & c \\ 0 & a & b \\ 0 & 0 & a \end{bmatrix}.$$

"The subspace of matrices that commute with the shift S has dimension _____."

38 Which of the following are bases for $\mathbf{R}^3$?

(a) $(1, 2, 0)$ and $(0, 1, -1)$
(b) $(1, 1, -1), (2, 3, 4), (4, 1, -1), (0, 1, -1)$
(c) $(1, 2, 2), (-1, 2, 1), (0, 8, 0)$
(d) $(1, 2, 2), (-1, 2, 1), (0, 8, 6)$

39 Suppose A is 5 by 4 with rank 4. Show that $A\boldsymbol{x} = \boldsymbol{b}$ has no solution when the 5 by 5 matrix $[\,A \;\; \boldsymbol{b}\,]$ is invertible. Show that $A\boldsymbol{x} = \boldsymbol{b}$ is solvable when $[\,A \;\; \boldsymbol{b}\,]$ is singular.

40 (a) Find a basis for all solutions to $d^4y/dx^4 = y(x)$.
(b) Find a particular solution to $d^4y/dx^4 = y(x) + 1$. Find the complete solution.

Challenge Problems

41 Write the 3 by 3 identity matrix as a combination of the other five permutation matrices! Then show that those five matrices are linearly independent. (Assume a combination gives $c_1P_1 + \cdots + c_5P_5 =$ zero matrix, and check entries to prove that c_1 to c_5 must all be zero.) The five permutations are a basis for the subspace of 3 by 3 matrices with row and column sums all equal.

42 Choose $\boldsymbol{x} = (x_1, x_2, x_3, x_4)$ in $\mathbf{R}^4$. It has 24 rearrangements like (x_2, x_1, x_3, x_4) and (x_4, x_3, x_1, x_2). Those 24 vectors, including $\boldsymbol{x}$ itself, span a subspace **S**. Find specific vectors $\boldsymbol{x}$ so that the dimension of **S** is: (a) zero, (b) one, (c) three, (d) four.

43 Intersections and sums have $\dim(\mathbf{V}) + \dim(\mathbf{W}) = \dim(\mathbf{V} \cap \mathbf{W}) + \dim(\mathbf{V} + \mathbf{W})$. Start with a basis $\boldsymbol{u}_1, \ldots, \boldsymbol{u}_r$ for the intersection $\mathbf{V} \cap \mathbf{W}$. Extend with $\boldsymbol{v}_1, \ldots, \boldsymbol{v}_s$ to a basis for $\mathbf{V}$, and separately with $\boldsymbol{w}_1, \ldots, \boldsymbol{w}_t$ to a basis for $\mathbf{W}$. Prove that the $\boldsymbol{u}$'s, $\boldsymbol{v}$'s and $\boldsymbol{w}$'s together are ***independent***. The dimensions have $(r + s) + (r + t) = (r) + (r + s + t)$ as desired.

44 Mike Artin suggested a neat higher-level proof of that dimension formula in Problem 43. From all inputs $\boldsymbol{v}$ in $\mathbf{V}$ and $\boldsymbol{w}$ in $\mathbf{W}$, the "sum transformation" produces $\boldsymbol{v} + \boldsymbol{w}$. Those outputs fill the space $\mathbf{V} + \mathbf{W}$. The nullspace contains all pairs $\boldsymbol{v} = \boldsymbol{u}$, $\boldsymbol{w} = -\boldsymbol{u}$ for vectors $\boldsymbol{u}$ in $\mathbf{V} \cap \mathbf{W}$. (Then $\boldsymbol{v} + \boldsymbol{w} = \boldsymbol{u} - \boldsymbol{u} = \boldsymbol{0}$.) So $\dim(\mathbf{V} + \mathbf{W}) + \dim(\mathbf{V} \cap \mathbf{W})$ equals $\dim(\mathbf{V}) + \dim(\mathbf{W})$ (*input dimension from* $\mathbf{V}$ *and* $\mathbf{W}$) by the Counting Theorem.

dimension of outputs + dimension of nullspace = dimension of inputs.

Problem For an m by n matrix of rank r, what are those 3 dimensions? Outputs = column space. This question will be answered in Section 3.5, can you do it now?

45 Inside $\mathbf{R}^n$, suppose dimension $(\mathbf{V})$ + dimension $(\mathbf{W}) > n$. Show that some nonzero vector is in both $\mathbf{V}$ and $\mathbf{W}$.

46 Suppose A is 10 by 10 and $A^2 = 0$ (zero matrix). So A multiplies each column of A to give the zero vector. This means that the column space of A is contained in the ______. If A has rank r, those subspaces have dimension $r \leq 10 - r$. So the rank is $r \leq 5$.

3.5 Dimensions of the Four Subspaces

1 The column space $\boldsymbol{C}(A)$ and the row space $\boldsymbol{C}(A^{\mathrm{T}})$ both have *dimension* r (the rank of A).

2 The nullspace $\boldsymbol{N}(A)$ has *dimension* $n-r$. The left nullspace $\boldsymbol{N}(A^{\mathrm{T}})$ has *dimension* $m-r$.

3 Elimination produces bases for the row space and nullspace of A: They are the same as for R.

4 Elimination often changes the column space and left nullspace (but dimensions don't change).

5 **Rank one matrices**: $A = \boldsymbol{u}\boldsymbol{v}^{\mathrm{T}} =$ column times row: $\boldsymbol{C}(A)$ has basis $\boldsymbol{u}$, $\boldsymbol{C}(A^{\mathrm{T}})$ has basis $\boldsymbol{v}$.

The main theorem in this chapter connects ***rank*** and ***dimension***. The ***rank*** of a matrix is the number of pivots. The ***dimension*** of a subspace is the number of vectors in a basis. We count pivots or we count basis vectors. *The rank of A reveals the dimensions of all four fundamental subspaces.* Here are the subspaces, including the new one.

Two subspaces come directly from A, and the other two from A^{T}:

Four Fundamental Subspaces

1. The ***row space*** is $\boldsymbol{C}(A^{\mathrm{T}})$, a subspace of $\mathbf{R}^n$.
2. The ***column space*** is $\boldsymbol{C}(A)$, a subspace of $\mathbf{R}^m$.
3. The ***nullspace*** is $\boldsymbol{N}(A)$, a subspace of $\mathbf{R}^n$.
4. The ***left nullspace*** is $\boldsymbol{N}(A^{\mathrm{T}})$, a subspace of $\mathbf{R}^m$. This is our new space.

In this book the column space and nullspace came first. We know $\boldsymbol{C}(A)$ and $\boldsymbol{N}(A)$ pretty well. Now the other two subspaces come forward. The row space contains all combinations of the rows. *This row space of A is the column space of* A^{T}.

For the left nullspace we solve $A^{\mathrm{T}}\boldsymbol{y} = \mathbf{0}$—that system is n by m. *This is the nullspace of* A^{T}. The vectors $\boldsymbol{y}$ go on the *left* side of A when the equation is written $\boldsymbol{y}^{\mathrm{T}}A = \mathbf{0}^{\mathrm{T}}$. The matrices A and A^{T} are usually different. So are their column spaces and their nullspaces. But those spaces are connected in an absolutely beautiful way.

Part 1 of the Fundamental Theorem finds the dimensions of the four subspaces. One fact stands out: ***The row space and column space have the same dimension*** r. This number r is the **rank** of the matrix. The other important fact involves the two nullspaces:

$\boldsymbol{N}(A)$ ***and*** $\boldsymbol{N}(A^{\mathrm{T}})$ ***have dimensions*** $n-r$ ***and*** $m-r$, ***to make up the full*** n ***and*** m.

Part 2 of the Fundamental Theorem will describe how the four subspaces fit together (two in $\mathbf{R}^n$ and two in $\mathbf{R}^m$). That completes the "right way" to understand every $A\boldsymbol{x} = \boldsymbol{b}$. Stay with it—you are doing real mathematics.

The Four Subspaces for R

Suppose A is reduced to its row echelon form R. For that special form, the four subspaces are easy to identify. We will find a basis for each subspace and check its dimension. Then we watch how the subspaces change (two of them don't change!) as we look back at A. The main point is that *the four dimensions are the same for A and R.*

As a specific 3 by 5 example, look at the four subspaces for this echelon matrix R:

$$\begin{matrix} m=3 \\ n=5 \\ r=2 \end{matrix} \qquad \boldsymbol{R} = \begin{bmatrix} \mathbf{1} & \mathbf{3} & \mathbf{5} & 0 & \mathbf{7} \\ 0 & 0 & 0 & \mathbf{1} & \mathbf{2} \\ 0 & 0 & 0 & 0 & 0 \end{bmatrix} \qquad \begin{matrix} \textbf{pivot rows } 1 \textbf{ and } 2 \\ \\ \textbf{pivot columns } 1 \textbf{ and } 4 \end{matrix}$$

The rank of this matrix is $r = 2$ (*two pivots*). Take the four subspaces in order.

1. The ***row space*** of R has dimension 2, matching the rank.

Reason: The first two rows are a basis. The row space contains combinations of all three rows, but the third row (the zero row) adds nothing new. So rows 1 and 2 span the row space $\boldsymbol{C}(R^{\mathrm{T}})$.

The pivot rows 1 and 2 are independent. That is obvious for this example, and it is always true. If we look only at the pivot columns, we see the r by r identity matrix. There is no way to combine its rows to give the zero row (except by the combination with all coefficients zero). So the r pivot rows are a basis for the row space.

The dimension of the row space is the rank r. The nonzero rows of R form a basis.

2. The ***column space*** of R also has dimension $r = 2$.

Reason: The pivot columns 1 and 4 form a basis for $\boldsymbol{C}(R)$. They are independent because they start with the r by r identity matrix. No combination of those pivot columns can give the zero column (except the combination with all coefficients zero). And they also span the column space. Every other (free) column is a combination of the pivot columns. Actually the combinations we need are the three special solutions !

Column 2 is 3 (column 1). The special solution is $(-3, 1, 0, 0, 0)$.

Column 3 is 5 (column 1). The special solution is $(-5, 0, 1, 0, 0,)$.

Column 5 is 7 (column 1) $+ 2$ (column 4). That solution is $(-7, 0, 0, -2, 1)$.

The pivot columns are independent, and they span, so they are a basis for $\boldsymbol{C}(R)$.

The dimension of the column space is the rank r. The pivot columns form a basis.

3. The ***nullspace*** of R has dimension $n - r = 5 - 2$. There are $n - r = 3$ free variables. Here x_2, x_3, x_5 are free (no pivots in those columns). They yield the three special solutions to $R\boldsymbol{x} = \mathbf{0}$. Set a free variable to 1, and solve for x_1 and x_4.

$$\boldsymbol{s}_2 = \begin{bmatrix} \mathbf{-3} \\ \mathbf{1} \\ 0 \\ 0 \\ 0 \end{bmatrix} \quad \boldsymbol{s}_3 = \begin{bmatrix} \mathbf{-5} \\ 0 \\ \mathbf{1} \\ 0 \\ 0 \end{bmatrix} \quad \boldsymbol{s}_5 = \begin{bmatrix} \mathbf{-7} \\ 0 \\ 0 \\ \mathbf{-2} \\ \mathbf{1} \end{bmatrix}$$

$R\boldsymbol{x} = \mathbf{0}$ has the complete solution
$\boldsymbol{x} = x_2\boldsymbol{s}_2 + x_3\boldsymbol{s}_3 + x_5\boldsymbol{s}_5$
The nullspace has dimension 3.

Reason: There is a special solution for each free variable. With n variables and r pivots, that leaves $n-r$ free variables and special solutions. The special solutions are independent, because they contain the identity matrix in rows 2, 3, 5. So $\boldsymbol{N}(R)$ has dimension $n - r$.

The nullspace has dimension $n - r$. The special solutions form a basis.

4. The ***nullspace of R^{T}*** (***left nullspace of R***) has dimension $m - r = 3 - 2$.

Reason: The equation $R^{\mathrm{T}}\boldsymbol{y} = \mathbf{0}$ looks for combinations of the columns of R^{T} (*the rows of* R) that produce zero. This equation $R^{\mathrm{T}}\boldsymbol{y} = \mathbf{0}$ or $\boldsymbol{y}^{\mathrm{T}}R = \mathbf{0}^{\mathrm{T}}$ is

$$\begin{array}{ll} \textbf{Left nullspace} & y_1\,[1,\ 3,\ 5,\ 0,\ 7] \\ \textbf{Combination} & +y_2\,[0,\ 0,\ 0,\ 1,\ 2] \\ \textbf{of rows is zero} & +y_3\,[0,\ 0,\ 0,\ 0,\ 0] \\ \hline & \quad\;\;\, [0,\ 0,\ 0,\ 0,\ 0] \end{array} \qquad (1)$$

The solutions y_1, y_2, y_3 are pretty clear. We need $y_1 = 0$ and $y_2 = 0$. The variable y_3 is free (it can be anything). **The nullspace of $\boldsymbol{R}^{\mathrm{T}}$ contains all vectors $\boldsymbol{y} = (0, 0, y_3)$.**

In all cases R ends with $m - r$ zero rows. Every combination of these $m - r$ rows gives zero. These are the *only* combinations of the rows of R that give zero, because the pivot rows are linearly independent. So $\boldsymbol{y}$ in the left nullspace has $y_1 = 0, \ldots, y_r = 0$.

If A is m by n of rank r, its left nullspace has dimension $m - r$.

Why is this a "*left* nullspace"? The reason is that $R^{\mathrm{T}}\boldsymbol{y} = \mathbf{0}$ can be transposed to $\boldsymbol{y}^{\mathrm{T}}R = \mathbf{0}^{\mathrm{T}}$. Now $\boldsymbol{y}^{\mathrm{T}}$ is a row vector to the *left* of R. You see the y's in equation (1) multiplying the rows. This subspace came fourth, and some linear algebra books omit it—but that misses the beauty of the whole subject.

In $\mathbf{R}^n$ the row space and nullspace have dimensions r and $n - r$ (adding to n).
In $\mathbf{R}^m$ the column space and left nullspace have dimensions r and $m - r$ (total m).

The Four Subspaces for A

We have a job still to do. ***The subspace dimensions for A are the same as for R.*** The job is to explain why. A is now any matrix that reduces to $R = \text{rref}(A)$.

This A reduces to R $\qquad A = \begin{bmatrix} 1 & 3 & 5 & 0 & 7 \\ 0 & 0 & 0 & 1 & 2 \\ 1 & 3 & 5 & 1 & 9 \end{bmatrix}$ $\qquad$ Notice $\boldsymbol{C}(A) \neq \boldsymbol{C}(R)!$ $\quad$ (2)

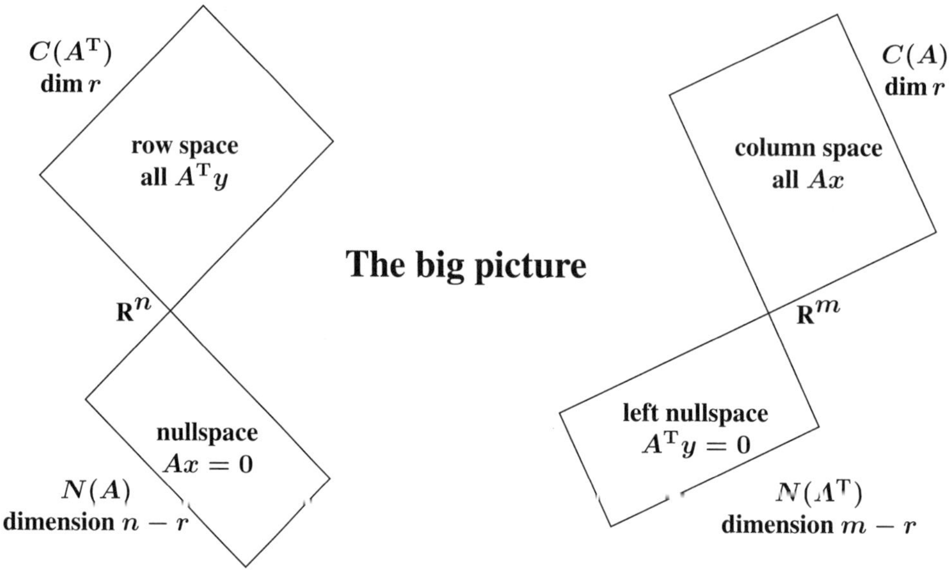

Figure 3.5: The dimensions of the Four Fundamental Subspaces (for R and for A).

1 ***A has the same row space as R.*** **Same dimension r and same basis.**

Reason: Every row of A is a combination of the rows of R. Also every row of R is a combination of the rows of A. Elimination changes rows, but not row *spaces*.

Since A has the same row space as R, we can choose the first r rows of R as a basis. Or we could choose r suitable rows of the original A. They might not always be the *first* r rows of A, because those could be dependent. The good r rows of A are the ones that end up as pivot rows in R.

2 ***The column space of A has dimension r. The column rank equals the row rank.***

Rank Theorem: The number of independent columns = the number of independent rows.

Wrong reason: "A and R have the same column space." This is false. *The columns of* R often end in zeros. The columns of A don't often end in zeros. Then $\boldsymbol{C}(A)$ is not $\boldsymbol{C}(R)$.

Right reason: The ***same*** combinations of the columns are zero (or nonzero) for A and R. Dependent in $A \Leftrightarrow$ dependent in R. Say that another way: $A\boldsymbol{x} = \mathbf{0}$ *exactly when* $R\boldsymbol{x} = \mathbf{0}$. The column spaces are different, but their *dimensions* are the same—equal to r.

Conclusion The r pivot columns of A are a basis for *its* column space $\boldsymbol{C}(A)$.

3 ***A has the same nullspace as R.*** **Same dimension $n - r$ and same basis.**

Reason: The elimination steps don't change the solutions. The special solutions are a basis for this nullspace (as we always knew). There are $n - r$ free variables, so the dimension of the nullspace is $n - r$. This is the **Counting Theorem**: $r + (n - r)$ equals n.

$$\textbf{(dimension of column space)} + \textbf{(dimension of nullspace)} = \textbf{dimension of } \mathbf{R}^n.$$

4 ***The left nullspace of A*** (the nullspace of A^{T}) ***has dimension $m - r$.***

Reason: A^{T} is just as good a matrix as A. When we know the dimensions for every A, we also know them for A^{T}. Its column space was proved to have dimension r. Since A^{T} is n by m, the "whole space" is now $\mathbf{R}^m$. The counting rule for A was $r + (n - r) = n$. The counting rule for A^{T} is $r + (m - r) = m$. We now have all details of a big theorem:

Fundamental Theorem of Linear Algebra, **Part 1**

The column space and row space both have dimension r.

The nullspaces have dimensions $n - r$ and $m - r$.

By concentrating on *spaces* of vectors, not on individual numbers or vectors, we get these clean rules. You will soon take them for granted—eventually they begin to look obvious. But if you write down an 11 by 17 matrix with 187 nonzero entries, I don't think most people would see why these facts are true:

Two key facts
$$\begin{aligned}&\text{dimension of } \boldsymbol{C}(A) = \text{dimension of } \boldsymbol{C}(A^{\mathrm{T}}) = \text{rank of } A\\ &\text{dimension of } \boldsymbol{C}(A) + \text{dimension of } \boldsymbol{N}(A) = 17.\end{aligned}$$

Example 1 $A = [\,1 \quad 2 \quad 3\,]$ has $m = 1$ and $n = 3$ and rank $r = 1$.

The row space is a line in $\mathbf{R}^3$. The nullspace is the plane $A\boldsymbol{x} = x_1 + 2x_2 + 3x_3 = 0$. This plane has dimension 2 (which is $3 - 1$). The dimensions add to $\mathbf{1} + \mathbf{2} = \mathbf{3}$.

The columns of this 1 by 3 matrix are in $\mathbf{R}^1$! The column space is all of $\mathbf{R}^1$. The left nullspace contains only the zero vector. The only solution to $A^{\mathrm{T}}\boldsymbol{y} = \mathbf{0}$ is $\boldsymbol{y} = \mathbf{0}$, no other multiple of $[\,1 \quad 2 \quad 3\,]$ gives the zero row. Thus $\boldsymbol{N}(A^{\mathrm{T}})$ is $\mathbf{Z}$, the zero space with dimension 0 (which is $m - r$). In $\mathbf{R}^m$ the dimensions of $\boldsymbol{C}(A)$ and $\boldsymbol{N}(A^{\mathrm{T}})$ add to $\mathbf{1} + \mathbf{0} = \mathbf{1}$.

Example 2 $A = \begin{bmatrix} 1 & 2 & 3 \\ 2 & 4 & 6 \end{bmatrix}$ has $m = 2$ with $n = 3$ and rank $r = 1$.

The row space is the same line through $(1, 2, 3)$. The nullspace must be the same plane $x_1 + 2x_2 + 3x_3 = 0$. The line and plane dimensions still add to $1 + 2 = 3$.

All columns are multiples of the first column $(1, 2)$. Twice the first row minus the second row is the zero row. Therefore $A^{\mathrm{T}}\boldsymbol{y} = \mathbf{0}$ has the solution $\boldsymbol{y} = (2, -1)$. The column space and left nullspace are **perpendicular lines** in $\mathbf{R}^2$. Dimensions $1 + 1 = 2$.

$$\text{Column space} = \text{line through } \begin{bmatrix} 1 \\ 2 \end{bmatrix} \qquad \text{Left nullspace} = \text{line through } \begin{bmatrix} 2 \\ -1 \end{bmatrix}.$$

If A has three equal rows, its rank is ______. What are two of the $\boldsymbol{y}$'s in its left nullspace?

The $\boldsymbol{y}$***'s in the left nullspace combine the rows to give the zero row.***

Example 3 You have nearly finished three chapters with made-up equations, and this can't continue forever. Here is a better example of five equations (one for every edge in Figure 3.6). The five equations have four unknowns (one for every node). The matrix in $A\boldsymbol{x} = \boldsymbol{b}$ is an **incidence matrix**. This matrix A has 1 and -1 on every row.

Differences $A\boldsymbol{x} = \boldsymbol{b}$
across edges 1, 2, 3, 4, 5
between nodes 1, 2, 3, 4

$$\begin{matrix} -x_1 & +x_2 & & & = b_1 \\ -x_1 & & +x_3 & & = b_2 \\ & -x_2 & +x_3 & & = b_3 \\ & -x_2 & & +x_4 & = b_4 \\ & & -x_3 & +x_4 & = b_5 \end{matrix} \qquad (3)$$

If you understand the four fundamental subspaces for this matrix (*the column spaces and the nullspaces for A and A^{T}*) you have captured the central ideas of linear algebra.

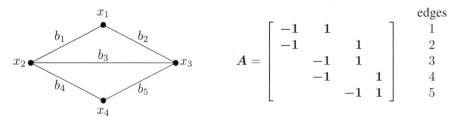

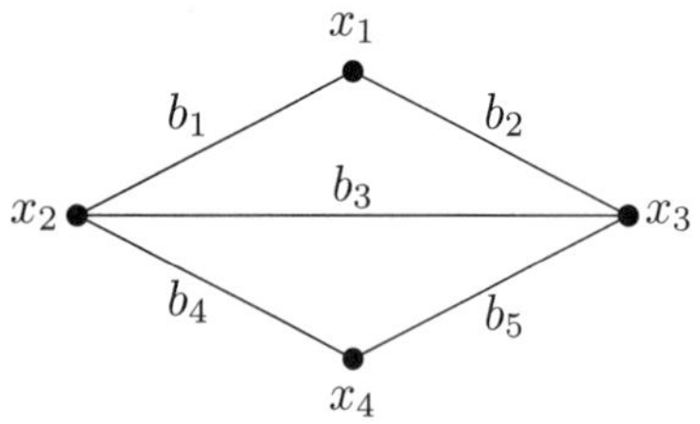

$$\boldsymbol{A} = \begin{bmatrix} -1 & 1 & & \\ -1 & & 1 & \\ & -1 & 1 & \\ & -1 & & 1 \\ & & -1 & 1 \end{bmatrix} \quad \begin{matrix} \text{edges} \\ 1 \\ 2 \\ 3 \\ 4 \\ 5 \end{matrix}$$

Figure 3.6: A "graph" with 5 edges and 4 nodes. A is its 5 by 4 incidence matrix.

The nullspace $\boldsymbol{N}(A)$ To find the nullspace we set $\boldsymbol{b} = \mathbf{0}$. Then the first equation says $x_1 = x_2$. The second equation is $x_3 = x_1$. Equation 4 is $x_2 = x_4$. *All four unknowns x_1, x_2, x_3, x_4 have the same value c.* The vectors $\boldsymbol{x} = (c, c, c, c)$ fill the nullspace of A.

That nullspace is a line in $\mathbf{R}^4$. The special solution $\boldsymbol{x} = (1, 1, 1, 1)$ is a basis for $\boldsymbol{N}(A)$. The dimension of $\boldsymbol{N}(A)$ is 1 (one vector in the basis). *The rank of A must be* 3, *since* $n - r = 4 - 3 = 1$. We now know the dimensions of all four subspaces.

The column space $C(A)$ There must be $r = 3$ independent columns. The fast way is to look at the first 3 columns. The systematic way is to find $R = \text{rref}(A)$.

$$\begin{array}{c} \textbf{Columns} \\ \mathbf{1,2,3} \\ \textbf{of } \boldsymbol{A} \end{array} \quad \begin{array}{rrr} -1 & 1 & 0 \\ -1 & 0 & 1 \\ 0 & -1 & 1 \\ 0 & -1 & 0 \\ 0 & 0 & -1 \end{array} \qquad \boldsymbol{R} = \begin{array}{c} \textbf{reduced row} \\ \textbf{echelon form} \end{array} = \begin{bmatrix} 1 & 0 & 0 & -1 \\ 0 & 1 & 0 & -1 \\ 0 & 0 & 1 & -1 \\ 0 & 0 & 0 & 0 \\ 0 & 0 & 0 & 0 \end{bmatrix}$$

From R we see again the special solution $\boldsymbol{x} = (1,1,1,1)$. The first 3 columns are basic, the fourth column is free. To produce a basis for $\boldsymbol{C}(A)$ and not $\boldsymbol{C}(R)$, we go back to columns $1,2,3$ of A. The column space has dimension $r = 3$.

The row space $C(A^{\mathrm{T}})$ The dimension must again be $r = 3$. But the first 3 rows of A are *not independent*: row $3 =$ row $2 -$ row 1. So row 3 became zero in elimination, and row 3 was exchanged with row 4. *The first three independent rows are rows* $1,2,4$. Those three rows are a basis (one possible basis) for the row space.

I notice that edges $1,2,3$ form a **loop** in the picture: Dependent rows $1,2,3$. Edges $1,2,4$ form a **tree** in the picture. **Trees have no loops**! Independent rows $1,2,4$.

The left nullspace $N(A^{\mathrm{T}})$ Now we solve $A^{\mathrm{T}}\boldsymbol{y} = \mathbf{0}$. Combinations of the rows give zero. We already noticed that row $3 =$ row $2 -$ row 1, so one solution is $\boldsymbol{y} = (1,-1,1,0,0)$. I would say: That $\boldsymbol{y}$ comes from following the upper loop in the picture. Another $\boldsymbol{y}$ comes from going around the lower loop and it is $\boldsymbol{y} = (0,0,\mathbf{-1},\mathbf{1},\mathbf{-1})$: *row* $3 =$ *row* $4 -$ *row* 5. Those two $\boldsymbol{y}$'s are independent, they solve $A^{\mathrm{T}}\boldsymbol{y} = \mathbf{0}$, and the dimension of $\boldsymbol{N}(A^{\mathrm{T}})$ is $\boldsymbol{m} - \boldsymbol{r} = 5 - 3 = \mathbf{2}$. So we have a basis for the left nullspace.

You may ask how "loops" and "trees" got into this problem. That didn't have to happen. We could have used elimination to solve $A^{\mathrm{T}}\boldsymbol{y} = \mathbf{0}$. The 4 by 5 matrix A^{T} would have three pivot columns $1,2,4$ and two free columns $3,5$. There are two special solutions and the nullspace of A^{T} has dimension two: $m - r = 5 - 3 = 2$. But *loops* and *trees* identify *dependent rows* and *independent rows* in a beautiful way. We use them in Section 10.1 for every incidence matrix like this A.

The equations $A\boldsymbol{x} = \boldsymbol{b}$ give "voltages" x_1, x_2, x_3, x_4 at the four nodes. The equations $A^{\mathrm{T}}\boldsymbol{y} = \mathbf{0}$ give "currents" y_1, y_2, y_3, y_4, y_5 on the five edges. These two equations are **Kirchhoff's Voltage Law** and **Kirchhoff's Current Law**. Those words apply to an electrical network. But the ideas behind the words apply all over engineering and science and economics and business.

Graphs are *the most important model in discrete applied mathematics*. You see graphs everywhere: roads, pipelines, blood flow, the brain, the Web, the economy of a country or the world. We can understand their matrices A and A^{T}.

Rank One Matrices (Review)

Suppose every row is a multiple of the first row. Here is a typical example:

$$\begin{bmatrix} 2 & 3 & 7 & 8 \\ 2a & 3a & 7a & 8a \\ 2b & 3b & 7b & 8b \end{bmatrix} = \begin{bmatrix} 1 \\ a \\ b \end{bmatrix} \begin{bmatrix} 2 & 3 & 7 & 8 \end{bmatrix} = \boldsymbol{u}\boldsymbol{v}^{\mathrm{T}}$$

On the left is a matrix with three rows. But its row *space* only has dimension $= 1$. The row vector $\boldsymbol{v}^{\mathrm{T}} = \begin{bmatrix} 2 & 3 & 7 & 8 \end{bmatrix}$ tells us a basis for that row space. *The row rank is* 1.

Now look at the columns. "The column rank equals the row rank which is 1." All columns of the matrix must be multiples of one column. Do you see that this key rule of linear algebra is true? The column vector $\boldsymbol{u} = (1, a, b)$ is multiplied by $2, 3, 7, 8$. That nonzero vector $\boldsymbol{u}$ is a basis for the column space. *The column rank is also* 1.

Every rank one matrix is one column times one row $\quad A = \boldsymbol{u}\boldsymbol{v}^{\mathrm{T}}$

Rank Two Matrices = Rank One plus Rank One

Here is a matrix A of rank $r = 2$. We can't see r immediately from A. So we reduce the matrix by row operations to $R = \mathrm{rref}(A)$. Some elimination matrix E simplifies A to $\boldsymbol{EA = R}$. Then the inverse matrix $C = E^{-1}$ connects R back to $\boldsymbol{A = CR}$.

You know the main point already: $\boldsymbol{R}$ **has the same row space as** $\boldsymbol{A}$.

Rank two
$$A = \begin{bmatrix} 1 & 0 & 3 \\ 1 & 1 & 7 \\ 4 & 2 & 20 \end{bmatrix} = \begin{bmatrix} \mathbf{1} & \mathbf{0} & 0 \\ \mathbf{1} & \mathbf{1} & 0 \\ \mathbf{4} & \mathbf{2} & 1 \end{bmatrix} \begin{bmatrix} \mathbf{1} & \mathbf{0} & \mathbf{3} \\ \mathbf{0} & \mathbf{1} & \mathbf{4} \\ 0 & 0 & 0 \end{bmatrix} = CR. \tag{4}$$

The row space of R clearly has two basis vectors $\boldsymbol{v}_1^{\mathrm{T}} = \begin{bmatrix} 1 & 0 & 3 \end{bmatrix}$ and $\boldsymbol{v}_2^{\mathrm{T}} = \begin{bmatrix} 0 & 1 & 4 \end{bmatrix}$. So the (same!) row space of A also has this basis: *row rank* $= 2$. Multiplying C times R says that row 3 of A is $4\boldsymbol{v}_1^{\mathrm{T}} + 2\boldsymbol{v}_2^{\mathrm{T}}$.

Now look at columns. The pivot columns of R are clearly $(1, 0, 0)$ and $(0, 1, 0)$. Then the pivot columns of A are also in columns 1 and 2: $\boldsymbol{u}_1 = (1, 1, 4)$ and $\boldsymbol{u}_2 = (0, 1, 2)$. Notice that C has those same first two columns! That was guaranteed since multiplying by two columns of the identity matrix (in R) won't change the pivot columns $\boldsymbol{u}_1$ and $\boldsymbol{u}_2$.

When you put in letters for the columns and rows, you see **rank 2 = rank 1 + rank 1**.

Matrix A Rank two
$$A = \begin{bmatrix} & & \\ \boldsymbol{u}_1 & \boldsymbol{u}_2 & \boldsymbol{u}_3 \\ & & \end{bmatrix} \begin{bmatrix} \boldsymbol{v}_1^{\mathrm{T}} \\ \boldsymbol{v}_2^{\mathrm{T}} \\ \textbf{zero row} \end{bmatrix} = \boldsymbol{u}_1\boldsymbol{v}_1^{\mathrm{T}} + \boldsymbol{u}_2\boldsymbol{v}_2^{\mathrm{T}} = (\text{rank } 1) + (\text{rank } 1).$$

Did you see that last step? I multiplied the matrices using **columns times rows**. That was perfect for this problem. *Every rank r matrix is a sum of r rank one matrices*: Pivot columns of A times nonzero rows of R. The row $\begin{bmatrix} 0 & 0 & 0 \end{bmatrix}$ simply disappeared.

The pivot columns $\boldsymbol{u}_1$ and $\boldsymbol{u}_2$ are a basis for the column space, which you knew.

■ REVIEW OF THE KEY IDEAS ■

1. The r pivot rows of R are a basis for the row spaces of R and A (same space).
2. The r pivot columns of A (!) are a basis for its column space $\boldsymbol{C}(A)$.
3. The $n - r$ special solutions are a basis for the nullspaces of A and R (same space).
4. If $EA = R$, the last $m - r$ rows of E are a basis for the left nullspace of A.

Note about the four subspaces The Fundamental Theorem looks like pure algebra, but it has very important applications. My favorites are the networks in Chapter 10 (often I go to 10.1 for my next lecture). The equation for $\boldsymbol{y}$ in the left nullspace is $A^{\mathrm{T}}\boldsymbol{y} = \mathbf{0}$:

Flow into a node equals flow out. Kirchhoff's Current Law is the "balance equation".

This must be the most important equation in applied mathematics. All models in science and engineering and economics involve a balance—of force or heat flow or charge or momentum or money. That balance equation, plus Hooke's Law or Ohm's Law or some law connecting "potentials" to "flows", gives a clear framework for applied mathematics.

My textbook on *Computational Science and Engineering* develops that framework, together with algorithms to solve the equations: Finite differences, finite elements, spectral methcds, iterative methods, and multigrid.

■ WORKED EXAMPLES ■

3.5 A Put four 1's into a 5 by 6 matrix of zeros, keeping the dimension of its *row space* as small as possible. Describe all the ways to make the dimension of its *column space* as small as possible. Describe all the ways to make the dimension of its *nullspace* as small as possible. How to make the *sum of the dimensions of all four subspaces small*?

Solution The rank is 1 if the four 1's go into the same row, or into the same column. They can also go into *two rows and two columns* (so $a_{ii} = a_{ij} = a_{ji} = a_{jj} = 1$). Since the column space and row space always have the same dimensions, this answers the first two questions: Dimension 1.

The nullspace has its smallest possible dimension $6 - 4 = 2$ when the rank is $r = 4$. To achieve rank 4, the 1's must go into four different rows and four different columns.

You can't do anything about the sum $r + (n-r) + r + (m-r) = \boldsymbol{n} + \boldsymbol{m}$. It will be $6 + 5 = 11$ no matter how the 1's are placed. The sum is 11 even if there aren't any 1's...

If all the other entries of A are 2's instead of 0's, how do these answers change?

3.5 B Fact: All the rows of AB are combinations of the rows of B. So the row space of AB is contained in (possibly equal to) the row space of B. **Rank** $(AB) \leq$ **rank** (B).

All columns of AB are combinations of the columns of A. So the column space of AB is contained in (possibly equal to) the column space of A. **Rank** $(AB) \leq$ **rank** (A).

If we multiply by an *invertible* matrix, the rank will not change. The rank can't drop, because when we multiply by the inverse matrix the rank can't jump back.

Problem Set 3.5

1 (a) If a 7 by 9 matrix has rank 5, what are the dimensions of the four subspaces? What is the sum of all four dimensions?

(b) If a 3 by 4 matrix has rank 3, what are its column space and left nullspace?

2 Find bases and dimensions for the four subspaces associated with A and B:

$$A = \begin{bmatrix} 1 & 2 & 4 \\ 2 & 4 & 8 \end{bmatrix} \quad \text{and} \quad B = \begin{bmatrix} 1 & 2 & 4 \\ 2 & 5 & 8 \end{bmatrix}.$$

3 Find a basis for each of the four subspaces associated with A:

$$A = \begin{bmatrix} 0 & 1 & 2 & 3 & 4 \\ 0 & 1 & 2 & 4 & 6 \\ 0 & 0 & 0 & 1 & 2 \end{bmatrix} = \begin{bmatrix} 1 & 0 & 0 \\ 1 & 1 & 0 \\ 0 & 1 & 1 \end{bmatrix} \begin{bmatrix} 0 & 1 & 2 & 3 & 4 \\ 0 & 0 & 0 & 1 & 2 \\ 0 & 0 & 0 & 0 & 0 \end{bmatrix}.$$

4 Construct a matrix with the required property or explain why this is impossible:

(a) Column space contains $\begin{bmatrix} 1 \\ 1 \\ 0 \end{bmatrix}, \begin{bmatrix} 0 \\ 0 \\ 1 \end{bmatrix}$, row space contains $\begin{bmatrix} 1 \\ 2 \end{bmatrix}, \begin{bmatrix} 2 \\ 5 \end{bmatrix}$.

(b) Column space has basis $\begin{bmatrix} 1 \\ 1 \\ 3 \end{bmatrix}$, nullspace has basis $\begin{bmatrix} 3 \\ 1 \\ 1 \end{bmatrix}$.

(c) Dimension of nullspace $= 1 +$ dimension of left nullspace.

(d) Nullspace contains $\begin{bmatrix} 1 \\ 3 \end{bmatrix}$, column space contains $\begin{bmatrix} 3 \\ 1 \end{bmatrix}$.

(e) Row space $=$ column space, nullspace $\neq$ left nullspace.

5 If $\mathbf{V}$ is the subspace spanned by $(1,1,1)$ and $(2,1,0)$, find a matrix A that has $\mathbf{V}$ as its row space. Find a matrix B that has $\mathbf{V}$ as its nullspace. Multiply AB.

6 Without using elimination, find dimensions and bases for the four subspaces for

$$A = \begin{bmatrix} 0 & 3 & 3 & 3 \\ 0 & 0 & 0 & 0 \\ 0 & 1 & 0 & 1 \end{bmatrix} \quad \text{and} \quad B = \begin{bmatrix} 1 \\ 4 \\ 5 \end{bmatrix}.$$

7 Suppose the 3 by 3 matrix A is invertible. Write down bases for the four subspaces for A, and also for the 3 by 6 matrix $B = [\, A \;\; A \,]$. (The basis for $\boldsymbol{Z}$ is empty.)

8 What are the dimensions of the four subspaces for A, B, and C, if I is the 3 by 3 identity matrix and 0 is the 3 by 2 zero matrix?

$$A = \begin{bmatrix} I & 0 \end{bmatrix} \quad \text{and} \quad B = \begin{bmatrix} I & I \\ 0^{\mathrm{T}} & 0^{\mathrm{T}} \end{bmatrix} \quad \text{and} \quad C = \begin{bmatrix} 0 \end{bmatrix}.$$

9 Which subspaces are the same for these matrices of different sizes?

(a) $\begin{bmatrix} A \end{bmatrix}$ and $\begin{bmatrix} A \\ A \end{bmatrix}$ (b) $\begin{bmatrix} A \\ A \end{bmatrix}$ and $\begin{bmatrix} A & A \\ A & A \end{bmatrix}$.

Prove that all three of those matrices have the *same rank* r.

10 If the entries of a 3 by 3 matrix are chosen randomly between 0 and 1, what are the most likely dimensions of the four subspaces? What if the random matrix is 3 by 5?

11 (Important) A is an m by n matrix of rank r. Suppose there are right sides $\boldsymbol{b}$ for which $A\boldsymbol{x} = \boldsymbol{b}$ has *no solution*.

(a) What are all inequalities ($<$ or $\leq$) that must be true between m, n, and r?

(b) How do you know that $A^{\mathrm{T}}\boldsymbol{y} = \mathbf{0}$ has solutions other than $\boldsymbol{y} = \mathbf{0}$?

12 Construct a matrix with $(1, 0, 1)$ and $(1, 2, 0)$ as a basis for its row space and its column space. Why can't this be a basis for the row space and nullspace?

13 True or false (with a reason or a counterexample):

(a) If $m = n$ then the row space of A equals the column space.

(b) The matrices A and $-A$ share the same four subspaces.

(c) If A and B share the same four subspaces then A is a multiple of B.

14 Without computing A, find bases for its four fundamental subspaces:

$$A = \begin{bmatrix} 1 & 0 & 0 \\ 6 & 1 & 0 \\ 9 & 8 & 1 \end{bmatrix} \begin{bmatrix} 1 & 2 & 3 & 4 \\ 0 & 1 & 2 & 3 \\ 0 & 0 & 1 & 2 \end{bmatrix}.$$

15 If you exchange the first two rows of A, which of the four subspaces stay the same? If $\boldsymbol{v} = (1, 2, 3, 4)$ is in the left nullspace of A, write down a vector in the left nullspace of the new matrix after the row exchange.

16 *Explain why* $\boldsymbol{v} = (1, 0, -1)$ *cannot be a row of A and also in the nullspace.*

17 Describe the four subspaces of $\mathbf{R}^3$ associated with

$$A = \begin{bmatrix} 0 & 1 & 0 \\ 0 & 0 & 1 \\ 0 & 0 & 0 \end{bmatrix} \quad \text{and} \quad I + A = \begin{bmatrix} 1 & 1 & 0 \\ 0 & 1 & 1 \\ 0 & 0 & 1 \end{bmatrix}.$$

18 (Left nullspace) Add the extra column $\boldsymbol{b}$ and reduce A to echelon form:

$$\begin{bmatrix} A & \boldsymbol{b} \end{bmatrix} = \begin{bmatrix} 1 & 2 & 3 & b_1 \\ 4 & 5 & 6 & b_2 \\ 7 & 8 & 9 & b_3 \end{bmatrix} \quad \to \quad \begin{bmatrix} 1 & 2 & 3 & b_1 \\ 0 & -3 & -6 & b_2 - 4b_1 \\ 0 & 0 & 0 & b_3 - 2b_2 + b_1 \end{bmatrix}.$$

A combination of the rows of A has produced the zero row. What combination is it? (Look at $b_3 - 2b_2 + b_1$ on the right side.) Which vectors are in the nullspace of A^{T} and which vectors are in the nullspace of A?

19 Following the method of Problem 18, reduce A to echelon form and look at zero rows. The $\boldsymbol{b}$ column tells which combinations you have taken of the rows:

$$\text{(a)} \quad \begin{bmatrix} 1 & 2 & b_1 \\ 3 & 4 & b_2 \\ 4 & 6 & b_3 \end{bmatrix} \qquad \text{(b)} \quad \begin{bmatrix} 1 & 2 & b_1 \\ 2 & 3 & b_2 \\ 2 & 4 & b_3 \\ 2 & 5 & b_4 \end{bmatrix}$$

From the $\boldsymbol{b}$ column after elimination, read off $m-r$ basis vectors in the left nullspace. Those $\boldsymbol{y}$'s are combinations of rows that give zero rows in the echelon form.

20 (a) Check that the solutions to $A\boldsymbol{x} = \boldsymbol{0}$ are perpendicular to the rows of A:

$$A = \begin{bmatrix} 1 & 0 & 0 \\ 2 & 1 & 0 \\ 3 & 4 & 1 \end{bmatrix} \begin{bmatrix} 4 & 2 & 0 & 1 \\ 0 & 0 & 1 & 3 \\ 0 & 0 & 0 & 0 \end{bmatrix} = ER.$$

(b) How many independent solutions to $A^{\mathrm{T}}\boldsymbol{y} = \boldsymbol{0}$? Why does $\boldsymbol{y}^{\mathrm{T}} =$ row 3 of E^{-1}?

21 Suppose A is the sum of two matrices of rank one: $A = \boldsymbol{u}\boldsymbol{v}^{\mathrm{T}} + \boldsymbol{w}\boldsymbol{z}^{\mathrm{T}}$.

(a) Which vectors span the column space of A?

(b) Which vectors span the row space of A?

(c) The rank is less than 2 if ______ or if ______.

(d) Compute A and its rank if $\boldsymbol{u} = \boldsymbol{z} = (1,0,0)$ and $\boldsymbol{v} = \boldsymbol{w} = (0,0,1)$.

22 Construct $A = \boldsymbol{u}\boldsymbol{v}^{\mathrm{T}} + \boldsymbol{w}\boldsymbol{z}^{\mathrm{T}}$ whose column space has basis $(1,2,4), (2,2,1)$ and whose row space has basis $(1,0), (1,1)$. Write A as (3 by 2) times (2 by 2).

23 Without multiplying matrices, find bases for the row and column spaces of A:

$$A = \begin{bmatrix} 1 & 2 \\ 4 & 5 \\ 2 & 7 \end{bmatrix} \begin{bmatrix} 3 & 0 & 3 \\ 1 & 1 & 2 \end{bmatrix}.$$

How do you know from these shapes that A cannot be invertible?

24 (Important) $A^{\mathrm{T}}\boldsymbol{y} = \boldsymbol{d}$ is solvable when $\boldsymbol{d}$ is in which of the four subspaces? The solution $\boldsymbol{y}$ is unique when the ______ contains only the zero vector.

25 True or false (with a reason or a counterexample):

(a) A and A^{T} have the same number of pivots.

(b) A and A^{T} have the same left nullspace.

(c) If the row space equals the column space then $A^{\mathrm{T}} = A$.

(d) If $A^{\mathrm{T}} = -A$ then the row space of A equals the column space.

26 If a, b, c are given with $a \neq 0$, how would you choose d so that $\begin{bmatrix} a & b \\ c & d \end{bmatrix}$ has rank 1? Find a basis for the row space and nullspace. Show they are perpendicular!

27 Find the ranks of the 8 by 8 checkerboard matrix B and the chess matrix C:

$$B = \begin{bmatrix} 1 & 0 & 1 & 0 & 1 & 0 & 1 & 0 \\ 0 & 1 & 0 & 1 & 0 & 1 & 0 & 1 \\ 1 & 0 & 1 & 0 & 1 & 0 & 1 & 0 \\ \cdot & \cdot & \cdot & \cdot & \cdot & \cdot & \cdot & \cdot \\ 0 & 1 & 0 & 1 & 0 & 1 & 0 & 1 \end{bmatrix} \quad \text{and} \quad C = \begin{bmatrix} r & n & b & q & k & b & n & r \\ p & p & p & p & p & p & p & p \\ & & & \text{four zero rows} & & & & \\ p & p & p & p & p & p & p & p \\ r & n & b & q & k & b & n & r \end{bmatrix}$$

The numbers r, n, b, q, k, p are all different. Find bases for the row space and left nullspace of B and C. Challenge problem: Find a basis for the nullspace of C.

28 Can tic-tac-toe be completed (5 ones and 4 zeros in A) so that rank $(A) = 2$ but neither side passed up a winning move?

Challenge Problems

29 If $A = \boldsymbol{u}\boldsymbol{v}^{\mathrm{T}}$ is a 2 by 2 matrix of rank 1, redraw Figure 3.5 to show clearly the Four Fundamental Subspaces. If B produces those same four subspaces, what is the exact relation of B to A?

30 $\mathbf{M}$ is the space of 3 by 3 matrices. Multiply every matrix X in $\mathbf{M}$ by

$$A = \begin{bmatrix} 1 & 0 & -1 \\ -1 & 1 & 0 \\ 0 & -1 & 1 \end{bmatrix}. \quad \text{Notice: } A \begin{bmatrix} 1 \\ 1 \\ 1 \end{bmatrix} = \begin{bmatrix} 0 \\ 0 \\ 0 \end{bmatrix}.$$

(a) Which matrices X lead to $AX =$ zero matrix?

(b) Which matrices have the form AX for some matrix X?

(a) finds the "nullspace" of that operation AX and (b) finds the "column space". What are the dimensions of those two subspaces of $\mathbf{M}$? Why do the dimensions add to $(n - r) + r = 9$?

31 Suppose the m by n matrices A and B have *the same four subspaces*. If they are both in row reduced echelon form, prove that F must equal G:

$$A = \begin{bmatrix} I & F \\ 0 & 0 \end{bmatrix} \qquad B = \begin{bmatrix} I & G \\ 0 & 0 \end{bmatrix}.$$

Chapter 4

Orthogonality

4.1 Orthogonality of the Four Subspaces

1 Orthogonal vectors have $\boldsymbol{v}^{\mathrm{T}}\boldsymbol{w} = 0$. Then $||\boldsymbol{v}||^2 + ||\boldsymbol{w}||^2 = ||\boldsymbol{v}+\boldsymbol{w}||^2 = ||\boldsymbol{v}-\boldsymbol{w}||^2$.

2 Subspaces $\boldsymbol{V}$ and $\boldsymbol{W}$ are orthogonal when $\boldsymbol{v}^{\mathrm{T}}\boldsymbol{w} = 0$ for every $\boldsymbol{v}$ in $\boldsymbol{V}$ and every $\boldsymbol{w}$ in $\boldsymbol{W}$.

3 The row space of A is orthogonal to the nullspace. The column space is orthogonal to $\boldsymbol{N}(A^{\mathrm{T}})$.

4 One pair of dimensions adds to $r + (n - r) = n$. The other pair has $r + (m - r) = m$.

5 Row space and nullspace are orthogonal *complements*. Every $\boldsymbol{x}$ in $\mathbf{R}^n$ splits into $\boldsymbol{x}_{\textbf{row}} + \boldsymbol{x}_{\textbf{null}}$.

6 Suppose a space $\boldsymbol{S}$ has dimension d. Then every basis for $\boldsymbol{S}$ consists of d vectors.

7 If d vectors in $\boldsymbol{S}$ are independent, they span $\boldsymbol{S}$. If d vectors span $\boldsymbol{S}$, they are independent.

Two vectors are orthogonal when their dot product is zero: $\boldsymbol{v} \cdot \boldsymbol{w} = \boldsymbol{v}^{\mathrm{T}}\boldsymbol{w} = 0$. This chapter moves to **orthogonal subspaces** and **orthogonal bases** and **orthogonal matrices**. The vectors in two subspaces, and the vectors in a basis, and the column vectors in Q, all pairs will be orthogonal. Think of $a^2 + b^2 = c^2$ for a *right triangle* with sides $\boldsymbol{v}$ and $\boldsymbol{w}$.

Orthogonal vectors	$\boldsymbol{v}^{\mathrm{T}}\boldsymbol{w} = 0$	and	$\|\boldsymbol{v}\|^2 + \|\boldsymbol{w}\|^2 = \|\boldsymbol{v}+\boldsymbol{w}\|^2$.

The right side is $(\boldsymbol{v}+\boldsymbol{w})^{\mathrm{T}}(\boldsymbol{v}+\boldsymbol{w})$. This equals $\boldsymbol{v}^{\mathrm{T}}\boldsymbol{v} + \boldsymbol{w}^{\mathrm{T}}\boldsymbol{w}$ when $\boldsymbol{v}^{\mathrm{T}}\boldsymbol{w} = \boldsymbol{w}^{\mathrm{T}}\boldsymbol{v} = 0$.

Subspaces entered Chapter 3 to throw light on $A\boldsymbol{x} = \boldsymbol{b}$. Right away we needed the column space and the nullspace. Then the light turned onto A^{T}, uncovering two more subspaces. Those four fundamental subspaces reveal what a matrix really does.

A matrix multiplies a vector: *A times* $\boldsymbol{x}$. At the first level this is only numbers. At the second level $A\boldsymbol{x}$ is a combination of column vectors. The third level shows subspaces. But I don't think you have seen the whole picture until you study Figure 4.2.

The subspaces fit together to show the hidden reality of A times $\boldsymbol{x}$. The $90°$ angles between subspaces are new—and we can say now what those right angles mean.

The row space is perpendicular to the nullspace. Every row of A is perpendicular to every solution of $A\boldsymbol{x} = \mathbf{0}$. That gives the $90°$ angle on the left side of the figure. This perpendicularity of subspaces is Part 2 of the Fundamental Theorem of Linear Algebra.

The column space is perpendicular to the nullspace of A^{T}. When $\boldsymbol{b}$ is outside the column space—when we want to solve $A\boldsymbol{x} = \boldsymbol{b}$ and can't do it—then this nullspace of A^{T} comes into its own. It contains the error $\boldsymbol{e} = \boldsymbol{b} - A\boldsymbol{x}$ in the "least-squares" solution. Least squares is the key application of linear algebra in this chapter.

Part 1 of the Fundamental Theorem gave the dimensions of the subspaces. The row and column spaces have the same dimension r (they are drawn the same size). The two nullspaces have the remaining dimensions $n - r$ and $m - r$. Now we will show that ***the row space and nullspace are orthogonal subspaces inside*** $\mathbf{R}^n$.

DEFINITION Two subspaces $\boldsymbol{V}$ and $\boldsymbol{W}$ of a vector space are ***orthogonal*** if every vector $\boldsymbol{v}$ in $\boldsymbol{V}$ is perpendicular to every vector $\boldsymbol{w}$ in $\boldsymbol{W}$:

Orthogonal subspaces $\quad \boldsymbol{v}^{\mathrm{T}}\boldsymbol{w} = 0$ ***for all*** $\boldsymbol{v}$ ***in*** $\boldsymbol{V}$ ***and all*** $\boldsymbol{w}$ ***in*** $\boldsymbol{W}$.

Example 1 The floor of your room (extended to infinity) is a subspace $\boldsymbol{V}$. The line where two walls meet is a subspace $\boldsymbol{W}$ (one-dimensional). Those subspaces are orthogonal. Every vector up the meeting line of the walls is perpendicular to every vector in the floor.

Example 2 Two walls look perpendicular but those two subspaces are not orthogonal! The meeting line is in both $\boldsymbol{V}$ and $\boldsymbol{W}$—and this line is not perpendicular to itself. Two planes (dimensions 2 and 2 in $\mathbf{R}^3$) cannot be orthogonal subspaces.

When a vector is in two orthogonal subspaces, it *must* be zero. It is perpendicular to itself. It is $\boldsymbol{v}$ and it is $\boldsymbol{w}$, so $\boldsymbol{v}^{\mathrm{T}}\boldsymbol{v} = 0$. This has to be the zero vector.

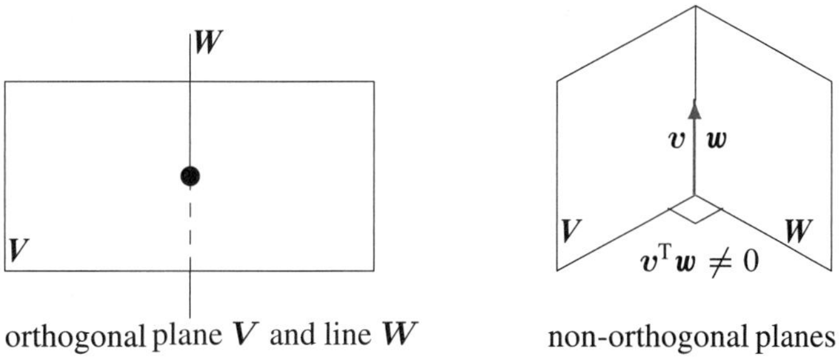

Figure 4.1: Orthogonality is impossible when $\dim \boldsymbol{V} + \dim \boldsymbol{W} > \dim$ (whole space).

The crucial examples for linear algebra come from the four fundamental subspaces. Zero is the only point where the nullspace meets the row space. More than that, the **nullspace and row space of** A **meet at** $\mathbf{90°}$. This key fact comes directly from $A\boldsymbol{x} = \mathbf{0}$:

Every vector $\boldsymbol{x}$ in the nullspace is perpendicular to every row of A, because $A\boldsymbol{x} = \mathbf{0}$. ***The nullspace*** $\boldsymbol{N}(A)$ ***and the row space*** $\boldsymbol{C}(A^{\mathrm{T}})$ ***are orthogonal subspaces of*** $\mathbf{R}^n$.

To see why $\boldsymbol{x}$ is perpendicular to the rows, look at $A\boldsymbol{x} = \mathbf{0}$. Each row multiplies $\boldsymbol{x}$:

$$A\boldsymbol{x} = \begin{bmatrix} \textbf{row 1} \\ \vdots \\ \textbf{row } m \end{bmatrix} \begin{bmatrix} \\ \boldsymbol{x} \\ \\ \end{bmatrix} = \begin{bmatrix} 0 \\ \vdots \\ 0 \end{bmatrix} \quad \begin{matrix} \longleftarrow & (\textbf{row 1}) \cdot \boldsymbol{x} \text{ is zero} \\ \\ \longleftarrow & (\textbf{row } m) \cdot \boldsymbol{x} \text{ is zero} \end{matrix} \tag{1}$$

The first equation says that row 1 is perpendicular to $\boldsymbol{x}$. The last equation says that row m is perpendicular to $\boldsymbol{x}$. *Every row has a zero dot product with* $\boldsymbol{x}$. Then $\boldsymbol{x}$ is also perpendicular to every *combination* of the rows. The whole row space $\boldsymbol{C}(A^{\mathrm{T}})$ is orthogonal to $\boldsymbol{N}(A)$.

Here is a second proof of that orthogonality for readers who like matrix shorthand. The vectors in the row space are combinations $A^{\mathrm{T}}\boldsymbol{y}$ of the rows. Take the dot product of $A^{\mathrm{T}}\boldsymbol{y}$ with any $\boldsymbol{x}$ in the nullspace. *These vectors are perpendicular*:

Nullspace orthogonal to row space $\qquad \boldsymbol{x}^{\mathrm{T}}(A^{\mathrm{T}}\boldsymbol{y}) = (A\boldsymbol{x})^{\mathrm{T}}\boldsymbol{y} = \mathbf{0}^{\mathrm{T}}\boldsymbol{y} = 0. \qquad (2)$

We like the first proof. You can see those rows of A multiplying $\boldsymbol{x}$ to produce zeros in equation (1). The second proof shows why A and A^{T} are both in the Fundamental Theorem.

Example 3 The rows of A are perpendicular to $\boldsymbol{x} = (1, 1, -1)$ in the nullspace:

$$A\boldsymbol{x} = \begin{bmatrix} 1 & 3 & 4 \\ 5 & 2 & 7 \end{bmatrix} \begin{bmatrix} 1 \\ 1 \\ -1 \end{bmatrix} = \begin{bmatrix} 0 \\ 0 \end{bmatrix} \quad \text{gives the dot products} \quad \begin{matrix} 1+3-4=0 \\ 5+2-7=0 \end{matrix}$$

Now we turn to the other two subspaces. In this example, the column space is all of $\mathbf{R}^2$. The nullspace of A^{T} is only the zero vector (orthogonal to every vector). The column space of A and the nullspace of A^{T} are always orthogonal subspaces.

Every vector $\boldsymbol{y}$ in the nullspace of A^{T} is perpendicular to every column of A. ***The left nullspace*** $\boldsymbol{N}(A^{\mathrm{T}})$ ***and the column space*** $\boldsymbol{C}(A)$ ***are orthogonal in*** $\mathbf{R}^m$.

Apply the original proof to A^{T}. The nullspace of A^{T} is orthogonal to the row space of A^{T}—and the row space of A^{T} is the column space of A. Q.E.D.

For a visual proof, look at $A^{\mathrm{T}}\boldsymbol{y} = \mathbf{0}$. Each column of A multiplies $\boldsymbol{y}$ to give 0:

$$\boldsymbol{C}(\boldsymbol{A}) \perp \boldsymbol{N}(\boldsymbol{A}^{\mathrm{T}}) \qquad A^{\mathrm{T}}\boldsymbol{y} = \begin{bmatrix} (\textbf{column 1})^{\mathrm{T}} \\ \cdots \\ (\textbf{column } n)^{\mathrm{T}} \end{bmatrix} \begin{bmatrix} \\ \boldsymbol{y} \\ \\ \end{bmatrix} = \begin{bmatrix} 0 \\ \cdot \\ 0 \end{bmatrix}. \tag{3}$$

The dot product of $\boldsymbol{y}$ with every column of A is zero. Then $\boldsymbol{y}$ in the left nullspace is perpendicular to each column of A—and to the whole column space.

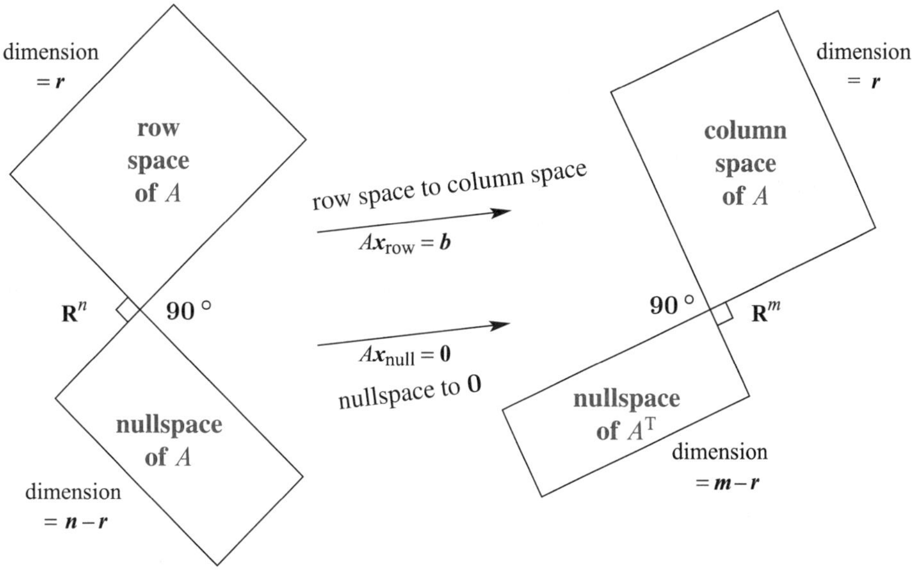

Figure 4.2: Two pairs of orthogonal subspaces. The dimensions add to n and add to m. **This is the Big Picture**—two subspaces in $\mathbf{R}^n$ and two subspaces in $\mathbf{R}^m$.

Orthogonal Complements

Important The fundamental subspaces are more than just orthogonal (in pairs). Their dimensions are also right. Two lines could be perpendicular in $\mathbf{R}^3$, **but those lines *could not be* the row space and nullspace of a 3 by 3 matrix**. The lines have dimensions 1 and 1, adding to 2. But the correct dimensions r and $n-r$ must add to $n=3$.

The fundamental subspaces of a 3 by 3 matrix have dimensions 2 and 1, or 3 and 0. Those pairs of subspaces are not only orthogonal, they are *orthogonal complements*.

DEFINITION The ***orthogonal complement*** of a subspace $\boldsymbol{V}$ contains ***every*** vector that is perpendicular to $\boldsymbol{V}$. This orthogonal subspace is denoted by $\boldsymbol{V}^{\perp}$ (pronounced "$\boldsymbol{V}$ perp").

By this definition, the nullspace is the orthogonal complement of the row space. ***Every*** $\boldsymbol{x}$ that is perpendicular to the rows satisfies $A\boldsymbol{x} = \mathbf{0}$, and lies in the nullspace.

The reverse is also true. ***If $\boldsymbol{v}$ is orthogonal to the nullspace, it must be in the row space***. Otherwise we could add this $\boldsymbol{v}$ as an extra row of the matrix, without changing its nullspace. The row space would grow, which breaks the law $r+(n-r)=n$. We conclude that the nullspace complement $\boldsymbol{N}(A)^{\perp}$ is exactly the row space $\boldsymbol{C}(A^{\mathrm{T}})$.

In the same way, the left nullspace and column space are orthogonal in $\mathbf{R}^m$, and they are orthogonal complements. Their dimensions r and $m-r$ add to the full dimension m.

Fundamental Theorem of Linear Algebra, Part 2

$N(A)$ is the orthogonal complement of the row space $C(A^{\mathrm{T}})$ (in $\mathbf{R}^n$).

$N(A^{\mathrm{T}})$ is the orthogonal complement of the column space $C(A)$ (in $\mathbf{R}^m$).

Part 1 gave the dimensions of the subspaces. Part 2 gives the 90° angles between them. The point of "complements" is that every $\boldsymbol{x}$ can be split into a *row space component* $\boldsymbol{x}_r$ and a *nullspace component* $\boldsymbol{x}_n$. When A multiplies $\boldsymbol{x} = \boldsymbol{x}_r + \boldsymbol{x}_n$, Figure 4.3 shows what happens to $A\boldsymbol{x} = A\boldsymbol{x}_r + A\boldsymbol{x}_n$:

The nullspace component goes to zero: $A\boldsymbol{x}_n = \mathbf{0}$.

The row space component goes to the column space: $A\boldsymbol{x}_r = A\boldsymbol{x}$.

Every vector goes to the column space! Multiplying by A cannot do anything else. More than that: *Every vector $\boldsymbol{b}$ in the column space comes from one and only one vector $\boldsymbol{x}_r$ in the row space*. Proof: If $A\boldsymbol{x}_r = A\boldsymbol{x}'_r$, the difference $\boldsymbol{x}_r - \boldsymbol{x}'_r$ is in the nullspace. It is also in the row space, where $\boldsymbol{x}_r$ and $\boldsymbol{x}'_r$ came from. This difference must be the zero vector, because the nullspace and row space are perpendicular. Therefore $\boldsymbol{x}_r = \boldsymbol{x}'_r$.

There is an r by r invertible matrix hiding inside A, if we throw away the two nullspaces. ***From the row space to the column space, A is invertible***. The "pseudoinverse" will invert that part of A in Section 7.4.

Example 4 Every matrix of rank r has an r by r invertible submatrix:

$$A = \begin{bmatrix} 3 & 0 & 0 & 0 & 0 \\ 0 & 5 & 0 & 0 & 0 \\ 0 & 0 & 0 & 0 & 0 \end{bmatrix} \quad \text{contains the submatrix} \quad \begin{bmatrix} 3 & 0 \\ 0 & 5 \end{bmatrix}.$$

The other eleven zeros are responsible for the nullspaces. The rank of B is also $r = 2$:

$$B = \begin{bmatrix} 1 & 2 & 3 & 4 & 5 \\ 1 & 2 & 4 & 5 & 6 \\ 1 & 2 & 4 & 5 & 6 \end{bmatrix} \quad \text{contains} \quad \begin{bmatrix} 1 & 3 \\ 1 & 4 \end{bmatrix} \quad \text{in the pivot rows and columns.}$$

Every matrix can be diagonalized, when we choose the right bases for $\mathbf{R}^n$ and $\mathbf{R}^m$. This ***Singular Value Decomposition*** has become extremely important in applications.

Let me repeat one clear fact. A row of A can't be in the nullspace of A (except for a zero row). The only vector in two orthogonal subspaces is the zero vector.

If a vector v is orthogonal to itself then v is the zero vector.

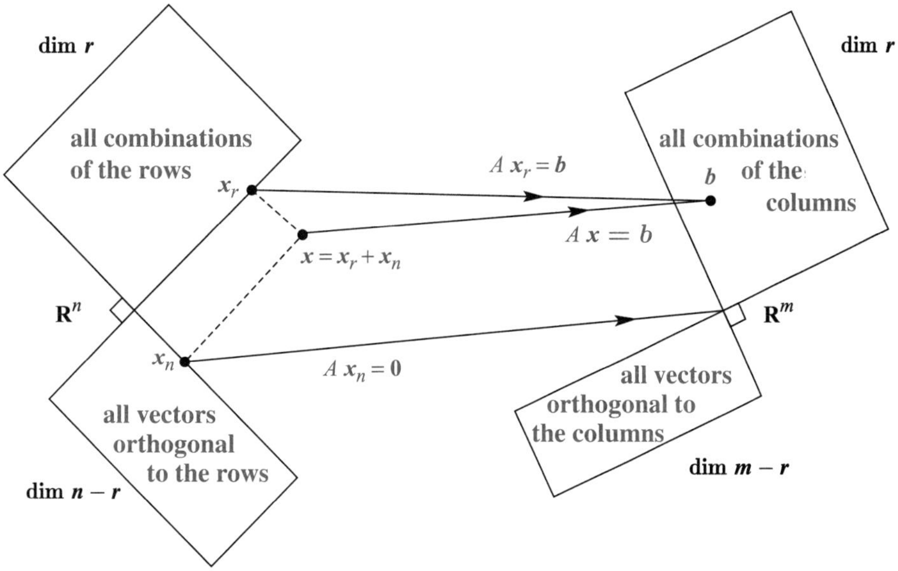

Figure 4.3: This update of Figure 4.2 shows the true action of A on $\boldsymbol{x} = \boldsymbol{x}_r + \boldsymbol{x}_n$. Row space vector $\boldsymbol{x}_r$ to column space, nullspace vector $\boldsymbol{x}_n$ to zero.

Drawing the Big Picture

I don't know the best way to draw the four subspaces in Figures 4.2 and 4.3. This big picture has to show the orthogonality of those subspaces. I can see a possible way to do it when a line meets a plane—maybe Figure 4.4 also shows that those spaces are infinite, more clearly than the rectangles in Figure 4.3. But how do I draw a pair of two-dimensional subspaces in $\mathbf{R}^4$, to show they are orthogonal to each other? Good ideas are welcome.

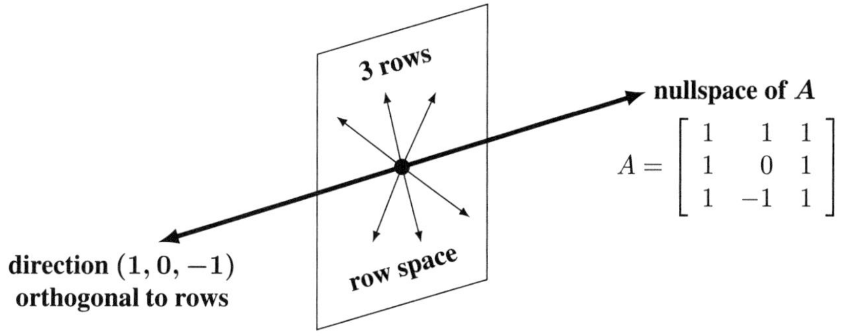

Figure 4.4: Row space of A = plane. Nullspace = orthogonal line. Dimensions $2 + 1 = 3$.

Combining Bases from Subspaces

What follows are some valuable facts about bases. They were saved until now—when we are ready to use them. After a week you have a clearer sense of what a basis is (*linearly independent* vectors that *span the space*). Normally we have to check both properties. When the count is right, one property implies the other:

Any n independent vectors in $\mathbf{R}^n$ must span $\mathbf{R}^n$. So they are a basis.
Any n vectors that span $\mathbf{R}^n$ must be independent. So they are a basis.

Starting with the correct number of vectors, one property of a basis produces the other. This is true in any vector space, but we care most about $\mathbf{R}^n$. When the vectors go into the columns of an n by n *square* matrix A, here are the same two facts:

If the n columns of A are independent, they span $\mathbf{R}^n$. So $A\boldsymbol{x} = \boldsymbol{b}$ is solvable.
If the n columns span $\mathbf{R}^n$, they are independent. So $A\boldsymbol{x} = \boldsymbol{b}$ has only one solution.

Uniqueness implies existence and existence implies uniqueness. ***Then A is invertible.*** If there are no free variables, the solution $\boldsymbol{x}$ is unique. There must be n pivot columns. Then back substitution solves $A\boldsymbol{x} = \boldsymbol{b}$ (the solution exists).

Starting in the opposite direction, suppose that $A\boldsymbol{x} = \boldsymbol{b}$ can be solved for every $\boldsymbol{b}$ (*existence of solutions*). Then elimination produced no zero rows. There are n pivots and no free variables. The nullspace contains only $\boldsymbol{x} = \mathbf{0}$ (*uniqueness* of solutions).

With bases for the row space and the nullspace, we have $r + (n - r) = n$ vectors. This is the right number. Those n vectors are independent.[2] *Therefore they span* $\mathbf{R}^n$.

Each $\boldsymbol{x}$ is the sum $\boldsymbol{x}_r + \boldsymbol{x}_n$ of a row space vector $\boldsymbol{x}_r$ and a nullspace vector $\boldsymbol{x}_n$.

The splitting in Figure 4.3 shows the key point of orthogonal complements—the dimensions add to n and all vectors are fully accounted for.

Example 5 For $A = \begin{bmatrix} 1 & 2 \\ 3 & 6 \end{bmatrix}$ split $\boldsymbol{x} = \begin{bmatrix} 4 \\ 3 \end{bmatrix}$ into $\boldsymbol{x}_r + \boldsymbol{x}_n = \begin{bmatrix} 2 \\ 4 \end{bmatrix} + \begin{bmatrix} 2 \\ -1 \end{bmatrix}$.

The vector $(2, 4)$ is in the row space. The orthogonal vector $(2, -1)$ is in the nullspace. The next section will compute this splitting for any A and $\boldsymbol{x}$, by a projection.

[2] If a combination of all n vectors gives $\boldsymbol{x}_r + \boldsymbol{x}_n = \mathbf{0}$, then $\boldsymbol{x}_r = -\boldsymbol{x}_n$ is in both subspaces. So $\boldsymbol{x}_r = \boldsymbol{x}_n = \mathbf{0}$. All coefficients of the row space basis and of the nullspace basis must be zero. This proves independence of the n vectors together.

■ REVIEW OF THE KEY IDEAS ■

1. Subspaces $\boldsymbol{V}$ and $\boldsymbol{W}$ are orthogonal if every $\boldsymbol{v}$ in $\boldsymbol{V}$ is orthogonal to every $\boldsymbol{w}$ in $\boldsymbol{W}$.
2. $\boldsymbol{V}$ and $\boldsymbol{W}$ are "orthogonal complements" if $\boldsymbol{W}$ contains **all** vectors perpendicular to $\boldsymbol{V}$ (and vice versa). Inside $\mathbf{R}^n$, the dimensions of complements $\boldsymbol{V}$ and $\boldsymbol{W}$ add to n.
3. The nullspace $\boldsymbol{N}(A)$ and the row space $\boldsymbol{C}(A^{\mathrm{T}})$ are orthogonal complements, with dimensions $(n-r)+r=n$. Similarly $\boldsymbol{N}(A^{\mathrm{T}})$ and $\boldsymbol{C}(A)$ are orthogonal complements with $(m-r)+r=m$.
4. Any n independent vectors in $\mathbf{R}^n$ span $\mathbf{R}^n$. Any n spanning vectors are independent.

■ WORKED EXAMPLES ■

4.1 A Suppose $\boldsymbol{S}$ is a six-dimensional subspace of nine-dimensional space $\mathbf{R}^9$.

(a) What are the possible dimensions of subspaces orthogonal to $\boldsymbol{S}$?

(b) What are the possible dimensions of the orthogonal complement $\boldsymbol{S}^{\perp}$ of $\boldsymbol{S}$?

(c) What is the smallest possible size of a matrix A that has row space $\boldsymbol{S}$?

(d) What is the smallest possible size of a matrix B that has nullspace $\boldsymbol{S}^{\perp}$?

Solution

(a) If $\boldsymbol{S}$ is six-dimensional in $\mathbf{R}^9$, subspaces orthogonal to $\boldsymbol{S}$ can have dimensions $0, 1, 2, 3$.

(b) The complement $\boldsymbol{S}^{\perp}$ is the largest orthogonal subspace, with dimension 3.

(c) The smallest matrix A is 6 by 9 (its six rows will be a basis for $\boldsymbol{S}$).

(d) This is the same as question (c) !

If a new row 7 of B is a combination of the six rows of A, then B has the same row space as A. It also has the same nullspace. The special solutions $\boldsymbol{s}_1, \boldsymbol{s}_2, \boldsymbol{s}_3$ to $A\boldsymbol{x}=\mathbf{0}$. will be the same for $B\boldsymbol{x}=\mathbf{0}$. Elimination will change row 7 of B to all zeros.

4.1 B The equation $x-3y-4z=0$ describes a plane $\boldsymbol{P}$ in $\mathbf{R}^3$ (actually a subspace).

(a) The plane $\boldsymbol{P}$ is the nullspace $\boldsymbol{N}(A)$ of what 1 by 3 matrix A? *Ans*: $A=[1\ -3\ -4]$.

(b) Find a basis $\boldsymbol{s}_1, \boldsymbol{s}_2$ of special solutions of $x-3y-4z=0$ (these would be the columns of the nullspace matrix N). *Answer*: $\boldsymbol{s}_1=(3,1,0)$ and $\boldsymbol{s}_2=(4,0,1)$.

(c) Find a basis for the line $\boldsymbol{P}^{\perp}$ that is perpendicular to $\boldsymbol{P}$. *Answer*: $(1,-3,-4)$**!**

Problem Set 4.1

Questions 1–12 grow out of Figures 4.2 and 4.3 with four subspaces.

1 Construct any 2 by 3 matrix of rank one. Copy Figure 4.2 and put one vector in each subspace (and put two in the nullspace). Which vectors are orthogonal?

2 Redraw Figure 4.3 for a 3 by 2 matrix of rank $r = 2$. Which subspace is $\boldsymbol{Z}$ (zero vector only)? The nullspace part of any vector $\boldsymbol{x}$ in $\mathbf{R}^2$ is $\boldsymbol{x}_n =$ ____.

3 Construct a matrix with the required property or say why that is impossible:

(a) Column space contains $\begin{bmatrix} 1 \\ 2 \\ -3 \end{bmatrix}$ and $\begin{bmatrix} 2 \\ -3 \\ 5 \end{bmatrix}$, nullspace contains $\begin{bmatrix} 1 \\ 1 \\ 1 \end{bmatrix}$

(b) Row space contains $\begin{bmatrix} 1 \\ 2 \\ -3 \end{bmatrix}$ and $\begin{bmatrix} 2 \\ -3 \\ 5 \end{bmatrix}$, nullspace contains $\begin{bmatrix} 1 \\ 1 \\ 1 \end{bmatrix}$

(c) $A\boldsymbol{x} = \begin{bmatrix} 1 \\ 1 \\ 1 \end{bmatrix}$ has a solution and $A^{\mathrm{T}} \begin{bmatrix} 1 \\ 0 \\ 0 \end{bmatrix} = \begin{bmatrix} 0 \\ 0 \\ 0 \end{bmatrix}$

(d) Every row is orthogonal to every column (A is not the zero matrix)

(e) Columns add up to a column of zeros, rows add to a row of 1's.

4 If $AB = 0$ then the columns of B are in the ____ of A. The rows of A are in the ____ of B. With $AB = 0$, why can't A and B be 3 by 3 matrices of rank 2?

5 (a) If $A\boldsymbol{x} = \boldsymbol{b}$ has a solution and $A^{\mathrm{T}}\boldsymbol{y} = \mathbf{0}$, *is* $(\boldsymbol{y}^{\mathrm{T}}\boldsymbol{x} = 0)$ *or* $(\boldsymbol{y}^{\mathrm{T}}\boldsymbol{b} = 0)$?

(b) If $A^{\mathrm{T}}\boldsymbol{y} = (1, 1, 1)$ has a solution and $A\boldsymbol{x} = \mathbf{0}$, then ____

6 This system of equations $A\boldsymbol{x} = \boldsymbol{b}$ has *no solution* (they lead to $0 = 1$):

$$\begin{aligned} x + 2y + 2z &= 5 \\ 2x + 2y + 3z &= 5 \\ 3x + 4y + 5z &= 9 \end{aligned}$$

Find numbers y_1, y_2, y_3 to multiply the equations so they add to $0 = 1$. You have found a vector $\boldsymbol{y}$ in which subspace? Its dot product $\boldsymbol{y}^{\mathrm{T}}\boldsymbol{b}$ is 1, so no solution $\boldsymbol{x}$.

7 Every system with no solution is like the one in Problem 6. There are numbers $y_1, \ldots, y_m$ that multiply the m equations so they add up to $0 = 1$. This is called **Fredholm's Alternative**:

Exactly one of these problems has a solution

$$A\boldsymbol{x} = \boldsymbol{b} \quad \textbf{OR} \quad A^{\mathrm{T}}\boldsymbol{y} = \mathbf{0} \ \text{ with } \ \boldsymbol{y}^{\mathrm{T}}\boldsymbol{b} = 1.$$

If $\boldsymbol{b}$ is not in the column space of A, it is not orthogonal to the nullspace of A^{T}. Multiply the equations $x_1 - x_2 = 1$ and $x_2 - x_3 = 1$ and $x_1 - x_3 = 1$ by numbers y_1, y_2, y_3 chosen so that the equations add up to $0 = 1$.

8 In Figure 4.3, how do we know that $A\boldsymbol{x}_r$ is equal to $A\boldsymbol{x}$? How do we know that this vector is in the column space? If $A = \left[\begin{smallmatrix}1 & 1\\ 1 & 1\end{smallmatrix}\right]$ and $\boldsymbol{x} = \left[\begin{smallmatrix}1\\ 0\end{smallmatrix}\right]$ what is $\boldsymbol{x}_r$?

9 If $A^{\mathrm{T}}A\boldsymbol{x} = \mathbf{0}$ then $A\boldsymbol{x} = \mathbf{0}$. Reason: $A\boldsymbol{x}$ is in the nullspace of A^{T} and also in the _____ of A and those spaces are _____. *Conclusion*: *$A^{\mathrm{T}}A$ has the same nullspace as A. This key fact is repeated in the next section.*

10 Suppose A is a symmetric matrix ($A^{\mathrm{T}} = A$).

(a) Why is its column space perpendicular to its nullspace?

(b) If $A\boldsymbol{x} = \mathbf{0}$ and $A\boldsymbol{z} = 5\boldsymbol{z}$, which subspaces contain these "eigenvectors" $\boldsymbol{x}$ and $\boldsymbol{z}$? **Symmetric matrices have perpendicular eigenvectors $\boldsymbol{x}^{\mathrm{T}}\boldsymbol{z} = 0$.**

11 (Recommended) Draw Figure 4.2 to show each subspace correctly for

$$A = \begin{bmatrix} 1 & 2 \\ 3 & 6 \end{bmatrix} \quad \text{and} \quad B = \begin{bmatrix} 1 & 0 \\ 3 & 0 \end{bmatrix}.$$

12 Find the pieces $\boldsymbol{x}_r$ and $\boldsymbol{x}_n$ and draw Figure 4.3 properly if

$$A = \begin{bmatrix} 1 & -1 \\ 0 & 0 \\ 0 & 0 \end{bmatrix} \quad \text{and} \quad \boldsymbol{x} = \begin{bmatrix} 2 \\ 0 \end{bmatrix}.$$

Questions 13–23 are about orthogonal subspaces.

13 Put bases for the subspaces $\boldsymbol{V}$ and $\boldsymbol{W}$ into the columns of matrices V and W. Explain why the test for orthogonal subspaces can be written $V^{\mathrm{T}}W =$ zero matrix. This matches $\boldsymbol{v}^{\mathrm{T}}\boldsymbol{w} = 0$ for orthogonal vectors.

14 The floor $\boldsymbol{V}$ and the wall $\boldsymbol{W}$ are not orthogonal subspaces, because they share a nonzero vector (along the line where they meet). No planes $\boldsymbol{V}$ and $\boldsymbol{W}$ in $\mathbf{R}^3$ can be orthogonal! Find a vector in the column spaces of both matrices:

$$A = \begin{bmatrix} 1 & 2 \\ 1 & 3 \\ 1 & 2 \end{bmatrix} \quad \text{and} \quad B = \begin{bmatrix} 5 & 4 \\ 6 & 3 \\ 5 & 1 \end{bmatrix}$$

This will be a vector $A\boldsymbol{x}$ and also $B\widehat{\boldsymbol{x}}$. Think 3 by 4 with the matrix $[\,A \;\; B\,]$.

15 Extend Problem 14 to a p-dimensional subspace $\boldsymbol{V}$ and a q-dimensional subspace $\boldsymbol{W}$ of $\mathbf{R}^n$. What inequality on $p + q$ guarantees that $\boldsymbol{V}$ intersects $\boldsymbol{W}$ in a nonzero vector? These subspaces cannot be orthogonal.

16 Prove that every $\boldsymbol{y}$ in $\boldsymbol{N}(A^{\mathrm{T}})$ is perpendicular to every $A\boldsymbol{x}$ in the column space, using the matrix shorthand of equation (2). Start from $A^{\mathrm{T}}\boldsymbol{y} = \mathbf{0}$.

17 If S is the subspace of $\mathbf{R}^3$ containing only the zero vector, what is $S^\perp$? If S is spanned by $(1,1,1)$, what is $S^\perp$? If S is spanned by $(1,1,1)$ and $(1,1,-1)$, what is a basis for $S^\perp$?

18 Suppose S only contains two vectors $(1,5,1)$ and $(2,2,2)$ (not a subspace). Then $S^\perp$ is the nullspace of the matrix $A =$ ____ . $S^\perp$ is a subspace even if S is not.

19 Suppose L is a one-dimensional subspace (a line) in $\mathbf{R}^3$. Its orthogonal complement $L^\perp$ is the ____ perpendicular to L. Then $(L^\perp)^\perp$ is a ____ perpendicular to $L^\perp$. In fact $(L^\perp)^\perp$ is the same as ____ .

20 Suppose V is the whole space $\mathbf{R}^4$. Then $V^\perp$ contains only the vector ____ . Then $(V^\perp)^\perp$ is ____ . So $(V^\perp)^\perp$ is the same as ____ .

21 Suppose S is spanned by the vectors $(1,2,2,3)$ and $(1,3,3,2)$. Find two vectors that span $S^\perp$. This is the same as solving $A\boldsymbol{x} = \mathbf{0}$ for which A?

22 If P is the plane of vectors in $\mathbf{R}^4$ satisfying $x_1 + x_2 + x_3 + x_4 = 0$, write a basis for $P^\perp$. Construct a matrix that has P as its nullspace.

23 If a subspace S is contained in a subspace V, prove that $S^\perp$ contains $V^\perp$.

Questions 24–30 are about perpendicular columns and rows.

24 Suppose an n by n matrix is invertible: $AA^{-1} = I$. Then the first column of A^{-1} is orthogonal to the space spanned by which rows of A?

25 Find $A^{\mathrm{T}}A$ if the columns of A are unit vectors, all mutually perpendicular.

26 Construct a 3 by 3 matrix A with no zero entries whose columns are mutually perpendicular. Compute $A^{\mathrm{T}}A$. Why is it a diagonal matrix?

27 The lines $3x + y = b_1$ and $6x + 2y = b_2$ are ____ . They are the same line if ____ . In that case (b_1, b_2) is perpendicular to the vector ____ . The nullspace of the matrix is the line $3x + y =$ ____ . One particular vector in that nullspace is ____ .

28 Why is each of these statements false?

(a) $(1,1,1)$ is perpendicular to $(1,1,-2)$ so the planes $x + y + z = 0$ and $x + y - 2z = 0$ are orthogonal subspaces.

(b) The subspace spanned by $(1,1,0,0,0)$ and $(0,0,0,1,1)$ is the orthogonal complement of the subspace spanned by $(1,-1,0,0,0)$ and $(2,-2,3,4,-4)$.

(c) Two subspaces that meet only in the zero vector are orthogonal.

29 Find a matrix with $\boldsymbol{v} = (1,2,3)$ in the row space and column space. Find another matrix with $\boldsymbol{v}$ in the nullspace and column space. Which pairs of subspaces can $\boldsymbol{v}$ *not* be in?

Challenge Problems

30 Suppose A is 3 by 4 and B is 4 by 5 and $AB = 0$. So $\boldsymbol{N}(A)$ contains $\boldsymbol{C}(B)$. Prove from the dimensions of $\boldsymbol{N}(A)$ and $\boldsymbol{C}(B)$ that $\text{rank}(A) + \text{rank}(B) \leq 4$.

31 The command $N = \mathsf{null}(A)$ will produce a basis for the nullspace of A. Then the command $B = \mathsf{null}(N')$ will produce a basis for the _____ of A.

32 Suppose I give you four nonzero vectors $\boldsymbol{r}, \boldsymbol{n}, \boldsymbol{c}, \boldsymbol{l}$ in $\mathbf{R}^2$.

(a) What are the conditions for those to be bases for the four fundamental subspaces $C(A^{\mathrm{T}}), N(A), C(A), N(A^{\mathrm{T}})$ of a 2 by 2 matrix?

(b) What is one possible matrix A?

33 Suppose I give you eight vectors $\boldsymbol{r}_1, \boldsymbol{r}_2, \boldsymbol{n}_1, \boldsymbol{n}_2, \boldsymbol{c}_1, \boldsymbol{c}_2, \boldsymbol{l}_1, \boldsymbol{l}_2$ in $\mathbf{R}^4$.

(a) What are the conditions for those pairs to be bases for the four fundamental subspaces of a 4 by 4 matrix?

(b) What is one possible matrix A?

4.2 Projections

1 The projection of a vector $\boldsymbol{b}$ onto the line through $\boldsymbol{a}$ is the closest point $\boldsymbol{p} = \boldsymbol{a}(\boldsymbol{a}^{\mathrm{T}}\boldsymbol{b} \,/\, \boldsymbol{a}^{\mathrm{T}}\boldsymbol{a})$.

2 The error $\boldsymbol{e} = \boldsymbol{b} - \boldsymbol{p}$ is perpendicular to $\boldsymbol{a}$: Right triangle $\boldsymbol{b}\,\boldsymbol{p}\,\boldsymbol{e}$ has $||\boldsymbol{p}||^2 + ||\boldsymbol{e}||^2 = ||\boldsymbol{b}||^2$.

3 The **projection** of $\boldsymbol{b}$ onto a subspace $\boldsymbol{S}$ is the closest vector $\boldsymbol{p}$ in $\boldsymbol{S}$; $\boldsymbol{b} - \boldsymbol{p}$ is orthogonal to $\boldsymbol{S}$.

4 $A^{\mathrm{T}}A$ is invertible (and symmetric) only if A has independent columns: $\boldsymbol{N}(A^{\mathrm{T}}A) = \boldsymbol{N}(A)$.

5 Then the projection of $\boldsymbol{b}$ onto the column space of A is the vector $\boldsymbol{p} = A(A^{\mathrm{T}}A)^{-1}A^{\mathrm{T}}\boldsymbol{b}$.

6 The **projection matrix** onto $\boldsymbol{C}(A)$ is $\boxed{P = A(A^{\mathrm{T}}A)^{-1}A^{\mathrm{T}}.}$ It has $\boldsymbol{p} = P\boldsymbol{b}$ and $P^2 = P = P^{\mathrm{T}}$.

May we start this section with two questions? (In addition to that one.) The first question aims to show that projections are easy to visualize. The second question is about "projection matrices"—symmetric matrices with $P^2 = P$. *The projection of* $\boldsymbol{b}$ *is* $P\boldsymbol{b}$.

1 What are the projections of $\boldsymbol{b} = (2, 3, 4)$ onto the z axis and the xy plane?

2 What matrices P_1 and P_2 produce those projections onto a line and a plane?

When $\boldsymbol{b}$ is projected onto a line, ***its projection p is the part of b along that line***. If $\boldsymbol{b}$ is projected onto a plane, $\boldsymbol{p}$ is the part in that plane. ***The projection p is*** $P\boldsymbol{b}$.

The projection matrix P multiplies $\boldsymbol{b}$ to give $\boldsymbol{p}$. This section finds $\boldsymbol{p}$ and also P.

The projection onto the z axis we call $\boldsymbol{p}_1$. The second projection drops straight down to the xy plane. The picture in your mind should be Figure 4.5. Start with $\boldsymbol{b} = (2, 3, 4)$. The projection across gives $\boldsymbol{p}_1 = (0, 0, 4)$. The projection down gives $\boldsymbol{p}_2 = (2, 3, 0)$. Those are the parts of $\boldsymbol{b}$ along the z axis and in the xy plane.

The projection matrices P_1 and P_2 are 3 by 3. They multiply $\boldsymbol{b}$ with 3 components to produce $\boldsymbol{p}$ with 3 components. Projection onto a line comes from a rank one matrix. Projection onto a plane comes from a rank two matrix:

Projection matrix
Onto the z axis:
$$P_1 = \begin{bmatrix} 0 & 0 & 0 \\ 0 & 0 & 0 \\ 0 & 0 & \mathbf{1} \end{bmatrix} \qquad \text{Onto the } xy \text{ plane:} \quad P_2 = \begin{bmatrix} \mathbf{1} & 0 & 0 \\ 0 & \mathbf{1} & 0 \\ 0 & 0 & 0 \end{bmatrix}.$$

P_1 picks out the z component of every vector. P_2 picks out the x and y components. To find the projections $\boldsymbol{p}_1$ and $\boldsymbol{p}_2$ of $\boldsymbol{b}$, multiply $\boldsymbol{b}$ by P_1 and P_2 (small $\boldsymbol{p}$ for the vector, capital P for the matrix that produces it):

$$\boldsymbol{p}_1 = P_1\boldsymbol{b} = \begin{bmatrix} 0 & 0 & 0 \\ 0 & 0 & 0 \\ 0 & 0 & 1 \end{bmatrix} \begin{bmatrix} x \\ y \\ z \end{bmatrix} = \begin{bmatrix} \mathbf{0} \\ \mathbf{0} \\ \boldsymbol{z} \end{bmatrix} \qquad \boldsymbol{p}_2 = P_2\boldsymbol{b} = \begin{bmatrix} 1 & 0 & 0 \\ 0 & 1 & 0 \\ 0 & 0 & 0 \end{bmatrix} \begin{bmatrix} x \\ y \\ z \end{bmatrix} = \begin{bmatrix} \boldsymbol{x} \\ \boldsymbol{y} \\ \mathbf{0} \end{bmatrix}.$$

In this case the projections p_1 and p_2 are perpendicular. The xy plane and the z axis are ***orthogonal subspaces***, like the floor of a room and the line between two walls.

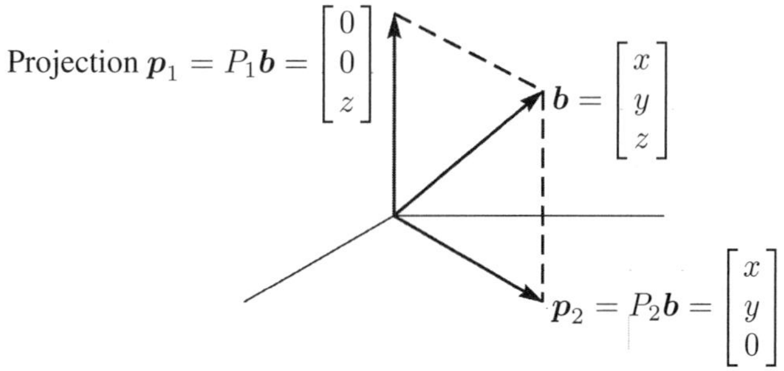

Figure 4.5: The projections $\boldsymbol{p}_1 = P_1\boldsymbol{b}$ and $\boldsymbol{p}_2 = P_2\boldsymbol{b}$ onto the z axis and the xy plane.

More than just orthogonal, the line and plane are orthogonal ***complements***. Their dimensions add to $1 + 2 = 3$. Every vector $\boldsymbol{b}$ in the whole space is the sum of its parts in the two subspaces. The projections $\boldsymbol{p}_1$ and $\boldsymbol{p}_2$ are exactly those two parts of $\boldsymbol{b}$:

$$\text{The vectors give } \boldsymbol{p}_1 + \boldsymbol{p}_2 = \boldsymbol{b}. \qquad \text{The matrices give } P_1 + P_2 = I. \tag{1}$$

This is perfect. Our goal is reached—for this example. We have the same goal for any line and any plane and any n-dimensional subspace. The object is to find the part $\boldsymbol{p}$ in each subspace, and the projection matrix P that produces that part $\boldsymbol{p} = P\boldsymbol{b}$. Every subspace of $\mathbf{R}^m$ has its own m by m projection matrix. To compute P, we absolutely need a good description of the subspace that it projects onto.

The best description of a subspace is a basis. We put the basis vectors into the columns of A. ***Now we are projecting onto the column space of A!*** Certainly the z axis is the column space of the 3 by 1 matrix A_1. The xy plane is the column space of A_2. That plane is *also* the column space of A_3 (a subspace has many bases). So $\boldsymbol{p}_2 = \boldsymbol{p}_3$ and $P_2 = P_3$.

$$A_1 = \begin{bmatrix} 0 \\ 0 \\ 1 \end{bmatrix} \quad \text{and} \quad A_2 = \begin{bmatrix} 1 & 0 \\ 0 & 1 \\ 0 & 0 \end{bmatrix} \quad \text{and} \quad A_3 = \begin{bmatrix} 1 & 2 \\ 2 & 3 \\ 0 & 0 \end{bmatrix}.$$

Our problem is ***to project any b onto the column space of any m by n matrix***. Start with a line (dimension $n = 1$). The matrix A will have only one column. Call it $\boldsymbol{a}$.

Projection Onto a Line

A line goes through the origin in the direction of $\boldsymbol{a} = (a_1, \ldots, a_m)$. Along that line, we want the point $\boldsymbol{p}$ closest to $\boldsymbol{b} = (b_1, \ldots, b_m)$. The key to projection is orthogonality: ***The line from b to p is perpendicular to the vector a***. This is the dotted line marked $\boldsymbol{e} = \boldsymbol{b} - \boldsymbol{p}$ for the error on the left side of Figure 4.6. We now compute $\boldsymbol{p}$ by algebra.

The projection p will be some multiple of a. Call it $p = \widehat{x}a$ = "x hat" times a. Computing this number $\widehat{x}$ will give the vector p. Then from the formula for p, we will read off the projection matrix P. These three steps will lead to all projection matrices: **find $\widehat{x}$, then find the vector p, then find the matrix P.**

The dotted line $b - p$ is the "error" $e = b - \widehat{x}a$. It is perpendicular to a—this will determine $\widehat{x}$. Use the fact that $b - \widehat{x}a$ **is perpendicular to** a when their dot product is zero:

Projecting b onto a with error $e = b - \widehat{x}a$

$a \cdot (b - \widehat{x}a) = 0$ or $a \cdot b - \widehat{x}a \cdot a = 0$

$$\widehat{x} = \frac{a \cdot b}{a \cdot a} = \frac{a^{\mathrm{T}}b}{a^{\mathrm{T}}a}. \tag{2}$$

The multiplication $a^{\mathrm{T}}b$ is the same as $a \cdot b$. Using the transpose is better, because it applies also to matrices. Our formula $\widehat{x} = a^{\mathrm{T}}b/a^{\mathrm{T}}a$ gives the projection $p = \widehat{x}a$.

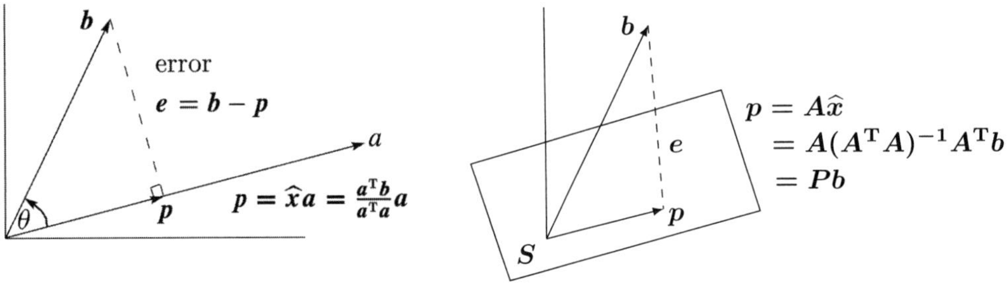

Figure 4.6: The projection p of b onto a line and onto S = column space of A.

The projection of b onto the line through a is the vector $p = \widehat{x}a = \dfrac{a^{\mathrm{T}}b}{a^{\mathrm{T}}a}a$.

Special case 1: If $b = a$ then $\widehat{x} = 1$. The projection of a onto a is itself. $Pa = a$.

Special case 2: If b is perpendicular to a then $a^{\mathrm{T}}b = 0$. The projection is $p = \mathbf{0}$.

Example 1 Project $b = \begin{bmatrix} 1 \\ 1 \\ 1 \end{bmatrix}$ onto $a = \begin{bmatrix} 1 \\ 2 \\ 2 \end{bmatrix}$ to find $p = \widehat{x}a$ in Figure 4.6.

Solution The number $\widehat{x}$ is the ratio of $a^{\mathrm{T}}b = 5$ to $a^{\mathrm{T}}a = 9$. So the projection is $p = \frac{5}{9}a$.

The error vector between $\boldsymbol{b}$ and $\boldsymbol{p}$ is $\boldsymbol{e} = \boldsymbol{b} - \boldsymbol{p}$. Those vectors $\boldsymbol{p}$ and $\boldsymbol{e}$ will add to $\boldsymbol{b} = (\mathbf{1}, \mathbf{1}, \mathbf{1})$:

$$\boldsymbol{p} = \frac{5}{9}\boldsymbol{a} = \left(\frac{5}{9}, \frac{10}{9}, \frac{10}{9}\right) \quad \text{and} \quad \boldsymbol{e} = \boldsymbol{b} - \boldsymbol{p} = \left(\frac{4}{9}, -\frac{1}{9}, -\frac{1}{9}\right).$$

The error $\boldsymbol{e}$ should be perpendicular to $\boldsymbol{a} = (1, 2, 2)$ and it is: $\boldsymbol{e}^{\mathrm{T}}\boldsymbol{a} = \frac{4}{9} - \frac{2}{9} - \frac{2}{9} = 0$.

Look at the right triangle of $\boldsymbol{b}$, $\boldsymbol{p}$, and $\boldsymbol{e}$. The vector $\boldsymbol{b}$ is split into two parts—its component along the line is $\boldsymbol{p}$, its perpendicular part is $\boldsymbol{e}$. Those two sides $\boldsymbol{p}$ and $\boldsymbol{e}$ have length $||\boldsymbol{p}|| = ||\boldsymbol{b}|| \cos\theta$ and $||\boldsymbol{e}|| = ||\boldsymbol{b}|| \sin\theta$. Trigonometry matches the dot product:

$$\boldsymbol{p} = \frac{\boldsymbol{a}^{\mathrm{T}}\boldsymbol{b}}{\boldsymbol{a}^{\mathrm{T}}\boldsymbol{a}}\boldsymbol{a} \quad \text{has length} \quad ||\boldsymbol{p}|| = \frac{||\boldsymbol{a}||\,||\boldsymbol{b}|| \cos\theta}{||\boldsymbol{a}||^2}||\boldsymbol{a}|| = ||\boldsymbol{b}|| \cos\theta. \tag{3}$$

The dot product is a lot simpler than getting involved with $\cos\theta$ and the length of $\boldsymbol{b}$. The example has square roots in $\cos\theta = 5/3\sqrt{3}$ and $||\boldsymbol{b}|| = \sqrt{3}$. There are no square roots in the projection $\boldsymbol{p} = 5\boldsymbol{a}/9$. The good way to $5/9$ is $\boldsymbol{a}^{\mathrm{T}}\boldsymbol{b}/\boldsymbol{a}^{\mathrm{T}}\boldsymbol{a}$.

Now comes the ***projection matrix***. In the formula for $\boldsymbol{p}$, what matrix is multiplying $\boldsymbol{b}$? You can see the matrix better if the number $\widehat{\boldsymbol{x}}$ is on the right side of $\boldsymbol{a}$:

Projection matrix P $\quad \boldsymbol{p} = \boldsymbol{a}\widehat{\boldsymbol{x}} = \boldsymbol{a}\dfrac{\boldsymbol{a}^{\mathrm{T}}\boldsymbol{b}}{\boldsymbol{a}^{\mathrm{T}}\boldsymbol{a}} = P\boldsymbol{b}$ ***when the matrix is*** $\quad P = \dfrac{\boldsymbol{a}\boldsymbol{a}^{\mathrm{T}}}{\boldsymbol{a}^{\mathrm{T}}\boldsymbol{a}}.$

P is a column times a row! The column is $\boldsymbol{a}$, the row is $\boldsymbol{a}^{\mathrm{T}}$. Then divide by the number $\boldsymbol{a}^{\mathrm{T}}\boldsymbol{a}$. The projection matrix P is m by m, but ***its rank is one***. We are projecting onto a one-dimensional subspace, the line through $\boldsymbol{a}$. *That line is the column space of P.*

Example 2 Find the projection matrix $P = \dfrac{\boldsymbol{a}\boldsymbol{a}^{\mathrm{T}}}{\boldsymbol{a}^{\mathrm{T}}\boldsymbol{a}}$ onto the line through $\boldsymbol{a} = \begin{bmatrix} \mathbf{1} \\ \mathbf{2} \\ \mathbf{2} \end{bmatrix}$.

Solution Multiply column $\boldsymbol{a}$ times row $\boldsymbol{a}^{\mathrm{T}}$ and divide by $\boldsymbol{a}^{\mathrm{T}}\boldsymbol{a} = 9$:

$$\text{Projection matrix} \qquad P = \frac{\boldsymbol{a}\boldsymbol{a}^{\mathrm{T}}}{\boldsymbol{a}^{\mathrm{T}}\boldsymbol{a}} = \frac{1}{9}\begin{bmatrix} 1 \\ 2 \\ 2 \end{bmatrix}\begin{bmatrix} 1 & 2 & 2 \end{bmatrix} = \frac{1}{9}\begin{bmatrix} 1 & 2 & 2 \\ 2 & 4 & 4 \\ 2 & 4 & 4 \end{bmatrix}.$$

This matrix projects *any* vector $\boldsymbol{b}$ onto $\boldsymbol{a}$. Check $\boldsymbol{p} = P\boldsymbol{b}$ for $\boldsymbol{b} = (1, 1, 1)$ in Example 1:

$$\boldsymbol{p} = P\boldsymbol{b} = \frac{1}{9}\begin{bmatrix} 1 & 2 & 2 \\ 2 & 4 & 4 \\ 2 & 4 & 4 \end{bmatrix}\begin{bmatrix} 1 \\ 1 \\ 1 \end{bmatrix} = \frac{1}{9}\begin{bmatrix} 5 \\ 10 \\ 10 \end{bmatrix} \quad \text{which is correct.}$$

If the vector $\boldsymbol{a}$ is doubled, the matrix P stays the same! It still projects onto the same line. If the matrix is squared, P^2 equals P. ***Projecting a second time doesn't change anything***, **so $\boldsymbol{P^2 = P}$**. The diagonal entries of P add up to $\frac{1}{9}(1 + 4 + 4) = 1$.

The matrix $I-P$ should be a projection too. It produces the other side $\boldsymbol{e}$ of the triangle—the perpendicular part of $\boldsymbol{b}$. Note that $(I-P)\boldsymbol{b}$ equals $\boldsymbol{b}-\boldsymbol{p}$ which is $\boldsymbol{e}$ in the left nullspace.

When P projects onto one subspace, $I-P$ projects onto the perpendicular subspace. Here $I-P$ projects onto the plane perpendicular to $\boldsymbol{a}$.

Now we move beyond projection onto a line. Projecting onto an n-dimensional subspace of $\mathbf{R}^m$ takes more effort. The crucial formulas will be collected in equations (5)–(6)–(7). Basically you need to remember those three equations.

Projection Onto a Subspace

Start with n vectors $\boldsymbol{a}_1, \ldots, \boldsymbol{a}_n$ in $\mathbf{R}^m$. Assume that these $\boldsymbol{a}$'s are linearly independent.

Problem: Find the combination $p = \widehat{x}_1 a_1 + \cdots + \widehat{x}_n a_n$ closest to a given vector b. We are projecting each $\boldsymbol{b}$ in $\mathbf{R}^m$ onto the subspace spanned by the $\boldsymbol{a}$'s.

With $n = 1$ (one vector $\boldsymbol{a}_1$) this is projection onto a line. The line is the column space of A, which has just one column. In general the matrix A has n columns $\boldsymbol{a}_1, \ldots, \boldsymbol{a}_n$.

The combinations in $\mathbf{R}^m$ are the vectors $A\boldsymbol{x}$ in the column space. We are looking for the particular combination $\boldsymbol{p} = A\widehat{\boldsymbol{x}}$ (***the projection***) that is closest to $\boldsymbol{b}$. The hat over $\widehat{\boldsymbol{x}}$ indicates the *best* choice $\widehat{\boldsymbol{x}}$, to give the closest vector in the column space. That choice is $\widehat{x} = \boldsymbol{a}^{\mathrm{T}}\boldsymbol{b}/\boldsymbol{a}^{\mathrm{T}}\boldsymbol{a}$ when $n = 1$. For $n > 1$, the best $\widehat{\boldsymbol{x}} = (\widehat{x}_1, \ldots, \widehat{x}_n)$ is to be found now.

We compute projections onto n-dimensional subspaces in three steps as before: ***Find the vector $\widehat{x}$, find the projection $p = A\widehat{x}$, find the projection matrix P.***

The key is in the geometry! The dotted line in Figure 4.6 goes from $\boldsymbol{b}$ to the nearest point $A\widehat{\boldsymbol{x}}$ in the subspace. ***This error vector $b - A\widehat{x}$ is perpendicular to the subspace.*** The error $\boldsymbol{b} - A\widehat{\boldsymbol{x}}$ makes a right angle with all the vectors $\boldsymbol{a}_1, \ldots, \boldsymbol{a}_n$ in the base. The n right angles give the n equations for $\widehat{\boldsymbol{x}}$:

$$\begin{matrix} \boldsymbol{a}_1^{\mathrm{T}}(\boldsymbol{b} - A\widehat{\boldsymbol{x}}) = 0 \\ \vdots \\ \boldsymbol{a}_n^{\mathrm{T}}(\boldsymbol{b} - A\widehat{\boldsymbol{x}}) = 0 \end{matrix} \qquad \text{or} \qquad \begin{bmatrix} - \boldsymbol{a}_1^{\mathrm{T}} - \\ \vdots \\ - \boldsymbol{a}_n^{\mathrm{T}} - \end{bmatrix} \begin{bmatrix} \\ \boldsymbol{b} - A\widehat{\boldsymbol{x}} \\ \\ \end{bmatrix} = \begin{bmatrix} \\ \mathbf{0} \\ \\ \end{bmatrix}. \tag{4}$$

The matrix with those rows $\boldsymbol{a}_i^{\mathrm{T}}$ is A^{T}. The n equations are exactly $A^{\mathrm{T}}(\boldsymbol{b} - A\widehat{\boldsymbol{x}}) = \mathbf{0}$.

Rewrite $A^{\mathrm{T}}(\boldsymbol{b} - A\widehat{\boldsymbol{x}}) = \mathbf{0}$ in its famous form $A^{\mathrm{T}}A\widehat{\boldsymbol{x}} = A^{\mathrm{T}}\boldsymbol{b}$. This is the equation for $\widehat{\boldsymbol{x}}$, and the coefficient matrix is $A^{\mathrm{T}}A$. Now we can find $\widehat{\boldsymbol{x}}$ and $\boldsymbol{p}$ and P, in that order.

The combination $\boldsymbol{p} = \widehat{x}_1\boldsymbol{a}_1 + \cdots + \widehat{x}_n\boldsymbol{a}_n = A\widehat{\boldsymbol{x}}$ that is closest to $\boldsymbol{b}$ comes from $\widehat{\boldsymbol{x}}$:

Find $\widehat{\boldsymbol{x}}$ $(n \times 1)$ $$A^{\mathrm{T}}(\boldsymbol{b} - A\widehat{\boldsymbol{x}}) = \boldsymbol{0} \quad \text{or} \quad A^{\mathrm{T}}A\widehat{\boldsymbol{x}} = A^{\mathrm{T}}\boldsymbol{b}. \tag{5}$$

This symmetric matrix $A^{\mathrm{T}}A$ is n by n. It is invertible if the $\boldsymbol{a}$'s are independent. The solution is $\widehat{\boldsymbol{x}} = (A^{\mathrm{T}}A)^{-1}A^{\mathrm{T}}\boldsymbol{b}$. The ***projection*** of $\boldsymbol{b}$ onto the subspace is $\boldsymbol{p}$:

Find $\boldsymbol{p}$ $(m \times 1)$ $$\boldsymbol{p} = A\widehat{\boldsymbol{x}} = A(A^{\mathrm{T}}A)^{-1}A^{\mathrm{T}}\boldsymbol{b}. \tag{6}$$

The next formula picks out the ***projection matrix*** that is multiplying $\boldsymbol{b}$ in (6):

Find $\boldsymbol{P}$ $(m \times m)$ $$P = A(A^{\mathrm{T}}A)^{-1}A^{\mathrm{T}}. \tag{7}$$

Compare with projection onto a line, when A has only one column: $A^{\mathrm{T}}A$ is $\boldsymbol{a}^{\mathrm{T}}\boldsymbol{a}$.

For $\boldsymbol{n = 1}$ $$\widehat{x} = \frac{\boldsymbol{a}^{\mathrm{T}}\boldsymbol{b}}{\boldsymbol{a}^{\mathrm{T}}\boldsymbol{a}} \quad \textbf{and} \quad \boldsymbol{p} = \boldsymbol{a}\frac{\boldsymbol{a}^{\mathrm{T}}\boldsymbol{b}}{\boldsymbol{a}^{\mathrm{T}}\boldsymbol{a}} \quad \textbf{and} \quad \boldsymbol{P} = \frac{\boldsymbol{a}\boldsymbol{a}^{\mathrm{T}}}{\boldsymbol{a}^{\mathrm{T}}\boldsymbol{a}}.$$

Those formulas are identical with (5) and (6) and (7). The number $\boldsymbol{a}^{\mathrm{T}}\boldsymbol{a}$ becomes the matrix $A^{\mathrm{T}}A$. When it is a number, we divide by it. When it is a matrix, we invert it. The new formulas contain $(A^{\mathrm{T}}A)^{-1}$ instead of $1/\boldsymbol{a}^{\mathrm{T}}\boldsymbol{a}$. The linear independence of the columns $\boldsymbol{a}_1, \ldots, \boldsymbol{a}_n$ will guarantee that this inverse matrix exists.

The key step was $A^{\mathrm{T}}(\boldsymbol{b} - A\widehat{\boldsymbol{x}}) = \boldsymbol{0}$. We used geometry ($\boldsymbol{e}$ is orthogonal to each $\boldsymbol{a}$). *Linear algebra gives this "normal equation" too*, in a very quick and beautiful way:

1. Our subspace is the column space of A.
2. The error vector $\boldsymbol{b} - A\widehat{\boldsymbol{x}}$ is perpendicular to that column space.
3. Therefore $\boldsymbol{b} - A\widehat{\boldsymbol{x}}$ is in the nullspace of A^{T}! This means $A^{\mathrm{T}}(\boldsymbol{b} - A\widehat{\boldsymbol{x}}) = \boldsymbol{0}$.

The left nullspace is important in projections. That nullspace of A^{T} contains the error vector $\boldsymbol{e} = \boldsymbol{b} - A\widehat{\boldsymbol{x}}$. The vector $\boldsymbol{b}$ is being split into the projection $\boldsymbol{p}$ and the error $\boldsymbol{e} = \boldsymbol{b} - \boldsymbol{p}$. Projection produces a right triangle with sides $\boldsymbol{p}$, $\boldsymbol{e}$, and $\boldsymbol{b}$.

Example 3 If $A = \begin{bmatrix} 1 & 0 \\ 1 & 1 \\ 1 & 2 \end{bmatrix}$ and $\boldsymbol{b} = \begin{bmatrix} 6 \\ 0 \\ 0 \end{bmatrix}$ find $\widehat{\boldsymbol{x}}$ and $\boldsymbol{p}$ and P.

Solution Compute the square matrix $A^{\mathrm{T}}A$ and also the vector $A^{\mathrm{T}}\boldsymbol{b}$:

$$A^{\mathrm{T}}A = \begin{bmatrix} 1 & 1 & 1 \\ 0 & 1 & 2 \end{bmatrix}\begin{bmatrix} 1 & 0 \\ 1 & 1 \\ 1 & 2 \end{bmatrix} = \begin{bmatrix} 3 & 3 \\ 3 & 5 \end{bmatrix} \quad \text{and} \quad A^{\mathrm{T}}\boldsymbol{b} = \begin{bmatrix} 1 & 1 & 1 \\ 0 & 1 & 2 \end{bmatrix}\begin{bmatrix} 6 \\ 0 \\ 0 \end{bmatrix} = \begin{bmatrix} 6 \\ 0 \end{bmatrix}.$$

Now solve the normal equation $A^{\mathrm{T}}A\widehat{\boldsymbol{x}} = A^{\mathrm{T}}\boldsymbol{b}$ to find $\widehat{\boldsymbol{x}}$:

$$\begin{bmatrix} 3 & 3 \\ 3 & 5 \end{bmatrix}\begin{bmatrix} \widehat{x}_1 \\ \widehat{x}_2 \end{bmatrix} = \begin{bmatrix} 6 \\ 0 \end{bmatrix} \quad \text{gives} \quad \widehat{\boldsymbol{x}} = \begin{bmatrix} \widehat{x}_1 \\ \widehat{x}_2 \end{bmatrix} = \begin{bmatrix} \mathbf{5} \\ \mathbf{-3} \end{bmatrix}. \tag{8}$$

The combination $\boldsymbol{p} = A\widehat{\boldsymbol{x}}$ is the projection of $\boldsymbol{b}$ onto the column space of A:

$$\boldsymbol{p} = \mathbf{5}\begin{bmatrix} 1 \\ 1 \\ 1 \end{bmatrix} - \mathbf{3}\begin{bmatrix} 0 \\ 1 \\ 2 \end{bmatrix} = \begin{bmatrix} 5 \\ 2 \\ -1 \end{bmatrix}. \quad \text{The error is} \quad \boldsymbol{e} = \boldsymbol{b} - \boldsymbol{p} = \begin{bmatrix} 1 \\ -2 \\ 1 \end{bmatrix}. \tag{9}$$

Two checks on the calculation. First, the error $\boldsymbol{e} = (1, -2, 1)$ is perpendicular to both columns $(1, 1, 1)$ and $(0, 1, 2)$. Second, the matrix P times $\boldsymbol{b} = (6, 0, 0)$ correctly gives $\boldsymbol{p} = (5, 2, -1)$. That solves the problem for one particular $\boldsymbol{b}$, as soon as we find P.

The projection matrix is $P = A(A^{\mathrm{T}}A)^{-1}A^{\mathrm{T}}$. The determinant of $A^{\mathrm{T}}A$ is $15 - 9 = 6$; then $(A^{\mathrm{T}}A)^{-1}$ is easy. Multiply A times $(A^{\mathrm{T}}A)^{-1}$ times A^{T} to reach P:

$$(A^{\mathrm{T}}A)^{-1} = \frac{1}{6}\begin{bmatrix} 5 & -3 \\ -3 & 3 \end{bmatrix} \quad \text{and} \quad P = \frac{1}{6}\begin{bmatrix} 5 & 2 & -1 \\ 2 & 2 & 2 \\ -1 & 2 & 5 \end{bmatrix}. \tag{10}$$

We must have $P^2 = P$, because a second projection doesn't change the first projection.

Warning The matrix $P = A(A^{\mathrm{T}}A)^{-1}A^{\mathrm{T}}$ is deceptive. You might try to split $(A^{\mathrm{T}}A)^{-1}$ into A^{-1} times $(A^{\mathrm{T}})^{-1}$. If you make that mistake, and substitute it into P, you will find $P = AA^{-1}(A^{\mathrm{T}})^{-1}A^{\mathrm{T}}$. Apparently everything cancels. This looks like $P = I$, the identity matrix. We want to say why this is wrong.

The matrix A is rectangular. It has no inverse matrix. We cannot split $(A^{\mathrm{T}}A)^{-1}$ into A^{-1} times $(A^{\mathrm{T}})^{-1}$ because there is no A^{-1} in the first place.

In our experience, a problem that involves a rectangular matrix almost always leads to $A^{\mathrm{T}}A$. When A has independent columns, $A^{\mathrm{T}}A$ is invertible. This fact is so crucial that we state it clearly and give a proof.

$A^{\mathrm{T}}A$ is invertible if and only if A has linearly independent columns.

Proof $A^{\mathrm{T}}A$ is a square matrix (n by n). For every matrix A, we will now show that $A^{\mathrm{T}}A$ ***has the same nullspace as*** A. When the columns of A are linearly independent, its nullspace contains only the zero vector. Then $A^{\mathrm{T}}A$, with this same nullspace, is invertible.

Let A be any matrix. If $\boldsymbol{x}$ is in its nullspace, then $A\boldsymbol{x} = \mathbf{0}$. Multiplying by A^{T} gives $A^{\mathrm{T}}A\boldsymbol{x} = \mathbf{0}$. So $\boldsymbol{x}$ is also in the nullspace of $A^{\mathrm{T}}A$.

Now start with the nullspace of $A^{\mathrm{T}}A$. **From** $A^{\mathrm{T}}A\boldsymbol{x} = \mathbf{0}$ **we must prove** $A\boldsymbol{x} = \mathbf{0}$. We can't multiply by $(A^{\mathrm{T}})^{-1}$, which generally doesn't exist. Just multiply by $\boldsymbol{x}^{\mathrm{T}}$:

$$(\boldsymbol{x}^{\mathrm{T}})A^{\mathrm{T}}A\boldsymbol{x} = 0 \quad \text{or} \quad (A\boldsymbol{x})^{\mathrm{T}}(A\boldsymbol{x}) = 0 \quad \text{or} \quad \|A\boldsymbol{x}\|^2 = 0. \tag{11}$$

We have shown: If $A^{\mathrm{T}}A\boldsymbol{x} = \mathbf{0}$ then $A\boldsymbol{x}$ has length zero. Therefore $A\boldsymbol{x} = \mathbf{0}$. Every vector $\boldsymbol{x}$ in one nullspace is in the other nullspace. If $A^{\mathrm{T}}A$ has dependent columns, so has A. If $A^{\mathrm{T}}A$ has independent columns, so has A. This is the good case: $A^{\mathrm{T}}A$ is invertible.

When A has independent columns, $A^{\mathrm{T}}A$ is square, symmetric, and invertible.

To repeat for emphasis: $A^{\mathrm{T}}A$ is (n by m) times (m by n). Then $A^{\mathrm{T}}A$ is square (n by n). It is symmetric, because its transpose is $(A^{\mathrm{T}}A)^{\mathrm{T}} = A^{\mathrm{T}}(A^{\mathrm{T}})^{\mathrm{T}}$ which equals $A^{\mathrm{T}}A$. We just proved that $A^{\mathrm{T}}A$ is invertible—provided A has independent columns. Watch the difference between dependent and independent columns:

$$\underset{A^{\mathrm{T}}}{\begin{bmatrix} 1 & 1 & 0 \\ 2 & 2 & 0 \end{bmatrix}} \underset{\text{dependent}}{\overset{A}{\begin{bmatrix} 1 & 2 \\ 1 & 2 \\ 0 & 0 \end{bmatrix}}} = \underset{\text{singular}}{\overset{A^{\mathrm{T}}A}{\begin{bmatrix} 2 & 4 \\ 4 & 8 \end{bmatrix}}} \qquad \underset{A^{\mathrm{T}}}{\begin{bmatrix} 1 & 1 & 0 \\ 2 & 2 & 1 \end{bmatrix}} \underset{\text{indep.}}{\overset{A}{\begin{bmatrix} 1 & 2 \\ 1 & 2 \\ 0 & 1 \end{bmatrix}}} = \underset{\text{invertible}}{\overset{A^{\mathrm{T}}A}{\begin{bmatrix} 2 & 4 \\ 4 & 9 \end{bmatrix}}}$$

Very brief summary To find the projection $\boldsymbol{p} = \widehat{x}_1\boldsymbol{a}_1 + \cdots + \widehat{x}_n\boldsymbol{a}_n$, solve $A^{\mathrm{T}}A\widehat{\boldsymbol{x}} = A^{\mathrm{T}}\boldsymbol{b}$. This gives $\widehat{\boldsymbol{x}}$. The projection is $\boldsymbol{p} = A\widehat{\boldsymbol{x}}$ and the error is $\boldsymbol{e} = \boldsymbol{b} - \boldsymbol{p} = \boldsymbol{b} - A\widehat{\boldsymbol{x}}$. The projection matrix $P = A(A^{\mathrm{T}}A)^{-1}A^{\mathrm{T}}$ gives $\boldsymbol{p} = P\boldsymbol{b}$.

This matrix satisfies $P^2 = P$. *The distance from b to the subspace $C(A)$ is $\|e\|$.*

■ REVIEW OF THE KEY IDEAS ■

1. The projection of $\boldsymbol{b}$ onto the line through $\boldsymbol{a}$ is $\boldsymbol{p} = \boldsymbol{a}\widehat{x} = \boldsymbol{a}(\boldsymbol{a}^{\mathrm{T}}\boldsymbol{b}/\boldsymbol{a}^{\mathrm{T}}\boldsymbol{a})$.
2. The rank one projection matrix $P = \boldsymbol{a}\boldsymbol{a}^{\mathrm{T}}/\boldsymbol{a}^{\mathrm{T}}\boldsymbol{a}$ multiplies $\boldsymbol{b}$ to produce $\boldsymbol{p}$.
3. Projecting $\boldsymbol{b}$ onto a subspace leaves $\boldsymbol{e} = \boldsymbol{b} - \boldsymbol{p}$ perpendicular to the subspace.
4. When A has full rank n, the equation $A^{\mathrm{T}}A\widehat{\boldsymbol{x}} = A^{\mathrm{T}}\boldsymbol{b}$ leads to $\widehat{\boldsymbol{x}}$ and $\boldsymbol{p} = A\widehat{\boldsymbol{x}}$.
5. The projection matrix $P = A(A^{\mathrm{T}}A)^{-1}A^{\mathrm{T}}$ has $P^{\mathrm{T}} = P$ and $P^2 = P$ and $P\boldsymbol{b} = \boldsymbol{p}$.

■ WORKED EXAMPLES ■

4.2 A Project the vector $\boldsymbol{b} = (3, 4, 4)$ onto the line through $\boldsymbol{a} = (2, 2, 1)$ and then onto the plane that also contains $\boldsymbol{a}^* = (1, 0, 0)$. Check that the first error vector $\boldsymbol{b} - \boldsymbol{p}$ is perpendicular to $\boldsymbol{a}$, and the second error vector $\boldsymbol{e}^* = \boldsymbol{b} - \boldsymbol{p}^*$ is also perpendicular to $\boldsymbol{a}^*$.

Find the 3 by 3 projection matrix P onto that plane of $\boldsymbol{a}$ and $\boldsymbol{a}^*$. Find a vector whose *projection onto the plane is the zero vector*. Why is it exactly the error $\boldsymbol{e}^*$?

Solution The projection of $\boldsymbol{b} = (3,4,4)$ onto the line through $\boldsymbol{a} = (2,2,1)$ is $\boldsymbol{p} = 2\boldsymbol{a}$:

Onto a line $$\boldsymbol{p} = \frac{\boldsymbol{a}^{\mathrm{T}}\boldsymbol{b}}{\boldsymbol{a}^{\mathrm{T}}\boldsymbol{a}}\boldsymbol{a} = \frac{18}{9}(2,2,1) = (4,4,2) = 2\boldsymbol{a}.$$

The error vector $\boldsymbol{e} = \boldsymbol{b} - \boldsymbol{p} = (-1,0,2)$ is perpendicular to $\boldsymbol{a} = (2,2,1)$. So $\boldsymbol{p}$ is correct.
The plane of $\boldsymbol{a} = (2,2,1)$ and $\boldsymbol{a}^* = (1,0,0)$ is the column space of $A = [\boldsymbol{a}\ \boldsymbol{a}^*]$:

$$A = \begin{bmatrix} 2 & 1 \\ 2 & 0 \\ 1 & 0 \end{bmatrix} \quad A^{\mathrm{T}}A = \begin{bmatrix} 9 & 2 \\ 2 & 1 \end{bmatrix} \quad (A^{\mathrm{T}}A)^{-1} = \frac{1}{5}\begin{bmatrix} 1 & -2 \\ -2 & 9 \end{bmatrix} \quad P = \begin{bmatrix} 1 & 0 & 0 \\ 0 & .8 & .4 \\ 0 & .4 & .2 \end{bmatrix}$$

Now $\boldsymbol{p}^* = P\boldsymbol{b} = (3, 4.8, 2.4)$. The error $\boldsymbol{e}^* = \boldsymbol{b} - \boldsymbol{p}^* = (0, -.8, 1.6)$ is perpendicular to $\boldsymbol{a}$ and $\boldsymbol{a}^*$. This $\boldsymbol{e}^*$ is in the nullspace of P and *its projection is zero*! Note $P^2 = P = P^{\mathrm{T}}$.

4.2 B Suppose your pulse is measured at $x = 70$ beats per minute, then at $x = 80$, then at $x = 120$. Those three equations $Ax = \boldsymbol{b}$ in one unknown have $A^{\mathrm{T}} = [1\ 1\ 1]$ and $\boldsymbol{b} = (70, 80, 120)$. ***The best $\widehat{x}$ is the _____ of 70, 80, 120.*** Use calculus and projection:

1. Minimize $E = (x-70)^2 + (x-80)^2 + (x-120)^2$ by solving $dE/dx = 0$.
2. Project $\boldsymbol{b} = (70, 80, 120)$ onto $\boldsymbol{a} = (1,1,1)$ to find $\widehat{x} = \boldsymbol{a}^{\mathrm{T}}\boldsymbol{b}/\boldsymbol{a}^{\mathrm{T}}\boldsymbol{a}$.

Solution The closest horizontal line to the heights 70, 80, 120 is the *average* $\widehat{x} = 90$:

$$\frac{dE}{dx} = 2(x-70) + 2(x-80) + 2(x-120) = 0 \quad \text{gives} \quad \widehat{x} = \frac{70+80+120}{3} = 90.$$

Also by projection : $$\widehat{x} = \frac{\boldsymbol{a}^{\mathrm{T}}\boldsymbol{b}}{\boldsymbol{a}^{\mathrm{T}}\boldsymbol{a}} = \frac{(1,1,1)^{\mathrm{T}}(70,80,120)}{(1,1,1)^{\mathrm{T}}(1,1,1)} = \frac{70+80+120}{3} = 90.$$

In *recursive* least squares, a fourth measurement 130 changes the average $\widehat{x}_{\text{old}} = 90$ to $\widehat{x}_{\text{new}} = 100$. Verify the *update formula* $\widehat{x}_{\text{new}} = \widehat{x}_{\text{old}} + \frac{1}{4}(130 - \widehat{x}_{\text{old}})$. When a new measurement arrives, we don't have to average all the old measurements again!

Problem Set 4.2

Questions 1–9 ask for projections $\boldsymbol{p}$ onto lines. Also errors $\boldsymbol{e} = \boldsymbol{b} - \boldsymbol{p}$ and matrices P.

1 Project the vector $\boldsymbol{b}$ onto the line through $\boldsymbol{a}$. Check that $\boldsymbol{e}$ is perpendicular to $\boldsymbol{a}$:

$$\text{(a)}\ \boldsymbol{b} = \begin{bmatrix} 1 \\ 2 \\ 2 \end{bmatrix} \text{ and } \boldsymbol{a} = \begin{bmatrix} 1 \\ 1 \\ 1 \end{bmatrix} \qquad \text{(b)}\ \boldsymbol{b} = \begin{bmatrix} 1 \\ 3 \\ 1 \end{bmatrix} \text{ and } \boldsymbol{a} = \begin{bmatrix} -1 \\ -3 \\ -1 \end{bmatrix}.$$

2 *Draw* the projection of $\boldsymbol{b}$ onto $\boldsymbol{a}$ and also compute it from $\boldsymbol{p} = \widehat{x}\boldsymbol{a}$:

$$\text{(a)}\ \boldsymbol{b} = \begin{bmatrix}\cos\theta\\ \sin\theta\end{bmatrix} \quad \text{and} \quad \boldsymbol{a} = \begin{bmatrix}1\\0\end{bmatrix} \qquad \text{(b)} \quad \boldsymbol{b} = \begin{bmatrix}1\\1\end{bmatrix} \quad \text{and} \quad \boldsymbol{a} = \begin{bmatrix}1\\-1\end{bmatrix}.$$

3 In Problem 1, find the projection matrix $P = \boldsymbol{a}\boldsymbol{a}^{\mathrm{T}}/\boldsymbol{a}^{\mathrm{T}}\boldsymbol{a}$ onto the line through each vector $\boldsymbol{a}$. Verify in both cases that $P^2 = P$. Multiply $P\boldsymbol{b}$ in each case to compute the projection $\boldsymbol{p}$.

4 Construct the projection matrices P_1 and P_2 onto the lines through the $\boldsymbol{a}$'s in Problem 2. Is it true that $(P_1 + P_2)^2 = P_1 + P_2$? This *would* be true if $P_1P_2 = 0$.

5 Compute the projection matrices $\boldsymbol{a}\boldsymbol{a}^{\mathrm{T}}/\boldsymbol{a}^{\mathrm{T}}\boldsymbol{a}$ onto the lines through $\boldsymbol{a}_1 = (-1, 2, 2)$ and $\boldsymbol{a}_2 = (2, 2, -1)$. Multiply those projection matrices and explain why their product P_1P_2 is what it is.

6 Project $\boldsymbol{b} = (1, 0, 0)$ onto the lines through $\boldsymbol{a}_1$ and $\boldsymbol{a}_2$ in Problem 5 and also onto $\boldsymbol{a}_3 = (2, -1, 2)$. Add up the three projections $\boldsymbol{p}_1 + \boldsymbol{p}_2 + \boldsymbol{p}_3$.

7 Continuing Problems 5–6, find the projection matrix P_3 onto $\boldsymbol{a}_3 = (2, -1, 2)$. Verify that $P_1 + P_2 + P_3 = I$. This is because the basis $\boldsymbol{a}_1, \boldsymbol{a}_2, \boldsymbol{a}_3$ is orthogonal!

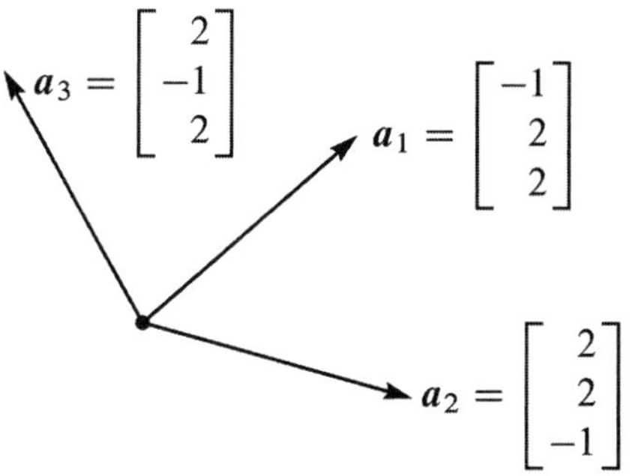

Questions 5–6–7: orthogonal

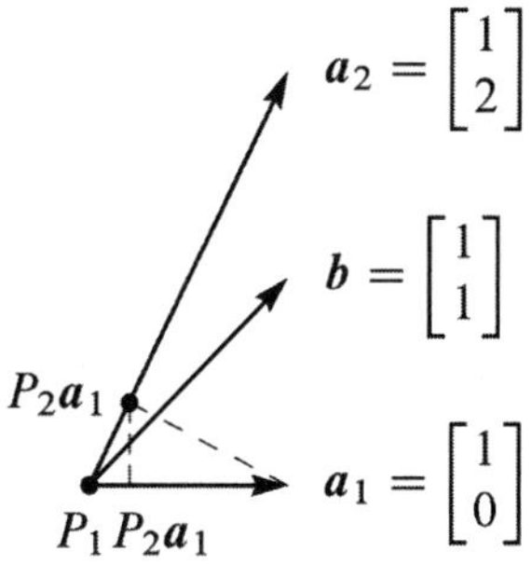

Questions 8–9–10: not orthogonal

8 Project the vector $\boldsymbol{b} = (1, 1)$ onto the lines through $\boldsymbol{a}_1 = (1, 0)$ and $\boldsymbol{a}_2 = (1, 2)$. Draw the projections $\boldsymbol{p}_1$ and $\boldsymbol{p}_2$ and add $\boldsymbol{p}_1 + \boldsymbol{p}_2$. The projections do not add to $\boldsymbol{b}$ because the $\boldsymbol{a}$'s are not orthogonal.

9 In Problem 8, the projection of $\boldsymbol{b}$ onto the *plane* of $\boldsymbol{a}_1$ and $\boldsymbol{a}_2$ will equal $\boldsymbol{b}$. Find $P = A(A^{\mathrm{T}}A)^{-1}A^{\mathrm{T}}$ for $A = \begin{bmatrix}\boldsymbol{a}_1 & \boldsymbol{a}_2\end{bmatrix} = \left[\begin{smallmatrix}1&1\\0&2\end{smallmatrix}\right] =$ invertible matrix.

10 Project $\boldsymbol{a}_1 = (1, 0)$ onto $\boldsymbol{a}_2 = (1, 2)$. Then project the result back onto $\boldsymbol{a}_1$. Draw these projections and multiply the projection matrices P_1P_2: Is this a projection?

Questions 11–20 ask for projections, and projection matrices, onto subspaces.

11 Project $\boldsymbol{b}$ onto the column space of A by solving $A^{\mathrm{T}}A\widehat{\boldsymbol{x}} = A^{\mathrm{T}}\boldsymbol{b}$ and $\boldsymbol{p} = A\widehat{\boldsymbol{x}}$:

$$\text{(a)}\ A = \begin{bmatrix} 1 & 1 \\ 0 & 1 \\ 0 & 0 \end{bmatrix} \text{ and } \boldsymbol{b} = \begin{bmatrix} 2 \\ 3 \\ 4 \end{bmatrix} \qquad \text{(b)}\quad A = \begin{bmatrix} 1 & 1 \\ 1 & 1 \\ 0 & 1 \end{bmatrix} \text{ and } \boldsymbol{b} = \begin{bmatrix} 4 \\ 4 \\ 6 \end{bmatrix}.$$

Find $\boldsymbol{e} = \boldsymbol{b} - \boldsymbol{p}$. It should be perpendicular to the columns of A.

12 Compute the projection matrices P_1 and P_2 onto the column spaces in Problem 11. Verify that $P_1\boldsymbol{b}$ gives the first projection $\boldsymbol{p}_1$. Also verify $P_2^2 = P_2$.

13 (Quick and Recommended) Suppose A is the 4 by 4 identity matrix with its last column removed. A is 4 by 3. Project $\boldsymbol{b} = (1,2,3,4)$ onto the column space of A. What shape is the projection matrix P and what is P?

14 Suppose $\boldsymbol{b}$ equals 2 times the first column of A. What is the projection of $\boldsymbol{b}$ onto the column space of A? Is $P = I$ for sure in this case? Compute $\boldsymbol{p}$ and P when $\boldsymbol{b} = (0,2,4)$ and the columns of A are $(0,1,2)$ and $(1,2,0)$.

15 If A is doubled, then $P = 2A(4A^{\mathrm{T}}A)^{-1}2A^{\mathrm{T}}$. This is the same as $A(A^{\mathrm{T}}A)^{-1}A^{\mathrm{T}}$. The column space of $2A$ is the same as ______. Is $\widehat{\boldsymbol{x}}$ the same for A and $2A$?

16 What linear combination of $(1,2,-1)$ and $(1,0,1)$ is closest to $\boldsymbol{b} = (2,1,1)$?

17 (*Important*) If $P^2 = P$ show that $(I-P)^2 = I - P$. When P projects onto the column space of A, $I - P$ projects onto the ______.

18 (a) If P is the 2 by 2 projection matrix onto the line through $(1,1)$, then $I - P$ is the projection matrix onto ______.

(b) If P is the 3 by 3 projection matrix onto the line through $(1,1,1)$, then $I - P$ is the projection matrix onto ______.

19 To find the projection matrix onto the plane $x - y - 2z = 0$, choose two vectors in that plane and make them the columns of A. The plane will be the column space of A! Then compute $P = A(A^{\mathrm{T}}A)^{-1}A^{\mathrm{T}}$.

20 To find the projection matrix P onto the same plane $x - y - 2z = 0$, write down a vector $\boldsymbol{e}$ that is perpendicular to that plane. Compute the projection $Q = \boldsymbol{e}\boldsymbol{e}^{\mathrm{T}}/\boldsymbol{e}^{\mathrm{T}}\boldsymbol{e}$ and then $P = I - Q$.

Questions 21–26 show that projection matrices satisfy $P^2 = P$ and $P^{\mathrm{T}} = P$.

21 Multiply the matrix $P = A(A^{\mathrm{T}}A)^{-1}A^{\mathrm{T}}$ by itself. Cancel to prove that $P^2 = P$. Explain why $P(P\boldsymbol{b})$ always equals $P\boldsymbol{b}$: The vector $P\boldsymbol{b}$ is in the column space of A so its projection onto that column space is ______.

22 Prove that $P = A(A^{\mathrm{T}}A)^{-1}A^{\mathrm{T}}$ is symmetric by computing P^{T}. Remember that the inverse of a symmetric matrix is symmetric.

23 If A is square and invertible, the warning against splitting $(A^{\mathrm{T}}A)^{-1}$ does not apply. It is true that $AA^{-1}(A^{\mathrm{T}})^{-1}A^{\mathrm{T}} = I$. *When A is invertible, why is $P = I$?* ***What is the error e?***

24 The nullspace of A^{T} is _____ to the column space $C(A)$. So if $A^{\mathrm{T}}\boldsymbol{b} = \mathbf{0}$, the projection of $\boldsymbol{b}$ onto $C(A)$ should be $\boldsymbol{p} =$ _____. Check that $P = A(A^{\mathrm{T}}A)^{-1}A^{\mathrm{T}}$ gives this answer.

25 The projection matrix P onto an n-dimensional subspace of $\mathbf{R}^m$ has rank $r = n$. ***Reason:*** The projections $P\boldsymbol{b}$ fill the subspace S. So S is the _____ of P.

26 If an m by m matrix has $A^2 = A$ and its rank is m, prove that $A = I$.

27 The important fact that ends the section is this: ***If*** $A^{\mathrm{T}}A\boldsymbol{x} = \mathbf{0}$ ***then*** $A\boldsymbol{x} = \mathbf{0}$. *New Proof*: The vector $A\boldsymbol{x}$ is in the nullspace of _____. $A\boldsymbol{x}$ is always in the column space of _____. To be in both of those perpendicular spaces, $A\boldsymbol{x}$ must be zero.

28 Use $P^{\mathrm{T}} = P$ and $P^2 = P$ to prove that the length squared of column 2 always equals the diagonal entry P_{22}. This number is $\frac{2}{6} = \frac{4}{36} + \frac{4}{36} + \frac{4}{36}$ for

$$P = \frac{1}{6}\begin{bmatrix} 5 & 2 & -1 \\ 2 & 2 & 2 \\ -1 & 2 & 5 \end{bmatrix}.$$

29 If B has rank m (full row rank, independent rows) show that BB^{T} is invertible.

Challenge Problems

30 (a) Find the projection matrix P_C onto the column space of A (after looking closely at the matrix!)

$$A = \begin{bmatrix} 3 & 6 & 6 \\ 4 & 8 & 8 \end{bmatrix}$$

(b) Find the 3 by 3 projection matrix P_R onto the row space of A. Multiply $B = P_C A P_R$. Your answer B should be a little surprising—can you explain it?

31 In $\mathbf{R}^m$, suppose I give you $\boldsymbol{b}$ and also a combination $\boldsymbol{p}$ of $\boldsymbol{a}_1, \ldots, \boldsymbol{a}_n$. How would you test to see if $\boldsymbol{p}$ is the projection of $\boldsymbol{b}$ onto the subspace spanned by the $\boldsymbol{a}$'s?

32 Suppose P_1 is the projection matrix onto the 1-dimensional subspace spanned by the first column of A. Suppose P_2 is the projection matrix onto the 2-dimensional column space of A. After thinking a little, compute the product P_2P_1.

$$A = \begin{bmatrix} 1 & 0 \\ 2 & 1 \\ 0 & 1 \end{bmatrix}.$$

33 Suppose you know the average $\widehat{x}_{\text{old}}$ of $b_1, b_2, \ldots, b_{999}$. When b_{1000} arrives, check that the new average is a combination of $\widehat{x}_{\text{old}}$ and the mismatch $b_{1000} - \widehat{x}_{\text{old}}$:

$$\widehat{\boldsymbol{x}}_{\textbf{new}} = \frac{b_1 + \cdots + b_{1000}}{1000} = \frac{b_1 + \cdots + b_{999}}{999} + \frac{1}{1000}\left(b_{1000} - \frac{b_1 + \cdots + b_{999}}{999}\right).$$

This is a "**Kalman filter**" $\widehat{x}_{\text{new}} = \widehat{x}_{\text{old}} + \frac{1}{1000}\,(b_{1000} - \widehat{x}_{\text{old}})$ with gain matrix $\frac{1}{1000}$. The last page of the book extends the Kalman filter to matrix updates.

34 (2017) Suppose P_1 and P_2 are projection matrices ($P_i^2 = P_i = P_i^{\mathrm{T}}$). Prove this fact:

$$P_1 P_2 \text{ is a projection matrix if and only if } P_1 P_2 = P_2 P_1.$$

4.3 Least Squares Approximations

1 Solving $\boxed{A^{\mathrm{T}}A\widehat{\boldsymbol{x}} = A^{\mathrm{T}}\boldsymbol{b}}$ gives the projection $\boldsymbol{p} = A\widehat{\boldsymbol{x}}$ of $\boldsymbol{b}$ onto the column space of A.

2 When $A\boldsymbol{x} = \boldsymbol{b}$ has no solution, $\widehat{\boldsymbol{x}}$ is the "least-squares solution": $||\boldsymbol{b} - A\widehat{\boldsymbol{x}}||^2 =$ minimum.

3 Setting partial derivatives of $E = ||A\boldsymbol{x} - \boldsymbol{b}||^2$ to zero $\left(\frac{\partial E}{\partial x_i} = 0\right)$ also produces $A^{\mathrm{T}}A\widehat{\boldsymbol{x}} = A^{\mathrm{T}}\boldsymbol{b}$.

4 To fit points $(t_1, b_1), \ldots, (t_m, b_m)$ by a straight line, A has columns $(1, \ldots, 1)$ and $(t_1, \ldots, t_m)$.

5 In that case $A^{\mathrm{T}}A$ is the 2 by 2 matrix $\begin{bmatrix} m & \Sigma\, t_i \\ \Sigma\, t_i & \Sigma\, t_i^2 \end{bmatrix}$ and $A^{\mathrm{T}}\boldsymbol{b}$ is the vector $\begin{bmatrix} \Sigma\, b_i \\ \Sigma\, t_i b_i \end{bmatrix}$.

It often happens that $A\boldsymbol{x} = \boldsymbol{b}$ has no solution. The usual reason is: *too many equations.* The matrix A has more rows than columns. There are more equations than unknowns (m is greater than n). The n columns span a small part of m-dimensional space. Unless all measurements are perfect, $\boldsymbol{b}$ is outside that column space of A. Elimination reaches an impossible equation and stops. But we can't stop just because measurements include noise!

To repeat: We cannot always get the error $\boldsymbol{e} = \boldsymbol{b} - A\boldsymbol{x}$ down to zero. When $\boldsymbol{e}$ is zero, $\boldsymbol{x}$ is an exact solution to $A\boldsymbol{x} = \boldsymbol{b}$. *When the length of* $\boldsymbol{e}$ *is as small as possible,* $\widehat{\boldsymbol{x}}$ *is a least squares solution.* Our goal in this section is to compute $\widehat{\boldsymbol{x}}$ and use it. These are real problems and they need an answer.

The previous section emphasized $\boldsymbol{p}$ (the projection). This section emphasizes $\widehat{\boldsymbol{x}}$ (the least squares solution). They are connected by $\boldsymbol{p} = A\widehat{\boldsymbol{x}}$. The fundamental equation is still $A^{\mathrm{T}}A\widehat{\boldsymbol{x}} = A^{\mathrm{T}}\boldsymbol{b}$. Here is a short unofficial way to reach this "*normal equation*":

When $A\boldsymbol{x} = \boldsymbol{b}$ has no solution, multiply by A^{T} and solve $A^{\mathrm{T}}A\widehat{\boldsymbol{x}} = A^{\mathrm{T}}\boldsymbol{b}$.

Example 1 A crucial application of least squares is fitting a straight line to m points. Start with three points: *Find the closest line to the points* $(0, 6)$, $(1, 0)$, *and* $(2, 0)$.

No straight line $b = C + Dt$ goes through those three points. We are asking for two numbers C and D that satisfy three equations: $n = 2$ and $m = 3$. Here are the three equations at $t = 0, 1, 2$ to match the given values $b = 6, 0, 0$:

$\boldsymbol{t = 0}$	The first point is on the line $b = C + Dt$ if	$C + D \cdot 0 = 6$
$\boldsymbol{t = 1}$	The second point is on the line $b = C + Dt$ if	$C + D \cdot 1 = 0$
$\boldsymbol{t = 2}$	The third point is on the line $b = C + Dt$ if	$C + D \cdot 2 = 0.$

This 3 by 2 system has *no solution*: $\boldsymbol{b} = (6,0,0)$ is not a combination of the columns $(1,1,1)$ and $(0,1,2)$. Read off A, $\boldsymbol{x}$, and $\boldsymbol{b}$ from those equations:

$$A = \begin{bmatrix} 1 & 0 \\ 1 & 1 \\ 1 & 2 \end{bmatrix} \quad \boldsymbol{x} = \begin{bmatrix} C \\ D \end{bmatrix} \quad \boldsymbol{b} = \begin{bmatrix} 6 \\ 0 \\ 0 \end{bmatrix} \quad A\boldsymbol{x} = \boldsymbol{b} \text{ is } \textit{not} \text{ solvable.}$$

The same numbers were in Example 3 in the last section. We computed $\widehat{\boldsymbol{x}} = (5, -3)$. **Those numbers are the best C and D, so $5 - 3t$ will be the best line for the 3 points.** We must connect projections to least squares, by explaining why $A^{\mathrm{T}}A\widehat{\boldsymbol{x}} = A^{\mathrm{T}}\boldsymbol{b}$.

In practical problems, there could easily be $m = 100$ points instead of $m = 3$. They don't exactly match any straight line $C + Dt$. Our numbers $6, 0, 0$ exaggerate the error so you can see e_1, e_2, and e_3 in Figure 4.6.

Minimizing the Error

How do we make the error $\boldsymbol{e} = \boldsymbol{b} - A\boldsymbol{x}$ as small as possible? This is an important question with a beautiful answer. The best $\boldsymbol{x}$ (called $\widehat{\boldsymbol{x}}$) can be found by geometry (the error $\boldsymbol{e}$ meets the column space of A at $90°$) and by algebra: $A^{\mathrm{T}}A\widehat{\boldsymbol{x}} = A^{\mathrm{T}}\boldsymbol{b}$. Calculus gives the same $\widehat{\boldsymbol{x}}$: the derivative of the error $||A\boldsymbol{x} - \boldsymbol{b}||^2$ is zero at $\widehat{\boldsymbol{x}}$.

By geometry Every $A\boldsymbol{x}$ lies in the plane of the columns $(1,1,1)$ and $(0,1,2)$. In that plane, we look for the point closest to $\boldsymbol{b}$. *The nearest point is the projection* $\boldsymbol{p}$.

The best choice for $A\widehat{\boldsymbol{x}}$ is $\boldsymbol{p}$. The smallest possible error is $\boldsymbol{e} = \boldsymbol{b} - \boldsymbol{p}$, perpendicular to the columns. *The three points at heights* (p_1, p_2, p_3) *do lie on a line*, because $\boldsymbol{p}$ is in the column space of A. In fitting a straight line, $\widehat{\boldsymbol{x}}$ is the best choice for (C, D).

By algebra Every vector $\boldsymbol{b}$ splits into two parts. The part in the column space is $\boldsymbol{p}$. The perpendicular part is $\boldsymbol{e}$. There is an equation we cannot solve ($A\boldsymbol{x} = \boldsymbol{b}$). There is an equation $A\widehat{\boldsymbol{x}} = \boldsymbol{p}$ we can and do solve (by removing $\boldsymbol{e}$ and solving $A^{\mathrm{T}}A\widehat{\boldsymbol{x}} = A^{\mathrm{T}}\boldsymbol{b}$):

$$A\boldsymbol{x} = \boldsymbol{b} = \boldsymbol{p} + \boldsymbol{e} \quad \text{is impossible} \qquad A\widehat{\boldsymbol{x}} = \boldsymbol{p} \quad \text{is solvable} \qquad \widehat{\boldsymbol{x}} \quad \text{is } (A^{\mathrm{T}}A)^{-1}A^{\mathrm{T}}\boldsymbol{b}. \tag{1}$$

The solution to $A\widehat{\boldsymbol{x}} = \boldsymbol{p}$ leaves the least possible error (which is $\boldsymbol{e}$):

$$\textbf{Squared length for any } \boldsymbol{x} \qquad ||A\boldsymbol{x} - \boldsymbol{b}||^2 = ||A\boldsymbol{x} - \boldsymbol{p}||^2 + ||\boldsymbol{e}||^2. \tag{2}$$

This is the law $c^2 = a^2 + b^2$ for a right triangle. The vector $A\boldsymbol{x} - \boldsymbol{p}$ in the column space is perpendicular to $\boldsymbol{e}$ in the left nullspace. We reduce $A\boldsymbol{x} - \boldsymbol{p}$ to **zero** by choosing $\boldsymbol{x} = \widehat{\boldsymbol{x}}$. That leaves the smallest possible error $\boldsymbol{e} = (e_1, e_2, e_3)$ which we can't reduce.

Notice what "smallest" means. The *squared length* of $A\boldsymbol{x} - \boldsymbol{b}$ is minimized:

> ***The least squares solution $\widehat{\boldsymbol{x}}$ makes $\boldsymbol{E} = ||A\boldsymbol{x} - \boldsymbol{b}||^2$ as small as possible.***

Figure 4.6a shows the closest line. It misses by distances $e_1, e_2, e_3 = 1, -2, 1$. *Those are vertical distances*. The least squares line minimizes $E = e_1^2 + e_2^2 + e_3^2$.

Figure 4.6b shows the same problem in 3-dimensional space ($\boldsymbol{b}\,\boldsymbol{p}\,\boldsymbol{e}$ space). The vector $\boldsymbol{b}$ is not in the column space of A. That is why we could not solve $A\boldsymbol{x} = \boldsymbol{b}$. No line goes through the three points. The smallest possible error is the perpendicular vector $\boldsymbol{e}$. This is $\boldsymbol{e} = \boldsymbol{b} - A\widehat{\boldsymbol{x}}$, the vector of errors $(1, -2, 1)$ in the three equations. Those are the distances from the best line. Behind both figures is the fundamental equation $A^{\mathrm{T}}A\widehat{\boldsymbol{x}} = A^{\mathrm{T}}\boldsymbol{b}$.

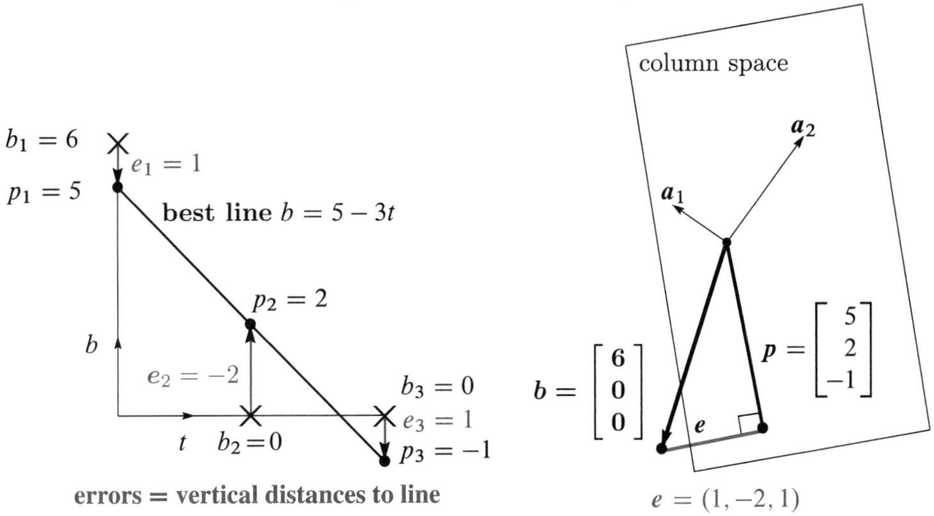

Figure 4.6: **Best line and projection: Two pictures, same problem.** The line has heights $\boldsymbol{p} = (5, 2, -1)$ with errors $\boldsymbol{e} = (1, -2, 1)$. The equations $A^{\mathrm{T}}A\widehat{\boldsymbol{x}} = A^{\mathrm{T}}\boldsymbol{b}$ give $\widehat{\boldsymbol{x}} = (5, -3)$. Same answer! The best line is $b = 5 - 3t$ and the closest point is $\boldsymbol{p} = 5\boldsymbol{a}_1 - 3\boldsymbol{a}_2$.

Notice that the errors $1, -2, 1$ add to zero. *Reason*: The error $\boldsymbol{e} = (e_1, e_2, e_3)$ is perpendicular to the first column $(1, 1, 1)$ in A. The dot product gives $e_1 + e_2 + e_3 = 0$.

By calculus Most functions are minimized by calculus! The graph bottoms out and the derivative in every direction is zero. Here the error function E to be minimized is a *sum of squares* $e_1^2 + e_2^2 + e_3^2$ (the square of the error in each equation):

$$E = \|A\boldsymbol{x} - \boldsymbol{b}\|^2 = (C + D \cdot 0 - 6)^2 + (C + D \cdot 1)^2 + (C + D \cdot 2)^2. \qquad (3)$$

The unknowns are C and D. With two unknowns there are *two derivatives*—both zero at the minimum. They are "partial derivatives" because $\partial E/\partial C$ treats D as constant and $\partial E/\partial D$ treats C as constant:

$$\partial E/\partial C = 2(C + D \cdot 0 - 6) \quad + 2(C + D \cdot 1) \quad + 2(C + D \cdot 2) \quad = 0$$

$$\partial E/\partial D = 2(C + D \cdot 0 - 6)(\mathbf{0}) + 2(C + D \cdot 1)(\mathbf{1}) + 2(C + D \cdot 2)(\mathbf{2}) = 0.$$

$\partial E/\partial D$ contains the extra factors $\mathbf{0}, \mathbf{1}, \mathbf{2}$ from the chain rule. (The last derivative from $(C + 2D)^2$ was 2 times $C + 2D$ times that extra 2.) Those factors are just $\mathbf{1}, \mathbf{1}, \mathbf{1}$ in $\partial E/\partial C$.

It is no accident that those factors 1, 1, 1 and 0, 1, 2 in the derivatives of $||A\boldsymbol{x} - \boldsymbol{b}||^2$ are the columns of A. Now cancel 2 from every term and collect all C's and all D's:

$$\begin{array}{ll}\text{The } C \text{ derivative is zero:} & 3C + 3D = 6 \\ \text{The } D \text{ derivative is zero:} & 3C + 5D = 0\end{array} \qquad \textbf{This matrix} \begin{bmatrix} 3 & 3 \\ 3 & 5 \end{bmatrix} \textbf{ is } \boldsymbol{A}^{\mathrm{T}}\boldsymbol{A} \qquad (4)$$

These equations are identical with $A^{\mathrm{T}}A\widehat{\boldsymbol{x}} = A^{\mathrm{T}}\boldsymbol{b}$. The best C and D are the components of $\widehat{\boldsymbol{x}}$. The equations from calculus are the same as the "normal equations" from linear algebra. These are the key equations of least squares:

The partial derivatives of $||A\boldsymbol{x} - \boldsymbol{b}||^2$ ***are zero when*** $\boldsymbol{A}^{\mathrm{T}}\boldsymbol{A}\widehat{\boldsymbol{x}} = \boldsymbol{A}^{\mathrm{T}}\boldsymbol{b}$.

The solution is $C = 5$ and $D = -3$. Therefore $b = 5 - 3t$ is the best line—it comes closest to the three points. At $t = 0, 1, 2$ this line goes through $\boldsymbol{p} = 5, 2, -1$. It could not go through $\boldsymbol{b} = 6, 0, 0$. The errors are $1, -2, 1$. This is the vector $\boldsymbol{e}$!

The Big Picture for Least Squares

The key figure of this book shows the four subspaces and the true action of a matrix. The vector $\boldsymbol{x}$ on the left side of Figure 4.3 went to $\boldsymbol{b} = A\boldsymbol{x}$ on the right side. In that figure $\boldsymbol{x}$ was split into $\boldsymbol{x}_r + \boldsymbol{x}_n$. There were *many* solutions to $A\boldsymbol{x} = \boldsymbol{b}$.

In this section the situation is just the opposite. There are *no* solutions to $A\boldsymbol{x} = \boldsymbol{b}$. *Instead of splitting up* $\boldsymbol{x}$ *we are splitting up* $\boldsymbol{b} = \boldsymbol{p} + \boldsymbol{e}$. Figure 4.7 shows the big picture for least squares. Instead of $A\boldsymbol{x} = \boldsymbol{b}$ we solve $A\widehat{\boldsymbol{x}} = \boldsymbol{p}$. The error $\boldsymbol{e} = \boldsymbol{b} - \boldsymbol{p}$ is unavoidable.

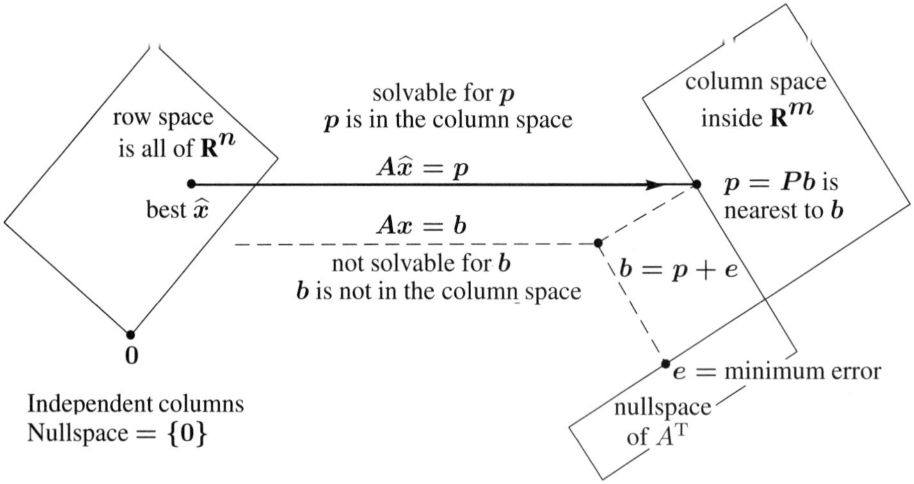

Figure 4.7: The projection $\boldsymbol{p} = A\widehat{\boldsymbol{x}}$ is closest to $\boldsymbol{b}$, so $\widehat{\boldsymbol{x}}$ minimizes $E = ||\boldsymbol{b} - A\boldsymbol{x}||^2$.

Notice how the nullspace $\boldsymbol{N}(A)$ is very small—just one point. With independent columns, the only solution to $A\boldsymbol{x} = \mathbf{0}$ is $\boldsymbol{x} = \mathbf{0}$. Then $A^{\mathrm{T}}A$ is invertible. The equation $A^{\mathrm{T}}A\widehat{\boldsymbol{x}} = A^{\mathrm{T}}\boldsymbol{b}$ fully determines the best vector $\widehat{\boldsymbol{x}}$. The error has $A^{\mathrm{T}}\boldsymbol{e} = \mathbf{0}$.

Chapter 7 will have the complete picture—all four subspaces included. Every $\boldsymbol{x}$ splits into $\boldsymbol{x}_r + \boldsymbol{x}_n$, and every $\boldsymbol{b}$ splits into $\boldsymbol{p} + \boldsymbol{e}$. The best solution is $\widehat{\boldsymbol{x}} = \widehat{\boldsymbol{x}}_r$ in the row space. We can't help $\boldsymbol{e}$ and we don't want $\boldsymbol{x}_n$ from the nullspace—this leaves $A\widehat{\boldsymbol{x}} = \boldsymbol{p}$.

Fitting a Straight Line

Fitting a line is the clearest application of least squares. It starts with $m > 2$ points, hopefully near a straight line. At times $t_1, \ldots, t_m$ those m points are at heights $b_1, \ldots, b_m$. The best line $C + Dt$ misses the points by vertical distances $e_1, \ldots, e_m$. No line is perfect, and the least squares line minimizes $E = e_1^2 + \cdots + e_m^2$.

The first example in this section had three points in Figure 4.6. Now we allow m points (and m can be large). The two components of $\widehat{\boldsymbol{x}}$ are still C and D.

A line goes through the m points when we exactly solve $A\boldsymbol{x} = \boldsymbol{b}$. Generally we can't do it. Two unknowns C and D determine a line, so A has only $n = 2$ columns. To fit the m points, we are trying to solve m equations (and we only have two unknowns!).

$$A\boldsymbol{x} = \boldsymbol{b} \quad \text{is} \quad \begin{matrix} C + Dt_1 = b_1 \\ C + Dt_2 = b_2 \\ \vdots \\ C + Dt_m = b_m \end{matrix} \qquad \text{with} \qquad A = \begin{bmatrix} 1 & t_1 \\ 1 & t_2 \\ \vdots & \vdots \\ 1 & t_m \end{bmatrix}. \tag{5}$$

The column space is so thin that almost certainly $\boldsymbol{b}$ is outside of it. When $\boldsymbol{b}$ happens to lie in the column space, the points happen to lie on a line. In that case $\boldsymbol{b} = \boldsymbol{p}$. Then $A\boldsymbol{x} = \boldsymbol{b}$ is solvable and the errors are $\boldsymbol{e} = (0, \ldots, 0)$.

The closest line $C + Dt$ has heights $p_1, \ldots, p_m$ with errors $e_1, \ldots, e_m$.

Solve $A^{\mathrm{T}} A\widehat{x} = A^{\mathrm{T}} b$ for $\widehat{x} = (C, D)$. The errors are $e_i = b_i - C - Dt_i$.

Fitting points by a straight line is so important that we give the two equations $A^{\mathrm{T}} A\widehat{\boldsymbol{x}} = A^{\mathrm{T}}\boldsymbol{b}$, once and for all. The two columns of A are independent (unless all times t_i are the same). So we turn to least squares and solve $A^{\mathrm{T}} A\widehat{\boldsymbol{x}} = A^{\mathrm{T}}\boldsymbol{b}$.

$$\textbf{Dot-product matrix} \;\; \boldsymbol{A}^{\mathrm{T}}\boldsymbol{A} = \begin{bmatrix} 1 & \cdots & 1 \\ t_1 & \cdots & t_m \end{bmatrix} \begin{bmatrix} 1 & t_1 \\ \vdots & \vdots \\ 1 & t_m \end{bmatrix} = \begin{bmatrix} m & \sum t_i \\ \sum t_i & \sum t_i^2 \end{bmatrix}. \tag{6}$$

On the right side of the normal equation is the 2 by 1 vector $A^{\mathrm{T}}\boldsymbol{b}$:

$$A^{\mathrm{T}}\boldsymbol{b} = \begin{bmatrix} 1 & \cdots & 1 \\ t_1 & \cdots & t_m \end{bmatrix} \begin{bmatrix} b_1 \\ \vdots \\ b_m \end{bmatrix} = \begin{bmatrix} \sum b_i \\ \sum t_i b_i \end{bmatrix}. \tag{7}$$

In a specific problem, these numbers are given. The best $\widehat{\boldsymbol{x}} = (C, D)$ is $(A^{\mathrm{T}}A)^{-1}A^{\mathrm{T}}\boldsymbol{b}$.

The line $C + Dt$ minimizes $e_1^2 + \cdots + e_m^2 = \|A\boldsymbol{x} - \boldsymbol{b}\|^2$ when $A^{\mathrm{T}}A\widehat{\boldsymbol{x}} = A^{\mathrm{T}}\boldsymbol{b}$:

$$A^{\mathrm{T}}A\widehat{\boldsymbol{x}} = A^{\mathrm{T}}\boldsymbol{b} \qquad \begin{bmatrix} m & \sum t_i \\ \sum t_i & \sum t_i^2 \end{bmatrix} \begin{bmatrix} C \\ D \end{bmatrix} = \begin{bmatrix} \sum b_i \\ \sum t_i b_i \end{bmatrix}. \tag{8}$$

The vertical errors at the m points on the line are the components of $\boldsymbol{e} = \boldsymbol{b} - \boldsymbol{p}$. This error vector (the *residual*) $\boldsymbol{b} - A\widehat{\boldsymbol{x}}$ is perpendicular to the columns of A (geometry). The error is in the nullspace of A^{T} (linear algebra). The best $\widehat{\boldsymbol{x}} = (C, D)$ minimizes the total error E, the sum of squares (calculus):

$$E(\boldsymbol{x}) = \|A\boldsymbol{x} - \boldsymbol{b}\|^2 = (C + Dt_1 - b_1)^2 + \cdots + (C + Dt_m - b_m)^2.$$

Calculus sets the derivatives $\partial E/\partial C$ and $\partial E/\partial D$ to zero, and produces $A^{\mathrm{T}}A\widehat{\boldsymbol{x}} = A^{\mathrm{T}}\boldsymbol{b}$.

Other least squares problems have more than two unknowns. Fitting by the best parabola has $n = 3$ coefficients C, D, E (see below). In general we are fitting m data points by n parameters $x_1, \ldots, x_n$. The matrix A has n columns and $n < m$. The derivatives of $\|A\boldsymbol{x} - \boldsymbol{b}\|^2$ give the n equations $A^{\mathrm{T}}A\widehat{\boldsymbol{x}} = A^{\mathrm{T}}\boldsymbol{b}$. **The derivative of a square is linear.** This is why the method of least squares is so popular.

Example 2 A has *orthogonal columns* when the measurement times t_i add to zero.

Suppose $b = 1, 2, 4$ at times $t = -2, 0, 2$. *Those times add to zero*. The columns of A have *zero dot product*: $(1, 1, 1)$ is orthogonal to $(-2, 0, 2)$:

$$\begin{aligned} C + D(-2) &= 1 \\ C + D(0) &= 2 \\ C + D(2) &= 4 \end{aligned} \qquad \text{or} \qquad A\boldsymbol{x} = \begin{bmatrix} 1 & -2 \\ 1 & 0 \\ 1 & 2 \end{bmatrix} \begin{bmatrix} C \\ D \end{bmatrix} = \begin{bmatrix} 1 \\ 2 \\ 4 \end{bmatrix}.$$

When the columns of A are orthogonal, $A^{\mathrm{T}}A$ will be a diagonal matrix:

$$A^{\mathrm{T}}A\widehat{\boldsymbol{x}} = A^{\mathrm{T}}\boldsymbol{b} \qquad \text{is} \qquad \begin{bmatrix} 3 & \mathbf{0} \\ \mathbf{0} & 8 \end{bmatrix} \begin{bmatrix} C \\ D \end{bmatrix} = \begin{bmatrix} 7 \\ 6 \end{bmatrix}. \tag{9}$$

Main point: Since $A^{\mathrm{T}}A$ *is diagonal*, we can solve separately for $C = \frac{7}{3}$ and $D = \frac{6}{8}$. The zeros in $A^{\mathrm{T}}A$ are dot products of perpendicular columns in A. The diagonal matrix $A^{\mathrm{T}}A$, with entries $m = 3$ and $t_1^2 + t_2^2 + t_3^2 = 8$, is virtually as good as the identity matrix.

Orthogonal columns are so helpful that it is worth *shifting the times by subtracting the average time* $\widehat{t} = (t_1 + \cdots + t_m)/m$. If the original times were $1, 3, 5$ then their average is $\widehat{t} = 3$. The shifted times $T = t - \widehat{t} = t - 3$ add up to zero!

$$\begin{aligned} T_1 &= 1 - 3 = -2 \\ T_2 &= 3 - 3 = 0 \\ T_3 &= 5 - 3 = 2 \end{aligned} \qquad A_{\text{new}} = \begin{bmatrix} 1 & T_1 \\ 1 & T_2 \\ 1 & T_3 \end{bmatrix} \qquad A_{\text{new}}^{\mathrm{T}}A_{\text{new}} = \begin{bmatrix} 3 & \mathbf{0} \\ \mathbf{0} & 8 \end{bmatrix}.$$

Now C and D come from the easy equation (9). Then the best straight line uses $C + DT$ which is $C + D(t - \widehat{t}\,) = C + D(t - 3)$. Problem 30 even gives a formula for C and D.

That was a perfect example of the "Gram-Schmidt idea" coming in the next section: *Make the columns orthogonal in advance*. Then $A_{\text{new}}^{\mathrm{T}} A_{\text{new}}$ is diagonal and $\widehat{\boldsymbol{x}}_{\text{new}}$ is easy.

Dependent Columns in A: What is $\widehat{x}$?

From the start, this chapter has assumed independent columns in A. Then $A^{\mathrm{T}} A$ is invertible. Then $A^{\mathrm{T}} A \widehat{\boldsymbol{x}} = A^{\mathrm{T}} \boldsymbol{b}$ produces the least squares solution to $A\boldsymbol{x} = \boldsymbol{b}$.

Which $\widehat{\boldsymbol{x}}$ is best if A has *dependent columns*? Here is a specific example.

$$\begin{bmatrix} 1 & 1 \\ 1 & 1 \end{bmatrix} \begin{bmatrix} x_1 \\ x_2 \end{bmatrix} = \begin{bmatrix} 3 \\ 1 \end{bmatrix} = \boldsymbol{b} \qquad \begin{bmatrix} 1 & 1 \\ 1 & 1 \end{bmatrix} \begin{bmatrix} \widehat{x}_1 \\ \widehat{x}_2 \end{bmatrix} = \begin{bmatrix} 2 \\ 2 \end{bmatrix} = \boldsymbol{p}$$

$$A\boldsymbol{x} = \boldsymbol{b} \qquad\qquad A\widehat{\boldsymbol{x}} = \boldsymbol{p}$$

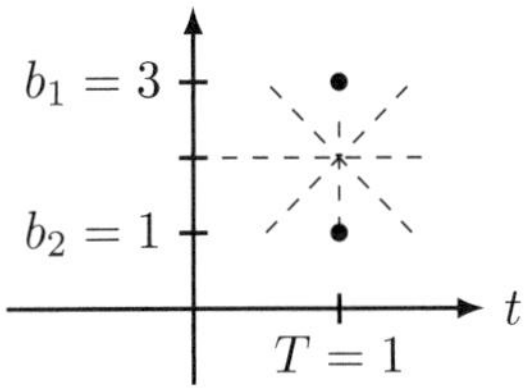

The measurements $b_1 = 3$ and $b_2 = 1$ are at the same time T! A straight line $C + Dt$ cannot go through both points. I think we are right to project $\boldsymbol{b} = (3, 1)$ to $\boldsymbol{p} = (2, 2)$ in the column space of A. That changes the equation $A\boldsymbol{x} = \boldsymbol{b}$ to the equation $A\widehat{\boldsymbol{x}} = \boldsymbol{p}$. An equation with no solution has become an equation with infinitely many solutions. The problem is that A has dependent columns and $(1, -1)$ is in its nullspace.

Which solution $\widehat{\boldsymbol{x}}$ should we choose? All the dashed lines in the figure have the same two errors 1 and -1 at time T. Those errors $(1, -1) = \boldsymbol{e} = \boldsymbol{b} - \boldsymbol{p}$ are as small as possible. But this doesn't tell us which dashed line is best.

My instinct is to go for the horizontal line at height 2. If the equation for the best line is $b = C + Dt$, then my choice will have $\widehat{\boldsymbol{x}}_1 = C = 2$ and $\widehat{\boldsymbol{x}}_2 = D = 0$. But what if the line had been written as $b = ct + d$? This is equally correct (just reversing C and D). Now the horizontal line has $\widehat{\boldsymbol{x}}_1 = c = 0$ and $\widehat{\boldsymbol{x}}_2 = d = 2$. I don't see any way out.

In Section 7.4, the *"pseudoinverse" of* A will choose the **shortest solution to** $A\widehat{x} = \boldsymbol{p}$. Here, that shortest solution will be $\boldsymbol{x}^+ = (1, 1)$. This is the particular solution in the row space of A, and $\boldsymbol{x}^+$ has length $\sqrt{2}$. (Both solutions $\widehat{\boldsymbol{x}} = (2, 0)$ and $(0, 2)$ have length 2.) We are arbitrarily choosing the nullspace component of the solution $\boldsymbol{x}^+$ to be zero.

When A has independent columns, the nullspace only contains the zero vector and the pseudoinverse is our usual left inverse $L = (A^{\mathrm{T}} A)^{-1} A^{\mathrm{T}}$. When I write it that way, the pseudoinverse sounds like the best way to choose $\boldsymbol{x}$.

Comment MATLAB experiments with singular matrices produced either **Inf** or **NaN** (Not a Number) or $\mathbf{10^{16}}$ (a bad number). There is a warning in every case! I believe that **Inf** and **NaN** and $\mathbf{10^{16}}$ come from the possibilities $\mathbf{0}\boldsymbol{x} = \boldsymbol{b}$ and $\mathbf{0}\boldsymbol{x} = \mathbf{0}$ and $\mathbf{10^{-16}}\,\boldsymbol{x} = \mathbf{1}$.

Those are three small examples of three big difficulties: singular with no solution, singular with many solutions, and very very close to singular.

Fitting by a Parabola

If we throw a ball, it would be crazy to fit the path by a straight line. A parabola $b = C + Dt + Et^2$ allows the ball to go up and come down again (b is the height at time t). The actual path is not a perfect parabola, but the whole theory of projectiles starts with that approximation.

When Galileo dropped a stone from the Leaning Tower of Pisa, it accelerated. The distance contains a quadratic term $\frac{1}{2}gt^2$. (Galileo's point was that the stone's mass is not involved.) Without that t^2 term we could never send a satellite into its orbit. But even with a nonlinear function like t^2, the unknowns C, D, E still appear linearly! Fitting points by the best parabola is still a problem in linear algebra.

Problem Fit heights $b_1, \ldots, b_m$ at times $t_1, \ldots, t_m$ by a parabola $C + Dt + Et^2$.

Solution With $m > 3$ points, the m equations for an exact fit are generally unsolvable:

$$\begin{aligned} C + Dt_1 + Et_1^2 &= b_1 \\ &\vdots \\ C + Dt_m + Et_m^2 &= b_m \end{aligned} \qquad \text{is } A\boldsymbol{x} = \boldsymbol{b} \text{ with the } m \text{ by 3 matrix} \qquad A = \begin{bmatrix} 1 & t_1 & t_1^2 \\ \vdots & \vdots & \vdots \\ 1 & t_m & t_m^2 \end{bmatrix}. \tag{10}$$

Least squares The closest parabola $C + Dt + Et^2$ chooses $\widehat{\boldsymbol{x}} = (C, D, E)$ to satisfy the three normal equations $A^{\mathrm{T}}A\widehat{\boldsymbol{x}} = A^{\mathrm{T}}\boldsymbol{b}$.

May I ask you to convert this to a problem of projection? The column space of A has dimension ______. The projection of $\boldsymbol{b}$ is $\boldsymbol{p} = A\widehat{\boldsymbol{x}}$, which combines the three columns using the coefficients C, D, E. The error at the first data point is $e_1 = b_1 - C - Dt_1 - Et_1^2$. The total squared error is $e_1^2 +$ ______. If you prefer to minimize by calculus, take the partial derivatives of E with respect to ______, ______, ______. These three derivatives will be zero when $\widehat{\boldsymbol{x}} = (C, D, E)$ solves the 3 by 3 system of equations $A^{\mathrm{T}}A\widehat{\boldsymbol{x}} = A^{\mathrm{T}}\boldsymbol{b}$.

Section 10.5 has more least squares applications. The big one is Fourier series—approximating functions instead of vectors. The function to be minimized changes from a sum of squared errors $e_1^2 + \cdots + e_m^2$ to an integral of the squared error.

Example 3 For a parabola $b = C + Dt + Et^2$ to go through the three heights $b = 6, 0, 0$ when $t = 0, 1, 2$, the equations for C, D, E are

$$\begin{aligned} C + D \cdot 0 + E \cdot 0^2 &= 6 \\ C + D \cdot 1 + E \cdot 1^2 &= 0 \\ C + D \cdot 2 + E \cdot 2^2 &= 0. \end{aligned} \tag{11}$$

This is $A\boldsymbol{x} = \boldsymbol{b}$. We can solve it exactly. Three data points give three equations and a square matrix. The solution is $\boldsymbol{x} = (C, D, E) = (\mathbf{6}, \mathbf{-9}, \mathbf{3})$. The parabola through the three points in Figure 4.8a is $b = \mathbf{6 - 9}\boldsymbol{t} + \mathbf{3}\boldsymbol{t}^{\mathbf{2}}$.

What does this mean for projection? The matrix has three columns, which span the whole space $\mathbf{R}^3$. The projection matrix is the identity. The projection of $\boldsymbol{b}$ is $\boldsymbol{b}$. The error is zero. We didn't need $A^{\mathrm{T}}A\widehat{\boldsymbol{x}} = A^{\mathrm{T}}\boldsymbol{b}$, because we solved $A\boldsymbol{x} = \boldsymbol{b}$. Of course we could multiply by A^{T}, but there is no reason to do it.

Figure 4.8 also shows a fourth point b_4 at time t_4. If that falls on the parabola, the new $A\boldsymbol{x} = \boldsymbol{b}$ (four equations) is still solvable. When the fourth point is not on the parabola, we turn to $A^{\mathrm{T}}A\widehat{\boldsymbol{x}} = A^{\mathrm{T}}\boldsymbol{b}$. Will the least squares parabola stay the same, with all the error at the fourth point? Not likely!

Least squares balances the four errors to get three equations for C, D, E.

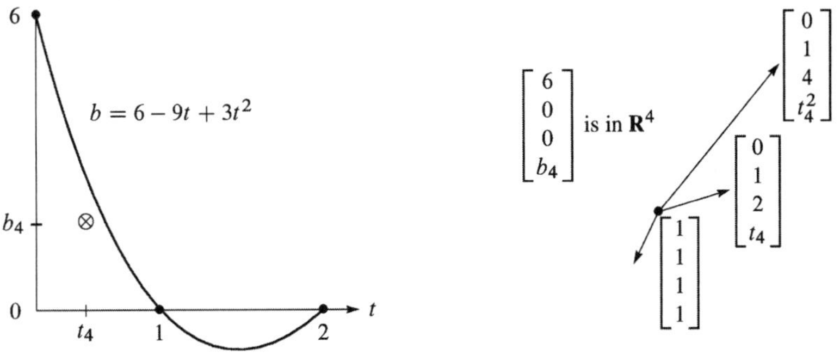

Figure 4.8: An exact fit of the parabola at $t = 0, 1, 2$ means that $\boldsymbol{p} = \boldsymbol{b}$ and $\boldsymbol{e} = \mathbf{0}$. The fourth point ⓧ off the parabola makes $m > n$ and we need least squares: project $\boldsymbol{b}$ on $\boldsymbol{C}(\boldsymbol{A})$. The figure on the right shows $\boldsymbol{b}$—not a combination of the three columns of A.

■ REVIEW OF THE KEY IDEAS ■

1. The least squares solution $\widehat{\boldsymbol{x}}$ minimizes $\|A\boldsymbol{x} - \boldsymbol{b}\|^2 = \boldsymbol{x}^{\mathrm{T}}A^{\mathrm{T}}A\boldsymbol{x} - 2\boldsymbol{x}^{\mathrm{T}}A^{\mathrm{T}}\boldsymbol{b} + \boldsymbol{b}^{\mathrm{T}}\boldsymbol{b}$. This is E, the sum of squares of the errors in the m equations $(m > n)$.

2. The best $\widehat{\boldsymbol{x}}$ comes from the normal equations $A^{\mathrm{T}}A\widehat{\boldsymbol{x}} = A^{\mathrm{T}}\boldsymbol{b}$. E is a minimum.

3. To fit m points by a line $b = C + Dt$, the normal equations give C and D.

4. The heights of the best line are $\boldsymbol{p} = (p_1, \ldots, p_m)$. The vertical distances to the data points are the errors $\boldsymbol{e} = (e_1, \ldots, e_m)$. A key equation is $A^{\mathrm{T}}\boldsymbol{e} = \mathbf{0}$.

5. If we try to fit m points by a combination of $n < m$ functions, the m equations $A\boldsymbol{x} = \boldsymbol{b}$ are generally unsolvable. The n equations $A^{\mathrm{T}}A\widehat{\boldsymbol{x}} = A^{\mathrm{T}}\boldsymbol{b}$ give the least squares solution—the combination with smallest MSE (mean square error).

■ WORKED EXAMPLES ■

4.3 A Start with nine measurements b_1 to b_9, *all zero*, at times $t = 1, \dots, 9$. The tenth measurement $b_{10} = 40$ is an outlier. Find the **best horizontal line $y = C$** to fit the ten points $(1,0), (2,0), \dots, (9,0), (10,40)$ using three options for the error E:

(1) Least *squares* $E_2 = e_1^2 + \cdots + e_{10}^2$ (then the normal equation for C is linear)

(2) Least *maximum* error $E_\infty = |e_{\max}|$ **(3)** Least *sum* of errors $E_1 = |e_1| + \cdots + |e_{10}|$.

Solution **(1)** The least squares fit to $0, 0, \dots, 0, 40$ by a horizontal line is $\boldsymbol{C = 4}$:

$$A = \text{column of 1's} \quad A^{\mathrm{T}}A = 10 \quad A^{\mathrm{T}}\boldsymbol{b} = \text{sum of } b_i = 40. \quad \text{So } 10\,C = 40.$$

(2) The least maximum error requires $\boldsymbol{C = 20}$, halfway between 0 and 40.

(3) The least sum requires $\boldsymbol{C = 0}$ (!!). The sum of errors $9|C| + |40 - C|$ would increase if C moves up from zero.

The least sum comes from the *median* measurement (the median of $0, \dots, 0, 40$ is zero). Many statisticians feel that the least squares solution is too heavily influenced by outliers like $b_{10} = 40$, and they prefer least sum. But the equations become *nonlinear*.

Now find the least squares line $C + Dt$ through those ten points $(1, 0)$ to $(10, 40)$:

$$A^{\mathrm{T}}A = \begin{bmatrix} 10 & \sum t_i \\ \sum t_i & \sum t_i^2 \end{bmatrix} = \begin{bmatrix} 10 & 55 \\ 55 & 385 \end{bmatrix} \qquad A^{\mathrm{T}}\boldsymbol{b} = \begin{bmatrix} \sum b_i \\ \sum t_i b_i \end{bmatrix} = \begin{bmatrix} 40 \\ 400 \end{bmatrix}$$

These come from equation (8). Then $A^{\mathrm{T}}A\widehat{\boldsymbol{x}} = A^{\mathrm{T}}\boldsymbol{b}$ gives $C = -8$ and $D = 24/11$.

What happens to C and D if you multiply $\boldsymbol{b} = (0, 0, \dots, 40)$ by 3 and then add 30 to get $\boldsymbol{b}_{\text{new}} = (30, 30, \dots, 150)$? Linearity allows us to rescale $\boldsymbol{b}$. Multiplying $\boldsymbol{b}$ by 3 will multiply C and D by 3. Adding 30 to all b_i will add 30 to C.

4.3 B Find the parabola $C + Dt + Et^2$ that comes closest (least squares error) to the values $\boldsymbol{b} = (0, 0, 1, 0, 0)$ at the times $t = -2, -1, 0, 1, 2$. First write down the five equations $A\boldsymbol{x} = \boldsymbol{b}$ in three unknowns $\boldsymbol{x} = (C, D, E)$ for a parabola to go through the five points. No solution because no such parabola exists. Solve $A^{\mathrm{T}}A\widehat{\boldsymbol{x}} = A^{\mathrm{T}}\boldsymbol{b}$.

I would predict $D = 0$. Why should the best parabola be symmetric around $t = 0$? In $A^{\mathrm{T}}A\widehat{\boldsymbol{x}} = A^{\mathrm{T}}\boldsymbol{b}$, equation 2 for D should uncouple from equations 1 and 3.

Solution The five equations $A\boldsymbol{x} = \boldsymbol{b}$ have a rectangular *Vandermonde matrix* A:

$$\begin{aligned} C + D\,(-2) + E\,(-2)^2 &= 0 \\ C + D\,(-1) + E\,(-1)^2 &= 0 \\ C + D\;\;(0) + E\;\;(0)^2 &= 1 \\ C + D\;\;(1) + E\;\;(1)^2 &= 0 \\ C + D\;\;(2) + E\;\;(2)^2 &= 0 \end{aligned} \qquad A = \begin{bmatrix} 1 & -2 & 4 \\ 1 & -1 & 1 \\ 1 & 0 & 0 \\ 1 & 1 & 1 \\ 1 & 2 & 4 \end{bmatrix} \qquad A^{\mathrm{T}}A = \begin{bmatrix} 5 & \mathbf{0} & 10 \\ \mathbf{0} & 10 & \mathbf{0} \\ 10 & \mathbf{0} & 34 \end{bmatrix}$$

Those zeros in $A^{\mathrm{T}}A$ mean that column 2 of A is orthogonal to columns 1 and 3. We see this directly in A (the times $-2,-1,0,1,2$ are symmetric). The best C, D, E in the parabola $C + Dt + Et^2$ come from $A^{\mathrm{T}}A\widehat{\boldsymbol{x}} = A^{\mathrm{T}}\boldsymbol{b}$, and D is uncoupled from C and E:

$$\begin{bmatrix} 5 & 0 & 10 \\ 0 & 10 & 0 \\ 10 & 0 & 34 \end{bmatrix}\begin{bmatrix} C \\ D \\ E \end{bmatrix} = \begin{bmatrix} 1 \\ 0 \\ 0 \end{bmatrix} \quad \text{leads to} \quad \begin{aligned} C &= 34/70 \\ D &= 0 \ \text{ as predicted} \\ E &= -10/70 \end{aligned}$$

Problem Set 4.3

Problems 1–11 use four data points $b = (0, 8, 8, 20)$ to bring out the key ideas.

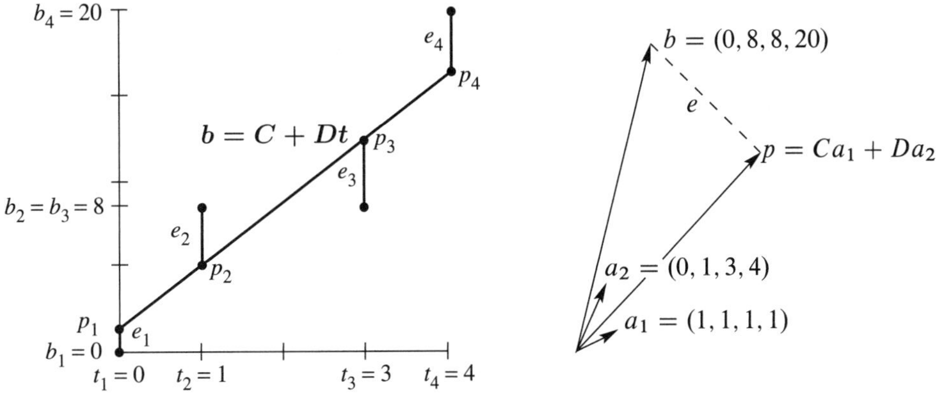

Figure 4.9: **Problems 1–11**: The closest line $C + Dt$ matches $C\boldsymbol{a}_1 + D\boldsymbol{a}_2$ in $\mathbf{R}^4$.

1 With $b = 0, 8, 8, 20$ at $t = 0, 1, 3, 4$, set up and solve the normal equations $A^{\mathrm{T}}A\widehat{\boldsymbol{x}} = A^{\mathrm{T}}\boldsymbol{b}$. For the best straight line in Figure 4.9a, find its four heights p_i and four errors e_i. What is the minimum value $E = e_1^2 + e_2^2 + e_3^2 + e_4^2$?

2 (Line $C + Dt$ does go through p's) With $b = 0, 8, 8, 20$ at times $t = 0, 1, 3, 4$, write down the four equations $A\boldsymbol{x} = \boldsymbol{b}$ (unsolvable). Change the measurements to $p = 1, 5, 13, 17$ and find an exact solution to $A\widehat{\boldsymbol{x}} = \boldsymbol{p}$.

3 Check that $\boldsymbol{e} = \boldsymbol{b} - \boldsymbol{p} = (-1, 3, -5, 3)$ is perpendicular to both columns of the same matrix A. What is the shortest distance $\|\boldsymbol{e}\|$ from $\boldsymbol{b}$ to the column space of A?

4 (By calculus) Write down $E = \|A\boldsymbol{x} - \boldsymbol{b}\|^2$ as a sum of four squares—the last one is $(C + 4D - 20)^2$. Find the derivative equations $\partial E/\partial C = 0$ and $\partial E/\partial D = 0$. Divide by 2 to obtain the normal equations $A^{\mathrm{T}}A\widehat{\boldsymbol{x}} = A^{\mathrm{T}}\boldsymbol{b}$.

5 Find the height C of the best *horizontal line* to fit $\boldsymbol{b} = (0, 8, 8, 20)$. An exact fit would solve the unsolvable equations $C = 0$, $C = 8$, $C = 8$, $C = 20$. Find the 4 by 1 matrix A in these equations and solve $A^{\mathrm{T}}A\widehat{x} = A^{\mathrm{T}}\boldsymbol{b}$. Draw the horizontal line at height $\widehat{x} = C$ and the four errors in $\boldsymbol{e}$.

6 Project $\boldsymbol{b} = (0, 8, 8, 20)$ onto the line through $\boldsymbol{a} = (1, 1, 1, 1)$. Find $\widehat{x} = \boldsymbol{a}^{\mathrm{T}}\boldsymbol{b}/\boldsymbol{a}^{\mathrm{T}}\boldsymbol{a}$ and the projection $\boldsymbol{p} = \widehat{x}\boldsymbol{a}$. Check that $\boldsymbol{e} = \boldsymbol{b} - \boldsymbol{p}$ is perpendicular to $\boldsymbol{a}$, and find the shortest distance $\|\boldsymbol{e}\|$ from $\boldsymbol{b}$ to the line through $\boldsymbol{a}$.

7 Find the closest line $b = Dt$, *through the origin*, to the same four points. An exact fit would solve $D \cdot 0 = 0, D \cdot 1 = 8, D \cdot 3 = 8, D \cdot 4 = 20$. Find the 4 by 1 matrix and solve $A^{\mathrm{T}}A\widehat{x} = A^{\mathrm{T}}\boldsymbol{b}$. Redraw Figure 4.9a showing the best line $b = Dt$ and the e's.

8 Project $\boldsymbol{b} = (0, 8, 8, 20)$ onto the line through $\boldsymbol{a} = (0, 1, 3, 4)$. Find $\widehat{x} = D$ and $\boldsymbol{p} = \widehat{x}\boldsymbol{a}$. The best C in Problems 5–6 and the best D in Problems 7–8 do *not* agree with the best (C, D) in Problems 1–4. That is because $(1, 1, 1, 1)$ and $(0, 1, 3, 4)$ are _____ perpendicular.

9 For the closest parabola $b = C + Dt + Et^2$ to the same four points, write down the unsolvable equations $A\boldsymbol{x} = \boldsymbol{b}$ in three unknowns $\boldsymbol{x} = (C, D, E)$. Set up the three normal equations $A^{\mathrm{T}}A\widehat{\boldsymbol{x}} = A^{\mathrm{T}}\boldsymbol{b}$ (solution not required). In Figure 4.9a you are now fitting a parabola to 4 points—what is happening in Figure 4.9b?

10 For the closest cubic $b = C + Dt + Et^2 + Ft^3$ to the same four points, write down the four equations $A\boldsymbol{x} = \boldsymbol{b}$. Solve them by elimination. In Figure 4.9a this cubic now goes exactly through the points. What are $\boldsymbol{p}$ and $\boldsymbol{e}$?

11 The average of the four times is $\widehat{t} = \frac{1}{4}(0 + 1 + 3 + 4) = 2$. The average of the four b's is $\widehat{b} = \frac{1}{4}(0 + 8 + 8 + 20) = 9$.

(a) Verify that the best line goes through the center point $(\widehat{t}, \widehat{b}) = (2, 9)$.

(b) Explain why $C + D\widehat{t} = \widehat{b}$ comes from the first equation in $A^{\mathrm{T}}A\widehat{\boldsymbol{x}} = A^{\mathrm{T}}\boldsymbol{b}$.

Questions 12–16 introduce basic ideas of statistics—the foundation for least squares.

12 (Recommended) This problem projects $\boldsymbol{b} = (b_1, \ldots, b_m)$ onto the line through $\boldsymbol{a} = (1, \ldots, 1)$. We solve m equations $\boldsymbol{a}x = \boldsymbol{b}$ in 1 unknown (by least squares).

(a) Solve $\boldsymbol{a}^{\mathrm{T}}\boldsymbol{a}\widehat{x} = \boldsymbol{a}^{\mathrm{T}}\boldsymbol{b}$ to show that $\widehat{x}$ is the *mean* (the average) of the $\boldsymbol{b}$'s.

(b) Find $\boldsymbol{e} = \boldsymbol{b} - \boldsymbol{a}\widehat{x}$ and the *variance* $\|\boldsymbol{e}\|^2$ and the *standard deviation* $\|\boldsymbol{e}\|$.

(c) The horizontal line $\widehat{\boldsymbol{b}} = 3$ is closest to $\boldsymbol{b} = (1, 2, 6)$. Check that $\boldsymbol{p} = (3, 3, 3)$ is perpendicular to $\boldsymbol{e}$ and find the 3 by 3 projection matrix P.

13 First assumption behind least squares: $A\boldsymbol{x} = \boldsymbol{b} -$ ***(noise e with mean zero)***. Multiply the error vectors $\boldsymbol{e} = \boldsymbol{b} - A\boldsymbol{x}$ by $(A^{\mathrm{T}}A)^{-1}A^{\mathrm{T}}$ to get $\widehat{\boldsymbol{x}} - \boldsymbol{x}$ on the right. The estimation errors $\widehat{\boldsymbol{x}} - \boldsymbol{x}$ also average to zero. The estimate $\widehat{\boldsymbol{x}}$ is *unbiased*.

14 Second assumption behind least squares: The m errors $\boldsymbol{e}_i$ are independent with variance σ^2, so the average of $(\boldsymbol{b} - A\boldsymbol{x})(\boldsymbol{b} - A\boldsymbol{x})^{\mathrm{T}}$ is $\sigma^2 I$. Multiply on the left by $(A^{\mathrm{T}}A)^{-1}A^{\mathrm{T}}$ and on the right by $A(A^{\mathrm{T}}A)^{-1}$ to show that the average matrix $(\widehat{\boldsymbol{x}} - \boldsymbol{x})(\widehat{\boldsymbol{x}} - \boldsymbol{x})^{\mathrm{T}}$ is $\sigma^2(A^{\mathrm{T}}A)^{-1}$. This is the ***covariance matrix*** W in Section 10.2.

15 A doctor takes 4 readings of your heart rate. The best solution to $x = b_1, \ldots, x = b_4$ is the average $\widehat{x}$ of $b_1, \ldots, b_4$. The matrix A is a column of 1's. Problem 14 gives the expected error $(\widehat{x} - x)^2$ as $\sigma^2(A^{\mathrm{T}}A)^{-1} =$ _____. ***By averaging, the variance drops from σ^2 to $\sigma^2/4$.***

16 If you know the average $\widehat{x}_9$ of 9 numbers $b_1, \ldots, b_9$, how can you quickly find the average $\widehat{x}_{10}$ with one more number b_{10} ? The idea of *recursive* least squares is to avoid adding 10 numbers. What number multiplies $\widehat{x}_9$ in computing $\widehat{x}_{10}$?

$$\widehat{x}_{10} = \tfrac{1}{10}b_{10} + \underline{\qquad}\,\widehat{x}_9 = \tfrac{1}{10}(b_1 + \cdots + b_{10}) \quad \text{as in Worked Example 4.2 C.}$$

Questions 17–24 give more practice with $\widehat{\boldsymbol{x}}$ and $\boldsymbol{p}$ and $\boldsymbol{e}$.

17 Write down three equations for the line $b = C + Dt$ to go through $b = 7$ at $t = -1$, $b = 7$ at $t = 1$, and $b = 21$ at $t = 2$. Find the least squares solution $\widehat{\boldsymbol{x}} = (C, D)$ and draw the closest line.

18 Find the projection $\boldsymbol{p} = A\widehat{\boldsymbol{x}}$ in Problem 17. This gives the three heights of the closest line. Show that the error vector is $\boldsymbol{e} = (2, -6, 4)$. Why is $P\boldsymbol{e} = \boldsymbol{0}$?

19 Suppose the measurements at $t = -1, 1, 2$ are the errors $2, -6, 4$ in Problem 18. Compute $\widehat{\boldsymbol{x}}$ and the closest line to these new measurements. Explain the answer: $\boldsymbol{b} = (2, -6, 4)$ is perpendicular to _____ so the projection is $\boldsymbol{p} = \boldsymbol{0}$.

20 Suppose the measurements at $t = -1, 1, 2$ are $\boldsymbol{b} = (5, 13, 17)$. Compute $\widehat{\boldsymbol{x}}$ and the closest line and $\boldsymbol{e}$. The error is $\boldsymbol{e} = \boldsymbol{0}$ because this $\boldsymbol{b}$ is _____.

21 Which of the four subspaces contains the error vector $\boldsymbol{e}$? Which contains $\boldsymbol{p}$? Which contains $\widehat{\boldsymbol{x}}$? What is the nullspace of A?

22 Find the best line $C + Dt$ to fit $b = 4, 2, -1, 0, 0$ at times $t = -2, -1, 0, 1, 2$.

23 Is the error vector $\boldsymbol{e}$ orthogonal to $\boldsymbol{b}$ or $\boldsymbol{p}$ or $\boldsymbol{e}$ or $\widehat{\boldsymbol{x}}$? Show that $\|\boldsymbol{e}\|^2$ equals $\boldsymbol{e}^{\mathrm{T}}\boldsymbol{b}$ which equals $\boldsymbol{b}^{\mathrm{T}}\boldsymbol{b} - \boldsymbol{p}^{\mathrm{T}}\boldsymbol{b}$. This is the smallest total error E.

24 The partial derivatives of $\|A\boldsymbol{x}\|^2$ with respect to $x_1, \ldots, x_n$ fill the vector $2A^{\mathrm{T}}A\boldsymbol{x}$. The derivatives of $2\boldsymbol{b}^{\mathrm{T}}A\boldsymbol{x}$ fill the vector $2A^{\mathrm{T}}\boldsymbol{b}$. So the derivatives of $\|A\boldsymbol{x} - \boldsymbol{b}\|^2$ are zero when _____.

Challenge Problems

25 *What condition on $(t_1, b_1), (t_2, b_2), (t_3, b_3)$ puts those three points onto a straight line*? A column space answer is: (b_1, b_2, b_3) must be a combination of $(1,\ 1,\ 1)$ and $(t_1,\ t_2,\ t_3)$. Try to reach a specific equation connecting the t's and b's. I should have thought of this question sooner!

26 Find the *plane* that gives the best fit to the 4 values $\boldsymbol{b} = (0, 1, 3, 4)$ at the corners $(1, 0)$ and $(0, 1)$ and $(-1, 0)$ and $(0, -1)$ of a square. The equations $C + Dx + Ey = b$ at those 4 points are $A\boldsymbol{x} = \boldsymbol{b}$ with 3 unknowns $\boldsymbol{x} = (C, D, E)$. What is A? At the center $(0, 0)$ of the square, show that $C + Dx + Ey =$ average of the b's.

27 (Distance between lines) The points $P = (x, x, x)$ and $Q = (y, 3y, -1)$ are on two lines in space that don't meet. Choose x and y to minimize the squared distance $\|P - Q\|^2$. The line connecting the closest P and Q is perpendicular to ______ .

28 Suppose the columns of A are not independent. How could you find a matrix B so that $P = B(B^{\mathrm{T}}B)^{-1}B^{\mathrm{T}}$ does give the projection onto the column space of A? (The usual formula will fail when $A^{\mathrm{T}}A$ is not invertible.)

29 Usually there will be exactly one hyperplane in $\mathbf{R}^n$ that contains the n given points $\boldsymbol{x} = \mathbf{0}, \boldsymbol{a}_1, \ldots, \boldsymbol{a}_{n-1}$. (Example for $n = 3$: There will be one plane containing $0, \boldsymbol{a}_1, \boldsymbol{a}_2$ unless ______ .) What is the test to have exactly one plane in $\mathbf{R}^n$?

30 Example 2 shifted the times t_i to make them add to zero. We subtracted away the average time $\widehat{t} = (t_1 + \cdots + t_m)/m$ to get $T_i = t_i - \widehat{t}$. Those T_i add to zero.

With the columns $(1, \ldots, 1)$ and $(T_1, \ldots, T_m)$ now orthogonal, $A^{\mathrm{T}}A$ is diagonal. Its entries are m and $T_1^2 + \cdots + T_m^2$. Show that the best C and D have direct formulas:

$$\boldsymbol{T} \textbf{ is } \boldsymbol{t} - \widehat{\boldsymbol{t}} \qquad C = \frac{b_1 + \cdots + b_m}{m} \qquad \text{and} \qquad D = \frac{b_1T_1 + \cdots + b_mT_m}{T_1^2 + \cdots + T_m^2}.$$

***The best line is** $C + DT$ **or** $C + D(t - \widehat{t})$*. The time shift that makes $A^{\mathrm{T}}A$ diagonal is an example of the Gram-Schmidt process: *orthogonalize the columns of A in advance.*

4.4 Orthonormal Bases and Gram-Schmidt

1 The columns $q_1, \ldots, q_n$ are orthonormal if $\boldsymbol{q}_i^{\mathrm{T}}\boldsymbol{q}_j = \begin{Bmatrix} 0 \text{ for } i \neq j \\ 1 \text{ for } i = j \end{Bmatrix}$. Then $\boxed{Q^{\mathrm{T}}Q = I\,.}$

2 If Q is also square, then $QQ^{\mathrm{T}} = I$ and $\boxed{Q^{\mathrm{T}} = Q^{-1}}$. Q is an "orthogonal matrix".

3 The least squares solution to $Q\boldsymbol{x} = \boldsymbol{b}$ is $\widehat{\boldsymbol{x}} = Q^{\mathrm{T}}\boldsymbol{b}$. Projection of $\boldsymbol{b}$: $\boldsymbol{p} = QQ^{\mathrm{T}}\boldsymbol{b} = P\boldsymbol{b}$.

4 The **Gram-Schmidt** process takes independent $\boldsymbol{a}_i$ to orthonormal $\boldsymbol{q}_i$. Start with $\boldsymbol{q}_1 = \boldsymbol{a}_1 / \,||\boldsymbol{a}_1||$.

5 $\boldsymbol{q}_i$ is $(\boldsymbol{a}_i -$ projection $\boldsymbol{p}_i) \,/\, ||\boldsymbol{a}_i - \boldsymbol{p}_i||$; projection $\boldsymbol{p}_i = (\boldsymbol{a}_i^{\mathrm{T}}\boldsymbol{q}_1)\boldsymbol{q}_1 + \cdots + (\boldsymbol{a}_i^{\mathrm{T}}\boldsymbol{q}_{i-1})\boldsymbol{q}_{i-1}$.

6 Each $\boldsymbol{a}_i$ will be a combination of $\boldsymbol{q}_1$ to $\boldsymbol{q}_i$. Then $\boldsymbol{A} = \boldsymbol{QR}$: orthogonal Q and triangular R.

This section has two goals, **why** and **how**. The first is to see why orthogonality is good. Dot products are zero, so $A^{\mathrm{T}}A$ will be diagonal. It becomes so easy to find $\widehat{\boldsymbol{x}}$ and $\boldsymbol{p} = A\widehat{\boldsymbol{x}}$. ***The second goal is to construct orthogonal vectors.*** You will see how Gram-Schmidt chooses combinations of the original basis vectors to produce right angles. Those original vectors are the columns of A, probably *not* orthogonal. ***The orthonormal basis vectors will be the columns of a new matrix Q.***

From Chapter 3, a basis consists of independent vectors that span the space. The basis vectors could meet at any angle (except 0° and 180°). But every time we visualize axes, they are perpendicular. *In our imagination, the coordinate axes are practically always orthogonal.* This simplifies the picture and it greatly simplifies the computations.

The vectors $\boldsymbol{q}_1, \ldots, \boldsymbol{q}_n$ are ***orthogonal*** when their dot products $\boldsymbol{q}_i \cdot \boldsymbol{q}_j$ are zero. More exactly $\boldsymbol{q}_i^{\mathrm{T}}\boldsymbol{q}_j = 0$ whenever $i \neq j$. With one more step—just *divide each vector by its length*—the vectors become ***orthogonal unit vectors***. Their lengths are all 1 (normal). Then the basis is called ***orthonormal***.

DEFINITION The vectors $\boldsymbol{q}_1, \ldots, \boldsymbol{q}_n$ are ***orthonormal*** if

$$\boldsymbol{q}_i^{\mathrm{T}}\boldsymbol{q}_j = \begin{cases} 0 & \text{when } i \neq j \quad (\textbf{\textit{orthogonal}} \text{ vectors}) \\ 1 & \text{when } i = j \quad (\textbf{\textit{unit}} \text{ vectors: } \|\boldsymbol{q}_i\| = 1) \end{cases}$$

A matrix with orthonormal columns is assigned the special letter Q.

The matrix Q is easy to work with because $Q^{\mathrm{T}}Q = I$. This repeats in matrix language that the columns $\boldsymbol{q}_1, \ldots, \boldsymbol{q}_n$ are orthonormal. Q is not required to be square.

A matrix Q with orthonormal columns satisfies $Q^{\mathrm{T}}Q = I$ **:**

$$Q^{\mathrm{T}}Q = \begin{bmatrix} -\,\boldsymbol{q}_1^{\mathrm{T}}\,- \\ -\,\boldsymbol{q}_2^{\mathrm{T}}\,- \\ \\ -\,\boldsymbol{q}_n^{\mathrm{T}}\,- \end{bmatrix} \begin{bmatrix} | & | & & | \\ \boldsymbol{q}_1 & \boldsymbol{q}_2 & & \boldsymbol{q}_n \\ | & | & & | \end{bmatrix} = \begin{bmatrix} 1 & 0 & \cdots & 0 \\ 0 & 1 & \cdots & 0 \\ \vdots & \vdots & \ddots & \vdots \\ 0 & 0 & \cdots & 1 \end{bmatrix} = \boldsymbol{I}. \quad (1)$$

When row i of Q^{T} multiplies column j of Q, the dot product is $\boldsymbol{q}_i^{\mathrm{T}}\boldsymbol{q}_j$. Off the diagonal $(i \neq j)$ that dot product is zero by orthogonality. On the diagonal $(i = j)$ the unit vectors give $\boldsymbol{q}_i^{\mathrm{T}}\boldsymbol{q}_i = \|\boldsymbol{q}_i\|^2 = 1$. Often Q is rectangular $(m > n)$. Sometimes $m = n$.

When Q is square, $Q^{\mathrm{T}}Q = I$ ***means that*** $Q^{\mathrm{T}} = Q^{-1}$***: transpose = inverse.***

If the columns are only orthogonal (not unit vectors), dot products still give a diagonal matrix (not the identity matrix). This diagonal matrix is almost as good as I. The important thing is orthogonality—then it is easy to produce unit vectors.

To repeat: $Q^{\mathrm{T}}Q = I$ even when Q is rectangular. In that case Q^{T} is only an inverse from the left. For square matrices we also have $QQ^{\mathrm{T}} = I$, so Q^{T} is the two-sided inverse of Q. The rows of a square Q are orthonormal like the columns. ***The inverse is the transpose***. In this square case we call Q an ***orthogonal matrix***.[1]

Here are three examples of orthogonal matrices—rotation and permutation and reflection. The quickest test is to check $Q^{\mathrm{T}}Q = I$.

Example 1 **(Rotation)** Q rotates every vector in the plane by the angle θ.

$$Q = \begin{bmatrix} \cos\theta & -\sin\theta \\ \sin\theta & \cos\theta \end{bmatrix} \quad \text{and} \quad Q^{\mathrm{T}} = Q^{-1} = \begin{bmatrix} \cos\theta & \sin\theta \\ -\sin\theta & \cos\theta \end{bmatrix}.$$

The columns of Q are orthogonal (take their dot product). They are unit vectors because $\sin^2\theta + \cos^2\theta = 1$. Those columns give an ***orthonormal basis*** for the plane $\mathbf{R}^2$.

The standard basis vectors $\boldsymbol{i}$ and $\boldsymbol{j}$ are rotated through θ (see Figure 4.10a). Q^{-1} rotates vectors back through $-\theta$. It agrees with Q^{T}, because the cosine of $-\theta$ equals the cosine of θ, and $\sin(-\theta) = -\sin\theta$. We have $Q^{\mathrm{T}}Q = I$ and $QQ^{\mathrm{T}} = I$.

Example 2 **(Permutation)** These matrices change the order to (y, z, x) and (y, x):

$$\begin{bmatrix} 0 & 1 & 0 \\ 0 & 0 & 1 \\ 1 & 0 & 0 \end{bmatrix} \begin{bmatrix} x \\ y \\ z \end{bmatrix} = \begin{bmatrix} y \\ z \\ x \end{bmatrix} \quad \text{and} \quad \begin{bmatrix} 0 & 1 \\ 1 & 0 \end{bmatrix} \begin{bmatrix} x \\ y \end{bmatrix} = \begin{bmatrix} y \\ x \end{bmatrix}.$$

All columns of these Q's are unit vectors (their lengths are obviously 1). They are also orthogonal (the 1's appear in different places). *The inverse of a permutation matrix is its transpose*: $Q^{-1} = Q^{\mathrm{T}}$. The inverse puts the components back into their original order:

[1]"Orthonormal matrix" would have been a better name for Q, but it's not used. Any matrix with orthonormal columns has the letter Q. But we only call it an **orthogonal matrix** when it is square.

$$\textbf{Inverse = transpose:} \quad \begin{bmatrix} 0 & 0 & 1 \\ 1 & 0 & 0 \\ 0 & 1 & 0 \end{bmatrix} \begin{bmatrix} \boldsymbol{y} \\ \boldsymbol{z} \\ \boldsymbol{x} \end{bmatrix} = \begin{bmatrix} \boldsymbol{x} \\ \boldsymbol{y} \\ \boldsymbol{z} \end{bmatrix} \quad \text{and} \quad \begin{bmatrix} 0 & 1 \\ 1 & 0 \end{bmatrix} \begin{bmatrix} \boldsymbol{y} \\ \boldsymbol{x} \end{bmatrix} = \begin{bmatrix} \boldsymbol{x} \\ \boldsymbol{y} \end{bmatrix}.$$

Every permutation matrix is an orthogonal matrix.

Example 3 **(Reflection)** If $\boldsymbol{u}$ is any unit vector, set $Q = I - 2\boldsymbol{u}\boldsymbol{u}^{\mathrm{T}}$. Notice that $\boldsymbol{u}\boldsymbol{u}^{\mathrm{T}}$ is a matrix while $\boldsymbol{u}^{\mathrm{T}}\boldsymbol{u}$ is the number $\|\boldsymbol{u}\|^2 = 1$. Then Q^{T} and Q^{-1} both equal Q:

$$Q^{\mathrm{T}} = I - 2\boldsymbol{u}\boldsymbol{u}^{\mathrm{T}} = Q \quad \text{and} \quad Q^{\mathrm{T}}Q = I - 4\boldsymbol{u}\boldsymbol{u}^{\mathrm{T}} + 4\boldsymbol{u}\boldsymbol{u}^{\mathrm{T}}\boldsymbol{u}\boldsymbol{u}^{\mathrm{T}} = I. \qquad (2)$$

Reflection matrices $I - 2\boldsymbol{u}\boldsymbol{u}^{\mathrm{T}}$ are symmetric and also orthogonal. If you square them, you get the identity matrix: $Q^2 = Q^{\mathrm{T}}Q = I$. Reflecting twice through a mirror brings back the original, like $(-1)^2 = 1$. Notice $\boldsymbol{u}^{\mathrm{T}}\boldsymbol{u} = 1$ inside $4\boldsymbol{u}\boldsymbol{u}^{\mathrm{T}}\boldsymbol{u}\boldsymbol{u}^{\mathrm{T}}$ in equation (2).

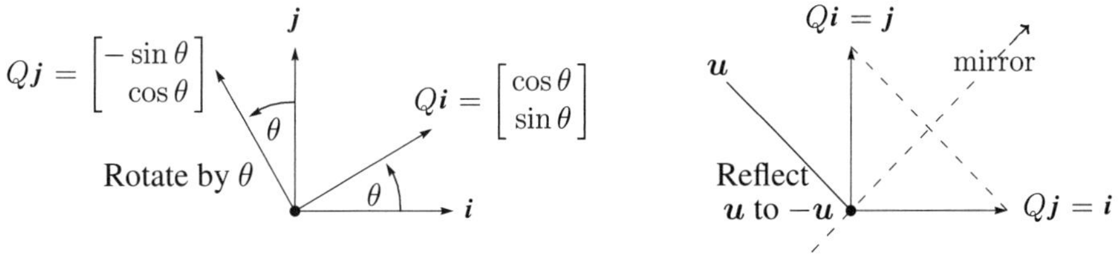

Figure 4.10: Rotation by $Q = \begin{bmatrix} c & -s \\ s & c \end{bmatrix}$ and reflection across $45°$ by $Q = \begin{bmatrix} 0 & 1 \\ 1 & 0 \end{bmatrix}$.

As example choose the direction $\boldsymbol{u} = (-1/\sqrt{2}, 1/\sqrt{2})$. Compute $2\boldsymbol{u}\boldsymbol{u}^{\mathrm{T}}$ (column times row) and subtract from I to get the reflection matrix Q in the direction of $\boldsymbol{u}$:

$$\textbf{Reflection} \quad Q = I - 2\begin{bmatrix} .5 & -.5 \\ -.5 & .5 \end{bmatrix} = \begin{bmatrix} 0 & 1 \\ 1 & 0 \end{bmatrix} \quad \text{and} \quad \begin{bmatrix} 0 & 1 \\ 1 & 0 \end{bmatrix}\begin{bmatrix} x \\ y \end{bmatrix} = \begin{bmatrix} y \\ x \end{bmatrix}.$$

When (x, y) goes to (y, x), a vector like $(3, 3)$ doesn't move. It is on the mirror line.

Rotations preserve the length of every vector. So do reflections. So do permutations. So does multiplication by any orthogonal matrix Q—***lengths and angles don't change***.

Proof $\|Q\boldsymbol{x}\|^2$ equals $\|\boldsymbol{x}\|^2$ because $(Q\boldsymbol{x})^{\mathrm{T}}(Q\boldsymbol{x}) = \boldsymbol{x}^{\mathrm{T}}Q^{\mathrm{T}}Q\boldsymbol{x} = \boldsymbol{x}^{\mathrm{T}}I\boldsymbol{x} = \boldsymbol{x}^{\mathrm{T}}\boldsymbol{x}$.

If Q has orthonormal columns ($Q^{\mathrm{T}}Q = I$), it leaves lengths unchanged:

Same length for $Q\boldsymbol{x}$ $\quad \|Q\boldsymbol{x}\| = \|\boldsymbol{x}\|$ **for every vector** $\boldsymbol{x}$. (3)

Q also preserves dot products: $(Q\boldsymbol{x})^{\mathrm{T}}(Q\boldsymbol{y}) = \boldsymbol{x}^{\mathrm{T}}Q^{\mathrm{T}}Q\boldsymbol{y} = \boldsymbol{x}^{\mathrm{T}}\boldsymbol{y}$. Just use $Q^{\mathrm{T}}Q = I$!

Projections Using Orthonormal Bases: Q Replaces A

Orthogonal matrices are excellent for computations—numbers can never grow too large when lengths of vectors are fixed. Stable computer codes use Q's as much as possible.

For projections onto subspaces, all formulas involve $A^{\mathrm{T}}A$. The entries of $A^{\mathrm{T}}A$ are the dot products $\boldsymbol{a}_i^{\mathrm{T}}\boldsymbol{a}_j$ of the basis vectors $\boldsymbol{a}_1, \ldots, \boldsymbol{a}_n$.

Suppose the basis vectors are actually orthonormal. The $\boldsymbol{a}$'s become the $\boldsymbol{q}$'s. Then $A^{\mathrm{T}}A$ *simplifies to* $Q^{\mathrm{T}}Q = I$. Look at the improvements in $\widehat{\boldsymbol{x}}$ and $\boldsymbol{p}$ and P. Instead of $Q^{\mathrm{T}}Q$ we print a blank for the identity matrix:

$$\underline{\qquad}\,\widehat{\boldsymbol{x}} = Q^{\mathrm{T}}\boldsymbol{b} \quad \text{and} \quad \boldsymbol{p} = Q\widehat{\boldsymbol{x}} \quad \text{and} \quad P = Q\,\underline{\qquad}\,Q^{\mathrm{T}}. \tag{4}$$

The least squares solution of $Qx = b$ is $\widehat{x} = Q^{\mathrm{T}}b$. The projection matrix is QQ^{T}.

There are no matrices to invert. This is the point of an orthonormal basis. The best $\widehat{\boldsymbol{x}} = Q^{\mathrm{T}}\boldsymbol{b}$ just has dot products of $\boldsymbol{q}_1, \ldots, \boldsymbol{q}_n$ with $\boldsymbol{b}$. We have 1-dimensional projections! The "coupling matrix" or "correlation matrix" $A^{\mathrm{T}}A$ is now $Q^{\mathrm{T}}Q = I$. There is no coupling. When A is Q, with orthonormal columns, here is $\boldsymbol{p} = Q\widehat{\boldsymbol{x}} = QQ^{\mathrm{T}}\boldsymbol{b}$:

Projection onto q's

$$\boldsymbol{p} = \begin{bmatrix} | & & | \\ \boldsymbol{q}_1 & \cdots & \boldsymbol{q}_n \\ | & & | \end{bmatrix} \begin{bmatrix} \boldsymbol{q}_1^{\mathrm{T}}\boldsymbol{b} \\ \vdots \\ \boldsymbol{q}_n^{\mathrm{T}}\boldsymbol{b} \end{bmatrix} = \boldsymbol{q}_1(\boldsymbol{q}_1^{\mathrm{T}}\boldsymbol{b}) + \cdots + \boldsymbol{q}_n(\boldsymbol{q}_n^{\mathrm{T}}\boldsymbol{b}). \tag{5}$$

Important case. When Q is square and $m = n$, the subspace is the whole space. Then $Q^{\mathrm{T}} = Q^{-1}$ and $\widehat{\boldsymbol{x}} = Q^{\mathrm{T}}\boldsymbol{b}$ is the same as $\boldsymbol{x} = Q^{-1}\boldsymbol{b}$. The solution is exact! The projection of $\boldsymbol{b}$ onto the whole space is $\boldsymbol{b}$ itself. In this case $\boldsymbol{p} = \boldsymbol{b}$ and $P = QQ^{\mathrm{T}} = I$.

You may think that projection onto the whole space is not worth mentioning. But when $\boldsymbol{p} = \boldsymbol{b}$, our formula assembles $\boldsymbol{b}$ out of its 1-dimensional projections. If $\boldsymbol{q}_1, \ldots, \boldsymbol{q}_n$ is an orthonormal basis for the whole space, then Q is square. Every $\boldsymbol{b} = QQ^{\mathrm{T}}\boldsymbol{b}$ *is the sum of its components along the* $\boldsymbol{q}$'s:

$$\boldsymbol{b} = \boldsymbol{q}_1(\boldsymbol{q}_1^{\mathrm{T}}\boldsymbol{b}) + \boldsymbol{q}_2(\boldsymbol{q}_2^{\mathrm{T}}\boldsymbol{b}) + \cdots + \boldsymbol{q}_n(\boldsymbol{q}_n^{\mathrm{T}}\boldsymbol{b}). \tag{6}$$

Transforms $QQ^{\mathrm{T}} = I$ is the foundation of Fourier series and all the great "transforms" of applied mathematics. They break vectors $\boldsymbol{b}$ or functions $f(x)$ into perpendicular pieces. Then by adding the pieces in (6), the inverse transform puts $\boldsymbol{b}$ and $f(x)$ back together.

Example 4 The columns of this orthogonal Q are orthonormal vectors $\boldsymbol{q}_1, \boldsymbol{q}_2, \boldsymbol{q}_3$:

$$\boldsymbol{m = n = 3} \qquad Q = \frac{1}{3}\begin{bmatrix} -1 & 2 & 2 \\ 2 & -1 & 2 \\ 2 & 2 & -1 \end{bmatrix} \quad \text{has} \quad \boldsymbol{Q^{\mathrm{T}}Q = QQ^{\mathrm{T}} = I}.$$

The separate projections of $\boldsymbol{b} = (0,0,1)$ onto $\boldsymbol{q}_1$ and $\boldsymbol{q}_2$ and $\boldsymbol{q}_3$ are $\boldsymbol{p}_1$ and $\boldsymbol{p}_2$ and $\boldsymbol{p}_3$:

$$\boldsymbol{q}_1(\boldsymbol{q}_1^{\mathrm{T}}\boldsymbol{b}) = \tfrac{2}{3}\boldsymbol{q}_1 \quad \text{and} \quad \boldsymbol{q}_2(\boldsymbol{q}_2^{\mathrm{T}}\boldsymbol{b}) = \tfrac{2}{3}\boldsymbol{q}_2 \quad \text{and} \quad \boldsymbol{q}_3(\boldsymbol{q}_3^{\mathrm{T}}\boldsymbol{b}) = -\tfrac{1}{3}\boldsymbol{q}_3.$$

The sum of the first two is the projection of $\boldsymbol{b}$ onto the *plane* of $\boldsymbol{q}_1$ and $\boldsymbol{q}_2$. The sum of all three is the projection of $\boldsymbol{b}$ onto the *whole space*—which is $\boldsymbol{p}_1 + \boldsymbol{p}_2 + \boldsymbol{p}_3 = \boldsymbol{b}$ itself:

Reconstruct $\boldsymbol{b}$
$\boldsymbol{b} = \boldsymbol{p}_1 + \boldsymbol{p}_2 + \boldsymbol{p}_3$

$$\frac{2}{3}\boldsymbol{q}_1 + \frac{2}{3}\boldsymbol{q}_2 - \frac{1}{3}\boldsymbol{q}_3 = \frac{1}{9}\begin{bmatrix} -2+4-2 \\ 4-2-2 \\ 4+4+1 \end{bmatrix} = \begin{bmatrix} 0 \\ 0 \\ 1 \end{bmatrix} = \boldsymbol{b}.$$

The Gram-Schmidt Process

The point of this section is that "orthogonal is good". Projections and least squares always involve $A^{\mathrm{T}}A$. When this matrix becomes $Q^{\mathrm{T}}Q = I$, the inverse is no problem. The one-dimensional projections are uncoupled. The best $\widehat{\boldsymbol{x}}$ is $Q^{\mathrm{T}}\boldsymbol{b}$ (just n separate dot products). For this to be true, we had to say "*If* the vectors are orthonormal". ***Now we explain the "Gram-Schmidt way" to create orthonormal vectors.***

Start with three independent vectors $\boldsymbol{a}, \boldsymbol{b}, \boldsymbol{c}$. We intend to construct three orthogonal vectors $\boldsymbol{A}, \boldsymbol{B}, \boldsymbol{C}$. Then (at the end may be easiest) we divide $\boldsymbol{A}, \boldsymbol{B}, \boldsymbol{C}$ by their lengths. That produces three orthonormal vectors $\boldsymbol{q}_1 = \boldsymbol{A}/\|\boldsymbol{A}\|$, $\boldsymbol{q}_2 = \boldsymbol{B}/\|\boldsymbol{B}\|$, $\boldsymbol{q}_3 = \boldsymbol{C}/\|\boldsymbol{C}\|$.

Gram-Schmidt Begin by choosing $\boldsymbol{A} = \boldsymbol{a}$. This first direction is accepted as it comes. The next direction $\boldsymbol{B}$ must be perpendicular to $\boldsymbol{A}$. ***Start with $\boldsymbol{b}$ and subtract its projection along $\boldsymbol{A}$.*** This leaves the perpendicular part, which is the orthogonal vector $\boldsymbol{B}$:

First Gram-Schmidt step

$$\boldsymbol{B} = \boldsymbol{b} - \frac{\boldsymbol{A}^{\mathrm{T}}\boldsymbol{b}}{\boldsymbol{A}^{\mathrm{T}}\boldsymbol{A}}\boldsymbol{A}. \tag{7}$$

$\boldsymbol{A}$ and $\boldsymbol{B}$ are orthogonal in Figure 4.11. Multiply equation (7) by $\boldsymbol{A}^{\mathrm{T}}$ to verify that $\boldsymbol{A}^{\mathrm{T}}\boldsymbol{B} = \boldsymbol{A}^{\mathrm{T}}\boldsymbol{b} - \boldsymbol{A}^{\mathrm{T}}\boldsymbol{b} = 0$. This vector $\boldsymbol{B}$ is what we have called the error vector $\boldsymbol{e}$, perpendicular to $\boldsymbol{A}$. Notice that $\boldsymbol{B}$ in equation (7) is not zero (otherwise $\boldsymbol{a}$ and $\boldsymbol{b}$ would be dependent). The directions $\boldsymbol{A}$ and $\boldsymbol{B}$ are now set.

The third direction starts with $\boldsymbol{c}$. This is not a combination of $\boldsymbol{A}$ and $\boldsymbol{B}$ (because $\boldsymbol{c}$ is not a combination of $\boldsymbol{a}$ and $\boldsymbol{b}$). But most likely $\boldsymbol{c}$ is not perpendicular to $\boldsymbol{A}$ and $\boldsymbol{B}$. So subtract off its components in those two directions to get a perpendicular direction $\boldsymbol{C}$:

Next Gram-Schmidt step

$$\boldsymbol{C} = \boldsymbol{c} - \frac{\boldsymbol{A}^{\mathrm{T}}\boldsymbol{c}}{\boldsymbol{A}^{\mathrm{T}}\boldsymbol{A}}\boldsymbol{A} - \frac{\boldsymbol{B}^{\mathrm{T}}\boldsymbol{c}}{\boldsymbol{B}^{\mathrm{T}}\boldsymbol{B}}\boldsymbol{B}. \tag{8}$$

This is the one and only idea of the Gram-Schmidt process. ***Subtract from every new vector its projections in the directions already set.*** That idea is repeated at every step.[2] If we had a fourth vector $\boldsymbol{d}$, we would subtract three projections onto $\boldsymbol{A}$, $\boldsymbol{B}, \boldsymbol{C}$ to get $\boldsymbol{D}$.

[2] I think Gram had the idea. I don't really know where Schmidt came in.

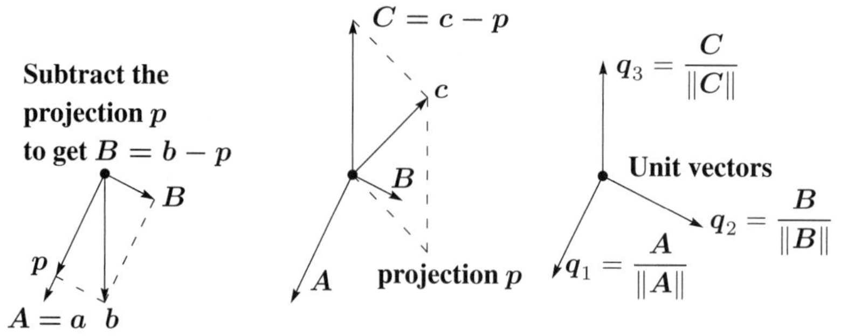

Figure 4.11: First project $\boldsymbol{b}$ onto the line through $\boldsymbol{a}$ and find the orthogonal $\boldsymbol{B}$ as $\boldsymbol{b} - \boldsymbol{p}$. Then project $\boldsymbol{c}$ onto the $\boldsymbol{AB}$ plane and find $\boldsymbol{C}$ as $\boldsymbol{c} - \boldsymbol{p}$. Divide by $\|\boldsymbol{A}\|$, $\|\boldsymbol{B}\|$, $\|\boldsymbol{C}\|$.

At the end, *or immediately when each one is found*, divide the orthogonal vectors $\boldsymbol{A}, \boldsymbol{B}, \boldsymbol{C}, \boldsymbol{D}$ by their lengths. The resulting vectors $\boldsymbol{q}_1, \boldsymbol{q}_2, \boldsymbol{q}_3, \boldsymbol{q}_4$ are orthonormal.

Example of **Gram-Schmidt** Suppose the independent non-orthogonal vectors $\boldsymbol{a}, \boldsymbol{b}, \boldsymbol{c}$ are

$$\boldsymbol{a} = \begin{bmatrix} 1 \\ -1 \\ 0 \end{bmatrix} \quad \text{and} \quad \boldsymbol{b} = \begin{bmatrix} 2 \\ 0 \\ -2 \end{bmatrix} \quad \text{and} \quad \boldsymbol{c} = \begin{bmatrix} 3 \\ -3 \\ 3 \end{bmatrix}.$$

Then $\boldsymbol{A} = \boldsymbol{a}$ has $\boldsymbol{A}^{\mathrm{T}}\boldsymbol{A} = 2$ and $A^{\mathrm{T}}\boldsymbol{b} = 2$. Subtract from $\boldsymbol{b}$ its projection $\boldsymbol{p}$ along $\boldsymbol{A}$:

First step $$\boldsymbol{B} = \boldsymbol{b} - \frac{\boldsymbol{A}^{\mathrm{T}}\boldsymbol{b}}{\boldsymbol{A}^{\mathrm{T}}\boldsymbol{A}}\boldsymbol{A} = \boldsymbol{b} - \frac{2}{2}\boldsymbol{A} = \begin{bmatrix} \mathbf{1} \\ \mathbf{1} \\ \mathbf{-2} \end{bmatrix}.$$

Check: $\boldsymbol{A}^{\mathrm{T}}\boldsymbol{B} = 0$ as required. Now subtract the projections of $\boldsymbol{c}$ on $\boldsymbol{A}$ and $\boldsymbol{B}$ to get $\boldsymbol{C}$:

Next step $$\boldsymbol{C} = \boldsymbol{c} - \frac{\boldsymbol{A}^{\mathrm{T}}\boldsymbol{c}}{\boldsymbol{A}^{\mathrm{T}}\boldsymbol{A}}\boldsymbol{A} - \frac{\boldsymbol{B}^{\mathrm{T}}\boldsymbol{c}}{\boldsymbol{B}^{\mathrm{T}}\boldsymbol{B}}\boldsymbol{B} = \boldsymbol{c} - \frac{6}{2}\boldsymbol{A} + \frac{6}{6}\boldsymbol{B} = \begin{bmatrix} \mathbf{1} \\ \mathbf{1} \\ \mathbf{1} \end{bmatrix}.$$

Check: $\boldsymbol{C} = (1, 1, 1)$ is perpendicular to both $\boldsymbol{A}$ and $\boldsymbol{B}$. Finally convert $\boldsymbol{A}, \boldsymbol{B}, \boldsymbol{C}$ to unit vectors (length 1, orthonormal). The lengths of $\boldsymbol{A}$, $\boldsymbol{B}$, $\boldsymbol{C}$ are $\sqrt{2}$ and $\sqrt{6}$ and $\sqrt{3}$. Divide by those lengths, for an orthonormal basis:

$$\boldsymbol{q}_1 = \frac{1}{\sqrt{2}}\begin{bmatrix} 1 \\ -1 \\ 0 \end{bmatrix} \quad \text{and} \quad \boldsymbol{q}_2 = \frac{1}{\sqrt{6}}\begin{bmatrix} 1 \\ 1 \\ -2 \end{bmatrix} \quad \text{and} \quad \boldsymbol{q}_3 = \frac{1}{\sqrt{3}}\begin{bmatrix} 1 \\ 1 \\ 1 \end{bmatrix}.$$

Usually $\boldsymbol{A}, \boldsymbol{B}, \boldsymbol{C}$ contain fractions. Almost always $\boldsymbol{q}_1, \boldsymbol{q}_2, \boldsymbol{q}_3$ contain square roots.

The Factorization $A = QR$

We started with a matrix A, whose columns were $\boldsymbol{a}, \boldsymbol{b}, \boldsymbol{c}$. We ended with a matrix Q, whose columns are $\boldsymbol{q}_1, \boldsymbol{q}_2, \boldsymbol{q}_3$. How are those matrices related? Since the vectors $\boldsymbol{a}, \boldsymbol{b}, \boldsymbol{c}$ are combinations of the $\boldsymbol{q}$'s (and vice versa), there must be a third matrix connecting A to Q. This third matrix is the triangular R in $A = QR$.

The first step was $\boldsymbol{q}_1 = \boldsymbol{a}/\|\boldsymbol{a}\|$ (other vectors not involved). The second step was equation (7), where $\boldsymbol{b}$ is a combination of $\boldsymbol{A}$ and $\boldsymbol{B}$. At that stage $\boldsymbol{C}$ and $\boldsymbol{q}_3$ were not involved. This non-involvement of later vectors is the key point of Gram-Schmidt:

- The vectors $\boldsymbol{a}$ and $\boldsymbol{A}$ and $\boldsymbol{q}_1$ are all along a single line.
- The vectors $\boldsymbol{a}, \boldsymbol{b}$ and $\boldsymbol{A}, \boldsymbol{B}$ and $\boldsymbol{q}_1, \boldsymbol{q}_2$ are all in the same plane.
- The vectors $\boldsymbol{a}, \boldsymbol{b}, \boldsymbol{c}$ and $\boldsymbol{A}, \boldsymbol{B}, \boldsymbol{C}$ and $\boldsymbol{q}_1, \boldsymbol{q}_2, \boldsymbol{q}_3$ are in one subspace (dimension 3).

At every step $\boldsymbol{a}_1, \ldots, \boldsymbol{a}_k$ are combinations of $\boldsymbol{q}_1, \ldots, \boldsymbol{q}_k$. Later $\boldsymbol{q}$'s are not involved. The connecting matrix R is ***triangular***, and we have $A = QR$:

$$\begin{bmatrix} & & \\ \boldsymbol{a} & \boldsymbol{b} & \boldsymbol{c} \\ & & \end{bmatrix} = \begin{bmatrix} & & \\ \boldsymbol{q}_1 & \boldsymbol{q}_2 & \boldsymbol{q}_3 \\ & & \end{bmatrix} \begin{bmatrix} \boldsymbol{q}_1^{\mathrm{T}}\boldsymbol{a} & \boldsymbol{q}_1^{\mathrm{T}}\boldsymbol{b} & \boldsymbol{q}_1^{\mathrm{T}}\boldsymbol{c} \\ & \boldsymbol{q}_2^{\mathrm{T}}\boldsymbol{b} & \boldsymbol{q}_2^{\mathrm{T}}\boldsymbol{c} \\ & & \boldsymbol{q}_3^{\mathrm{T}}\boldsymbol{c} \end{bmatrix} \quad \text{or} \quad \boldsymbol{A = QR}. \tag{9}$$

$A = QR$ is Gram-Schmidt in a nutshell. Multiply by Q^{T} to recognize $\boldsymbol{R = Q^{\mathrm{T}}A}$ above.

(Gram-Schmidt) From independent vectors $\boldsymbol{a}_1, \ldots, \boldsymbol{a}_n$, Gram-Schmidt constructs orthonormal vectors $\boldsymbol{q}_1, \ldots, \boldsymbol{q}_n$. The matrices with these columns satisfy $A = QR$. Then $R = Q^{\mathrm{T}}A$ is ***upper triangular*** because later $\boldsymbol{q}$'s are orthogonal to earlier $\boldsymbol{a}$'s.

Here are the original $\boldsymbol{a}$'s and the final $\boldsymbol{q}$'s from the example. The i, j entry of $R = Q^{\mathrm{T}}A$ is row i of Q^{T} times column j of A. The dot products $\boldsymbol{q}_i^{\mathrm{T}}\boldsymbol{a}_j$ go into R. **Then** $\boldsymbol{A = QR}$:

$$A = \begin{bmatrix} 1 & 2 & 3 \\ -1 & 0 & -3 \\ 0 & -2 & 3 \end{bmatrix} = \begin{bmatrix} 1/\sqrt{2} & 1/\sqrt{6} & 1/\sqrt{3} \\ -1/\sqrt{2} & 1/\sqrt{6} & 1/\sqrt{3} \\ 0 & -2/\sqrt{6} & 1/\sqrt{3} \end{bmatrix} \begin{bmatrix} \boldsymbol{\sqrt{2}} & \sqrt{2} & \sqrt{18} \\ \boldsymbol{0} & \boldsymbol{\sqrt{6}} & -\sqrt{6} \\ \boldsymbol{0} & \boldsymbol{0} & \boldsymbol{\sqrt{3}} \end{bmatrix} = QR.$$

Look closely at Q and R. The lengths of $\boldsymbol{A}, \boldsymbol{B}, \boldsymbol{C}$ are $\sqrt{2}, \sqrt{6}, \sqrt{3}$ on the diagonal of R. The columns of Q are orthonormal. Because of the square roots, QR might look harder than LU. Both factorizations are absolutely central to calculations in linear algebra.

Any m by n matrix A with independent columns can be factored into $A = QR$. The m by n matrix Q has orthonormal columns, and the square matrix R is upper triangular with positive diagonal. We must not forget why this is useful for least squares: $\boldsymbol{A^{\mathrm{T}}A = (QR)^{\mathrm{T}}QR = R^{\mathrm{T}}Q^{\mathrm{T}}QR = R^{\mathrm{T}}R}$. The least squares equation $A^{\mathrm{T}}A\widehat{\boldsymbol{x}} = A^{\mathrm{T}}\boldsymbol{b}$ simplifies to $R^{\mathrm{T}}R\widehat{\boldsymbol{x}} = R^{\mathrm{T}}Q^{\mathrm{T}}\boldsymbol{b}$. Then finally we reach $R\widehat{\boldsymbol{x}} = Q^{\mathrm{T}}\boldsymbol{b}$: good.

Least squares $\quad R^{\mathrm{T}}R\widehat{\boldsymbol{x}} = R^{\mathrm{T}}Q^{\mathrm{T}}\boldsymbol{b}$ or $R\widehat{\boldsymbol{x}} = Q^{\mathrm{T}}\boldsymbol{b}$ or $\widehat{\boldsymbol{x}} = R^{-1}Q^{\mathrm{T}}\boldsymbol{b}$ (10)

Instead of solving $A\boldsymbol{x} = \boldsymbol{b}$, which is impossible, we solve $R\widehat{\boldsymbol{x}} = Q^{\mathrm{T}}\boldsymbol{b}$ by back substitution—which is very fast. The real cost is the mn^2 multiplications in the Gram-Schmidt process, which are needed to construct the orthogonal Q and the triangular R with $A = QR$.

Below is an informal code. It executes equations (11) for $j = 1$ then $j = 2$ and eventually $j = n$. The important lines 4-5 subtract from $\boldsymbol{v} = \boldsymbol{a}_j$ its projection onto each $\boldsymbol{q}_i, i < j$. The last line of that code normalizes $\boldsymbol{v}$ (divides by $r_{jj} = ||\boldsymbol{v}||$) to get the unit vector $\boldsymbol{q}_j$:

$$r_{kj} = \sum_{i=1}^{m} q_{ik}v_{ij} \text{ and } v_{ij} = v_{ij} - q_{ik}r_{kj} \text{ and } r_{jj} = \left(\sum_{i=1}^{m} v_{ij}^2\right)^{1/2} \text{ and } q_{ij} = \frac{v_{ij}}{r_{jj}}. \quad (11)$$

Starting from $\boldsymbol{a}, \boldsymbol{b}, \boldsymbol{c} = \boldsymbol{a}_1, \boldsymbol{a}_2, \boldsymbol{a}_3$ this code will construct $\boldsymbol{q}_1$, then $\boldsymbol{B}$, $\boldsymbol{q}_2$, then $\boldsymbol{C}$, $\boldsymbol{q}_3$:

$$\boldsymbol{q}_1 = \boldsymbol{a}_1/\|\boldsymbol{a}_1\| \qquad \boldsymbol{B} = \boldsymbol{a}_2 - (\boldsymbol{q}_1^{\mathrm{T}}\boldsymbol{a}_2)\boldsymbol{q}_1 \qquad \boldsymbol{q}_2 = \boldsymbol{B}/\|\boldsymbol{B}\|$$

$$\boldsymbol{C}^* = \boldsymbol{a}_3 - (\boldsymbol{q}_1^{\mathrm{T}}\boldsymbol{a}_3)\boldsymbol{q}_1 \quad \boldsymbol{C} = \boldsymbol{C}^* - (\boldsymbol{q}_2^{\mathrm{T}}\boldsymbol{C}^*)\boldsymbol{q}_2 \quad \boldsymbol{q}_3 = \boldsymbol{C}/\|\boldsymbol{C}\|$$

Equation (11) subtracts **one projection at a time** as in $\boldsymbol{C}^*$ and $\boldsymbol{C}$. That change is called ***modified Gram-Schmidt***. This code is numerically more stable than equation (8) which subtracts all projections at once.

```
for j = 1:n                          % modified Gram-Schmidt
   v = A(:, j);                      % v begins as column j of the original A
   for i = 1:j-1                     % columns q_1 to q_{j-1} are already settled in Q
      R(i, j) = Q(:, i)'*v;          % compute R_ij = q_i^T a_j which is q_i^T v
      v = v-R(i, j)*Q(:, i);         % subtract the projection (q_i^T v) q_i
   end                               % v is now perpendicular to all of q_1, ..., q_{j-1}
   R(j, j) = norm(v);                % the diagonal entries R_jj are lengths
   Q(:, j) = v/R(j, j);              % divide v by its length to get the next q_j
end                                  % the "for j = 1 : n loop" produces all of the q_j
```

To recover column j of A, undo the last step and the middle steps of the code:

$$R(j,j)\boldsymbol{q}_j = (\boldsymbol{v} \textit{ minus its projections}) = (\text{column } j \text{ of } A) - \sum_{i=1}^{j-1} R(i,j)\boldsymbol{q}_i\,. \quad (12)$$

Moving the sum to the far left, this is column j in the multiplication $QR = A$.

Confession Good software like LAPACK, used in good systems like MATLAB and Julia and Python, will not use this Gram-Schmidt code. There is now a better way. "Householder reflections" act on A to produce the upper triangular R. This happens one column at a time in the same way that elimination produces the upper triangular U in LU.

Those reflection matrices $I - 2\boldsymbol{u}\boldsymbol{u}^{\mathrm{T}}$ will be described in Chapter 11 on numerical linear algebra. If A is tridiagonal we can simplify even more to use 2 by 2 rotations. The result is always $A = QR$ and the MATLAB command to orthogonalize A is $[Q, R] = \mathsf{qr}(A)$. I believe that Gram-Schmidt is still the good process to understand, even if the reflections or rotations lead to a more perfect Q.

■ REVIEW OF THE KEY IDEAS ■

1. If the orthonormal vectors $\boldsymbol{q}_1, \ldots, \boldsymbol{q}_n$ are the columns of Q, then $\boldsymbol{q}_i^{\mathrm{T}}\boldsymbol{q}_j = 0$ and $\boldsymbol{q}_i^{\mathrm{T}}\boldsymbol{q}_i = 1$ translate into the matrix multiplication $Q^{\mathrm{T}}Q = I$.
2. If Q is square (an ***orthogonal matrix***) then $Q^{\mathrm{T}} = Q^{-1}$: ***transpose*** = ***inverse***.
3. The length of $Q\boldsymbol{x}$ equals the length of $\boldsymbol{x}$: $\|Q\boldsymbol{x}\| = \|\boldsymbol{x}\|$.
4. The projection onto the column space of Q spanned by the $\boldsymbol{q}$'s is $P = QQ^{\mathrm{T}}$.
5. If Q is square then $P = QQ^{\mathrm{T}} = I$ and every $\boldsymbol{b} = \boldsymbol{q}_1(\boldsymbol{q}_1^{\mathrm{T}}\boldsymbol{b}) + \cdots + \boldsymbol{q}_n(\boldsymbol{q}_n^{\mathrm{T}}\boldsymbol{b})$.
6. Gram-Schmidt produces orthonormal vectors $\boldsymbol{q}_1, \boldsymbol{q}_2, \boldsymbol{q}_3$ from independent $\boldsymbol{a}, \boldsymbol{b}, \boldsymbol{c}$. In matrix form this is the factorization $A = QR =$ (orthogonal Q)(triangular R).

■ WORKED EXAMPLES ■

4.4 A Add two more columns with all entries 1 or -1, so the columns of this 4 by 4 "Hadamard matrix" are orthogonal. How do you turn H_4 into an *orthogonal matrix* Q?

$$H_2 = \begin{bmatrix} 1 & 1 \\ 1 & -1 \end{bmatrix} \qquad H_4 = \begin{bmatrix} 1 & 1 & x & x \\ 1 & -1 & x & x \\ 1 & 1 & x & x \\ 1 & -1 & x & x \end{bmatrix} \qquad \text{and} \qquad Q_4 = \begin{bmatrix} & & & \\ & & & \\ & & & \\ & & & \end{bmatrix}$$

The block matrix $H_8 = \begin{bmatrix} H_4 & H_4 \\ H_4 & -H_4 \end{bmatrix}$ is the next Hadamard matrix with 1's and -1's. What is the product $H_8^{\mathrm{T}} H_8$?

The projection of $\boldsymbol{b} = (6, 0, 0, 2)$ onto the first column of H_4 is $\boldsymbol{p}_1 = (2, 2, 2, 2)$. The projection onto the second column is $\boldsymbol{p}_2 = (1, -1, 1, -1)$. What is the projection $\boldsymbol{p}_{1,2}$ of $\boldsymbol{b}$ onto the 2-dimensional space spanned by the first two columns?

Solution H_4 can be built from H_2 just as H_8 is built from H_4:

$$H_4 = \begin{bmatrix} H_2 & H_2 \\ H_2 & -H_2 \end{bmatrix} = \begin{bmatrix} 1 & 1 & 1 & 1 \\ 1 & -1 & 1 & -1 \\ 1 & 1 & -1 & -1 \\ 1 & -1 & -1 & 1 \end{bmatrix} \text{ has orthogonal columns.}$$

Then $Q = H/2$ has orthonormal columns. Dividing by 2 gives unit vectors in Q. A 5 by 5 Hadamard matrix is impossible because the dot product of columns would have five 1's and/or -1's and could not add to zero. H_8 has orthogonal columns of length $\sqrt{8}$.

$$H_8^{\mathrm{T}} H_8 = \begin{bmatrix} H^{\mathrm{T}} & H^{\mathrm{T}} \\ H^{\mathrm{T}} & -H^{\mathrm{T}} \end{bmatrix} \begin{bmatrix} H & H \\ H & -H \end{bmatrix} = \begin{bmatrix} 2H^{\mathrm{T}}H & 0 \\ 0 & 2H^{\mathrm{T}}H \end{bmatrix} = \begin{bmatrix} 8I & 0 \\ 0 & 8I \end{bmatrix}. \ Q_8 = \frac{H_8}{\sqrt{8}}$$

4.4 B What is the key point of orthogonal columns? Answer: $A^{\mathrm{T}}A$ is diagonal and easy to invert. **We can project onto lines and just add**. The axes are orthogonal.

Add *p*'s Projection $\boldsymbol{p}_{1,2}$ onto a plane equals $\boldsymbol{p}_1 + \boldsymbol{p}_2$ onto orthogonal lines.

Problem Set 4.4

Problems 1–12 are about orthogonal vectors and orthogonal matrices.

1 Are these pairs of vectors orthonormal or only orthogonal or only independent?

(a) $\begin{bmatrix} 1 \\ 0 \end{bmatrix}$ and $\begin{bmatrix} 1 \\ 1 \end{bmatrix}$ (b) $\begin{bmatrix} .6 \\ .8 \end{bmatrix}$ and $\begin{bmatrix} .4 \\ -.3 \end{bmatrix}$ (c) $\begin{bmatrix} \cos\theta \\ \sin\theta \end{bmatrix}$ and $\begin{bmatrix} -\sin\theta \\ \cos\theta \end{bmatrix}$.

Change the second vector when necessary to produce orthonormal vectors.

2 The vectors $(2, 2, -1)$ and $(-1, 2, 2)$ are orthogonal. Divide them by their lengths to find orthonormal vectors $\boldsymbol{q}_1$ and $\boldsymbol{q}_2$. Put those into the columns of Q and multiply $Q^{\mathrm{T}}Q$ and QQ^{T}.

3 (a) If A has three orthogonal columns each of length 4, what is $A^{\mathrm{T}}A$?

(b) If A has three orthogonal columns of lengths $1, 2, 3$, what is $A^{\mathrm{T}}A$?

4 Give an example of each of the following:

(a) A matrix Q that has orthonormal columns but $QQ^{\mathrm{T}} \neq I$.

(b) Two orthogonal vectors that are not linearly independent.

(c) An orthonormal basis for $\mathbf{R}^3$, including the vector $\boldsymbol{q}_1 = (1, 1, 1)/\sqrt{3}$.

5 Find two orthogonal vectors in the plane $x + y + 2z = 0$. Make them orthonormal.

6 If Q_1 and Q_2 are orthogonal matrices, show that their product Q_1Q_2 is also an orthogonal matrix. (Use $Q^{\mathrm{T}}Q = I$.)

7 If Q has orthonormal columns, what is the least squares solution $\widehat{\boldsymbol{x}}$ to $Q\boldsymbol{x} = \boldsymbol{b}$?

8 If $\boldsymbol{q}_1$ and $\boldsymbol{q}_2$ are orthonormal vectors in $\mathbf{R}^5$, what combination ____ $\boldsymbol{q}_1 +$ ____ $\boldsymbol{q}_2$ is closest to a given vector $\boldsymbol{b}$?

9 (a) Compute $P = QQ^{\mathrm{T}}$ when $\boldsymbol{q}_1 = (.8, .6, 0)$ and $\boldsymbol{q}_2 = (-.6, .8, 0)$. Verify that $P^2 = P$.

(b) Prove that always $(QQ^{\mathrm{T}})^2 = QQ^{\mathrm{T}}$ by using $Q^{\mathrm{T}}Q = I$. Then $P = QQ^{\mathrm{T}}$ is the projection matrix onto the column space of Q.

10 Orthonormal vectors are automatically linearly independent.

(a) Vector proof: When $c_1\boldsymbol{q}_1 + c_2\boldsymbol{q}_2 + c_3\boldsymbol{q}_3 = \boldsymbol{0}$, what dot product leads to $c_1 = 0$? Similarly $c_2 = 0$ and $c_3 = 0$. Thus the $\boldsymbol{q}$'s are independent.

(b) Matrix proof: Show that $Q\boldsymbol{x} = \boldsymbol{0}$ leads to $\boldsymbol{x} = \boldsymbol{0}$. Since Q may be rectangular, you can use Q^{T} but not Q^{-1}.

11 (a) Gram-Schmidt: Find orthonormal vectors $\boldsymbol{q}_1$ and $\boldsymbol{q}_2$ in the plane spanned by $\boldsymbol{a} = (1, 3, 4, 5, 7)$ and $\boldsymbol{b} = (-6, 6, 8, 0, 8)$.

(b) Which vector in this plane is closest to $(1, 0, 0, 0, 0)$?

12 If $\boldsymbol{a}_1, \boldsymbol{a}_2, \boldsymbol{a}_3$ is a basis for $\mathbf{R}^3$, any vector $\boldsymbol{b}$ can be written as

$$\boldsymbol{b} = x_1\boldsymbol{a}_1 + x_2\boldsymbol{a}_2 + x_3\boldsymbol{a}_3 \qquad \text{or} \qquad \begin{bmatrix} & & \\ \boldsymbol{a}_1 & \boldsymbol{a}_2 & \boldsymbol{a}_3 \\ & & \end{bmatrix} \begin{bmatrix} x_1 \\ x_2 \\ x_3 \end{bmatrix} = \boldsymbol{b}.$$

(a) Suppose the $\boldsymbol{a}$'s are orthonormal. Show that $x_1 = \boldsymbol{a}_1^{\mathrm{T}}\boldsymbol{b}$.

(b) Suppose the $\boldsymbol{a}$'s are orthogonal. Show that $x_1 = \boldsymbol{a}_1^{\mathrm{T}}\boldsymbol{b}/\boldsymbol{a}_1^{\mathrm{T}}\boldsymbol{a}_1$.

(c) If the $\boldsymbol{a}$'s are independent, x_1 is the first component of ____ times $\boldsymbol{b}$.

Problems 13–25 are about the Gram-Schmidt process and $A = QR$.

13 What multiple of $\boldsymbol{a} = \left[\begin{smallmatrix}1\\1\end{smallmatrix}\right]$ should be subtracted from $\boldsymbol{b} = \left[\begin{smallmatrix}4\\0\end{smallmatrix}\right]$ to make the result $\boldsymbol{B}$ orthogonal to $\boldsymbol{a}$? Sketch a figure to show $\boldsymbol{a}$, $\boldsymbol{b}$, and $\boldsymbol{B}$.

14 Complete the Gram-Schmidt process in Problem 13 by computing $\boldsymbol{q}_1 = \boldsymbol{a}/\|\boldsymbol{a}\|$ and $\boldsymbol{q}_2 = \boldsymbol{B}/\|\boldsymbol{B}\|$ and factoring into QR:

$$\begin{bmatrix} 1 & 4 \\ 1 & 0 \end{bmatrix} = \begin{bmatrix} \boldsymbol{q}_1 & \boldsymbol{q}_2 \end{bmatrix} \begin{bmatrix} \|\boldsymbol{a}\| & ? \\ 0 & \|\boldsymbol{B}\| \end{bmatrix}.$$

15 (a) Find orthonormal vectors $\boldsymbol{q}_1, \boldsymbol{q}_2, \boldsymbol{q}_3$ such that $\boldsymbol{q}_1, \boldsymbol{q}_2$ span the column space of

$$A = \begin{bmatrix} 1 & 1 \\ 2 & -1 \\ -2 & 4 \end{bmatrix}.$$

(b) Which of the four fundamental subspaces contains $\boldsymbol{q}_3$?

(c) Solve $A\boldsymbol{x} = (1, 2, 7)$ by least squares.

16 What multiple of $\boldsymbol{a} = (4, 5, 2, 2)$ is closest to $\boldsymbol{b} = (1, 2, 0, 0)$? Find orthonormal vectors $\boldsymbol{q}_1$ and $\boldsymbol{q}_2$ in the plane of $\boldsymbol{a}$ and $\boldsymbol{b}$.

17 Find the projection of $\boldsymbol{b}$ onto the line through $\boldsymbol{a}$:

$$\boldsymbol{a} = \begin{bmatrix} 1 \\ 1 \\ 1 \end{bmatrix} \quad \text{and} \quad \boldsymbol{b} = \begin{bmatrix} 1 \\ 3 \\ 5 \end{bmatrix} \quad \text{and} \quad \boldsymbol{p} = ? \quad \text{and} \quad \boldsymbol{e} = \boldsymbol{b} - \boldsymbol{p} = ?$$

Compute the orthonormal vectors $\boldsymbol{q}_1 = \boldsymbol{a}/\|\boldsymbol{a}\|$ and $\boldsymbol{q}_2 = \boldsymbol{e}/\|\boldsymbol{e}\|$.

18 (Recommended) Find orthogonal vectors $\boldsymbol{A}, \boldsymbol{B}, \boldsymbol{C}$ by Gram-Schmidt from $\boldsymbol{a}, \boldsymbol{b}, \boldsymbol{c}$:

$$\boldsymbol{a} = (1, -1, 0, 0) \qquad \boldsymbol{b} = (0, 1, -1, 0) \qquad \boldsymbol{c} = (0, 0, 1, -1).$$

$\boldsymbol{A}, \boldsymbol{B}, \boldsymbol{C}$ and $\boldsymbol{a}, \boldsymbol{b}, \boldsymbol{c}$ are bases for the vectors perpendicular to $\boldsymbol{d} = (1, 1, 1, 1)$.

19 If $A = QR$ then $A^{\mathrm{T}}A = R^{\mathrm{T}}R =$ _____ triangular times _____ triangular. *Gram-Schmidt on A corresponds to elimination on $A^{\mathrm{T}}A$.* The pivots for $A^{\mathrm{T}}A$ must be the squares of diagonal entries of R. Find Q and R by Gram-Schmidt for this A:

$$A = \begin{bmatrix} -1 & 1 \\ 2 & 1 \\ 2 & 4 \end{bmatrix} \quad \text{and} \quad A^{\mathrm{T}}A = \begin{bmatrix} 9 & 9 \\ 9 & 18 \end{bmatrix} = \begin{bmatrix} 1 & 0 \\ 1 & 1 \end{bmatrix} \begin{bmatrix} 9 & \\ & 9 \end{bmatrix} \begin{bmatrix} 1 & 1 \\ 0 & 1 \end{bmatrix}.$$

20 True or false (give an example in either case):

(a) Q^{-1} is an orthogonal matrix when Q is an orthogonal matrix.

(b) If Q (3 by 2) has orthonormal columns then $\|Q\boldsymbol{x}\|$ always equals $\|\boldsymbol{x}\|$.

21 Find an orthonormal basis for the column space of A:

$$A = \begin{bmatrix} 1 & -2 \\ 1 & 0 \\ 1 & 1 \\ 1 & 3 \end{bmatrix} \quad \text{and} \quad \boldsymbol{b} = \begin{bmatrix} -4 \\ -3 \\ 3 \\ 0 \end{bmatrix}.$$

Then compute the projection of $\boldsymbol{b}$ onto that column space.

22 Find orthogonal vectors $\boldsymbol{A}, \boldsymbol{B}, \boldsymbol{C}$ by Gram-Schmidt from

$$\boldsymbol{a} = \begin{bmatrix} 1 \\ 1 \\ 2 \end{bmatrix} \quad \text{and} \quad \boldsymbol{b} = \begin{bmatrix} 1 \\ -1 \\ 0 \end{bmatrix} \quad \text{and} \quad \boldsymbol{c} = \begin{bmatrix} 1 \\ 0 \\ 4 \end{bmatrix}.$$

23 Find $\boldsymbol{q}_1, \boldsymbol{q}_2, \boldsymbol{q}_3$ (orthonormal) as combinations of $\boldsymbol{a}, \boldsymbol{b}, \boldsymbol{c}$ (independent columns). Then write A as QR:

$$A = \begin{bmatrix} 1 & 2 & 4 \\ 0 & 0 & 5 \\ 0 & 3 & 6 \end{bmatrix}.$$

24 (a) Find a basis for the subspace $\boldsymbol{S}$ in $\mathbf{R}^4$ spanned by all solutions of

$$x_1 + x_2 + x_3 - x_4 = 0.$$

(b) Find a basis for the orthogonal complement $\boldsymbol{S}^\perp$.

(c) Find $\boldsymbol{b}_1$ in $\boldsymbol{S}$ and $\boldsymbol{b}_2$ in $\boldsymbol{S}^\perp$ so that $\boldsymbol{b}_1 + \boldsymbol{b}_2 = \boldsymbol{b} = (1, 1, 1, 1)$.

25 If $ad - bc > 0$, the entries in $A = QR$ are

$$\begin{bmatrix} a & b \\ c & d \end{bmatrix} = \frac{\begin{bmatrix} a & -c \\ c & a \end{bmatrix}}{\sqrt{a^2 + c^2}} \frac{\begin{bmatrix} a^2 + c^2 & ab + cd \\ 0 & ad - bc \end{bmatrix}}{\sqrt{a^2 + c^2}}.$$

Write $A = QR$ when $a, b, c, d = 2, 1, 1, 1$ and also $1, 1, 1, 1$. Which entry of R becomes zero when the columns are dependent and Gram-Schmidt breaks down?

Problems 26–29 use the QR code in equation (11). It executes Gram-Schmidt.

26 Show why $\boldsymbol{C}$ (found via $\boldsymbol{C}^*$ in the steps after (11)) is equal to $\boldsymbol{C}$ in equation (8).

27 Equation (8) subtracts from $\boldsymbol{c}$ its components along $\boldsymbol{A}$ and $\boldsymbol{B}$. Why not subtract the components along $\boldsymbol{a}$ and along $\boldsymbol{b}$?

28 Where are the mn^2 multiplications in equation (11)?

29 Apply the MATLAB qr code to $\boldsymbol{a} = (2, 2, -1)$, $\boldsymbol{b} = (0, -3, 3)$, $\boldsymbol{c} = (1, 0, 0)$. What are the $\boldsymbol{q}$'s?

Problems 30–35 involve orthogonal matrices that are special.

30 The first four ***wavelets*** are in the columns of this wavelet matrix W:

$$W = \frac{1}{2} \begin{bmatrix} 1 & 1 & \sqrt{2} & 0 \\ 1 & 1 & -\sqrt{2} & 0 \\ 1 & -1 & 0 & \sqrt{2} \\ 1 & -1 & 0 & -\sqrt{2} \end{bmatrix}.$$

What is special about the columns? Find the inverse wavelet transform W^{-1}.

31 (a) Choose c so that Q is an orthogonal matrix:

$$Q = c\begin{bmatrix} 1 & -1 & -1 & -1 \\ -1 & 1 & -1 & -1 \\ -1 & -1 & 1 & -1 \\ -1 & -1 & -1 & 1 \end{bmatrix}.$$

Project $\boldsymbol{b} = (1,1,1,1)$ onto the first column. Then project $\boldsymbol{b}$ onto the plane of the first two columns.

32 If $\boldsymbol{u}$ is a unit vector, then $Q = I - 2\boldsymbol{u}\boldsymbol{u}^{\mathrm{T}}$ is a reflection matrix (Example 3). Find Q_1 from $\boldsymbol{u} = (0,1)$ and Q_2 from $\boldsymbol{u} = (0, \sqrt{2}/2, \sqrt{2}/2)$. Draw the reflections when Q_1 and Q_2 multiply the vectors $(1,2)$ and $(1,1,1)$.

33 Find all matrices that are both orthogonal and lower triangular.

34 $Q = I - 2\boldsymbol{u}\boldsymbol{u}^{\mathrm{T}}$ is a reflection matrix when $\boldsymbol{u}^{\mathrm{T}}\boldsymbol{u} = 1$. Two reflections give $Q^2 = I$.

(a) Show that $Q\boldsymbol{u} = -\boldsymbol{u}$. The mirror is perpendicular to $\boldsymbol{u}$.

(b) Find $Q\boldsymbol{v}$ when $\boldsymbol{u}^{\mathrm{T}}\boldsymbol{v} = 0$. The mirror contains $\boldsymbol{v}$. It reflects to itself.

Challenge Problems

35 (MATLAB) Factor $[Q, R] = \mathbf{qr}(A)$ for $A = \mathbf{eye}(4) - \mathbf{diag}([1\ \ 1\ \ 1], -1)$. You are orthogonalizing the columns $(1,-1,0,0)$ and $(0,1,-1,0)$ and $(0,0,1,-1)$ and $(0,0,0,1)$ of A. Can you scale the orthogonal columns of Q to get nice integer components?

36 If A is m by n with rank n, $\mathbf{qr}(A)$ produces a *square* Q and zeros below R:

The factors from MATLAB are $(m \text{ by } m)(m \text{ by } n)$ $\qquad A = [Q_1 \quad Q_2]\begin{bmatrix} R \\ 0 \end{bmatrix}.$

The n columns of Q_1 are an orthonormal basis for which fundamental subspace? The $m-n$ columns of Q_2 are an orthonormal basis for which fundamental subspace?

37 We know that $P = QQ^{\mathrm{T}}$ is the projection onto the column space of $Q(m \text{ by } n)$. Now add another column $\boldsymbol{a}$ to produce $A = [Q \quad \boldsymbol{a}]$. Gram-Schmidt replaces $\boldsymbol{a}$ by what vector $\boldsymbol{q}$? Start with $\boldsymbol{a}$, subtract ______, divide by ______ to find $\boldsymbol{q}$.

Chapter 5

Determinants

1 The **determinant** of $A = \begin{bmatrix} a & b \\ c & d \end{bmatrix}$ is $\boldsymbol{ad - bc}$. Singular matrix $A = \begin{bmatrix} a & xa \\ c & xc \end{bmatrix}$ has $\mathbf{det} = \mathbf{0}$.

2 **Row exchange reverses signs** $PA = \begin{bmatrix} 0 & 1 \\ 1 & 0 \end{bmatrix} \begin{bmatrix} a & b \\ c & d \end{bmatrix} = \begin{bmatrix} c & d \\ a & b \end{bmatrix}$ has $\det PA = bc - ad = -\,\mathbf{det}\, \boldsymbol{A}$.

3 The determinant of $\begin{bmatrix} xa + yA & xb + yB \\ c & d \end{bmatrix}$ is $x(ad - bc) + y(Ad - Bc)$. **Det is linear in row 1 by itself.**

4 **Elimination** $EA = \begin{bmatrix} a & b \\ 0 & d - \frac{c}{a}b \end{bmatrix}$ $\mathbf{det}\, \boldsymbol{EA} = a\left(d - \frac{c}{a}b\right) =$ **product of pivots** $=$ **det** $\boldsymbol{A}$.

5 If A is n by n then $\mathbf{1, 2, 3, 4}$ remain true: $\mathbf{det} = \mathbf{0}$ when A is singular, **det reverses sign** when rows are exchanged, **det is linear in row 1 by itself**, det $=$ **product of the pivots**. Always $\mathbf{det}\, \boldsymbol{BA} = (\mathbf{det}\, \boldsymbol{B})(\mathbf{det}\, \boldsymbol{A})$ and $\mathbf{det}\, \boldsymbol{A}^{\mathrm{T}} = \mathbf{det}\, \boldsymbol{A}$. This is an amazing number.

5.1 The Properties of Determinants

The determinant of a square matrix is a single number. That number contains an amazing amount of information about the matrix. It tells immediately whether the matrix is invertible. ***The determinant is zero when the matrix has no inverse***. When A is invertible, the determinant of A^{-1} is $1/(\det A)$. If $\det A = 2$ then $\det A^{-1} = \frac{1}{2}$. In fact the determinant leads to a formula for every entry in A^{-1}.

This is one use for determinants—to find formulas for inverse matrices and pivots and solutions $A^{-1}\boldsymbol{b}$. For a large matrix we seldom use those formulas, because elimination is faster. For a 2 by 2 matrix with entries a, b, c, d, its determinant $ad - bc$ shows how A^{-1} changes as A changes. Notice the division by the determinant!

$$A = \begin{bmatrix} a & b \\ c & d \end{bmatrix} \quad \text{has inverse} \quad A^{-1} = \frac{\mathbf{1}}{\boldsymbol{ad - bc}} \begin{bmatrix} d & -b \\ -c & a \end{bmatrix}. \tag{1}$$

Multiply those matrices to get I. When the determinant is $ad - bc = 0$, we are asked to divide by zero and we can't—then A has no inverse. (The rows are parallel when $a/c = b/d$. This gives $ad = bc$ and $\det A = 0$.) Dependent rows always lead to $\det A = 0$.

The determinant is also connected to the pivots. For a 2 by 2 matrix the pivots are a and $d - (c/a)b$. ***The product of the pivots is the determinant***:

$$\textbf{Product of pivots} \qquad a\left(d - \frac{c}{a}b\right) = ad - bc \quad \textbf{which is} \quad \det A.$$

After a row exchange the pivots change to c and $b - (a/c)d$. Those new pivots multiply to give $bc - ad$. The row exchange to $\left[\begin{smallmatrix} c & d \\ a & b \end{smallmatrix}\right]$ reversed the sign of the determinant.

Looking ahead The determinant of an n by n matrix can be found in three ways:

1 Multiply the n pivots (times 1 or -1)	This is the **pivot formula**.
2 Add up $n!$ terms (times 1 or -1)	This is the **"big" formula**.
3 Combine n smaller determinants (times 1 or -1)	This is the **cofactor formula**.

You see that *plus or minus signs*—the decisions between 1 and -1—play a big part in determinants. That comes from the following rule for n by n matrices:

The determinant changes sign when two rows* (or two columns) *are exchanged.

The identity matrix has determinant $+1$. Exchange two rows and $\det P = -1$. Exchange two more rows and the new permutation has $\det P = +1$. Half of all permutations are *even* ($\det P = 1$) and half are *odd* ($\det P = -1$). Starting from I, half of the P's involve an even number of exchanges and half require an odd number. In the 2 by 2 case, ad has a plus sign and bc has minus—coming from the row exchange:

$$\det \begin{bmatrix} 1 & 0 \\ 0 & 1 \end{bmatrix} = 1 \quad \text{and} \quad \det \begin{bmatrix} 0 & 1 \\ 1 & 0 \end{bmatrix} = -1.$$

The other essential rule is linearity—but a warning comes first. Linearity does not mean that $\det(A + B) = \det A + \det B$. ***This is absolutely false***. That kind of linearity is not even true when $A = I$ and $B = I$. The false rule would say that $\det(I + I) = 1 + 1 = 2$. The true rule is $\det 2I = 2^n$. Determinants are multiplied by 2^n (not just by 2) when matrices are multiplied by 2.

We don't intend to define the determinant by its formulas. It is better to start with its properties—*sign reversal and linearity*. The properties are simple (Section 5.1). They prepare for the formulas (Section 5.2). Then come the applications, including these three:

(1) Determinants give A^{-1} and $A^{-1}\boldsymbol{b}$ (this formula is called **Cramer's Rule**).

(2) When the edges of a box are the rows of A, the **volume** is $|\det A|$.

(3) For n special numbers λ, called **eigenvalues**, the determinant of $A - \lambda I$ is zero. This is a truly important application and it fills Chapter 6.

The Properties of the Determinant

Determinants have three basic properties (rules 1, 2, 3). By using those rules we can compute the determinant of any square matrix A. ***This number is written in two ways,* det *A and* $|A|$.** Notice: Brackets for the matrix, straight bars for its determinant. When A is a 2 by 2 matrix, the rules 1, 2, 3 lead to the answer we expect:

$$\text{The determinant of} \quad \begin{bmatrix} a & b \\ c & d \end{bmatrix} \quad \text{is} \quad \begin{vmatrix} a & b \\ c & d \end{vmatrix} = ad - bc.$$

From rules 1–3 we will reach rules 4–10. The last two are $\det(AB) = (\det A)(\det B)$ and $\det A^{\mathrm{T}} = \det A$. We will check all rules with the 2 by 2 formula, but do not forget: The rules apply to any n by n matrix A.

Rule 1 (the easiest) matches $\det I = 1$ with volume $= 1$ for a unit cube.

1 ***The determinant of the n by n identity matrix is* 1.**

$$\begin{vmatrix} 1 & 0 \\ 0 & 1 \end{vmatrix} = 1 \qquad \text{and} \qquad \begin{vmatrix} 1 & & \\ & \ddots & \\ & & 1 \end{vmatrix} = 1.$$

2 ***The determinant changes sign when two rows are exchanged*** (sign reversal):

$$\text{Check:} \quad \begin{vmatrix} c & d \\ a & b \end{vmatrix} = -\begin{vmatrix} a & b \\ c & d \end{vmatrix} \quad \text{(both sides equal } bc - ad\text{)}.$$

Because of this rule, we can find $\det P$ for any permutation matrix. Just exchange rows of I until you reach P. Then $\det P = +1$ for an ***even*** number of row exchanges and $\det P = -1$ for an ***odd*** number.

The third rule has to make the big jump to the determinants of all matrices.

3 ***The determinant is a linear function of each row separately*** (all other rows stay fixed). If the first row is multiplied by t, the determinant is multiplied by t. If first rows are added, determinants are added. This rule only applies when the other rows do not change! Notice how c and d stay the same:

multiply row 1 by any number t **det is multiplied by t**	$\begin{vmatrix} ta & tb \\ c & d \end{vmatrix} = t\begin{vmatrix} a & b \\ c & d \end{vmatrix}$
add row 1 of A to row 1 of A': **then determinants add**	$\begin{vmatrix} a+a' & b+b' \\ c & d \end{vmatrix} = \begin{vmatrix} a & b \\ c & d \end{vmatrix} + \begin{vmatrix} a' & b' \\ c & d \end{vmatrix}.$

In the first case, both sides are $tad - tbc$. Then t factors out. In the second case, both sides are $ad + a'd - bc - b'c$. These rules still apply when A is n by n, and **one row changes**.

$$A = \begin{vmatrix} \mathbf{4} & \mathbf{8} & \mathbf{8} \\ 0 & 1 & 1 \\ 0 & 0 & 1 \end{vmatrix} = \mathbf{4}\begin{vmatrix} \mathbf{1} & \mathbf{2} & \mathbf{2} \\ 0 & 1 & 1 \\ 0 & 0 & 1 \end{vmatrix} \quad \text{and} \quad \begin{vmatrix} \mathbf{4} & \mathbf{8} & \mathbf{8} \\ 0 & 1 & 1 \\ 0 & 0 & 1 \end{vmatrix} = \begin{vmatrix} \mathbf{4} & \mathbf{0} & \mathbf{0} \\ 0 & 1 & 1 \\ 0 & 0 & 1 \end{vmatrix} + \begin{vmatrix} \mathbf{0} & \mathbf{8} & \mathbf{8} \\ 0 & 1 & 1 \\ 0 & 0 & 1 \end{vmatrix}.$$

By itself, rule 3 does not say what those determinants are ($\det A$ is 4).

Combining multiplication and addition, we get *any linear combination in one row*. Rule 2 for row exchanges can put that row into the first row and back again.

This rule does not mean that $\det 2I = 2 \det I$. To obtain $2I$ we have to multiply *both* rows by 2, and the factor 2 comes out both times:

$$\begin{vmatrix} 2 & 0 \\ 0 & 2 \end{vmatrix} = 2^2 = 4 \qquad \text{and} \qquad \begin{vmatrix} t & 0 \\ 0 & t \end{vmatrix} = t^2.$$

This is just like area and volume. Expand a rectangle by 2 and its area increases by 4. Expand an n-dimensional box by t and its volume increases by t^n. The connection is no accident—we will see how *determinants equal volumes*.

Pay special attention to rules 1–3. They completely determine the number $\det A$. We could stop here to find a formula for n by n determinants (a little complicated). We prefer to go gradually, because rules $4 - 10$ make determinants much easier to work with.

4 ***If two rows of A are equal, then* $\det A = 0$.**

Equal rows Check 2 by 2 : $\begin{vmatrix} a & b \\ a & b \end{vmatrix} = 0.$

Rule 4 follows from rule 2. (Remember we must use the rules and not the 2 by 2 formula.) *Exchange the two equal rows.* The determinant D is supposed to change sign. But also D has to stay the same, because the matrix is not changed. The only number with $-D = D$ is $D = 0$—this must be the determinant. (Note: In Boolean algebra the reasoning fails, because $-1 = 1$. Then D is defined by rules 1, 3, 4.)

A matrix with two equal rows has no inverse. Rule 4 makes $\det A = 0$. But matrices can be singular and determinants can be zero without having equal rows! Rule 5 will be the key. We can do row operations (like elimination) without changing $\det A$.

5 ***Subtracting a multiple of one row from another row leaves* $\det A$ *unchanged.***

ℓ times row 1
from row 2

$$\begin{vmatrix} a & b \\ c - \ell a & d - \ell b \end{vmatrix} = \begin{vmatrix} a & b \\ c & d \end{vmatrix}.$$

Rule 3 (linearity) splits the left side into the right side plus another term $-\ell \begin{vmatrix} a & b \\ a & b \end{vmatrix}$. This extra term is zero by rule 4: equal rows. Therefore rule 5 is correct (not just 2 by 2).

Conclusion *The determinant is not changed by the usual elimination steps from A to U.* Thus $\det A$ *equals* $\det U$. If we can find determinants of triangular matrices U, we can find determinants of all matrices A. Every row exchange reverses the sign, so always $\det A = \pm \det U$. Rule 5 has narrowed the problem to triangular matrices.

6 ***A matrix with a row of zeros has* $\det A = 0$.**

Row of zeros $\begin{vmatrix} 0 & 0 \\ c & d \end{vmatrix} = 0 \quad \text{and} \quad \begin{vmatrix} a & b \\ 0 & 0 \end{vmatrix} = 0.$

For an easy proof, add some other row to the zero row. The determinant is not changed (rule 5). But the matrix now has two equal rows. So $\det A = 0$ by rule 4.

7 ***If A is triangular then*** $\det A = a_{11}a_{22}\cdots a_{nn}$ ***= product of diagonal entries.***

$$\textbf{Triangular} \qquad \begin{vmatrix} a & b \\ 0 & d \end{vmatrix} = ad \quad \text{and also} \quad \begin{vmatrix} a & 0 \\ c & d \end{vmatrix} = ad.$$

Suppose all diagonal entries are nonzero. Remove the off-diagonal entries by elimination! (If A is lower triangular, subtract multiples of each row from lower rows. If A is upper triangular, subtract from higher rows.) By rule 5 the determinant is not changed—and now the matrix is diagonal:

$$\textbf{Diagonal matrix} \qquad \det \begin{bmatrix} a_{11} & & & 0 \\ & a_{22} & & \\ & & \ddots & \\ 0 & & & a_{nn} \end{bmatrix} = (a_{11})(a_{22})\cdots(a_{nn}).$$

Factor a_{11} from the first row by rule 3. Then factor a_{22} from the second row. Eventually factor a_{nn} from the last row. The determinant is a_{11} times a_{22} times $\cdots$ times a_{nn} times $\det I$. Then rule 1 (used at last!) is $\det I = 1$.

What if a diagonal entry a_{ii} is zero? Then the triangular A is singular. Elimination produces a *zero row*. By rule 5 the determinant is unchanged, and by rule 6 a zero row means $\det A = 0$. We reach the great test for **singular or invertible** matrices.

8 ***If A is singular then*** $\det A = 0$. ***If A is invertible then*** $\det A \neq 0$.

$$\textbf{Singular} \qquad \begin{bmatrix} a & b \\ c & d \end{bmatrix} \quad \text{is singular if and only if} \quad ad - bc = 0.$$

Proof Elimination goes from A to U. If A is singular then U has a zero row. The rules give $\det A = \det U = 0$. If A is invertible then U has the pivots along its diagonal. The product of nonzero pivots (using rule 7) gives a nonzero determinant:

$$\textbf{Multiply pivots} \qquad \det A = \pm \det U = \pm\ \textbf{(product of the pivots)}. \tag{2}$$

The pivots of a 2 by 2 matrix (if $a \neq 0$) are a and $d - (c/a)b$:

$$\textbf{The determinant is} \qquad \begin{vmatrix} a & b \\ c & d \end{vmatrix} = \begin{vmatrix} a & b \\ 0 & d - (c/a)b \end{vmatrix} = ad - bc.$$

This is the first formula for the determinant. MATLAB multiplies the pivots to find $\det A$. The sign in $\pm \det U$ depends on whether the number of row exchanges is even or odd: $+1$ or -1 is the determinant of the permutation P that exchanges rows.

With no row exchanges, $P = I$ and $\det A = \det U =$ *product of pivots*. And $\det L = 1$:

$$\text{If} \quad PA = LU \quad \text{then} \quad \det P \, \det A = \det L \, \det U \quad \text{and} \quad \det A = \pm \det U. \tag{3}$$

9 ***The determinant of AB is* $\det A$ *times* $\det B$*:* $|AB| = |A|\,|B|$.**

Product rule $$\begin{vmatrix} a & b \\ c & d \end{vmatrix}\begin{vmatrix} p & q \\ r & s \end{vmatrix} = \begin{vmatrix} ap+br & aq+bs \\ cp+dr & cq+ds \end{vmatrix}.$$

When the matrix B is A^{-1}, *this rule says that the determinant of* A^{-1} is $\mathbf{1/\det A}$:

A times A^{-1} $$\boldsymbol{AA^{-1} = I} \quad \textbf{so} \quad \boldsymbol{(\det A)(\det A^{-1}) = \det I = 1.}$$

This product rule is the most intricate so far. Even the 2 by 2 case needs some algebra:

$$|A|\,|B| = (ad-bc)(ps-qr) = (ap+br)(cq+ds) - (aq+bs)(cp+dr) = |AB|.$$

For the n by n case, here is a snappy proof that $|AB| = |A|\,|B|$. When $|B|$ is not zero, consider the ratio $\boldsymbol{D(A) = |AB|/|B|}$. *Check that this ratio $D(A)$ has properties* 1,2,3. Then $D(A)$ has to be the determinant and we have $|AB|/|B| = |A|$. Good.

Property 1 (*Determinant of I*) If $A = I$ then the ratio $D(A)$ becomes $|B|/|B| = 1$.

Property 2 (*Sign reversal*) When two rows of A are exchanged, so are the same two rows of AB. Therefore $|AB|$ changes sign and so does the ratio $|AB|/|B|$.

Property 3 (*Linearity*) When row 1 of A is multiplied by t, so is row 1 of AB. This multiplies the determinant $|AB|$ by t. So the ratio $|AB|/|B|$ is multiplied by t.

Add row 1 of A to row 1 of A'. Then row 1 of AB adds to row 1 of $A'B$. By rule 3, determinants add. After dividing by $|B|$, the ratios add—as desired.

Conclusion This ratio $|AB|/|B|$ has the same three properties that define $|A|$. Therefore it equals $|A|$. This proves the product rule $|AB| = |A|\,|B|$. The case $|B| = 0$ is separate and easy, because AB is singular when B is singular. Then $|AB| = |A|\,|B|$ is $0 = 0$.

10 ***The transpose A^{T} has the same determinant as A.***

Transpose $$\begin{vmatrix} a & b \\ c & d \end{vmatrix} = \begin{vmatrix} a & c \\ b & d \end{vmatrix} \quad \text{since both sides equal} \quad ad-bc.$$

The equation $|A^{\mathrm{T}}| = |A|$ becomes $0 = 0$ when A is singular (we know that A^{T} is also singular). Otherwise A has the usual factorization $PA = LU$. Transposing both sides gives $A^{\mathrm{T}}P^{\mathrm{T}} = U^{\mathrm{T}}L^{\mathrm{T}}$. The proof of $|A| = |A^{\mathrm{T}}|$ comes by using rule 9 for products:

$$\text{Compare} \quad \det P \det A = \det L \det U \quad \text{with} \quad \det A^{\mathrm{T}} \det P^{\mathrm{T}} = \det U^{\mathrm{T}} \det L^{\mathrm{T}}.$$

First, $\det L = \det L^{\mathrm{T}} = 1$ (both have 1's on the diagonal). Second, $\det U = \det U^{\mathrm{T}}$ (those triangular matrices have the same diagonal). Third, $\det P = \det P^{\mathrm{T}}$ (permutations have $P^{\mathrm{T}}P = I$, so $|P^{\mathrm{T}}|\,|P| = 1$ by rule 9; thus $|P|$ and $|P^{\mathrm{T}}|$ both equal 1 or both equal -1). So L, U, P have the same determinants as $L^{\mathrm{T}}, U^{\mathrm{T}}, P^{\mathrm{T}}$ and this leaves $\det A = \det A^{\mathrm{T}}$.

Important comment on columns Every rule for the rows can apply to the columns (just by transposing, since $|A| = |A^{\mathrm{T}}|$). The determinant changes sign when two columns are exchanged. *A zero column or two equal columns will make the determinant zero.* If a column is multiplied by t, so is the determinant. The determinant is a linear function of each column separately.

It is time to stop. The list of properties is long enough. Next we find and use an explicit formula for the determinant.

■ REVIEW OF THE KEY IDEAS ■

1. The determinant is defined by $\det I = 1$, sign reversal, and linearity in each row.
2. After elimination $\det A$ is $\pm$ (product of the pivots).
3. The determinant is zero exactly when A is not invertible.
4. Two remarkable properties are $\det AB = (\det A)(\det B)$ and $\det A^{\mathrm{T}} = \det A$.

■ WORKED EXAMPLES ■

5.1 A Apply these operations to A and find the determinants of M_1, M_2, M_3, M_4:
In M_1, multiplying each a_{ij} by $(-1)^{i+j}$ gives a checkerboard sign pattern.
In M_2, rows $1, 2, 3$ of A are *subtracted* from rows $2, 3, 1$.
In M_3, rows $1, 2, 3$ of A are *added* to rows $2, 3, 1$.
How are the determinants of M_1, M_2, M_3 related to the determinant of A?

$$\begin{bmatrix} a_{11} & -a_{12} & a_{13} \\ -a_{21} & a_{22} & -a_{23} \\ a_{31} & -a_{32} & a_{33} \end{bmatrix} \qquad \begin{bmatrix} \text{row } 1 - \text{row } 3 \\ \text{row } 2 - \text{row } 1 \\ \text{row } 3 - \text{row } 2 \end{bmatrix} \qquad \begin{bmatrix} \text{row } 1 + \text{row } 3 \\ \text{row } 2 + \text{row } 1 \\ \text{row } 3 + \text{row } 2 \end{bmatrix}$$

Solution The three determinants are $\det A$, 0, and $2 \det A$. Here are reasons:

$$M_1 = \begin{bmatrix} 1 & & \\ & -1 & \\ & & 1 \end{bmatrix} \begin{bmatrix} a_{11} & a_{12} & a_{13} \\ a_{21} & a_{22} & a_{23} \\ a_{31} & a_{32} & a_{33} \end{bmatrix} \begin{bmatrix} 1 & & \\ & -1 & \\ & & 1 \end{bmatrix} \quad \text{so } \det M_1 = (-1)(\det A)(-1).$$

M_2 is singular because its rows add to the zero row. Its determinant is zero.
M_3 can be split into *eight matrices* by Rule 3 (linearity in each row separately):

$$\begin{vmatrix} \text{row } 1 + \text{row } 3 \\ \text{row } 2 + \text{row } 1 \\ \text{row } 3 + \text{row } 3 \end{vmatrix} = \begin{vmatrix} \text{row } 1 \\ \text{row } 2 \\ \text{row } 3 \end{vmatrix} + \begin{vmatrix} \text{row } 3 \\ \text{row } 2 \\ \text{row } 3 \end{vmatrix} + \begin{vmatrix} \text{row } 1 \\ \text{row } 1 \\ \text{row } 3 \end{vmatrix} + \cdots + \begin{vmatrix} \text{row } 3 \\ \text{row } 1 \\ \text{row } 2 \end{vmatrix}.$$

All but the first and last have repeated rows and zero determinant. The first is A and the last has *two* row exchanges. So $\det M_3 = \det A + \det A$. (Try $A = I$.)

5.1 B Explain how to reach this determinant by row operations:

$$\det\begin{bmatrix} 1-a & 1 & 1 \\ 1 & 1-a & 1 \\ 1 & 1 & 1-a \end{bmatrix} = a^2(3-a). \tag{4}$$

Solution Subtract row 3 from row 1 and then from row 2. This leaves

$$\det\begin{bmatrix} -a & 0 & a \\ 0 & -a & a \\ 1 & 1 & 1-a \end{bmatrix}.$$

Now add column 1 to column 3, and also column 2 to column 3. This leaves a lower triangular matrix with $-a, -a, 3-a$ on the diagonal: $\det = (-a)(-a)(3-a)$.

The determinant is zero if $a = 0$ or $a = 3$. For $a = 0$ we have the *all-ones matrix*—certainly singular. For $a = 3$, each row adds to zero—again singular. Those numbers 0 and 3 are the **eigenvalues** of the all-ones matrix. This example is revealing and important, leading toward Chapter 6.

Problem Set 5.1

Questions 1–12 are about the rules for determinants.

1 If a 4 by 4 matrix has $\det A = \frac{1}{2}$, find $\det(2A)$ and $\det(-A)$ and $\det(A^2)$ and $\det(A^{-1})$.

2 If a 3 by 3 matrix has $\det A = -1$, find $\det(\frac{1}{2}A)$ and $\det(-A)$ and $\det(A^2)$ and $\det(A^{-1})$.

3 True or false, with a reason if true or a counterexample if false:

(a) The determinant of $I + A$ is $1 + \det A$.

(b) The determinant of ABC is $|A|\,|B|\,|C|$.

(c) The determinant of $4A$ is $4|A|$.

(d) The determinant of $AB - BA$ is zero. Try an example with $A = \begin{bmatrix} 0 & 0 \\ 0 & 1 \end{bmatrix}$.

4 Which row exchanges show that these "reverse identity matrices" J_3 and J_4 have $|J_3| = -1$ but $|J_4| = +1$?

$$\det\begin{bmatrix} 0 & 0 & 1 \\ 0 & 1 & 0 \\ 1 & 0 & 0 \end{bmatrix} = -1 \quad \text{but} \quad \det\begin{bmatrix} 0 & 0 & 0 & 1 \\ 0 & 0 & 1 & 0 \\ 0 & 1 & 0 & 0 \\ 1 & 0 & 0 & 0 \end{bmatrix} = +1.$$

5 For $n = 5, 6, 7$, count the row exchanges to permute the reverse identity J_n to the identity matrix I_n. Propose a rule for every size n and predict whether J_{101} has determinant $+1$ or -1.

6 Show how Rule 6 (determinant $= 0$ if a row is all zero) comes from Rule 3.

7 Find the determinants of rotations and reflections:

$$Q = \begin{bmatrix} \cos\theta & -\sin\theta \\ \sin\theta & \cos\theta \end{bmatrix} \quad \text{and} \quad Q = \begin{bmatrix} 1-2\cos^2\theta & -2\cos\theta\sin\theta \\ -2\cos\theta\sin\theta & 1-2\sin^2\theta \end{bmatrix}.$$

8 Prove that every orthogonal matrix ($Q^{\mathrm{T}}Q = I$) has determinant 1 or -1.

(a) Use the product rule $|AB| = |A|\,|B|$ and the transpose rule $|Q| = |Q^{\mathrm{T}}|$.

(b) Use only the product rule. If $|\det Q| > 1$ then $\det Q^n = (\det Q)^n$ blows up. How do you know this can't happen to Q^n?

9 Do these matrices have determinant $0, 1, 2$, or 3?

$$A = \begin{bmatrix} 0 & 0 & 1 \\ 1 & 0 & 0 \\ 0 & 1 & 0 \end{bmatrix} \quad B = \begin{bmatrix} 0 & 1 & 1 \\ 1 & 0 & 1 \\ 1 & 1 & 0 \end{bmatrix} \quad C = \begin{bmatrix} 1 & 1 & 1 \\ 1 & 1 & 1 \\ 1 & 1 & 1 \end{bmatrix}.$$

10 If the entries in every row of A add to zero, solve $A\boldsymbol{x} = \mathbf{0}$ to prove $\det A = 0$. If those entries add to one, show that $\det(A - I) = 0$. Does this mean $\det A = 1$?

11 Suppose that $CD = -DC$ and find the flaw in this reasoning: Taking determinants gives $|C|\,|D| = -|D|\,|C|$. Therefore $|C| = 0$ or $|D| = 0$. One or both of the matrices must be singular. (That is not true.)

12 The inverse of a 2 by 2 matrix seems to have determinant $= 1$:

$$\det A^{-1} = \det \frac{1}{ad-bc}\begin{bmatrix} d & -b \\ -c & a \end{bmatrix} = \frac{ad-bc}{ad-bc} = 1.$$

What is wrong with this calculation? What is the correct $\det A^{-1}$?

Questions 13–27 use the rules to compute specific determinants.

13 Reduce A to U and find $\det A =$ product of the pivots:

$$A = \begin{bmatrix} 1 & 1 & 1 \\ 1 & 2 & 2 \\ 1 & 2 & 3 \end{bmatrix} \qquad A = \begin{bmatrix} 1 & 2 & 3 \\ 2 & 2 & 3 \\ 3 & 3 & 3 \end{bmatrix}.$$

14 By applying row operations to produce an upper triangular U, compute

$$\det \begin{bmatrix} 1 & 2 & 3 & 0 \\ 2 & 6 & 6 & 1 \\ -1 & 0 & 0 & 3 \\ 0 & 2 & 0 & 7 \end{bmatrix} \quad \text{and} \quad \det \begin{bmatrix} 2 & -1 & 0 & 0 \\ -1 & 2 & -1 & 0 \\ 0 & -1 & 2 & -1 \\ 0 & 0 & -1 & 2 \end{bmatrix}.$$

15 Use row operations to simplify and compute these determinants:

$$\det\begin{bmatrix}101 & 201 & 301\\ 102 & 202 & 302\\ 103 & 203 & 303\end{bmatrix} \quad\text{and}\quad \det\begin{bmatrix}1 & t & t^2\\ t & 1 & t\\ t^2 & t & 1\end{bmatrix}.$$

16 Find the determinants of a rank one matrix and a skew-symmetric matrix:

$$A=\begin{bmatrix}1\\2\\3\end{bmatrix}\begin{bmatrix}1 & -4 & 5\end{bmatrix} \quad\text{and}\quad A=\begin{bmatrix}0 & 1 & 3\\ -1 & 0 & 4\\ -3 & -4 & 0\end{bmatrix}.$$

17 A skew-symmetric matrix has $A^{\mathrm{T}}=-A$. Insert a, b, c for $1, 3, 4$ in Question 16 and show that $|A|=0$. Write down a 4 by 4 example with $|A|=1$.

18 Use row operations to show that the 3 by 3 "Vandermonde determinant" is

$$\det\begin{bmatrix}1 & a & a^2\\ 1 & b & b^2\\ 1 & c & c^2\end{bmatrix}=(b-a)(c-a)(c-b).$$

19 Find the determinants of U and U^{-1} and U^2:

$$U=\begin{bmatrix}1 & 4 & 6\\ 0 & 2 & 5\\ 0 & 0 & 3\end{bmatrix} \quad\text{and}\quad U=\begin{bmatrix}a & b\\ 0 & d\end{bmatrix}.$$

20 Suppose you do two row operations at once, going from

$$\begin{bmatrix}a & b\\ c & d\end{bmatrix} \quad\text{to}\quad \begin{bmatrix}a-Lc & b-Ld\\ c-la & d-lb\end{bmatrix}.$$

Find the second determinant. Does it equal $ad-bc$?

21 *Row exchange*: Add row 1 of A to row 2, then subtract row 2 from row 1. Then add row 1 to row 2 and multiply row 1 by -1 to reach B. Which rules show

$$\det B=\begin{vmatrix}c & d\\ a & b\end{vmatrix} \quad\text{equals}\quad -\det A=-\begin{vmatrix}a & b\\ c & d\end{vmatrix}?$$

Those rules could replace Rule 2 in the definition of the determinant.

22 From $ad-bc$, find the determinants of A and A^{-1} and $A-\lambda I$:

$$A=\begin{bmatrix}2 & 1\\ 1 & 2\end{bmatrix} \quad\text{and}\quad A^{-1}=\frac{1}{3}\begin{bmatrix}2 & -1\\ -1 & 2\end{bmatrix} \quad\text{and}\quad A-\lambda I=\begin{bmatrix}2-\lambda & 1\\ 1 & 2-\lambda\end{bmatrix}.$$

Which two numbers λ lead to $\det(A-\lambda I)=0$? Write down the matrix $A-\lambda I$ for each of those numbers λ—it should not be invertible.

23 From $A = \begin{bmatrix} 4 & 1 \\ 2 & 3 \end{bmatrix}$ find A^2 and A^{-1} and $A - \lambda I$ and their determinants. Which two numbers λ lead to $\det(A - \lambda I) = 0$?

24 Elimination reduces A to U. Then $A = LU$:

$$A = \begin{bmatrix} 3 & 3 & 4 \\ 6 & 8 & 7 \\ -3 & 5 & -9 \end{bmatrix} = \begin{bmatrix} 1 & 0 & 0 \\ 2 & 1 & 0 \\ -1 & 4 & 1 \end{bmatrix} \begin{bmatrix} 3 & 3 & 4 \\ 0 & 2 & -1 \\ 0 & 0 & -1 \end{bmatrix} = LU.$$

Find the determinants of L, U, A, $U^{-1}L^{-1}$, and $U^{-1}L^{-1}A$.

25 If the i, j entry of A is i times j, show that $\det A = 0$. (Exception when $A = [\,1\,]$.)

26 If the i, j entry of A is $i + j$, show that $\det A = 0$. (Exception when $n = 1$ or 2.)

27 Compute the determinants of these matrices by row operations:

$$A = \begin{bmatrix} 0 & a & 0 \\ 0 & 0 & b \\ c & 0 & 0 \end{bmatrix} \quad \text{and} \quad B = \begin{bmatrix} 0 & a & 0 & 0 \\ 0 & 0 & b & 0 \\ 0 & 0 & 0 & c \\ d & 0 & 0 & 0 \end{bmatrix} \quad \text{and} \quad C = \begin{bmatrix} a & a & a \\ a & b & b \\ a & b & c \end{bmatrix}.$$

28 True or false (give a reason if true or a 2 by 2 example if false):

(a) If A is not invertible then AB is not invertible.

(b) The determinant of A is always the product of its pivots.

(c) The determinant of $A - B$ equals $\det A - \det B$.

(d) AB and BA have the same determinant.

29 What is wrong with this proof that projection matrices have $\det P = 1$?

$$P = A(A^{\mathrm{T}}A)^{-1}A^{\mathrm{T}} \qquad \text{so} \qquad |P| = |A|\frac{1}{|A^{\mathrm{T}}||A|}|A^{\mathrm{T}}| = 1.$$

30 (Calculus question) Show that the partial derivatives of $\ln(\det A)$ give A^{-1}!

$$f(a,b,c,d) = \ln(ad - bc) \qquad \text{leads to} \qquad \begin{bmatrix} \partial f/\partial a & \partial f/\partial c \\ \partial f/\partial b & \partial f/\partial d \end{bmatrix} = A^{-1}.$$

31 (MATLAB) The Hilbert matrix **hilb**(n) has i, j entry equal to $1/(i + j - 1)$. Print the determinants of **hilb**(1), **hilb**$(2), \ldots,$ **hilb**(10). Hilbert matrices are hard to work with! What are the pivots of **hilb** (5)?

32 (MATLAB) What is a typical determinant (experimentally) of **rand**(n) and **randn**(n) for $n = 50, 100, 200, 400$? (And what does "Inf" mean in MATLAB?)

33 (MATLAB) Find the largest determinant of a 6 by 6 matrix of 1's and -1's.

34 If you know that $\det A = 6$, what is the determinant of B?

$$\text{From } \det A = \begin{vmatrix} \text{row 1} \\ \text{row 2} \\ \text{row 3} \end{vmatrix} = 6 \text{ find } \det B = \begin{vmatrix} \text{row 3 + row 2 + row 1} \\ \text{row 2 + row 1} \\ \text{row 1} \end{vmatrix}.$$

5.2 Permutations and Cofactors

1 **2 by 2** : $ad - bc$ has $2!$ terms with $\pm$ signs. **n by n : det A adds $n!$ terms with $\pm$ signs.**

2 For $n = 3$, $\det A$ adds $3! = 6$ terms. Two terms are $+a_{12}a_{23}a_{31}$ and $-a_{13}a_{22}a_{31}$. **Rows 1, 2, 3 and columns 1, 2, 3 appear once in each term.**

3 That minus sign came because the column order $3, 2, 1$ needs one exchange to recover $1, 2, 3$.

4 The six terms include $+a_{11}a_{22}a_{33} - a_{11}a_{23}a_{32} = a_{11}(\boldsymbol{a_{22}a_{33} - a_{23}a_{32}}) = a_{11}(\textbf{cofactor } \boldsymbol{C_{11}})$.

5 Always $\det A = a_{11}C_{11} + a_{12}C_{12} + \cdots + a_{1n}C_{1n}$. Cofactors are determinants of size $n - 1$.

A computer finds the determinant from the pivots. This section explains two other ways to do it. There is a "big formula" using all $n!$ permutations. There is a "cofactor formula" using determinants of size $n - 1$. The best example is my favorite 4 by 4 matrix:

$$A = \begin{bmatrix} 2 & -1 & 0 & 0 \\ -1 & 2 & -1 & 0 \\ 0 & -1 & 2 & -1 \\ 0 & 0 & -1 & 2 \end{bmatrix} \quad \text{has} \quad \det A = 5.$$

We can find this determinant in all three ways: ***pivots, big formula, cofactors***.

1. The product of the pivots is $2 \cdot \frac{3}{2} \cdot \frac{4}{3} \cdot \frac{5}{4}$. Cancellation produces 5.

2. The "big formula" in equation (8) has $4! = 24$ terms. Only five terms are nonzero:

$$\det A = 16 - 4 - 4 - 4 + 1 = 5.$$

   The 16 comes from $2 \cdot 2 \cdot 2 \cdot 2$ on the diagonal of A. Where do -4 and $+1$ come from? When you can find those five terms, you have understood formula (8).

3. The numbers $2, -1, 0, 0$ in the first row multiply their cofactors $4, 3, 2, 1$ from the other rows. That gives $2 \cdot 4 - 1 \cdot 3 = 5$. Those cofactors are 3 by 3 determinants. Cofactors use the rows and columns that are *not* used by the entry in the first row. ***Every term in a determinant uses each row and column once!***

The Pivot Formula

When elimination leads to $A = LU$, the pivots $d_1, \ldots, d_n$ are on the diagonal of the upper triangular U. If no row exchanges are involved, ***multiply those pivots*** to find the determinant:

$$\det A = (\det L)(\det U) = (1)(d_1 d_2 \cdots d_n). \tag{1}$$

This formula for $\det A$ appeared in Section 5.1, with the further possibility of row exchanges. Then a permutation enters $PA = LU$. The determinant of P is -1 or $+1$.

$$(\det \boldsymbol{P})(\det \boldsymbol{A}) = (\det \boldsymbol{L})(\det \boldsymbol{U}) \quad \text{gives} \quad \det A = \pm(d_1 d_2 \cdots d_n). \tag{2}$$

Example 1 A row exchange produces pivots 4, 2, 1 and that important minus sign:

$$A = \begin{bmatrix} 0 & 0 & 1 \\ 0 & 2 & 3 \\ 4 & 5 & 6 \end{bmatrix} \qquad PA = \begin{bmatrix} 4 & 5 & 6 \\ 0 & 2 & 3 \\ 0 & 0 & 1 \end{bmatrix} \qquad \det A = -(4)(2)(1) = -8.$$

The odd number of row exchanges (namely one exchange) means that $\det P = -1$.

The next example has no row exchanges. It may be the first matrix we factored into LU (when it was 3 by 3). What is remarkable is that we can go directly to n by n. Pivots give the determinant. We will also see how determinants give the pivots.

Example 2 The first pivots of this tridiagonal matrix A are $2, \frac{3}{2}, \frac{4}{3}$. The next are $\frac{5}{4}$ and $\frac{6}{5}$ and eventually $\frac{n+1}{n}$. Factoring this n by n matrix reveals its determinant:

$$\begin{bmatrix} 2 & -1 & & & \\ -1 & 2 & -1 & & \\ & -1 & 2 & \cdot & \\ & & \cdot & \cdot & -1 \\ & & & -1 & 2 \end{bmatrix} = \begin{bmatrix} 1 & & & & \\ -\frac{1}{2} & 1 & & & \\ & -\frac{2}{3} & 1 & & \\ & & \cdot & \cdot & \\ & & & -\frac{n-1}{n} & 1 \end{bmatrix} \begin{bmatrix} \mathbf{2} & -1 & & & \\ & \mathbf{\frac{3}{2}} & -1 & & \\ & & \mathbf{\frac{4}{3}} & -1 & \\ & & & \cdot & \cdot \\ & & & & \boldsymbol{\frac{n+1}{n}} \end{bmatrix}$$

The pivots are on the diagonal of U (the last matrix). When 2 and $\frac{3}{2}$ and $\frac{4}{3}$ and $\frac{5}{4}$ are multiplied, the fractions cancel. The determinant of the 4 by 4 matrix is 5. The 3 by 3 determinant is 4. *The n by n determinant is* $n+1$:

$$\mathbf{-1, 2, -1 \text{ matrix}} \qquad \det A = (\mathbf{2})\left(\mathbf{\tfrac{3}{2}}\right)\left(\mathbf{\tfrac{4}{3}}\right)\cdots\left(\mathbf{\tfrac{n+1}{n}}\right) = \boldsymbol{n+1}.$$

Important point: The first pivots depend only on the *upper left corner* of the original matrix A. This is a rule for all matrices without row exchanges.

> The first k pivots come from the k by k matrix A_k in the top left corner of A.
> ***The determinant of that corner submatrix*** A_k ***is*** $d_1 d_2 \cdots d_k$ ***(first k pivots)***.

The 1 by 1 matrix A_1 contains the very first pivot d_1. This is $\det A_1$. The 2 by 2 matrix in the corner has $\det A_2 = d_1 d_2$. Eventually the n by n determinant multiplies all n pivots.

Elimination deals with the matrix A_k in the upper left corner while starting on the whole matrix. We assume no row exchanges—then $A = LU$ and $A_k = L_k U_k$. Dividing one determinant by the previous determinant ($\det A_k$ divided by $\det A_{k-1}$) cancels everything but the latest pivot d_k. ***Each pivot is a ratio of determinants***:

$$\textbf{Pivots from determinants} \qquad \textbf{The } k\textbf{th pivot is } \boldsymbol{d_k} = \frac{d_1 d_2 \cdots d_k}{d_1 d_2 \cdots d_{k-1}} = \frac{\det A_k}{\det A_{k-1}}. \tag{3}$$

We don't need row exchanges when all the upper left submatrices have $\det A_k \neq 0$.

The Big Formula for Determinants

Pivots are good for computing. They concentrate a lot of information—enough to find the determinant. But it is hard to connect them to the original a_{ij}. That part will be clearer if we go back to rules 1-2-3, linearity and sign reversal and $\det I = 1$. We want to derive a single explicit formula for the determinant, directly from the entries a_{ij}.

The formula has $n!$ terms. Its size grows fast because $n! = 1, 2, 6, 24, 120, \ldots$. For $n = 11$ there are about forty million terms. For $n = 2$, the two terms are ad and bc. Half the terms have minus signs (as in $-bc$). The other half have plus signs (as in ad). For $n = 3$ there are $3! = (3)(2)(1)$ terms. Here are those six terms:

$$\text{3 by 3 determinant} \quad \begin{vmatrix} a_{11} & a_{12} & \boldsymbol{a_{13}} \\ \boldsymbol{a_{21}} & a_{22} & a_{23} \\ a_{31} & \boldsymbol{a_{32}} & a_{33} \end{vmatrix} = \begin{matrix} +a_{11}a_{22}a_{33} + a_{12}a_{23}a_{31} + \boldsymbol{a_{13}a_{21}a_{32}} \\ -a_{11}a_{23}a_{32} - a_{12}a_{21}a_{33} - a_{13}a_{22}a_{31}. \end{matrix} \tag{4}$$

Notice the pattern. Each product like $a_{11}a_{23}a_{32}$ has ***one entry from each row***. It also has ***one entry from each column***. The column order 1, 3, 2 means that this particular term comes with a minus sign. The column order 3, 1, 2 in $a_{13}a_{21}a_{32}$ has a plus sign (boldface). It will be "permutations" that tell us the sign.

The next step ($n = 4$) brings $4! = 24$ terms. There are 24 ways to choose one entry from each row and column. Down the main diagonal, $a_{11}a_{22}a_{33}a_{44}$ with column order $1, 2, 3, 4$ always has a plus sign. That is the "identity permutation".

To derive the big formula I start with $n = 2$. The goal is to reach $ad - bc$ in a systematic way. Break each row into two simpler rows:

$$\begin{bmatrix} a & b \end{bmatrix} = \begin{bmatrix} a & 0 \end{bmatrix} + \begin{bmatrix} 0 & b \end{bmatrix} \quad \text{and} \quad \begin{bmatrix} c & d \end{bmatrix} = \begin{bmatrix} c & 0 \end{bmatrix} + \begin{bmatrix} 0 & d \end{bmatrix}.$$

Now apply linearity, first in row 1 (with row 2 fixed) and then in row 2 (with row 1 fixed):

$$\begin{aligned} \begin{vmatrix} a & b \\ c & d \end{vmatrix} &= \begin{vmatrix} a & 0 \\ c & d \end{vmatrix} + \begin{vmatrix} 0 & b \\ c & d \end{vmatrix} && \textit{(break up row 1)} \\ &= \begin{vmatrix} a & 0 \\ c & 0 \end{vmatrix} + \begin{vmatrix} a & 0 \\ 0 & d \end{vmatrix} + \begin{vmatrix} 0 & b \\ c & 0 \end{vmatrix} + \begin{vmatrix} 0 & b \\ 0 & d \end{vmatrix} && \textit{(break up row 2).} \end{aligned} \tag{5}$$

The last line has $2^2 = 4$ determinants. The first and fourth are zero because one row is a multiple of the other row. We are left with $2! = 2$ determinants to compute:

$$\begin{vmatrix} a & 0 \\ 0 & d \end{vmatrix} + \begin{vmatrix} 0 & b \\ c & 0 \end{vmatrix} = ad \begin{vmatrix} 1 & 0 \\ 0 & 1 \end{vmatrix} + bc \begin{vmatrix} 0 & 1 \\ 1 & 0 \end{vmatrix} = ad - bc.$$

The splitting led to permutation matrices. Their determinants give a plus or minus sign. The permutation tells the column sequence. In this case the column order is $(1, 2)$ or $(2, 1)$.

Now try $n = 3$. Each row splits into 3 simpler rows like $\begin{bmatrix} a_{11} & 0 & 0 \end{bmatrix}$. Using linearity in each row, $\det A$ splits into $3^3 = 27$ simple determinants. If a column choice is repeated—for example if we also choose the row $\begin{bmatrix} a_{21} & 0 & 0 \end{bmatrix}$—then the simple determinant is zero.

We pay attention only when ***the entries a_{ij} come from different columns***, like $(\mathbf{3},\mathbf{1},\mathbf{2})$:

$$\begin{vmatrix} a_{11} & a_{12} & a_{13} \\ a_{21} & a_{22} & a_{23} \\ a_{31} & a_{32} & a_{33} \end{vmatrix} = \begin{vmatrix} a_{11} & & \\ & a_{22} & \\ & & a_{33} \end{vmatrix} + \begin{vmatrix} & a_{12} & \\ & & a_{23} \\ a_{31} & & \end{vmatrix} + \begin{vmatrix} & & \boldsymbol{a_{13}} \\ \boldsymbol{a_{21}} & & \\ & \boldsymbol{a_{32}} & \end{vmatrix}$$

Six terms

$$+ \begin{vmatrix} a_{11} & & \\ & & a_{23} \\ & a_{32} & \end{vmatrix} + \begin{vmatrix} & a_{12} & \\ a_{21} & & \\ & & a_{33} \end{vmatrix} + \begin{vmatrix} & & a_{13} \\ & a_{22} & \\ a_{31} & & \end{vmatrix}.$$

There are $3! = 6$ ***ways to order the columns, so six determinants***. The six permutations of $(1,2,3)$ include the identity permutation $(1,2,3)$ from $P = I$.

$$\textbf{Column numbers} = (1,2,3),(2,3,1),(\mathbf{3},\mathbf{1},\mathbf{2}),(1,3,2),(2,1,3),(3,2,1). \tag{6}$$

The last three are *odd permutations* (one exchange). The first three are *even permutations* (0 or 2 exchanges). When the column sequence is $(\mathbf{3},\mathbf{1},\mathbf{2})$, we have chosen the entries $a_{1\mathbf{3}}a_{2\mathbf{1}}a_{3\mathbf{2}}$—that particular column sequence comes with a plus sign (2 exchanges). The determinant of A is now split into six simple terms. Factor out the a_{ij}:

$$\begin{aligned} \det A = {} & a_{11}a_{22}a_{33} \begin{vmatrix} 1 & & \\ & 1 & \\ & & 1 \end{vmatrix} + a_{12}a_{23}a_{31} \begin{vmatrix} & 1 & \\ & & 1 \\ 1 & & \end{vmatrix} + a_{1\mathbf{3}}a_{2\mathbf{1}}a_{3\mathbf{2}} \begin{vmatrix} & & 1 \\ 1 & & \\ & 1 & \end{vmatrix} \\ & + a_{11}a_{23}a_{32} \begin{vmatrix} 1 & & \\ & & 1 \\ & 1 & \end{vmatrix} + a_{12}a_{21}a_{33} \begin{vmatrix} & 1 & \\ 1 & & \\ & & 1 \end{vmatrix} + a_{13}a_{22}a_{31} \begin{vmatrix} & & 1 \\ & 1 & \\ 1 & & \end{vmatrix}. \end{aligned} \tag{7}$$

The first three (even) permutations have $\det P = +1$, the last three (odd) permutations have $\det P = -1$. We have proved the 3 by 3 formula in a systematic way.

Now you can see the n by n formula. There are $n!$ orderings of the columns. The columns $(1, 2, \ldots, n)$ go in each possible order $(\alpha, \beta, \ldots, \omega)$. Taking $a_{1\alpha}$ from row 1 and $a_{2\beta}$ from row 2 and eventually $a_{n\omega}$ from row n, the determinant contains the product $a_{1\alpha}a_{2\beta}\cdots a_{n\omega}$ times $+1$ or -1. Half the column orderings have sign -1.

The determinant of A is the sum of these $n!$ simple determinants, times 1 or -1. The simple determinants $a_{1\alpha}a_{2\beta}\cdots a_{n\omega}$ choose ***one entry from every row and column***. For 5 by 5, the term $a_{15}a_{22}a_{33}a_{44}a_{51}$ would have $\det P = -1$ from exchanging 5 and 1.

$$\begin{aligned} \det A &= \text{sum over all } \mathbf{n!} \text{ column permutations } P = (\alpha, \beta, \ldots, \omega) \\ &= \boxed{\sum (\det P) a_{1\alpha}a_{2\beta}\cdots a_{n\omega}} = \textbf{BIG FORMULA}. \end{aligned} \tag{8}$$

The 2 by 2 case is $+a_{11}a_{22} - a_{12}a_{21}$ (which is $ad - bc$). Here P is $(1,2)$ or $(2,1)$.

The 3 by 3 case has three products "down to the right" (see Problem 28) and three products "down to the left". Warning: Many people believe they should follow this pattern in the 4 by 4 case. They only take 8 products—but we need 24.

Example 3 (Determinant of U) When U is upper triangular, only one of the $n!$ products can be nonzero. This one term comes from the diagonal: $\det U = +u_{11}u_{22}\cdots u_{nn}$. All other column orderings pick at least one entry below the diagonal, where U has zeros. As soon as we pick a number like $u_{21} = 0$, that term in equation (8) is sure to be zero.

Of course $\det I = 1$. The only nonzero term is $+(1)(1)\cdots(1)$ from the diagonal.

Example 4 Suppose Z is the identity matrix except for column 3. Then

$$\text{The determinant of } Z = \begin{vmatrix} 1 & 0 & \boldsymbol{a} & 0 \\ 0 & 1 & \boldsymbol{b} & 0 \\ 0 & 0 & \boldsymbol{c} & 0 \\ 0 & 0 & \boldsymbol{d} & 1 \end{vmatrix} \text{ is } \boldsymbol{c}. \tag{9}$$

The term $(1)(1)(c)(1)$ comes from the main diagonal with a plus sign. There are $4! = 24$ products (choosing one factor from each row and column) but the other 23 products are zero. Reason: If we pick a, b, or d from column 3, that column is used up. Then the only available choice from row 3 is zero.

Here is a different reason for the same answer. If $c = 0$, then Z has a row of zeros and $\det Z = c = 0$ is correct. If c is not zero, *use elimination*. Subtract multiples of row 3 from the other rows, to knock out a, b, d. That leaves a diagonal matrix and $\det Z = c$.

This example will soon be used for "Cramer's Rule". If we move a, b, c, d into the first column of Z, the determinant is $\det Z = a$. (*Why*?) Changing one column of I leaves Z with an easy determinant, coming from its main diagonal only.

Example 5 Suppose A has 1's just above and below the main diagonal. Here $n = 4$:

$$A = \begin{bmatrix} 0 & \mathbf{1} & 0 & 0 \\ \mathbf{1} & 0 & \mathbf{1} & 0 \\ 0 & \mathbf{1} & 0 & \mathbf{1} \\ 0 & 0 & \mathbf{1} & 0 \end{bmatrix} \quad \text{and} \quad P = \begin{bmatrix} 0 & \mathbf{1} & 0 & 0 \\ \mathbf{1} & 0 & 0 & 0 \\ 0 & 0 & 0 & \mathbf{1} \\ 0 & 0 & \mathbf{1} & 0 \end{bmatrix} \quad \text{have } \textbf{determinant 1}.$$

The only nonzero choice in the first row is column 2. The only nonzero choice in row 4 is column 3. Then rows 2 and 3 *must* choose columns 1 and 4. In other words $\det P = \det A$. The determinant of P is $+1$ (two exchanges to reach $2, 1, 4, 3$). Therefore $\det A = +1$.

Determinant by Cofactors

Formula (8) is a direct definition of the determinant. It gives you everything at once—but you have to digest it. Somehow this sum of $n!$ terms must satisfy rules 1-2-3 (then all the other properties 4-10 will follow). The easiest is $\det I = 1$, already checked.

When you separate out the factor a_{11} ***or*** a_{12} ***or*** $a_{1\alpha}$ ***that comes from the first row***, you see linearity. For 3 by 3, separate the usual 6 terms of the determinant into 3 pairs:

$$\det A = \boldsymbol{a_{11}}(a_{22}a_{33} - a_{23}a_{32}) + \boldsymbol{a_{12}}(a_{23}a_{31} - a_{21}a_{33}) + \boldsymbol{a_{13}}(a_{21}a_{32} - a_{22}a_{31}). \quad (10)$$

Those three quantities in parentheses are called ***"cofactors"***. They are **2 by 2 determinants**, from rows 2 and 3. The first row contributes the factors a_{11}, a_{12}, a_{13}. *The lower rows contribute the cofactors* C_{11}, C_{12}, C_{13}. Certainly the determinant $a_{11}C_{11} + a_{12}C_{12} + a_{13}C_{13}$ depends linearly on a_{11}, a_{12}, a_{13}—this is Rule 3.

The cofactor of a_{11} is $C_{11} = a_{22}a_{33} - a_{23}a_{32}$. You can see it in this splitting:

$$\begin{vmatrix} a_{11} & a_{12} & a_{13} \\ a_{21} & a_{22} & a_{23} \\ a_{31} & a_{32} & a_{33} \end{vmatrix} = \begin{vmatrix} a_{11} & & \\ & a_{22} & a_{23} \\ & a_{32} & a_{33} \end{vmatrix} + \begin{vmatrix} & a_{12} & \\ a_{21} & & a_{23} \\ a_{31} & & a_{33} \end{vmatrix} + \begin{vmatrix} & & a_{13} \\ a_{21} & a_{22} & \\ a_{31} & a_{32} & \end{vmatrix}.$$

We are still choosing ***one entry from each row and column***. Since a_{11} uses up row 1 and column 1, that leaves a 2 by 2 determinant as its cofactor.

As always, we have to watch signs. The 2 by 2 determinant that goes with a_{12} looks like $a_{21}a_{33} - a_{23}a_{31}$. But in the cofactor C_{12}, *its sign is reversed.* Then $a_{12}C_{12}$ is the correct 3 by 3 determinant. The sign pattern for cofactors along the first row is *plus-minus-plus-minus*. ***You cross out row*** 1 ***and column*** j ***to get a submatrix*** M_{1j} ***of size*** $n - 1$. Multiply its determinant by the sign $(-1)^{1+j}$ to get the cofactor:

$$\text{The cofactors along row 1 are } C_{1j} = (-1)^{1+j} \det M_{1j}.$$

$$\textbf{The cofactor expansion is } \det \boldsymbol{A} = \boldsymbol{a_{11}C_{11} + a_{12}C_{12} + \cdots + a_{1n}C_{1n}}. \quad (11)$$

In the big formula (8), the terms that multiply a_{11} combine to give $C_{11} = \det M_{11}$. The sign is $(-1)^{1+1}$, meaning *plus*. Equation (11) is another form of equation (8) and also equation (10), with factors from row 1 multiplying cofactors that use only the other rows.

Note Whatever is possible for row 1 is possible for row i. The entries a_{ij} in that row also have cofactors C_{ij}. Those are determinants of order $n - 1$, multiplied by $(-1)^{i+j}$. Since a_{ij} accounts for row i and column j, ***the submatrix*** M_{ij} ***throws out row*** i ***and column*** j. The display shows a_{43} and M_{43} (with row 4 and column 3 removed). The sign $(-1)^{4+3}$ multiplies the determinant of M_{43} to give C_{43}. The sign matrix shows the $\pm$ pattern:

$$A = \begin{bmatrix} \bullet & \bullet & & \bullet \\ \bullet & \bullet & & \bullet \\ \bullet & \bullet & & \bullet \\ & & a_{43} & \end{bmatrix} \qquad \textbf{signs } (-1)^{i+j} = \begin{bmatrix} + & - & + & - \\ - & + & - & + \\ + & - & + & - \\ - & + & - & + \end{bmatrix}.$$

The determinant is the dot product of any row i of A with its cofactors using other rows:

COFACTOR FORMULA $\qquad \det A = a_{i1}C_{i1} + a_{i2}C_{i2} + \cdots + a_{in}C_{in}.$ (12)

Each cofactor C_{ij} (order $n-1$, without row i and column j) includes its correct sign:

Cofactor $\qquad C_{ij} = (-1)^{i+j} \det M_{ij}.$

A determinant of order n is a combination of determinants of order $n-1$. A recursive person would keep going. Each subdeterminant breaks into determinants of order $n-2$. *We could define all determinants via equation* (12). This rule goes from order n to $n-1$ to $n-2$ and eventually to order 1. Define the 1 by 1 determinant $|a|$ to be the number a. Then the cofactor method is complete.

We preferred to construct $\det A$ from its properties (linearity, sign reversal, $\det I = 1$). The big formula (8) and the cofactor formulas (10)–(12) follow from those rules. One last formula comes from the rule that $\det A = \det A^{\mathrm{T}}$. We can expand in cofactors, *down a column* instead of across a row. Down column j the entries are a_{1j} to a_{nj}. The cofactors are C_{1j} to C_{nj}. The determinant is the dot product:

Cofactors down column j $\qquad \det A = a_{1j}C_{1j} + a_{2j}C_{2j} + \cdots + a_{nj}C_{nj}.$ (13)

Cofactors are useful when matrices have many zeros—as in the next examples.

Example 6 The $-1, 2, -1$ matrix has only two nonzeros in its first row. So only two cofactors C_{11} and C_{12} are involved in the determinant. I will highlight C_{12} :

$$\begin{vmatrix} 2 & -1 & & \\ \mathbf{-1} & 2 & \mathbf{-1} & \\ & -1 & \mathbf{2} & \mathbf{-1} \\ & & \mathbf{-1} & \mathbf{2} \end{vmatrix} = 2\begin{vmatrix} 2 & -1 & \\ -1 & 2 & -1 \\ & -1 & 2 \end{vmatrix} - (-1)\begin{vmatrix} \mathbf{-1} & \mathbf{-1} & \\ & \mathbf{2} & \mathbf{-1} \\ & \mathbf{-1} & \mathbf{2} \end{vmatrix}. \qquad (14)$$

You see 2 times C_{11} first on the right, from crossing out row 1 and column 1. This cofactor C_{11} has exactly the same $-1, 2, -1$ pattern as the original A—but one size smaller.

To compute the boldface C_{12}, *use cofactors down its first column.* The only nonzero is at the top. That contributes another -1 (so we are back to minus). Its cofactor is the $-1, 2, -1$ determinant which is 2 by 2, *two sizes smaller* than the original A.

Summary ***Each determinant D_n of order n comes from D_{n-1} and D_{n-2}***:

$$D_4 = 2D_3 - D_2 \qquad \text{and generally} \qquad D_n = 2D_{n-1} - D_{n-2}. \qquad (15)$$

Direct calculation gives $D_2 = 3$ and $D_3 = 4$. Equation (14) has $D_4 = 2(4) - 3 = 5$. These determinants 3, 4, 5 fit the formula $\boldsymbol{D_n = n+1}$. Then D_n equals $2n - (n-1)$. That "special tridiagonal answer" also came from the product of pivots in Example 2.

Example 7 This is the same matrix, except the first entry (upper left) is now 1:

$$B_4 = \begin{bmatrix} 1 & -1 & & \\ -1 & 2 & -1 & \\ & -1 & 2 & -1 \\ & & -1 & 2 \end{bmatrix}.$$

All pivots of this matrix turn out to be 1. So its determinant is 1. How does that come from cofactors? Expanding on row 1, the cofactors all agree with Example 6. Just change $a_{11} = 2$ to $b_{11} = 1$:

$$\det B_4 = D_3 - D_2 \qquad \text{instead of} \qquad \det A_4 = 2D_3 - D_2.$$

The determinant of B_4 is $4 - 3 = 1$. The determinant of every B_n is $n - (n-1) = 1$.
If you also change the last 2 into 1, why is $\det = 0$?

■ REVIEW OF THE KEY IDEAS ■

1. With no row exchanges, $\det A$ = (*product of pivots*). In the upper left corner of A, $\det A_k$ = (*product of the first* k *pivots*).

2. Every term in the big formula (8) uses each row and column once. Half of the $n!$ terms have plus signs (when $\det P = +1$) and half have minus signs.

3. The cofactor C_{ij} is $(-1)^{i+j}$ times the smaller determinant that omits row i and column j (because a_{ij} uses that row and column).

4. The determinant is the dot product of any row of A with its row of cofactors. When a row of A has a lot of zeros, we only need a few cofactors.

■ WORKED EXAMPLES ■

5.2 A A *Hessenberg matrix* is a triangular matrix with one extra diagonal. Use cofactors of row 1 to show that the 4 by 4 determinant satisfies Fibonacci's rule $|H_4| = |H_3| + |H_2|$. The same rule will continue for all sizes, $|H_n| = |H_{n-1}| + |H_{n-2}|$. Which Fibonacci number is $|H_n|$?

$$H_2 = \begin{bmatrix} 2 & 1 \\ 1 & 2 \end{bmatrix} \qquad H_3 = \begin{bmatrix} 2 & 1 & \\ 1 & 2 & 1 \\ 1 & 1 & 2 \end{bmatrix} \qquad H_4 = \begin{bmatrix} 2 & 1 & & \\ \mathbf{1} & 2 & \mathbf{1} & \\ \mathbf{1} & 1 & \mathbf{2} & \mathbf{1} \\ \mathbf{1} & 1 & \mathbf{1} & \mathbf{2} \end{bmatrix}$$

Solution The cofactor C_{11} for H_4 is the determinant $|H_3|$. We also need C_{12} (in boldface):

$$C_{12} = -\begin{vmatrix} \mathbf{1} & \mathbf{1} & \mathbf{0} \\ \mathbf{1} & \mathbf{2} & \mathbf{1} \\ \mathbf{1} & \mathbf{1} & \mathbf{2} \end{vmatrix} = -\begin{vmatrix} 2 & 1 & 0 \\ 1 & 2 & 1 \\ 1 & 1 & 2 \end{vmatrix} + \begin{vmatrix} 1 & 0 & 0 \\ 1 & 2 & 1 \\ 1 & 1 & 2 \end{vmatrix}$$

Rows 2 and 3 stayed the same and we used linearity in row 1. The two determinants on the right are $-|H_3|$ and $+|H_2|$. Then the 4 by 4 determinant is

$$|H_4| = 2C_{11} + 1C_{12} = 2|H_3| - |H_3| + |H_2| = |H_3| + |H_2|.$$

The actual numbers are $|H_2| = 3$ and $|H_3| = 5$ (and of course $|H_1| = 2$). Since $|H_n| = 2, 3, 5, 8, \ldots$ follows Fibonacci's rule $|H_{n-1}| + |H_{n-2}|$, it must be $|H_n| = F_{n+2}$.

5.2 B These questions use the $\pm$ signs (even and odd P's) in the big formula for $\det A$:

1. If A is the 10 by 10 all-ones matrix, how does the big formula give $\det A = 0$?
2. If you multiply all $n!$ permutations together into a single P, is P odd or even?
3. If you multiply each a_{ij} by the fraction i/j, why is $\det A$ unchanged?

Solution In Question **1**, with all $a_{ij} = 1$, all the products in the big formula (8) will be 1. Half of them come with a plus sign, and half with minus. So they cancel to leave $\det A = 0$. (Of course the all-ones matrix is singular. I am assuming $n > 1$.)

In Question **2**, multiplying $\left[\begin{smallmatrix} 1 & 0 \\ 0 & 1 \end{smallmatrix}\right]\left[\begin{smallmatrix} 0 & 1 \\ 1 & 0 \end{smallmatrix}\right]$ gives an odd permutation. Also for 3 by 3, the three odd permutations multiply (in any order) to give *odd*. But for $n > 3$ the product of all permutations will be *even*. There are $n!/2$ odd permutations and that is an even number as soon as $n!$ includes the factor 4.

In Question **3**, each a_{ij} is multiplied by i/j. So each product $a_{1\alpha}a_{2\beta}\cdots a_{n\omega}$ in the big formula is multiplied by all the row numbers $i = 1, 2, \ldots, n$ and divided by all the column numbers $j = 1, 2, \ldots, n$. (The columns come in some permuted order!) Then each product is unchanged and $\det A$ stays the same.

Another approach to Question **3**: We are multiplying the matrix A by the diagonal matrix $D = \mathbf{diag}(1:n)$ when row i is multiplied by i. And we are postmultiplying by D^{-1} when column j is divided by j. The determinant of DAD^{-1} is the same as $\det A$ by the product rule.

Problem Set 5.2

Problems 1–10 use the big formula with $n!$ terms: $|A| = \sum \pm a_{1\alpha}a_{2\beta}\cdots a_{n\omega}$. Every term uses each row and each column once.

1 Compute the determinants of A, B, C from six terms. Are their rows independent?

$$A = \begin{bmatrix} 1 & 2 & 3 \\ 3 & 1 & 2 \\ 3 & 2 & 1 \end{bmatrix} \qquad B = \begin{bmatrix} 1 & 2 & 3 \\ 4 & 4 & 4 \\ 5 & 6 & 7 \end{bmatrix} \qquad C = \begin{bmatrix} 1 & 1 & 1 \\ 1 & 1 & 0 \\ 1 & 0 & 0 \end{bmatrix}.$$

2 Compute the determinants of A, B, C, D. Are their columns independent?

$$A = \begin{bmatrix} 1 & 1 & 0 \\ 1 & 0 & 1 \\ 0 & 1 & 1 \end{bmatrix} \quad B = \begin{bmatrix} 1 & 2 & 3 \\ 4 & 5 & 6 \\ 7 & 8 & 9 \end{bmatrix} \quad C = \begin{bmatrix} A & 0 \\ 0 & A \end{bmatrix} \quad D = \begin{bmatrix} A & 0 \\ 0 & B \end{bmatrix}.$$

3 Show that $\det A = 0$, regardless of the five nonzeros marked by x's:

$$A = \begin{bmatrix} x & x & x \\ 0 & 0 & x \\ 0 & 0 & x \end{bmatrix}.$$

What are the cofactors of row 1?
What is the rank of A?
What are the 6 terms in $\det A$?

4 Find two ways to choose nonzeros from four different rows and columns:

$$A = \begin{bmatrix} 1 & 0 & 0 & 1 \\ 0 & 1 & 1 & 1 \\ 1 & 1 & 0 & 1 \\ 1 & 0 & 0 & 1 \end{bmatrix} \qquad B = \begin{bmatrix} 1 & 0 & 0 & 2 \\ 0 & 3 & 4 & 5 \\ 5 & 4 & 0 & 3 \\ 2 & 0 & 0 & 1 \end{bmatrix} \quad (B \text{ has the same zeros as } A).$$

Is $\det A$ equal to $1 + 1$ or $1 - 1$ or $-1 - 1$? What is $\det B$?

5 Place the smallest number of zeros in a 4 by 4 matrix that will guarantee $\det A = 0$. Place as many zeros as possible while still allowing $\det A \neq 0$.

6 (a) If $a_{11} = a_{22} = a_{33} = 0$, how many of the six terms in $\det A$ will be zero?

(b) If $a_{11} = a_{22} = a_{33} = a_{44} = 0$, how many of the 24 products $a_{1j}a_{2k}a_{3l}a_{4m}$ are sure to be zero?

7 How many 5 by 5 permutation matrices have $\det P = +1$? Those are even permutations. Find one that needs four exchanges to reach the identity matrix.

8 If $\det A$ is not zero, at least one of the $n!$ terms in formula (8) is not zero. Deduce from the big formula that some ordering of the rows of A leaves no zeros on the diagonal. (Don't use P from elimination; that PA can have zeros on the diagonal.)

9 Show that 4 is the largest determinant for a 3 by 3 matrix of 1's and -1's.

10 How many permutations of $(1, 2, 3, 4)$ are even and what are they? Extra credit: What are all the possible 4 by 4 determinants of $I + P_{\text{even}}$?

Problems 11–22 use cofactors $C_{ij} = (-1)^{i+j} \det M_{ij}$. Remove row i and column j.

11 Find all cofactors and put them into cofactor matrices C, D. Find AC and $\det B$.

$$A = \begin{bmatrix} a & b \\ c & d \end{bmatrix} \qquad B = \begin{bmatrix} 1 & 2 & 3 \\ 4 & 5 & 6 \\ 7 & 0 & 0 \end{bmatrix}.$$

12 Find the cofactor matrix C and multiply A times C^{T}. Compare AC^{T} with A^{-1}:

$$A = \begin{bmatrix} 2 & -1 & 0 \\ -1 & 2 & -1 \\ 0 & -1 & 2 \end{bmatrix} \qquad A^{-1} = \frac{1}{4}\begin{bmatrix} 3 & 2 & 1 \\ 2 & 4 & 2 \\ 1 & 2 & 3 \end{bmatrix}.$$

13 The n by n determinant C_n has 1's above and below the main diagonal:

$$C_1 = |0| \quad C_2 = \begin{vmatrix} 0 & 1 \\ 1 & 0 \end{vmatrix} \quad C_3 = \begin{vmatrix} 0 & 1 & 0 \\ 1 & 0 & 1 \\ 0 & 1 & 0 \end{vmatrix} \quad C_4 = \begin{vmatrix} 0 & 1 & 0 & 0 \\ 1 & 0 & 1 & 0 \\ 0 & 1 & 0 & 1 \\ 0 & 0 & 1 & 0 \end{vmatrix}.$$

(a) What are these determinants C_1, C_2, C_3, C_4?

(b) By cofactors find the relation between C_n and C_{n-1} and C_{n-2}. Find C_{10}.

14 The matrices in Problem 13 have 1's just above and below the main diagonal. Going down the matrix, which order of columns (if any) gives all 1's? Explain why that permutation is *even* for $n = 4, 8, 12, \ldots$ and *odd* for $n = 2, 6, 10, \ldots$. Then

$$C_n = 0 \text{ (odd } n) \qquad C_n = 1 \; (n = 4, 8, \cdots) \qquad C_n = -1 \; (n = 2, 6, \cdots).$$

15 The tridiagonal $1, 1, 1$ matrix of order n has determinant E_n :

$$E_1 = |1| \quad E_2 = \begin{vmatrix} 1 & 1 \\ 1 & 1 \end{vmatrix} \quad E_3 = \begin{vmatrix} 1 & 1 & 0 \\ 1 & 1 & 1 \\ 0 & 1 & 1 \end{vmatrix} \quad E_4 = \begin{vmatrix} 1 & 1 & 0 & 0 \\ 1 & 1 & 1 & 0 \\ 0 & 1 & 1 & 1 \\ 0 & 0 & 1 & 1 \end{vmatrix}.$$

(a) By cofactors show that $E_n = E_{n-1} - E_{n-2}$.

(b) Starting from $E_1 = 1$ and $E_2 = 0$ find $E_3, E_4, \ldots, E_8$.

(c) By noticing how these numbers eventually repeat, find E_{100}.

16 F_n is the determinant of the $1, 1, -1$ tridiagonal matrix of order n:

$$F_2 = \begin{vmatrix} 1 & -1 \\ 1 & 1 \end{vmatrix} = 2 \quad F_3 = \begin{vmatrix} 1 & -1 & 0 \\ 1 & 1 & -1 \\ 0 & 1 & 1 \end{vmatrix} = 3 \quad F_4 = \begin{vmatrix} 1 & -1 & & \\ 1 & 1 & -1 & \\ & 1 & 1 & -1 \\ & & 1 & 1 \end{vmatrix} \neq 4.$$

Expand in cofactors to show that $F_n = F_{n-1} + F_{n-2}$. These determinants are *Fibonacci numbers* $1, 2, 3, 5, 8, 13, \ldots$. The sequence usually starts $1, 1, 2, 3$ (with two 1's) so our F_n is the usual F_{n+1}.

17 The matrix B_n is the $-1, 2, -1$ matrix A_n except that $b_{11} = 1$ instead of $a_{11} = 2$. Using cofactors of the *last* row of B_4 show that $|B_4| = 2|B_3| - |B_2| = 1$.

$$B_4 = \begin{bmatrix} \mathbf{1} & -1 & & \\ -1 & 2 & -1 & \\ & -1 & 2 & -1 \\ & & -1 & 2 \end{bmatrix} \qquad B_3 = \begin{bmatrix} \mathbf{1} & -1 & \\ -1 & 2 & -1 \\ & -1 & 2 \end{bmatrix} \qquad B_2 = \begin{bmatrix} \mathbf{1} & -1 \\ -1 & 2 \end{bmatrix}.$$

The recursion $|B_n| = 2|B_{n-1}| - |B_{n-2}|$ is satisfied when every $|B_n| = 1$. This recursion is the same as for the A's in Example 6. The difference is in the starting values $1, 1, 1$ for the determinants of sizes $n = 1, 2, 3$.

18 Go back to B_n in Problem 17. It is the same as A_n except for $b_{11} = 1$. So use linearity in the first row, where $[1 \;\; -1 \;\; 0]$ equals $[2 \;\; -1 \;\; 0]$ minus $[1 \;\; 0 \;\; 0]$:

$$|B_n| = \begin{vmatrix} 1 & -1 & & 0 \\ -1 & & & \\ & & A_{n-1} & \\ 0 & & & \end{vmatrix} = \begin{vmatrix} 2 & -1 & & 0 \\ -1 & & & \\ & & A_{n-1} & \\ 0 & & & \end{vmatrix} - \begin{vmatrix} 1 & 0 & & 0 \\ -1 & & & \\ & & A_{n-1} & \\ 0 & & & \end{vmatrix}.$$

Linearity gives $|B_n| = |A_n| - |A_{n-1}| = $ _____.

19 Explain why the 4 by 4 Vandermonde determinant contains x^3 but not x^4 or x^5:

$$V_4 = \det \begin{bmatrix} 1 & a & a^2 & a^3 \\ 1 & b & b^2 & b^3 \\ 1 & c & c^2 & c^3 \\ 1 & x & x^2 & x^3 \end{bmatrix}.$$

The determinant is zero at $x = $ _____, _____, and _____. The cofactor of x^3 is $V_3 = (b-a)(c-a)(c-b)$. Then $V_4 = (b-a)(c-a)(c-b)(x-a)(x-b)(x-c)$.

20 Find G_2 and G_3 and then by row operations G_4. Can you predict G_n?

$$G_2 = \begin{vmatrix} 0 & 1 \\ 1 & 0 \end{vmatrix} \qquad G_3 = \begin{vmatrix} 0 & 1 & 1 \\ 1 & 0 & 1 \\ 1 & 1 & 0 \end{vmatrix} \qquad G_4 = \begin{vmatrix} 0 & 1 & 1 & 1 \\ 1 & 0 & 1 & 1 \\ 1 & 1 & 0 & 1 \\ 1 & 1 & 1 & 0 \end{vmatrix}.$$

21 Compute S_1, S_2, S_3 for these $1, 3, 1$ matrices. By Fibonacci guess and check S_4.

$$S_1 = |3| \qquad S_2 = \begin{vmatrix} 3 & 1 \\ 1 & 3 \end{vmatrix} \qquad S_3 = \begin{vmatrix} 3 & 1 & 0 \\ 1 & 3 & 1 \\ 0 & 1 & 3 \end{vmatrix}$$

22 Change 3 to 2 in the upper left corner of the matrices in Problem 21. Why does that subtract S_{n-1} from the determinant S_n? Show that the determinants of the new matrices become the Fibonacci numbers $2, 5, 13$ (always F_{2n+1}).

Problems 23–26 are about block matrices and block determinants.

23 With 2 by 2 blocks in 4 by 4 matrices, you cannot always use block determinants:

$$\begin{vmatrix} A & B \\ 0 & D \end{vmatrix} = |A|\,|D| \qquad \text{but} \qquad \begin{vmatrix} A & B \\ C & D \end{vmatrix} \neq |A|\,|D| - |C|\,|B|.$$

(a) Why is the first statement true? Somehow B doesn't enter.

(b) Show by example that equality fails (as shown) when C enters.

(c) Show by example that the answer $\det(AD - CB)$ is also wrong.

24 With block multiplication, $A = LU$ has $A_k = L_k U_k$ in the top left corner:

$$A = \begin{bmatrix} A_k & * \\ * & * \end{bmatrix} = \begin{bmatrix} L_k & 0 \\ * & * \end{bmatrix} \begin{bmatrix} U_k & * \\ 0 & * \end{bmatrix}.$$

(a) Suppose the first three pivots of A are $2, 3, -1$. What are the determinants of L_1, L_2, L_3 (with diagonal 1's) and U_1, U_2, U_3 and A_1, A_2, A_3?

(b) If A_1, A_2, A_3 have determinants $5, 6, 7$ find the three pivots from equation (3).

25 Block elimination subtracts CA^{-1} times the first row $[\,A \;\; B\,]$ from the second row $[C \;\; D]$. This leaves the *Schur complement* $D - CA^{-1}B$ in the corner:

$$\begin{bmatrix} I & 0 \\ -CA^{-1} & I \end{bmatrix} \begin{bmatrix} A & B \\ C & D \end{bmatrix} = \begin{bmatrix} A & B \\ 0 & D - CA^{-1}B \end{bmatrix}.$$

Take determinants of these block matrices to prove correct rules if A^{-1} exists:

$$\begin{vmatrix} A & B \\ C & D \end{vmatrix} = |A|\,|D - CA^{-1}B| = |AD - CB| \ \text{ provided } AC = CA.$$

26 If A is m by n and B is n by m, block multiplication gives $\det M = \det AB$:

$$M = \begin{bmatrix} 0 & A \\ -B & I \end{bmatrix} = \begin{bmatrix} AB & A \\ 0 & I \end{bmatrix} \begin{bmatrix} I & 0 \\ -B & I \end{bmatrix}.$$

If A is a single row and B is a single column what is $\det M$? If A is a column and B is a row what is $\det M$? Do a 3 by 3 example of each.

27 (A calculus question) Show that the derivative of $\det A$ with respect to a_{11} is the cofactor C_{11}. The other entries are fixed—we are only changing a_{11}.

28 A 3 by 3 determinant has three products "down to the right" and three "down to the left" with minus signs. Compute the six terms like $(1)(5)(9) = 45$ to find D.

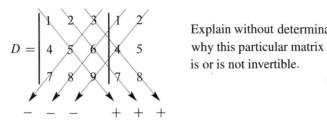

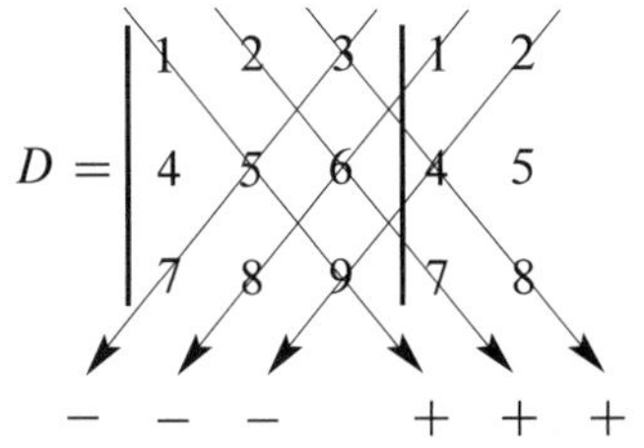

Explain without determinants why this particular matrix is or is not invertible.

29 For E_4 in Problem 15, five of the $4! = 24$ terms in the big formula (8) are nonzero. Find those five terms to show that $E_4 = -1$.

30 For the 4 by 4 tridiagonal second difference matrix (entries -1, 2, -1) find the five terms in the big formula that give $\det A = 16 - 4 - 4 - 4 + 1$.

31 Find the determinant of this cyclic P by cofactors of row 1 and then the "big formula". How many exchanges reorder $4, 1, 2, 3$ into $1, 2, 3, 4$? Is $|P^2| = 1$ or -1?

$$P = \begin{bmatrix} 0 & 0 & 0 & 1 \\ 1 & 0 & 0 & 0 \\ 0 & 1 & 0 & 0 \\ 0 & 0 & 1 & 0 \end{bmatrix} \qquad P^2 = \begin{bmatrix} 0 & 0 & 1 & 0 \\ 0 & 0 & 0 & 1 \\ 1 & 0 & 0 & 0 \\ 0 & 1 & 0 & 0 \end{bmatrix} = \begin{bmatrix} 0 & I \\ I & 0 \end{bmatrix}.$$

Challenge Problems

32 Cofactors of the $1, 3, 1$ matrices in Problem 21 give a recursion $S_n = 3S_{n-1} - S_{n-2}$. Amazingly that recursion produces every second Fibonacci number. Here is the challenge.

Show that S_n is the Fibonacci number F_{2n+2} by proving $F_{2n+2} = 3F_{2n} - F_{2n-2}$. Keep using Fibonacci's rule $F_k = F_{k-1} + F_{k-2}$ starting with $k = 2n + 2$.

33 The symmetric Pascal matrices have determinant 1. If I subtract 1 from the n, n entry, why does the determinant become zero? (Use rule 3 or cofactors.)

$$\det \begin{bmatrix} 1 & 1 & 1 & 1 \\ 1 & 2 & 3 & 4 \\ 1 & 3 & 6 & 10 \\ 1 & 4 & 10 & 20 \end{bmatrix} = 1 \text{ (known)} \qquad \det \begin{bmatrix} 1 & 1 & 1 & 1 \\ 1 & 2 & 3 & 4 \\ 1 & 3 & 6 & 10 \\ 1 & 4 & 10 & \mathbf{19} \end{bmatrix} = \mathbf{0} \text{ (to explain)}.$$

34 This problem shows in two ways that $\det A = 0$ (the x's are any numbers):

$$A = \begin{bmatrix} x & x & x & x & x \\ x & x & x & x & x \\ 0 & 0 & 0 & x & x \\ 0 & 0 & 0 & x & x \\ 0 & 0 & 0 & x & x \end{bmatrix}.$$

(a) How do you know that the rows are linearly dependent?

(b) Explain why all 120 terms are zero in the big formula for $\det A$.

35 If $|\det(A)| > 1$, prove that the powers A^n cannot stay bounded. But if $|\det(A)| \leq 1$, show that some entries of A^n might still grow large. Eigenvalues will give the right test for stability, determinants tell us only one number.

5.3 Cramer's Rule, Inverses, and Volumes

1 $\boldsymbol{A^{-1}}$ **equals** $\boldsymbol{C^{\mathrm{T}}}$ **/ det** $\boldsymbol{A}$. Then $(A^{-1})_{ij}$ = cofactor $\boldsymbol{C_{ji}}$ divided by the determinant of A.

2 Cramer's Rule computes $\boldsymbol{x} = A^{-1}\boldsymbol{b}$ from $x_j = \det(A$ with column j changed to $\boldsymbol{b}) / \det A$.

3 Area of parallelogram $= |ad-bc|$ if the four corners are $(0,0), (a,b), (c,d)$, and $(a+c, b+d)$.

4 Volume of box $= |\det \boldsymbol{A}|$ if the rows of A (or the columns of A) give the sides of the box.

5 The **cross product** $\boldsymbol{w} = \boldsymbol{u} \times \boldsymbol{v}$ is det $\begin{bmatrix} \boldsymbol{i} & \boldsymbol{j} & \boldsymbol{k} \\ u_1 & u_2 & u_3 \\ v_1 & v_2 & v_3 \end{bmatrix}$. Notice $\boldsymbol{v} \times \boldsymbol{u} = -(\boldsymbol{u} \times \boldsymbol{v})$. w_1, w_2, w_3 are cofactors of row 1. Notice $\boldsymbol{w}^{\mathrm{T}}\boldsymbol{u} = 0$ and $\boldsymbol{w}^{\mathrm{T}}\boldsymbol{v} = 0$.

This section solves $A\boldsymbol{x} = \boldsymbol{b}$ and also finds A^{-1}—by algebra and not by elimination. In all formulas you will see a division by $\det A$. Each entry in A^{-1} and $A^{-1}\boldsymbol{b}$ is a determinant divided by the determinant of A. Let me start with Cramer's Rule.

Cramer's Rule solves $A\boldsymbol{x} = \boldsymbol{b}$. A neat idea gives the first component x_1. Replacing the first column of I by $\boldsymbol{x}$ gives a matrix with determinant x_1. When you multiply it by A, *the first column becomes $A\boldsymbol{x}$ which is* $\boldsymbol{b}$. The other columns of B_1 are copied from A:

$$\textbf{Key idea} \qquad \begin{bmatrix} & & \\ & A & \\ & & \end{bmatrix} \begin{bmatrix} \boldsymbol{x_1} & 0 & 0 \\ \boldsymbol{x_2} & 1 & 0 \\ \boldsymbol{x_3} & 0 & 1 \end{bmatrix} = \begin{bmatrix} \boldsymbol{b_1} & a_{12} & a_{13} \\ \boldsymbol{b_2} & a_{22} & a_{23} \\ \boldsymbol{b_3} & a_{32} & a_{33} \end{bmatrix} = B_1. \qquad (1)$$

We multiplied a column at a time. *Take determinants of the three matrices to find* x_1 :

$$\textbf{Product rule} \qquad (\det \boldsymbol{A})(\boldsymbol{x_1}) = \det \boldsymbol{B_1} \qquad \text{or} \qquad \boldsymbol{x_1} = \frac{\det \boldsymbol{B_1}}{\det \boldsymbol{A}}. \qquad (2)$$

This is the first component of $\boldsymbol{x}$ in Cramer's Rule! Changing a column of A gave B_1. To find x_2 and B_2, put the vectors $\boldsymbol{x}$ and $\boldsymbol{b}$ into the *second* columns of I and A:

$$\textbf{Same idea} \qquad \begin{bmatrix} & & \\ \boldsymbol{a}_1 & \boldsymbol{a}_2 & \boldsymbol{a}_3 \\ & & \end{bmatrix} \begin{bmatrix} 1 & x_1 & 0 \\ 0 & x_2 & 0 \\ 0 & x_3 & 1 \end{bmatrix} = \begin{bmatrix} & & \\ \boldsymbol{a}_1 & \boldsymbol{b} & \boldsymbol{a}_3 \\ & & \end{bmatrix} = B_2. \qquad (3)$$

Take determinants to find $(\det A)(x_2) = \det B_2$. This gives $\boldsymbol{x_2} = \boldsymbol{(\det B_2)/(\det A)}$.

Example 1 Solving $3x_1 + 4x_2 = \boldsymbol{2}$ and $5x_1 + 6x_2 = \boldsymbol{4}$ needs three determinants:

$$\det \boldsymbol{A} = \begin{vmatrix} 3 & 4 \\ 5 & 6 \end{vmatrix} \qquad \det \boldsymbol{B_1} = \begin{vmatrix} \boldsymbol{2} & 4 \\ \boldsymbol{4} & 6 \end{vmatrix} \qquad \det \boldsymbol{B_2} = \begin{vmatrix} 3 & \boldsymbol{2} \\ 5 & \boldsymbol{4} \end{vmatrix}$$

Those determinants of A, B_1, B_2 are -2 and -4 and 2. All ratios divide by $\det A = -2$:

$$\textbf{Find } \boldsymbol{x} = A^{-1}\boldsymbol{b} \qquad x_1 = \frac{-4}{-2} = 2 \qquad x_2 = \frac{2}{-2} = -1 \qquad \textbf{Check} \begin{bmatrix} 3 & 4 \\ 5 & 6 \end{bmatrix} \begin{bmatrix} 2 \\ -1 \end{bmatrix} = \begin{bmatrix} 2 \\ 4 \end{bmatrix}$$

CRAMER's RULE If $\det A$ is not zero, $A\boldsymbol{x} = \boldsymbol{b}$ is solved by determinants:

$$x_1 = \frac{\det B_1}{\det A} \qquad x_2 = \frac{\det B_2}{\det A} \qquad \ldots \qquad x_n = \frac{\det B_n}{\det A} \tag{4}$$

The matrix B_j has the jth column of A replaced by the vector $\boldsymbol{b}$.

To solve an n by n system, Cramer's Rule evaluates $n+1$ determinants (of A and the n different B's). When each one is the sum of $n!$ terms—applying the "big formula" with all permutations—this makes a total of $(n+1)!$ terms. *It would be crazy to solve equations that way.* But we do finally have an explicit formula for the solution $\boldsymbol{x}$.

Example 2 Cramer's Rule is inefficient for numbers but it is well suited to letters. For $n = 2$, find the columns of $A^{-1} = [\boldsymbol{x}\ \ \boldsymbol{y}]$ by solving $AA^{-1} = I$:

$$\begin{array}{l}\textbf{Columns of } A^{-1} \\ \textbf{are } \boldsymbol{x} \textbf{ and } \boldsymbol{y}\end{array} \qquad \begin{bmatrix} a & b \\ c & d \end{bmatrix} \begin{bmatrix} x_1 \\ x_2 \end{bmatrix} = \begin{bmatrix} \mathbf{1} \\ \mathbf{0} \end{bmatrix} \qquad \begin{bmatrix} a & b \\ c & d \end{bmatrix} \begin{bmatrix} y_1 \\ y_2 \end{bmatrix} = \begin{bmatrix} \mathbf{0} \\ \mathbf{1} \end{bmatrix}$$

Those share the same matrix A. We need $|A|$ and four determinants for $x_1,\ x_2,\ y_1,\ y_2$:

$$\begin{vmatrix} a & b \\ c & d \end{vmatrix} \text{ and } \begin{vmatrix} \mathbf{1} & b \\ \mathbf{0} & d \end{vmatrix} \quad \begin{vmatrix} a & \mathbf{1} \\ c & \mathbf{0} \end{vmatrix} \quad \begin{vmatrix} \mathbf{0} & b \\ \mathbf{1} & d \end{vmatrix} \quad \begin{vmatrix} a & \mathbf{0} \\ c & \mathbf{1} \end{vmatrix}$$

The last four determinants are $d, -c, -b$, and a. (They are the cofactors!) Here is A^{-1}:

$$x_1 = \frac{d}{|A|},\ x_2 = \frac{-c}{|A|},\ y_1 = \frac{-b}{|A|},\ y_2 = \frac{a}{|A|} \text{ and then } A^{-1} = \frac{1}{ad-bc}\begin{bmatrix} d & -b \\ -c & a \end{bmatrix}.$$

I chose 2 by 2 so that the main points could come through clearly. The new idea is: $\boldsymbol{A^{-1}}$ **involves the cofactors**. When the right side is a column of the identity matrix I, as in $AA^{-1} = I$, **the determinant of each $\boldsymbol{B_j}$ in Cramer's Rule is a cofactor of $\boldsymbol{A}$.**

You can see those cofactors for $n = 3$. Solve $A\boldsymbol{x} = (1, 0, 0)$ to find column 1 of A^{-1}:

$$\begin{array}{l}\textbf{Determinants of } B\textbf{'s} \\ = \textbf{Cofactors of } A\end{array} \quad \begin{vmatrix} \mathbf{1} & a_{12} & a_{13} \\ \mathbf{0} & a_{22} & a_{23} \\ \mathbf{0} & a_{32} & a_{33} \end{vmatrix} \quad \begin{vmatrix} a_{11} & \mathbf{1} & a_{13} \\ a_{21} & \mathbf{0} & a_{23} \\ a_{31} & \mathbf{0} & a_{33} \end{vmatrix} \quad \begin{vmatrix} a_{11} & a_{12} & \mathbf{1} \\ a_{21} & a_{22} & \mathbf{0} \\ a_{31} & a_{32} & \mathbf{0} \end{vmatrix} \tag{5}$$

That first determinant $|B_1|$ is the cofactor $C_{11} = a_{22}a_{33} - a_{23}a_{32}$. Then $|B_2|$ is the cofactor C_{12}. Notice that the correct minus sign appears in $-(a_{21}a_{33} - a_{23}a_{31})$. This cofactor C_{12} goes into column 1 of A^{-1}. When we divide by $\det A$, we have the inverse matrix!

***The* i, j *entry of* A^{-1} *is the cofactor* C_{ji} *(not* C_{ij}*) divided by* $\det A$:**

FORMULA FOR A^{-1} $$(A^{-1})_{ij} = \frac{C_{ji}}{\det A} \quad \text{and} \quad A^{-1} = \frac{C^{\mathrm{T}}}{\det A}. \tag{6}$$

The cofactors C_{ij} go into the *"cofactor matrix"* C. ***The transpose of C leads to*** A^{-1}. To compute the i, j entry of A^{-1}, cross out row j and column i of A. Multiply the determinant by $(-1)^{i+j}$ to get the cofactor C_{ji}, and divide by $\det A$.

Check this rule for the $3, 1$ entry of A^{-1}. For column 1 we solve $Ax = (1, 0, 0)$. The third component x_3 needs the third determinant in equation (5), divided by $\det A$. That determinant is exactly the cofactor $C_{13} = a_{21}a_{32} - a_{22}a_{31}$. So $(A^{-1})_{31} = C_{13}/\det A$.

Summary In solving $AA^{-1} = I$, each column of I leads to a column of A^{-1}. Every entry of A^{-1} is a ratio: determinant of size $n-1$ / determinant of size n.

Direct proof of the formula $\boldsymbol{A^{-1} = C^{\mathrm{T}}/\det A}$ This means $\boldsymbol{AC^{\mathrm{T}} = (\det A)I}$:

$$\begin{bmatrix} a_{11} & a_{12} & a_{13} \\ a_{21} & a_{22} & a_{23} \\ a_{31} & a_{32} & a_{33} \end{bmatrix} \begin{bmatrix} C_{11} & C_{21} & C_{31} \\ C_{12} & C_{22} & C_{32} \\ C_{13} & C_{23} & C_{33} \end{bmatrix} = \begin{bmatrix} \det A & 0 & 0 \\ 0 & \det A & 0 \\ 0 & 0 & \det A \end{bmatrix}. \tag{7}$$

(Row 1 of A) times (column 1 of C^{T}) yields the first $\det A$ on the right:

$$a_{11}C_{11} + a_{12}C_{12} + a_{13}C_{13} = \det A \quad \text{This is exactly the cofactor rule!}$$

Similarly row 2 of A times column 2 of C^{T} (*notice the transpose*) also yields $\det A$. The entries a_{2j} are multiplying cofactors C_{2j} as they should, to give the determinant.

How to explain the zeros off the main diagonal in equation (7)? The rows of A are multiplying cofactors from *different* rows. Why is the answer zero?

Row 2 of A
Row 1 of C
$$a_{21}C_{11} + a_{22}C_{12} + a_{23}C_{13} = 0. \tag{8}$$

Answer: This is the cofactor rule for a new matrix, when the second row of A is copied into its first row. The new matrix A^* has two equal rows, so $\det A^* = 0$ in equation (8). Notice that A^* has the same cofactors C_{11}, C_{12}, C_{13} as A—because all rows agree after the first row. Thus the remarkable multiplication (7) is correct:

$$AC^{\mathrm{T}} = (\det A)I \quad \text{or} \quad A^{-1} = \frac{C^{\mathrm{T}}}{\det A}.$$

Example 3 The "sum matrix" A has determinant 1. Then A^{-1} contains cofactors:

$$A = \begin{bmatrix} 1 & 0 & 0 & 0 \\ 1 & 1 & 0 & 0 \\ 1 & 1 & 1 & 0 \\ 1 & 1 & 1 & 1 \end{bmatrix} \quad \text{has inverse} \quad A^{-1} = \frac{C^{\mathrm{T}}}{1} = \begin{bmatrix} 1 & 0 & 0 & 0 \\ -1 & 1 & 0 & 0 \\ 0 & -1 & 1 & 0 \\ 0 & 0 & -1 & 1 \end{bmatrix}.$$

Cross out row 1 and column 1 of A to see the 3 by 3 cofactor $C_{11} = 1$. Now cross out row 1 and column 2 for C_{12}. The 3 by 3 submatrix is still triangular with determinant 1. But the cofactor C_{12} is -1 because of the sign $(-1)^{1+2}$. This number -1 goes into the $(2,1)$ entry of A^{-1}—don't forget to transpose C.

The inverse of a triangular matrix is triangular. Cofactors give a reason why.

Example 4 If all cofactors are nonzero, is A sure to be invertible? *No way.*

Area of a Triangle

Everybody knows the area of a rectangle—base times height. The area of a triangle is *half* the base times the height. But here is a question that those formulas don't answer. ***If we know the corners (x_1, y_1) and (x_2, y_2) and (x_3, y_3) of a triangle, what is the area?*** Using the corners to find the base and height is not a good way to compute area.

Determinants are the best way to find area. ***The area of a triangle is half of a 3 by 3 determinant.*** The square roots in the base and height cancel out in the good formula. If one corner is at the origin, say $(x_3, y_3) = (0,0)$, the determinant is only 2 by 2.

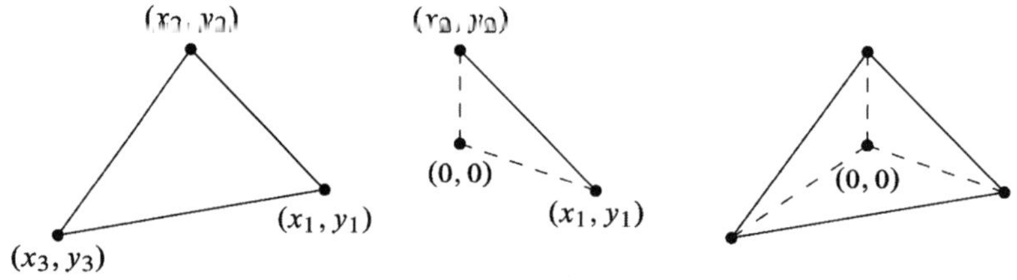

Figure 5.1: General triangle; special triangle from $(0,0)$; general from three specials.

The triangle with corners (x_1, y_1) and (x_2, y_2) and (x_3, y_3) has **area** $= \dfrac{\textbf{determinant}}{\textbf{2}}$:

Area of triangle $\dfrac{1}{2}\begin{vmatrix} x_1 & y_1 & 1 \\ x_2 & y_2 & 1 \\ x_3 & y_3 & 1 \end{vmatrix}$ Area $= \dfrac{1}{2}\begin{vmatrix} x_1 & y_1 \\ x_2 & y_2 \end{vmatrix}$ when $(x_3, y_3) = (0,0)$.

When you set $x_3 = y_3 = 0$ in the 3 by 3 determinant, you get the 2 by 2 determinant. These formulas have no square roots—they are reasonable to memorize. The 3 by 3 determinant breaks into a sum of three 2 by 2's (cofactors), just as the third triangle in Figure 5.1 breaks into three special triangles from $(0, 0)$:

$$\textbf{Area} = \frac{1}{2}\begin{vmatrix} x_1 & y_1 & 1 \\ x_2 & y_2 & 1 \\ x_3 & y_3 & 1 \end{vmatrix} = \begin{array}{l} +\frac{1}{2}(x_1y_2 - x_2y_1) \\ +\frac{1}{2}(x_2y_3 - x_3y_2) \\ +\frac{1}{2}(x_3y_1 - x_1y_3). \end{array} \tag{9}$$

If $(0, 0)$ is outside the triangle, two of the special areas can be negative—but the sum is still correct. The real problem is to explain the area of a triangle with corner $(0, 0)$.

Why is $\frac{1}{2}|x_1y_2 - x_2y_1|$ the area of this triangle? We can remove the factor $\frac{1}{2}$ for a parallelogram (twice as big, because the parallelogram contains two equal triangles). We now prove that the parallelogram area is the determinant $x_1y_2 - x_2y_1$. This area in Figure 5.2 is 11, and therefore the triangle has area $\frac{11}{2}$.

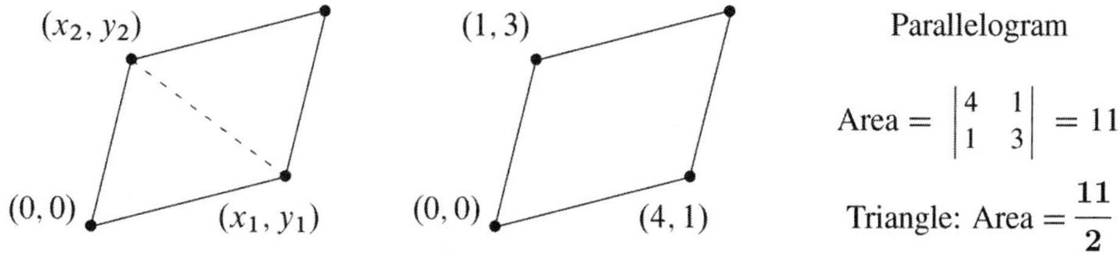

Figure 5.2: A triangle is half of a parallelogram. Area is half of a determinant.

Proof that a parallelogram starting from $(0, 0)$ has area = 2 by 2 determinant.

There are many proofs but this one fits with the book. We show that the area has the same properties 1-2-3 as the determinant. Then area = determinant! Remember that those three rules defined the determinant and led to all its other properties.

1 When $A = I$, the parallelogram becomes the unit square. Its area is $\det I = 1$.

2 When rows are exchanged, the determinant reverses sign. The absolute value (positive area) stays the same—it is the same parallelogram.

3 If row 1 is multiplied by t, Figure 5.3a shows that the area is also multiplied by t. Suppose a new row (x_1', y_1') is added to (x_1, y_1) (keeping row 2 fixed). Figure 5.3b shows that the solid parallelogram areas add to the dotted parallelogram area (because the two triangles completed by dotted lines are the same).

That is an exotic proof, when we could use plane geometry. But the proof has a major attraction—it applies in n dimensions. The n edges going out from the origin are given by the *rows of an n by n matrix*. The box is completed by more edges, like the parallelogram.

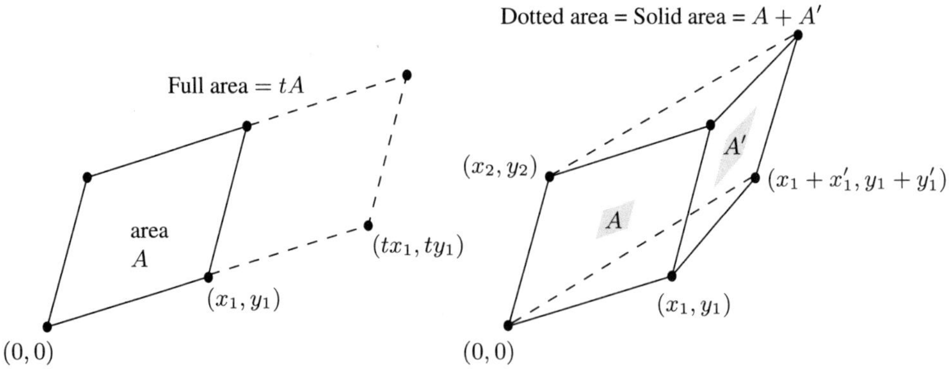

Figure 5.3: Areas obey the rule of linearity in side 1 (keeping the side (x_2, y_2) constant).

Figure 5.4 shows a three-dimensional box—whose edges are not at right angles. ***The volume equals the absolute value of* det *A*.** Our proof checks again that rules 1–3 for determinants are also obeyed by volumes. When an edge is stretched by a factor t, the volume is multiplied by t. When edge 1 is added to edge $1'$, the volume is the sum of the two original volumes. This is Figure 5.3b lifted into three dimensions or n dimensions. I would draw the boxes but this paper is only two-dimensional.

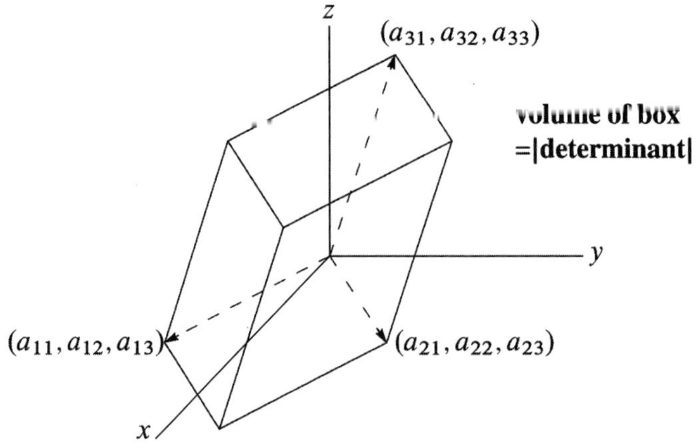

Figure 5.4: Three-dimensional box formed from the three rows of A.

The unit cube has volume $= 1$, which is $\det I$. Row exchanges or edge exchanges leave the same box and the same absolute volume. The determinant changes sign, to indicate whether the edges are a *right-handed triple* $(\det A > 0)$ or a *left-handed triple* $(\det A < 0)$. The box volume follows the rules for determinants, so volume of $\det A =$ absolute value.

Example 5 Suppose a rectangular box $(90°$ angles) has side lengths r, s, and t. Its volume is r times s times t. The diagonal matrix A with entries r, s, and t produces those three sides. Then $\det A$ also equals the volume $r\,s\,t$.

Example 6 In calculus, the box is infinitesimally small! To integrate over a circle, we might change x and y to r and θ. Those are polar coordinates: $x = r\cos\theta$ and $y = r\sin\theta$. The area of a "polar box" is a determinant J times $dr\, d\theta$:

$$\textbf{Area } r\,dr\,d\theta \textbf{ in calculus} \qquad J = \begin{vmatrix} \partial x/\partial r & \partial x/\partial\theta \\ \partial y/\partial r & \partial y/\partial\theta \end{vmatrix} = \begin{vmatrix} \cos\theta & -r\sin\theta \\ \sin\theta & r\cos\theta \end{vmatrix} = r.$$

This determinant is the r in the small area $dA = r\,dr\,d\theta$. The stretching factor J goes into double integrals just as dx/du goes into an ordinary integral $\int dx = \int (dx/du)\,du$. For triple integrals the Jacobian matrix J with nine derivatives will be 3 by 3.

The Cross Product

The *cross product* is an extra (and optional) application, special for three dimensions. Start with vectors $\boldsymbol{u} = (u_1, u_2, u_3)$ and $\boldsymbol{v} = (v_1, v_2, v_3)$. Unlike the dot product, which is a number, **the cross product is a vector**—also in three dimensions. It is written $\boldsymbol{u} \times \boldsymbol{v}$ and pronounced "$\boldsymbol{u}$ cross $\boldsymbol{v}$." *The components of this cross product are* 2 *by* 2 *cofactors.* We will explain the properties that make $\boldsymbol{u} \times \boldsymbol{v}$ useful in geometry and physics.

This time we bite the bullet, and write down the formula before the properties.

DEFINITION The ***cross product*** of $\boldsymbol{u} = (u_1, u_2, u_3)$ and $\boldsymbol{v} = (v_1, v_2, v_3)$ is a vector

$$\boldsymbol{u} \times \boldsymbol{v} = \begin{vmatrix} \boldsymbol{i} & \boldsymbol{j} & \boldsymbol{k} \\ u_1 & u_2 & u_3 \\ v_1 & v_2 & v_3 \end{vmatrix} = (u_2v_3 - u_3v_2)\boldsymbol{i} + (u_3v_1 - u_1v_3)\boldsymbol{j} + (u_1v_2 - u_2v_1)\boldsymbol{k}. \qquad (10)$$

This vector $\boldsymbol{u} \times \boldsymbol{v}$ is perpendicular to $\boldsymbol{u}$ and $\boldsymbol{v}$. The cross product $\boldsymbol{v} \times \boldsymbol{u}$ is $-(\boldsymbol{u} \times \boldsymbol{v})$.

Comment The 3 by 3 determinant is the easiest way to remember $\boldsymbol{u} \times \boldsymbol{v}$. It is not especially legal, because the first row contains vectors $\boldsymbol{i}, \boldsymbol{j}, \boldsymbol{k}$ and the other rows contain numbers. In the determinant, the vector $\boldsymbol{i} = (1, 0, 0)$ multiplies u_2v_3 and $-u_3v_2$. The result is $(u_2v_3 - u_3v_2, 0, 0)$, which displays the first component of the cross product.

Notice the cyclic pattern of the subscripts: 2 and 3 give component 1 of $\boldsymbol{u} \times \boldsymbol{v}$, then 3 and 1 give component 2, then 1 and 2 give component 3. This completes the definition of $\boldsymbol{u} \times \boldsymbol{v}$. Now we list the properties of the cross product:

Property 1 $\boldsymbol{v} \times \boldsymbol{u}$ reverses rows 2 and 3 in the determinant so it equals $-(\boldsymbol{u} \times \boldsymbol{v})$.

Property 2 The cross product $\boldsymbol{u} \times \boldsymbol{v}$ is perpendicular to $\boldsymbol{u}$ (and also to $\boldsymbol{v}$). The direct proof is to watch terms cancel, producing a zero dot product:

$$\boldsymbol{u} \cdot (\boldsymbol{u} \times \boldsymbol{v}) = u_1(u_2v_3 - u_3v_2) + u_2(u_3v_1 - u_1v_3) + u_3(u_1v_2 - u_2v_1) = 0. \qquad (11)$$

The determinant for $\boldsymbol{u} \cdot (\boldsymbol{u} \times \boldsymbol{v})$ has rows $\boldsymbol{u}$, $\boldsymbol{u}$ and $\boldsymbol{v}$ (2 equal rows) so it is zero.

Property 3 The cross product of any vector with itself (two equal rows) is $\boldsymbol{u} \times \boldsymbol{u} = \mathbf{0}$.

When $\boldsymbol{u}$ and $\boldsymbol{v}$ are parallel, the cross product is zero. When $\boldsymbol{u}$ and $\boldsymbol{v}$ are perpendicular, the dot product is zero. One involves $\sin\theta$ and the other involves $\cos\theta$:

$$\|\boldsymbol{u} \times \boldsymbol{v}\| = \|\boldsymbol{u}\|\,\|\boldsymbol{v}\|\,|\sin\theta| \qquad \text{and} \qquad |\boldsymbol{u} \cdot \boldsymbol{v}| = \|\boldsymbol{u}\|\,\|\boldsymbol{v}\|\,|\cos\theta|. \tag{12}$$

Example 7 $\boldsymbol{u} = (3, 2, 0)$ and $\boldsymbol{v} = (1, 4, 0)$ are in the xy plane, $\boldsymbol{u} \times \boldsymbol{v}$ goes up the z axis:

$$\boldsymbol{u} \times \boldsymbol{v} = \begin{vmatrix} \boldsymbol{i} & \boldsymbol{j} & \boldsymbol{k} \\ 3 & 2 & 0 \\ 1 & 4 & 0 \end{vmatrix} = 10\boldsymbol{k}. \quad \text{The cross product is } \boldsymbol{u} \times \boldsymbol{v} = (\mathbf{0}, \mathbf{0}, \mathbf{10}).$$

***The length of* $\boldsymbol{u} \times \boldsymbol{v}$ *equals the area of the parallelogram with sides* $\boldsymbol{u}$ *and* $\boldsymbol{v}$.** This will be important: In this example the area is 10.

Example 8 The cross product of $\boldsymbol{u} = (1, 1, 1)$ and $\boldsymbol{v} = (1, 1, 2)$ is $(1, -1, 0)$:

$$\begin{vmatrix} \boldsymbol{i} & \boldsymbol{j} & \boldsymbol{k} \\ 1 & 1 & 1 \\ 1 & 1 & 2 \end{vmatrix} = \boldsymbol{i}\begin{vmatrix} 1 & 1 \\ 1 & 2 \end{vmatrix} - \boldsymbol{j}\begin{vmatrix} 1 & 1 \\ 1 & 2 \end{vmatrix} + \boldsymbol{k}\begin{vmatrix} 1 & 1 \\ 1 & 1 \end{vmatrix} = \boldsymbol{i} - \boldsymbol{j}.$$

This vector $(1, -1, 0)$ is perpendicular to $(1, 1, 1)$ and $(1, 1, 2)$ as predicted. Area $= \sqrt{2}$.

Example 9 The cross product of $\boldsymbol{i} = (1, 0, 0)$ and $\boldsymbol{j} = (0, 1, 0)$ obeys the *right hand rule*. That cross product $\boldsymbol{k} = \boldsymbol{i} \times \boldsymbol{j}$ goes up not down:

$$\begin{vmatrix} \boldsymbol{i} & \boldsymbol{j} & \boldsymbol{k} \\ 1 & 0 & 0 \\ 0 & 1 & 0 \end{vmatrix} = \boldsymbol{k}$$

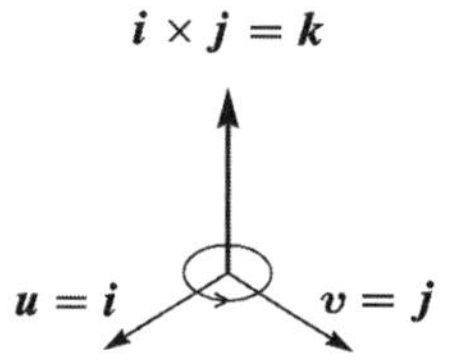

Rule $\boldsymbol{u} \times \boldsymbol{v}$ points along your right thumb when the fingers curl from $\boldsymbol{u}$ to $\boldsymbol{v}$.

Thus $\boldsymbol{i} \times \boldsymbol{j} = \boldsymbol{k}$. The right hand rule also gives $\boldsymbol{j} \times \boldsymbol{k} = \boldsymbol{i}$ and $\boldsymbol{k} \times \boldsymbol{i} = \boldsymbol{j}$. Note the cyclic order. In the opposite order (anti-cyclic) the thumb is reversed and the cross product goes the other way: $\boldsymbol{k} \times \boldsymbol{j} = -\boldsymbol{i}$ and $\boldsymbol{i} \times \boldsymbol{k} = -\boldsymbol{j}$ and $\boldsymbol{j} \times \boldsymbol{i} = -\boldsymbol{k}$. You see the three plus signs and three minus signs from a 3 by 3 determinant.

The definition of $\boldsymbol{u} \times \boldsymbol{v}$ can be based on vectors instead of their components:

DEFINITION The ***cross product*** is a vector with length $\|\boldsymbol{u}\|\,\|\boldsymbol{v}\|\,|\sin\theta|$. Its direction is perpendicular to $\boldsymbol{u}$ and $\boldsymbol{v}$. It points "up" or "down" by the right hand rule.

This definition appeals to physicists, who hate to choose axes and coordinates. They see (u_1, u_2, u_3) as the position of a mass and (F_x, F_y, F_z) as a force acting on it. If $\boldsymbol{F}$ is parallel to $\boldsymbol{u}$, then $\boldsymbol{u} \times \boldsymbol{F} = \boldsymbol{0}$—there is no turning. The cross product $\boldsymbol{u} \times \boldsymbol{F}$ is the turning force or *torque*. It points along the turning axis (perpendicular to $\boldsymbol{u}$ and $\boldsymbol{F}$). Its length $\|\boldsymbol{u}\|\,\|\boldsymbol{F}\|\sin\theta$ measures the "moment" that produces turning.

Triple Product = Determinant = Volume

Since $\boldsymbol{u} \times \boldsymbol{v}$ is a vector, we can take its dot product with a third vector $\boldsymbol{w}$. That produces the ***triple product*** $(\boldsymbol{u} \times \boldsymbol{v}) \cdot \boldsymbol{w}$. It is called a "scalar" triple product, because it is a number. In fact it is a determinant—it gives the volume of the $\boldsymbol{u}, \boldsymbol{v}, \boldsymbol{w}$ box:

Triple product
$$(\boldsymbol{u} \times \boldsymbol{v}) \cdot \boldsymbol{w} = \begin{vmatrix} w_1 & w_2 & w_3 \\ u_1 & u_2 & u_3 \\ v_1 & v_2 & v_3 \end{vmatrix} = \begin{vmatrix} u_1 & u_2 & u_3 \\ v_1 & v_2 & v_3 \\ w_1 & w_2 & w_3 \end{vmatrix}. \qquad (13)$$

We can put $\boldsymbol{w}$ in the top or bottom row. The two determinants are the same because _____ row exchanges go from one to the other. Notice when this determinant is zero:

$$(\boldsymbol{u} \times \boldsymbol{v}) \cdot \boldsymbol{w} = 0 \quad \text{exactly when the vectors } \boldsymbol{u}, \boldsymbol{v}, \boldsymbol{w} \text{ lie in the } \textit{same plane}.$$

First reason $\boldsymbol{u} \times \boldsymbol{v}$ is perpendicular to that plane so its dot product with $\boldsymbol{w}$ is zero.

Second reason Three vectors in a plane are dependent. The matrix is singular ($\det = 0$).

Third reason Zero volume when the $\boldsymbol{u}, \boldsymbol{v}, \boldsymbol{w}$ box is squashed onto a plane.

It is remarkable that $(\boldsymbol{u} \times \boldsymbol{v}) \cdot \boldsymbol{w}$ equals the volume of the box with sides $\boldsymbol{u}, \boldsymbol{v}, \boldsymbol{w}$. This 3 by 3 determinant carries tremendous information. Like $ad - bc$ for a 2 by 2 matrix, it separates invertible from singular. Chapter 6 will be looking for singular.

■ REVIEW OF THE KEY IDEAS ■

1. Cramer's Rule solves $A\boldsymbol{x} = \boldsymbol{b}$ by ratios like $x_1 = |B_1|/|A| = |\boldsymbol{b}\,\boldsymbol{a}_2 \cdots \boldsymbol{a}_n|/|A|$.

2. When C is the cofactor matrix for A, the inverse is $A^{-1} = C^{\mathrm{T}}/\det A$.

3. The volume of a box is $|\det A|$, when the box edges are the rows of A.

4. Area and volume are needed to change variables in double and triple integrals.

5. In $\mathbf{R}^3$, the cross product $\boldsymbol{u} \times \boldsymbol{v}$ is perpendicular to $\boldsymbol{u}$ and $\boldsymbol{v}$. Notice $\boldsymbol{i} \times \boldsymbol{j} = \boldsymbol{k}$.

■ WORKED EXAMPLES ■

5.3 A If A is singular, the equation $AC^{\mathrm{T}} = (\det A)I$ becomes $\boldsymbol{AC^{\mathrm{T}}} =$ **zero matrix**. *Then each column of* C^{T} *is in the nullspace of* A. Those columns contain cofactors along rows of A. So the cofactors quickly find the nullspace for a 3 by 3 matrix of rank 2. My apologies that this comes so late!

Solve $A\boldsymbol{x} = \mathbf{0}$ by $\boldsymbol{x} =$ cofactors along a row, for these singular matrices of rank 2:

Cofactors give nullspace $\quad A = \begin{bmatrix} 1 & 4 & 7 \\ 2 & 3 & 9 \\ 2 & 2 & 8 \end{bmatrix} \qquad A = \begin{bmatrix} 1 & 1 & 2 \\ 1 & 1 & 1 \\ 1 & 1 & 1 \end{bmatrix}$

Solution The first matrix has these cofactors along its top row (note each minus sign):

$$\begin{vmatrix} 3 & 9 \\ 2 & 8 \end{vmatrix} = 6 \qquad -\begin{vmatrix} 2 & 9 \\ 2 & 8 \end{vmatrix} = 2 \qquad \begin{vmatrix} 2 & 3 \\ 2 & 2 \end{vmatrix} = -2$$

Then $\boldsymbol{x} = (6, 2, -2)$ solves $A\boldsymbol{x} = \mathbf{0}$. The cofactors along the second row are $(-18, -6, 6)$ which is just $-3\boldsymbol{x}$. This is also in the one-dimensional nullspace of A.

The second matrix has *zero cofactors* along its first row. The nullvector $\boldsymbol{x} = (0, 0, 0)$ is not interesting. The cofactors of row 2 give $\boldsymbol{x} = (1, -1, 0)$ which solves $A\boldsymbol{x} = \mathbf{0}$.

Every n by n matrix of rank $n-1$ has at least one nonzero cofactor by Problem 3.3.12. But for rank $n-2$, all cofactors are zero and we only find $\boldsymbol{x} = \mathbf{0}$.

5.3 B Use Cramer's Rule with ratios $\det B_j / \det A$ to solve $A\boldsymbol{x} = \boldsymbol{b}$. Also find the inverse matrix $A^{-1} = C^{\mathrm{T}} / \det A$. For this $\boldsymbol{b} = (0, 0, 1)$ the solution $\boldsymbol{x}$ is column 3 of A^{-1}! Which cofactors are involved in computing that column $\boldsymbol{x} = (x, y, z)$?

Column 3 of $\boldsymbol{A^{-1}}$ $\quad \begin{bmatrix} 2 & 6 & 2 \\ 1 & 4 & 2 \\ 5 & 9 & 0 \end{bmatrix} \begin{bmatrix} x \\ y \\ z \end{bmatrix} = \begin{bmatrix} 0 \\ 0 \\ 1 \end{bmatrix}.$

Find the volumes of two boxes: edges are *columns* of A and edges are rows of A^{-1}.

Solution The determinants of the B_j (with right side $\boldsymbol{b}$ placed in column j) are

$$|B_1| = \begin{vmatrix} \mathbf{0} & 6 & 2 \\ \mathbf{0} & 4 & 2 \\ \mathbf{1} & 9 & 0 \end{vmatrix} = 4 \qquad |B_2| = \begin{vmatrix} 2 & \mathbf{0} & 2 \\ 1 & \mathbf{0} & 2 \\ 5 & \mathbf{1} & 0 \end{vmatrix} = -2 \qquad |B_3| = \begin{vmatrix} 2 & 6 & \mathbf{0} \\ 1 & 4 & \mathbf{0} \\ 5 & 9 & \mathbf{1} \end{vmatrix} = 2.$$

Those are cofactors C_{31}, C_{32}, C_{33} of row 3. Their dot product with row 3 is $\det A = 2$:

$$\det A = a_{31}C_{31} + a_{32}C_{32} + a_{33}C_{33} = (5, 9, 0) \cdot (4, -2, 2) = 2.$$

The three ratios $\det B_j / \det A$ give the three components of $\boldsymbol{x} = (2, -1, 1)$. This $\boldsymbol{x}$ is the third column of A^{-1} because $\boldsymbol{b} = (0, 0, 1)$ is the third column of I.

The cofactors along the other *rows* of A, divided by $\det A$, give the other *columns* of A^{-1}:

$$A^{-1} = \frac{C^{\mathrm{T}}}{\det A} = \frac{1}{2}\begin{bmatrix} -18 & 18 & 4 \\ 10 & -10 & -2 \\ -11 & 12 & 2 \end{bmatrix}. \quad \text{Multiply to check} \quad AA^{-1} = I$$

The box from the columns of A has volume $= \det A = 2$. The box from the rows also has volume 2, since $|A^{\mathrm{T}}| = |A|$. The box from the rows of A^{-1} has volume $1/|A| = \frac{1}{2}$.

Problem Set 5.3

Problems 1–5 are about Cramer's Rule for $\boldsymbol{x} = A^{-1}\boldsymbol{b}$.

1 Solve these linear equations by Cramer's Rule $x_j = \det B_j / \det A$:

(a) $\begin{aligned} 2x_1 + 5x_2 &= 1 \\ x_1 + 4x_2 &= 2 \end{aligned}$ (b) $\begin{aligned} 2x_1 + x_2 &= 1 \\ x_1 + 2x_2 + x_3 &= 0 \\ x_2 + 2x_3 &= 0. \end{aligned}$

2 Use Cramer's Rule to solve for y (only). Call the 3 by 3 determinant D:

(a) $\begin{aligned} ax + by &= 1 \\ cx + dy &= 0 \end{aligned}$ (b) $\begin{aligned} ax + by + cz &= 1 \\ dx + ey + fz &= 0 \\ gx + hy + iz &= 0. \end{aligned}$

3 Cramer's Rule breaks down when $\det A = 0$. Example (a) has no solution while (b) has infinitely many. What are the ratios $x_j = \det B_j / \det A$ in these two cases?

(a) $\begin{aligned} 2x_1 + 3x_2 &= 1 \\ 4x_1 + 6x_2 &= 1 \end{aligned}$ (parallel lines) (b) $\begin{aligned} 2x_1 + 3x_2 &= 1 \\ 4x_1 + 6x_2 &= 2 \end{aligned}$ (same line)

4 *Quick proof of Cramer's rule.* The determinant is a linear function of column 1. It is zero if two columns are equal. When $\boldsymbol{b} = A\boldsymbol{x} = x_1\boldsymbol{a}_1 + x_2\boldsymbol{a}_2 + x_3\boldsymbol{a}_3$ goes into the first column of A, the determinant of this matrix B_1 is

$$|\boldsymbol{b} \quad \boldsymbol{a}_2 \quad \boldsymbol{a}_3| = |x_1\boldsymbol{a}_1 + x_2\boldsymbol{a}_2 + x_3\boldsymbol{a}_3 \quad \boldsymbol{a}_2 \quad \boldsymbol{a}_3| = x_1|\boldsymbol{a}_1 \quad \boldsymbol{a}_2 \quad \boldsymbol{a}_3| = x_1 \det A.$$

(a) What formula for x_1 comes from left side = right side?

(b) What steps lead to the middle equation?

5 If the right side $\boldsymbol{b}$ is the first column of A, solve the 3 by 3 system $A\boldsymbol{x} = \boldsymbol{b}$. How does each determinant in Cramer's Rule lead to this solution $\boldsymbol{x}$?

Problems 6–15 are about $A^{-1} = C^{\mathrm{T}}/\det A$. Remember to transpose C.

6 Find A^{-1} from the cofactor formula $C^{\mathrm{T}}/\det A$. Use symmetry in part (b).

$$\text{(a)}\ A = \begin{bmatrix} 1 & 2 & 0 \\ 0 & 3 & 0 \\ 0 & 7 & 1 \end{bmatrix} \qquad \text{(b)}\quad A = \begin{bmatrix} 2 & -1 & 0 \\ -1 & 2 & -1 \\ 0 & -1 & 2 \end{bmatrix}.$$

7 If all the cofactors are zero, how do you know that A has no inverse? If none of the cofactors are zero, is A sure to be invertible?

8 Find the cofactors of A and multiply AC^{T} to find $\det A$:

$$A = \begin{bmatrix} 1 & 1 & 4 \\ 1 & 2 & 2 \\ 1 & 2 & 5 \end{bmatrix} \quad \text{and} \quad C = \begin{bmatrix} 6 & -3 & 0 \\ \cdot & \cdot & \cdot \\ \cdot & \cdot & \cdot \end{bmatrix} \quad \text{and } AC^{\mathrm{T}} = \underline{\qquad}.$$

If you change that 4 to 100, why is $\det A$ unchanged?

9 Suppose $\det A = 1$ and you know all the cofactors in C. How can you find A?

10 From the formula $AC^{\mathrm{T}} = (\det A)I$ show that $\det C = (\det A)^{n-1}$.

11 If all entries of A are integers, and $\det A = 1$ or -1, prove that all entries of A^{-1} are integers. Give a 2 by 2 example with no zero entries.

12 If all entries of A and A^{-1} are integers, prove that $\det A = 1$ or -1. Hint: What is $\det A$ times $\det A^{-1}$?

13 Complete the calculation of A^{-1} by cofactors that was started in Example 5.

14 L is lower triangular and S is symmetric. Assume they are invertible:

To invert triangular L symmetric S
$$L = \begin{bmatrix} a & 0 & 0 \\ b & c & 0 \\ d & e & f \end{bmatrix} \qquad S = \begin{bmatrix} a & b & d \\ b & c & e \\ d & e & f \end{bmatrix}.$$

(a) Which three cofactors of L are zero? Then L^{-1} is also lower triangular.

(b) Which three pairs of cofactors of S are equal? Then S^{-1} is also symmetric.

(c) The cofactor matrix C of an orthogonal Q will be _____. *Why?*

15 For $n = 5$ the matrix C contains _____ cofactors. Each 4 by 4 cofactor contains _____ terms and each term needs _____ multiplications. Compare with $5^3 = 125$ for the Gauss-Jordan computation of A^{-1} in Section 2.4.

Problems 16–26 are about area and volume by determinants.

16 (a) Find the area of the parallelogram with edges $\boldsymbol{v} = (3, 2)$ and $\boldsymbol{w} = (1, 4)$.

(b) Find the area of the triangle with sides $\boldsymbol{v}$, $\boldsymbol{w}$, and $\boldsymbol{v} + \boldsymbol{w}$. Draw it.

(c) Find the area of the triangle with sides $\boldsymbol{v}$, $\boldsymbol{w}$, and $\boldsymbol{w} - \boldsymbol{v}$. Draw it.

17 A box has edges from $(0,0,0)$ to $(3,1,1)$ and $(1,3,1)$ and $(1,1,3)$. Find its volume. Also find the area of each parallelogram face using $\|\boldsymbol{u}\times\boldsymbol{v}\|$.

18 (a) The corners of a triangle are $(2,1)$ and $(3,4)$ and $(0,5)$. What is the area?

(b) Add a corner at $(-1,0)$ to make a lopsided region (four sides). Find the area.

19 The parallelogram with sides $(2,1)$ and $(2,3)$ has the same area as the parallelogram with sides $(2,2)$ and $(1,3)$. Find those areas from 2 by 2 determinants and say why they must be equal. (I can't see why from a picture. Please write to me if you do.)

20 The Hadamard matrix H has orthogonal rows. The box is a hypercube!

$$\text{What is} \quad |H| = \begin{vmatrix} 1 & 1 & 1 & 1 \\ 1 & 1 & -1 & -1 \\ 1 & -1 & -1 & 1 \\ 1 & -1 & 1 & -1 \end{vmatrix} = \text{volume of a hypercube in } \mathbf{R}^4?$$

21 If the columns of a 4 by 4 matrix have lengths L_1, L_2, L_3, L_4, what is the largest possible value for the determinant (based on volume)? If all entries of the matrix are 1 or -1, what are those lengths and the maximum determinant?

22 Show by a picture how a rectangle with area x_1y_2 minus a rectangle with area x_2y_1 produces the same area as our parallelogram.

23 When the edge vectors $\boldsymbol{a}, \boldsymbol{b}, \boldsymbol{c}$ are perpendicular, the volume of the box is $\|\boldsymbol{a}\|$ times $\|\boldsymbol{b}\|$ times $\|\boldsymbol{c}\|$. The matrix $A^{\mathrm{T}}A$ is _____. Find $\det A^{\mathrm{T}}A$ and $\det A$.

24 The box with edges $\boldsymbol{i}$ and $\boldsymbol{j}$ and $\boldsymbol{w} = 2\boldsymbol{i}+3\boldsymbol{j}+4\boldsymbol{k}$ has height _____. What is the volume? What is the matrix with this determinant? What is $\boldsymbol{i}\times\boldsymbol{j}$ and what is its dot product with $\boldsymbol{w}$?

25 An n-dimensional cube has how many corners? How many edges? How many $(n-1)$-dimensional faces? The cube in $\mathbf{R}^n$ whose edges are the rows of $2I$ has volume _____. A hypercube computer has parallel processors at the corners with connections along the edges.

26 The triangle with corners $(0,0), (1,0), (0,1)$ has area $\frac{1}{2}$. The pyramid in $\mathbf{R}^3$ with four corners $(0,0,0), (1,0,0), (0,1,0), (0,0,1)$ has volume _____. What is the volume of a pyramid in $\mathbf{R}^4$ with five corners at $(0,0,0,0)$ and the rows of I?

Problems 27–30 are about areas dA and volumes dV in calculus.

27 Polar coordinates satisfy $x = r\cos\theta$ and $y = r\sin\theta$. Polar area is $J\,dr\,d\theta$:

$$J = \begin{vmatrix} \partial x/\partial r & \partial x/\partial\theta \\ \partial y/\partial r & \partial y/\partial\theta \end{vmatrix} = \begin{vmatrix} \cos\theta & -r\sin\theta \\ \sin\theta & r\cos\theta \end{vmatrix}.$$

The two columns are orthogonal. Their lengths are _____. Thus $J =$ _____.

28 Spherical coordinates ρ, ϕ, θ satisfy $x = \rho \sin\phi \cos\theta$ and $y = \rho \sin\phi \sin\theta$ and $z = \rho \cos\phi$. Find the 3 by 3 matrix of partial derivatives: $\partial x/\partial \rho, \partial x/\partial \phi, \partial x/\partial \theta$ in row 1. Simplify its determinant to $J = \rho^2 \sin\phi$. Then dV in spherical coordinates is $\rho^2 \sin\phi\, d\rho\, d\phi d\theta$, the volume of an infinitesimal "coordinate box".

29 The matrix that connects r, θ to x, y is in Problem 27. Invert that 2 by 2 matrix:

$$J^{-1} = \begin{vmatrix} \partial r/\partial x & \partial r/\partial y \\ \partial \theta/\partial x & \partial \theta/\partial y \end{vmatrix} = \begin{vmatrix} \cos\theta & ? \\ ? & ? \end{vmatrix} = ?$$

It is surprising that $\partial r/\partial x = \partial x/\partial r$ (***Calculus***, Gilbert Strang, p. 501). Multiplying the matrices J and J^{-1} gives the chain rule $\frac{\partial x}{\partial x} = \frac{\partial x}{\partial r}\frac{\partial r}{\partial x} + \frac{\partial x}{\partial \theta}\frac{\partial \theta}{\partial x} = 1$.

30 The triangle with corners $(0,0)$, $(6,0)$, and $(1,4)$ has area ______. When you rotate it by $\theta = 60°$ the area is ______. The determinant of the rotation matrix is

$$J = \begin{vmatrix} \cos\theta & -\sin\theta \\ \sin\theta & \cos\theta \end{vmatrix} = \begin{vmatrix} \frac{1}{2} & ? \\ ? & ? \end{vmatrix} = ?$$

Problems 31–38 are about the triple product $(u \times v) \cdot w$ in three dimensions.

31 A box has base area $\|\boldsymbol{u} \times \boldsymbol{v}\|$. Its perpendicular height is $\|\boldsymbol{w}\| \cos\theta$. Base area times height $=$ volume $= \|\boldsymbol{u} \times \boldsymbol{v}\|\, \|\boldsymbol{w}\| \cos\theta$ which is $(\boldsymbol{u} \times \boldsymbol{v}) \cdot \boldsymbol{w}$. Compute base area, height, and volume for $\boldsymbol{u} = (2,4,0)$, $\boldsymbol{v} = (-1,3,0)$, $\boldsymbol{w} = (1,2,2)$.

32 The volume of the same box is given more directly by a 3 by 3 determinant. Evaluate that determinant.

33 Expand the 3 by 3 determinant in equation (13) in cofactors of its row u_1, u_2, u_3. This expansion is the dot product of $\boldsymbol{u}$ with the vector ______.

34 Which of the triple products $(\boldsymbol{u} \times \boldsymbol{w}) \cdot \boldsymbol{v}$ and $(\boldsymbol{w} \times \boldsymbol{u}) \cdot \boldsymbol{v}$ and $(\boldsymbol{v} \times \boldsymbol{w}) \cdot \boldsymbol{u}$ are the same as $(\boldsymbol{u} \times \boldsymbol{v}) \cdot \boldsymbol{w}$? Which orders of the rows $\boldsymbol{u}$, $\boldsymbol{v}$, $\boldsymbol{w}$ give the correct determinant?

35 Let $P = (1,0,-1)$ and $Q = (1,1,1)$ and $R = (2,2,1)$. Choose S so that $PQRS$ is a parallelogram and compute its area. Choose T, U, V so that $OPQRSTUV$ is a tilted box and compute its volume.

36 Suppose (x,y,z) and $(1,1,0)$ and $(1,2,1)$ lie on a plane through the origin. What determinant is zero? What equation does this give for the plane?

37 Suppose (x,y,z) is a linear combination of $(2,3,1)$ and $(1,2,3)$. What determinant is zero? What equation does this give for the plane of all combinations?

38 (a) Explain from volumes why $\det 2A = 2^n \det A$ for n by n matrices.

(b) For what size matrix is the false statement $\det A + \det A = \det(A + A)$ true?

Challenge Problems

39 If you know all 16 cofactors of a 4 by 4 invertible matrix A, how would you find A?

40 Suppose A is a 5 by 5 matrix. Its entries in row 1 multiply determinants (cofactors) in rows 2–5 to give the determinant. Can you guess a "Jacobi formula" for $\det A$ using 2 by 2 determinants from rows 1–2 *times* 3 by 3 determinants from rows 3–5?

Test your formula on the $-1, 2, -1$ tridiagonal matrix that has determinant $= 6$.

41 The 2 by 2 matrix AB =(2 by 3)(3 by 2) has a "Cauchy-Binet formula" for $\det AB$:

$$\det AB = \text{sum of (2 by 2 determinants in } A) \text{ (2 by 2 determinants in } B)$$

(a) Guess which 2 by 2 determinants to use from A and B.

(b) Test your formula when the rows of A are $1, 2, 3$ and $1, 4, 7$ with $B = A^{\mathrm{T}}$.

42 The big formula has $n!$ terms. But if an entry of A is zero, $(n-1)!$ terms disappear. If A has only *three diagonals*, how many terms are left ?

For $n = 1, 2, 3, 4$ the tridiagonal determinant has $1, 2, 3, 5$ terms. Those are Fibonacci numbers in Section 6.2 ! Show why a tridiagonal 5 by 5 determinant has $5 + 3 = 8$ nonzero terms (Fibonacci again). Use the cofactors of a_{11} and a_{12}.

Chapter 6

Eigenvalues and Eigenvectors

6.1 Introduction to Eigenvalues

1 An **eigenvector** $\boldsymbol{x}$ lies along the same line as $A\boldsymbol{x}$: $\boxed{A\boldsymbol{x} = \lambda\boldsymbol{x}.}$ The **eigenvalue** is λ.

2 If $A\boldsymbol{x} = \lambda\boldsymbol{x}$ then $A^2\boldsymbol{x} = \lambda^2\boldsymbol{x}$ and $A^{-1}\boldsymbol{x} = \lambda^{-1}\boldsymbol{x}$ and $(A + cI)\boldsymbol{x} = (\lambda + c)\boldsymbol{x}$: the same $\boldsymbol{x}$.

3 If $A\boldsymbol{x} = \lambda\boldsymbol{x}$ then $(A–\lambda I)\boldsymbol{x} = \mathbf{0}$ and $A–\lambda I$ is singular and $\boxed{\det(\boldsymbol{A}-\boldsymbol{\lambda I}) = \mathbf{0}.}$ n eigenvalues.

4 Check λ's by $\det A = (\lambda_1)(\lambda_2)\cdots(\lambda_n)$ and diagonal sum $a_{11} + a_{22} + \cdots + a_{nn} =$ sum of λ's.

5 Projections have $\lambda = 1$ and 0. Reflections have 1 and -1. Rotations have $e^{i\theta}$ and $e^{-i\theta}$: *complex!*

This chapter enters a new part of linear algebra. The first part was about $A\boldsymbol{x} = \boldsymbol{b}$: balance and equilibrium and steady state. Now the second part is about **change**. Time enters the picture—continuous time in a differential equation $d\boldsymbol{u}/dt = A\boldsymbol{u}$ or time steps in a difference equation $\boldsymbol{u}_{k+1} = A\boldsymbol{u}_k$. Those equations are NOT solved by elimination.

The key idea is to avoid all the complications presented by the matrix A. Suppose the solution vector $\boldsymbol{u}(t)$ stays in the direction of a fixed vector $\boldsymbol{x}$. Then we only need to find the number (changing with time) that multiplies $\boldsymbol{x}$. A number is easier than a vector. **We want "eigenvectors" $\boldsymbol{x}$ that don't change direction when you multiply by A.**

A good model comes from the powers $A, A^2, A^3, \ldots$ of a matrix. Suppose you need the hundredth power A^{100}. Its columns are very close to the *eigenvector* $(.6,\ .4)$:

$$A, A^2, A^3 = \begin{bmatrix} .8 & .3 \\ .2 & .7 \end{bmatrix} \quad \begin{bmatrix} .70 & .45 \\ .30 & .55 \end{bmatrix} \quad \begin{bmatrix} .650 & .525 \\ .350 & .475 \end{bmatrix} \qquad \boldsymbol{A^{100}} \approx \begin{bmatrix} \mathbf{.6000} & \mathbf{.6000} \\ \mathbf{.4000} & \mathbf{.4000} \end{bmatrix}$$

A^{100} was found by using the *eigenvalues* of A, not by multiplying 100 matrices. Those eigenvalues (here they are $\lambda = 1$ and $1/2$) are a new way to see into the heart of a matrix.

To explain eigenvalues, we first explain eigenvectors. Almost all vectors change direction, when they are multiplied by A. ***Certain exceptional vectors $\boldsymbol{x}$ are in the same direction as $A\boldsymbol{x}$. Those are the "eigenvectors"***. Multiply an eigenvector by A, and the vector $A\boldsymbol{x}$ is a number λ times the original $\boldsymbol{x}$.

The basic equation is $Ax = \lambda x$. The number λ is an eigenvalue of A.

The eigenvalue λ tells whether the special vector $\boldsymbol{x}$ is stretched or shrunk or reversed or left unchanged—when it is multiplied by A. We may find $\lambda = 2$ or $\frac{1}{2}$ or -1 or 1. The eigenvalue λ could be zero! Then $A\boldsymbol{x} = 0\boldsymbol{x}$ means that this eigenvector $\boldsymbol{x}$ is in the nullspace.

If A is the identity matrix, every vector has $A\boldsymbol{x} = \boldsymbol{x}$. All vectors are eigenvectors of I. All eigenvalues "lambda" are $\lambda = 1$. This is unusual to say the least. Most 2 by 2 matrices have *two* eigenvector directions and *two* eigenvalues. We will show that $\det(A - \lambda I) = 0$.

This section will explain how to compute the $\boldsymbol{x}$'s and λ's. It can come early in the course because we only need the determinant of a 2 by 2 matrix. Let me use $\det(A - \lambda I) = 0$ to find the eigenvalues for this first example, and then derive it properly in equation (3).

Example 1 The matrix A has two eigenvalues $\lambda = 1$ and $\lambda = 1/2$. Look at $\det(A - \lambda I)$:

$$A = \begin{bmatrix} .8 & .3 \\ .2 & .7 \end{bmatrix} \quad \det \begin{bmatrix} .8-\lambda & .3 \\ .2 & .7-\lambda \end{bmatrix} = \lambda^2 - \frac{3}{2}\lambda + \frac{1}{2} = (\lambda - 1)\left(\lambda - \frac{1}{2}\right).$$

I factored the quadratic into $\lambda - 1$ times $\lambda - \frac{1}{2}$, to see the two eigenvalues $\boldsymbol{\lambda = 1}$ and $\boldsymbol{\lambda = \frac{1}{2}}$. For those numbers, the matrix $A - \lambda I$ becomes *singular* (zero determinant). The eigenvectors $\boldsymbol{x}_1$ and $\boldsymbol{x}_2$ are in the nullspaces of $A - I$ and $A - \frac{1}{2}I$.

$(A - I)\boldsymbol{x}_1 = 0$ is $A\boldsymbol{x}_1 = \boldsymbol{x}_1$ and the first eigenvector is $(\boldsymbol{.6},\ \boldsymbol{.4})$.

$(A - \frac{1}{2}I)\boldsymbol{x}_2 = 0$ is $A\boldsymbol{x}_2 = \boldsymbol{\frac{1}{2}}\boldsymbol{x}_2$ and the second eigenvector is $(\boldsymbol{1},\ \boldsymbol{-1})$:

$$\boldsymbol{x}_1 = \begin{bmatrix} .6 \\ .4 \end{bmatrix} \quad \text{and} \quad A\boldsymbol{x}_1 = \begin{bmatrix} .8 & .3 \\ .2 & .7 \end{bmatrix}\begin{bmatrix} .6 \\ .4 \end{bmatrix} = \boldsymbol{x}_1 \quad (A\boldsymbol{x} = \boldsymbol{x} \text{ means that } \lambda_1 = 1)$$

$$\boldsymbol{x}_2 = \begin{bmatrix} 1 \\ -1 \end{bmatrix} \quad \text{and} \quad A\boldsymbol{x}_2 = \begin{bmatrix} .8 & .3 \\ .2 & .7 \end{bmatrix}\begin{bmatrix} 1 \\ -1 \end{bmatrix} = \begin{bmatrix} .5 \\ -.5 \end{bmatrix} \quad (\text{this is } \tfrac{1}{2}\,\boldsymbol{x}_2 \text{ so } \lambda_2 = \tfrac{1}{2}).$$

If $\boldsymbol{x}_1$ is multiplied again by A, we still get $\boldsymbol{x}_1$. Every power of A will give $A^n\boldsymbol{x}_1 = \boldsymbol{x}_1$. Multiplying $\boldsymbol{x}_2$ by A gave $\frac{1}{2}\boldsymbol{x}_2$, and if we multiply again we get $(\frac{1}{2})^2$ times $\boldsymbol{x}_2$.

When A is squared, the eigenvectors stay the same. The eigenvalues are squared.

This pattern keeps going, because the eigenvectors stay in their own directions (Figure 6.1) and never get mixed. The eigenvectors of A^{100} are the same $\boldsymbol{x}_1$ and $\boldsymbol{x}_2$. The eigenvalues of A^{100} are $1^{100} = 1$ and $(\frac{1}{2})^{100} =$ very small number.

Other vectors do change direction. But all other vectors are combinations of the two eigenvectors. The first column of A is the combination $\boldsymbol{x}_1 + (.2)\boldsymbol{x}_2$:

Separate into eigenvectors
Then multiply by A

$$\begin{bmatrix} .8 \\ .2 \end{bmatrix} = \boldsymbol{x}_1 + (.2)\boldsymbol{x}_2 = \begin{bmatrix} .6 \\ .4 \end{bmatrix} + \begin{bmatrix} .2 \\ -.2 \end{bmatrix}. \tag{1}$$

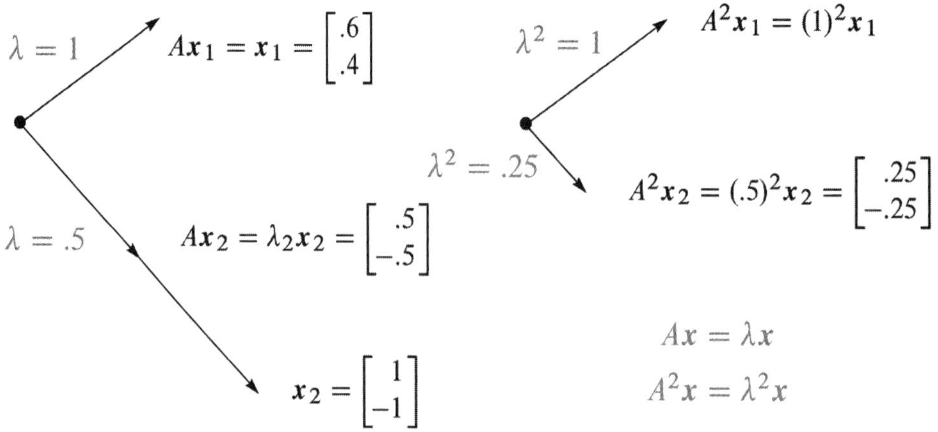

Figure 6.1: The eigenvectors keep their directions. $A^2\boldsymbol{x} = \lambda^2 \boldsymbol{x}$ with $\lambda^2 = 1^2$ and $(.5)^2$.

When we multiply separately for $\boldsymbol{x}_1$ and $(.2)\boldsymbol{x}_2$, A multiplies $\boldsymbol{x}_2$ by its eigenvalue $\frac{1}{2}$:

$$\textbf{Multiply each } \boldsymbol{x}_i \textbf{ by } \boldsymbol{\lambda}_i \qquad A\begin{bmatrix} .8 \\ .2 \end{bmatrix} \quad \textbf{is} \quad \boldsymbol{x}_1 + \frac{1}{2}(.2)\boldsymbol{x}_2 = \begin{bmatrix} .6 \\ .4 \end{bmatrix} + \begin{bmatrix} .1 \\ -.1 \end{bmatrix} = \begin{bmatrix} .7 \\ .3 \end{bmatrix}.$$

Each eigenvector is multiplied by its eigenvalue, when we multiply by A. At every step $\boldsymbol{x}_1$ is unchanged and $\boldsymbol{x}_2$ is multiplied by $\left(\frac{1}{2}\right)$, so 99 steps give the small number $\left(\frac{1}{2}\right)^{99}$:

$$A^{99}\begin{bmatrix} .8 \\ .2 \end{bmatrix} \quad \text{is really} \quad \boldsymbol{x}_1 + (.2)\left(\tfrac{1}{2}\right)^{99}\boldsymbol{x}_2 = \begin{bmatrix} .6 \\ .4 \end{bmatrix} + \begin{bmatrix} \text{very} \\ \text{small} \\ \text{vector} \end{bmatrix}.$$

This is the first column of A^{100}. The number we originally wrote as .6000 was not exact. We left out $(.2)(\frac{1}{2})^{99}$ which wouldn't show up for 30 decimal places.

The eigenvector $\boldsymbol{x}_1$ is a "steady state" that doesn't change (because $\lambda_1 = 1$). The eigenvector $\boldsymbol{x}_2$ is a "decaying mode" that virtually disappears (because $\lambda_2 = .5$). The higher the power of A, the more closely its columns approach the steady state.

This particular A is a ***Markov matrix***. Its largest eigenvalue is $\lambda = 1$. Its eigenvector $\boldsymbol{x}_1 = (.6, .4)$ is the *steady state*—which all columns of A^k will approach. Section 10.3 shows how Markov matrices appear when you search with Google.

For projection matrices P, we can see when $P\boldsymbol{x}$ is parallel to $\boldsymbol{x}$. The eigenvectors for $\lambda = 1$ and $\lambda = 0$ fill the column space and nullspace. The column space doesn't move $(P\boldsymbol{x} = \boldsymbol{x})$. The nullspace goes to zero $(P\boldsymbol{x} = 0\,\boldsymbol{x})$.

Example 2 **The projection matrix** $P = \begin{bmatrix} .5 & .5 \\ .5 & .5 \end{bmatrix}$ has eigenvalues $\lambda = 1$ and $\lambda = 0$.

Its eigenvectors are $\boldsymbol{x}_1 = (1,1)$ and $\boldsymbol{x}_2 = (1,-1)$. For those vectors, $P\boldsymbol{x}_1 = \boldsymbol{x}_1$ (steady state) and $P\boldsymbol{x}_2 = \mathbf{0}$ (nullspace). This example illustrates Markov matrices and singular matrices and (most important) symmetric matrices. All have special λ's and $\boldsymbol{x}$'s:

1. **Markov matrix**: Each column of P adds to 1, so $\lambda = 1$ is an eigenvalue.
2. P is **singular**, so $\lambda = 0$ is an eigenvalue.
3. P is **symmetric**, so its eigenvectors $(1,1)$ and $(1,-1)$ are perpendicular.

The only eigenvalues of a projection matrix are 0 and 1. The eigenvectors for $\lambda = 0$ (which means $P\boldsymbol{x} = 0\boldsymbol{x}$) fill up the nullspace. The eigenvectors for $\lambda = 1$ (which means $P\boldsymbol{x} = \boldsymbol{x}$) fill up the column space. The nullspace is projected to zero. The column space projects onto itself. The projection keeps the column space and destroys the nullspace:

$$\textbf{Project each part} \quad \boldsymbol{v} = \begin{bmatrix} 1 \\ -1 \end{bmatrix} + \begin{bmatrix} 2 \\ 2 \end{bmatrix} \quad \textbf{projects onto} \quad P\boldsymbol{v} = \begin{bmatrix} \mathbf{0} \\ \mathbf{0} \end{bmatrix} + \begin{bmatrix} \mathbf{2} \\ \mathbf{2} \end{bmatrix}.$$

Projections have $\lambda = 0$ and 1. Permutations have all $|\lambda| = 1$. The next matrix R is a reflection and at the same time a permutation. R also has special eigenvalues.

Example 3 **The reflection matrix** $R = \left[\begin{smallmatrix} 0 & 1 \\ 1 & 0 \end{smallmatrix}\right]$ **has eigenvalues 1 and** -1.

The eigenvector $(1,1)$ is unchanged by R. The second eigenvector is $(1,-1)$—its signs are reversed by R. A matrix with no negative entries can still have a negative eigenvalue! The eigenvectors for R are the same as for P, because *reflection* $= 2$(*projection*) $- I$:

$$\boldsymbol{R = 2P - I} \qquad \begin{bmatrix} 0 & 1 \\ 1 & 0 \end{bmatrix} = 2\begin{bmatrix} .5 & .5 \\ .5 & .5 \end{bmatrix} - \begin{bmatrix} 1 & 0 \\ 0 & 1 \end{bmatrix}. \tag{2}$$

***When a matrix is shifted by I, each λ is shifted by* 1**. No change in eigenvectors.

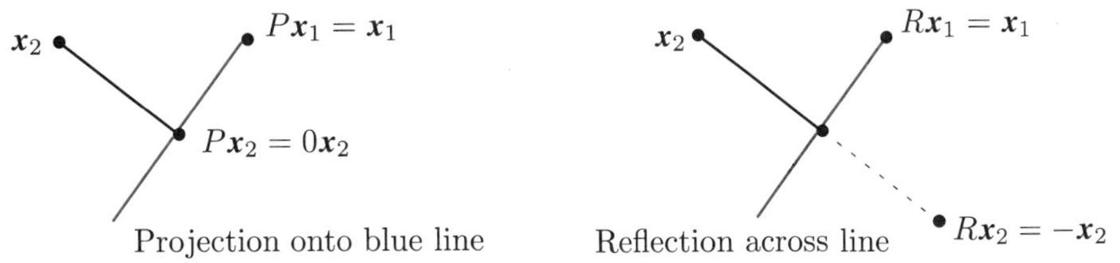

Figure 6.2: Projections P have eigenvalues 1 and 0. Reflections R have $\lambda = 1$ and -1. A typical $\boldsymbol{x}$ changes direction, but an eigenvector stays along the same line.

The Equation for the Eigenvalues

For projection matrices we found λ's and $\boldsymbol{x}$'s by geometry: $P\boldsymbol{x} = \boldsymbol{x}$ and $P\boldsymbol{x} = \mathbf{0}$. For other matrices we use determinants and linear algebra. *This is the key calculation in the chapter*—almost every application starts by solving $A\boldsymbol{x} = \lambda\boldsymbol{x}$.

First move $\lambda\boldsymbol{x}$ to the left side. Write the equation $A\boldsymbol{x} = \lambda\boldsymbol{x}$ as $(A - \lambda I)\boldsymbol{x} = \mathbf{0}$. The matrix $A - \lambda I$ times the eigenvector $\boldsymbol{x}$ is the zero vector. ***The eigenvectors make up the nullspace of $A - \lambda I$.*** When we know an eigenvalue λ, we find an eigenvector by solving $(A - \lambda I)\boldsymbol{x} = \mathbf{0}$.

Eigenvalues first. If $(A - \lambda I)\boldsymbol{x} = \mathbf{0}$ has a nonzero solution, $A - \lambda I$ is not invertible. ***The determinant of $A - \lambda I$ must be zero.*** This is how to recognize an eigenvalue λ:

Eigenvalues The number λ is an eigenvalue of A if and only if $A - \lambda I$ is singular.

Equation for the eigenvalues $\quad \det(A - \lambda I) = 0. \quad (3)$

This "*characteristic polynomial*" $\det(A - \lambda I)$ involves only λ, not $\boldsymbol{x}$. When A is n by n, equation (3) has degree n. Then A has n eigenvalues (repeats possible!) Each λ leads to $\boldsymbol{x}$:

For each eigenvalue λ solve $(A - \lambda I)\boldsymbol{x} = \mathbf{0}$ or $A\boldsymbol{x} = \lambda\boldsymbol{x}$ to find an eigenvector $\boldsymbol{x}$.

Example 4 $A = \begin{bmatrix} \mathbf{1} & \mathbf{2} \\ \mathbf{2} & \mathbf{4} \end{bmatrix}$ is already singular (zero determinant). Find its λ's and $\boldsymbol{x}$'s.

When A is singular, $\lambda = 0$ is one of the eigenvalues. The equation $A\boldsymbol{x} = 0\boldsymbol{x}$ has solutions. They are the eigenvectors for $\lambda = 0$. But $\det(A - \lambda I) = 0$ is the way to find *all* λ's and $\boldsymbol{x}$'s. Always subtract λI from A:

$$\textit{Subtract } \lambda \textit{ from the diagonal to find} \quad A - \lambda I = \begin{bmatrix} 1-\lambda & 2 \\ 2 & 4-\lambda \end{bmatrix}. \tag{4}$$

Take the determinant "$ad - bc$" of this 2 by 2 matrix. From $1 - \lambda$ times $4 - \lambda$, the "ad" part is $\lambda^2 - 5\lambda + 4$. The "bc" part, not containing λ, is 2 times 2.

$$\det \begin{bmatrix} 1-\lambda & 2 \\ 2 & 4-\lambda \end{bmatrix} = (1-\lambda)(4-\lambda) - (2)(2) = \lambda^2 - 5\lambda. \tag{5}$$

Set this determinant $\boldsymbol{\lambda^2 - 5\lambda}$ to zero. One solution is $\lambda = 0$ (as expected, since A is singular). Factoring into λ times $\lambda - 5$, the other root is $\lambda = 5$:

$\det(A - \lambda I) = \lambda^2 - 5\lambda = 0$ yields the eigenvalues $\lambda_1 = 0$ and $\lambda_2 = 5$.

Now find the eigenvectors. Solve $(A - \lambda I)\boldsymbol{x} = \mathbf{0}$ separately for $\lambda_1 = 0$ and $\lambda_2 = 5$:

$$(A - 0I)\boldsymbol{x} = \begin{bmatrix} 1 & 2 \\ 2 & 4 \end{bmatrix} \begin{bmatrix} y \\ z \end{bmatrix} = \begin{bmatrix} 0 \\ 0 \end{bmatrix} \text{ yields an eigenvector } \begin{bmatrix} y \\ z \end{bmatrix} = \begin{bmatrix} 2 \\ -1 \end{bmatrix} \text{ for } \lambda_1 = 0$$

$$(A - 5I)\boldsymbol{x} = \begin{bmatrix} -4 & 2 \\ 2 & -1 \end{bmatrix} \begin{bmatrix} y \\ z \end{bmatrix} = \begin{bmatrix} 0 \\ 0 \end{bmatrix} \text{ yields an eigenvector } \begin{bmatrix} y \\ z \end{bmatrix} = \begin{bmatrix} 1 \\ 2 \end{bmatrix} \text{ for } \lambda_2 = 5.$$

The matrices $A - 0I$ and $A - 5I$ are singular (because 0 and 5 are eigenvalues). The eigenvectors $(2, -1)$ and $(1, 2)$ are in the nullspaces: $(A - \lambda I)\boldsymbol{x} = \mathbf{0}$ is $A\boldsymbol{x} = \lambda\boldsymbol{x}$.

We need to emphasize: *There is nothing exceptional about* $\lambda = 0$. Like every other number, zero might be an eigenvalue and it might not. If A is singular, the eigenvectors for $\lambda = 0$ fill the nullspace: $A\boldsymbol{x} = 0\boldsymbol{x} = \mathbf{0}$. If A is invertible, zero is not an eigenvalue. We shift A by a multiple of I to *make it singular.*

In the example, the shifted matrix $A - 5I$ is singular and 5 is the other eigenvalue.

Summary To solve the eigenvalue problem for an n by n matrix, follow these steps:

1. ***Compute the determinant of*** $A - \lambda I$. With λ subtracted along the diagonal, this determinant starts with λ^n or $-\lambda^n$. It is a polynomial in λ of degree n.

2. ***Find the roots of this polynomial,*** by solving $\det(A - \lambda I) = 0$. The n roots are the n eigenvalues of A. They make $A - \lambda I$ singular.

3. For each eigenvalue λ, ***solve*** $(A - \lambda I)\boldsymbol{x} = \mathbf{0}$ ***to find an eigenvector*** $\boldsymbol{x}$.

A note on the eigenvectors of 2 by 2 matrices. When $A - \lambda I$ is singular, both rows are multiples of a vector (a, b). *The eigenvector is any multiple of* $(b, -a)$. The example had

$\lambda = 0$: rows of $A - 0I$ in the direction $(1, 2)$; eigenvector in the direction $(2, -1)$

$\lambda = 5$: rows of $A - 5I$ in the direction $(-4, 2)$; eigenvector in the direction $(2, 4)$.

Previously we wrote that last eigenvector as $(1, 2)$. Both $(1, 2)$ and $(2, 4)$ are correct. There is a whole *line of eigenvectors*—any nonzero multiple of $\boldsymbol{x}$ is as good as $\boldsymbol{x}$. MATLAB's **eig**(A) divides by the length, to make the eigenvector into a unit vector.

We must add a warning. Some 2 by 2 matrices have only *one* line of eigenvectors. This can only happen when two eigenvalues are equal. (On the other hand $A = I$ has equal eigenvalues and plenty of eigenvectors.) Without a full set of eigenvectors, we don't have a basis. We can't write every $\boldsymbol{v}$ as a combination of eigenvectors. In the language of the next section, *we can't diagonalize a matrix without n independent eigenvectors.*

Determinant and Trace

Bad news first: If you add a row of A to another row, or exchange rows, the eigenvalues usually change. *Elimination does not preserve the* λ's. The triangular U has *its* eigenvalues sitting along the diagonal—they are the pivots. But they are not the eigenvalues of A! Eigenvalues are changed when row 1 is added to row 2:

$$U = \begin{bmatrix} 1 & 3 \\ 0 & 0 \end{bmatrix} \quad \text{has } \lambda = 0 \text{ and } \lambda = 1; \quad A = \begin{bmatrix} 1 & 3 \\ 2 & 6 \end{bmatrix} \quad \text{has } \lambda = 0 \text{ and } \lambda = 7.$$

Good news second: The *product* λ_1 *times* λ_2 *and the sum* $\lambda_1 + \lambda_2$ *can be found quickly from the matrix.* For this A, the product is 0 times 7. That agrees with the determinant (which is 0). The sum of eigenvalues is $0 + 7$. That agrees with the sum down the main diagonal (the **trace** is $1 + 6$). These quick checks always work:

> ***The product of the n eigenvalues** equals **the determinant.***
> ***The sum of the n eigenvalues** equals **the sum of the n diagonal entries.***

The sum of the entries along the main diagonal is called the ***trace*** of A:

$$\lambda_1 + \lambda_2 + \cdots + \lambda_n = \textbf{\textit{trace}} = a_{11} + a_{22} + \cdots + a_{nn}. \tag{6}$$

Those checks are very useful. They are proved in Problems 16–17 and again in the next section. They don't remove the pain of computing λ's. But when the computation is wrong, they generally tell us so. To compute the correct λ's, go back to $\det(A - \lambda I) = 0$.

The trace and determinant *do* tell everything when the matrix is 2 by 2. We never want to get those wrong! Here trace $=$ **3** and det $=$ **2**, so the eigenvalues are $\lambda =$ **1** and **2**:

$$A = \begin{bmatrix} 1 & 9 \\ 0 & 2 \end{bmatrix} \quad \text{or} \quad \begin{bmatrix} 3 & 1 \\ -2 & 0 \end{bmatrix} \quad \text{or} \quad \begin{bmatrix} 7 & -3 \\ 10 & -4 \end{bmatrix}. \tag{7}$$

And here is a question about the best matrices for finding eigenvalues: *triangular.*

Why do the eigenvalues of a triangular matrix lie along its diagonal?

Imaginary Eigenvalues

One more bit of news (not too terrible). The eigenvalues might not be real numbers.

> **Example 5** *The* 90° *rotation* $Q = \begin{bmatrix} 0 & -1 \\ 1 & 0 \end{bmatrix}$ *has no real eigenvectors. Its eigenvalues are* $\lambda_1 = i$ *and* $\lambda_2 = -i$. *Then* $\lambda_1 + \lambda_2 =$ **trace** $= 0$ *and* $\lambda_1\lambda_2 =$ *determinant* $= 1$.

After a rotation, *no real vector* Qx *stays in the same direction as* x ($x = \mathbf{0}$ is useless). There cannot be an eigenvector, unless we go to ***imaginary numbers***. Which we do.

To see how $i = \sqrt{-1}$ can help, look at Q^2 which is $-I$. If Q is rotation through 90°, then Q^2 is rotation through 180°. Its eigenvalues are -1 and -1. (Certainly $-I\boldsymbol{x} = -1\boldsymbol{x}$.) Squaring Q will square each λ, so we must have $\lambda^2 = -1$. *The eigenvalues of the* 90° *rotation matrix* Q *are* $+i$ *and* $-i$, because $i^2 = -1$.

Those λ's come as usual from $\det(Q - \lambda I) = 0$. This equation gives $\lambda^2 + 1 = 0$. Its roots are i and $-i$. We meet the imaginary number i also in the eigenvectors:

Complex eigenvectors $\quad \begin{bmatrix} 0 & -1 \\ 1 & 0 \end{bmatrix} \begin{bmatrix} 1 \\ i \end{bmatrix} = -i \begin{bmatrix} 1 \\ i \end{bmatrix}$ and $\begin{bmatrix} 0 & -1 \\ 1 & 0 \end{bmatrix} \begin{bmatrix} i \\ 1 \end{bmatrix} = i \begin{bmatrix} i \\ 1 \end{bmatrix}.$

Somehow these complex vectors $\boldsymbol{x}_1 = (1, i)$ and $\boldsymbol{x}_2 = (i, 1)$ keep their direction as they are rotated. Don't ask me how. This example makes the all-important point that real matrices can easily have complex eigenvalues and eigenvectors. The particular eigenvalues i and $-i$ also illustrate two special properties of Q:

1. Q is an orthogonal matrix so the absolute value of each λ is $|\lambda| = 1$.

2. Q is a skew-symmetric matrix so each λ is pure imaginary.

A symmetric matrix ($S^{\mathrm{T}} = S$) can be compared to a real number. A skew-symmetric matrix ($A^{\mathrm{T}} = -A$) can be compared to an imaginary number. An orthogonal matrix ($Q^{\mathrm{T}}Q = I$) corresponds to a complex number with $|\lambda| = 1$. For the eigenvalues of S and A and Q, those are more than analogies—they are facts to be proved in Section 6.4.

The eigenvectors for all these special matrices are perpendicular. Somehow $(i, 1)$ and $(1, i)$ are perpendicular (Chapter 9 explains the dot product of complex vectors).

Eigenvalues of AB and $A+B$

The first guess about the eigenvalues of AB is not true. An eigenvalue λ of A times an eigenvalue β of B usually does *not* give an eigenvalue of AB:

False proof $$AB\boldsymbol{x} = A\beta\boldsymbol{x} = \beta A\boldsymbol{x} = \beta\lambda\boldsymbol{x}. \tag{8}$$

It seems that β times λ is an eigenvalue. When $\boldsymbol{x}$ is an eigenvector for A and B, this proof is correct. ***The mistake is to expect that A and B automatically share the same eigenvector $\boldsymbol{x}$.*** Usually they don't. Eigenvectors of A are not generally eigenvectors of B. A and B could have all zero eigenvalues while 1 is an eigenvalue of AB:

$$A = \begin{bmatrix} 0 & 1 \\ 0 & 0 \end{bmatrix} \text{ and } B = \begin{bmatrix} 0 & 0 \\ 1 & 0 \end{bmatrix}; \text{ then } AB = \begin{bmatrix} 1 & 0 \\ 0 & 0 \end{bmatrix} \text{ and } A+B = \begin{bmatrix} 0 & 1 \\ 1 & 0 \end{bmatrix}.$$

For the same reason, the eigenvalues of $A + B$ are generally not $\lambda + \beta$. Here $\lambda + \beta = 0$ while $A + B$ has eigenvalues 1 and -1. (At least they add to zero.)

The false proof suggests what is true. Suppose $\boldsymbol{x}$ really is an eigenvector for both A and B. Then we do have $AB\boldsymbol{x} = \lambda\beta\boldsymbol{x}$ and $BA\boldsymbol{x} = \lambda\beta\boldsymbol{x}$. When all n eigenvectors are shared, we *can* multiply eigenvalues. The test $AB = BA$ for shared eigenvectors is important in quantum mechanics—time out to mention this application of linear algebra:

A and B share the same $\boldsymbol{n}$ *independent* eigenvectors if and only if $\boldsymbol{AB = BA}$.

Heisenberg's uncertainty principle In quantum mechanics, the position matrix P and the momentum matrix Q do not commute. In fact $QP - PQ = I$ (these are infinite matrices). To have $P\boldsymbol{x} = \boldsymbol{0}$ at the same time as $Q\boldsymbol{x} = \boldsymbol{0}$ would require $\boldsymbol{x} = I\boldsymbol{x} = \boldsymbol{0}$. If we knew the position exactly, we could not also know the momentum exactly. Problem 36 derives Heisenberg's uncertainty principle $\|P\boldsymbol{x}\|\,\|Q\boldsymbol{x}\| \geq \frac{1}{2}\|\boldsymbol{x}\|^2$.

■ REVIEW OF THE KEY IDEAS ■

1. $A\boldsymbol{x} = \lambda\boldsymbol{x}$ says that eigenvectors $\boldsymbol{x}$ keep the same direction when multiplied by A.
2. $A\boldsymbol{x} = \lambda\boldsymbol{x}$ also says that $\det(A - \lambda I) = 0$. This determines n eigenvalues.
3. The eigenvalues of A^2 and A^{-1} are λ^2 and λ^{-1}, with the same eigenvectors.
4. The sum of the λ's equals the sum down the main diagonal of A (*the trace*). The product of the λ's equals the determinant of A.
5. Projections P, reflections R, $90°$ rotations Q have special eigenvalues $1, 0, -1, i, -i$. Singular matrices have $\lambda = 0$. Triangular matrices have λ's on their diagonal.
6. *Special properties of a matrix lead to special eigenvalues and eigenvectors.* That is a major theme of this chapter (it is captured in a table at the very end).

■ WORKED EXAMPLES ■

6.1 A Find the eigenvalues and eigenvectors of A and A^2 and A^{-1} and $A + 4I$:

$$A = \begin{bmatrix} 2 & -1 \\ -1 & 2 \end{bmatrix} \quad \text{and} \quad A^2 = \begin{bmatrix} 5 & -4 \\ -4 & 5 \end{bmatrix}.$$

Check the trace $\lambda_1 + \lambda_2 = 4$ and the determinant $\lambda_1\lambda_2 = 3$.

Solution The eigenvalues of A come from $\det(A - \lambda I) = 0$:

$$A = \begin{bmatrix} 2 & -1 \\ -1 & 2 \end{bmatrix} \qquad \det(A - \lambda I) = \begin{vmatrix} \mathbf{2 - \lambda} & \mathbf{-1} \\ \mathbf{-1} & \mathbf{2 - \lambda} \end{vmatrix} = \lambda^2 - 4\lambda + 3 = 0.$$

This factors into $(\lambda - 1)(\lambda - 3) = 0$ so the eigenvalues of A are $\lambda_1 = \mathbf{1}$ and $\lambda_2 = \mathbf{3}$. For the trace, the sum $2 + 2$ agrees with $1 + 3$. The determinant 3 agrees with the product $\lambda_1\lambda_2$.

The eigenvectors come separately by solving $(A - \lambda I)\boldsymbol{x} = \mathbf{0}$ which is $A\boldsymbol{x} = \lambda\boldsymbol{x}$:

$$\boldsymbol{\lambda = 1:}\quad (A - I)\boldsymbol{x} = \begin{bmatrix} 1 & -1 \\ -1 & 1 \end{bmatrix}\begin{bmatrix} x \\ y \end{bmatrix} = \begin{bmatrix} 0 \\ 0 \end{bmatrix} \text{ gives the eigenvector } \boldsymbol{x}_1 = \begin{bmatrix} 1 \\ 1 \end{bmatrix}$$

$$\boldsymbol{\lambda = 3:}\quad (A - 3I)\boldsymbol{x} = \begin{bmatrix} -1 & -1 \\ -1 & -1 \end{bmatrix}\begin{bmatrix} x \\ y \end{bmatrix} = \begin{bmatrix} 0 \\ 0 \end{bmatrix} \text{ gives the eigenvector } \boldsymbol{x}_2 = \begin{bmatrix} 1 \\ -1 \end{bmatrix}$$

A^2 and A^{-1} and $A + 4I$ keep the *same eigenvectors as* A. Their eigenvalues are λ^2 and λ^{-1} and $\lambda + 4$:

$$A^2 \text{ has eigenvalues } 1^2 = 1 \text{ and } 3^2 = 9 \qquad A^{-1} \text{ has } \frac{1}{1} \text{ and } \frac{1}{3} \qquad A + 4I \text{ has } \begin{matrix} 1 + 4 = 5 \\ 3 + 4 = 7 \end{matrix}$$

Notes for later sections: A has *orthogonal eigenvectors* (Section 6.4 on symmetric matrices). A can be *diagonalized* since $\lambda_1 \neq \lambda_2$ (Section 6.2). A is *similar* to any 2 by 2 matrix with eigenvalues 1 and 3 (Section 6.2). A is a *positive definite matrix* (Section 6.5) since $A = A^{\mathrm{T}}$ and the λ's are positive.

6.1 B How can you estimate the eigenvalues of any A? Gershgorin gave this answer.

Every eigenvalue of A must be "near" at least one of the entries a_{ii} on the main diagonal. For λ to be "near a_{ii}" means that $|a_{ii} - \lambda|$ is no more than **the sum R_i of all other $|a_{ij}|$ in that row i of the matrix**. Then $R_i = \Sigma_{j \neq i}|a_{ij}|$ is the radius of a circle centered at a_{ii}.

Every λ is in the circle around one or more diagonal entries a_{ii}: $|a_{ii} - \lambda| \leq R_i$.

Here is the reasoning. If λ is an eigenvalue, then $A - \lambda I$ is not invertible. Then $A - \lambda I$ cannot be diagonally dominant (see Section 2.5). So at least one diagonal entry $a_{ii} - \lambda$ is *not larger* than the sum R_i of all other entries $|a_{ij}|$ (we take absolute values!) in row i.

Example 1. Every eigenvalue λ of this A falls into one or both of the **Gershgorin circles**: The centers are a and d, the radii are $R_1 = |b|$ and $R_2 = |c|$.

$$A = \begin{bmatrix} a & b \\ c & d \end{bmatrix} \qquad \begin{matrix} \text{First circle:} & |\lambda - a| \leq |b| \\ \text{Second circle:} & |\lambda - d| \leq |c| \end{matrix}$$

Those are circles in the complex plane, since λ could certainly be complex.

Example 2. All eigenvalues of this A lie in a circle of radius $R = 3$ around *one or more* of the diagonal entries d_1, d_2, d_3:

$$A = \begin{bmatrix} d_1 & 1 & 2 \\ 2 & d_2 & 1 \\ -1 & 2 & d_3 \end{bmatrix} \qquad \begin{matrix} |\lambda - d_1| \leq 1 + 2 = R_1 \\ |\lambda - d_2| \leq 2 + 1 = R_2 \\ |\lambda - d_3| \leq 1 + 2 = R_3 \end{matrix}$$

You see that "near" means not more than 3 away from d_1 or d_2 or d_3, for this example.

6.1 C Find the eigenvalues and eigenvectors of this symmetric 3 by 3 matrix S:

Symmetric matrix
Singular matrix
Trace $1+2+1=4$

$$S=\begin{bmatrix}1 & -1 & 0\\ -1 & 2 & -1\\ 0 & -1 & 1\end{bmatrix}$$

Solution Since all rows of S add to zero, the vector $\boldsymbol{x}=(1,1,1)$ gives $S\boldsymbol{x}=\mathbf{0}$. This is an eigenvector for $\lambda=0$. To find λ_2 and λ_3 I will compute the 3 by 3 determinant:

$$\det(S-\lambda I)=\begin{vmatrix}1-\lambda & -1 & 0\\ -1 & 2-\lambda & -1\\ 0 & -1 & 1-\lambda\end{vmatrix}\begin{aligned}&=(1-\lambda)(2-\lambda)(1-\lambda)-2(1-\lambda)\\ &=(1-\lambda)[(2-\lambda)(1-\lambda)-2]\\ &=\boldsymbol{(1-\lambda)(-\lambda)(3-\lambda)}.\end{aligned}$$

Those three factors give $\lambda=0,1,3$. Each eigenvalue corresponds to an eigenvector (or a line of eigenvectors):

$$\boldsymbol{x}_1=\begin{bmatrix}1\\1\\1\end{bmatrix}\quad S\boldsymbol{x}_1=\mathbf{0}\boldsymbol{x}_1\qquad \boldsymbol{x}_2=\begin{bmatrix}1\\0\\-1\end{bmatrix}\quad S\boldsymbol{x}_2=\mathbf{1}\boldsymbol{x}_2\qquad \boldsymbol{x}_3=\begin{bmatrix}1\\-2\\1\end{bmatrix}\quad S\boldsymbol{x}_3=\mathbf{3}\boldsymbol{x}_3\,.$$

I notice again that eigenvectors are perpendicular when S is symmetric. We were lucky to find $\lambda=0,1,3$. For a larger matrix I would use **eig**(A), and never touch determinants.

The full command $[\boldsymbol{X},\boldsymbol{E}]=$**eig**$(A)$ will produce unit eigenvectors in the columns of X.

Problem Set 6.1

1 The example at the start of the chapter has powers of this matrix A.

$$A=\begin{bmatrix}.8 & .3\\ .2 & .7\end{bmatrix}\quad\text{and}\quad A^2=\begin{bmatrix}.70 & .45\\ .30 & .55\end{bmatrix}\quad\text{and}\quad A^\infty=\begin{bmatrix}.6 & .6\\ .4 & .4\end{bmatrix}.$$

Find the eigenvalues of these matrices. All powers have the same eigenvectors.

(a) Show from A how a row exchange can produce different eigenvalues.

(b) Why is a zero eigenvalue *not* changed by the steps of elimination?

2 Find the eigenvalues and the eigenvectors of these two matrices:

$$A=\begin{bmatrix}1 & 4\\ 2 & 3\end{bmatrix}\quad\text{and}\quad A+I=\begin{bmatrix}2 & 4\\ 2 & 4\end{bmatrix}.$$

$A+I$ has the ______ eigenvectors as A. Its eigenvalues are ______ by 1.

3 Compute the eigenvalues and eigenvectors of A and A^{-1}. Check the trace !

$$A=\begin{bmatrix}0 & 2\\ 1 & 1\end{bmatrix}\quad\text{and}\quad A^{-1}=\begin{bmatrix}-1/2 & 1\\ 1/2 & 0\end{bmatrix}.$$

A^{-1} has the ______ eigenvectors as A. When A has eigenvalues λ_1 and λ_2, its inverse has eigenvalues ______.

4 Compute the eigenvalues and eigenvectors of A and A^2:

$$A = \begin{bmatrix} -1 & 3 \\ 2 & 0 \end{bmatrix} \quad \text{and} \quad A^2 = \begin{bmatrix} 7 & -3 \\ -2 & 6 \end{bmatrix}.$$

A^2 has the same ______ as A. When A has eigenvalues λ_1 and λ_2, A^2 has eigenvalues ______. In this example, why is $\lambda_1^2 + \lambda_2^2 = 13$?

5 Find the eigenvalues of A and B (easy for triangular matrices) and $A + B$:

$$A = \begin{bmatrix} 3 & 0 \\ 1 & 1 \end{bmatrix} \quad \text{and} \quad B = \begin{bmatrix} 1 & 1 \\ 0 & 3 \end{bmatrix} \quad \text{and} \quad A + B = \begin{bmatrix} 4 & 1 \\ 1 & 4 \end{bmatrix}.$$

Eigenvalues of $A + B$ (*are equal to*)(*are not equal to*) eigenvalues of A plus eigenvalues of B.

6 Find the eigenvalues of A and B and AB and BA:

$$A = \begin{bmatrix} 1 & 0 \\ 1 & 1 \end{bmatrix} \quad \text{and} \quad B = \begin{bmatrix} 1 & 2 \\ 0 & 1 \end{bmatrix} \quad \text{and} \quad AB = \begin{bmatrix} 1 & 2 \\ 1 & 3 \end{bmatrix} \quad \text{and} \quad BA = \begin{bmatrix} 3 & 2 \\ 1 & 1 \end{bmatrix}.$$

(a) Are the eigenvalues of AB equal to eigenvalues of A times eigenvalues of B?

(b) Are the eigenvalues of AB equal to the eigenvalues of BA?

7 Elimination produces $A = LU$. The eigenvalues of U are on its diagonal; they are the ______. The eigenvalues of L are on its diagonal; they are all ______. The eigenvalues of A are not the same as ______.

8 (a) If you know that $\boldsymbol{x}$ is an eigenvector, the way to find λ is to ______.

(b) If you know that λ is an eigenvalue, the way to find $\boldsymbol{x}$ is to ______.

9 What do you do to the equation $A\boldsymbol{x} = \lambda\boldsymbol{x}$, in order to prove (a), (b), and (c)?

(a) λ^2 is an eigenvalue of A^2, as in Problem 4.

(b) λ^{-1} is an eigenvalue of A^{-1}, as in Problem 3.

(c) $\lambda + 1$ is an eigenvalue of $A + I$, as in Problem 2.

10 Find the eigenvalues and eigenvectors for both of these Markov matrices A and A^∞. Explain from those answers why A^{100} is close to A^∞:

$$A = \begin{bmatrix} .6 & .2 \\ .4 & .8 \end{bmatrix} \quad \text{and} \quad A^\infty = \begin{bmatrix} 1/3 & 1/3 \\ 2/3 & 2/3 \end{bmatrix}.$$

11 Here is a strange fact about 2 by 2 matrices with eigenvalues $\lambda_1 \neq \lambda_2$: The columns of $A - \lambda_1 I$ are multiples of the eigenvector $\boldsymbol{x}_2$. Any idea why this should be?

12 Find three eigenvectors for this matrix P (projection matrices have $\lambda = 1$ and 0):

$$\textbf{Projection matrix} \qquad P = \begin{bmatrix} .2 & .4 & 0 \\ .4 & .8 & 0 \\ 0 & 0 & 1 \end{bmatrix}.$$

If two eigenvectors share the same λ, so do all their linear combinations. Find an eigenvector of P with no zero components.

13 From the unit vector $\boldsymbol{u} = \left(\frac{1}{6}, \frac{1}{6}, \frac{3}{6}, \frac{5}{6}\right)$ construct the rank one projection matrix $P = \boldsymbol{u}\boldsymbol{u}^{\mathrm{T}}$. This matrix has $P^2 = P$ because $\boldsymbol{u}^{\mathrm{T}}\boldsymbol{u} = 1$.

(a) $P\boldsymbol{u} = \boldsymbol{u}$ comes from $(\boldsymbol{u}\boldsymbol{u}^{\mathrm{T}})\boldsymbol{u} = \boldsymbol{u}(\_\_\_\_)$. Then $\boldsymbol{u}$ is an eigenvector with $\lambda = 1$.

(b) If $\boldsymbol{v}$ is perpendicular to $\boldsymbol{u}$ show that $P\boldsymbol{v} = \mathbf{0}$. Then $\lambda = 0$.

(c) Find three independent eigenvectors of P all with eigenvalue $\lambda = 0$.

14 Solve $\det(Q - \lambda I) = 0$ by the quadratic formula to reach $\lambda = \cos\theta \pm i \sin\theta$:

$$Q = \begin{bmatrix} \cos\theta & -\sin\theta \\ \sin\theta & \cos\theta \end{bmatrix} \quad \text{rotates the } xy \text{ plane by the angle } \theta. \text{ No real } \lambda\text{'s.}$$

Find the eigenvectors of Q by solving $(Q - \lambda I)\boldsymbol{x} = \mathbf{0}$. Use $i^2 = -1$.

15 Every permutation matrix leaves $\boldsymbol{x} = (1, 1, \ldots, 1)$ unchanged. Then $\lambda = 1$. Find two more λ's (possibly complex) for these permutations, from $\det(P - \lambda I) = 0$:

$$P = \begin{bmatrix} 0 & 1 & 0 \\ 0 & 0 & 1 \\ 1 & 0 & 0 \end{bmatrix} \quad \text{and} \quad P = \begin{bmatrix} 0 & 0 & 1 \\ 0 & 1 & 0 \\ 1 & 0 & 0 \end{bmatrix}.$$

16 **The determinant of A equals the product $\lambda_1 \lambda_2 \cdots \lambda_n$.** Start with the polynomial $\det(A - \lambda I)$ separated into its n factors (always possible). Then set $\lambda = 0$:

$$\det(A - \lambda I) = (\lambda_1 - \lambda)(\lambda_2 - \lambda) \cdots (\lambda_n - \lambda) \quad \text{so} \quad \det A = \_\_\_\_.$$

Check this rule in Example 1 where the Markov matrix has $\lambda = 1$ and $\frac{1}{2}$.

17 The sum of the diagonal entries (the *trace*) equals the sum of the eigenvalues:

$$A = \begin{bmatrix} a & b \\ c & d \end{bmatrix} \quad \text{has} \quad \det(A - \lambda I) = \lambda^2 - (a + d)\lambda + ad - bc = 0.$$

The quadratic formula gives the eigenvalues $\lambda = (a + d + \sqrt{\quad})/2$ and $\lambda = \_\_\_\_$. Their sum is $\_\_\_\_$. If A has $\lambda_1 = 3$ and $\lambda_2 = 4$ then $\det(A - \lambda I) = \_\_\_\_$.

18 If A has $\lambda_1 = 4$ and $\lambda_2 = 5$ then $\det(A - \lambda I) = (\lambda - 4)(\lambda - 5) = \lambda^2 - 9\lambda + 20$. Find three matrices that have trace $a + d = 9$ and determinant 20 and $\lambda = 4, 5$.

19 A 3 by 3 matrix B is known to have eigenvalues $0, 1, 2$. This information is enough to find three of these (give the answers where possible):

(a) the rank of B

(b) the determinant of $B^{\mathrm{T}}B$

(c) the eigenvalues of $B^{\mathrm{T}}B$

(d) the eigenvalues of $(B^2 + I)^{-1}$.

20 Choose the last rows of A and C to give eigenvalues $4, 7$ and $1, 2, 3$:

Companion matrices $$A = \begin{bmatrix} 0 & 1 \\ * & * \end{bmatrix} \quad C = \begin{bmatrix} 0 & 1 & 0 \\ 0 & 0 & 1 \\ * & * & * \end{bmatrix}.$$

21 ***The eigenvalues of A equal the eigenvalues of*** A^{T}. This is because $\det(A - \lambda I)$ equals $\det(A^{\mathrm{T}} - \lambda I)$. That is true because ______. Show by an example that the eigenvectors of A and A^{T} are *not* the same.

22 Construct any 3 by 3 Markov matrix M: positive entries down each column add to 1. Show that $M^{\mathrm{T}}(1, 1, 1) = (1, 1, 1)$. By Problem 21, $\lambda = 1$ is also an eigenvalue of M. Challenge: A 3 by 3 singular Markov matrix with trace $\frac{1}{2}$ has what λ's ?

23 Find three 2 by 2 matrices that have $\lambda_1 = \lambda_2 = 0$. The trace is zero and the determinant is zero. A might not be the zero matrix but check that $A^2 = 0$.

24 This matrix is singular with rank one. Find three λ's and three eigenvectors:

$$A = \begin{bmatrix} 1 \\ 2 \\ 1 \end{bmatrix} \begin{bmatrix} 2 & 1 & 2 \end{bmatrix} = \begin{bmatrix} 2 & 1 & 2 \\ 4 & 2 & 4 \\ 2 & 1 & 2 \end{bmatrix}.$$

25 Suppose A and B have the same eigenvalues $\lambda_1, \ldots, \lambda_n$ with the same independent eigenvectors $\boldsymbol{x}_1, \ldots, \boldsymbol{x}_n$. Then $A = B$. *Reason*: Any vector $\boldsymbol{x}$ is a combination $c_1\boldsymbol{x}_1 + \cdots + c_n\boldsymbol{x}_n$. What is $A\boldsymbol{x}$? What is $B\boldsymbol{x}$?

26 The block B has eigenvalues $1, 2$ and C has eigenvalues $3, 4$ and D has eigenvalues $5, 7$. Find the eigenvalues of the 4 by 4 matrix A:

$$A = \begin{bmatrix} B & C \\ 0 & D \end{bmatrix} = \begin{bmatrix} 0 & 1 & 3 & 0 \\ -2 & 3 & 0 & 4 \\ 0 & 0 & 6 & 1 \\ 0 & 0 & 1 & 6 \end{bmatrix}.$$

27 Find the rank and the four eigenvalues of A and C:

$$A = \begin{bmatrix} 1 & 1 & 1 & 1 \\ 1 & 1 & 1 & 1 \\ 1 & 1 & 1 & 1 \\ 1 & 1 & 1 & 1 \end{bmatrix} \quad \text{and} \quad C = \begin{bmatrix} 1 & 0 & 1 & 0 \\ 0 & 1 & 0 & 1 \\ 1 & 0 & 1 & 0 \\ 0 & 1 & 0 & 1 \end{bmatrix}.$$

28 Subtract I from the previous A. Find the λ's and then the determinants of

$$B = A - I = \begin{bmatrix} 0 & 1 & 1 & 1 \\ 1 & 0 & 1 & 1 \\ 1 & 1 & 0 & 1 \\ 1 & 1 & 1 & 0 \end{bmatrix} \quad \text{and} \quad C = I - A = \begin{bmatrix} 0 & -1 & -1 & -1 \\ -1 & 0 & -1 & -1 \\ -1 & -1 & 0 & -1 \\ -1 & -1 & -1 & 0 \end{bmatrix}.$$

29 (Review) Find the eigenvalues of A, B, and C:

$$A = \begin{bmatrix} 1 & 2 & 3 \\ 0 & 4 & 5 \\ 0 & 0 & 6 \end{bmatrix} \quad \text{and} \quad B = \begin{bmatrix} 0 & 0 & 1 \\ 0 & 2 & 0 \\ 3 & 0 & 0 \end{bmatrix} \quad \text{and} \quad C = \begin{bmatrix} 2 & 2 & 2 \\ 2 & 2 & 2 \\ 2 & 2 & 2 \end{bmatrix}.$$

30 When $a + b = c + d$ show that $(1, 1)$ is an eigenvector and find both eigenvalues:

$$A = \begin{bmatrix} a & b \\ c & d \end{bmatrix}.$$

31 If we exchange rows 1 and 2 *and* columns 1 and 2, the eigenvalues don't change. Find eigenvectors of A and B for $\lambda = 11$. *Rank one gives* $\lambda_2 = \lambda_3 = 0$.

$$A = \begin{bmatrix} 1 & 2 & 1 \\ 3 & 6 & 3 \\ 4 & 8 & 4 \end{bmatrix} \quad \text{and} \quad B = PAP^{\mathrm{T}} = \begin{bmatrix} 6 & 3 & 3 \\ 2 & 1 & 1 \\ 8 & 4 & 4 \end{bmatrix}.$$

32 Suppose A has eigenvalues $0, 3, 5$ with independent eigenvectors $\boldsymbol{u}, \boldsymbol{v}, \boldsymbol{w}$.

(a) Give a basis for the nullspace and a basis for the column space.

(b) Find a particular solution to $A\boldsymbol{x} = \boldsymbol{v} + \boldsymbol{w}$. Find all solutions.

(c) $A\boldsymbol{x} = \boldsymbol{u}$ has no solution. If it did then _____ would be in the column space.

Challenge Problems

33 Show that $\boldsymbol{u}$ is an eigenvector of the rank one 2×2 matrix $A = \boldsymbol{u}\boldsymbol{v}^{\mathrm{T}}$. Find both eigenvalues of A. Check that $\lambda_1 + \lambda_2$ agrees with the trace $u_1 v_1 + u_2 v_2$.

34 Find the eigenvalues of this permutation matrix P from $\det(P - \lambda I) = 0$. Which vectors are not changed by the permutation? They are eigenvectors for $\lambda = 1$. Can you find three more eigenvectors?

$$P = \begin{bmatrix} 0 & 0 & 0 & 1 \\ 1 & 0 & 0 & 0 \\ 0 & 1 & 0 & 0 \\ 0 & 0 & 1 & 0 \end{bmatrix}.$$

35 There are six 3 by 3 permutation matrices P. What numbers can be the *determinants* of P? What numbers can be *pivots*? What numbers can be the *trace* of P? What *four numbers* can be eigenvalues of P, as in Problem 15?

36 (**Heisenberg's Uncertainty Principle**) $AB - BA = I$ can happen for infinite matrices with $A = A^{\mathrm{T}}$ and $B = -B^{\mathrm{T}}$. Then

$$\boldsymbol{x}^{\mathrm{T}}\boldsymbol{x} = \boldsymbol{x}^{\mathrm{T}}AB\boldsymbol{x} - \boldsymbol{x}^{\mathrm{T}}BA\boldsymbol{x} \leq 2\|A\boldsymbol{x}\|\,\|B\boldsymbol{x}\|.$$

Explain that last step by using the Schwarz inequality $|\boldsymbol{u}^{\mathrm{T}}\boldsymbol{v}| \leq \|\boldsymbol{u}\|\,\|\boldsymbol{v}\|$. Then Heisenberg's inequality says that $\|A\boldsymbol{x}\|/\|\boldsymbol{x}\|$ times $\|B\boldsymbol{x}\|/\|\boldsymbol{x}\|$ is at least $\frac{1}{2}$. It is impossible to get the position error and momentum error both very small.

37 *Find a* 2 *by* 2 *rotation matrix* (other than I) *with* $A^3 = I$. Its eigenvalues must satisfy $\lambda^3 = 1$. They can be $e^{2\pi i/3}$ and $e^{-2\pi i/3}$. What are the trace and determinant?

38 (a) Find the eigenvalues and eigenvectors of A. They depend on c:

$$A = \begin{bmatrix} .4 & 1-c \\ .6 & c \end{bmatrix}.$$

(b) Show that A has just one line of eigenvectors when $c = 1.6$.

(c) This is a Markov matrix when $c = .8$. Then A^n will approach what matrix A^{∞}?

Eigshow in MATLAB

There is a MATLAB demo (just type **eigshow**), displaying the eigenvalue problem for a 2 by 2 matrix. It starts with the unit vector $\boldsymbol{x} = (1,0)$. *The mouse makes this vector move around the unit circle.* At the same time the screen shows $A\boldsymbol{x}$, in color and also moving. Possibly $A\boldsymbol{x}$ is ahead of $\boldsymbol{x}$. Possibly $A\boldsymbol{x}$ is behind $\boldsymbol{x}$. *Sometimes $A\boldsymbol{x}$ is parallel to $\boldsymbol{x}$.*

At that parallel moment, $A\boldsymbol{x} = \lambda\boldsymbol{x}$ (at $\boldsymbol{x}_1$ and $\boldsymbol{x}_2$ in the second figure).

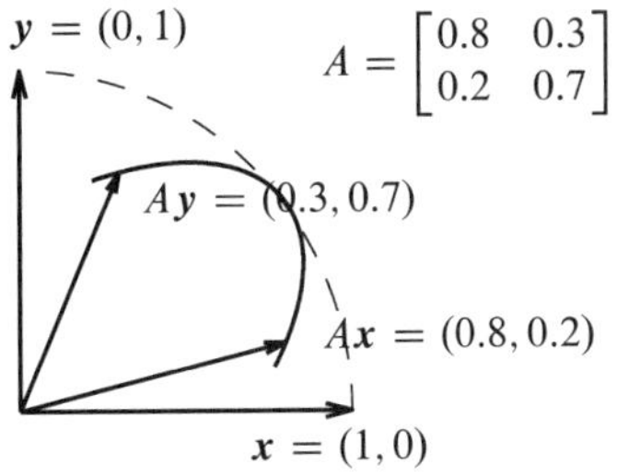

These are not eigenvectors

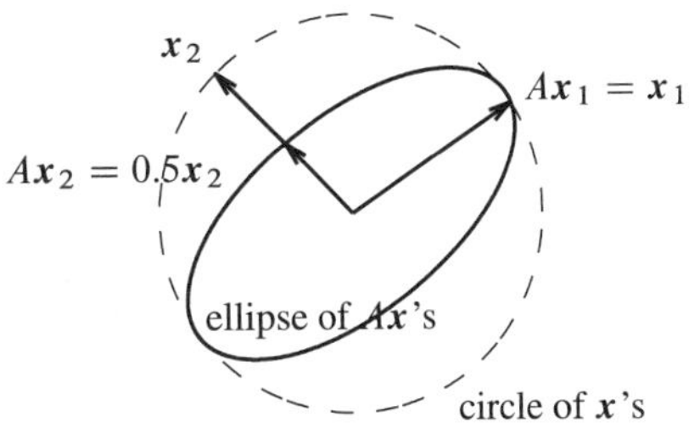

$A\boldsymbol{x}$ lines up with $\boldsymbol{x}$ at eigenvectors

The eigenvalue λ is the length of $A\boldsymbol{x}$, when the unit eigenvector $\boldsymbol{x}$ lines up. The built-in choices for A illustrate three possibilities: $0, 1$, or 2 real vectors where $A\boldsymbol{x}$ crosses $\boldsymbol{x}$.

The axes of the ellipse are **singular vectors** in 7.4—and eigenvectors if $A^{\mathrm{T}} = A$.

6.2 Diagonalizing a Matrix

1 The columns of $AX = X\Lambda$ are $A\boldsymbol{x}_k = \lambda_k \boldsymbol{x}_k$. The eigenvalue matrix Λ is diagonal.

2 $\boldsymbol{n}$ independent eigenvectors in X diagonalize A $\boxed{\boldsymbol{A = X\Lambda X^{-1}} \text{ and } \boldsymbol{\Lambda = X^{-1}AX}}$

3 The eigenvector matrix X also diagonalizes all powers A^k: $\boxed{\boldsymbol{A^k = X\Lambda^k X^{-1}}}$

4 Solve $\boldsymbol{u}_{k+1} = A\boldsymbol{u}_k$ by $\boldsymbol{u}_k = A^k\boldsymbol{u}_0 = X\Lambda^k X^{-1}\boldsymbol{u}_0 = \boxed{c_1(\lambda_1)^k\boldsymbol{x}_1 + \cdots + c_n(\lambda_n)^k\boldsymbol{x}_n}$

5 **No equal eigenvalues** $\Rightarrow$ X is invertible and A can be diagonalized.

Equal eigenvalues $\Rightarrow$ A *might* have too few independent eigenvectors. Then X^{-1} fails.

6 Every matrix $C = B^{-1}AB$ has the **same eigenvalues** as A. These C's are "**similar**" to A.

When $\boldsymbol{x}$ is an eigenvector, multiplication by A is just multiplication by a number λ: $A\boldsymbol{x} = \lambda\boldsymbol{x}$. All the difficulties of matrices are swept away. Instead of an interconnected system, we can follow the eigenvectors separately. It is like having a *diagonal matrix*, with no off-diagonal interconnections. The 100th power of a diagonal matrix is easy.

The point of this section is very direct. ***The matrix A turns into a diagonal matrix*** $\boldsymbol{\Lambda}$ ***when we use the eigenvectors properly***. This is the matrix form of our key idea. We start right off with that one essential computation. The next page explains why $\boldsymbol{AX = X\Lambda}$.

Diagonalization Suppose the n by n matrix A has n linearly independent eigenvectors $\boldsymbol{x}_1, \ldots, \boldsymbol{x}_n$. Put them into the columns of an ***eigenvector matrix X***. Then $X^{-1}AX$ is the ***eigenvalue matrix*** $\boldsymbol{\Lambda}$:

Eigenvector matrix X
Eigenvalue matrix Λ

$$\boldsymbol{X^{-1}AX = \Lambda} = \begin{bmatrix} \lambda_1 & & \\ & \ddots & \\ & & \lambda_n \end{bmatrix}. \qquad (1)$$

The matrix A is "diagonalized." We use capital lambda for the eigenvalue matrix, because the small λ's (the eigenvalues) are on its diagonal.

Example 1 This A is triangular so its eigenvalues are on the diagonal: $\lambda = 1$ and $\lambda = 6$.

Eigenvectors go into X $\begin{bmatrix} 1 \\ 0 \end{bmatrix} \begin{bmatrix} 1 \\ 1 \end{bmatrix}$

$$\underset{X^{-1}}{\begin{bmatrix} 1 & -1 \\ 0 & 1 \end{bmatrix}} \underset{A}{\begin{bmatrix} \mathbf{1} & \mathbf{5} \\ 0 & \mathbf{6} \end{bmatrix}} \underset{X}{\begin{bmatrix} 1 & 1 \\ 0 & 1 \end{bmatrix}} = \underset{\Lambda}{\begin{bmatrix} \mathbf{1} & 0 \\ 0 & \mathbf{6} \end{bmatrix}}$$

In other words $A = X\Lambda X^{-1}$. Then watch $A^2 = X\Lambda X^{-1}X\Lambda X^{-1}$. So $\boldsymbol{A^2}$ **is** $\boldsymbol{X\Lambda^2 X^{-1}}$.

A^2 has the same eigenvectors in X and squared eigenvalues in Λ^2.

Why is $AX = X\Lambda$? A multiplies its eigenvectors, which are the columns of X. The first column of AX is $A\boldsymbol{x}_1$. That is $\lambda_1\boldsymbol{x}_1$. Each column of X is multiplied by its eigenvalue:

$$\textbf{A times X} \qquad AX = A\begin{bmatrix} \boldsymbol{x}_1 & \cdots & \boldsymbol{x}_n \end{bmatrix} = \begin{bmatrix} \lambda_1\boldsymbol{x}_1 & \cdots & \lambda_n\boldsymbol{x}_n \end{bmatrix}.$$

The trick is to split this matrix AX into X times Λ:

$$\textbf{X times }\boldsymbol{\Lambda} \qquad \begin{bmatrix} \lambda_1\boldsymbol{x}_1 & \cdots & \lambda_n\boldsymbol{x}_n \end{bmatrix} = \begin{bmatrix} \boldsymbol{x}_1 & \cdots & \boldsymbol{x}_n \end{bmatrix}\begin{bmatrix} \lambda_1 & & \\ & \ddots & \\ & & \lambda_n \end{bmatrix} = X\Lambda.$$

Keep those matrices in the right order! Then λ_1 multiplies the first column $\boldsymbol{x}_1$, as shown. The diagonalization is complete, and we can write $AX = X\Lambda$ in two good ways:

$$AX = X\Lambda \quad \text{is} \quad \boldsymbol{X^{-1}AX = \Lambda} \quad \text{or} \quad \boldsymbol{A = X\Lambda X^{-1}}. \tag{2}$$

The matrix X has an inverse, because its columns (the eigenvectors of A) were assumed to be linearly independent. *Without n independent eigenvectors, we can't diagonalize.*

A and Λ have the same eigenvalues $\lambda_1, \ldots, \lambda_n$. The eigenvectors are different. The job of the original eigenvectors $\boldsymbol{x}_1, \ldots, \boldsymbol{x}_n$ was to diagonalize A. Those eigenvectors in X produce $A = X\Lambda X^{-1}$. You will soon see their simplicity and importance and meaning. The kth power will be $A^k = X\Lambda^k X^{-1}$ which is easy to compute:

$$\boldsymbol{A^k} = (\boldsymbol{X\Lambda X^{-1}})(\boldsymbol{X\Lambda X^{-1}})\ldots(\boldsymbol{X\Lambda X^{-1}}) = \boldsymbol{X\Lambda^k X^{-1}}.$$

Powers of A
Example 1

$$\begin{bmatrix} 1 & 5 \\ 0 & 6 \end{bmatrix}^k = \begin{bmatrix} 1 & 1 \\ 0 & 1 \end{bmatrix}\begin{bmatrix} 1 & \\ & 6^k \end{bmatrix}\begin{bmatrix} 1 & -1 \\ 0 & 1 \end{bmatrix} = \begin{bmatrix} \mathbf{1} & \boldsymbol{6^k - 1} \\ 0 & \boldsymbol{6^k} \end{bmatrix} = A^k.$$

With $k = 1$ we get A. With $k = 0$ we get $A^0 = I$ (and $\lambda^0 = 1$). *With $k = -1$ we get A^{-1}.* You can see how $A^2 = [1 \;\; 35; \;\; 0 \;\; 36]$ fits that formula when $k = 2$.

Here are four small remarks before we use Λ again in Example 2.

Remark 1 Suppose the eigenvalues $\lambda_1, \ldots, \lambda_n$ are all different. Then it is automatic that the eigenvectors $\boldsymbol{x}_1, \ldots, \boldsymbol{x}_n$ are independent. The eigenvector matrix X will be *invertible*. ***Any matrix that has no repeated eigenvalues can be diagonalized.***

Remark 2 ***We can multiply eigenvectors by any nonzero constants.*** $A(c\boldsymbol{x}) = \lambda(c\boldsymbol{x})$ is still true. In Example 1, we can divide $\boldsymbol{x} = (1, 1)$ by $\sqrt{2}$ to produce a unit vector.

MATLAB and virtually all other codes produce eigenvectors of length $||\boldsymbol{x}|| = 1$.

Remark 3 The eigenvectors in X come in the same order as the eigenvalues in Λ. To reverse the order in Λ, put the eigenvector $(1,1)$ before $(1,0)$ in X:

$$\textbf{New order } 6,1 \qquad \begin{bmatrix} 0 & 1 \\ 1 & -1 \end{bmatrix} \begin{bmatrix} 1 & 5 \\ 0 & 6 \end{bmatrix} \begin{bmatrix} 1 & 1 \\ 1 & 0 \end{bmatrix} = \begin{bmatrix} 6 & 0 \\ 0 & 1 \end{bmatrix} = \Lambda_{\text{new}}$$

To diagonalize A we *must* use an eigenvector matrix. From $X^{-1}AX = \Lambda$ we know that $AX = X\Lambda$. Suppose the first column of X is $\boldsymbol{x}$. Then the first columns of AX and $X\Lambda$ are $A\boldsymbol{x}$ and $\lambda_1 \boldsymbol{x}$. For those to be equal, $\boldsymbol{x}$ must be an eigenvector.

Remark 4 (repeated warning for repeated eigenvalues) Some matrices have too few eigenvectors. *Those matrices cannot be diagonalized.* Here are two examples:

$$\textbf{Not diagonalizable} \qquad A = \begin{bmatrix} 1 & -1 \\ 1 & -1 \end{bmatrix} \quad \textbf{and} \quad B = \begin{bmatrix} 0 & 1 \\ 0 & 0 \end{bmatrix}.$$

Their eigenvalues happen to be 0 and 0. Nothing is special about $\lambda = 0$, the problem is the repetition of λ. All eigenvectors of the first matrix are multiples of $(1,1)$:

$$\begin{array}{l}\textbf{Only one line} \\ \textbf{of eigenvectors}\end{array} \qquad A\boldsymbol{x} = 0\boldsymbol{x} \quad \text{means} \quad \begin{bmatrix} 1 & -1 \\ 1 & -1 \end{bmatrix} \begin{bmatrix} \boldsymbol{x} \end{bmatrix} = \begin{bmatrix} 0 \\ 0 \end{bmatrix} \quad \text{and} \quad \boldsymbol{x} = c\begin{bmatrix} 1 \\ 1 \end{bmatrix}.$$

There is no second eigenvector, so this unusual matrix A cannot be diagonalized.

Those matrices are the best examples to test any statement about eigenvectors. In many true-false questions, non-diagonalizable matrices lead to *false*.

Remember that there is no connection between invertibility and diagonalizability:

- ***Invertibility*** is concerned with the ***eigenvalues*** ($\lambda = 0$ or $\lambda \neq 0$).
- ***Diagonalizability*** is concerned with the ***eigenvectors*** (too few or enough for X).

Each eigenvalue has at least one eigenvector! $A - \lambda I$ is singular. If $(A - \lambda I)\boldsymbol{x} = \boldsymbol{0}$ leads you to $\boldsymbol{x} = \boldsymbol{0}$, λ is *not* an eigenvalue. Look for a mistake in solving $\det(A - \lambda I) = 0$.

Eigenvectors for n different λ's are independent. Then we can diagonalize A.

> **Independent $\boldsymbol{x}$ from different λ** Eigenvectors $\boldsymbol{x}_1, \ldots, \boldsymbol{x}_j$ that correspond to distinct (all different) eigenvalues are linearly independent. An n by n matrix that has n different eigenvalues (no repeated λ's) must be diagonalizable.

Proof Suppose $c_1\boldsymbol{x}_1 + c_2\boldsymbol{x}_2 = \boldsymbol{0}$. Multiply by A to find $c_1\lambda_1\boldsymbol{x}_1 + c_2\lambda_2\boldsymbol{x}_2 = \boldsymbol{0}$. Multiply by λ_2 to find $c_1\lambda_2\boldsymbol{x}_1 + c_2\lambda_2\boldsymbol{x}_2 = \boldsymbol{0}$. Now subtract one from the other:

$$\text{Subtraction leaves} \quad (\lambda_1 - \lambda_2)c_1\boldsymbol{x}_1 = \boldsymbol{0}. \quad \text{Therefore } c_1 = 0.$$

Since the λ's are different and $\boldsymbol{x}_1 \neq \boldsymbol{0}$, we are forced to the conclusion that $c_1 = 0$. Similarly $c_2 = 0$. Only the combination with $c_1 = c_2 = 0$ gives $c_1\boldsymbol{x}_1 + c_2\boldsymbol{x}_2 = \boldsymbol{0}$. So the eigenvectors $\boldsymbol{x}_1$ and $\boldsymbol{x}_2$ must be independent.

This proof extends directly to j eigenvectors. Suppose that $c_1\boldsymbol{x}_1 + \cdots + c_j\boldsymbol{x}_j = \mathbf{0}$. Multiply by A, multiply by λ_j, and subtract. This multiplies $\boldsymbol{x}_j$ by $\lambda_j - \lambda_j = 0$, and $\boldsymbol{x}_j$ is gone. Now multiply by A and by λ_{j-1} and subtract. This removes $\boldsymbol{x}_{j-1}$. Eventually only $\boldsymbol{x}_1$ is left:

$$\text{We reach} \quad (\lambda_1 - \lambda_2)\cdots(\lambda_1 - \lambda_j)c_1\boldsymbol{x}_1 = \mathbf{0} \quad \text{which forces} \quad c_1 = 0. \tag{3}$$

Similarly every $c_i = 0$. When the λ's are all different, the eigenvectors are independent. A full set of eigenvectors can go into the columns of the eigenvector matrix X.

Example 2 **Powers of A** The Markov matrix $A = \begin{bmatrix} .8 & .3 \\ .2 & .7 \end{bmatrix}$ in the last section had $\lambda_1 = 1$ and $\lambda_2 = .5$. Here is $A = X\Lambda X^{-1}$ with those eigenvalues in the diagonal Λ:

$$\textbf{Markov example} \qquad \begin{bmatrix} .8 & .3 \\ .2 & .7 \end{bmatrix} = \begin{bmatrix} .6 & 1 \\ .4 & -1 \end{bmatrix} \begin{bmatrix} 1 & 0 \\ 0 & .5 \end{bmatrix} \begin{bmatrix} 1 & 1 \\ .4 & -.6 \end{bmatrix} = X\Lambda X^{-1}.$$

The eigenvectors $(.6, .4)$ and $(1, -1)$ are in the columns of X. They are also the eigenvectors of A^2. Watch how A^2 has the same X, and ***the eigenvalue matrix of A^2 is Λ^2***:

$$\textbf{Same } X \textbf{ for } A^2 \qquad A^2 = X\Lambda X^{-1}X\Lambda X^{-1} = X\Lambda^2 X^{-1}. \tag{4}$$

Just keep going, and you see why the high powers A^k approach a "steady state":

$$\textbf{Powers of } A \qquad A^k = X\Lambda^k X^{-1} = \begin{bmatrix} .6 & 1 \\ .4 & -1 \end{bmatrix} \begin{bmatrix} 1^k & 0 \\ 0 & (.5)^k \end{bmatrix} \begin{bmatrix} 1 & 1 \\ .4 & -.6 \end{bmatrix}.$$

As k gets larger, $(.5)^k$ gets smaller. In the limit it disappears completely. That limit is A^∞:

$$\textbf{Limit } k \to \infty \qquad A^\infty = \begin{bmatrix} .6 & 1 \\ .4 & -1 \end{bmatrix} \begin{bmatrix} 1 & 0 \\ 0 & 0 \end{bmatrix} \begin{bmatrix} 1 & 1 \\ .4 & -.6 \end{bmatrix} = \begin{bmatrix} .6 & .6 \\ .4 & .4 \end{bmatrix}.$$

The limit has the eigenvector $\boldsymbol{x}_1$ in both columns. We saw this A^∞ on the very first page of Chapter 6. Now we see it coming from powers like $A^{100} = X\Lambda^{100}X^{-1}$.

Question ***When does $A^k \to$ zero matrix?*** Answer ***All $|\lambda| < 1$.***

Similar Matrices: Same Eigenvalues

Suppose the eigenvalue matrix Λ is fixed. As we change the eigenvector matrix X, we get a whole family of different matrices $A = X\Lambda X^{-1}$—*all with the same eigenvalues in* Λ. All those matrices A (with the same Λ) are called **similar**.

This idea extends to matrices that can't be diagonalized. Again we choose one constant matrix C (not necessarily Λ). And we look at the whole family of matrices $\boldsymbol{A = BCB^{-1}}$, allowing all invertible matrices B. Again those matrices A and C are called **similar**.

We are using C instead of Λ because C might not be diagonal. We are using B instead of X because the columns of B might not be eigenvectors. We only require that B is invertible—its columns can contain any basis for $\mathbf{R}^n$. The key fact about similar matrices stays true. **Similar matrices A and C have the same eigenvalues.**

All the matrices $A = BCB^{-1}$ are "similar." They all share the eigenvalues of C.

Proof Suppose $C\boldsymbol{x} = \lambda\boldsymbol{x}$. Then BCB^{-1} has the same eigenvalue λ with the new eigenvector $B\boldsymbol{x}$:

$$\textbf{\textit{Same } }\boldsymbol{\lambda} \qquad (BCB^{-1})(B\boldsymbol{x}) = BC\boldsymbol{x} = B\lambda\boldsymbol{x} = \lambda(B\boldsymbol{x}). \tag{5}$$

A fixed matrix C produces a family of similar matrices BCB^{-1}, allowing all B. When C is the identity matrix, the "family" is very small. The only member is $BIB^{-1} = I$. The identity matrix is the only diagonalizable matrix with all eigenvalues $\lambda = 1$.

The family is larger when $\lambda = 1$ and 1 *with only one eigenvector* (not diagonalizable). The simplest C is the *Jordan form*—to be developed in Section 8.3. All the similar A's have two parameters r and s, not both zero: always determinant $= 1$ and trace $= 2$.

$$C = \begin{bmatrix} \mathbf{1} & \mathbf{1} \\ 0 & \mathbf{1} \end{bmatrix} = \text{Jordan form gives } A = BCB^{-1} = \begin{bmatrix} 1 - rs & r^2 \\ -s^2 & 1 + rs \end{bmatrix}. \tag{6}$$

For an important example I will take eigenvalues $\lambda = 1$ and 0 (not repeated!). Now the whole family is diagonalizable with the same eigenvalue matrix Λ. We get every 2 by 2 matrix that has eigenvalues 1 and 0. The trace is 1 and the determinant is zero:

$$\begin{matrix}\textbf{All} \\ \textbf{similar}\end{matrix} \quad \Lambda = \begin{bmatrix} \mathbf{1} & 0 \\ 0 & \mathbf{0} \end{bmatrix} \quad A = \begin{bmatrix} 1 & 1 \\ 0 & 0 \end{bmatrix} \quad \text{or} \quad A = \begin{bmatrix} .5 & .5 \\ .5 & .5 \end{bmatrix} \quad \text{or any } A = \frac{\boldsymbol{x}\boldsymbol{y}^{\mathrm{T}}}{\boldsymbol{x}^{\mathrm{T}}\boldsymbol{y}}.$$

The family contains all matrices with $A^2 = A$, including $A = \Lambda$ when $B = I$. When A is symmetric these are also projection matrices. Eigenvalues 1 and 0 make life easy.

Fibonacci Numbers

We present a famous example, where eigenvalues tell how fast the Fibonacci numbers grow. ***Every new Fibonacci number is the sum of the two previous F's***:

The sequence $0, 1, 1, 2, 3, 5, 8, 13, \ldots$ ***comes from*** $F_{k+2} = F_{k+1} + F_k$.

These numbers turn up in a fantastic variety of applications. Plants and trees grow in a spiral pattern, and a pear tree has 8 growths for every 3 turns. For a willow those numbers can be 13 and 5. The champion is a sunflower of Daniel O'Connell, which had 233 seeds in 144 loops. Those are the Fibonacci numbers F_{13} and F_{12}. Our problem is more basic.

Problem: Find the Fibonacci number F_{100} The slow way is to apply the rule $F_{k+2} = F_{k+1} + F_k$ one step at a time. By adding $F_6 = 8$ to $F_7 = 13$ we reach $F_8 = 21$. Eventually we come to F_{100}. Linear algebra gives a better way.

The key is to begin with a matrix equation $\mathbf{u}_{k+1} = A\boldsymbol{u}_k$. That is a *one-step* rule for vectors, while Fibonacci gave a two-step rule for scalars. We match those rules by putting two Fibonacci numbers into a vector. Then you will see the matrix A.

$$\text{Let } \boldsymbol{u}_k = \begin{bmatrix} F_{k+1} \\ F_k \end{bmatrix}. \text{ The rule } \begin{array}{l} F_{k+2} = F_{k+1} + F_k \\ F_{k+1} = F_{k+1} \end{array} \text{ is } \boldsymbol{u}_{k+1} = \begin{bmatrix} \mathbf{1} & \mathbf{1} \\ \mathbf{1} & \mathbf{0} \end{bmatrix} \boldsymbol{u}_k. \tag{7}$$

Every step multiplies by $A = \left[\begin{smallmatrix} 1 & 1 \\ 1 & 0 \end{smallmatrix}\right]$. After 100 steps we reach $\boldsymbol{u}_{100} = A^{100}\boldsymbol{u}_0$:

$$\boldsymbol{u}_0 = \begin{bmatrix} 1 \\ 0 \end{bmatrix}, \quad \boldsymbol{u}_1 = \begin{bmatrix} 1 \\ 1 \end{bmatrix}, \quad \boldsymbol{u}_2 = \begin{bmatrix} 2 \\ 1 \end{bmatrix}, \quad \boldsymbol{u}_3 = \begin{bmatrix} 3 \\ 2 \end{bmatrix}, \quad \ldots, \quad \boldsymbol{u}_{100} = \begin{bmatrix} F_{101} \\ F_{100} \end{bmatrix}.$$

This problem is just right for eigenvalues. Subtract λ from the diagonal of A:

$$A - \lambda I = \begin{bmatrix} 1-\lambda & 1 \\ 1 & -\lambda \end{bmatrix} \quad \text{leads to} \quad \det(A - \lambda I) = \lambda^2 - \lambda - 1.$$

The equation $\lambda^2 - \lambda - 1 = 0$ is solved by the quadratic formula $\left(-b \pm \sqrt{b^2 - 4ac}\right)/2a$:

$$\textbf{Eigenvalues} \qquad \lambda_1 = \frac{1+\sqrt{5}}{2} \approx 1.618 \quad \text{and} \quad \lambda_2 = \frac{1-\sqrt{5}}{2} \approx -.618.$$

These eigenvalues lead to eigenvectors $\boldsymbol{x}_1 = (\lambda_1, 1)$ and $\boldsymbol{x}_2 = (\lambda_2, 1)$. Step 2 finds the combination of those eigenvectors that gives $\boldsymbol{u}_0 = (1, 0)$:

$$\begin{bmatrix} 1 \\ 0 \end{bmatrix} = \frac{1}{\lambda_1 - \lambda_2}\left(\begin{bmatrix} \lambda_1 \\ 1 \end{bmatrix} - \begin{bmatrix} \lambda_2 \\ 1 \end{bmatrix}\right) \quad \text{or} \quad \boldsymbol{u}_0 = \frac{\boldsymbol{x}_1 - \boldsymbol{x}_2}{\lambda_1 - \lambda_2}. \tag{8}$$

Step 3 multiplies $\boldsymbol{u}_0$ by A^{100} to find $\boldsymbol{u}_{100}$. The eigenvectors $\boldsymbol{x}_1$ and $\boldsymbol{x}_2$ stay separate! They are multiplied by $(\lambda_1)^{100}$ and $(\lambda_2)^{100}$:

$$\textbf{100 steps from } \boldsymbol{u}_0 \qquad \boldsymbol{u}_{100} = \frac{(\lambda_1)^{100}\boldsymbol{x}_1 - (\lambda_2)^{100}\boldsymbol{x}_2}{\lambda_1 - \lambda_2}. \tag{9}$$

We want F_{100} = second component of $\boldsymbol{u}_{100}$. The second components of $\boldsymbol{x}_1$ and $\boldsymbol{x}_2$ are 1. The difference between $\lambda_1 = (1+\sqrt{5})/2$ and $\lambda_2 = (1-\sqrt{5})/2$ is $\sqrt{5}$. And $\lambda_2^{100} \approx 0$.

$$\text{100th Fibonacci number} = \frac{\lambda_1^{100} - \lambda_2^{100}}{\lambda_1 - \lambda_2} = \text{nearest integer to } \frac{1}{\sqrt{5}}\left(\frac{1+\sqrt{5}}{2}\right)^{100}. \tag{10}$$

Every F_k is a whole number. The ratio F_{101}/F_{100} must be very close to the limiting ratio $\left(1 + \sqrt{5}\right)/2$. The Greeks called this number the *"golden mean"*. For some reason a rectangle with sides 1.618 and 1 looks especially graceful.

Matrix Powers A^k

Fibonacci's example is a typical difference equation $\boldsymbol{u}_{k+1} = A\mathbf{u}_k$. ***Each step multiplies by A.*** The solution is $\boldsymbol{u}_k = A^k\boldsymbol{u}_0$. We want to make clear how diagonalizing the matrix gives a quick way to compute A^k and find $\boldsymbol{u}_k$ in three steps.

The eigenvector matrix X produces $A = X\Lambda X^{-1}$. This is a factorization of the matrix, like $A = LU$ or $A = QR$. The new factorization is perfectly suited to computing powers, because ***every time*** $\boldsymbol{X^{-1}}$ ***multiplies X we get I***:

Powers of A $$A^k\boldsymbol{u}_0 = (X\Lambda X^{-1})\cdots(X\Lambda X^{-1})\boldsymbol{u}_0 = X\Lambda^k X^{-1}\boldsymbol{u}_0$$

I will split $X\Lambda^k X^{-1}\boldsymbol{u}_0$ into three steps that show how eigenvalues work:

1. Write $\boldsymbol{u}_0$ as a combination $c_1\boldsymbol{x}_1 + \cdots + c_n\boldsymbol{x}_n$ of the eigenvectors. Then $\boldsymbol{c} = X^{-1}\boldsymbol{u}_0$.

2. Multiply each eigenvector $\boldsymbol{x}_i$ by $(\lambda_i)^k$. Now we have $\Lambda^k X^{-1}\boldsymbol{u}_0$.

3. Add up the pieces $c_i(\lambda_i)^k\boldsymbol{x}_i$ to find the solution $\boldsymbol{u}_k = A^k\boldsymbol{u}_0$. This is $X\Lambda^k X^{-1}\boldsymbol{u}_0$.

Solution for $\boldsymbol{u}_{k+1} = A\boldsymbol{u}_k$ $$\boldsymbol{u}_k = A^k\boldsymbol{u}_0 = c_1(\lambda_1)^k\boldsymbol{x}_1 + \cdots + c_n(\lambda_n)^k\boldsymbol{x}_n. \tag{11}$$

In matrix language A^k equals $(X\Lambda X^{-1})^k$ which is X times Λ^k times X^{-1}. In Step 1, the eigenvectors in X lead to the c's in the combination $\boldsymbol{u}_0 = c_1\boldsymbol{x}_1 + \cdots + c_n\boldsymbol{x}_n$:

Step 1 $$\boldsymbol{u}_0 = \begin{bmatrix} & & \\ \boldsymbol{x}_1 & \cdots & \boldsymbol{x}_n \\ & & \end{bmatrix}\begin{bmatrix} c_1 \\ \vdots \\ c_n \end{bmatrix}. \quad \text{This says that} \quad \boldsymbol{u}_0 = X\mathbf{c}. \tag{12}$$

The coefficients in Step 1 are $\boldsymbol{c} = X^{-1}\boldsymbol{u}_0$. Then Step 2 multiplies by Λ^k. The final result $\boldsymbol{u}_k = \sum c_i(\lambda_i)^k\boldsymbol{x}_i$ in Step 3 is the product of X and Λ^k and $X^{-1}\boldsymbol{u}_0$:

$$A^k\boldsymbol{u}_0 = X\Lambda^k X^{-1}\boldsymbol{u}_0 = X\Lambda^k\boldsymbol{c} = \begin{bmatrix} & & \\ \boldsymbol{x}_1 & \cdots & \boldsymbol{x}_n \\ & & \end{bmatrix}\begin{bmatrix} (\lambda_1)^k & & \\ & \ddots & \\ & & (\lambda_n)^k \end{bmatrix}\begin{bmatrix} c_1 \\ \vdots \\ c_n \end{bmatrix}. \tag{13}$$

This result is exactly $\boldsymbol{u}_k = c_1(\lambda_1)^k\boldsymbol{x}_1 + \cdots + c_n(\lambda_n)^k\boldsymbol{x}_n$. It solves $\boldsymbol{u}_{k+1} = A\boldsymbol{u}_k$.

Example 3 Start from $\boldsymbol{u}_0 = (1, 0)$. Compute $A^k\boldsymbol{u}_0$ for this faster Fibonacci:

$$A = \begin{bmatrix} 1 & 2 \\ 1 & 0 \end{bmatrix} \quad \text{has} \quad \lambda_1 = \mathbf{2} \quad \text{and} \quad \boldsymbol{x}_1 = \begin{bmatrix} 2 \\ 1 \end{bmatrix}, \quad \lambda_2 = \mathbf{-1} \quad \text{and} \quad \boldsymbol{x}_2 = \begin{bmatrix} 1 \\ -1 \end{bmatrix}.$$

This matrix is like Fibonacci except the rule is changed to $F_{k+2} = F_{k+1} + \mathbf{2}F_k$. The new numbers start with $0, 1, 1, 3$. They grow faster because of $\lambda = 2$.

Find $u_k = A^k u_0$ in 3 steps $\quad u_0 = c_1 x_1 + c_2 x_2$ and $u_k = c_1(\lambda_1)^k x_1 + c_2(\lambda_2)^k x_2$

Step 1 $\quad u_0 = \begin{bmatrix}1\\0\end{bmatrix} = \frac{1}{3}\begin{bmatrix}2\\1\end{bmatrix} + \frac{1}{3}\begin{bmatrix}1\\-1\end{bmatrix}$ so $c_1 = c_2 = \frac{1}{3}$

Step 2 $\quad$ Multiply the two parts by $(\lambda_1)^k = 2^k$ and $(\lambda_2)^k = (-1)^k$

Step 3 $\quad$ Combine eigenvectors $c_1(\lambda_1)^k x_1$ and $c_2(\lambda_2)^k x_2$ into u_k:

$$\boldsymbol{u_k = A^k u_0} \qquad u_k = \frac{1}{3}2^k\begin{bmatrix}2\\1\end{bmatrix} + \frac{1}{3}(-1)^k\begin{bmatrix}1\\-1\end{bmatrix} = \begin{bmatrix}F_{k+1}\\F_k\end{bmatrix}.$$

The new number is $F_k = \boldsymbol{(2^k - (-1)^k)/3}$. After $0, 1, 1, 3$ comes $F_4 = 15/3 = 5$.

Behind these numerical examples lies a fundamental idea: ***Follow the eigenvectors***. In Section 6.3 this is the crucial link from linear algebra to differential equations (λ^k will become $e^{\lambda t}$). Chapter 8 sees the same idea as "transforming to an eigenvector basis." The best example of all is a ***Fourier series***, built from the eigenvectors e^{ikx} of d/dx.

Nondiagonalizable Matrices (Optional)

Suppose λ is an eigenvalue of A. We discover that fact in two ways:

1. **Eigenvectors** (geometric) There are nonzero solutions to $Ax = \lambda x$.
2. **Eigenvalues** (algebraic) The determinant of $A - \lambda I$ is zero.

The number λ may be a simple eigenvalue or a multiple eigenvalue, and we want to know its ***multiplicity***. Most eigenvalues have multiplicity $M = 1$ (simple eigenvalues). Then there is a single line of eigenvectors, and $\det(A - \lambda I)$ does not have a double factor.

For exceptional matrices, an eigenvalue can be ***repeated***. Then there are two different ways to count its multiplicity. Always GM $\leq$ AM for each λ:

1. (Geometric Multiplicity = GM) Count the **independent eigenvectors** for λ. Then GM is the dimension of the nullspace of $A - \lambda I$.

2. (Algebraic Multiplicity = AM) AM counts the **repetitions of $\boldsymbol{\lambda}$** among the eigenvalues. Look at the n roots of $\det(A - \lambda I) = 0$.

If A has $\lambda = 4, 4, 4$, then that eigenvalue has AM $= 3$ and GM $= 1, 2,$ or 3.

The following matrix A is the standard example of trouble. Its eigenvalue $\lambda = 0$ is repeated. It is a double eigenvalue (AM $= 2$) with only one eigenvector (GM $= 1$).

AM = 2
GM = 1 $\quad A = \begin{bmatrix}0 & 1\\0 & 0\end{bmatrix}$ has $\det(A - \lambda I) = \begin{vmatrix}-\lambda & 1\\0 & -\lambda\end{vmatrix} = \lambda^2.$ $\quad$ **$\boldsymbol{\lambda = 0, 0}$ but 1 eigenvector**

There "should" be two eigenvectors, because $\lambda^2 = 0$ has a double root. The double factor λ^2 makes AM $= 2$. But there is only one eigenvector $\boldsymbol{x} = (1, 0)$ and GM $= 1$. ***This shortage of eigenvectors when*** GM ***is below*** AM ***means that*** A ***is not diagonalizable.***

These three matrices all have the same shortage of eigenvectors. Their repeated eigenvalue is $\lambda = 5$. Traces are 10 and determinants are 25:

$$A = \begin{bmatrix} 5 & 1 \\ 0 & 5 \end{bmatrix} \quad \text{and} \quad A = \begin{bmatrix} 6 & -1 \\ 1 & 4 \end{bmatrix} \quad \text{and} \quad A = \begin{bmatrix} 7 & 2 \\ -2 & 3 \end{bmatrix}.$$

Those all have $\det(A - \lambda I) = (\lambda - 5)^2$. The algebraic multiplicity is AM $= 2$. But each $A - 5I$ has rank $r = 1$. The geometric multiplicity is GM $= 1$. There is only one line of eigenvectors for $\lambda = 5$, and these matrices are not diagonalizable.

■ REVIEW OF THE KEY IDEAS ■

1. If A has n independent eigenvectors $\boldsymbol{x}_1, \ldots, \boldsymbol{x}_n$, they go into the columns of X.

   A is diagonalized by X $\qquad X^{-1}AX = \Lambda \quad \text{and} \quad A = X\Lambda X^{-1}.$

2. The powers of A are $A^k = X\Lambda^k X^{-1}$. The eigenvectors in X are unchanged.

3. The eigenvalues of A^k are $(\lambda_1)^k, \ldots, (\lambda_n)^k$ in the matrix Λ^k.

4. The solution to $\boldsymbol{u}_{k+1} = A\boldsymbol{u}_k$ starting from $\boldsymbol{u}_0$ is $\boldsymbol{u}_k = A^k\boldsymbol{u}_0 = X\Lambda^k X^{-1}\boldsymbol{u}_0$:

   $$\boldsymbol{u}_k = c_1(\lambda_1)^k\boldsymbol{x}_1 + \cdots + c_n(\lambda_n)^k\boldsymbol{x}_n \quad \text{provided} \quad \boldsymbol{u}_0 = c_1\boldsymbol{x}_1 + \cdots + c_n\boldsymbol{x}_n.$$

   That shows Steps $1, 2, 3$ (c's from $X^{-1}\boldsymbol{u}_0$, λ^k from Λ^k, and $\boldsymbol{x}$'s from X)

5. A is diagonalizable if every eigenvalue has enough eigenvectors (GM = AM).

■ WORKED EXAMPLES ■

6.2 A The **Lucas numbers** are like the Fibonacci numbers except they start with $L_1 = 1$ and $L_2 = 3$. Using the same rule $L_{k+2} = L_{k+1} + L_k$, the next Lucas numbers are $4, 7, 11, 18$. Show that the Lucas number L_{100} is $\lambda_1^{100} + \lambda_2^{100}$.

Solution $u_{k+1} = \begin{bmatrix} 1 & 1 \\ 1 & 0 \end{bmatrix} u_k$ is the same as for Fibonacci, because $L_{k+2} = L_{k+1} + L_k$ is the same rule (with different starting values). The equation becomes a 2 by 2 system:

$$\text{Let } \ u_k = \begin{bmatrix} L_{k+1} \\ L_k \end{bmatrix}. \qquad \text{The rule } \begin{array}{l} L_{k+2} = L_{k+1} + L_k \\ L_{k+1} = L_{k+1} \end{array} \ \text{ is } \ u_{k+1} = \begin{bmatrix} 1 & 1 \\ 1 & 0 \end{bmatrix} u_k.$$

The eigenvalues and eigenvectors of $A = \begin{bmatrix} 1 & 1 \\ 1 & 0 \end{bmatrix}$ still come from $\lambda^2 = \lambda + 1$:

$$\lambda_1 = \frac{1+\sqrt{5}}{2} \quad \text{and} \quad x_1 = \begin{bmatrix} \lambda_1 \\ 1 \end{bmatrix} \qquad \lambda_2 = \frac{1-\sqrt{5}}{2} \quad \text{and} \quad x_2 = \begin{bmatrix} \lambda_2 \\ 1 \end{bmatrix}.$$

Now solve $c_1 x_1 + c_2 x_2 = u_1 = (3, 1)$. The solution is $c_1 = \lambda_1$ and $c_2 = \lambda_2$. Check:

$$\lambda_1 x_1 + \lambda_2 x_2 = \begin{bmatrix} \lambda_1^2 + \lambda_2^2 \\ \lambda_1 + \lambda_2 \end{bmatrix} = \begin{bmatrix} \text{trace of } A^2 \\ \text{trace of } A \end{bmatrix} = \begin{bmatrix} 3 \\ 1 \end{bmatrix} = u_1$$

$u_{100} = A^{99} u_1$ tells us the Lucas numbers (L_{101}, L_{100}). The second components of the eigenvectors x_1 and x_2 are 1, so the second component of u_{100} is the answer we want:

$$\textbf{Lucas number} \qquad L_{100} = c_1 \lambda_1^{99} + c_2 \lambda_2^{99} = \boldsymbol{\lambda_1^{100} + \lambda_2^{100}}.$$

Lucas starts faster than Fibonacci, and ends up larger by a factor near $\sqrt{5}$.

6.2 B Find the inverse and the eigenvalues and the determinant of this matrix A:

$$A = 5 * \textbf{eye}(4) - \textbf{ones}(4) = \begin{bmatrix} 4 & -1 & -1 & -1 \\ -1 & 4 & -1 & -1 \\ -1 & -1 & 4 & -1 \\ -1 & -1 & -1 & 4 \end{bmatrix}.$$

Describe an eigenvector matrix X that gives $X^{-1} A X = \Lambda$.

Solution What are the eigenvalues of the all-ones matrix? Its rank is certainly 1, so three eigenvalues are $\lambda = 0, 0, 0$. Its trace is 4, so the other eigenvalue is $\lambda = 4$. Subtract this all-ones matrix from $5I$ to get our matrix A:

Subtract the eigenvalues $4, 0, 0, 0$ from $5, 5, 5, 5$. The eigenvalues of A are $1, 5, 5, 5$.

The determinant of A is 125, the product of those four eigenvalues. The eigenvector for $\lambda = 1$ is $x = (1, 1, 1, 1)$ or (c, c, c, c). The other eigenvectors are perpendicular to x (since A is symmetric). The nicest eigenvector matrix X is the symmetric orthogonal **Hadamard matrix H** The factor $\frac{1}{2}$ produces unit column vectors.

$$\textbf{Orthonormal eigenvectors} \qquad X = H = \frac{1}{2} \begin{bmatrix} 1 & 1 & 1 & 1 \\ 1 & -1 & 1 & -1 \\ 1 & 1 & -1 & -1 \\ 1 & -1 & -1 & 1 \end{bmatrix} = H^{\mathrm{T}} = H^{-1}.$$

The eigenvalues of A^{-1} are $1, \frac{1}{5}, \frac{1}{5}, \frac{1}{5}$. The eigenvectors are not changed so $A^{-1} = H\Lambda^{-1}H^{-1}$. The inverse matrix is surprisingly neat:

$$A^{-1} = \frac{1}{5} * (\mathbf{eye}(4) + \mathbf{ones}(4)) = \frac{1}{5}\begin{bmatrix} 2 & 1 & 1 & 1 \\ 1 & 2 & 1 & 1 \\ 1 & 1 & 2 & 1 \\ 1 & 1 & 1 & 2 \end{bmatrix}$$

A is a rank-one change from $5I$. So A^{-1} is a rank-one change from $I/5$.

In a graph with 5 nodes, the determinant 125 counts the "spanning trees" (trees that touch all nodes). *Trees have no loops* (graphs and trees are in Section 10.1).

With 6 nodes, the matrix $6 * \mathbf{eye}(5) - \mathbf{ones}(5)$ has the five eigenvalues $1, 6, 6, 6, 6$.

Problem Set 6.2

Questions 1–7 are about the eigenvalue and eigenvector matrices Λ and X.

1 (a) Factor these two matrices into $A = X\Lambda X^{-1}$:

$$A = \begin{bmatrix} 1 & 2 \\ 0 & 3 \end{bmatrix} \quad \text{and} \quad A = \begin{bmatrix} 1 & 1 \\ 3 & 3 \end{bmatrix}.$$

(b) If $A = X\Lambda X^{-1}$ then $A^3 = (\quad)(\quad)(\quad)$ and $A^{-1} = (\quad)(\quad)(\quad)$.

2 If A has $\lambda_1 = 2$ with eigenvector $\boldsymbol{x}_1 = \begin{bmatrix} 1 \\ 0 \end{bmatrix}$ and $\lambda_2 = 5$ with $\boldsymbol{x}_2 = \begin{bmatrix} 1 \\ 1 \end{bmatrix}$, use $X\Lambda X^{-1}$ to find A. No other matrix has the same λ's and $\boldsymbol{x}$'s.

3 Suppose $A = X\Lambda X^{-1}$. What is the eigenvalue matrix for $A + 2I$? What is the eigenvector matrix? Check that $A + 2I = (\quad)(\quad)(\quad)^{-1}$.

4 True or false: If the columns of X (eigenvectors of A) are linearly independent, then

(a) A is invertible (b) A is diagonalizable

(c) X is invertible (d) X is diagonalizable.

5 If the eigenvectors of A are the columns of I, then A is a _____ matrix. If the eigenvector matrix X is triangular, then X^{-1} is triangular. Prove that A is also triangular.

6 Describe all matrices X that diagonalize this matrix A (find all eigenvectors):

$$A = \begin{bmatrix} 4 & 0 \\ 1 & 2 \end{bmatrix}.$$

Then describe all matrices that diagonalize A^{-1}.

7 Write down the most general matrix that has eigenvectors $\begin{bmatrix} 1 \\ 1 \end{bmatrix}$ and $\begin{bmatrix} 1 \\ -1 \end{bmatrix}$.

Questions 8–10 are about Fibonacci and Gibonacci numbers.

8 Diagonalize the Fibonacci matrix by completing X^{-1}:

$$\begin{bmatrix} 1 & 1 \\ 1 & 0 \end{bmatrix} = \begin{bmatrix} \lambda_1 & \lambda_2 \\ 1 & 1 \end{bmatrix} \begin{bmatrix} \lambda_1 & 0 \\ 0 & \lambda_2 \end{bmatrix} \begin{bmatrix} & \\ & \end{bmatrix}.$$

Do the multiplication $X\Lambda^k X^{-1}\left[\begin{smallmatrix} 1 \\ 0 \end{smallmatrix}\right]$ to find its second component. This is the kth Fibonacci number $F_k = \left(\lambda_1^k - \lambda_2^k\right)/\left(\lambda_1 - \lambda_2\right)$.

9 Suppose G_{k+2} is the *average* of the two previous numbers G_{k+1} and G_k:

$$\begin{aligned} G_{k+2} &= \tfrac{1}{2}G_{k+1} + \tfrac{1}{2}G_k \\ G_{k+1} &= G_{k+1} \end{aligned} \quad \text{is} \quad \begin{bmatrix} G_{k+2} \\ G_{k+1} \end{bmatrix} = \begin{bmatrix} & A & \end{bmatrix} \begin{bmatrix} G_{k+1} \\ G_k \end{bmatrix}.$$

(a) Find the eigenvalues and eigenvectors of A.

(b) Find the limit as $n \to \infty$ of the matrices $A^n = X\Lambda^n X^{-1}$.

(c) If $G_0 = 0$ and $G_1 = 1$ show that the Gibonacci numbers approach $\frac{2}{3}$.

10 Prove that every third Fibonacci number in $0, 1, 1, 2, 3, \ldots$ is even.

Questions 11–14 are about diagonalizability.

11 True or false: If the eigenvalues of A are $2, 2, 5$ then the matrix is certainly

(a) invertible (b) diagonalizable (c) not diagonalizable.

12 True or false: If the only eigenvectors of A are multiples of $(1, 4)$ then A has

(a) no inverse (b) a repeated eigenvalue (c) no diagonalization $X\Lambda X^{-1}$.

13 Complete these matrices so that $\det A = 25$. Then check that $\lambda = 5$ is repeated—the trace is 10 so the determinant of $A - \lambda I$ is $(\lambda - 5)^2$. Find an eigenvector with $A\boldsymbol{x} = 5\boldsymbol{x}$. These matrices will not be diagonalizable because there is no second line of eigenvectors.

$$A = \begin{bmatrix} 8 & \\ & 2 \end{bmatrix} \quad \text{and} \quad A = \begin{bmatrix} 9 & 4 \\ & 1 \end{bmatrix} \quad \text{and} \quad A = \begin{bmatrix} 10 & 5 \\ -5 & \end{bmatrix}$$

14 The matrix $A = \left[\begin{smallmatrix} 3 & 1 \\ 0 & 3 \end{smallmatrix}\right]$ is not diagonalizable because the rank of $A - 3I$ is _____. Change one entry to make A diagonalizable. Which entries could you change?

Questions 15–19 are about powers of matrices.

15 $A^k = X\Lambda^k X^{-1}$ approaches the zero matrix as $k \to \infty$ if and only if every λ has absolute value less than _____. Which of these matrices has $A^k \to 0$?

$$A_1 = \begin{bmatrix} .6 & .9 \\ .4 & .1 \end{bmatrix} \quad \text{and} \quad A_2 = \begin{bmatrix} .6 & .9 \\ .1 & .6 \end{bmatrix}.$$

16 (Recommended) Find Λ and X to diagonalize A_1 in Problem 15. What is the limit of Λ^k as $k \to \infty$? What is the limit of $X\Lambda^k X^{-1}$? In the columns of this limiting matrix you see the ______.

17 Find Λ and X to diagonalize A_2 in Problem 15. What is $(A_2)^{10}\boldsymbol{u}_0$ for these $\boldsymbol{u}_0$?

$$\boldsymbol{u}_0 = \begin{bmatrix} 3 \\ 1 \end{bmatrix} \quad \text{and} \quad \boldsymbol{u}_0 = \begin{bmatrix} 3 \\ -1 \end{bmatrix} \quad \text{and} \quad \boldsymbol{u}_0 = \begin{bmatrix} 6 \\ 0 \end{bmatrix}.$$

18 Diagonalize A and compute $X\Lambda^k X^{-1}$ to prove this formula for A^k:

$$A = \begin{bmatrix} 2 & -1 \\ -1 & 2 \end{bmatrix} \quad \text{has} \quad A^k = \frac{1}{2}\begin{bmatrix} 1+3^k & 1-3^k \\ 1-3^k & 1+3^k \end{bmatrix}.$$

19 Diagonalize B and compute $X\Lambda^k X^{-1}$ to prove this formula for B^k:

$$B = \begin{bmatrix} 5 & 1 \\ 0 & 4 \end{bmatrix} \quad \text{has} \quad B^k = \begin{bmatrix} 5^k & 5^k - 4^k \\ 0 & 4^k \end{bmatrix}.$$

20 Suppose $A = X\Lambda X^{-1}$. Take determinants to prove $\det A = \det \Lambda = \lambda_1\lambda_2 \cdots \lambda_n$. This quick proof only works when A can be ______.

21 Show that trace XY = trace YX, by adding the diagonal entries of XY and YX:

$$X = \begin{bmatrix} a & b \\ c & d \end{bmatrix} \quad \text{and} \quad Y = \begin{bmatrix} q & r \\ s & t \end{bmatrix}.$$

Now choose Y to be ΛX^{-1}. Then $X\Lambda X^{-1}$ has the same trace as $\Lambda X^{-1}X = \Lambda$. This proves that *the trace of A equals the trace of Λ = sum of the eigenvalues.*

22 $AB - BA = I$ is impossible since the left side has trace = ______. But find an elimination matrix so that $A = E$ and $B = E^{\mathrm{T}}$ give

$$AB - BA = \begin{bmatrix} -1 & 0 \\ 0 & 1 \end{bmatrix} \quad \text{which has trace zero.}$$

23 If $A = X\Lambda X^{-1}$, diagonalize the block matrix $B = \begin{bmatrix} A & 0 \\ 0 & 2A \end{bmatrix}$. Find its eigenvalue and eigenvector (block) matrices.

24 Consider all 4 by 4 matrices A that are diagonalized by the same fixed eigenvector matrix X. Show that the A's form a subspace (cA and $A_1 + A_2$ have this same X). What is this subspace when $X = I$? What is its dimension?

25 Suppose $A^2 = A$. On the left side A multiplies each column of A. Which of our four subspaces contains eigenvectors with $\lambda = 1$? Which subspace contains eigenvectors with $\lambda = 0$? From the dimensions of those subspaces, A has a full set of independent eigenvectors. So a matrix with $A^2 = A$ can be diagonalized.

26 (Recommended) Suppose $A\boldsymbol{x} = \lambda\boldsymbol{x}$. If $\lambda = 0$ then $\boldsymbol{x}$ is in the nullspace. If $\lambda \neq 0$ then $\boldsymbol{x}$ is in the column space. Those spaces have dimensions $(n - r) + r = n$. So why doesn't every square matrix have n linearly independent eigenvectors?

27 The eigenvalues of A are 1 and 9, and the eigenvalues of B are -1 and 9:

$$A = \begin{bmatrix} 5 & 4 \\ 4 & 5 \end{bmatrix} \quad \text{and} \quad B = \begin{bmatrix} 4 & 5 \\ 5 & 4 \end{bmatrix}.$$

Find a matrix square root of A from $R = X\sqrt{\Lambda}\,X^{-1}$. Why is there no real matrix square root of B?

28 If A and B have the same λ's with the same independent eigenvectors, their factorizations into _____ are the same. So $A = B$.

29 Suppose the same X diagonalizes both A and B. They have the *same eigenvectors* in $A = X\Lambda_1X^{-1}$ and $B = X\Lambda_2X^{-1}$. Prove that $AB = BA$.

30 (a) If $A = \left[\begin{smallmatrix} \mathbf{a} & \mathbf{b} \\ \mathbf{0} & \mathbf{d} \end{smallmatrix}\right]$ then the determinant of $A - \lambda I$ is $(\lambda - a)(\lambda - d)$. Check the "Cayley-Hamilton Theorem" that $(A - aI)(A - dI) = \textit{zero matrix}$.

(b) Test the Cayley-Hamilton Theorem on Fibonacci's $A = \left[\begin{smallmatrix} 1 & 1 \\ 1 & 0 \end{smallmatrix}\right]$. The theorem predicts that $A^2 - A - I = 0$, since the polynomial $\det(A - \lambda I)$ is $\lambda^2 - \lambda - 1$.

31 Substitute $A = X\Lambda X^{-1}$ into the product $(A - \lambda_1 I)(A - \lambda_2 I)\cdots(A - \lambda_n I)$ and explain why this produces the zero matrix. We are substituting the matrix A for the number λ in the polynomial $p(\lambda) = \det(A - \lambda I)$. The ***Cayley-Hamilton Theorem*** says that this product is always $p(A) = \textit{zero matrix}$, even if A is not diagonalizable.

32 If $A = \left[\begin{smallmatrix} 1 & 0 \\ 0 & 2 \end{smallmatrix}\right]$ and $AB = BA$, show that $B = \left[\begin{smallmatrix} \mathbf{a} & \mathbf{b} \\ \mathbf{c} & \mathbf{d} \end{smallmatrix}\right]$ is also a diagonal matrix. B has the same eigen_____ as A but different eigen_____ . These diagonal matrices B form a two-dimensional subspace of matrix space. $AB - BA = 0$ gives four equations for the unknowns $\mathbf{a}, \mathbf{b}, \mathbf{c}, \mathbf{d}$—find the rank of the 4 by 4 matrix.

33 The powers A^k approach zero if all $|\lambda_i| < 1$ and they blow up if any $|\lambda_i| > 1$. Peter Lax gives these striking examples in his book *Linear Algebra*:

$$A = \begin{bmatrix} 3 & 2 \\ 1 & 4 \end{bmatrix} \qquad B = \begin{bmatrix} 3 & 2 \\ -5 & -3 \end{bmatrix} \qquad C = \begin{bmatrix} 5 & 7 \\ -3 & -4 \end{bmatrix} \qquad D = \begin{bmatrix} 5 & 6.9 \\ -3 & -4 \end{bmatrix}$$

$$\|\boldsymbol{A}^{1024}\| > \mathbf{10^{700}} \qquad \boldsymbol{B}^{1024} = \boldsymbol{I} \qquad \boldsymbol{C}^{1024} = -\boldsymbol{C} \qquad \|\boldsymbol{D}^{1024}\| < \mathbf{10^{-78}}$$

Find the eigenvalues $\lambda = e^{i\theta}$ of B and C to show $B^4 = I$ and $C^3 = -I$.

Challenge Problems

34 The nth power of rotation through θ is rotation through $n\theta$:

$$A^n = \begin{bmatrix} \cos\theta & -\sin\theta \\ \sin\theta & \cos\theta \end{bmatrix}^n = \begin{bmatrix} \cos n\theta & -\sin n\theta \\ \sin n\theta & \cos n\theta \end{bmatrix}.$$

Prove that neat formula by diagonalizing $A = X\Lambda X^{-1}$. The eigenvectors (columns of X) are $(1, i)$ and $(i, 1)$. You need to know Euler's formula $e^{i\theta} = \cos\theta + i\sin\theta$.

35 The transpose of $A = X\Lambda X^{-1}$ is $A^{\mathrm{T}} = (X^{-1})^{\mathrm{T}}\Lambda X^{\mathrm{T}}$. The eigenvectors in $A^{\mathrm{T}}\boldsymbol{y} = \lambda\boldsymbol{y}$ are the columns of that matrix $(X^{-1})^{\mathrm{T}}$. They are often called ***left eigenvectors of A***, because $\boldsymbol{y}^{\mathrm{T}}A = \lambda\boldsymbol{y}^{\mathrm{T}}$. How do you multiply matrices to find this formula for A?

Sum of rank-1 matrices $\quad A = X\Lambda X^{-1} = \lambda_1\boldsymbol{x}_1\boldsymbol{y}_1^{\mathrm{T}} + \cdots + \lambda_n\boldsymbol{x}_n\boldsymbol{y}_n^{\mathrm{T}}.$

36 The inverse of $A = \textbf{eye}(n) + \textbf{ones}(n)$ is $A^{-1} = \textbf{eye}(n) + C * \textbf{ones}(n)$. Multiply AA^{-1} to find that number C (depending on n).

37 Suppose A_1 and A_2 are n by n invertible matrices. What matrix B shows that $A_2A_1 = B(A_1A_2)B^{-1}$? Then A_2A_1 is similar to A_1A_2: *same eigenvalues.*

38 **When is a matrix A similar to its eigenvalue matrix Λ?**

A and Λ always have the same eigenvalues. But similarity requires a matrix B with $A = B\Lambda B^{-1}$. Then B is the ______ matrix and A must have n independent ______.

39 (Pavel Grinfeld) Without writing down any calculations, can you find the eigenvalues of this matrix? Can you find the 2017th power A^{2017}?

$$A = \begin{bmatrix} 110 & 55 & -164 \\ 42 & 21 & -62 \\ 88 & 44 & -131 \end{bmatrix}.$$

If A is m by n and B is n by m, then AB and BA have same nonzero eigenvalues.

Proof. Start with this identity between square matrices (easily checked). The first and third matrices are inverses. The "size matrix" shows the shapes of all blocks.

$$\begin{bmatrix} I & -A \\ 0 & I \end{bmatrix}\begin{bmatrix} AB & 0 \\ B & 0 \end{bmatrix}\begin{bmatrix} I & A \\ 0 & I \end{bmatrix} = \begin{bmatrix} 0 & 0 \\ B & BA \end{bmatrix} \qquad \begin{bmatrix} m\times m & m\times n \\ n\times m & n\times n \end{bmatrix}$$

This equation $D^{-1}ED = F$ says $\boldsymbol{F}$ **is similar to** $\boldsymbol{E}$—they have the *same* $m+n$ eigenvalues.

$E = \begin{bmatrix} AB & 0 \\ B & 0 \end{bmatrix}$ has the m eigenvalues of AB, plus n zeros

$F = \begin{bmatrix} 0 & 0 \\ B & BA \end{bmatrix}$ has the n eigenvalues of BA, plus m zeros

So AB and BA have the **same eigenvalues** except for $|n - m|$ zeros. Wow.

If $A = [1\ \ 1]$ and $B = A^{\mathrm{T}}$ then $A^{\mathrm{T}}A = \begin{bmatrix} \mathbf{1} & \mathbf{1} \\ \mathbf{1} & \mathbf{1} \end{bmatrix}$ (notice $\lambda = 2$ and 0) and $AA^{\mathrm{T}} = [\ 2\ ]$.

6.3 Systems of Differential Equations

1 If $A\boldsymbol{x} = \lambda\boldsymbol{x}$ then $\boldsymbol{u}(t) = e^{\lambda t}\boldsymbol{x}$ will solve $\dfrac{d\boldsymbol{u}}{dt} = A\boldsymbol{u}$. Each λ and $\boldsymbol{x}$ give a solution $e^{\lambda t}\boldsymbol{x}$.

2 If $A = X\Lambda X^{-1}$ then $\boxed{\boldsymbol{u}(t) = e^{At}\boldsymbol{u}(0) = Xe^{\Lambda t}X^{-1}\boldsymbol{u}(0) = c_1 e^{\lambda_1 t}\boldsymbol{x}_1 + \cdots + c_n e^{\lambda_n t}\boldsymbol{x}_n.}$

3 A is **stable** and $\boldsymbol{u}(t) \to \boldsymbol{0}$ and $e^{At} \to 0$ when all eigenvalues of A have real part < 0.

4 **Matrix exponential** $e^{At} = I + At + \cdots + (At)^n/n! + \cdots = Xe^{\Lambda t}X^{-1}$ if A is diagonalizable.

5 **Second order equation** / **First order system** $u'' + Bu' + Cu = 0$ is equivalent to $\begin{bmatrix} u \\ u' \end{bmatrix}' = \begin{bmatrix} 0 & 1 \\ -C & -B \end{bmatrix}\begin{bmatrix} u \\ u' \end{bmatrix}$.

Eigenvalues and eigenvectors and $A = X\Lambda X^{-1}$ are perfect for matrix powers A^k. They are also perfect for differential equations $d\boldsymbol{u}/dt = A\boldsymbol{u}$. This section is mostly linear algebra, but to read it you need one fact from calculus: *The derivative of* $e^{\lambda t}$ *is* $\lambda e^{\lambda t}$. The whole point of the section is this: **To convert constant-coefficient differential equations into linear algebra.**

The ordinary equations $\dfrac{du}{dt} = u$ and $\dfrac{du}{dt} = \lambda u$ are solved by exponentials:

$$\frac{du}{dt} = u \;\text{ produces }\; u(t) = \boldsymbol{Ce^t} \qquad \frac{du}{dt} = \lambda u \;\text{ produces }\; u(t) = \boldsymbol{Ce^{\lambda t}} \tag{1}$$

At time $t = 0$ those solutions include $e^0 = 1$. So they both reduce to $u(0) = C$. This "initial value" tells us the right choice for C. **The solutions that start from the number $\boldsymbol{u(0)}$ at time $\boldsymbol{t = 0}$ are $\boldsymbol{u(t) = u(0)e^t}$ and $\boldsymbol{u(t) = u(0)e^{\lambda t}}$.**

We just solved a 1 by 1 problem. Linear algebra moves to n by n. The unknown is a vector $\boldsymbol{u}$ (now boldface). It starts from the initial vector $\boldsymbol{u}(0)$, which is given. The n equations contain a square matrix A. We expect n exponents $e^{\lambda t}$ in $\boldsymbol{u}(t)$, from n λ's:

System of $\boldsymbol{n}$ equations $\quad \dfrac{d\boldsymbol{u}}{dt} = A\boldsymbol{u}$ starting from the vector $\boldsymbol{u}(0) = \begin{bmatrix} u_1(0) \\ \cdots \\ u_n(0) \end{bmatrix}$ at $t = 0$. (2)

These differential equations are *linear*. If $\boldsymbol{u}(t)$ and $\boldsymbol{v}(t)$ are solutions, so is $C\boldsymbol{u}(t) + D\boldsymbol{v}(t)$. We will need n constants like C and D to match the n components of $\boldsymbol{u}(0)$. Our first job is to find n "pure exponential solutions" $\boldsymbol{u} = e^{\lambda t}\boldsymbol{x}$ by using $A\boldsymbol{x} = \lambda\boldsymbol{x}$.

Notice that A is a *constant* matrix. In other linear equations, A changes as t changes. In nonlinear equations, A changes as $\boldsymbol{u}$ changes. We don't have those difficulties, $d\boldsymbol{u}/dt = A\boldsymbol{u}$ is "linear with constant coefficients". Those and only those are the differential equations that we will convert directly to linear algebra. Here is the key:

Solve linear constant coefficient equations by exponentials $e^{\lambda t}x$, ***when*** $Ax = \lambda x$.

Solution of $du/dt = Au$

Our pure exponential solution will be $e^{\lambda t}$ times a fixed vector $\boldsymbol{x}$. You may guess that λ is an eigenvalue of A, and ***$\boldsymbol{x}$ is the eigenvector***. Substitute $\boldsymbol{u}(t) = e^{\lambda t}\boldsymbol{x}$ into the equation $d\boldsymbol{u}/dt = A\boldsymbol{u}$ to prove you are right. The factor $e^{\lambda t}$ will cancel to leave $\lambda\boldsymbol{x} = A\boldsymbol{x}$:

Choose $\boldsymbol{u} = e^{\lambda t}\boldsymbol{x}$ when $A\boldsymbol{x} = \lambda\boldsymbol{x}$
$$\frac{d\boldsymbol{u}}{dt} = \lambda e^{\lambda t}\boldsymbol{x} \quad \text{agrees with} \quad A\boldsymbol{u} = Ae^{\lambda t}\boldsymbol{x} \tag{3}$$

All components of this special solution $\boldsymbol{u} = e^{\lambda t}\boldsymbol{x}$ share the same $e^{\lambda t}$. The solution grows when $\lambda > 0$. It decays when $\lambda < 0$. If λ is a complex number, its real part decides growth or decay. The imaginary part ω gives oscillation $e^{i\omega t}$ like a sine wave.

Example 1 Solve $\dfrac{d\boldsymbol{u}}{dt} = A\boldsymbol{u} = \begin{bmatrix} \mathbf{0} & \mathbf{1} \\ \mathbf{1} & \mathbf{0} \end{bmatrix}\boldsymbol{u}$ starting from $\boldsymbol{u}(0) = \begin{bmatrix} \mathbf{4} \\ \mathbf{2} \end{bmatrix}$.

This is a vector equation for $\boldsymbol{u}$. It contains two scalar equations for the components y and z. They are "coupled together" because the matrix A is not diagonal:

$$\frac{d\boldsymbol{u}}{dt} = A\boldsymbol{u} \qquad \frac{d}{dt}\begin{bmatrix} y \\ z \end{bmatrix} = \begin{bmatrix} 0 & 1 \\ 1 & 0 \end{bmatrix}\begin{bmatrix} y \\ z \end{bmatrix} \quad \text{means that} \quad \frac{\boldsymbol{dy}}{\boldsymbol{dt}} = \boldsymbol{z} \text{ and } \frac{\boldsymbol{dz}}{\boldsymbol{dt}} = \boldsymbol{y}.$$

The idea of eigenvectors is to combine those equations in a way that gets back to 1 by 1 problems. The combinations $y + z$ and $y - z$ will do it. Add and subtract equations:

$$\frac{d}{dt}(y + z) = z + y \qquad \text{and} \qquad \frac{d}{dt}(y - z) = -(y - z).$$

The combination $y + z$ grows like e^t, because it has $\lambda = 1$. The combination $y - z$ decays like e^{-t}, because it has $\lambda = -1$. Here is the point: We don't have to juggle the original equations $d\boldsymbol{u}/dt = A\boldsymbol{u}$, looking for these special combinations. The eigenvectors and eigenvalues of A will do it for us.

This matrix A has eigenvalues 1 and -1. The eigenvectors $\boldsymbol{x}$ are $(1, 1)$ and $(1, -1)$. The pure exponential solutions $\boldsymbol{u}_1$ and $\boldsymbol{u}_2$ take the form $e^{\lambda t}\boldsymbol{x}$ with $\lambda_1 = 1$ and $\lambda_2 = -1$:

$$\boldsymbol{u}_1(t) = e^{\lambda_1 t}\boldsymbol{x}_1 = e^t\begin{bmatrix} 1 \\ 1 \end{bmatrix} \quad \text{and} \quad \boldsymbol{u}_2(t) = e^{\lambda_2 t}\boldsymbol{x}_2 = e^{-t}\begin{bmatrix} 1 \\ -1 \end{bmatrix}. \tag{4}$$

Notice: These $\boldsymbol{u}$'s satisfy $A\boldsymbol{u}_1 = \boldsymbol{u}_1$ and $A\boldsymbol{u}_2 = -\boldsymbol{u}_2$, just like $\boldsymbol{x}_1$ and $\boldsymbol{x}_2$. The factors e^t and e^{-t} change with time. Those factors give $d\boldsymbol{u}_1/dt = \boldsymbol{u}_1 = A\boldsymbol{u}_1$ and $d\boldsymbol{u}_2/dt = -\boldsymbol{u}_2 = A\boldsymbol{u}_2$. **We have two solutions to $d\boldsymbol{u}/dt = A\boldsymbol{u}$.** To find all other solutions, **multiply those special solutions by any numbers C and D and add:**

Complete solution
$$\boldsymbol{u}(t) = \boldsymbol{C}e^{\boldsymbol{t}}\begin{bmatrix} \mathbf{1} \\ \mathbf{1} \end{bmatrix} + \boldsymbol{D}e^{-\boldsymbol{t}}\begin{bmatrix} \mathbf{1} \\ \mathbf{-1} \end{bmatrix} = \begin{bmatrix} Ce^t + De^{-t} \\ Ce^t - De^{-t} \end{bmatrix}. \tag{5}$$

With these two constants C and D, we can match any starting vector $\boldsymbol{u}(0) = (u_1(0), u_2(0))$. Set $t = 0$ and $e^0 = 1$. Example 1 asked for the initial value to be $\boldsymbol{u}(0) = (4, 2)$:

$$\boldsymbol{u}(0) \textbf{ decides } \boldsymbol{C}, \boldsymbol{D} \qquad C\begin{bmatrix}1\\1\end{bmatrix} + D\begin{bmatrix}1\\-1\end{bmatrix} = \begin{bmatrix}4\\2\end{bmatrix} \quad \text{yields} \quad \boldsymbol{C = 3} \quad \text{and} \quad \boldsymbol{D = 1}.$$

With $C = 3$ and $D = 1$ in the solution (5), the initial value problem is completely solved.

The same three steps that solved $\boldsymbol{u}_{k+1} = A\boldsymbol{u}_k$ now solve $d\boldsymbol{u}/dt = A\boldsymbol{u}$:

1. Write $\boldsymbol{u}(0)$ as a **combination** $c_1\boldsymbol{x}_1 + \cdots + c_n\boldsymbol{x}_n$ **of the eigenvectors of** $\boldsymbol{A}$.
2. Multiply each eigenvector $\boldsymbol{x}_i$ by **its growth factor** $e^{\lambda_i t}$.
3. The solution is the same combination of those pure solutions $e^{\lambda t}\boldsymbol{x}$:

$$\frac{d\boldsymbol{u}}{dt} = A\boldsymbol{u} \qquad \boldsymbol{u(t) = c_1 e^{\lambda_1 t}x_1 + \cdots + c_n e^{\lambda_n t}x_n.} \tag{6}$$

Not included: If two λ's are equal, with only one eigenvector, another solution is needed. (It will be $te^{\lambda t}\boldsymbol{x}$.) Step 1 needs to diagonalize $A = X\Lambda X^{-1}$: a basis of n eigenvectors.

Example 2 Solve $d\boldsymbol{u}/dt = A\boldsymbol{u}$ knowing the eigenvalues $\lambda = 1, 2, 3$ of A:

Typical example
Equation for $\boldsymbol{u}$
Initial condition $\boldsymbol{u}(0)$

$$\frac{d\boldsymbol{u}}{dt} = \begin{bmatrix}1 & 1 & 1\\0 & 2 & 1\\0 & 0 & 3\end{bmatrix}\boldsymbol{u} \quad \text{starting from} \quad \boldsymbol{u}(0) = \begin{bmatrix}9\\7\\4\end{bmatrix}.$$

The eigenvectors are $\boldsymbol{x}_1 = (1, 0, 0)$ and $\boldsymbol{x}_2 = (1, 1, 0)$ and $\boldsymbol{x}_3 = (1, 1, 1)$.

Step 1 The vector $\boldsymbol{u}(0) = (9, 7, 4)$ is $\boldsymbol{2x_1 + 3x_2 + 4x_3}$. Thus $(c_1, c_2, c_3) = \boldsymbol{(2, 3, 4)}$.

Step 2 The factors $e^{\lambda t}$ give exponential solutions $e^t\boldsymbol{x}_1$ and $e^{2t}\boldsymbol{x}_2$ and $e^{3t}\boldsymbol{x}_3$.

Step 3 The combination that starts from $\boldsymbol{u}(0)$ is $\boldsymbol{u}(t) = \boldsymbol{2}e^t\boldsymbol{x}_1 + \boldsymbol{3}e^{2t}\boldsymbol{x}_2 + \boldsymbol{4}e^{3t}\boldsymbol{x}_3$.

The coefficients 2, 3, 4 came from solving the linear equation $c_1\boldsymbol{x}_1 + c_2\boldsymbol{x}_2 + c_3\boldsymbol{x}_3 = \boldsymbol{u}(0)$:

$$\begin{bmatrix} & & \\ \boldsymbol{x}_1 & \boldsymbol{x}_2 & \boldsymbol{x}_3 \\ & & \end{bmatrix}\begin{bmatrix}c_1\\c_2\\c_3\end{bmatrix} = \begin{bmatrix}1 & 1 & 1\\0 & 1 & 1\\0 & 0 & 1\end{bmatrix}\begin{bmatrix}2\\3\\4\end{bmatrix} = \begin{bmatrix}9\\7\\4\end{bmatrix} \quad \text{which is} \quad X\boldsymbol{c} = \boldsymbol{u}(0). \tag{7}$$

You now have the basic idea—how to solve $d\boldsymbol{u}/dt = A\boldsymbol{u}$. The rest of this section goes further. We solve equations that contain *second* derivatives, because they arise so often in applications. We also decide whether $\boldsymbol{u}(t)$ approaches zero or blows up or just oscillates.

At the end comes the ***matrix exponential*** e^{At}. The short formula $e^{At}\boldsymbol{u}(0)$ solves the equation $d\boldsymbol{u}/dt = A\boldsymbol{u}$ in the same way that $A^k\boldsymbol{u}_0$ solves the equation $\boldsymbol{u}_{k+1} = A\boldsymbol{u}_k$. Example 3 will show how "difference equations" help to solve differential equations.

All these steps use the λ's and the $\boldsymbol{x}$'s. This section solves the constant coefficient problems that turn into linear algebra. It clarifies these simplest but most important differential equations—whose solution is completely based on growth factors $e^{\lambda t}$.

Second Order Equations

The most important equation in mechanics is $my'' + by' + ky = 0$. The first term is the mass m times the acceleration $a = y''$. This term ma balances the force F (that is *Newton's Law*). The force includes the damping $-by'$ and the elastic force $-ky$, proportional to distance moved. This is a second-order equation because it contains the second derivative $y'' = d^2y/dt^2$. It is still linear with constant coefficients m, b, k.

In a differential equations course, the method of solution is to substitute $y = e^{\lambda t}$. Each derivative of y brings down a factor λ. We want $y = e^{\lambda t}$ to solve the equation:

$$m\frac{d^2y}{dt^2} + b\frac{dy}{dt} + ky = 0 \quad \text{becomes} \quad (m\lambda^2 + b\lambda + k)\,e^{\lambda t} = 0. \tag{8}$$

Everything depends on $m\lambda^2 + b\lambda + k = 0$. This equation for λ has two roots λ_1 and λ_2. Then the equation for y has two pure solutions $y_1 = e^{\lambda_1 t}$ and $y_2 = e^{\lambda_2 t}$. Their combinations $c_1y_1 + c_2y_2$ give the complete solution unless $\lambda_1 = \lambda_2$.

In a linear algebra course we expect matrices and eigenvalues. Therefore we turn the scalar equation (with y'') into a *vector equation for y and y'*: first derivative only. Suppose the mass is $m = 1$. Two equations for $\boldsymbol{u} = (y, y')$ give $d\boldsymbol{u}/dt = A\boldsymbol{u}$:

$$\begin{aligned} dy/dt &= y' \\ dy'/dt &= -ky - by' \end{aligned} \quad \text{converts to} \quad \frac{d}{dt}\begin{bmatrix} y \\ y' \end{bmatrix} = \begin{bmatrix} 0 & 1 \\ -k & -b \end{bmatrix}\begin{bmatrix} y \\ y' \end{bmatrix} = A\boldsymbol{u}. \tag{9}$$

The first equation $dy/dt = y'$ is trivial (but true). The second is equation (8) connecting y'' to y' and y. Together they connect $\boldsymbol{u}'$ to $\boldsymbol{u}$. So we solve $\boldsymbol{u}' = A\boldsymbol{u}$ by eigenvalues of A:

$$A - \lambda I = \begin{bmatrix} -\lambda & 1 \\ -k & -b-\lambda \end{bmatrix} \quad \text{has determinant} \quad \lambda^2 + b\lambda + k = 0.$$

The equation for the λ's is the same as (8)! **It is still $\lambda^2 + b\lambda + k = 0$, since $m = 1$.** The roots λ_1 and λ_2 are now *eigenvalues of A*. The eigenvectors and the solution are

$$\boldsymbol{x}_1 = \begin{bmatrix} 1 \\ \lambda_1 \end{bmatrix} \qquad \boldsymbol{x}_2 = \begin{bmatrix} 1 \\ \lambda_2 \end{bmatrix} \qquad \boldsymbol{u}(t) = c_1 e^{\lambda_1 t}\begin{bmatrix} 1 \\ \lambda_1 \end{bmatrix} + c_2 e^{\lambda_2 t}\begin{bmatrix} 1 \\ \lambda_2 \end{bmatrix}.$$

The first component of $\boldsymbol{u}(t)$ has $y = c_1e^{\lambda_1 t} + c_2e^{\lambda_2 t}$—the same solution as before. It can't be anything else. In the second component of $\boldsymbol{u}(t)$ you see the velocity dy/dt. The vector problem is completely consistent with the scalar problem. The 2 by 2 matrix A is called a *companion matrix*—a companion to the second order equation with y''.

Example 3 *Motion around a circle with* $y'' + y = 0$ *and* $y = \cos t$

This is our master equation with mass $m = 1$ and stiffness $k = 1$ and $d = 0$: no damping. Substitute $y = e^{\lambda t}$ into $y'' + y = 0$ to reach $\boldsymbol{\lambda^2 + 1 = 0}$. *The roots are* $\boldsymbol{\lambda = i}$ *and* $\boldsymbol{\lambda = -i}$. Then half of $e^{it} + e^{-it}$ gives the solution $y = \cos t$.

As a first-order system, the initial values $y(0) = 1$, $y'(0) = 0$ go into $\boldsymbol{u}(0) = (1, 0)$:

$$\textbf{Use } \boldsymbol{y'' = -y} \qquad \frac{d\boldsymbol{u}}{dt} = \frac{d}{dt}\begin{bmatrix} y \\ y' \end{bmatrix} = \begin{bmatrix} \mathbf{0} & \mathbf{1} \\ \mathbf{-1} & \mathbf{0} \end{bmatrix}\begin{bmatrix} y \\ y' \end{bmatrix} = A\boldsymbol{u}. \tag{10}$$

The eigenvalues of A are again the same $\lambda = i$ and $\lambda = -i$ (no surprise). A is antisymmetric with eigenvectors $\boldsymbol{x}_1 = (1, i)$ and $\boldsymbol{x}_2 = (1, -i)$. The combination that matches $\boldsymbol{u}(0) = (1, 0)$ is $\frac{1}{2}(\boldsymbol{x}_1 + \boldsymbol{x}_2)$. Step 2 multiplies the x's by e^{it} and e^{-it}. Step 3 combines the pure oscillations into $\boldsymbol{u}(t)$ to find $y = \cos t$ as expected:

$$\boldsymbol{u}(t) = \frac{1}{2}e^{it}\begin{bmatrix} 1 \\ i \end{bmatrix} + \frac{1}{2}e^{-it}\begin{bmatrix} 1 \\ -i \end{bmatrix} = \begin{bmatrix} \cos t \\ -\sin t \end{bmatrix}. \qquad \text{This is } \begin{bmatrix} y(t) \\ y'(t) \end{bmatrix}.$$

All good. The vector $\boldsymbol{u} = (\cos t, -\sin t)$ goes around a circle (Figure 6.3). The radius is 1 because $\cos^2 t + \sin^2 t = 1$.

Difference Equations (optional)

To display a circle on a screen, replace $y'' = -y$ by a ***difference equation***. Here are three choices using $\boldsymbol{Y}(t + \Delta t) - 2\boldsymbol{Y}(t) + \boldsymbol{Y}(t - \Delta t)$. Divide by $(\Delta t)^2$ to approximate y''.

F	**Forward from $n - 1$**		$-\boldsymbol{Y}_{n-1}$	(11 **F**)
C	**Centered at time n**	$\dfrac{Y_{n+1} - 2Y_n + Y_{n-1}}{(\Delta t)^2} =$	$-\boldsymbol{Y}_n$	(11 **C**)
B	**Backward from $n + 1$**		$-\boldsymbol{Y}_{n+1}$	(11 **B**)

Figure 6.3 shows the exact $y(t) = \cos t$ completing a circle at $t = 2\pi$. The three difference methods *don't* complete a perfect circle in 32 time steps of length $\Delta t = 2\pi/32$. Those pictures will be explained by eigenvalues:

Forward $|\lambda| > 1$ **(spiral out)** **Centered** $|\lambda| = 1$ **(best)** **Backward** $|\lambda| < 1$ **(spiral in)**

The 2-step equations (11) reduce to 1-step systems $\boldsymbol{U}_{n+1} = A\boldsymbol{U}_n$. Instead of $\boldsymbol{u} = (y, y')$ the discrete unknown is $\boldsymbol{U}_n = (Y_n, Z_n)$. We take n time steps Δt starting from $\boldsymbol{U}_0$:

$$\textbf{Forward (11F)} \quad \begin{aligned} Y_{n+1} &= Y_n + \Delta t\, Z_n \\ Z_{n+1} &= Z_n - \Delta t\, Y_n \end{aligned} \quad \text{becomes} \quad \boldsymbol{U}_{n+1} = \begin{bmatrix} 1 & \Delta t \\ -\Delta t & 1 \end{bmatrix}\begin{bmatrix} Y_n \\ Z_n \end{bmatrix} = A\boldsymbol{U}_n. \tag{12}$$

Those are like $Y' = Z$ and $Z' = -Y$. They are first order equations involving times n and $n + 1$. Eliminating Z would bring back the "forward" second order equation (11 **F**).

My question is simple. *Do the points* (Y_n, Z_n) *stay on the circle* $Y^2 + Z^2 = 1$? No, they are growing to infinity in Figure 6.3. ***We are taking powers*** $\boldsymbol{A^n}$ ***and not*** $\boldsymbol{e^{At}}$***, so we test the magnitude*** $\boldsymbol{|\lambda|}$ ***and not the real parts of the eigenvalues.***

Eigenvalues of A $\boldsymbol{\lambda = 1 \pm i\Delta t}$ **Then** $\boldsymbol{|\lambda| > 1}$ **and** $\boldsymbol{(Y_n, Z_n)}$ **spirals out**

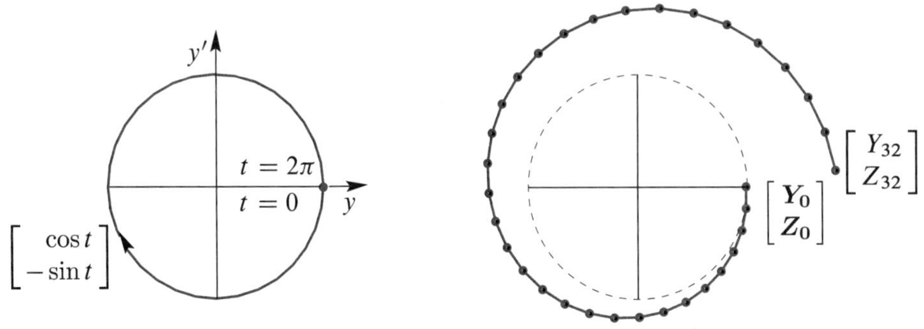

Figure 6.3: Exact $\boldsymbol{u} = (\cos t, -\sin t)$ on a circle. **Forward Euler spirals out** (32 steps).

The backward choice in (11 **B**) will do the opposite in Figure 6.4. Notice the new A:

$$\textbf{Backward} \quad \begin{array}{l} Y_{n+1} = Y_n + \Delta t\, Z_{n+1} \\ Z_{n+1} = Z_n - \Delta t\, \mathbf{Y_{n+1}} \end{array} \quad \text{is} \quad \begin{bmatrix} 1 & -\Delta t \\ \Delta t & 1 \end{bmatrix} \begin{bmatrix} Y_{n+1} \\ Z_{n+1} \end{bmatrix} = \begin{bmatrix} Y_n \\ Z_n \end{bmatrix} = \boldsymbol{U}_n. \quad (13)$$

That matrix has eigenvalues $1 \pm i\Delta t$. But we *invert* it to reach $\boldsymbol{U}_{n+1}$ from $\boldsymbol{U}_n$. Then $|\lambda| < 1$ explains why *the solution spirals in* to $(0, 0)$ for backward differences.

On the right side of Figure 6.4 you see 32 steps with the *centered* choice. The solution stays close to the circle (Problem 28) if $\Delta t < 2$. This is the **leapfrog method**, constantly used. The second difference $Y_{n+1} - 2Y_n + Y_{n-1}$ "leaps over" the center value Y_n in (11).

This is the way a chemist follows the motion of molecules (molecular dynamics leads to giant computations). Computational science is lively because one differential equation can be replaced by many difference equations—some unstable, some stable, some neutral. Problem 30 has a fourth (very good) method that stays right on the circle.

Real engineering and real physics deal with systems (not just a single mass at one point). The unknown $\boldsymbol{y}$ is a vector. The coefficient of $\boldsymbol{y}''$ is a *mass matrix* M, with n masses. The coefficient of $\boldsymbol{y}$ is a *stiffness matrix* K, not a number k. The coefficient of $\boldsymbol{y}'$ is a damping matrix which might be zero.

The vector equation $My'' + Ky = \boldsymbol{f}$ is a major part of computational mechanics. It is controlled by the eigenvalues of $M^{-1}K$ in $K\boldsymbol{x} = \lambda M\boldsymbol{x}$.

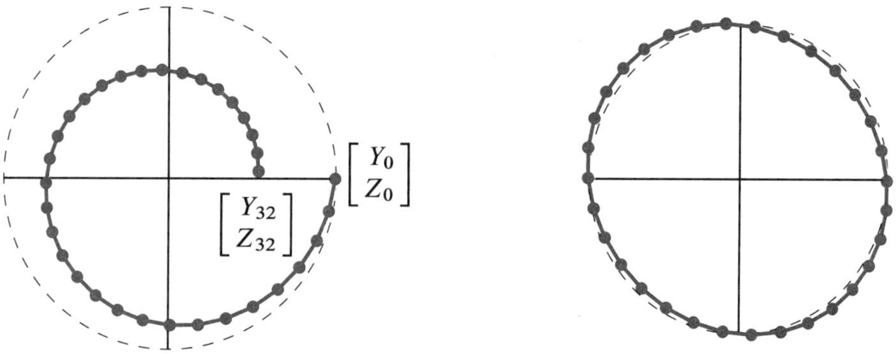

Figure 6.4: Backward differences spiral in. Leapfrog stays near the correct circle.

Stability of 2 by 2 Matrices

For the solution of $d\boldsymbol{u}/dt = A\boldsymbol{u}$, there is a fundamental question. *Does the solution approach $\boldsymbol{u} = \boldsymbol{0}$ as $t \to \infty$?* Is the problem *stable*, by dissipating energy? A solution that includes e^t is unstable. Stability depends on the eigenvalues of A.

The complete solution $\boldsymbol{u}(t)$ is built from pure solutions $e^{\lambda t}\boldsymbol{x}$. If the eigenvalue λ is real, we know exactly when $e^{\lambda t}$ will approach zero: *The number λ must be negative.* If the eigenvalue is a complex number $\lambda = r + is$, *the real part r must be negative.* When $e^{\lambda t}$ splits into $e^{rt}e^{ist}$, the factor e^{ist} has absolute value fixed at 1:

$$e^{ist} = \cos st + i \sin st \quad \text{has} \quad |e^{ist}|^2 = \cos^2 st + \sin^2 st = 1.$$

The real part of λ controls the growth ($r > 0$) or the decay ($r < 0$).

The question is: ***Which matrices have negative eigenvalues?*** More accurately, when are the ***real parts of the λ's all negative***? 2 by 2 matrices allow a clear answer.

Stability A is ***stable*** and $\boldsymbol{u}(t) \to \boldsymbol{0}$ when all eigenvalues λ have ***negative real parts***. The 2 by 2 matrix $A = \begin{bmatrix} a & b \\ c & d \end{bmatrix}$ must pass two tests:

$\boldsymbol{\lambda_1 + \lambda_2 < 0}$	The trace $T = a + d$ must be negative.
$\boldsymbol{\lambda_1 \lambda_2 > 0}$	The determinant $D = ad - bc$ must be positive.

Reason If the λ's are real and negative, their sum is negative. This is the trace T. Their product is positive. This is the determinant D. The argument also goes in the reverse direction. If $D = \lambda_1\lambda_2$ is positive, then λ_1 and λ_2 have the same sign. If $T = \lambda_1 + \lambda_2$ is negative, that sign will be negative. We can test T and D.

If the λ's are complex numbers, they must have the form $r + is$ and $r - is$. Otherwise T and D will not be real. The determinant D is automatically positive, since $(r + is)(r - is) = r^2 + s^2$. The trace T is $r + is + r - is = 2r$. So a negative trace T means that the real part r is negative and the matrix is stable. Q.E.D.

Figure 6.5 shows the parabola $T^2 = 4D$ separating real λ's from complex λ's. Solving $\lambda^2 - T\lambda + D = 0$ involves the square root $\sqrt{T^2 - 4D}$. This is real below the parabola and imaginary above it. The stable region is the *upper left quarter* of the figure—where the trace T is negative and the determinant D is positive.

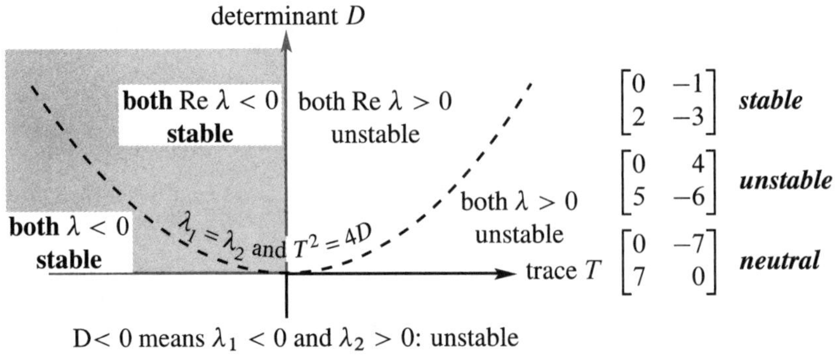

Figure 6.5: A 2 by 2 matrix is stable ($\boldsymbol{u}(t) \to \mathbf{0}$) when **trace** $<$ **0** and **det** $>$ **0**.

The Exponential of a Matrix

We want to write the solution $\boldsymbol{u}(t)$ ***in a new form*** $e^{At}\boldsymbol{u}(0)$. First we have to say what e^{At} means, with a matrix in the exponent. To define e^{At} for matrices, we copy e^x for numbers.

The direct definition of e^x is by the infinite series $1 + x + \frac{1}{2}x^2 + \frac{1}{6}x^3 + \cdots$. When you change x to a square matrix At, this series defines the matrix exponential e^{At}:

Matrix exponential e^{At}
$$e^{At} = I + At + \tfrac{1}{2}(At)^2 + \tfrac{1}{6}(At)^3 + \cdots \tag{14}$$

Its t derivative is Ae^{At}
$$A + A^2t + \tfrac{1}{2}A^3t^2 + \cdots = Ae^{At}$$

Its eigenvalues are $e^{\lambda t}$
$$(I + At + \tfrac{1}{2}(At)^2 + \cdots)\boldsymbol{x} = (1 + \lambda t + \tfrac{1}{2}(\lambda t)^2 + \cdots)\,\boldsymbol{x}$$

The number that divides $(At)^n$ is "n factorial". This is $n! = (1)(2)\cdots(n-1)(n)$. The factorials after $1, 2, 6$ are $4! = 24$ and $5! = 120$. They grow quickly. The series always converges and its derivative is always Ae^{At}. Therefore $e^{At}\boldsymbol{u}(0)$ solves the differential equation with one quick formula—*even if there is a shortage of eigenvectors.*

I will use this series in Example 4, to see it work with a missing eigenvector. **It will produce $te^{\lambda t}$.** First let me reach $Xe^{\Lambda t}X^{-1}$ in the good (diagonalizable) case.

This chapter emphasizes how to find $\boldsymbol{u}(t) = e^{At}\boldsymbol{u}(0)$ by diagonalization. Assume A does have n independent eigenvectors, so it is diagonalizable. Substitute $A = X\Lambda X^{-1}$ into the series for e^{At}. Whenever $X\Lambda X^{-1}X\Lambda X^{-1}$ appears, cancel $X^{-1}X$ in the middle:

Use the series $$\boldsymbol{e}^{\boldsymbol{At}} = I + X\Lambda X^{-1}t + \tfrac{1}{2}(X\Lambda X^{-1}t)(X\Lambda X^{-1}t) + \cdots$$

Factor out X and X^{-1} $$= X\left[I + \Lambda t + \tfrac{1}{2}(\Lambda t)^2 + \cdots\right]X^{-1} \tag{15}$$

e^{At} is diagonalized! $$\boxed{\boldsymbol{e}^{\boldsymbol{At}} = \boldsymbol{X}\,\boldsymbol{e}^{\boldsymbol{\Lambda t}}\,\boldsymbol{X}^{-\boldsymbol{1}}.}$$

e^{At} has the same eigenvector matrix X as A. Then Λ is a diagonal matrix and so is $e^{\Lambda t}$. The numbers $e^{\lambda_i t}$ are on the diagonal. Multiply $Xe^{\Lambda t}X^{-1}\boldsymbol{u}(0)$ to recognize $\boldsymbol{u}(t)$:

$$e^{At}\boldsymbol{u}(0) = Xe^{\Lambda t}X^{-1}\boldsymbol{u}(0) = \begin{bmatrix} \\ \boldsymbol{x}_1 & \cdots & \boldsymbol{x}_n \\ \\ \end{bmatrix}\begin{bmatrix} e^{\lambda_1 t} & & \\ & \ddots & \\ & & e^{\lambda_n t} \end{bmatrix}\begin{bmatrix} c_1 \\ \vdots \\ c_n \end{bmatrix}. \tag{16}$$

This solution $e^{At}\boldsymbol{u}(0)$ is the same answer that came in equation (6) from three steps:

> **1**. $\boldsymbol{u}(0) = c_1\boldsymbol{x}_1 + \cdots + c_n\boldsymbol{x}_n = X\boldsymbol{c}$. Here we need n independent eigenvectors.
>
> **2**. Multiply each $\boldsymbol{x}_i$ by its growth factor $e^{\lambda_i t}$ to follow it forward in time.
>
> **3**. The best form of $e^{At}\boldsymbol{u}(0)$ is $\quad \boldsymbol{u}(t) = \boldsymbol{c_1 e^{\lambda_1 t}x_1 + \cdots + c_n e^{\lambda_n t}x_n}$. (17)

Example 4 When you substitute $y = e^{\lambda t}$ into $y'' - 2y' + y = 0$, you get an equation with **repeated roots**: $\lambda^2 - 2\lambda + 1 = 0$ is $(\lambda - 1)^2 = 0$ with $\boldsymbol{\lambda = 1, 1}$. A differential equations course would propose e^t and te^t as two independent solutions. Here we discover why.

Linear algebra reduces $y'' - 2y' + y = 0$ to a vector equation for $\boldsymbol{u} = (y, y')$:

$$\frac{d}{dt}\begin{bmatrix} y \\ y' \end{bmatrix} = \begin{bmatrix} y' \\ 2y' - y \end{bmatrix} \quad \text{is} \quad \frac{d\boldsymbol{u}}{dt} = A\boldsymbol{u} = \begin{bmatrix} 0 & 1 \\ -1 & 2 \end{bmatrix}\boldsymbol{u}. \tag{18}$$

A has a **repeated eigenvalue $\boldsymbol{\lambda = 1, 1}$** (with trace $= 2$ and $\det A = 1$). The only eigenvectors are multiples of $\boldsymbol{x} = (1, 1)$. *Diagonalization is not possible*, A has only one line of eigenvectors. So we compute e^{At} from its definition as a series:

Short series $$\boldsymbol{e}^{\boldsymbol{At}} = e^{It}\,e^{(A-I)t} = e^t\left[I + (A - I)t\right]. \tag{19}$$

That "infinite" series for $e^{(A-I)t}$ ended quickly because $(A - I)^2$ is the zero matrix! You can see $\boldsymbol{te^t}$ in equation (19). The first component of $\boldsymbol{e}^{At}\,\boldsymbol{u}(0)$ is our answer $y(t)$:

$$\begin{bmatrix} y \\ y' \end{bmatrix} = e^t\left[I + \begin{bmatrix} -1 & 1 \\ -1 & 1 \end{bmatrix}t\right]\begin{bmatrix} y(0) \\ y'(0) \end{bmatrix} \qquad y(t) = e^t y(0) - \boldsymbol{te^t}y(0) + \boldsymbol{te^t}y'(0).$$

Example 5 Use the infinite series to find e^{At} for $A = \left[\begin{smallmatrix} 0 & 1 \\ -1 & 0 \end{smallmatrix}\right]$. Notice that $A^4 = I$:

$$A = \begin{bmatrix} & 1 \\ -1 & \end{bmatrix} \quad A^2 = \begin{bmatrix} -1 & \\ & -1 \end{bmatrix} \quad A^3 = \begin{bmatrix} & -1 \\ 1 & \end{bmatrix} \quad A^4 = \begin{bmatrix} 1 & \\ & 1 \end{bmatrix}.$$

A^5, A^6, A^7, A^8 will be a repeat of A, A^2, A^3, A^4. The top right corner has $1, 0, -1, 0$ repeating over and over in powers of A. Then $t - \frac{1}{6}t^3$ starts the infinite series for e^{At} in that top right corner, and $1 - \frac{1}{2}t^2$ starts the top left corner:

$$e^{At} = I + At + \tfrac{1}{2}(At)^2 + \tfrac{1}{6}(At)^3 + \cdots = \begin{bmatrix} 1 - \frac{1}{2}t^2 + \cdots & t - \frac{1}{6}t^3 + \cdots \\ -t + \frac{1}{6}t^3 - \cdots & 1 - \frac{1}{2}t^2 + \cdots \end{bmatrix}.$$

The top row of that matrix e^{At} shows the infinite series for the cosine and sine!

$$\boldsymbol{A = \begin{bmatrix} 0 & 1 \\ -1 & 0 \end{bmatrix}} \qquad \boldsymbol{e^{At} = \begin{bmatrix} \cos t & \sin t \\ -\sin t & \cos t \end{bmatrix}}. \tag{20}$$

A is an antisymmetric matrix ($A^{\mathrm{T}} = -A$). Its exponential e^{At} is an orthogonal matrix. The eigenvalues of A are i and $-i$. The eigenvalues of e^{At} are e^{it} and e^{-it}. Three rules:

1 $\boldsymbol{e^{At}}$ ***always has the inverse*** $\boldsymbol{e^{-At}}$**.**

2 ***The eigenvalues of*** $\boldsymbol{e^{At}}$ ***are always*** $\boldsymbol{e^{\lambda t}}$**.**

3 ***When A is antisymmetric,*** $\boldsymbol{e^{At}}$ ***is orthogonal. Inverse = transpose =*** $\boldsymbol{e^{-At}}$**.**

Antisymmetric is the same as "skew-symmetric". Those matrices have pure imaginary eigenvalues like i and $-i$. Then e^{At} has eigenvalues like e^{it} and e^{-it}. Their absolute value is 1: neutral stability, pure oscillation, energy is conserved. So $||\boldsymbol{u}(t)|| = ||\boldsymbol{u}(0)||$.

Our final example has a triangular matrix A. Then the eigenvector matrix X is triangular. So are X^{-1} and e^{At}. You will see the two forms of the solution: a combination of eigenvectors and the short form $e^{At}\boldsymbol{u}(0)$.

Example 6 Solve $\dfrac{d\boldsymbol{u}}{dt} = A\boldsymbol{u} = \begin{bmatrix} 1 & 1 \\ 0 & 2 \end{bmatrix}\boldsymbol{u}$ starting from $\boldsymbol{u}(0) = \begin{bmatrix} 2 \\ 1 \end{bmatrix}$ at $t = 0$.

Solution The eigenvalues 1 and 2 are on the diagonal of A (since A is triangular). The eigenvectors are $(1, 0)$ and $(1, 1)$. The starting $\boldsymbol{u}(0)$ is $\boldsymbol{x}_1 + \boldsymbol{x}_2$ so $c_1 = c_2 = 1$. Then $\boldsymbol{u}(t)$ is the same combination of pure exponentials (*no* $te^{\lambda t}$ *when* $\lambda = 1$ and 2):

Solution to $\boldsymbol{u}' = A\boldsymbol{u}$ $\qquad \boldsymbol{u}(t) = e^t \begin{bmatrix} 1 \\ 0 \end{bmatrix} + e^{2t} \begin{bmatrix} 1 \\ 1 \end{bmatrix}.$

That is the clearest form. But the matrix form with e^{At} produces $\boldsymbol{u}(t)$ for every $\boldsymbol{u}(0)$:

$$\boldsymbol{u}(t) = Xe^{\Lambda t}X^{-1}\boldsymbol{u}(0) \text{ is } \begin{bmatrix} 1 & 1 \\ 0 & 1 \end{bmatrix} \begin{bmatrix} e^t & \\ & e^{2t} \end{bmatrix} \begin{bmatrix} 1 & -1 \\ 0 & 1 \end{bmatrix} \boldsymbol{u}(0) = \begin{bmatrix} e^t & e^{2t} + e^t \\ 0 & e^{2t} \end{bmatrix} \boldsymbol{u}(0).$$

That last matrix is $\boldsymbol{e^{At}}$**.** It is nice because A is triangular. The situation is the same as for $A\boldsymbol{x} = \boldsymbol{b}$ and inverses. We don't need A^{-1} to find $\boldsymbol{x}$, and we don't need e^{At} to solve $d\boldsymbol{u}/dt = A\boldsymbol{u}$. But as quick formulas for the answers, $A^{-1}\boldsymbol{b}$ and $e^{At}\boldsymbol{u}(0)$ are unbeatable.

■ REVIEW OF THE KEY IDEAS ■

1. The equation $\boldsymbol{u}' = A\boldsymbol{u}$ is linear with constant coefficients in A. Start from $\boldsymbol{u}(0)$.

2. Its solution is usually a combination of exponentials, involving every λ and $\boldsymbol{x}$:

   Independent eigenvectors $\quad \boldsymbol{u}(t) = c_1 e^{\lambda_1 t}\boldsymbol{x}_1 + \cdots + c_n e^{\lambda_n t}\boldsymbol{x}_n.$

3. The constants $c_1, \ldots, c_n$ are determined by $\boldsymbol{u}(0) = c_1\boldsymbol{x}_1 + \cdots + c_n\boldsymbol{x}_n = X\boldsymbol{c}$.

4. $\boldsymbol{u}(t)$ approaches zero (**stability**) if every λ has negative real part: All $e^{\lambda t} \to 0$.

5. Solutions have the short form $\boldsymbol{u}(t) = e^{At}\boldsymbol{u}(0)$, with the matrix exponential e^{At}.

6. Equations with y'' reduce to $\boldsymbol{u}' = A\boldsymbol{u}$ by combining y and y' into the vector $\boldsymbol{u}$.

■ WORKED EXAMPLES ■

6.3 A Solve $y'' + 4y' + 3y = 0$ by substituting $e^{\lambda t}$ and also by linear algebra.

Solution Substituting $y = e^{\lambda t}$ yields $(\lambda^2 + 4\lambda + 3)e^{\lambda t} = 0$. That quadratic factors into $\lambda^2 + 4\lambda + 3 = (\lambda+1)(\lambda+3) = 0$. Therefore $\boldsymbol{\lambda_1 = -1}$ and $\boldsymbol{\lambda_2 = -3}$. The pure solutions are $y_1 = e^{-t}$ and $y_2 = e^{-3t}$. The complete solution $\boldsymbol{y} = c_1y_1 + c_2y_2$ approaches zero.

To use linear algebra we set $\boldsymbol{u} = (y, y')$. Then the vector equation is $\boldsymbol{u}' = A\boldsymbol{u}$:

$$\begin{aligned} dy/dt &= y' \\ dy'/dt &= -3y - 4y' \end{aligned} \quad \text{converts to} \quad \frac{d\boldsymbol{u}}{dt} = \begin{bmatrix} 0 & 1 \\ -3 & -4 \end{bmatrix} \boldsymbol{u}.$$

This A is a "companion matrix" and its eigenvalues are again -1 and -3:

$$\textbf{Same quadratic} \qquad \det(A - \lambda I) = \begin{vmatrix} -\lambda & 1 \\ -3 & -4-\lambda \end{vmatrix} = \lambda^2 + 4\lambda + 3 = 0.$$

The eigenvectors of A are $(1, \lambda_1)$ and $(1, \lambda_2)$. Either way, the decay in $y(t)$ comes from e^{-t} and e^{-3t}. With constant coefficients, calculus leads to linear algebra $A\boldsymbol{x} = \lambda\boldsymbol{x}$.

Note In linear algebra the serious danger is a shortage of eigenvectors. Our eigenvectors $(1, \lambda_1)$ and $(1, \lambda_2)$ are the same if $\lambda_1 = \lambda_2$. Then we can't diagonalize A. In this case we don't yet have two independent solutions to $d\boldsymbol{u}/dt = A\boldsymbol{u}$.

In differential equations the danger is also a repeated λ. After $y = e^{\lambda t}$, a second solution has to be found. It turns out to be $y = \boldsymbol{te^{\lambda t}}$. This "impure" solution (with an extra $\boldsymbol{t}$) appears in the matrix exponential e^{At}. Example 4 showed how.

6.3 B Find the eigenvalues and eigenvectors of A. Then write $\boldsymbol{u}(0) = (0, 2\sqrt{2}, 0)$ as a combination of the eigenvectors. Solve both equations $\boldsymbol{u}' = A\boldsymbol{u}$ and $\boldsymbol{u}'' = A\boldsymbol{u}$:

$$\frac{d\boldsymbol{u}}{dt} = \begin{bmatrix} -2 & 1 & 0 \\ 1 & -2 & 1 \\ 0 & 1 & -2 \end{bmatrix} \boldsymbol{u} \quad \text{and} \quad \frac{d^2\boldsymbol{u}}{dt^2} = \begin{bmatrix} -2 & 1 & 0 \\ 1 & -2 & 1 \\ 0 & 1 & -2 \end{bmatrix} \boldsymbol{u} \quad \text{with} \quad \frac{d\boldsymbol{u}}{dt}(0) = \mathbf{0}.$$

$\boldsymbol{u}' = A\boldsymbol{u}$ is like the heat equation $\partial u/\partial t = \partial^2 u/\partial x^2$.
Its solution $u(t)$ will decay (A has negative eigenvalues).
$\boldsymbol{u}'' = A\boldsymbol{u}$ is like the wave equation $\partial^2 u/\partial t^2 = \partial^2 u/\partial x^2$.
Its solution will oscillate (the square roots of λ are imaginary).

Solution The eigenvalues and eigenvectors come from $\det(A - \lambda I) = 0$:

$$\det(A - \lambda I) = \begin{vmatrix} -2-\lambda & 1 & 0 \\ 1 & -2-\lambda & 1 \\ 0 & 1 & -2-\lambda \end{vmatrix} = (-2-\lambda)[(-2-\lambda)^2 - 2] = 0.$$

One eigenvalue is $\lambda = -2$, when $-2 - \lambda$ is zero. The other factor is $\lambda^2 + 4\lambda + 2$, so the other eigenvalues (also real and negative) are $\lambda = -2 \pm \sqrt{2}$. Find the eigenvectors:

$$\boldsymbol{\lambda = -2} \qquad (A + 2I)\boldsymbol{x} = \begin{bmatrix} 0 & 1 & 0 \\ 1 & 0 & 1 \\ 0 & 1 & 0 \end{bmatrix} \begin{bmatrix} x \\ y \\ z \end{bmatrix} = \begin{bmatrix} 0 \\ 0 \\ 0 \end{bmatrix} \quad \text{for } \boldsymbol{x}_1 = \begin{bmatrix} 1 \\ 0 \\ -1 \end{bmatrix}$$

$$\boldsymbol{\lambda = -2 - \sqrt{2}} \quad (A - \lambda I)\boldsymbol{x} = \begin{bmatrix} \sqrt{2} & 1 & 0 \\ 1 & \sqrt{2} & 1 \\ 0 & 1 & \sqrt{2} \end{bmatrix} \begin{bmatrix} x \\ y \\ z \end{bmatrix} = \begin{bmatrix} 0 \\ 0 \\ 0 \end{bmatrix} \quad \text{for } \boldsymbol{x}_2 = \begin{bmatrix} 1 \\ -\sqrt{2} \\ 1 \end{bmatrix}$$

$$\boldsymbol{\lambda = -2 + \sqrt{2}} \quad (A - \lambda I)\boldsymbol{x} = \begin{bmatrix} -\sqrt{2} & 1 & 0 \\ 1 & -\sqrt{2} & 1 \\ 0 & 1 & -\sqrt{2} \end{bmatrix} \begin{bmatrix} x \\ y \\ z \end{bmatrix} = \begin{bmatrix} 0 \\ 0 \\ 0 \end{bmatrix} \quad \text{for } \boldsymbol{x}_3 = \begin{bmatrix} 1 \\ \sqrt{2} \\ 1 \end{bmatrix}$$

The eigenvectors are *orthogonal* (proved in Section 6.4 for all symmetric matrices). All three λ_i are negative. This A is *negative definite* and e^{At} decays to zero (stability).

The starting $\boldsymbol{u}(0) = (0, 2\sqrt{2}, 0)$ is $\boldsymbol{x}_3 - \boldsymbol{x}_2$. The solution is $\boldsymbol{u}(t) = \boldsymbol{e^{\lambda_3 t} x_3 - e^{\lambda_2 t} x_2}$.

Heat equation In Figure 6.6a, the temperature at the center starts at $2\sqrt{2}$. Heat diffuses into the neighboring boxes and then to the outside boxes (frozen at $0°$). The rate of heat flow between boxes is the temperature difference. From box 2, heat flows left and right at the rate $u_1 - u_2$ and $u_3 - u_2$. So the flow out is $u_1 - 2u_2 + u_3$ in the second row of $A\boldsymbol{u}$.

Wave equation $d^2\boldsymbol{u}/dt^2 = A\boldsymbol{u}$ has the same eigenvectors $\boldsymbol{x}$. But now the eigenvalues λ lead to **oscillations** $e^{i\omega t}\boldsymbol{x}$ and $e^{-i\omega t}\boldsymbol{x}$. The frequencies come from $\omega^2 = -\lambda$:

$$\frac{d^2}{dt^2}(e^{i\omega t}\boldsymbol{x}) = A(e^{i\omega t}\boldsymbol{x}) \qquad \text{becomes} \qquad (i\omega)^2 e^{i\omega t}\boldsymbol{x} = \lambda e^{i\omega t}\boldsymbol{x} \quad \text{and} \quad \boldsymbol{\omega^2 = -\lambda}.$$

There are two square roots of $-\lambda$, so we have $e^{i\omega t}\boldsymbol{x}$ and $e^{-i\omega t}\boldsymbol{x}$. With three eigenvectors this makes *six* solutions to $\boldsymbol{u}'' = A\boldsymbol{u}$. A combination will match the six components of $\boldsymbol{u}(0)$ and $\boldsymbol{u}'(0)$. Since $\boldsymbol{u}' = \mathbf{0}$ in this problem, $e^{i\omega t}\boldsymbol{x}$ and $e^{-i\omega t}\boldsymbol{x}$ produce $2\cos\omega t\, \boldsymbol{x}$.

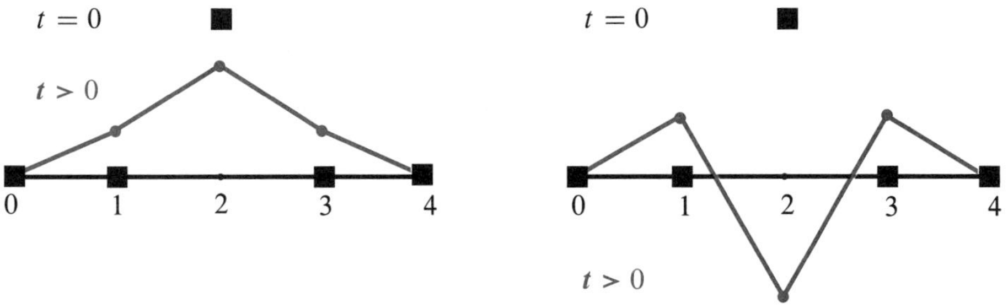

Figure 6.6: Heat diffuses away from box 2 (left). Wave travels from box 2 (right).

6.3 C Solve the four equations $da/dt = 0, db/dt = a, dc/dt = 2b, dz/dt = 3c$ in that order starting from $\boldsymbol{u}(0) = (a(0), b(0), c(0), z(0))$. Solve the same equations by the matrix exponential in $\boldsymbol{u}(t) = e^{At}\boldsymbol{u}(0)$.

Four equations
$\boldsymbol{\lambda = 0, 0, 0, 0}$
Eigenvalues on the diagonal

$$\frac{d}{dt}\begin{bmatrix} a \\ b \\ c \\ z \end{bmatrix} = \begin{bmatrix} 0 & 0 & 0 & 0 \\ \mathbf{1} & 0 & 0 & 0 \\ 0 & \mathbf{2} & 0 & 0 \\ 0 & 0 & \mathbf{3} & 0 \end{bmatrix}\begin{bmatrix} a \\ b \\ c \\ z \end{bmatrix} \quad \text{is} \quad \frac{d\boldsymbol{u}}{dt} = A\boldsymbol{u}.$$

First find A^2, A^3, A^4 and $e^{At} = I + At + \frac{1}{2}(At)^2 + \frac{1}{6}(At)^3$. Why does the series stop? Why is it true that $(e^A)(e^A) = (e^{2A})$? ***Always*** $\boldsymbol{e^{As}}$ ***times*** $\boldsymbol{e^{At}}$ ***is*** $\boldsymbol{e^{A(s+t)}}$.

Solution 1 Integrate $da/dt = 0$, then $db/dt = a$, then $dc/dt = 2b$ and $dz/dt = 3c$:

$$\begin{aligned} a(t) &= a(0) \\ b(t) &= ta(0) + b(0) \\ c(t) &= t^2a(0) + 2tb(0) + c(0) \\ z(t) &= t^3a(0) + 3t^2b(0) + 3tc(0) + z(0) \end{aligned}$$

The 4 by 4 matrix which is multiplying $a(0), b(0), c(0), d(0)$ to produce $a(t), b(t), c(t), d(t)$ must be the same e^{At} as below

Solution 2 *The powers of A (strictly triangular) are all zero after A^3.*

$$\boldsymbol{A} = \begin{bmatrix} 0 & 0 & 0 & 0 \\ \mathbf{1} & 0 & 0 & 0 \\ 0 & \mathbf{2} & 0 & 0 \\ 0 & 0 & \mathbf{3} & 0 \end{bmatrix} \quad \boldsymbol{A^2} = \begin{bmatrix} 0 & 0 & 0 & 0 \\ 0 & 0 & 0 & 0 \\ \mathbf{2} & 0 & 0 & 0 \\ 0 & \mathbf{6} & 0 & 0 \end{bmatrix} \quad \boldsymbol{A^3} = \begin{bmatrix} 0 & 0 & 0 & 0 \\ 0 & 0 & 0 & 0 \\ 0 & 0 & 0 & 0 \\ \mathbf{6} & 0 & 0 & 0 \end{bmatrix} \quad \boldsymbol{A^4 = 0}$$

The diagonals move down at each step. So the series for e^{At} stops after four terms:

Same e^{At} as in Solution 1

$$e^{At} = I + At + \frac{(At)^2}{2} + \frac{(At)^3}{6} = \begin{bmatrix} 1 & & & \\ t & 1 & & \\ t^2 & 2t & 1 & \\ t^3 & 3t^2 & 3t & 1 \end{bmatrix}$$

The square of e^A is e^{2A}. But e^Ae^B and e^Be^A and e^{A+B} can be all different.

Problem Set 6.3

1 Find two λ's and $\boldsymbol{x}$'s so that $\boldsymbol{u} = e^{\lambda t}\boldsymbol{x}$ solves

$$\frac{d\boldsymbol{u}}{dt} = \begin{bmatrix} 4 & 3 \\ 0 & 1 \end{bmatrix} \boldsymbol{u}.$$

What combination $\boldsymbol{u} = c_1 e^{\lambda_1 t}\boldsymbol{x}_1 + c_2 e^{\lambda_2 t}\boldsymbol{x}_2$ starts from $\boldsymbol{u}(0) = (5, -2)$?

2 Solve Problem 1 for $\boldsymbol{u} = (y, z)$ by back substitution, z before y:

$$\text{Solve } \frac{dz}{dt} = z \text{ from } z(0) = -2. \qquad \text{Then solve } \frac{dy}{dt} = 4y + 3z \text{ from } y(0) = 5.$$

The solution for y will be a combination of e^{4t} and e^t. The λ's are 4 and 1.

3 (a) If every column of A adds to zero, why is $\lambda = 0$ an eigenvalue?

(b) With negative diagonal and positive off-diagonal adding to zero, $\boldsymbol{u}' = A\boldsymbol{u}$ will be a "continuous" Markov equation. Find the eigenvalues and eigenvectors, and the *steady state* as $t \to \infty$

$$\text{Solve } \frac{d\boldsymbol{u}}{dt} = \begin{bmatrix} -2 & 3 \\ 2 & -3 \end{bmatrix} \boldsymbol{u} \text{ with } \boldsymbol{u}(0) = \begin{bmatrix} 4 \\ 1 \end{bmatrix}. \quad \text{What is } \boldsymbol{u}(\infty)?$$

4 A door is opened between rooms that hold $v(0) = 30$ people and $w(0) = 10$ people. The movement between rooms is proportional to the difference $v - w$:

$$\frac{dv}{dt} = w - v \qquad \text{and} \qquad \frac{dw}{dt} = v - w.$$

Show that the total $v + w$ is constant (40 people). Find the matrix in $d\boldsymbol{u}/dt = A\boldsymbol{u}$ and its eigenvalues and eigenvectors. What are v and w at $t = 1$ and $t = \infty$?

5 Reverse the diffusion of people in Problem 4 to $d\boldsymbol{u}/dt = -A\boldsymbol{u}$:

$$\frac{dv}{dt} = v - w \qquad \text{and} \qquad \frac{dw}{dt} = w - v.$$

The total $v + w$ still remains constant. How are the λ's changed now that A is changed to $-A$? But show that $v(t)$ grows to infinity from $v(0) = 30$.

6 A has real eigenvalues but B has complex eigenvalues:

$$A = \begin{bmatrix} a & 1 \\ 1 & a \end{bmatrix} \quad B = \begin{bmatrix} b & -1 \\ 1 & b \end{bmatrix} \quad (a \text{ and } b \text{ are real})$$

Find the conditions on a and b so that all solutions of $d\boldsymbol{u}/dt = A\boldsymbol{u}$ and $d\boldsymbol{v}/dt = B\boldsymbol{v}$ approach zero as $t \to \infty$: $\mathbf{Re}\,\lambda < 0$ for all eigenvalues.

7 Suppose P is the projection matrix onto the $45°$ line $y = x$ in $\mathbf{R}^2$. What are its eigenvalues? If $d\boldsymbol{u}/dt = -P\boldsymbol{u}$ (notice minus sign) can you find the limit of $\boldsymbol{u}(t)$ at $t = \infty$ starting from $\boldsymbol{u}(0) = (3, 1)$?

8 The rabbit population shows fast growth (from $6r$) but loss to wolves (from $-2w$). The wolf population always grows in this model ($-w^2$ would control wolves):

$$\frac{dr}{dt} = 6r - 2w \qquad \text{and} \qquad \frac{dw}{dt} = 2r + w.$$

Find the eigenvalues and eigenvectors. If $r(0) = w(0) = 30$ what are the populations at time t? After a long time, what is the ratio of rabbits to wolves?

9 (a) Write $(4, 0)$ as a combination $c_1\boldsymbol{x}_1 + c_2\boldsymbol{x}_2$ of these two eigenvectors of A:

$$\begin{bmatrix} 0 & 1 \\ -1 & 0 \end{bmatrix}\begin{bmatrix} 1 \\ i \end{bmatrix} = i\begin{bmatrix} 1 \\ i \end{bmatrix} \qquad \begin{bmatrix} 0 & 1 \\ -1 & 0 \end{bmatrix}\begin{bmatrix} 1 \\ -i \end{bmatrix} = -i\begin{bmatrix} 1 \\ -i \end{bmatrix}.$$

(b) The solution to $d\boldsymbol{u}/dt = A\boldsymbol{u}$ starting from $(4, 0)$ is $c_1e^{it}\boldsymbol{x}_1 + c_2e^{-it}\boldsymbol{x}_2$. Substitute $e^{it} = \cos t + i\sin t$ and $e^{-it} = \cos t - i\sin t$ to find $\boldsymbol{u}(t)$.

Questions 10–13 reduce second-order equations to first-order systems for (y, y').

10 Find A to change the scalar equation $y'' = 5y' + 4y$ into a vector equation for $\boldsymbol{u} = (y, y')$:

$$\frac{d\boldsymbol{u}}{dt} = \begin{bmatrix} y' \\ y'' \end{bmatrix} = \begin{bmatrix} \quad & \quad \\ \quad & \quad \end{bmatrix}\begin{bmatrix} y \\ y' \end{bmatrix} = A\boldsymbol{u}.$$

What are the eigenvalues of A? Find them also by substituting $y = e^{\lambda t}$ into $y'' = 5y' + 4y$.

11 The solution to $y'' = 0$ is a straight line $y = C + Dt$. Convert to a matrix equation:

$$\frac{d}{dt}\begin{bmatrix} y \\ y' \end{bmatrix} = \begin{bmatrix} 0 & 1 \\ 0 & 0 \end{bmatrix}\begin{bmatrix} y \\ y' \end{bmatrix} \text{ has the solution } \begin{bmatrix} y \\ y' \end{bmatrix} = e^{At}\begin{bmatrix} y(0) \\ y'(0) \end{bmatrix}.$$

This matrix A has $\lambda = 0, 0$ and it cannot be diagonalized. Find A^2 and compute $e^{At} = I + At + \frac{1}{2}A^2t^2 + \cdots$. Multiply your e^{At} times $\big(y(0), y'(0)\big)$ to check the straight line $y(t) = y(0) + y'(0)t$.

12 Substitute $y = e^{\lambda t}$ into $y'' = 6y' - 9y$ to show that $\lambda = 3$ is a repeated root. This is trouble; we need a second solution after e^{3t}. The matrix equation is

$$\frac{d}{dt}\begin{bmatrix} y \\ y' \end{bmatrix} = \begin{bmatrix} 0 & 1 \\ -9 & 6 \end{bmatrix}\begin{bmatrix} y \\ y' \end{bmatrix}.$$

Show that this matrix has $\lambda = 3, 3$ and only one line of eigenvectors. *Trouble here too.* Show that the second solution to $y'' = 6y' - 9y$ is $y = te^{3t}$.

13 (a) Write down two familiar functions that solve the equation $d^2y/dt^2 = -9y$. Which one starts with $y(0) = 3$ and $y'(0) = 0$?

(b) This second-order equation $y'' = -9y$ produces a vector equation $\boldsymbol{u}' = A\boldsymbol{u}$:

$$\boldsymbol{u} = \begin{bmatrix} y \\ y' \end{bmatrix} \qquad \frac{d\boldsymbol{u}}{dt} = \begin{bmatrix} y' \\ y'' \end{bmatrix} = \begin{bmatrix} 0 & 1 \\ -9 & 0 \end{bmatrix} \begin{bmatrix} y \\ y' \end{bmatrix} = A\boldsymbol{u}.$$

Find $\boldsymbol{u}(t)$ by using the eigenvalues and eigenvectors of A: $\boldsymbol{u}(0) = (3, 0)$.

14 The matrix in this question is skew-symmetric ($A^{\mathrm{T}} = -A$):

$$\frac{d\boldsymbol{u}}{dt} = \begin{bmatrix} 0 & c & -b \\ -c & 0 & a \\ b & -a & 0 \end{bmatrix} \boldsymbol{u} \qquad \text{or} \qquad \begin{aligned} u_1' &= cu_2 - bu_3 \\ u_2' &= au_3 - cu_1 \\ u_3' &= bu_1 - au_2. \end{aligned}$$

(a) The derivative of $\|\boldsymbol{u}(t)\|^2 = u_1^2 + u_2^2 + u_3^2$ is $2u_1u_1' + 2u_2u_2' + 2u_3u_3'$. Substitute u_1', u_2', u_3' to get *zero*. Then $\|\boldsymbol{u}(t)\|^2$ stays equal to $\|\boldsymbol{u}(0)\|^2$.

(b) *When A is skew-symmetric, $Q = e^{At}$ is orthogonal.* Prove $Q^{\mathrm{T}} = e^{-At}$ from the series for $Q = e^{At}$. Then $Q^{\mathrm{T}}Q = I$.

15 A particular solution to $d\boldsymbol{u}/dt = A\boldsymbol{u} - \boldsymbol{b}$ is $\boldsymbol{u}_p = A^{-1}\boldsymbol{b}$, if A is invertible. The usual solutions to $d\boldsymbol{u}/dt = A\boldsymbol{u}$ give $\boldsymbol{u}_n$. Find the complete solution $\boldsymbol{u} = \boldsymbol{u}_p + \boldsymbol{u}_n$:

(a) $\dfrac{du}{dt} = u - 4$ (b) $\dfrac{d\boldsymbol{u}}{dt} = \begin{bmatrix} 1 & 0 \\ 1 & 1 \end{bmatrix} \boldsymbol{u} - \begin{bmatrix} 4 \\ 6 \end{bmatrix}$.

16 If c is not an eigenvalue of A, substitute $\boldsymbol{u} = e^{ct}\boldsymbol{v}$ and find a particular solution to $d\boldsymbol{u}/dt = A\boldsymbol{u} - e^{ct}\boldsymbol{b}$. How does it break down when c is an eigenvalue of A? The "nullspace" of $d\boldsymbol{u}/dt = A\boldsymbol{u}$ contains the usual solutions $e^{\lambda_i t}\boldsymbol{x}_i$.

17 Find a matrix A to illustrate each of the unstable regions in Figure 6.5:

(a) $\lambda_1 < 0$ and $\lambda_2 > 0$ (b) $\lambda_1 > 0$ and $\lambda_2 > 0$ (c) $\lambda = a \pm ib$ with $a > 0$.

Questions 18–27 are about the matrix exponential e^{At}.

18 Write five terms of the infinite series for e^{At}. Take the t derivative of each term. Show that you have four terms of Ae^{At}. Conclusion: $e^{At}\boldsymbol{u}_0$ solves $\boldsymbol{u}' = A\boldsymbol{u}$.

19 The matrix $B = \begin{bmatrix} 0 & -4 \\ 0 & 0 \end{bmatrix}$ has $B^2 = 0$. Find e^{Bt} from a (short) infinite series. Check that the derivative of e^{Bt} is Be^{Bt}.

20 Starting from $\boldsymbol{u}(0)$ the solution at time T is $e^{AT}\boldsymbol{u}(0)$. Go an additional time t to reach $e^{At}\, e^{AT}\boldsymbol{u}(0)$. This solution at time $t + T$ can also be written as _____. Conclusion: e^{At} times e^{AT} equals _____.

21 Write $A = \begin{bmatrix} 1 & 4 \\ 0 & 0 \end{bmatrix}$ in the form $X\Lambda X^{-1}$. Find e^{At} from $Xe^{\Lambda t}X^{-1}$.

22 If $A^2 = A$ show that the infinite series produces $e^{At} = I + (e^t - 1)A$. For $A = \left[\begin{smallmatrix}1 & 4\\ 0 & 0\end{smallmatrix}\right]$ in Problem 21 this gives $e^{At} =$ _____.

23 Generally $e^A e^B$ is different from $e^B e^A$. They are both different from e^{A+B}. Check this using Problems 21–22 and 19. (If $AB = BA$, all three are the same.)

$$A = \begin{bmatrix}1 & 4\\ 0 & 0\end{bmatrix} \qquad B = \begin{bmatrix}0 & -4\\ 0 & 0\end{bmatrix} \qquad A + B = \begin{bmatrix}1 & 0\\ 0 & 0\end{bmatrix}.$$

24 Write $A = \left[\begin{smallmatrix}1 & 1\\ 0 & 3\end{smallmatrix}\right]$ as $X\Lambda X^{-1}$. Multiply $Xe^{\Lambda t}X^{-1}$ to find the matrix exponential e^{At}. Check e^{At} and the derivative of e^{At} when $t = 0$.

25 Put $A = \left[\begin{smallmatrix}1 & 3\\ 0 & 0\end{smallmatrix}\right]$ into the infinite series to find e^{At}. First compute A^2 and A^n:

$$e^{At} = \begin{bmatrix}1 & 0\\ 0 & 1\end{bmatrix} + \begin{bmatrix}t & 3t\\ 0 & 0\end{bmatrix} + \frac{1}{2}\begin{bmatrix} & \\ & \end{bmatrix} + \cdots = \begin{bmatrix}e^t & \\ 0 & \end{bmatrix}.$$

26 (Recommended) Give two reasons why the matrix exponential e^{At} is never singular:

(a) Write down its inverse.

(b) Why are these eigenvalues nonzero? If $A\boldsymbol{x} = \lambda\boldsymbol{x}$ then $e^{At}\boldsymbol{x} =$ _____ $\boldsymbol{x}$.

27 Find a solution $x(t), y(t)$ that gets large as $t \to \infty$. To avoid this instability a scientist exchanged the two equations:

$$\begin{aligned} dx/dt &= 0x - 4y \\ dy/dt &= -2x + 2y \end{aligned} \qquad \text{becomes} \qquad \begin{aligned} dy/dt &= -2x + 2y \\ dx/dt &= 0x - 4y. \end{aligned}$$

Now the matrix $\left[\begin{smallmatrix}-2 & 2\\ 0 & -4\end{smallmatrix}\right]$ is stable. It has negative eigenvalues. How can this be?

Challenge Problems

28 Centering $y'' = -y$ in Example 3 will produce $Y_{n+1} - 2Y_n + Y_{n-1} = -(\Delta t)^2 Y_n$. This can be written as a one-step difference equation for $\boldsymbol{U} = (Y, Z)$:

$$\begin{aligned} Y_{n+1} &= Y_n + \Delta t\, Z_n \\ Z_{n+1} &= Z_n - \Delta t\, Y_{n+1} \end{aligned} \qquad \begin{bmatrix}1 & 0\\ \Delta t & 1\end{bmatrix}\begin{bmatrix}Y_{n+1}\\ Z_{n+1}\end{bmatrix} = \begin{bmatrix}1 & \Delta t\\ 0 & 1\end{bmatrix}\begin{bmatrix}Y_n\\ Z_n\end{bmatrix}$$

Invert the matrix on the left side to write this as $\boldsymbol{U}_{n+1} = A\boldsymbol{U}_n$. Show that $\det A = 1$. Choose the large time step $\Delta t = 1$ and find the eigenvalues λ_1 and $\lambda_2 = \overline{\lambda_1}$ of A:

$$A = \begin{bmatrix}1 & 1\\ -1 & 0\end{bmatrix} \text{ has } |\lambda_1| = |\lambda_2| = 1. \text{ Show that } A^6 \text{ is exactly } I.$$

After 6 steps to $t = 6$, $\boldsymbol{U}_6$ equals $\boldsymbol{U}_0$. The exact $y = \cos t$ returns to 1 at $t = 2\pi$.

29 That centered choice (*leapfrog method*) in Problem 28 is very successful for small time steps Δt. But find the eigenvalues of A for $\Delta t = \sqrt{2}$ and 2:

$$A = \begin{bmatrix} 1 & \sqrt{2} \\ -\sqrt{2} & -1 \end{bmatrix} \quad \text{and} \quad A = \begin{bmatrix} 1 & 2 \\ -2 & -3 \end{bmatrix}.$$

Both matrices have $|\lambda| = 1$. Compute A^4 in both cases and find the eigenvectors of A. That second value $\Delta t = 2$ is at the border of instability. Any time step $\Delta t > 2$ will lead to $|\lambda| > 1$, and the powers in $\boldsymbol{U}_n = A^n \boldsymbol{U}_0$ will explode.

Note You might say that nobody would compute with $\Delta t > 2$. But if an atom vibrates with $y'' = -1000000y$, then $\Delta t > .0002$ will give instability. Leapfrog has a very strict stability limit. $Y_{n+1} = Y_n + 3Z_n$ and $Z_{n+1} = Z_n - 3Y_{n+1}$ will explode because $\Delta t = 3$ is too large. The matrix has $|\lambda| > 1$.

30 Another good idea for $y'' = -y$ is the trapezoidal method (half forward/half back). ***This may be the best way to keep (Y_n, Z_n) exactly on a circle.***

$$\textbf{Trapezoidal} \quad \begin{bmatrix} 1 & -\Delta t/2 \\ \Delta t/2 & 1 \end{bmatrix} \begin{bmatrix} Y_{n+1} \\ Z_{n+1} \end{bmatrix} = \begin{bmatrix} 1 & \Delta t/2 \\ -\Delta t/2 & 1 \end{bmatrix} \begin{bmatrix} Y_n \\ Z_n \end{bmatrix}.$$

(a) Invert the left matrix to write this equation as $\boldsymbol{U}_{n+1} = A\boldsymbol{U}_n$. *Show that A is an orthogonal matrix*: $A^{\mathrm{T}}A = I$. **These points $\boldsymbol{U}_n$ never leave the circle.** $A = (I - B)^{-1}(I + B)$ is always an orthogonal matrix if $B^{\mathrm{T}} = -B$.

(b) (Optional MATLAB) Take 32 steps from $\boldsymbol{U}_0 = (1, 0)$ to $\boldsymbol{U}_{32}$ with $\Delta t = 2\pi/32$. Is $\boldsymbol{U}_{32} = \boldsymbol{U}_0$? I think there is a small error.

31 The ***cosine of a matrix*** is defined like e^A, by copying the series for $\cos t$:

$$\cos t = 1 - \frac{1}{2!}t^2 + \frac{1}{4!}t^4 - \cdots \qquad \cos A = I - \frac{1}{2!}A^2 + \frac{1}{4!}A^4 - \cdots$$

(a) If $A\boldsymbol{x} = \lambda\boldsymbol{x}$, multiply each term times $\boldsymbol{x}$ to find the eigenvalue of $\cos A$.

(b) Find the eigenvalues of $A = \begin{bmatrix} \pi & \pi \\ \pi & \pi \end{bmatrix}$ with eigenvectors $(1, 1)$ and $(1, -1)$. From the eigenvalues and eigenvectors of $\cos A$, find that matrix $C = \cos A$.

(c) The second derivative of $\cos(At)$ is $-A^2 \cos(At)$.

$$\boldsymbol{u}(t) = \mathbf{cos}(\boldsymbol{At})\,\boldsymbol{u}(0) \ \text{ solves } \ \frac{d^2\boldsymbol{u}}{dt^2} = -A^2\boldsymbol{u} \ \text{ starting from } \ \boldsymbol{u}'(0) = 0.$$

Construct $\boldsymbol{u}(t) = \cos(At)\,\boldsymbol{u}(0)$ by the usual three steps for that specific A:

1. Expand $\boldsymbol{u}(0) = (4, 2) = c_1\boldsymbol{x}_1 + c_2\boldsymbol{x}_2$ in the eigenvectors.

2. Multiply those eigenvectors by ______ and ______ (instead of $e^{\lambda t}$).

3. Add up the solution $\boldsymbol{u}(t) = c_1$ ______ $\boldsymbol{x}_1 + c_2$ ______ $\boldsymbol{x}_2$.

32 Explain one of these three proofs that the square of e^A is e^{2A}.

1. Solving with e^A from $t = 0$ to 1 and then 1 to 2 agrees with e^{2A} from 0 to 2.

2. The squared series $(I + A + \frac{A^2}{2} + \cdots)^2$ matches $I + 2A + \frac{(2A)^2}{2} + \cdots = e^{2A}$.

3. If A can be diagonalized then $(Xe^{\Lambda}X^{-1})(Xe^{\Lambda}X^{-1}) = Xe^{2\Lambda}X^{-1}$.

Notes on a Differential Equations Course

Certainly constant-coefficient linear equations are the simplest to solve. This section 6.3 of the book shows you part of a differential equations course, but there is more:

1. The second order equation $mu'' + bu' + ku = 0$ has major importance in applications. The exponents λ in the solutions $u = e^{\lambda t}$ solve $m\lambda^2 + b\lambda + k = 0$. The damping coefficient b is crucial:

   Underdamping $b^2 < 4mk$ **Critical damping** $b^2 = 4mk$ **Overdamping** $b^2 > 4mk$

   This decides whether λ_1 and λ_2 are real roots or repeated roots or complex roots. With complex λ's the solution $u(t)$ oscillates as it decays.

2. Our equations had no forcing term $f(t)$. We were finding the "nullspace solution". To $\boldsymbol{u}_n(t)$ we need to add a particular solution $u_p(t)$ that balances the force $f(t)$:

   Input $f(s)$ at time s
   Growth factor $e^{A(t-s)}$
   Add up outputs at time t

   $$\boldsymbol{u}_{\text{particular}} = \int_0^t e^{A(t-s)} f(s)\, ds.$$

   This solution can also be discovered and studied by *Laplace transform*—that is the established way to convert linear differential equations to linear algebra.

In real applications, nonlinear differential equations are solved numerically. A standard method with good accuracy is "Runge-Kutta"—named after its discoverers. Analysis can find the constant solutions to $du/dt = f(u)$. Those are solutions $u(t) = Y$ with $f(Y) = 0$ and $du/dt = 0$: *no movement*. We can also understand stability or instability near $u = Y$. Far from Y, the computer takes over.

This basic course is the subject of my textbook (companion to this one) on *Differential Equations and Linear Algebra*: **math.mit.edu/dela**.

6.4 Symmetric Matrices

1 A symmetric matrix S has n **real eigenvalues** λ_i and n **orthonormal eigenvectors** $\boldsymbol{q}_1, \ldots, \boldsymbol{q}_n$.

2 Every real symmetric S can be diagonalized: $\boxed{S = Q\Lambda Q^{-1} = \boldsymbol{Q\Lambda Q^{\mathrm{T}}}}$

3 The number of positive eigenvalues of S equals the number of positive pivots.

4 Antisymmetric matrices $A = -A^{\mathrm{T}}$ have *imaginary* λ's and *orthonormal* (*complex*) $\boldsymbol{q}$'s.

5 Section 9.2 explains why the test $S = S^{\mathrm{T}}$ becomes $\boldsymbol{S} = \overline{\boldsymbol{S}}^{\mathrm{T}}$ for *complex matrices*.

$S = \begin{bmatrix} 0 & i \\ -i & 0 \end{bmatrix} = \overline{S}^{\mathrm{T}}$ has real $\lambda = 1, -1$. $\qquad A = \begin{bmatrix} 0 & i \\ i & 0 \end{bmatrix} = -\overline{A}^{\mathrm{T}}$ has $\lambda = i, -i$.

It is no exaggeration to say that symmetric matrices S are the most important matrices the world will ever see—in the theory of linear algebra and also in the applications. We come immediately to the key question about symmetry. Not only the question, but also the two-part answer.

What is special about $\boldsymbol{Sx = \lambda x}$ ***when S is symmetric***?

We look for special properties of the eigenvalues λ and eigenvectors $\boldsymbol{x}$ when $S = S^{\mathrm{T}}$.

The diagonalization $S = X\Lambda X^{-1}$ will reflect the symmetry of S. We get some hint by transposing to $S^{\mathrm{T}} = (X^{-1})^{\mathrm{T}}\Lambda X^{\mathrm{T}}$. Those are the same since $S = S^{\mathrm{T}}$. Possibly X^{-1} in the first form equals X^{T} in the second form? Then $X^{\mathrm{T}}X = I$. That makes each eigenvector in X orthogonal to the other eigenvectors when $S = S^{\mathrm{T}}$. Here are the key facts:

1. **A symmetric matrix has only** ***real eigenvalues***.
2. **The** ***eigenvectors*** **can be chosen** ***orthonormal***.

Those n orthonormal eigenvectors go into the columns of X. Every symmetric matrix can be diagonalized. ***Its eigenvector matrix*** $\boldsymbol{X}$ ***becomes an orthogonal matrix*** $\boldsymbol{Q}$. Orthogonal matrices have $Q^{-1} = Q^{\mathrm{T}}$—what we suspected about the eigenvector matrix is true. To remember it we write Q instead of X, when we choose orthonormal eigenvectors.

Why do we use the word "choose"? Because the eigenvectors do not *have* to be unit vectors. Their lengths are at our disposal. We will choose unit vectors—eigenvectors of length one, which are orthonormal and not just orthogonal. Then $A = X\Lambda X^{-1}$ is in its special and particular form $S = Q\Lambda Q^{\mathrm{T}}$ for symmetric matrices.

(Spectral Theorem) Every symmetric matrix has the factorization $S = Q\Lambda Q^{\mathrm{T}}$ with real eigenvalues in Λ and orthonormal eigenvectors in the columns of Q:

Symmetric diagonalization $\quad S = Q\Lambda Q^{-1} = Q\Lambda Q^{\mathrm{T}} \quad \text{with} \quad Q^{-1} = Q^{\mathrm{T}}. \quad (1)$

It is easy to see that $Q\Lambda Q^{\mathrm{T}}$ is symmetric. Take its transpose. You get $(Q^{\mathrm{T}})^{\mathrm{T}}\Lambda^{\mathrm{T}}Q^{\mathrm{T}}$, which is $Q\Lambda Q^{\mathrm{T}}$ again. The harder part is to prove that every symmetric matrix has real λ's and orthonormal $\boldsymbol{x}$'s. This is the *"spectral theorem"* in mathematics and the *"principal axis theorem"* in geometry and physics. We have to prove it! No choice. I will approach the proof in three steps:

1. By an example, showing real λ's in Λ and orthonormal $\boldsymbol{x}$'s in Q.

2. By a proof of those facts when no eigenvalues are repeated.

3. By a proof that allows repeated eigenvalues (at the end of this section).

Example 1 Find the λ's and $\boldsymbol{x}$'s when $S = \begin{bmatrix} 1 & 2 \\ 2 & 4 \end{bmatrix}$ and $S - \lambda I = \begin{bmatrix} 1-\lambda & 2 \\ 2 & 4-\lambda \end{bmatrix}$.

Solution The determinant of $S - \lambda I$ is $\lambda^2 - 5\lambda$. The eigenvalues are 0 and 5 (*both real*). We can see them directly: $\lambda = 0$ is an eigenvalue because S is singular, and $\lambda = 5$ matches the *trace* down the diagonal of S: $0 + 5$ agrees with $1 + 4$.

Two eigenvectors are $(\mathbf{2}, \mathbf{-1})$ and $(\mathbf{1}, \mathbf{2})$—orthogonal but not yet orthonormal. The eigenvector for $\lambda = 0$ is in the *nullspace* of A. The eigenvector for $\lambda = 5$ is in the *column space*. We ask ourselves, why are the nullspace and column space perpendicular? The Fundamental Theorem says that the nullspace is perpendicular to the ***row space***—not the column space. But our matrix is *symmetric*! Its row and column spaces are the same. Its eigenvectors $(2, -1)$ and $(1, 2)$ must be (and are) perpendicular.

These eigenvectors have length $\sqrt{5}$. Divide them by $\sqrt{5}$ to get unit vectors. Put those unit eigenvectors into the columns of Q. Then $Q^{-1}SQ$ is Λ and $Q^{-1} = Q^{\mathrm{T}}$:

$$Q^{-1}SQ = \frac{1}{\sqrt{5}}\begin{bmatrix} 2 & -1 \\ 1 & 2 \end{bmatrix}\begin{bmatrix} 1 & 2 \\ 2 & 4 \end{bmatrix}\frac{1}{\sqrt{5}}\begin{bmatrix} 2 & 1 \\ -1 & 2 \end{bmatrix} = \begin{bmatrix} 0 & 0 \\ 0 & 5 \end{bmatrix} = \Lambda.$$

Now comes the n by n case. The λ's are real when $S = S^{\mathrm{T}}$ and $S\boldsymbol{x} = \lambda\boldsymbol{x}$.

Real Eigenvalues All the eigenvalues of a real symmetric matrix are real.

Proof Suppose that $S\boldsymbol{x} = \lambda\boldsymbol{x}$. Until we know otherwise, λ might be a complex number $a + ib$ (a and b real). *Its complex conjugate is* $\overline{\lambda} = a - ib$. Similarly the components of $\boldsymbol{x}$ may be complex numbers, and switching the signs of their imaginary parts gives $\overline{\boldsymbol{x}}$.

The good thing is that $\overline{\lambda}$ times $\overline{\boldsymbol{x}}$ is always the conjugate of λ times $\boldsymbol{x}$. So we can take conjugates of $S\boldsymbol{x} = \lambda\boldsymbol{x}$, remembering that S is real:

$$S\,\boldsymbol{x} = \lambda\,\boldsymbol{x} \quad \text{leads to} \quad S\,\overline{\boldsymbol{x}} = \overline{\lambda}\,\overline{\boldsymbol{x}}. \qquad \text{Transpose to} \quad \overline{\boldsymbol{x}}^{\mathrm{T}} S = \overline{\boldsymbol{x}}^{\mathrm{T}}\,\overline{\lambda}.$$

Now take the dot product of the first equation with $\overline{\boldsymbol{x}}$ and the last equation with $\boldsymbol{x}$:

$$\overline{\boldsymbol{x}}^{\mathrm{T}} S\,\boldsymbol{x} = \overline{\boldsymbol{x}}^{\mathrm{T}} \lambda\,\boldsymbol{x} \qquad \text{and also} \qquad \overline{\boldsymbol{x}}^{\mathrm{T}} S\,\boldsymbol{x} = \overline{\boldsymbol{x}}^{\mathrm{T}}\,\overline{\lambda}\,\boldsymbol{x}. \tag{2}$$

The left sides are the same so the right sides are equal. One equation has λ, the other has $\overline{\lambda}$. They multiply $\overline{\boldsymbol{x}}^{\mathrm{T}}\boldsymbol{x} = |x_1|^2 + |x_2|^2 + \cdots =$ length squared which is not zero. *Therefore* λ *must equal* $\overline{\lambda}$, and $a + ib$ equals $a - ib$. So $b = 0$ and $\lambda = a =$ *real*. Q.E.D.

The eigenvectors come from solving the real equation $(S - \lambda I)\boldsymbol{x} = \mathbf{0}$. So the $\boldsymbol{x}$'s are also real. The important fact is that they are perpendicular.

Orthogonal Eigenvectors Eigenvectors of a real symmetric matrix (when they correspond to different λ's) are always perpendicular.

Proof Suppose $S\boldsymbol{x} = \lambda_1\boldsymbol{x}$ and $S\boldsymbol{y} = \lambda_2\boldsymbol{y}$. We are assuming here that $\lambda_1 \neq \lambda_2$. Take dot products of the first equation with $\boldsymbol{y}$ and the second with $\boldsymbol{x}$:

$$\textbf{Use } S^{\mathrm{T}} = S \qquad (\lambda_1\boldsymbol{x})^{\mathrm{T}}\boldsymbol{y} = (S\boldsymbol{x})^{\mathrm{T}}\boldsymbol{y} = \boldsymbol{x}^{\mathrm{T}} S^{\mathrm{T}}\boldsymbol{y} = \boldsymbol{x}^{\mathrm{T}} S\boldsymbol{y} = \boldsymbol{x}^{\mathrm{T}}\lambda_2\boldsymbol{y}. \tag{3}$$

The left side is $\boldsymbol{x}^{\mathrm{T}}\lambda_1\boldsymbol{y}$, the right side is $\boldsymbol{x}^{\mathrm{T}}\lambda_2\boldsymbol{y}$. Since $\lambda_1 \neq \lambda_2$, this proves that $\boldsymbol{x}^{\mathrm{T}}\boldsymbol{y} = 0$. The eigenvector $\boldsymbol{x}$ (for λ_1) is perpendicular to the eigenvector $\boldsymbol{y}$ (for λ_2).

Example 2 The eigenvectors of a 2 by 2 symmetric matrix have a special form:

$$\textbf{Not widely known} \quad S = \begin{bmatrix} a & b \\ b & c \end{bmatrix} \quad \text{has} \quad \boldsymbol{x}_1 = \begin{bmatrix} b \\ \lambda_1 - a \end{bmatrix} \quad \text{and} \quad \boldsymbol{x}_2 = \begin{bmatrix} \lambda_2 - c \\ b \end{bmatrix}. \tag{4}$$

This is in the Problem Set. The point here is that $\boldsymbol{x}_1$ is perpendicular to $\boldsymbol{x}_2$:

$$\boldsymbol{x}_1^{\mathrm{T}}\boldsymbol{x}_2 = b(\lambda_2 - c) + (\lambda_1 - a)b = b(\lambda_1 + \lambda_2 - a - c) = 0.$$

This is zero because $\lambda_1 + \lambda_2$ equals the trace $a + c$. Thus $\boldsymbol{x}_1^{\mathrm{T}}\boldsymbol{x}_2 = 0$. Eagle eyes might notice the special case $S = I$, when b and $\lambda_1 - a$ and $\lambda_2 - c$ and $\boldsymbol{x}_1$ and $\boldsymbol{x}_2$ are all zero. Then $\lambda_1 = \lambda_2 = 1$ is repeated. But of course $S = I$ has perpendicular eigenvectors.

Symmetric matrices S have orthogonal eigenvector matrices Q. Look at this again:

$$\textbf{Symmetry} \qquad S = X\Lambda X^{-1} \quad \text{becomes} \quad S = Q\Lambda Q^{\mathrm{T}} \quad \text{with} \quad Q^{\mathrm{T}}Q = I.$$

This says that every 2 by 2 symmetric matrix is (**rotation**)(**stretch**)(**rotate back**)

$$S = Q\Lambda Q^{\mathrm{T}} = \begin{bmatrix} \boldsymbol{q}_1 & \boldsymbol{q}_2 \end{bmatrix} \begin{bmatrix} \lambda_1 & \\ & \lambda_2 \end{bmatrix} \begin{bmatrix} \boldsymbol{q}_1^{\mathrm{T}} \\ \boldsymbol{q}_2^{\mathrm{T}} \end{bmatrix}. \tag{5}$$

Columns $\boldsymbol{q}_1$ ***and*** $\boldsymbol{q}_2$ ***multiply rows*** $\lambda_1\boldsymbol{q}_1^{\mathrm{T}}$ ***and*** $\lambda_2\boldsymbol{q}_2^{\mathrm{T}}$ ***to produce*** $\boldsymbol{S = \lambda_1 q_1 q_1^{\mathrm{T}} + \lambda_2 q_2 q_2^{\mathrm{T}}}$.

Every symmetric matrix $\quad S = Q\Lambda Q^{\mathrm{T}} = \lambda_1 \boldsymbol{q}_1 \boldsymbol{q}_1^{\mathrm{T}} + \cdots + \lambda_n \boldsymbol{q}_n \boldsymbol{q}_n^{\mathrm{T}}. \quad (6)$

Remember the steps to this great result (the spectral theorem).

Section 6.2	Write $A\boldsymbol{x}_i = \lambda_i \boldsymbol{x}_i$ in matrix form	$AX = X\Lambda$ or $A = X\Lambda X^{-1}$
Section 6.4	Orthonormal $\boldsymbol{x}_i = \boldsymbol{q}_i$ gives $X = Q$	$S = Q\Lambda Q^{-1} = Q\Lambda Q^{\mathrm{T}}$

$Q\Lambda Q^{\mathrm{T}}$ in equation (6) has columns of $Q\Lambda$ times rows of Q^{T}. Here is a direct proof.

S has correct eigenvectors
Those q's are orthonormal $\quad S\boldsymbol{q}_i = (\lambda_1 \boldsymbol{q}_1 \boldsymbol{q}_1^{\mathrm{T}} + \cdots + \lambda_n \boldsymbol{q}_n \boldsymbol{q}_n^{\mathrm{T}})\, \boldsymbol{q}_i = \lambda_i \boldsymbol{q}_i\,. \quad (7)$

Complex Eigenvalues of Real Matrices

For any real matrix, $S\,x = \lambda\,\boldsymbol{x}$ gives $S\,\overline{\boldsymbol{x}} = \overline{\lambda}\,\overline{\boldsymbol{x}}$. For a symmetric matrix, λ and $\boldsymbol{x}$ turn out to be real. Those two equations become the same. But a *non*symmetric matrix can easily produce λ and $\boldsymbol{x}$ that are complex. Then $A\,\overline{\boldsymbol{x}} = \overline{\lambda}\,\overline{\boldsymbol{x}}$ is true but different from $A\,\boldsymbol{x} = \lambda\,\boldsymbol{x}$. We get another complex eigenvalue (which is $\overline{\lambda}$) and a new eigenvector (which is $\overline{\boldsymbol{x}}$):

For real matrices, complex λ's and x's come in "conjugate pairs."

$$\begin{aligned}\lambda &= a + ib \\ \overline{\lambda} &= a - ib\end{aligned} \qquad \textit{If} \quad A\,\boldsymbol{x} = \lambda\,\boldsymbol{x} \quad \textit{then} \quad A\,\overline{\boldsymbol{x}} = \overline{\lambda}\,\overline{\boldsymbol{x}}. \tag{8}$$

Example 3 $\quad A = \begin{bmatrix} \cos\boldsymbol{\theta} & -\sin\boldsymbol{\theta} \\ \sin\boldsymbol{\theta} & \cos\boldsymbol{\theta} \end{bmatrix}$ has $\lambda_1 = \cos\theta + i\sin\theta$ and $\lambda_2 = \cos\theta - i\sin\theta$.

Those eigenvalues are conjugate to each other. They are λ and $\overline{\lambda}$. The eigenvectors must be $\boldsymbol{x}$ and $\overline{\boldsymbol{x}}$, because A is real:

$$\begin{aligned} &\textbf{This is } \lambda\,\boldsymbol{x} \qquad A\boldsymbol{x} = \begin{bmatrix} \cos\theta & -\sin\theta \\ \sin\theta & \cos\theta \end{bmatrix} \begin{bmatrix} 1 \\ -i \end{bmatrix} = (\cos\theta + i\sin\theta) \begin{bmatrix} 1 \\ -i \end{bmatrix} \\ &\textbf{This is } \overline{\lambda}\,\overline{\boldsymbol{x}} \qquad A\overline{\boldsymbol{x}} = \begin{bmatrix} \cos\theta & -\sin\theta \\ \sin\theta & \cos\theta \end{bmatrix} \begin{bmatrix} 1 \\ i \end{bmatrix} = (\cos\theta - i\sin\theta) \begin{bmatrix} 1 \\ i \end{bmatrix}. \end{aligned} \tag{9}$$

Those eigenvectors $(1, -i)$ and $(1, i)$ are complex conjugates because A is real.

For this rotation matrix the absolute value is $|\lambda| = 1$, because $\cos^2\theta + \sin^2\theta = 1$. ***This fact*** $|\lambda| = 1$ ***holds for the eigenvalues of every orthogonal matrix Q.***

We apologize that a touch of complex numbers slipped in. They are unavoidable even when the matrix is real. Chapter 9 goes beyond complex numbers λ and complex eigenvectors $\boldsymbol{x}$ to complex matrices A. Then you have the whole picture.

We end with two optional discussions.

Eigenvalues versus Pivots

The eigenvalues of A are very different from the pivots. For eigenvalues, we solve $\det(A - \lambda I) = 0$. For pivots, we use elimination. The only connection so far is this:

$$\textit{product of pivots} = \textit{determinant} = \textit{product of eigenvalues.}$$

We are assuming a full set of pivots $d_1, \ldots, d_n$. There are n real eigenvalues $\lambda_1, \ldots, \lambda_n$. The d's and λ's are not the same, but they come from the same symmetric matrix. Those d's and λ's have a hidden relation. ***For symmetric matrices the pivots and the eigenvalues have the same signs***:

> ***The number of positive eigenvalues of $S = S^{\mathrm{T}}$ equals the number of positive pivots.***
> Special case: S has all $\lambda_i > 0$ if and only if all pivots are positive.

That special case is an all-important fact for **positive definite matrices** in Section 6.5.

Example 4 This symmetric matrix has one positive eigenvalue and one positive pivot:

$$\textbf{Matching signs} \qquad S = \begin{bmatrix} 1 & 3 \\ 3 & 1 \end{bmatrix} \qquad \begin{array}{l} \text{has pivots } 1 \text{ and } -8 \\ \text{eigenvalues } 4 \text{ and } -2. \end{array}$$

The signs of the pivots match the signs of the eigenvalues, one plus and one minus. This could be false when the matrix is not symmetric:

$$\textbf{Opposite signs} \quad B = \begin{bmatrix} 1 & 6 \\ -1 & -4 \end{bmatrix} \qquad \begin{array}{l} \text{has pivots } 1 \text{ and } 2 \\ \text{eigenvalues } -1 \text{ and } -2. \end{array}$$

The diagonal entries are a third set of numbers and we say nothing about them.

Here is a proof that the pivots and eigenvalues have matching signs, when $S = S^{\mathrm{T}}$.

You see it best when the pivots are divided out of the rows of U. Then S is LDL^{T}. The diagonal pivot matrix D goes between triangular matrices L and L^{T}:

$$\begin{bmatrix} 1 & 3 \\ 3 & 1 \end{bmatrix} = \begin{bmatrix} 1 & 0 \\ \mathbf{3} & 1 \end{bmatrix} \begin{bmatrix} 1 & \\ & -8 \end{bmatrix} \begin{bmatrix} 1 & \mathbf{3} \\ 0 & 1 \end{bmatrix} \qquad \textbf{This is } S = LDL^{\mathrm{T}}\textbf{. It is symmetric.}$$

Watch the eigenvalues of LDL^{T} when L moves to I. S changes to D.

The eigenvalues of LDL^{T} are 4 and -2. The eigenvalues of IDI^{T} are 1 and -8 (the pivots!). The eigenvalues are changing, as the "3" in L moves to zero. But to change *sign*, a real eigenvalue would have to cross zero. The matrix would at that moment be singular. Our changing matrix always has pivots 1 and -8, so it is *never* singular. The signs cannot change, as the λ's move to the d's.

We repeat the proof for any $S = LDL^{\mathrm{T}}$. Move L toward I, by moving the off-diagonal entries to zero. The pivots are not changing and not zero. The eigenvalues λ of LDL^{T} change to the eigenvalues d of IDI^{T}. Since these eigenvalues cannot cross zero as they move into the pivots, their signs cannot change. **Same signs for the λ's and d's.**

This connects the two halves of applied linear algebra—pivots and eigenvalues.

All Symmetric Matrices are Diagonalizable

When no eigenvalues of A are repeated, the eigenvectors are sure to be independent. Then A can be diagonalized. But a repeated eigenvalue can produce a shortage of eigenvectors. This *sometimes* happens for nonsymmetric matrices. It *never* happens for symmetric matrices. ***There are always enough eigenvectors to diagonalize*** $S = S^{\mathrm{T}}$.

Here is one idea for a proof. Change S slightly by a diagonal matrix **diag**$(c, 2c, \ldots, nc)$. If c is very small, the new symmetric matrix will have no repeated eigenvalues. Then we know it has a full set of orthonormal eigenvectors. As $c \to 0$ we obtain n orthonormal eigenvectors of the original S—even if some eigenvalues of that S are repeated.

Every mathematician knows that this argument is incomplete. How do we guarantee that the small diagonal matrix will separate the eigenvalues? (I am sure this is true.)

A different proof comes from a useful new factorization that applies to *all square matrices* A, symmetric or not. This new factorization quickly produces $S = Q\Lambda Q^{\mathrm{T}}$ with a full set of real orthonormal eigenvectors when S is any real symmetric matrix.

Every square A ***factors into*** QTQ^{-1} ***where*** T ***is upper triangular and*** $\overline{Q}^{\mathrm{T}} = Q^{-1}$.
If A ***has real eigenvalues then*** Q ***and*** T ***can be chosen real:*** $Q^{\mathrm{T}}Q = I$.

This is Schur's Theorem. Its proof will go onto the website **math.mit.edu/linearalgebra**. Here I will show how T is diagonal ($T = \Lambda$) when S is symmetric. Then S is $Q\Lambda Q^{\mathrm{T}}$.

We know that every symmetric S has real eigenvalues, and Schur allows repeated λ's:

Schur's $S = QTQ^{-1}$ means that $T = Q^{\mathrm{T}}SQ$. The transpose is again $Q^{\mathrm{T}}SQ$.
The triangular T *is symmetric when* $S^{\mathrm{T}} = S$. Then T must be diagonal and $T = \Lambda$.

This proves that $S = Q\Lambda Q^{-1}$. The symmetric S has n orthonormal eigenvectors in Q.

Note. I have added another proof in Section 7.2 of this book. That proof shows how the eigenvalues λ can be described *one at a time*. The largest λ_1 is the maximum of $\boldsymbol{x}^{\mathrm{T}}S\boldsymbol{x}/\boldsymbol{x}^{\mathrm{T}}\boldsymbol{x}$. Then λ_2 (second largest) is again the same maximum, if we only allow vectors $\boldsymbol{x}$ that are perpendicular to the first eigenvector. The third eigenvalue λ_3 comes by requiring $\boldsymbol{x}^{\mathrm{T}}\boldsymbol{q}_1 = 0$ and $\boldsymbol{x}^{\mathrm{T}}\boldsymbol{q}_2 = 0\ldots$

This proof is placed in Chapter 7 because the same one-at-a-time idea succeeds for the *singular values of any matrix* A. **Singular values come from** $\boldsymbol{A^{\mathrm{T}}A}$ **and** $\boldsymbol{AA^{\mathrm{T}}}$.

■ REVIEW OF THE KEY IDEAS ■

1. Every symmetric matrix S has *real eigenvalues* and *perpendicular eigenvectors.*
2. Diagonalization becomes $\boldsymbol{S = Q\Lambda Q^{\mathrm{T}}}$ with an orthogonal eigenvector matrix Q.
3. All symmetric matrices are diagonalizable, even with repeated eigenvalues.
4. The signs of the eigenvalues match the signs of the pivots, when $S = S^{\mathrm{T}}$.
5. Every square matrix can be "triangularized" by $A = QTQ^{-1}$. If $A = S$ then $T = \Lambda$.

■ WORKED EXAMPLES ■

6.4 A What matrix A has eigenvalues $\lambda = 1, -1$ and eigenvectors $\boldsymbol{x}_1 = (\cos\theta, \sin\theta)$ and $\boldsymbol{x}_2 = (-\sin\theta, \cos\theta)$? Which of these properties can be predicted in advance?

$$A = A^{\mathrm{T}} \qquad A^2 = I \qquad \det A = -1 \qquad \text{pivot are} + \text{and} - \qquad A^{-1} = A$$

Solution All those properties can be predicted! With real eigenvalues $1, -1$ and orthonormal $\boldsymbol{x}_1$ and $\boldsymbol{x}_2$, the matrix $A = Q\Lambda Q^{\mathrm{T}}$ must be symmetric. The eigenvalues 1 and -1 tell us that $A^2 = I$ (since $\lambda^2 = 1$) and $A^{-1} = A$ (same thing) and $\det A = -1$. The two pivots must be positive and negative like the eigenvalues, since A is symmetric.

The matrix will be a reflection. Vectors in the direction of $\boldsymbol{x}_1$ are unchanged by A (since $\lambda = 1$). Vectors in the perpendicular direction are reversed (since $\lambda = -1$). The reflection $A = Q\Lambda Q^{\mathrm{T}}$ is across the "θ-line". Write c for $\cos\theta$ and s for $\sin\theta$:

$$A = \begin{bmatrix} c & -s \\ s & c \end{bmatrix} \begin{bmatrix} 1 & 0 \\ 0 & -1 \end{bmatrix} \begin{bmatrix} c & s \\ -s & c \end{bmatrix} = \begin{bmatrix} c^2 - s^2 & 2cs \\ 2cs & s^2 - c^2 \end{bmatrix} = \begin{bmatrix} \cos 2\theta & \sin 2\theta \\ \sin 2\theta & -\cos 2\theta \end{bmatrix}.$$

Notice that $\boldsymbol{x} = (1, 0)$ goes to $A\boldsymbol{x} = (\cos 2\theta, \sin 2\theta)$ on the 2θ-line. And $(\cos 2\theta, \sin 2\theta)$ goes back across the θ-line to $\boldsymbol{x} = (1, 0)$.

6.4 B Find the eigenvalues and eigenvectors (discrete sines and cosines) of A_3 and B_4.

$$A_3 = \begin{bmatrix} 2 & -1 & 0 \\ -1 & 2 & -1 \\ 0 & -1 & 2 \end{bmatrix} \qquad B_4 = \begin{bmatrix} 1 & -1 & & \\ -1 & 2 & -1 & \\ & -1 & 2 & -1 \\ & & -1 & 1 \end{bmatrix}$$

The $-1, 2, -1$ pattern in both matrices is a "second difference". This is like a second derivative. Then $Ax = \lambda x$ and $Bx = \lambda x$ are like $d^2x/dt^2 = \lambda x$. This has eigenvectors $x = \sin kt$ and $x = \cos kt$ that are the bases for Fourier series.

A_n and B_n lead to "discrete sines" and "discrete cosines" that are the bases for the *Discrete Fourier Transform*. This DFT is absolutely central to all areas of digital signal processing. The favorite choice for JPEG in image processing has been B_8 of size $n = 8$.

Solution The eigenvalues of A_3 are $\lambda = 2 - \sqrt{2}$ and 2 and $2 + \sqrt{2}$ (see **6.3 B**). Their sum is 6 (the trace of A_3) and their product is 4 (the determinant). The eigenvector matrix gives the "Discrete Sine Transform" and the eigenvectors fall onto sine curves.

$$\textbf{Sines} = \begin{bmatrix} 1 & \sqrt{2} & 1 \\ \sqrt{2} & 0 & -\sqrt{2} \\ 1 & -\sqrt{2} & 1 \end{bmatrix} \qquad \textbf{Cosines} = \begin{bmatrix} 1 & 1 & 1 & 1 \\ 1 & \sqrt{2}-1 & -1 & 1-\sqrt{2} \\ 1 & 1-\sqrt{2} & -1 & \sqrt{2}-1 \\ 1 & -1 & 1 & -1 \end{bmatrix}$$

Sine matrix = Eigenvectors of A_3 **Cosine matrix = Eigenvectors of B_4**

The eigenvalues of B_4 are $\lambda = 2 - \sqrt{2}$ and 2 and $2 + \sqrt{2}$ and 0 (the same as for A_3, plus the zero eigenvalue). The trace is still 6, but the determinant is now zero. The eigenvector matrix C gives the 4-point "Discrete Cosine Transform". The graph on the Web shows how the first two eigenvectors fall onto cosine curves. (So do all the eigenvectors of B.) These eigenvectors match cosines at the *halfway points* $\pi/8, 3\pi/8, 5\pi/8, 7\pi/8$.

Problem Set 6.4

1 Which of these matrices ASB will be symmetric with eigenvalues 1 and -1?

$$\begin{bmatrix}1&0\\1&1\end{bmatrix}\begin{bmatrix}1&\\&-1\end{bmatrix}\begin{bmatrix}1&1\\0&1\end{bmatrix}\quad\begin{bmatrix}1&0\\1&1\end{bmatrix}\begin{bmatrix}1&\\&-1\end{bmatrix}\begin{bmatrix}1&0\\-1&1\end{bmatrix}\quad\begin{bmatrix}0&-1\\1&0\end{bmatrix}\begin{bmatrix}1&\\&-1\end{bmatrix}\begin{bmatrix}0&1\\-1&0\end{bmatrix}$$

$B = A^{\mathrm{T}}$ doesn't do it. $B = A^{-1}$ doesn't do it. $B =$ ___ $=$ ___ will succeed. So B must be an _____ matrix.

2 Suppose $S = S^{\mathrm{T}}$. When is ASB also symmetric with the same eigenvalues as S?

(a) Transpose ASB to see that it stays symmetric when $B =$ _____.

(b) ASB is similar to S (same eigenvalues) when $B =$ _____.

Put (a) and (b) together. The symmetric matrices similar to S look like (___)S(___).

3 Write A as $S + N$, symmetric matrix S plus skew-symmetric matrix N:

$$A = \begin{bmatrix}1&2&4\\4&3&0\\8&6&5\end{bmatrix} = S + N \qquad (S^{\mathrm{T}} = S \text{ and } N^{\mathrm{T}} = -N).$$

For any square matrix, $S = \frac{1}{2}(A + A^{\mathrm{T}})$ and $N =$ _____ add up to A.

4 If C is symmetric prove that $A^{\mathrm{T}}CA$ is also symmetric. (Transpose it.) When A is 6 by 3, what are the shapes of C and $A^{\mathrm{T}}CA$?

5 Find the eigenvalues and the unit eigenvectors of

$$S = \begin{bmatrix}2&2&2\\2&0&0\\2&0&0\end{bmatrix}.$$

6 Find an orthogonal matrix Q that diagonalizes $S = \begin{bmatrix}-2&6\\6&7\end{bmatrix}$. What is Λ?

7 Find an orthogonal matrix Q that diagonalizes this symmetric matrix:

$$S = \begin{bmatrix}1&0&2\\0&-1&-2\\2&-2&0\end{bmatrix}.$$

8 Find *all* orthogonal matrices that diagonalize $S = \begin{bmatrix}9&12\\12&16\end{bmatrix}$.

9 (a) Find a symmetric matrix $\begin{bmatrix}1&b\\b&1\end{bmatrix}$ that has a negative eigenvalue.

(b) How do you know it must have a negative pivot?

(c) How do you know it can't have two negative eigenvalues?

10 If $A^3 = 0$ then the eigenvalues of A must be ____. Give an example that has $A \neq 0$. But if A is symmetric, diagonalize it to prove that A must be a zero matrix.

11 If $\lambda = a + ib$ is an eigenvalue of a real matrix A, then its conjugate $\overline{\lambda} = a - ib$ is also an eigenvalue. (If $A\boldsymbol{x} = \lambda\boldsymbol{x}$ then also $A\overline{\boldsymbol{x}} = \overline{\lambda}\overline{\boldsymbol{x}}$: a conjugate pair λ and $\overline{\lambda}$.) Explain why every real 3 by 3 matrix has at least one real eigenvalue.

12 Here is a quick "proof" that the eigenvalues of every real matrix A are real:

$$\textbf{False proof} \quad A\boldsymbol{x} = \lambda\boldsymbol{x} \quad \text{gives} \quad \boldsymbol{x}^{\mathrm{T}}A\boldsymbol{x} = \lambda\boldsymbol{x}^{\mathrm{T}}\boldsymbol{x} \quad \text{so} \quad \lambda = \frac{\boldsymbol{x}^{\mathrm{T}}A\boldsymbol{x}}{\boldsymbol{x}^{\mathrm{T}}\boldsymbol{x}} = \frac{\text{real}}{\text{real}}.$$

Find the flaw in this reasoning—a hidden assumption that is not justified. You could test those steps on the 90° rotation matrix $[\,0 \;\; -1; \;\; 1 \;\; 0\,]$ with $\lambda = i$ and $\boldsymbol{x} = (i, 1)$.

13 Write S and B in the form $\lambda_1\boldsymbol{x}_1\boldsymbol{x}_1^{\mathrm{T}} + \lambda_2\boldsymbol{x}_2\boldsymbol{x}_2^{\mathrm{T}}$ of the spectral theorem $Q\Lambda Q^T$:

$$S = \begin{bmatrix} 3 & 1 \\ 1 & 3 \end{bmatrix} \qquad B = \begin{bmatrix} 9 & 12 \\ 12 & 16 \end{bmatrix} \qquad (\text{keep } \|\boldsymbol{x}_1\| = \|\boldsymbol{x}_2\| = 1).$$

14 Every 2 by 2 symmetric matrix is $\lambda_1\boldsymbol{x}_1\boldsymbol{x}_1^{\mathrm{T}} + \lambda_2\boldsymbol{x}_2\boldsymbol{x}_2^{\mathrm{T}} = \lambda_1 P_1 + \lambda_2 P_2$. Explain $P_1 + P_2 = \boldsymbol{x}_1\boldsymbol{x}_1^{\mathrm{T}} + \boldsymbol{x}_2\boldsymbol{x}_2^{\mathrm{T}} = I$ from columns times rows of Q. Why is $P_1P_2 = 0$?

15 What are the eigenvalues of $A = \left[\begin{smallmatrix} 0 & b \\ -b & 0 \end{smallmatrix}\right]$? Create a 4 by 4 antisymmetric matrix $(A^{\mathrm{T}} = -A)$ and verify that all its eigenvalues are imaginary.

16 (Recommended) This matrix M is antisymmetric and also ____. Then all its eigenvalues are pure imaginary and they also have $|\lambda| = 1$. ($\|M\boldsymbol{x}\| = \|\boldsymbol{x}\|$ for every $\boldsymbol{x}$ so $\|\lambda\boldsymbol{x}\| = \|\boldsymbol{x}\|$ for eigenvectors.) Find all four eigenvalues from the trace of M:

$$M = \frac{1}{\sqrt{3}} \begin{bmatrix} 0 & 1 & 1 & 1 \\ -1 & 0 & -1 & 1 \\ -1 & 1 & 0 & -1 \\ -1 & -1 & 1 & 0 \end{bmatrix} \quad \text{can only have eigenvalues } i \text{ or } -i.$$

17 Show that this A (**symmetric but complex**) has only one line of eigenvectors:

$$A = \begin{bmatrix} i & 1 \\ 1 & -i \end{bmatrix} \quad \text{is not even diagonalizable: eigenvalues } \lambda = 0, 0.$$

$A^{\mathrm{T}} = A$ is not such a special property for complex matrices. The good property is $\overline{A}^{\mathrm{T}} = A$ (Section 9.2). Then all λ's are real and the eigenvectors are orthogonal.

18 Even if A is rectangular, the block matrix $S = \begin{bmatrix} 0 & A \\ A^{\mathrm{T}} & 0 \end{bmatrix}$ is symmetric:

$$S\boldsymbol{x} = \lambda\boldsymbol{x} \quad \text{is} \quad \begin{bmatrix} 0 & A \\ A^{\mathrm{T}} & 0 \end{bmatrix} \begin{bmatrix} \boldsymbol{y} \\ \boldsymbol{z} \end{bmatrix} = \lambda \begin{bmatrix} \boldsymbol{y} \\ \boldsymbol{z} \end{bmatrix} \quad \text{which is} \quad \begin{matrix} A\boldsymbol{z} = \lambda\boldsymbol{y} \\ A^{\mathrm{T}}\boldsymbol{y} = \lambda\boldsymbol{z}. \end{matrix}$$

(a) Show that $-\lambda$ is also an eigenvalue, with the eigenvector $(\boldsymbol{y}, -\boldsymbol{z})$.

(b) Show that $A^{\mathrm{T}} A\boldsymbol{z} = \lambda^2 \boldsymbol{z}$, so that λ^2 is an eigenvalue of $A^{\mathrm{T}} A$.

(c) If $A = I$ (2 by 2) find all four eigenvalues and eigenvectors of S.

19 If $A = \begin{bmatrix} 1 \\ 1 \end{bmatrix}$ in Problem 18, find all three eigenvalues and eigenvectors of S.

20 ***Another proof that eigenvectors are perpendicular when $S = S^{\mathrm{T}}$.*** Two steps:

1. Suppose $S\boldsymbol{x} = \lambda\boldsymbol{x}$ and $S\boldsymbol{y} = 0\boldsymbol{y}$ and $\lambda \neq 0$. Then $\boldsymbol{y}$ is in the nullspace and $\boldsymbol{x}$ is in the column space. They are perpendicular because ____. Go carefully—why are these subspaces orthogonal?
2. If $S\boldsymbol{y} = \beta\boldsymbol{y}$, apply that argument to $S - \beta I$. One eigenvalue of $S - \beta I$ moves to zero. The eigenvectors $\boldsymbol{x}, \boldsymbol{y}$ stay the same—so they are perpendicular.

21 Find the eigenvector matrices Q for S and X for B. Show that X doesn't collapse at $d = 1$, even though $\lambda = 1$ is repeated. Are those eigenvectors perpendicular?

$$S = \begin{bmatrix} 0 & d & 0 \\ d & 0 & 0 \\ 0 & 0 & 1 \end{bmatrix} \qquad B = \begin{bmatrix} -d & 0 & 1 \\ 0 & 1 & 0 \\ 0 & 0 & d \end{bmatrix} \qquad \text{have} \quad \lambda = 1, d, -d.$$

22 Write a 2 by 2 *complex* matrix with $\overline{S}^{\mathrm{T}} = S$ (a "Hermitian matrix"). Find λ_1 and λ_2 for your complex matrix. Check that $\overline{\boldsymbol{x}}_1^{\mathrm{T}} \boldsymbol{x}_2 = 0$ (this is complex orthogonality).

23 ***True*** (with reason) ***or false*** (with example).

(a) A matrix with real eigenvalues and n real eigenvectors is symmetric.

(b) A matrix with real eigenvalues and n orthonormal eigenvectors is symmetric.

(c) The inverse of an invertible symmetric matrix is symmetric.

(d) The eigenvector matrix Q of a symmetric matrix is symmetric.

24 (A paradox for instructors) If $AA^{\mathrm{T}} = A^{\mathrm{T}} A$ then A and A^{T} share the same eigenvectors (true). A and A^{T} always share the same eigenvalues. Find the flaw in this conclusion: A and A^{T} must have the same X and same Λ. Therefore A equals A^{T}.

25 (Recommended) Which of these classes of matrices do A and B belong to: Invertible, orthogonal, projection, permutation, diagonalizable, Markov?

$$A = \begin{bmatrix} 0 & 0 & 1 \\ 0 & 1 & 0 \\ 1 & 0 & 0 \end{bmatrix} \qquad B = \frac{1}{3} \begin{bmatrix} 1 & 1 & 1 \\ 1 & 1 & 1 \\ 1 & 1 & 1 \end{bmatrix}.$$

Which of these factorizations are possible for A and B: $\boldsymbol{LU}$, $\boldsymbol{QR}$, $\boldsymbol{X\Lambda X^{-1}}$, $\boldsymbol{Q\Lambda Q^{\mathrm{T}}}$?

26 What number b in $A = \begin{bmatrix} 2 & b \\ 1 & 0 \end{bmatrix}$ makes $A = Q\Lambda Q^{\mathrm{T}}$ possible? What number will make it impossible to diagonalize A? What number makes A singular?

27 Find all 2 by 2 matrices that are orthogonal and also symmetric. Which two numbers can be eigenvalues of those two matrices?

28 This A is nearly symmetric. But its eigenvectors are far from orthogonal:

$$A = \begin{bmatrix} 1 & 10^{-15} \\ 0 & 1+10^{-15} \end{bmatrix} \quad \text{has eigenvectors} \quad \begin{bmatrix} 1 \\ 0 \end{bmatrix} \quad \text{and} \quad [?]$$

What is the angle between the eigenvectors?

29 (MATLAB) Take two symmetric matrices with different eigenvectors, say $A = \left[\begin{smallmatrix} \mathbf{1} & \mathbf{0} \\ \mathbf{0} & \mathbf{2} \end{smallmatrix}\right]$ and $B = \left[\begin{smallmatrix} \mathbf{8} & \mathbf{1} \\ \mathbf{1} & \mathbf{0} \end{smallmatrix}\right]$. Graph the eigenvalues $\lambda_1(A+tB)$ and $\lambda_2(A+tB)$ for $-8 < t < 8$. Peter Lax says on page 113 of *Linear Algebra* that λ_1 and λ_2 appear to be on a collision course at certain values of t. "Yet at the last minute they turn aside." How close does λ_1 come to λ_2 ?

Challenge Problems

30 *For complex matrices*, the symmetry $S^{\mathrm{T}} = S$ that produces real eigenvalues must change in Section 9.2 to $\overline{\boldsymbol{S}}^{\mathbf{T}} = \boldsymbol{S}$. From $\det(S - \lambda I) = 0$, find the eigenvalues of the 2 by 2 **Hermitian matrix** $\boldsymbol{S} = [4 \quad 2+i; \quad 2-i \quad 0] = \overline{\boldsymbol{S}}^{\mathbf{T}}$.

31 **Normal matrices** have $\overline{N}^{\mathrm{T}} N = N\overline{N}^{\mathrm{T}}$. For real matrices, this is $N^{\mathrm{T}}N = NN^{\mathrm{T}}$. Normal includes symmetric, skew-symmetric, and orthogonal (with real λ, imaginary λ, and $|\lambda| = 1$). Other normal matrices can have any complex eigenvalues.

Key point: ***Normal matrices have*** n ***orthonormal eigenvectors***. Those vectors $\boldsymbol{x}_i$ probably will have complex components. In that complex case (Chapter 9) orthogonality means $\overline{\boldsymbol{x}}_i^{\mathrm{T}} \boldsymbol{x}_j = 0$. Inner products (dot products) $\boldsymbol{x}^{\mathrm{T}}\boldsymbol{y}$ become $\overline{\boldsymbol{x}}^{\mathrm{T}}\boldsymbol{y}$.

The test for n ***orthonormal columns in*** Q ***becomes*** $\overline{Q}^{\mathrm{T}}Q = I$ ***instead of*** $Q^{\mathrm{T}}Q = I$.

N has n **orthonormal eigenvectors** $(N = Q\Lambda\overline{Q}^{\mathrm{T}})$ if and only if N is **normal**.

(a) Start from $N = Q\Lambda\overline{Q}^{\mathrm{T}}$ with $\overline{Q}^{\mathrm{T}}Q = I$. Show that $\overline{N}^{\mathrm{T}}N = N\overline{N}^{\mathrm{T}}$: N is normal.

(b) Now start from $\overline{N}^{\mathrm{T}}N = N\overline{N}^{\mathrm{T}}$. Schur found $A = QT\overline{Q}^{\mathrm{T}}$ for every matrix A, with a *triangular* T. For normal matrices $A = N$ we must show (in 3 steps) that this triangular matrix T will actually be diagonal. Then $T = \Lambda$.

Step 1. Put $N = QT\overline{Q}^{\mathrm{T}}$ into $\overline{N}^{\mathrm{T}}N = N\overline{N}^{\mathrm{T}}$ to find $\overline{T}^{\mathrm{T}}T = T\overline{T}^{\mathrm{T}}$.

Step 2. Suppose $T = \begin{bmatrix} a & b \\ 0 & d \end{bmatrix}$ has $\overline{T}^{\mathrm{T}}T = T\overline{T}^{\mathrm{T}}$. Prove that $b = 0$.

Step 3. Extend Step 2 to size n. *Any normal triangular* T *must be diagonal.*

32 If $\lambda_{\max}$ is the largest eigenvalue of a symmetric matrix S, no diagonal entry can be larger than $\lambda_{\max}$. What is the first entry a_{11} of $S = Q\Lambda Q^{\mathrm{T}}$? Show why $a_{11} \leq \lambda_{\max}$.

33 Suppose $A^{\mathrm{T}} = -A$ (real *antisymmetric* matrix). Explain these facts about A:

(a) $\boldsymbol{x}^{\mathrm{T}} A \boldsymbol{x} = 0$ for every real vector $\boldsymbol{x}$.

(b) The eigenvalues of A are pure imaginary.

(c) The determinant of A is positive or zero (not negative).

For (a), multiply out an example of $\boldsymbol{x}^{\mathrm{T}} A \boldsymbol{x}$ and watch terms cancel. Or reverse $\boldsymbol{x}^{\mathrm{T}}(A\boldsymbol{x})$ to $-(A\boldsymbol{x})^{\mathrm{T}}\boldsymbol{x}$. For (b), $A\boldsymbol{z} = \lambda\boldsymbol{z}$ leads to $\overline{\boldsymbol{z}}^{\mathrm{T}} A\boldsymbol{z} = \lambda\overline{\boldsymbol{z}}^{\mathrm{T}}\boldsymbol{z} = \lambda\|\boldsymbol{z}\|^2$. Part (a) shows that $\overline{\boldsymbol{z}}^{\mathrm{T}} A\boldsymbol{z} = (\boldsymbol{x} - i\boldsymbol{y})^{\mathrm{T}} A(\boldsymbol{x} + i\boldsymbol{y})$ has zero real part. Then (b) helps with (c).

34 If S is symmetric and all its eigenvalues are $\lambda = 2$, how do you know that S must be $2I$? Key point: Symmetry guarantees that $S = Q\Lambda Q^{\mathrm{T}}$. What is that Λ?

35 *Which symmetric matrices S are also orthogonal*? Then $S^{\mathrm{T}} = S^{-1}$.

(a) Show how symmetry and orthogonality lead to $S^2 = I$.

(b) What are the possible eigenvalues of this S?

(c) What are the possible eigenvalue matrices Λ? Then S must be $Q\Lambda Q^{\mathrm{T}}$ for those Λ and any orthogonal Q.

36 If S is symmetric, show that $A^{\mathrm{T}}SA$ is also symmetric (take the transpose of $A^{\mathrm{T}}SA$). Here A is m by n and S is m by m. Are eigenvalues of S = eigenvalues of $A^{\mathrm{T}}SA$?

In case A is square and invertible, $A^{\mathrm{T}}SA$ is called *congruent* to S. They have the same number of positive, negative, and zero eigenvalues: ***Law of Inertia***.

37 Here is a way to show that a is *in between* the eigenvalues λ_1 and λ_2 of S:

$$S = \begin{bmatrix} a & b \\ b & c \end{bmatrix} \qquad \begin{array}{l} \det(S - \lambda I) = \lambda^2 - a\lambda - c\lambda + ac - b^2 \\ \text{is a parabola opening upwards (because of } \lambda^2\text{)} \end{array}$$

Show that $\det(S - \lambda I)$ is negative at $\lambda = a$. So the parabola crosses the axis left and right of $\lambda = a$. It crosses at the two eigenvalues of S so they must enclose a.

The $n-1$ eigenvalues of A always fall between the n eigenvalues of $S = \begin{bmatrix} A & \boldsymbol{b} \\ \boldsymbol{b}^{\mathrm{T}} & c \end{bmatrix}$.

6.5 Positive Definite Matrices

1 Symmetric S: all eigenvalues $> \mathbf{0} \Leftrightarrow$ all pivots $> \mathbf{0} \Leftrightarrow$ all upper left determinants $> \mathbf{0}$.

2 The matrix S is then **positive definite**. The energy test is $\boldsymbol{x}^{\mathrm{T}}S\boldsymbol{x} > 0$ for all vectors $\boldsymbol{x} \neq \mathbf{0}$.

3 One more test for positive definiteness: $S = A^{\mathrm{T}}A$ with independent columns in A.

4 Positive semidefinite S allows $\lambda = 0$, pivot $= 0$, determinant $= 0$, energy $\boldsymbol{x}^{\mathrm{T}}S\boldsymbol{x} = 0$.

5 The equation $\boldsymbol{x}^{\mathrm{T}}S\boldsymbol{x} = 1$ gives an ellipse in $\mathbf{R}^n$ when S is symmetric positive definite.

This section concentrates on ***symmetric matrices that have positive eigenvalues***. If symmetry makes a matrix important, this extra property (*all* $\lambda > 0$) makes it truly special. When we say special, we don't mean rare. Symmetric matrices with positive eigenvalues are at the center of all kinds of applications. They are called ***positive definite***.

The first problem is to recognize positive definite matrices. You may say, just find all the eigenvalues and test $\lambda > 0$. That is exactly what we want to avoid. Calculating eigenvalues is work. When the λ's are needed, we can compute them. But if we just want to know that all the λ's are positive, there are faster ways. Here are two goals of this section:

- To find *quick tests* on a symmetric matrix that guarantee *positive eigenvalues*.
- To explain important applications of positive definiteness.

Every eigenvalue is real because the matrix is symmetric.

Start with 2 by 2. When does $S = \begin{bmatrix} a & b \\ b & c \end{bmatrix}$ ***have*** $\lambda_1 > 0$ ***and*** $\lambda_2 > 0$?

Test: The eigenvalues of S are positive if and only if $\boldsymbol{a > 0}$ ***and*** $\boldsymbol{ac - b^2 > 0}$.

$$S_1 = \begin{bmatrix} 1 & 2 \\ 2 & 1 \end{bmatrix} \text{ is **not** positive definite because } ac - b^2 = 1 - 4 < 0$$

$$S_2 = \begin{bmatrix} 1 & -2 \\ -2 & 6 \end{bmatrix} \text{ is positive definite because } a = 1 \text{ and } ac - b^2 = 6 - 4 > 0$$

$$S_3 = \begin{bmatrix} -1 & 2 \\ 2 & -6 \end{bmatrix} \text{ is **not** positive definite (even with } \det A = +2) \text{ because } a = -1$$

The eigenvalues 3 and -1 of S_1 confirm that S_1 is *not* positive definite. Positive trace $3 - 1 = 2$, but negative determinant $(3)(-1) = -3$. And $S_3 = -S_2$ is *negative* definite. Two positive eigenvalues for S_2, two negative eigenvalues for S_3.

Proof that the 2 *by* 2 *test is passed when* $\lambda_1 > 0$ *and* $\lambda_2 > 0$. Their product $\lambda_1\lambda_2$ is the determinant so $ac - b^2 > 0$. Their sum $\lambda_1 + \lambda_2$ is the trace so $a + c > 0$. Then a and c are both positive (if a or c is not positive, $ac - b^2 > 0$ will fail). Problem 1 reverses the reasoning to show that the tests $a > 0$ and $ac > b^2$ guarantee $\lambda_1 > 0$ and $\lambda_2 > 0$.

This test uses the 1 by 1 determinant a and the 2 by 2 determinant $ac - b^2$. When S is 3 by 3, det $S > 0$ is the third part of the test. The next test requires *positive pivots.*

***Test: The eigenvalues of S are positive if and only if the pivots are positive*:**

$$a > 0 \qquad \text{and} \qquad \frac{ac - b^2}{a} > 0.$$

$a > 0$ is required in both tests. So $ac > b^2$ is also required, for the determinant test and now the pivot test. The point is to recognize that ratio as the *second pivot* of S:

$$\begin{bmatrix} a & b \\ b & c \end{bmatrix} \quad \underset{\text{The multiplier is } b/a}{\xrightarrow{\text{The first pivot is } a}} \quad \begin{bmatrix} a & b \\ 0 & c - \frac{b}{a}b \end{bmatrix} \qquad \begin{matrix} \textbf{\textit{The second pivot is}} \\ \boldsymbol{c - \dfrac{b^2}{a} = \dfrac{ac - b^2}{a}} \end{matrix}$$

This connects two big parts of linear algebra. ***Positive eigenvalues mean positive pivots and vice versa***. Each pivot is a ratio of upper left determinants. The pivots give a quick test for $\lambda > 0$, and they are a lot faster to compute than the eigenvalues. It is very satisfying to see pivots and determinants and eigenvalues come together in this course.

3 by 3 example $\quad S = \begin{bmatrix} 2 & 1 & 1 \\ 1 & 2 & 1 \\ 1 & 1 & 2 \end{bmatrix}$ is positive definite $\qquad$ **eigenvalues** $1, 1, 4$; **determinants** 2 and 3 and 4; **pivots** 2 and $3/2$ and $4/3$

$S - I$ will be *semidefinite*: eigenvalues $0, 0, 3$. $\quad S - 2I$ is *indefinite* because $\lambda = -1, -1, 2$.

Now comes a different way to look at symmetric matrices with positive eigenvalues.

Energy-based Definition

From $S\boldsymbol{x} = \lambda\boldsymbol{x}$, multiply by $\boldsymbol{x}^{\mathrm{T}}$ to get $\boldsymbol{x}^{\mathrm{T}}S\boldsymbol{x} = \lambda\boldsymbol{x}^{\mathrm{T}}\boldsymbol{x}$. The right side is a positive λ times a positive number $\boldsymbol{x}^{\mathrm{T}}\boldsymbol{x} = ||\boldsymbol{x}||^2$. So the left side $\boldsymbol{x}^{\mathrm{T}}S\boldsymbol{x}$ is positive for any eigenvector.

Important point: The new idea is that ***$\boldsymbol{x}^{\mathrm{T}}S\boldsymbol{x}$ is positive for all nonzero vectors $\boldsymbol{x}$***, not just the eigenvectors. In many applications this number $\boldsymbol{x}^{\mathrm{T}}S\boldsymbol{x}$ (or $\frac{1}{2}\boldsymbol{x}^{\mathrm{T}}S\boldsymbol{x}$) is the **energy** in the system. The requirement of positive energy gives *another definition* of a positive definite matrix. I think this energy-based definition is the fundamental one.

Eigenvalues and pivots are two equivalent ways to test the new requirement $\boldsymbol{x}^{\mathrm{T}}S\boldsymbol{x} > 0$.

Definition *S is positive definite if $\boldsymbol{x}^{\mathrm{T}}S\boldsymbol{x} > 0$ for every nonzero vector $\boldsymbol{x}$:*

$$\textbf{2 by 2} \qquad \boldsymbol{x}^{\mathrm{T}}S\boldsymbol{x} = \begin{bmatrix} x & y \end{bmatrix}\begin{bmatrix} a & b \\ b & c \end{bmatrix}\begin{bmatrix} x \\ y \end{bmatrix} = ax^2 + 2bxy + cy^2 > 0. \qquad (1)$$

The four entries a, b, b, c give the four parts of $\boldsymbol{x}^{\mathrm{T}}S\boldsymbol{x}$. From a and c come the pure squares ax^2 and cy^2. From b and b off the diagonal come the cross terms bxy and byx (the same). Adding those four parts gives $\boldsymbol{x}^{\mathrm{T}}S\boldsymbol{x}$. This energy-based definition leads to a basic fact:

If S and T are symmetric positive definite, so is $S + T$.

Reason: $\boldsymbol{x}^{\mathrm{T}}(S+T)\boldsymbol{x}$ is simply $\boldsymbol{x}^{\mathrm{T}}S\boldsymbol{x}+\boldsymbol{x}^{\mathrm{T}}T\boldsymbol{x}$. Those two terms are positive (for $\boldsymbol{x} \neq \boldsymbol{0}$) so $S + T$ is also positive definite. The pivots and eigenvalues are not easy to follow when matrices are added, but the energies just add.

$\boldsymbol{x}^{\mathrm{T}}S\boldsymbol{x}$ also connects with our final way to recognize a positive definite matrix. Start with any matrix A, possibly rectangular. We know that $S = A^{\mathrm{T}}A$ is square and symmetric. More than that, S will be positive definite when A has independent columns:

Test: If the columns of A are independent, then $S = A^{\mathrm{T}}A$ is positive definite.

Again eigenvalues and pivots are not easy. But the number $\boldsymbol{x}^{\mathrm{T}}S\boldsymbol{x}$ is the same as $\boldsymbol{x}^{\mathrm{T}}A^{\mathrm{T}}A\boldsymbol{x}$. $\boldsymbol{x}^{\mathrm{T}}A^{\mathrm{T}}A\boldsymbol{x}$ is exactly $(A\boldsymbol{x})^{\mathrm{T}}(A\boldsymbol{x}) = \|A\boldsymbol{x}\|^2$—another important proof by parenthesis! That vector $A\boldsymbol{x}$ is not zero when $\boldsymbol{x} \neq \boldsymbol{0}$ (this is the meaning of independent columns). Then $\boldsymbol{x}^{\mathrm{T}}S\boldsymbol{x}$ is the positive number $\|A\boldsymbol{x}\|^2$ and the matrix S is positive definite.

Let me collect this theory together, into *five equivalent statements* of positive definiteness. You will see how that key idea connects the whole subject of linear algebra: pivots, determinants, eigenvalues, and least squares (from $A^{\mathrm{T}}A$). Then come the applications.

When a symmetric matrix S has one of these five properties, it has them all :

1. All ***n pivots*** of S are positive.
2. All ***n upper left determinants*** are positive.
3. All ***n eigenvalues*** of S are positive.
4. $\boldsymbol{x}^{\mathrm{T}}S\boldsymbol{x}$ is positive except at $\boldsymbol{x} = \boldsymbol{0}$. This is the ***energy-based*** definition.
5. S equals $A^{\mathrm{T}}A$ for a matrix A with ***independent columns***.

The "upper left determinants" are 1 by 1, 2 by 2, . . ., n by n. The last one is the determinant of the complete matrix S. This theorem ties together the whole linear algebra course.

Example 1 Test these symmetric matrices S and T for positive definiteness:

$$S = \begin{bmatrix} 2 & -1 & 0 \\ -1 & 2 & -1 \\ 0 & -1 & 2 \end{bmatrix} \quad \text{and} \quad T = \begin{bmatrix} 2 & -1 & b \\ -1 & 2 & -1 \\ b & -1 & 2 \end{bmatrix}.$$

Solution The pivots of S are 2 and $\frac{3}{2}$ and $\frac{4}{3}$, all positive. Its upper left determinants are 2 and 3 and 4, all positive. The eigenvalues of S are $2 - \sqrt{2}$ and 2 and $2 + \sqrt{2}$, all positive. That completes tests **1**, **2**, and **3**. Any one test is decisive!

I have three candidates A_1, A_2, A_3 to suggest for $S = A^{\mathrm{T}}A$. They all show that S is positive definite. A_1 is a first difference matrix, 4 by 3, to produce $-1, 2, -1$ in S:

$$\boldsymbol{S = A_1^{\mathrm{T}} A_1} \qquad \begin{bmatrix} 2 & -1 & 0 \\ -1 & 2 & -1 \\ 0 & -1 & 2 \end{bmatrix} = \begin{bmatrix} 1 & -1 & 0 & 0 \\ 0 & 1 & -1 & 0 \\ 0 & 0 & 1 & -1 \end{bmatrix} \begin{bmatrix} 1 & 0 & 0 \\ -1 & 1 & 0 \\ 0 & -1 & 1 \\ 0 & 0 & -1 \end{bmatrix}$$

The three columns of A_1 are independent. Therefore S is positive definite.

A_2 comes from $S = LDL^{\mathrm{T}}$ (the symmetric version of $S = LU$). Elimination gives the pivots $2, \frac{3}{2}, \frac{4}{3}$ in D and the multipliers $-\frac{1}{2}, 0, -\frac{2}{3}$ in L. **Just put** $\boldsymbol{A_2 = L\sqrt{D}}$**.**

$$LDL^{\mathrm{T}} = \begin{bmatrix} 1 & & \\ -\frac{1}{2} & 1 & \\ 0 & -\frac{2}{3} & 1 \end{bmatrix} \begin{bmatrix} 2 & & \\ & \frac{3}{2} & \\ & & \frac{4}{3} \end{bmatrix} \begin{bmatrix} 1 & -\frac{1}{2} & 0 \\ & 1 & -\frac{2}{3} \\ & & 1 \end{bmatrix} = (L\sqrt{D})(L\sqrt{D})^{\mathrm{T}} = A_2^{\mathrm{T}} A_2.$$

$\boldsymbol{A_2}$ ***is the Cholesky factor of S***

This triangular choice of A has square roots (not so beautiful). It is the "Cholesky factor" of S and the MATLAB command is $A = \mathrm{chol}(S)$. In applications, the rectangular A_1 is how we build S and this Cholesky A_2 is how we break it apart.

Eigenvalues give the symmetric choice $\boldsymbol{A_3 = Q\sqrt{\Lambda}Q^{\mathrm{T}}}$. This is also successful with $A_3^{\mathrm{T}} A_3 = Q\Lambda Q^{\mathrm{T}} = S$. All tests show that the $-1, 2, -1$ matrix S is positive definite.

To see that the energy $\boldsymbol{x}^{\mathrm{T}} S \boldsymbol{x}$ is positive, we can write it as a sum of squares. The three choices A_1, A_2, A_3 give three different ways to split up $\boldsymbol{x}^{\mathrm{T}} S \boldsymbol{x}$:

$$\boldsymbol{x}^{\mathrm{T}} S \boldsymbol{x} = 2x_1^2 - 2x_1x_2 + 2x_2^2 - 2x_2x_3 + 2x_3^2 \qquad \textbf{Rewrite with squares}$$

$$||A_1\boldsymbol{x}||^2 = x_1^2 + (x_2 - x_1)^2 + (x_3 - x_2)^2 + x_3^2 \qquad \textbf{Using differences in } \boldsymbol{A_1}$$

$$||A_2\boldsymbol{x}||^2 = 2\left(x_1 - \tfrac{1}{2}x_2\right)^2 + \tfrac{3}{2}\left(x_2 - \tfrac{2}{3}x_3\right)^2 + \tfrac{4}{3}\,x_3^2 \qquad \textbf{Using } \boldsymbol{S = LDL^{\mathrm{T}}}$$

$$||A_3\boldsymbol{x}||^2 = \lambda_1(\boldsymbol{q}_1^{\mathrm{T}}\boldsymbol{x})^2 + \lambda_2(\boldsymbol{q}_2^{\mathrm{T}}\boldsymbol{x})^2 + \lambda_3(\boldsymbol{q}_3^{\mathrm{T}}\boldsymbol{x})^2 \qquad \textbf{Using } \boldsymbol{S = Q\Lambda Q^{\mathrm{T}}}$$

Now turn to T (top of this page). The $(1,3)$ and $(3,1)$ entries move away from 0 to b. This b must not be too large! *The determinant test is easiest.* The 1 by 1 determinant is 2, the 2 by 2 determinant T is still 3. The 3 by 3 determinant involves b:

$$\textbf{Test on } \boldsymbol{T} \qquad \det T = 4 + 2b - 2b^2 = (1+b)(4-2b) \quad \text{must be positive.}$$

At $b = -1$ and $b = 2$ we get $\det T = 0$. *Between $b = -1$ and $b = 2$ this matrix T is positive definite*. The corner entry $b = 0$ in the matrix S was safely between -1 and 2.

Positive Semidefinite Matrices

Often we are at the edge of positive definiteness. The determinant is zero. The smallest eigenvalue is zero. The energy in its eigenvector is $\boldsymbol{x}^{\mathrm{T}}S\boldsymbol{x} = \boldsymbol{x}^{\mathrm{T}}0\boldsymbol{x} = 0$. These matrices on the edge are called *positive semidefinite*. Here are two examples (not invertible):

$$S = \begin{bmatrix} 1 & 2 \\ 2 & 4 \end{bmatrix} \text{ and } T = \begin{bmatrix} 2 & -1 & -1 \\ -1 & 2 & -1 \\ -1 & -1 & 2 \end{bmatrix} \text{ are positive semidefinite.}$$

S has eigenvalues 5 and 0. Its upper left determinants are 1 and 0. Its rank is only 1. This matrix S factors into $A^{\mathrm{T}}A$ with **dependent columns** in A:

$$\begin{matrix}\textbf{Dependent columns in } \boldsymbol{A} \\ \textbf{Positive semidefinite } \boldsymbol{S}\end{matrix} \qquad \begin{bmatrix} 1 & 2 \\ 2 & 4 \end{bmatrix} = \begin{bmatrix} 1 & 0 \\ 2 & 0 \end{bmatrix} \begin{bmatrix} 1 & 2 \\ 0 & 0 \end{bmatrix} = A^{\mathrm{T}}A.$$

If 4 is increased by any small number, the matrix S will become positive definite.

The cyclic T also has zero determinant (computed above when $b = -1$). T is singular. The eigenvector $\boldsymbol{x} = (1, 1, 1)$ has $T\boldsymbol{x} = \boldsymbol{0}$ and energy $\boldsymbol{x}^{\mathrm{T}}T\boldsymbol{x} = 0$. Vectors $\boldsymbol{x}$ in all other directions do give positive energy. This T can be written as $A^{\mathrm{T}}A$ in many ways, but A will always have *dependent* columns, with $(1, 1, 1)$ in its nullspace:

$$\begin{matrix}\textbf{Second differences } \boldsymbol{T} \\ \textbf{from first differences } \boldsymbol{A} \\ \textbf{Cyclic } \boldsymbol{T} \textbf{ from cyclic } \boldsymbol{A}\end{matrix} \quad \begin{bmatrix} 2 & -1 & -1 \\ -1 & 2 & -1 \\ -1 & -1 & 2 \end{bmatrix} = \begin{bmatrix} 1 & -1 & 0 \\ 0 & 1 & -1 \\ -1 & 0 & 1 \end{bmatrix} \begin{bmatrix} 1 & 0 & -1 \\ -1 & 1 & 0 \\ 0 & -1 & 1 \end{bmatrix}.$$

Positive semidefinite matrices have all $\lambda \geq 0$ and all $\boldsymbol{x}^{\mathrm{T}}S\boldsymbol{x} \geq 0$. Those weak inequalities ($\geq$ **instead of** $>$) include positive definite S and also the singular matrices at the edge.

The Ellipse $ax^2 + 2bxy + cy^2 = 1$

Think of a tilted ellipse $\boldsymbol{x}^{\mathrm{T}}S\boldsymbol{x} = 1$. Its center is $(0, 0)$, as in Figure 6.7a. Turn it to line up with the coordinate axes (X and Y axes). That is Figure 6.7b. These two pictures show the geometry behind the factorization $S = Q\Lambda Q^{-1} = Q\Lambda Q^{\mathrm{T}}$:

1. The tilted ellipse is associated with S. Its equation is $\boldsymbol{x}^{\mathrm{T}}S\boldsymbol{x} = 1$.

2. The lined-up ellipse is associated with Λ. Its equation is $\boldsymbol{X}^{\mathrm{T}}\Lambda\boldsymbol{X} = 1$.

3. The rotation matrix that lines up the ellipse is the eigenvector matrix Q.

Example 2 Find the axes of this tilted ellipse $5x^2 + 8xy + 5y^2 = 1$.

Solution Start with the positive definite matrix that matches this equation:

$$\text{The equation is } \begin{bmatrix} x & y \end{bmatrix} \begin{bmatrix} 5 & 4 \\ 4 & 5 \end{bmatrix} \begin{bmatrix} x \\ y \end{bmatrix} = 1. \qquad \text{The matrix is } \quad S = \begin{bmatrix} 5 & 4 \\ 4 & 5 \end{bmatrix}.$$

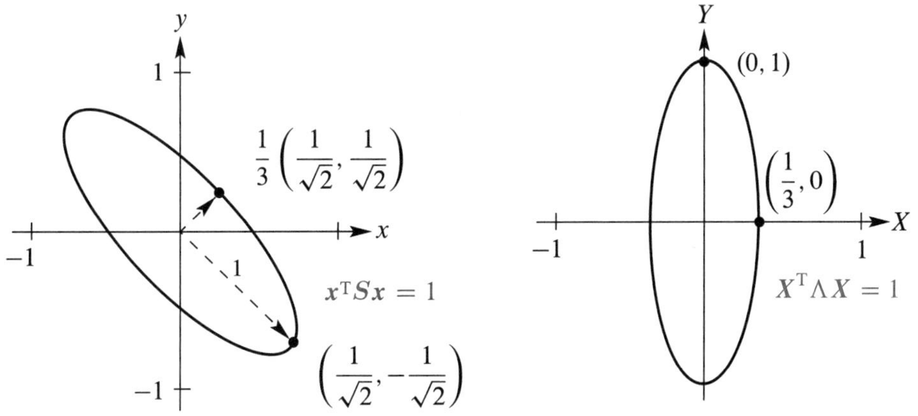

Figure 6.7: The tilted ellipse $5x^2 + 8xy + 5y^2 = 1$. Lined up it is $9X^2 + Y^2 = 1$.

The eigenvectors are $\left[\begin{smallmatrix}1\\1\end{smallmatrix}\right]$ and $\left[\begin{smallmatrix}1\\-1\end{smallmatrix}\right]$. Divide by $\sqrt{2}$ for unit vectors. Then $S = Q\Lambda Q^{\mathrm{T}}$:

Eigenvectors in Q
Eigenvalues 9 and 1
$$\begin{bmatrix} 5 & 4 \\ 4 & 5 \end{bmatrix} = \frac{1}{\sqrt{2}} \begin{bmatrix} 1 & 1 \\ 1 & -1 \end{bmatrix} \begin{bmatrix} \mathbf{9} & 0 \\ 0 & \mathbf{1} \end{bmatrix} \frac{1}{\sqrt{2}} \begin{bmatrix} 1 & 1 \\ 1 & -1 \end{bmatrix}.$$

Now multiply by $\begin{bmatrix} x & y \end{bmatrix}$ on the left and $\begin{bmatrix} x \\ y \end{bmatrix}$ on the right to get $\boldsymbol{x}^{\mathrm{T}} S \boldsymbol{x} = (\boldsymbol{x}^{\mathrm{T}} Q)\Lambda(Q^{\mathrm{T}}\boldsymbol{x})$:

$$x^{\mathrm{T}} S x = \textbf{ sum of squares } \quad 5x^2 + 8xy + 5y^2 = 9\left(\frac{x+y}{\sqrt{2}}\right)^2 + 1\left(\frac{x-y}{\sqrt{2}}\right)^2. \tag{2}$$

The coefficients are not the pivots 5 and 9/5 from D, they are the eigenvalues 9 and 1 from Λ. Inside the squares are the eigenvectors $\boldsymbol{q}_1 = (1, 1)/\sqrt{2}$ and $\boldsymbol{q}_2 = (1, -1)/\sqrt{2}$.

The axes of the tilted ellipse point along those eigenvectors. This explains why $S = Q\Lambda Q^{\mathrm{T}}$ is called the "principal axis theorem"—it displays the axes. Not only the axis directions (from the eigenvectors) but also the axis lengths (from the eigenvalues). To see it all, use capital letters for the new coordinates that line up the ellipse:

$$\textbf{Lined up} \qquad \frac{x+y}{\sqrt{2}} = X \quad \text{and} \quad \frac{x-y}{\sqrt{2}} = Y \quad \text{and} \quad \boldsymbol{9X^2 + Y^2 = 1}.$$

The largest value of X^2 is $1/9$. The endpoint of the shorter axis has $X = 1/3$ and $Y = 0$. Notice: The *bigger* eigenvalue λ_1 gives the *shorter* axis, of half-length $1/\sqrt{\lambda_1} = 1/3$. The smaller eigenvalue $\lambda_2 = 1$ gives the greater length $1/\sqrt{\lambda_2} = 1$.

In the xy system, the axes are along the eigenvectors of S. In the XY system, the **axes are along the eigenvectors of $\boldsymbol{\Lambda}$**—the coordinate axes. All comes from $S = Q\Lambda Q^{\mathrm{T}}$.

$\boldsymbol{S} = Q\Lambda Q^{\mathrm{T}}$ is positive definite when all $\lambda_i > 0$. The graph of $\boldsymbol{x}^{\mathrm{T}} S \boldsymbol{x} = 1$ is an ellipse:

$$\textbf{Ellipse} \begin{bmatrix} x & y \end{bmatrix} Q\Lambda Q^{\mathrm{T}} \begin{bmatrix} x \\ y \end{bmatrix} = \begin{bmatrix} X & Y \end{bmatrix} \Lambda \begin{bmatrix} X \\ Y \end{bmatrix} = \boldsymbol{\lambda_1 X^2 + \lambda_2 Y^2 = 1}. \qquad (3)$$

The axes point along eigenvectors of S. The half-lengths are $1/\sqrt{\lambda_1}$ and $1/\sqrt{\lambda_2}$.

$S = I$ gives the circle $x^2 + y^2 = 1$. If one eigenvalue is negative (exchange 4's and 5's in S), the ellipse changes to a *hyperbola*. The sum of squares becomes a *difference of squares*: $9X^2 - Y^2 = 1$. For a negative definite matrix like $S = -I$, with both λ's negative, the graph of $-x^2 - y^2 = 1$ has no points at all.

If S is n by n, $\boldsymbol{x}^{\mathrm{T}} S \boldsymbol{x} = 1$ is an "ellipsoid" in $\mathbf{R}^n$. Its axes are the eigenvectors of S.

Important Application: Test for a Minimum

Does $F(x, y)$ have a minimum if $\partial F/\partial x = 0$ and $\partial F/\partial y = 0$ at the point $(x, y) = (0, 0)$?

For $f(x)$, the test for a minimum comes from calculus: df/dx is zero and $d^2f/dx^2 > 0$. Two variables in $F(x, y)$ produce a symmetric matrix S. It contains *four second derivatives*. **Positive d^2f/dx^2 changes to positive definite S:**

Second derivatives $\quad S = \begin{bmatrix} \partial^2 F/\partial x^2 & \partial^2 F/\partial x \partial y \\ \partial^2 F/\partial y \partial x & \partial^2 F/\partial y^2 \end{bmatrix}$

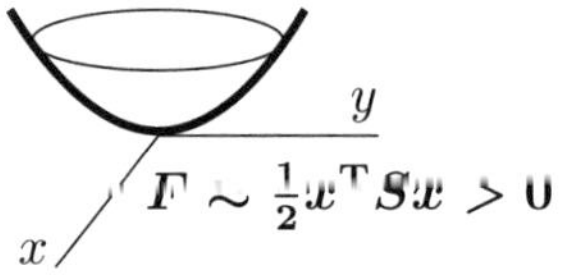

$F(x, y)$ has a minimum if $\partial F/\partial x = \partial F/\partial y = 0$ and S is positive definite.

Reason: S reveals the all-important terms $ax^2 + 2bxy + cy^2$ near $(x, y) = (0, 0)$. The second derivatives of F are $2a, 2b, 2b, 2c$. For $F(x, y, z)$ the matrix S will be 3 by 3.

■ REVIEW OF THE KEY IDEAS ■

1. Positive definite matrices have positive eigenvalues and positive pivots.
2. A quick test is given by the upper left determinants: $a > 0$ and $ac - b^2 > 0$.
3. The graph of the energy $\boldsymbol{x}^{\mathrm{T}} S \boldsymbol{x}$ is then a "bowl" going up from $\boldsymbol{x} = \mathbf{0}$:

   $\boldsymbol{x}^{\mathrm{T}} S \boldsymbol{x} = ax^2 + 2bxy + cy^2$ is positive except at $(x, y) = (0, 0)$.
4. $S = A^{\mathrm{T}} A$ is automatically positive definite if A has independent columns.
5. The ellipsoid $\boldsymbol{x}^{\mathrm{T}} S \boldsymbol{x} = 1$ has its axes along the eigenvectors of S. Lengths $1/\sqrt{\lambda}$.
6. Minimum of $F(x, y)$ if $\dfrac{\partial F}{\partial x} = \dfrac{\partial F}{\partial y} = 0$ and 2nd derivative matrix is positive definite.

■ WORKED EXAMPLES ■

6.5 A The great factorizations of a symmetric matrix are $S = LDL^{\mathrm{T}}$ from pivots and multipliers, and $S = Q\Lambda Q^{\mathrm{T}}$ from eigenvalues and eigenvectors. Try these n by n tests on pascal(6) and ones(6) and hilb(6) and other matrices in MATLAB's gallery.

pascal(6) is positive *definite* because all its pivots are 1 (Worked Example **2.6 A**).

ones(6) is positive *semidefinite* because its eigenvalues are $0, 0, 0, 0, 0, 6$.

H=hilb(6) is positive *definite* even though eig(H) shows eigenvalues very near zero.

Hilbert matrix $\boldsymbol{x}^{\mathrm{T}} H \boldsymbol{x} = \int_0^1 (x_1 + x_2 s + \cdots + x_6 s^5)^2 \, ds > 0$, $H_{ij} = 1/(i+j-1)$.

rand(6) + rand(6)′ can be positive definite or not. *Experiments gave only* 2 *in* 20000.

$n = 20000$; $p = 0$; for $k = 1:n$, $A =$ rand(6); $p = p +$ all(eig$(A + A') > 0$); end, $p\,/\,n$

6.5 B **When is the symmetric block matrix** $M = \begin{bmatrix} A & B \\ B^{\mathrm{T}} & C \end{bmatrix}$ **positive definite?**

Solution Multiply the first row of M by $B^{\mathrm{T}}A^{-1}$ and subtract from the second row, to get a block of zeros. The *Schur complement* $S = C - B^{\mathrm{T}}A^{-1}B$ appears in the corner:

$$\begin{bmatrix} I & 0 \\ -B^{\mathrm{T}}A^{-1} & I \end{bmatrix} \begin{bmatrix} A & B \\ B^{\mathrm{T}} & C \end{bmatrix} = \begin{bmatrix} A & B \\ 0 & C - B^{\mathrm{T}}A^{-1}B \end{bmatrix} = \begin{bmatrix} \boldsymbol{A} & B \\ 0 & \boldsymbol{S} \end{bmatrix} \tag{4}$$

Those two blocks A and S must be positive definite. Their pivots are the pivots of M.

6.5 C Find the eigenvalues of the $-1, 2, -1$ tridiagonal n by n matrix S (my favorite).

Solution The best way is to guess λ and $\boldsymbol{x}$. Then check $S\boldsymbol{x} = \lambda\boldsymbol{x}$. Guessing could not work for most matrices, but special cases are a big part of mathematics (pure and applied).

The key is hidden in a differential equation. The second difference matrix S is like a *second derivative*, and those eigenvalues are much easier to see:

Eigenvalues $\boldsymbol{\lambda_1, \lambda_2, \ldots}$ **Eigenfunctions** $\boldsymbol{y_1, y_2, \ldots}$	$\dfrac{d^2y}{dx^2} = \lambda y(x)$ **with**	$y(0) = 0$ $y(1) = 0$	(5)

Try $y = \sin cx$. Its second derivative is $y'' = -c^2 \sin cx$. So the eigenvalue in (5) will be $\lambda = -c^2$, provided $y = \sin cx$ satisfies the end point conditions $y(0) = 0 = y(1)$.

Certainly $\sin 0 = 0$ (this is where cosines are eliminated). At the other end $x = 1$, we need $y(1) = \sin c = 0$. The number c must be $k\pi$, a multiple of π. Then λ is $-k^2\pi^2$:

Eigenvalues $\boldsymbol{\lambda = -k^2\pi^2}$

Eigenfunctions $\boldsymbol{y = \sin k\pi x}$

$$\frac{d^2}{dx^2} \sin k\pi x = -k^2\pi^2 \sin k\pi x. \tag{6}$$

Now we go back to the matrix S and guess its eigenvectors. They come from $\sin k\pi x$ at n points $x = h, 2h, \ldots, nh$, equally spaced between 0 and 1. The spacing Δx is $h = 1/(n+1)$, so the $(n+1)$st point has $(n+1)h = 1$. Multiply that sine vector $\boldsymbol{x}$ by S:

Eigenvalue of S is positive $\quad S\boldsymbol{x} = \lambda_k \boldsymbol{x} = (2 - 2\cos k\pi h)\, \boldsymbol{x}$

Eigenvector of S is sine vector $\quad \boldsymbol{x} = (\sin k\pi h, \ldots, \sin nk\pi h)$. $\quad$ (7)

Problem Set 6.5

Problems 1–13 are about tests for positive definiteness.

1 Suppose the 2 by 2 tests $a > 0$ and $ac - b^2 > 0$ are passed. Then $c > b^2/a > 0$.

(i) λ_1 and λ_2 have the *same sign* because their product $\lambda_1\lambda_2$ equals _____.

(i) That sign is positive because $\lambda_1 + \lambda_2$ equals _____.

Conclusion: The tests $a > 0, ac - b^2 > 0$ guarantee positive eigenvalues λ_1, λ_2.

2 Which of S_1, S_2, S_3, S_4 has two positive eigenvalues? Use a test, don't compute the λ's. Also find an $\boldsymbol{x}$ so that $\boldsymbol{x}^{\mathrm{T}} S_1 \boldsymbol{x} < 0$, so S_1 is not positive definite.

$$S_1 = \begin{bmatrix} 5 & 6 \\ 6 & 7 \end{bmatrix} \quad S_2 = \begin{bmatrix} -1 & -2 \\ -2 & -5 \end{bmatrix} \quad S_3 = \begin{bmatrix} 1 & 10 \\ 10 & 100 \end{bmatrix} \quad S_4 = \begin{bmatrix} 1 & 10 \\ 10 & 101 \end{bmatrix}.$$

3 For which numbers b and c are these matrices positive definite?

$$S = \begin{bmatrix} 1 & b \\ b & 9 \end{bmatrix} \qquad S = \begin{bmatrix} 2 & 4 \\ 4 & c \end{bmatrix} \qquad S = \begin{bmatrix} c & b \\ b & c \end{bmatrix}.$$

With the pivots in D and multiplier in L, factor each A into LDL^{T}.

4 What is the function $f = ax^2 + 2bxy + cy^2$ for each of these matrices? Complete the square to write each f as a sum of one or two squares $f = d_1(\quad)^2 + d_2(\quad)^2$.

$$S_1 = \begin{bmatrix} 1 & 2 \\ 2 & 9 \end{bmatrix} \qquad S_2 = \begin{bmatrix} 1 & 3 \\ 3 & 9 \end{bmatrix} \qquad f = [x \ \ y] \begin{bmatrix} & S & \end{bmatrix} \begin{bmatrix} x \\ y \end{bmatrix}$$

5 Write $f(x, y) = x^2 + 4xy + 3y^2$ as a *difference* of squares and find a point (x, y) where f is negative. No minimum at $(0, 0)$ even though f has positive coefficients.

6 The function $f(x, y) = 2xy$ certainly has a saddle point and not a minimum at $(0, 0)$. What symmetric matrix S produces this f? What are its eigenvalues?

7 Test to see if $A^{\mathrm{T}}A$ is positive definite in each case: A needs independent columns.

$$A = \begin{bmatrix} 1 & 2 \\ 0 & 3 \end{bmatrix} \quad \text{and} \quad A = \begin{bmatrix} 1 & 1 \\ 1 & 2 \\ 2 & 1 \end{bmatrix} \quad \text{and} \quad A = \begin{bmatrix} 1 & 1 & 2 \\ 1 & 2 & 1 \end{bmatrix}.$$

8 The function $f(x, y) = 3(x + 2y)^2 + 4y^2$ is positive except at $(0, 0)$. What is the matrix in $f = [\,x \ \ y\,]S[\,x \ \ y\,]^{\mathrm{T}}$? Check that the pivots of A are 3 and 4.

9 Find the 3 by 3 matrix S and its pivots, rank, eigenvalues, and determinant:

$$\begin{bmatrix} x_1 & x_2 & x_3 \end{bmatrix} \begin{bmatrix} & & \\ & S & \\ & & \end{bmatrix} \begin{bmatrix} x_1 \\ x_2 \\ x_3 \end{bmatrix} = 4(x_1 - x_2 + 2x_3)^2.$$

10 Which 3 by 3 symmetric matrices S and T produce these quadratics?

$$\boldsymbol{x}^{\mathrm{T}} S \boldsymbol{x} = 2\left(x_1^2 + x_2^2 + x_3^2 - x_1 x_2 - x_2 x_3\right). \quad \text{Why is } S \text{ positive definite?}$$
$$\boldsymbol{x}^{\mathrm{T}} T \boldsymbol{x} = 2\left(x_1^2 + x_2^2 + x_3^2 - x_1 x_2 - x_1 x_3 - x_2 x_3\right). \quad \text{Why is } T \text{ semidefinite?}$$

11 Compute the three upper left determinants of S to establish positive definiteness. Verify that their ratios give the second and third pivots.

$$\textbf{Pivots = ratios of determinants} \qquad S = \begin{bmatrix} 2 & 2 & 0 \\ 2 & 5 & 3 \\ 0 & 3 & 8 \end{bmatrix}.$$

12 For what numbers c and d are S and T positive definite? Test their 3 determinants:

$$S = \begin{bmatrix} c & 1 & 1 \\ 1 & c & 1 \\ 1 & 1 & c \end{bmatrix} \quad \text{and} \quad T = \begin{bmatrix} 1 & 2 & 3 \\ 2 & d & 4 \\ 3 & 4 & 5 \end{bmatrix}.$$

13 Find a matrix with $a > 0$ and $c > 0$ and $a + c > 2b$ that has a negative eigenvalue.

Problems 14–20 are about applications of the tests.

14 *If S is positive definite then S^{-1} is positive definite.* Best proof: The eigenvalues of S^{-1} are positive because _____. *Second proof* (only for 2 by 2):

$$\text{The entries of } S^{-1} = \frac{1}{ac - b^2} \begin{bmatrix} c & -b \\ -b & a \end{bmatrix} \quad \text{pass the determinant tests} \quad \_\_\_\_\_.$$

15 *If S and T are positive definite, their sum $S + T$ is positive definite.* Pivots and eigenvalues are not convenient for $S + T$. Better to use $\boldsymbol{x}^{\mathrm{T}}(S + T)\boldsymbol{x} > 0$. Also $S = A^{\mathrm{T}} A$ and $T = B^{\mathrm{T}} B$ give $S + T = \begin{bmatrix} \mathrm{A} & \mathrm{B} \end{bmatrix}^{\mathrm{T}} \begin{bmatrix} \mathrm{A} \\ \mathrm{B} \end{bmatrix}$ with independent columns.

16 A positive definite matrix cannot have a zero (or even worse, a negative number) on its main diagonal. Show that this matrix fails to have $\boldsymbol{x}^{\mathrm{T}} S \boldsymbol{x} > 0$:

$$\begin{bmatrix} x_1 & x_2 & x_3 \end{bmatrix} \begin{bmatrix} 4 & 1 & 1 \\ 1 & 0 & 2 \\ 1 & 2 & 5 \end{bmatrix} \begin{bmatrix} x_1 \\ x_2 \\ x_3 \end{bmatrix} \quad \text{is not positive when } (x_1, x_2, x_3) = (\quad,\quad,\quad).$$

17 A diagonal entry s_{jj} of a symmetric matrix cannot be smaller than all the λ's. If it were, then $S - s_{jj} I$ would have _____ eigenvalues and would be positive definite. But $S - s_{jj} I$ has a _____ on the main diagonal.

18 If $S\boldsymbol{x} = \lambda\boldsymbol{x}$ then $\boldsymbol{x}^{\mathrm{T}}S\boldsymbol{x} =$ _____. Why is this number positive when $\lambda > 0$?

19 Reverse Problem 18 to show that *if all* $\lambda > 0$ *then* $\boldsymbol{x}^{\mathrm{T}}S\boldsymbol{x} > 0$. We must do this for **every nonzero** $\boldsymbol{x}$, not just the eigenvectors. So write $\boldsymbol{x}$ as a combination of the eigenvectors and *explain why all "cross terms" are* $\boldsymbol{x}_i^{\mathrm{T}}\boldsymbol{x}_j = 0$. Then $\boldsymbol{x}^{\mathrm{T}}S\boldsymbol{x}$ is

$$(c_1\boldsymbol{x}_1+\cdots+c_n\boldsymbol{x}_n)^{\mathrm{T}}(c_1\lambda_1\boldsymbol{x}_1+\cdots+c_n\lambda_n x_n) = c_1^2\lambda_1\boldsymbol{x}_1^{\mathrm{T}}\boldsymbol{x}_1+\cdots+c_n^2\lambda_n\boldsymbol{x}_n^{\mathrm{T}}\boldsymbol{x}_n > 0.$$

20 Give a quick reason why each of these statements is true:

(a) Every positive definite matrix is invertible.

(b) The only positive definite projection matrix is $P = I$.

(c) A diagonal matrix with positive diagonal entries is positive definite.

(d) A symmetric matrix with a positive determinant might not be positive definite!

Problems 21–24 use the eigenvalues; Problems 25–27 are based on pivots.

21 For which s and t do S and T have all $\lambda > 0$ (therefore positive definite)?

$$S = \begin{bmatrix} s & -4 & -4 \\ -4 & s & -4 \\ -4 & -4 & s \end{bmatrix} \quad \text{and} \quad T = \begin{bmatrix} t & 3 & 0 \\ 3 & t & 4 \\ 0 & 4 & t \end{bmatrix}.$$

22 From $S = Q\Lambda Q^{\mathrm{T}}$ compute the positive definite symmetric square root $Q\sqrt{\Lambda}Q^{\mathrm{T}}$ of each matrix. Check that this square root gives $A^{\mathrm{T}}A = S$:

$$S = \begin{bmatrix} 5 & 4 \\ 4 & 5 \end{bmatrix} \quad \text{and} \quad S = \begin{bmatrix} 10 & 6 \\ 6 & 10 \end{bmatrix}$$

23 You may have seen the equation for an ellipse as $x^2/a^2 + y^2/b^2 = 1$. What are a and b when the equation is written $\lambda_1 x^2 + \lambda_2 y^2 = 1$? The ellipse $9x^2 + 4y^2 = 1$ has axes with half-lengths $a =$ _____ and $b =$ _____.

24 Draw the tilted ellipse $x^2 + xy + y^2 = 1$ and find the half-lengths of its axes from the eigenvalues of the corresponding matrix S.

25 With positive pivots in D, the factorization $S = LDL^{\mathrm{T}}$ becomes $L\sqrt{D}\sqrt{D}L^{\mathrm{T}}$. (Square roots of the pivots give $D = \sqrt{D}\sqrt{D}$.) Then $C = \sqrt{D}L^{\mathrm{T}}$ yields the ***Cholesky factorization*** $A = C^{\mathrm{T}}C$ which is "symmetrized LU":

$$\text{From} \quad C = \begin{bmatrix} 3 & 1 \\ 0 & 2 \end{bmatrix} \quad \text{find } S. \qquad \text{From} \quad S = \begin{bmatrix} 4 & 8 \\ 8 & 25 \end{bmatrix} \quad \text{find } C = \textbf{chol}(S).$$

26 In the Cholesky factorization $S = C^{\mathrm{T}}C$, with $C = \sqrt{D}L^{\mathrm{T}}$, the square roots of the pivots are on the diagonal of C. Find C (upper triangular) for

$$S = \begin{bmatrix} 9 & 0 & 0 \\ 0 & 1 & 2 \\ 0 & 2 & 8 \end{bmatrix} \quad \text{and} \quad S = \begin{bmatrix} 1 & 1 & 1 \\ 1 & 2 & 2 \\ 1 & 2 & 7 \end{bmatrix}.$$

27 The symmetric factorization $S = LDL^{\mathrm{T}}$ means that $\boldsymbol{x}^{\mathrm{T}}S\boldsymbol{x} = \boldsymbol{x}^{\mathrm{T}}LDL^{\mathrm{T}}\boldsymbol{x}$:

$$\begin{bmatrix} x & y \end{bmatrix}\begin{bmatrix} a & b \\ b & c \end{bmatrix}\begin{bmatrix} x \\ y \end{bmatrix} = \begin{bmatrix} x & y \end{bmatrix}\begin{bmatrix} 1 & 0 \\ b/a & 1 \end{bmatrix}\begin{bmatrix} a & 0 \\ 0 & (ac-b^2)/a \end{bmatrix}\begin{bmatrix} 1 & b/a \\ 0 & 1 \end{bmatrix}\begin{bmatrix} x \\ y \end{bmatrix}.$$

The left side is $ax^2 + 2bxy + cy^2$. The right side is $a\left(x + \frac{b}{a}y\right)^2 +$ _____ y^2. The second pivot completes the square! Test with $a = 2$, $b = 4$, $c = 10$.

28 Without multiplying $S = \begin{bmatrix} \cos\theta & -\sin\theta \\ \sin\theta & \cos\theta \end{bmatrix}\begin{bmatrix} 2 & 0 \\ 0 & 5 \end{bmatrix}\begin{bmatrix} \cos\theta & \sin\theta \\ -\sin\theta & \cos\theta \end{bmatrix}$, find

(a) the determinant of S (b) the eigenvalues of S
(c) the eigenvectors of S (d) a reason why S is symmetric positive definite.

29 For $F_1(x,y) = \frac{1}{4}x^4 + x^2y + y^2$ and $F_2(x,y) = x^3 + xy - x$ find the second derivative matrices S_1 and S_2:

Test for minimum $S = \begin{bmatrix} \partial^2F/\partial x^2 & \partial^2F/\partial x\partial y \\ \partial^2F/\partial y\partial x & \partial^2F/\partial y^2 \end{bmatrix}$ is positive definite

S_1 is positive definite so F_1 is concave up (= convex). Find the minimum point of F_1. *Find the saddle point of* F_2 (look only where first derivatives are zero).

30 The graph of $z = x^2 + y^2$ is a bowl opening upward. *The graph* of $z = x^2 - y^2$ *is a saddle.* The graph of $z = -x^2 - y^2$ is a bowl opening downward. What is a test on a, b, c for $z = ax^2 + 2bxy + cy^2$ to have a saddle point at $(x,y) = (0,0)$?

31 Which values of c give a bowl and which c give a saddle point for the graph of $z = 4x^2 + 12xy + cy^2$? Describe this graph at the borderline value of c.

The Minimum of a Function $F(x, y, z)$

What tests would you expect for a minimum point? First come zero slopes:

First derivatives are zero $\dfrac{\partial F}{\partial x} = \dfrac{\partial F}{\partial y} = \dfrac{\partial F}{\partial z} = \mathbf{0}$ at the minimum point.

Next comes the linear algebra version of the usual calculus test $d^2f/dx^2 > 0$:

Second derivative matrix S is positive definite $S = \begin{bmatrix} F_{xx} & F_{xy} & F_{xz} \\ F_{yx} & F_{yy} & F_{yz} \\ F_{zx} & F_{zy} & F_{zz} \end{bmatrix}$

Here $\boldsymbol{F_{xy}} = \dfrac{\partial}{\partial x}\left(\dfrac{\partial F}{\partial y}\right) = \dfrac{\partial}{\partial y}\left(\dfrac{\partial F}{\partial x}\right) = \boldsymbol{F_{yx}}$ is a 'mixed" second derivative.

Challenge Problems

32 A *group* of nonsingular matrices includes AB and A^{-1} if it includes A and B. "Products and inverses stay in the group." Which of these are groups (as in 2.7.37)?

Invent a "subgroup" of two of these groups (not I by itself $=$ the smallest group).

(a) Positive definite symmetric matrices S.

(b) Orthogonal matrices Q.

(c) All exponentials e^{tA} of a fixed matrix A.

(d) Matrices P with positive eigenvalues.

(e) Matrices D with determinant 1.

33 When S and T are symmetric positive definite, ST might not even be symmetric. But its eigenvalues are still positive. Start from $ST\boldsymbol{x} = \lambda\boldsymbol{x}$ and take dot products with $T\boldsymbol{x}$. Then prove $\lambda > 0$.

34 Write down the 5 by 5 sine matrix Q from Worked Example **6.5 C**, containing the eigenvectors of S when $n = 5$ and $h = 1/6$. Multiply SQ to see the five λ's.

The sum of λ's should equal the trace 10. Their product should be $\det S = 6$.

35 Suppose C is positive definite (so $\boldsymbol{y}^{\mathrm{T}}C\boldsymbol{y} > 0$ whenever $\boldsymbol{y} \neq \mathbf{0}$) and A has independent columns (so $A\boldsymbol{x} \neq \mathbf{0}$ whenever $\boldsymbol{x} \neq \mathbf{0}$). Apply the energy test to $\boldsymbol{x}^{\mathrm{T}}A^{\mathrm{T}}CA\boldsymbol{x}$ to show that ***$S = A^{\mathrm{T}}CA$* is positive definite: the crucial matrix in engineering**.

36 **Important!** Suppose S is positive definite with eigenvalues $\lambda_1 \geq \lambda_2 \geq \ldots \geq \lambda_n$.

(a) What are the eigenvalues of the matrix $\lambda_1 I - S$? Is it positive semidefinite?

(b) How does it follow that $\lambda_1 \boldsymbol{x}^{\mathrm{T}}\boldsymbol{x} \geq \boldsymbol{x}^{\mathrm{T}}S\boldsymbol{x}$ for every $\boldsymbol{x}$?

(c) Draw this conclusion: **The maximum value of $\boldsymbol{x}^{\mathrm{T}}S\boldsymbol{x}/\boldsymbol{x}^{\mathrm{T}}\boldsymbol{x}$ is** _____ .

37 For which a and c is this matrix positive definite? For which a and c is it positive semidefinite (this includes definite)?

$$S = \begin{bmatrix} a & a & a \\ a & a+c & a-c \\ a & a-c & a+c \end{bmatrix}$$

All 5 tests are possible.
The energy $\boldsymbol{x}^{\mathrm{T}}S\boldsymbol{x}$ equals
$a\,(x_1 + x_2 + x_3)^2 + c\,(x_2 - x_3)^2$.

Table of Eigenvalues and Eigenvectors

How are the properties of a matrix reflected in its eigenvalues and eigenvectors? This question is fundamental throughout Chapter 6. A table that organizes the key facts may be helpful. Here are the special properties of the eigenvalues λ_i and the eigenvectors $\boldsymbol{x}_i$.

Symmetric: $S^{\mathrm{T}} = S = Q\Lambda Q^{\mathrm{T}}$	real eigenvalues	orthogonal $\boldsymbol{x}_i^{\mathrm{T}}\boldsymbol{x}_j = 0$
Orthogonal: $Q^{\mathrm{T}} = Q^{-1}$	all $\lvert\lambda\rvert = 1$	orthogonal $\overline{\boldsymbol{x}}_i^{\mathrm{T}}\boldsymbol{x}_j = 0$
Skew-symmetric: $A^{\mathrm{T}} = -A$	imaginary λ's	orthogonal $\overline{\boldsymbol{x}}_i^{\mathrm{T}}\boldsymbol{x}_j = 0$
Complex Hermitian: $\overline{S}^{\mathrm{T}} = S$	real λ's	orthogonal $\overline{\boldsymbol{x}}_i^{\mathrm{T}}\boldsymbol{x}_j = 0$
Positive Definite: $\boldsymbol{x}^{\mathrm{T}}S\boldsymbol{x} > 0$	all $\lambda > 0$	orthogonal since $S^{\mathrm{T}} = S$
Markov: $m_{ij} > 0, \sum_{i=1}^{n} m_{ij} = 1$	$\lambda_{\max} = 1$	steady state $\boldsymbol{x} > 0$
Similar: $A = BCB^{-1}$	$\lambda(A) = \lambda(C)$	B times eigenvector of C
Projection: $P = P^2 = P^{\mathrm{T}}$	$\lambda = 1;\ 0$	column space; nullspace
Plane Rotation : cosine-sine	$e^{i\theta}$ and $e^{-i\theta}$	$\boldsymbol{x} = (1, i)$ and $(1, -i)$
Reflection: $I - 2\boldsymbol{u}\boldsymbol{u}^{\mathrm{T}}$	$\lambda = -1;\ 1, .., 1$	$\boldsymbol{u}$; whole plane $\boldsymbol{u}^{\perp}$
Rank One: $\boldsymbol{u}\boldsymbol{v}^{\mathrm{T}}$	$\lambda = \boldsymbol{v}^{\mathrm{T}}\boldsymbol{u};\ 0, .., 0$	$\boldsymbol{u}$; whole plane $\boldsymbol{v}^{\perp}$
Inverse: A^{-1}	$1/\lambda(A)$	keep eigenvectors of A
Shift: $A + cI$	$\lambda(A) + c$	keep eigenvectors of A
Stable Powers: $A^n \to 0$	all $\lvert\lambda\rvert < 1$	any eigenvectors
Stable Exponential: $e^{At} \to 0$	all Re $\lambda < 0$	any eigenvectors
Cyclic Permutation: $P_{i,i+1} = 1, P_{n1} = 1$	$\lambda_k = e^{2\pi ik/n} =$ roots of 1	$\boldsymbol{x}_k = (1, \lambda_k, \ldots, \lambda_k^{n-1})$
Circulant: $c_0 I + c_1 P + \cdots$	$\lambda_k = c_0 + c_1\, e^{2\pi ik/n} + \cdots$	$\boldsymbol{x}_k = (1, \lambda_k, \ldots, \lambda_k^{n-1})$
Tridiagonal: $-1, 2, -1$ on diagonals	$\lambda_k = 2 - 2\cos\frac{k\pi}{n+1}$	$\boldsymbol{x}_k = \left(\sin\frac{k\pi}{n+1}, \sin\frac{2k\pi}{n+1}, \ldots\right)$
Diagonalizable: $A = X\Lambda X^{-1}$	diagonal of Λ	columns of X are independent
Schur: $A = QTQ^{-1}$	diagonal of triangular T	columns of Q *if* $A^{\mathrm{T}}A = AA^{\mathrm{T}}$
Jordan: $A = BJB^{-1}$	diagonal of J	each block gives 1 eigenvector
SVD: $A = U\Sigma V^{\mathrm{T}}$	r singular values in Σ	eigenvectors of $A^{\mathrm{T}}A$, AA^{T} in V, U

Chapter 7

The Singular Value Decomposition (SVD)

7.1 Image Processing by Linear Algebra

1 An image is a large matrix of grayscale values, one for each pixel and color.

2 When nearby pixels are correlated (not random) the image can be compressed.

3 The SVD separates any matrix A into rank one pieces $\boldsymbol{u}\boldsymbol{v}^{\mathrm{T}}$ = (**column**)(**row**).

4 The columns and rows are eigenvectors of symmetric matrices AA^{T} and $A^{\mathrm{T}}A$.

The singular value theorem for A is the eigenvalue theorem for $A^{\mathrm{T}}A$ and AA^{T}.

That is a quick preview of what you will see in this chapter. A has *two* sets of singular vectors (the eigenvectors of $A^{\mathrm{T}}A$ and AA^{T}). There is *one* set of positive singular values (because $A^{\mathrm{T}}A$ has the same positive eigenvalues as AA^{T}). A is often rectangular, but $A^{\mathrm{T}}A$ and AA^{T} are square, symmetric, and positive semidefinite.

The Singular Value Decomposition (SVD) separates any matrix into simple pieces.

Each piece is a column vector times a row vector. An m by n matrix has m times n entries (a big number when the matrix represents an image). But a column and a row only have $\boldsymbol{m+n}$ **components, far less than** $\boldsymbol{m}$ **times** $\boldsymbol{n}$. Those (column)(row) pieces are full size matrices that can be processed with extreme speed—they need only m *plus* n numbers.

Unusually, this image processing application of the SVD is coming before the matrix algebra it depends on. I will start with simple images that only involve one or two pieces. Right now I am thinking of an image as a large rectangular matrix. The entries a_{ij} tell the grayscales of all the pixels in the image. Think of a pixel as a small square, i steps across and j steps up from the lower left corner. Its grayscale is a number (often a whole number in the range $0 \leq a_{ij} < 256 = 2^8$). An all-white pixel has $a_{ij} = 255 = 11111111$. That number has eight 1's when the computer writes 255 in binary notation.

You see how an image that has m times n pixels, with each pixel using 8 bits (0 or 1) for its grayscale, becomes an m by n matrix with 256 possible values for each entry a_{ij}.

In short, an image is a large matrix. To copy it perfectly, we need $8\,(m)(n)$ bits of information. High definition television typically has $m = 1080$ and $n = 1920$. Often there are 24 frames each second and you probably like to watch in color (3 color scales). This requires transmitting $(3)(8)(48,470,400)$ bits per second. That is too expensive and it is not done. The transmitter can't keep up with the show.

When compression is well done, you can't see the difference from the original. *Edges in the image* (sudden changes in the grayscale) are the hard parts to compress.

Major success in compression will be impossible if every a_{ij} is an independent random number. We totally depend on the fact that *nearby pixels generally have similar grayscales*. An edge produces a sudden jump when you cross over it. Cartoons are more compressible than real-world images, with edges everywhere.

For a video, the numbers a_{ij} don't change much between frames. **We only transmit the small changes**. This is *difference coding* in the H.264 video compression standard (on this book's website). We compress each change matrix by linear algebra (and by nonlinear "quantization" for an efficient step to integers in the computer).

The natural images that we see every day are absolutely ready and open for compression—but that doesn't make it easy to do.

Low Rank Images (Examples)

The easiest images to compress are all black or all white or all a constant grayscale g. The matrix A has the same number g in every entry : $a_{ij} = g$. When $g = 1$ and $m = n = 6$, here is an extreme example of the central SVD dogma of image processing :

Example 1 **Don't send** $A = \begin{bmatrix} 1&1&1&1&1&1\\ 1&1&1&1&1&1\\ 1&1&1&1&1&1\\ 1&1&1&1&1&1\\ 1&1&1&1&1&1\\ 1&1&1&1&1&1 \end{bmatrix}$ **Send this** $A = \begin{bmatrix} 1\\1\\1\\1\\1\\1 \end{bmatrix} \begin{bmatrix} 1&1&1&1&1&1 \end{bmatrix}$

36 numbers become 12 numbers. With 300 by 300 pixels, $90,000$ numbers become 600. And if we define the all-ones vector $\boldsymbol{x}$ in advance, we only have to send **one number**. That number would be the constant grayscale g that multiplies $\boldsymbol{x}\boldsymbol{x}^{\mathrm{T}}$ to produce the matrix.

Of course this first example is extreme. But it makes an important point. If there are special vectors like $\boldsymbol{x}$ = **ones** that can usefully be defined in advance, then image processing can be extremely fast. The battle is between **preselected bases** (the Fourier basis allows speed-up from the FFT) and **adaptive bases** determined by the image. The SVD produces bases from the image itself—this is adaptive and it can be expensive.

I am not saying that the SVD always or usually gives the most effective algorithm in practice. The purpose of these next examples is instruction and not production.

Example 2
"ace flag"
French flag $\boldsymbol{A}$
Italian flag $\boldsymbol{A}$
German flag $\boldsymbol{A}^{\mathrm{T}}$

$$\textbf{Don't send } A = \begin{bmatrix} a & a & c & c & e & e \\ a & a & c & c & e & e \\ a & a & c & c & e & e \\ a & a & c & c & e & e \\ a & a & c & c & e & e \\ a & a & c & c & e & e \end{bmatrix} \qquad \textbf{Send } A = \begin{bmatrix} 1 \\ 1 \\ 1 \\ 1 \\ 1 \\ 1 \end{bmatrix} \begin{bmatrix} a & a & c & c & e & e \end{bmatrix}$$

This flag has 3 colors but it still has rank 1. We still have one column times one row. The 36 entries could even be all different, provided they keep that rank 1 pattern $A = \boldsymbol{u}_1\boldsymbol{v}_1^{\mathrm{T}}$. But when the rank moves up to $r = 2$, we need $\boldsymbol{u}_1\boldsymbol{v}_1^{\mathrm{T}} + \boldsymbol{u}_2\boldsymbol{v}_2^{\mathrm{T}}$. Here is one choice:

Example 3
Embedded square

$$A = \begin{bmatrix} 1 & 0 \\ 1 & 1 \end{bmatrix} \text{ is equal to } A = \begin{bmatrix} 1 \\ 1 \end{bmatrix} \begin{bmatrix} 1 & 1 \end{bmatrix} - \begin{bmatrix} 1 \\ 0 \end{bmatrix} \begin{bmatrix} 0 & 1 \end{bmatrix}$$

The 1's and the 0 in A could be blocks of 1's and a block of 0's. *We would still have rank* 2. We would still only need two terms $\boldsymbol{u}_1\boldsymbol{v}_1^{\mathrm{T}}$ and $\boldsymbol{u}_2\boldsymbol{v}_2^{\mathrm{T}}$. A 6 by 6 image would be compressed into 24 numbers. An N by N image (N^2 numbers) would be compressed into $4N$ numbers from the four vectors $\boldsymbol{u}_1, \boldsymbol{v}_1, \boldsymbol{u}_2, \boldsymbol{v}_2$.

Have I made the best choice for the $\boldsymbol{u}$'s and $\boldsymbol{v}$'s ? This is *not* the choice from the SVD! I notice that $\boldsymbol{u}_1 = (1, 1)$ is not orthogonal to $\boldsymbol{u}_2 = (1, 0)$. And $\boldsymbol{v}_1 = (1, 1)$ is not orthogonal to $\boldsymbol{v}_2 = (0, 1)$. The theory says that orthogonality will produce a smaller second piece $c_2\boldsymbol{u}_2\boldsymbol{v}_2^{\mathrm{T}}$. (**The SVD chooses rank one pieces in order of importance.**)

If the rank of A is much higher than 2, as we expect for real images, then A will add up many rank one pieces. We want the small ones to be really small—they can be discarded with no loss to visual quality. Image compression becomes lossy, but good image compression is virtually undetectable by the human visual system.

The question becomes: **What are the orthogonal choices from the SVD**?

Eigenvectors for the SVD

I want to introduce the use of eigenvectors. But the eigenvectors of most images are not orthogonal. Furthermore the eigenvectors $\boldsymbol{x}_1, \boldsymbol{x}_2$ give only one set of vectors, and we want two sets ($\boldsymbol{u}$'s and $\boldsymbol{v}$'s). The answer to both of those difficulties is the SVD idea:

Use the eigenvectors $\boldsymbol{u}$ of $\boldsymbol{A}\boldsymbol{A}^{\mathrm{T}}$ and the eigenvectors $\boldsymbol{v}$ of $\boldsymbol{A}^{\mathrm{T}}\boldsymbol{A}$.

Since AA^{T} and $A^{\mathrm{T}}A$ are automatically symmetric (but not usually equal!) the $\boldsymbol{u}$'s will be one orthogonal set and the eigenvectors $\boldsymbol{v}$ will be another orthogonal set. We can and will make them all unit vectors: $||\boldsymbol{u}_i|| = 1$ and $||\boldsymbol{v}_i|| = 1$. Then our rank 2 matrix will be $A = \sigma_1\boldsymbol{u}_1\boldsymbol{v}_1^{\mathrm{T}} + \sigma_2\boldsymbol{u}_2\boldsymbol{v}_2^{\mathrm{T}}$. The size of those numbers σ_1 and σ_2 will decide whether they can be ignored in compression. *We keep larger σ's, we discard small σ's.*

The $\boldsymbol{u}$'s from the SVD are called **left singular vectors** (unit eigenvectors of AA^{T}). The $\boldsymbol{v}$'s are **right singular vectors** (unit eigenvectors of $A^{\mathrm{T}}A$). The σ's are **singular values**, square roots of the equal eigenvalues of AA^{T} and $A^{\mathrm{T}}A$:

Choices from the SVD $\quad AA^{\mathrm{T}}\boldsymbol{u}_i = \sigma_i^2\boldsymbol{u}_i \quad A^{\mathrm{T}}A\boldsymbol{v}_i = \sigma_i^2\boldsymbol{v}_i \quad A\boldsymbol{v}_i = \sigma_i\boldsymbol{u}_i \qquad (1)$

In Example 3 (the embedded square), here are the symmetric matrices AA^{T} and $A^{\mathrm{T}}A$:

$$AA^{\mathrm{T}} = \begin{bmatrix} 1 & 0 \\ 1 & 1 \end{bmatrix}\begin{bmatrix} 1 & 1 \\ 0 & 1 \end{bmatrix} = \begin{bmatrix} \mathbf{1} & \mathbf{1} \\ \mathbf{1} & \mathbf{2} \end{bmatrix} \qquad A^{\mathrm{T}}A = \begin{bmatrix} 1 & 1 \\ 0 & 1 \end{bmatrix}\begin{bmatrix} 1 & 0 \\ 1 & 1 \end{bmatrix} = \begin{bmatrix} \mathbf{2} & \mathbf{1} \\ \mathbf{1} & \mathbf{1} \end{bmatrix}.$$

Their determinants are 1, so $\lambda_1\lambda_2 = 1$. Their traces (diagonal sums) are 3:

$$\det\begin{bmatrix} 1-\lambda & 1 \\ 1 & 2-\lambda \end{bmatrix} = \lambda^2 - 3\lambda + 1 = 0 \quad \text{gives} \quad \lambda_1 = \frac{3+\sqrt{5}}{2} \quad \text{and} \quad \lambda_2 = \frac{3-\sqrt{5}}{2}.$$

The square roots of λ_1 and λ_2 are $\sigma_1 = \dfrac{\sqrt{5}+1}{2}$ and $\sigma_2 = \dfrac{\sqrt{5}-1}{2}$ with $\sigma_1\,\sigma_2 = 1$.

The nearest rank 1 matrix to A will be $\sigma_1\boldsymbol{u}_1\boldsymbol{v}_1^{\mathrm{T}}$. The error is only $\sigma_2 \approx 0.6 =$ best possible.

The orthonormal eigenvectors of AA^{T} and $A^{\mathrm{T}}A$ are

$$\boldsymbol{u}_1 = \begin{bmatrix} 1 \\ \sigma_1 \end{bmatrix} \quad \boldsymbol{u}_2 = \begin{bmatrix} \sigma_1 \\ -1 \end{bmatrix} \quad \boldsymbol{v}_1 = \begin{bmatrix} \sigma_1 \\ 1 \end{bmatrix} \quad \boldsymbol{v}_2 = \begin{bmatrix} 1 \\ -\sigma_1 \end{bmatrix} \quad \text{all divided by } \sqrt{1+\sigma_1^2}. \qquad (2)$$

Every reader understands that in real life those calculations are done by computers! (Certainly not by unreliable professors. I corrected myself using svd(A) in MATLAB.) And we can check that the matrix A is correctly recovered from $\sigma_1\boldsymbol{u}_1\boldsymbol{v}_1^{\mathrm{T}} + \sigma_2\boldsymbol{u}_2\boldsymbol{v}_2^{\mathrm{T}}$:

$$A = \begin{bmatrix} \boldsymbol{u}_1 & \boldsymbol{u}_2 \end{bmatrix}\begin{bmatrix} \sigma_1 & \\ & \sigma_2 \end{bmatrix}\begin{bmatrix} \boldsymbol{v}_1^{\mathrm{T}} \\ \boldsymbol{v}_2^{\mathrm{T}} \end{bmatrix} \quad \text{or more simply} \quad A\begin{bmatrix} \boldsymbol{v}_1 & \boldsymbol{v}_2 \end{bmatrix} = \begin{bmatrix} \sigma_1\,\boldsymbol{u}_1 & \sigma_2\,\boldsymbol{u}_2 \end{bmatrix} \qquad (3)$$

Important The key point is not that images tend to have low rank. **No**: Images mostly have full rank. But they do have **low effective rank**. This means: Many singular values are small and can be set to zero. *We transmit a low rank approximation.*

Example 4 Suppose the flag has two triangles of different colors. The lower left triangle has 1's and the upper right triangle has 0's. The main diagonal is included with the 1's. Here is the image matrix when $n = 4$. It has full rank $r = 4$ so it is invertible:

Triangular flag matrix

$$A = \begin{bmatrix} \mathbf{1} & 0 & 0 & 0 \\ \mathbf{1} & \mathbf{1} & 0 & 0 \\ \mathbf{1} & \mathbf{1} & \mathbf{1} & 0 \\ \mathbf{1} & \mathbf{1} & \mathbf{1} & \mathbf{1} \end{bmatrix} \quad \text{and} \quad A^{-1} = \begin{bmatrix} 1 & 0 & 0 & 0 \\ -1 & 1 & 0 & 0 \\ 0 & -1 & 1 & 0 \\ 0 & 0 & -1 & 1 \end{bmatrix}$$

With full rank, A has a full set of n singular values σ (all positive). The SVD will produce n pieces $\sigma_i\, \boldsymbol{u}_i\, \boldsymbol{v}_i^{\mathrm{T}}$ of rank one. Perfect reproduction needs all n pieces.

In compression *small* σ's can be discarded with no serious loss in image quality. We want to understand and plot the σ's for $n = 4$ and also for large n. Notice that Example 3 was the special case $n = 2$ of this triangular Example 4.

Working by hand, we begin with AA^{T} (a computer would proceed differently):

$$AA^{\mathrm{T}} = \begin{bmatrix} 1 & 1 & 1 & 1 \\ 1 & 2 & 2 & 2 \\ 1 & 2 & 3 & 3 \\ 1 & 2 & 3 & 4 \end{bmatrix} \quad \text{and} \quad (AA^{\mathrm{T}})^{-1} = (A^{-1})^{\mathrm{T}} A^{-1} = \begin{bmatrix} \mathbf{2} & -1 & 0 & 0 \\ -1 & \mathbf{2} & -1 & 0 \\ 0 & -1 & \mathbf{2} & -1 \\ 0 & 0 & -1 & \mathbf{1} \end{bmatrix}. \tag{4}$$

That $-1, 2, -1$ inverse matrix is included because its eigenvalues all have the form $2 - 2\cos\theta$. So we know the λ's for AA^{T} and the σ's for A:

$$\lambda = \frac{1}{2 - 2\cos\theta} = \frac{1}{4\sin^2(\theta/2)} \quad \text{gives} \quad \sigma = \sqrt{\lambda} = \frac{1}{2\sin(\theta/2)}. \tag{5}$$

The n different angles θ are equally spaced, which makes this example so exceptional:

$$\theta = \frac{\pi}{2n+1}, \frac{3\pi}{2n+1}, \ldots, \frac{(2n-1)\,\pi}{2n+1} \qquad \left(n = 4 \text{ includes } \theta = \frac{3\pi}{9} \text{ with } 2\sin\frac{\theta}{2} = 1 \right).$$

That special case gives $\lambda = 1$ as an eigenvalue of AA^{T} when $n = 4$. So $\sigma = \sqrt{\lambda} = 1$ is a singular value of A. You can check that the vector $\boldsymbol{u} = (1, 1, 0, -1)$ has $AA^{\mathrm{T}}\boldsymbol{u} = \boldsymbol{u}$ (a truly special case).

The important point is to graph the n singular values of A. Those numbers drop off (unlike the eigenvalues of A, which are all 1). But the dropoff is not steep. So the SVD gives only moderate compression of this triangular flag. *Great compression for Hilbert.*

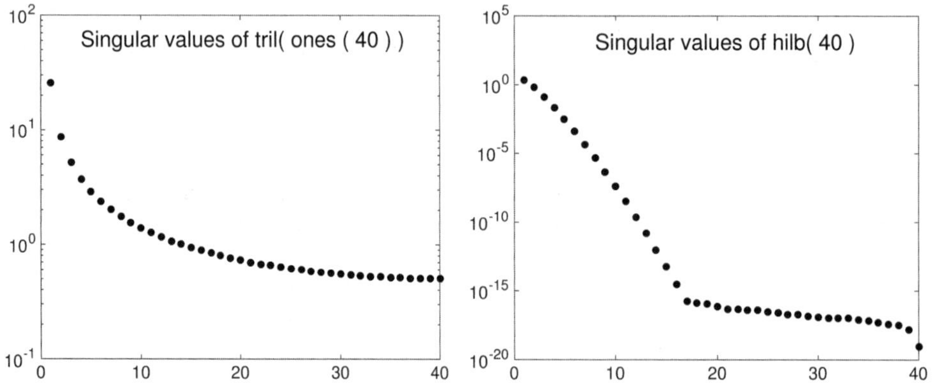

Figure 7.1: Singular values of the triangle of 1's in Examples 3-4 (not compressible) and the evil Hilbert matrix $H(i, j) = (i + j - 1)^{-1}$ in Section 8.3: compress it to work with it.

Your faithful author has continued research on the ranks of flags. Quite a few are based on horizontal or vertical stripes. Those have *rank one*—all rows or all columns are multiples of the *ones* vector $(1, 1, \ldots, 1)$. Armenia, Austria, Belgium, Bulgaria, Chad, Colombia, Ireland, Madagascar, Mali, Netherlands, Nigeria, Romania, Russia (and more) have three stripes. Indonesia and Poland have two ! Libya was the extreme case in the Gadaffi years 1977 to 2011 (*the whole flag was green*).

At the other extreme, many flags include diagonal lines. Those could be long diagonals as in the British flag. Or they could be short diagonals coming from the edges of a star—as in the US flag. The text example of a triangle of ones shows how those flag matrices will have large rank. The rank increases to infinity as the pixel sizes get small.

Other flags have circles or crescents or various curved shapes. Their ranks are large and also increasing to infinity. These are still compressible! The compressed image won't be perfect but our eyes won't see the difference (with enough terms $\sigma_i \boldsymbol{u}_i \boldsymbol{v}_i^{\mathrm{T}}$ from the SVD). Those examples actually bring out the main purpose of image compression:

Visual quality can be preserved even with a big reduction in the rank.

For fun I looked back at the flags with finite rank. They can have stripes and they can also have crosses—provided the edges of the cross are horizontal or vertical. Some flags have a thin outline around the cross. This artistic touch will increase the rank. Right now my champion is the flag of Greece shown below, with a cross and also stripes. Its rank is **three** by my counting (three different columns). I see no US State Flags of finite rank !

The reader could google "national flags" to see the variety of designs and colors. I would be glad to know any finite rank examples with rank > 3. Good examples of all kinds will go on the book's website **math.mit.edu/linearalgebra** (and flags in full color).

Problem Set 7.1

1 What are the ranks r for these matrices with entries i times j and i plus j? Write A and B as the sum of r pieces $\boldsymbol{u}\boldsymbol{v}^{\mathrm{T}}$ of rank one. Not requiring $\boldsymbol{u}_1^{\mathrm{T}}\boldsymbol{u}_2 = \boldsymbol{v}_1^{\mathrm{T}}\boldsymbol{v}_2 = \mathbf{0}$.

$$A = \begin{bmatrix} 1 & 2 & 3 & 4 \\ 2 & 4 & 6 & 8 \\ 3 & 6 & 9 & 12 \\ 4 & 8 & 12 & 16 \end{bmatrix} \qquad B = \begin{bmatrix} 2 & 3 & 4 & 5 \\ 3 & 4 & 5 & 6 \\ 4 & 5 & 6 & 7 \\ 5 & 6 & 7 & 8 \end{bmatrix}$$

2 We usually think that the identity matrix I is as simple as possible. But why is I completely incompressible? *Draw a rank* 5 *flag with a cross.*

3 These flags have rank 2. Write A and B in any way as $\boldsymbol{u}_1\boldsymbol{v}_1^{\mathrm{T}} + \boldsymbol{u}_2\boldsymbol{v}_2^{\mathrm{T}}$.

$$\boldsymbol{A}_{\textbf{Sweden}} = \boldsymbol{A}_{\textbf{Finland}} = \begin{bmatrix} 1 & 2 & 1 & 1 \\ 2 & 2 & 2 & 2 \\ 1 & 2 & 1 & 1 \end{bmatrix} \qquad \boldsymbol{B}_{\textbf{Benin}} = \begin{bmatrix} 1 & 2 & 2 \\ 1 & 3 & 3 \end{bmatrix}$$

4 Now find the trace and determinant of BB^{T} and $B^{\mathrm{T}}B$ in Problem 3. The singular values of B are close to $\sigma_1^2 = 28 - \frac{1}{14}$ and $\sigma_2^2 = \frac{1}{14}$. Is B compressible or not?

5 Use $[U, S, V] =$ svd (A) to find two orthogonal pieces $\sigma\boldsymbol{u}\boldsymbol{v}^{\mathrm{T}}$ of $\boldsymbol{A}_{\textbf{Sweden}}$.

6 Find the eigenvalues and the singular values of this 2 by 2 matrix A.

$$A = \begin{bmatrix} 2 & 1 \\ 4 & 2 \end{bmatrix} \quad \text{with} \quad A^{\mathrm{T}}A = \begin{bmatrix} 20 & 10 \\ 10 & 5 \end{bmatrix} \quad \text{and} \quad AA^{\mathrm{T}} = \begin{bmatrix} 5 & 10 \\ 10 & 20 \end{bmatrix}.$$

The eigenvectors $(1, 2)$ and $(1, -2)$ of A are not orthogonal. How do you know the eigenvectors $\boldsymbol{v}_1, \boldsymbol{v}_2$ of $A^{\mathrm{T}}A$ are orthogonal? Notice that $A^{\mathrm{T}}A$ and AA^{T} have the same eigenvalues (25 and 0).

7 How does the second form $\boldsymbol{AV} = \boldsymbol{U\Sigma}$ in equation (3) follow from the first form $\boldsymbol{A} = \boldsymbol{U\Sigma V}^{\mathbf{T}}$? That is the most famous form of the SVD.

8 The two columns of $AV = U\Sigma$ are $A\boldsymbol{v}_1 = \sigma_1\boldsymbol{u}_1$ and $A\boldsymbol{v}_2 = \sigma_2\boldsymbol{u}_2$. So we hope that

$$A\boldsymbol{v}_1 = \begin{bmatrix} 1 & 0 \\ 1 & 1 \end{bmatrix}\begin{bmatrix} \sigma_1 \\ 1 \end{bmatrix} = \sigma_1\begin{bmatrix} 1 \\ \sigma_1 \end{bmatrix} \quad \text{and} \quad \begin{bmatrix} 1 & 0 \\ 1 & 1 \end{bmatrix}\begin{bmatrix} 1 \\ -\sigma_1 \end{bmatrix} = \sigma_2\begin{bmatrix} \sigma_1 \\ -1 \end{bmatrix}$$

The first needs $\sigma_1 + 1 = \sigma_1^2$ and the second needs $1 - \sigma_1 = -\sigma_2$. Are those true?

9 The MATLAB commands $A =$ rand (20, 40) and $B =$ randn (20, 40) produce 20 by 40 random matrices. The entries of A are between 0 and 1 with uniform probability. The entries of B have a normal "bell-shaped" probability distribution. Using an svd command, find and graph their singular values σ_1 to σ_{20}. Why do they have 20 σ's?

7.2 Bases and Matrices in the SVD

1 The SVD produces **orthonormal basis** of $\boldsymbol{v}$'s and $\boldsymbol{u}$'s for the four fundamental subspaces.

2 Using those bases, A becomes a diagonal matrix Σ and $\boldsymbol{Av_i = \sigma_i u_i}$: $\boldsymbol{\sigma_i}$ = **singular value**.

3 The two-bases diagonalization $\boldsymbol{A = U\Sigma V^{\mathrm{T}}}$ often has more information than $A = X\Lambda X^{-1}$.

4 $U\Sigma V^{\mathrm{T}}$ separates A into rank-1 matrices $\sigma_1 \boldsymbol{u}_1 \boldsymbol{v}_1^{\mathrm{T}} + \cdots + \sigma_r \boldsymbol{u}_r \boldsymbol{v}_r^{\mathrm{T}}$. $\sigma_1 \boldsymbol{u}_1 \boldsymbol{v}_1^{\mathrm{T}}$ is the largest!

The Singular Value Decomposition is a highlight of linear algebra. A is any m by n matrix, square or rectangular. Its rank is r. We will diagonalize this A, but not by $X^{-1}AX$. The eigenvectors in X have three big problems: They are usually not orthogonal, there are not always enough eigenvectors, and $A\boldsymbol{x} = \lambda \boldsymbol{x}$ requires A to be a square matrix. The ***singular vectors*** of A solve all those problems in a perfect way.

Let me describe what we want from the SVD : **the right bases for the four subspaces.** Then I will write about the steps to find those basis vectors **in order of importance**.

The price we pay is to have **two sets of singular vectors**, $\boldsymbol{u}$'s and $\boldsymbol{v}$'s. The $\boldsymbol{u}$'s are in $\mathbf{R}^m$ and the $\boldsymbol{v}$'s are in $\mathbf{R}^n$. They will be the columns of an m by m matrix $\boldsymbol{U}$ and an n by n matrix $\boldsymbol{V}$. I will first describe the SVD in terms of those basis vectors. Then I can also describe the SVD in terms of the orthogonal matrices U and V.

(using vectors) The $\boldsymbol{u}$'s and $\boldsymbol{v}$'s give bases for the four fundamental subspaces :

$\boldsymbol{u}_1, \ldots, \boldsymbol{u}_r$	is an orthonormal basis for the **column space**
$\boldsymbol{u}_{r+1}, \ldots, \boldsymbol{u}_m$	is an orthonormal basis for the **left nullspace** $\boldsymbol{N}(A^{\mathrm{T}})$
$\boldsymbol{v}_1, \ldots, \boldsymbol{v}_r$	is an orthonormal basis for the **row space**
$\boldsymbol{v}_{r+1}, \ldots, \boldsymbol{v}_n$	is an orthonormal basis for the **nullspace** $\boldsymbol{N}(A)$.

More than just orthogonality, these basis vectors diagonalize the matrix A :

"*A* is diagonalized" $\quad A\boldsymbol{v}_1 = \sigma_1 \boldsymbol{u}_1 \quad A\boldsymbol{v}_2 = \sigma_2 \boldsymbol{u}_2 \quad \ldots \quad A\boldsymbol{v}_r = \sigma_r \boldsymbol{u}_r \qquad (1)$

Those **singular values $\boldsymbol{\sigma_1}$ to $\boldsymbol{\sigma_r}$** will be positive numbers: σ_i *is the length of* $A\boldsymbol{v}_i$. The σ's go into a diagonal matrix that is otherwise zero. That matrix is Σ.

(using matrices) Since the $\boldsymbol{u}$'s are orthonormal, the matrix U_r with those r columns has $U_r^{\mathrm{T}} U_r = I$. Since the $\boldsymbol{v}$'s are orthonormal, the matrix V_r has $V_r^{\mathrm{T}} V_r = I$. Then the equations $A\boldsymbol{v}_i = \sigma_i \boldsymbol{u}_i$ tell us column by column that $\boldsymbol{AV_r = U_r \Sigma_r}$:

$$\begin{matrix} (m \text{ by } n)(n \text{ by } r) \\ \boldsymbol{AV_r = U_r\Sigma_r} \\ (m \text{ by } r)(r \text{ by } r) \end{matrix} \qquad A\begin{bmatrix} \\ \boldsymbol{v}_1 & \cdot\cdot & \boldsymbol{v}_r \\ \\ \end{bmatrix} = \begin{bmatrix} \\ \boldsymbol{u}_1 & \cdot\cdot & \boldsymbol{u}_r \\ \\ \end{bmatrix} \begin{bmatrix} \sigma_1 & & \\ & \ddots & \\ & & \sigma_r \end{bmatrix}. \qquad (2)$$

This is the heart of the SVD, but there is more. Those $\boldsymbol{v}$'s and $\boldsymbol{u}$'s account for the row space and column space of A. We have $n - r$ more $\boldsymbol{v}$'s and $m - r$ more $\boldsymbol{u}$'s, from the nullspace $\boldsymbol{N}(A)$ and the left nullspace $\boldsymbol{N}(A^{\mathrm{T}})$. They are automatically orthogonal to the first $\boldsymbol{v}$'s and $\boldsymbol{u}$'s (because the whole nullspaces are orthogonal). We now include all the $\boldsymbol{v}$'s and $\boldsymbol{u}$'s in V and U, so these matrices become *square*. ***We still have*** $\boldsymbol{AV = U\Sigma}$.

$$\begin{array}{c}(m \text{ by } n)(n \text{ by } n)\\ \boldsymbol{AV} \text{ equals } \boldsymbol{U\Sigma}\\ (m \text{ by } m)(m \text{ by } n)\end{array} \quad \boldsymbol{A}\begin{bmatrix} \boldsymbol{v}_1 & \cdot\cdot & \boldsymbol{v}_r & \cdot\cdot & \boldsymbol{v}_n \end{bmatrix} = \begin{bmatrix} \boldsymbol{u}_1 & \cdot\cdot & \boldsymbol{u}_r & \cdot\cdot & \boldsymbol{u}_m \end{bmatrix} \begin{bmatrix} \boldsymbol{\sigma_1} & & \\ & \ddots & \\ & & \boldsymbol{\sigma_r} & \\ & & & \end{bmatrix} \tag{3}$$

The new Σ is m by n. It is just the r by r matrix in equation (2) with $m - r$ extra zero rows and $n - r$ new zero columns. The real change is in the shapes of U and V. Those are square matrices and $V^{-1} = V^{\mathrm{T}}$. So $AV = U\Sigma$ becomes $\boldsymbol{A = U\Sigma V^{\mathrm{T}}}$. This is the ***Singular Value Decomposition***. I can multiply columns $\boldsymbol{u}_i\sigma_i$ from $U\Sigma$ by rows of V^{T}:

SVD $$\boldsymbol{A} = \boldsymbol{U\Sigma V^{\mathrm{T}}} = \boldsymbol{u_1\sigma_1 v_1^{\mathrm{T}}} + \cdots + \boldsymbol{u_r\sigma_r v_r^{\mathrm{T}}}. \tag{4}$$

Equation (2) was a "reduced SVD" with bases for the row space and column space. Equation (3) is the full SVD with nullspaces included. They both split up A into the same r matrices $\boldsymbol{u}_i\sigma_i\boldsymbol{v}_i^{\mathrm{T}}$ of rank one. Column times row is the fourth way to multiply matrices.

We will see that each σ_i^2 is an eigenvalue of $A^{\mathrm{T}}A$ and also AA^{T}. When we put the singular values in descending order, $\sigma_1 \geq \sigma_2 \geq \ldots \sigma_r > 0$, the splitting in equation (4) gives the r rank-one pieces of A ***in order of importance***. This is crucial.

Example 1 When is $A = U\Sigma V^{\mathrm{T}}$ (singular values) the *same* as $X\Lambda X^{-1}$ (eigenvalues)?

Solution A needs orthonormal eigenvectors to allow $X = U = V$. A also needs eigenvalues $\lambda \geq 0$ if $\Lambda = \Sigma$. So A must be a ***positive semidefinite*** (***or definite***) ***symmetric matrix***. Only then will $A = X\Lambda X^{-1}$ which is also $Q\Lambda Q^{\mathrm{T}}$ coincide with $A = U\Sigma V^{\mathrm{T}}$.

Example 2 If $A = \boldsymbol{xy}^{\mathrm{T}}$ (rank 1) with unit vectors $\boldsymbol{x}$ and $\boldsymbol{y}$, what is the SVD of A?

Solution The reduced SVD in (2) is exactly $\boldsymbol{xy}^{\mathrm{T}}$, with rank $r = 1$. It has $\boldsymbol{u}_1 = \boldsymbol{x}$ and $\boldsymbol{v}_1 = \boldsymbol{y}$ and $\sigma_1 = 1$. For the full SVD, complete $\boldsymbol{u}_1 = \boldsymbol{x}$ to an orthonormal basis of $\boldsymbol{u}$'s, and complete $\boldsymbol{v}_1 = \boldsymbol{y}$ to an orthonormal basis of $\boldsymbol{v}$'s. No new σ's, only $\sigma_1 = 1$.

Proof of the SVD

We need to show how those amazing $\boldsymbol{u}$'s and $\boldsymbol{v}$'s can be constructed. The $\boldsymbol{v}$'s will be **orthonormal eigenvectors of** $\boldsymbol{A^{\mathrm{T}}A}$. This must be true because we are aiming for

$$\boldsymbol{A^{\mathrm{T}}A} = (U\Sigma V^{\mathrm{T}})^{\mathrm{T}}(U\Sigma V^{\mathrm{T}}) = V\Sigma^{\mathrm{T}}U^{\mathrm{T}}U\Sigma V^{\mathrm{T}} = \boldsymbol{V\Sigma^{\mathrm{T}}\Sigma V^{\mathrm{T}}}. \tag{5}$$

On the right you see the eigenvector matrix V for the symmetric positive (semi) definite matrix $A^{\mathrm{T}}A$. And $(\Sigma^{\mathrm{T}}\Sigma)$ must be the eigenvalue matrix of $(A^{\mathrm{T}}A)$: *Each* σ^2 *is* $\lambda(A^{\mathrm{T}}A)$!

Now $Av_i = \sigma_i u_i$ tells us the unit vectors u_1 to u_r. This is the key equation (1). The essential point—the whole reason that the SVD succeeds—is that those unit vectors u_1 to u_r are automatically orthogonal to each other (*because the* v's *are orthogonal*):

$$\textbf{Key step } i \neq j \qquad u_i^{\mathrm{T}} u_j = \left(\frac{Av_i}{\sigma_i}\right)^{\mathrm{T}} \left(\frac{Av_j}{\sigma_j}\right) = \frac{v_i^{\mathrm{T}} A^{\mathrm{T}} A v_j}{\sigma_i \sigma_j} = \frac{\sigma_j^2}{\sigma_i \sigma_j} v_i^{\mathrm{T}} v_j = \textbf{zero}. \quad (6)$$

The v's are eigenvectors of $A^{\mathrm{T}}A$ (symmetric). They are orthogonal and now the u's are also orthogonal. *Actually those* u's *will be eigenvectors of* AA^{T}.

Finally we complete the v's and u's to n v's and m u's with any orthonormal bases for the nullspaces $N(A)$ and $N(A^{\mathrm{T}})$. We have found V and Σ and U in $A = U\Sigma V^{\mathrm{T}}$.

An Example of the SVD

Here is an example to show the computation of all three matrices in $A = U\Sigma V^{\mathrm{T}}$.

Example 3 Find the matrices U, Σ, V for $A = \begin{bmatrix} \mathbf{3} & \mathbf{0} \\ \mathbf{4} & \mathbf{5} \end{bmatrix}$. The rank is $r = \mathbf{2}$.

With rank 2, this A has positive singular values σ_1 and σ_2. We will see that σ_1 is larger than $\lambda_{\max} = 5$, and σ_2 is smaller than $\lambda_{\min} = 3$. Begin with $A^{\mathrm{T}}A$ and AA^{T}:

$$A^{\mathrm{T}}A = \begin{bmatrix} 25 & 20 \\ 20 & 25 \end{bmatrix} \qquad AA^{\mathrm{T}} = \begin{bmatrix} 9 & 12 \\ 12 & 41 \end{bmatrix}$$

Those have the same trace (50) and the same eigenvalues $\sigma_1^2 = 45$ and $\sigma_2^2 = 5$. The square roots are $\sigma_1 = \sqrt{\mathbf{45}}$ and $\sigma_2 = \sqrt{\mathbf{5}}$. Then $\sigma_1\sigma_2 = 15$ and this is the determinant of A.

A key step is to find the eigenvectors of $A^{\mathrm{T}}A$ (with eigenvalues 45 and 5):

$$\begin{bmatrix} 25 & 20 \\ 20 & 25 \end{bmatrix} \begin{bmatrix} 1 \\ 1 \end{bmatrix} = \mathbf{45} \begin{bmatrix} 1 \\ 1 \end{bmatrix} \qquad \begin{bmatrix} 25 & 20 \\ 20 & 25 \end{bmatrix} \begin{bmatrix} -1 \\ 1 \end{bmatrix} = \mathbf{5} \begin{bmatrix} -1 \\ 1 \end{bmatrix}$$

Then v_1 and v_2 are those orthogonal eigenvectors rescaled to length 1. Divide by $\sqrt{2}$.

$$\textbf{Right singular vectors } v_1 = \frac{1}{\sqrt{2}}\begin{bmatrix} 1 \\ 1 \end{bmatrix} \quad v_2 = \frac{1}{\sqrt{2}}\begin{bmatrix} -1 \\ 1 \end{bmatrix} \quad \textbf{Left singular vectors } u_i = \frac{Av_i}{\sigma_i}$$

Now compute Av_1 and Av_2 which will be $\sigma_1 u_1 = \sqrt{45}\, u_1$ and $\sigma_2 u_2 = \sqrt{5}\, u_2$:

$$Av_1 = \frac{3}{\sqrt{2}}\begin{bmatrix} 1 \\ 3 \end{bmatrix} = \sqrt{45}\frac{\mathbf{1}}{\sqrt{\mathbf{10}}}\begin{bmatrix} \mathbf{1} \\ \mathbf{3} \end{bmatrix} = \sigma_1 u_1$$

$$Av_2 = \frac{1}{\sqrt{2}}\begin{bmatrix} -3 \\ 1 \end{bmatrix} = \sqrt{5}\frac{\mathbf{1}}{\sqrt{\mathbf{10}}}\begin{bmatrix} \mathbf{-3} \\ \mathbf{1} \end{bmatrix} = \sigma_2 u_2$$

The division by $\sqrt{10}$ makes u_1 and u_2 orthonormal. Then $\sigma_1 = \sqrt{45}$ and $\sigma_2 = \sqrt{5}$ as expected. The Singular Value Decomposition of A is U times Σ times V^{T}.

$$\boldsymbol{U} = \frac{1}{\sqrt{10}}\begin{bmatrix} 1 & -3 \\ 3 & 1 \end{bmatrix} \quad \boldsymbol{\Sigma} = \begin{bmatrix} \sqrt{45} & \\ & \sqrt{5} \end{bmatrix} \quad \boldsymbol{V} = \frac{1}{\sqrt{2}}\begin{bmatrix} 1 & -1 \\ 1 & 1 \end{bmatrix}. \qquad (7)$$

U and V contain orthonormal bases for the column space and the row space (both spaces are just $\mathbf{R}^2$). The real achievement is that those two bases diagonalize A: AV equals $U\Sigma$. The matrix A splits into a combination of two rank-one matrices, columns times rows:

$$\sigma_1 \boldsymbol{u}_1 \boldsymbol{v}_1^{\mathrm{T}} + \sigma_2 \boldsymbol{u}_2 \boldsymbol{v}_2^{\mathrm{T}} = \frac{\sqrt{45}}{\sqrt{20}}\begin{bmatrix} \mathbf{1} & \mathbf{1} \\ \mathbf{3} & \mathbf{3} \end{bmatrix} + \frac{\sqrt{5}}{\sqrt{20}}\begin{bmatrix} \mathbf{3} & \mathbf{-3} \\ \mathbf{-1} & \mathbf{1} \end{bmatrix} = \begin{bmatrix} 3 & 0 \\ 4 & 5 \end{bmatrix} = A.$$

An Extreme Matrix

Here is a larger example, when the $\boldsymbol{u}$'s and the $\boldsymbol{v}$'s are just columns of the identity matrix. So the computations are easy, but keep your eye on the *order of the columns*. The matrix A is badly lopsided (strictly triangular). All its eigenvalues are zero. AA^{T} is not close to $A^{\mathrm{T}}A$. The matrices U and V will be permutations that fix these problems properly.

$$A = \begin{bmatrix} 0 & \mathbf{1} & 0 & 0 \\ 0 & 0 & \mathbf{2} & 0 \\ 0 & 0 & 0 & \mathbf{3} \\ 0 & 0 & 0 & 0 \end{bmatrix} \quad \begin{array}{l} \text{eigenvalues } \lambda = 0,0,0,0 \text{ all zero!} \\ \text{only one eigenvector } (1,0,0,0) \\ \text{singular values } \sigma = \mathbf{3,2,1} \\ \text{singular vectors are columns of } I \end{array}$$

$A^{\mathrm{T}}A$ and AA^{T} are diagonal (with easy eigenvectors, but in different orders):

$$A^{\mathrm{T}}A = \begin{bmatrix} \mathbf{0} & 0 & 0 & 0 \\ 0 & \mathbf{1} & 0 & 0 \\ 0 & 0 & \mathbf{4} & 0 \\ 0 & 0 & 0 & \mathbf{9} \end{bmatrix} \qquad AA^{\mathrm{T}} = \begin{bmatrix} \mathbf{1} & 0 & 0 & 0 \\ 0 & \mathbf{4} & 0 & 0 \\ 0 & 0 & \mathbf{9} & 0 \\ 0 & 0 & 0 & \mathbf{0} \end{bmatrix}$$

Their eigenvectors ($\boldsymbol{u}$'s for AA^{T} and $\boldsymbol{v}$'s for $A^{\mathrm{T}}A$) go in decreasing order $\sigma_1^2 > \sigma_2^2 > \sigma_3^2$ of the eigenvalues. Those eigenvalues are $\sigma^2 = 9, 4, 1$.

$$\boldsymbol{U} = \begin{bmatrix} 0 & 0 & \mathbf{1} & 0 \\ 0 & \mathbf{1} & 0 & 0 \\ \mathbf{1} & 0 & 0 & 0 \\ 0 & 0 & 0 & \mathbf{1} \end{bmatrix} \qquad \Sigma = \begin{bmatrix} \mathbf{3} & & & \\ & \mathbf{2} & & \\ & & \mathbf{1} & \\ & & & 0 \end{bmatrix} \qquad \boldsymbol{V} = \begin{bmatrix} 0 & 0 & 0 & \mathbf{1} \\ 0 & 0 & \mathbf{1} & 0 \\ 0 & \mathbf{1} & 0 & 0 \\ \mathbf{1} & 0 & 0 & 0 \end{bmatrix}$$

Those first columns $\boldsymbol{u}_1$ and $\boldsymbol{v}_1$ have 1's in positions 3 and 4. Then $\boldsymbol{u}_1\sigma_1\boldsymbol{v}_1^{\mathrm{T}}$ picks out the biggest number $A_{34} = 3$ in the original matrix A. The three rank-one matrices in the SVD come (for this extreme example) exactly from the numbers $3, 2, 1$ in A.

$$\boldsymbol{A} = \boldsymbol{U\Sigma V}^{\mathbf{T}} = \mathbf{3}\boldsymbol{u}_1\boldsymbol{v}_1^{\mathrm{T}} + \mathbf{2}\boldsymbol{u}_2\boldsymbol{v}_2^{\mathrm{T}} + \mathbf{1}\boldsymbol{u}_3\boldsymbol{v}_3^{\mathrm{T}}.$$

Note Suppose I remove the last row of A (all zeros). Then A is a 3 by 4 matrix and AA^{T} is 3 by 3—its fourth row and column will disappear. We still have eigenvalues $\lambda = 1, 4, 9$ in $A^{\mathrm{T}}A$ and AA^{T}, producing the same singular values $\sigma = 3, 2, 1$ in Σ.

Removing the zero row of A (now 3×4) just removes the last row of Σ and also the last row and column of U. Then $(3 \times 4) = U\Sigma V^{\mathrm{T}} = (3 \times 3)(3 \times 4)(4 \times 4)$. The SVD is totally adapted to rectangular matrices.

A good thing, because the rows and columns of a data matrix A often have completely different meanings (like a spreadsheet). If we have the grades for all courses, there would be a column for each student and a row for each course: The entry a_{ij} would be the grade. Then $\sigma_1 \boldsymbol{u}_1 \boldsymbol{v}_1^{\mathrm{T}}$ could have $\boldsymbol{u}_1$ = **combination course** and $\boldsymbol{v}_1$ = **combination student**. And σ_1 would be the grade for those combinations: the highest grade.

The matrix A could count the frequency of key words in a journal: A different article for each column of A and a different word for each row. The whole journal is indexed by the matrix A and the most important information is in $\sigma_1 \boldsymbol{u}_1 \boldsymbol{v}_1^{\mathrm{T}}$. Then σ_1 is the largest frequency for a hyperword (the word combination $\boldsymbol{u}_1$) in the hyperarticle $\boldsymbol{v}_1$.

Section 7.3 will apply the SVD to finance and genetics and search engines.

Singular Value Stability versus Eigenvalue Instability

The 4 by 4 example A provides an example (an extreme case) of the instability of eigenvalues. **Suppose the 4,1 entry barely changes** from zero to $1/60,000$. The rank is now 4.

$$A = \begin{bmatrix} 0 & 1 & 0 & 0 \\ 0 & 0 & 2 & 0 \\ 0 & 0 & 0 & 3 \\ \dfrac{1}{60,000} & 0 & 0 & 0 \end{bmatrix}$$

That change by only $1/60,000$ produces a much bigger jump in the eigenvalues of A

$$\lambda = 0,0,0,0 \text{ to } \boldsymbol{\lambda} = \frac{\mathbf{1}}{\mathbf{10}}, \frac{\boldsymbol{i}}{\mathbf{10}}, \frac{\mathbf{-1}}{\mathbf{10}}, \frac{\boldsymbol{-i}}{\mathbf{10}}$$

The four eigenvalues moved from zero onto a circle around zero. The circle has radius $\frac{1}{10}$ when the new entry is only $1/60,000$. This shows serious instability of eigenvalues when AA^{T} is far from $A^{\mathrm{T}}A$. At the other extreme, if $A^{\mathrm{T}}A = AA^{\mathrm{T}}$ (a "normal matrix") the eigenvectors of A are orthogonal and the eigenvalues of A are totally stable.

By contrast, **the singular values of any matrix are stable**. They don't change more than the change in A. In this example, the new singular values are $\mathbf{3, 2, 1}$, and $\mathbf{1/60,000}$. The matrices U and V stay the same. The new fourth piece of A is $\sigma_4 \boldsymbol{u}_4 \boldsymbol{v}_4^{\mathrm{T}}$, with fifteen zeros and that small entry $\sigma_4 = 1/60,000$.

Singular Vectors of A and Eigenvectors of $S = A^{\mathrm{T}}A$

Equations (5–6) "proved" the SVD *all at once*. The singular vectors $\boldsymbol{v}_i$ are the eigenvectors $\boldsymbol{q}_i$ of $S = A^{\mathrm{T}}A$. The eigenvalues λ_i of S are the same as σ_i^2 for A. The rank r of S equals the rank of A. The expansions in eigenvectors and singular vectors are perfectly parallel.

Symmetric *S*	$S = Q\Lambda Q^{\mathrm{T}}$	$= \lambda_1 \boldsymbol{q}_1 \boldsymbol{q}_1^{\mathrm{T}} + \lambda_2 \boldsymbol{q}_2 \boldsymbol{q}_2^{\mathrm{T}} + \cdots + \lambda_r \boldsymbol{q}_r \boldsymbol{q}_r^{\mathrm{T}}$
Any matrix *A*	$A = U\Sigma V^{\mathrm{T}}$	$= \sigma_1 \boldsymbol{u}_1 \boldsymbol{v}_1^{\mathrm{T}} + \sigma_2 \boldsymbol{u}_2 \boldsymbol{v}_2^{\mathrm{T}} + \cdots + \sigma_r \boldsymbol{u}_r \boldsymbol{v}_r^{\mathrm{T}}$

The $\boldsymbol{q}$'s are orthonormal, the $\boldsymbol{u}$'s are orthonormal, the $\boldsymbol{v}$'s are orthonormal. Beautiful.

But I want to look again, for two good reasons. One is to fix a weak point in the eigenvalue part, where Chapter 6 was not complete. If λ is a *double* eigenvalue of S, we can and must find *two* orthonormal eigenvectors. The other reason is to see how the SVD picks off the largest term $\sigma_1\boldsymbol{u}_1\boldsymbol{v}_1^{\mathrm{T}}$ before $\sigma_2\boldsymbol{u}_2\boldsymbol{v}_2^{\mathrm{T}}$. We want to understand the eigenvalues λ (of S) and the singular values σ (of A) **one at a time instead of all at once**.

Start with the largest eigenvalue λ_1 of S. It solves this problem:

$$\boldsymbol{\lambda_1} = \textbf{maximum ratio } \frac{\boldsymbol{x}^{\mathrm{T}}S\boldsymbol{x}}{\boldsymbol{x}^{\mathrm{T}}\boldsymbol{x}}. \quad \text{The winning vector is } \boldsymbol{x} = \boldsymbol{q}_1 \text{ with } S\boldsymbol{q}_1 = \lambda_1\boldsymbol{q}_1. \quad (8)$$

Compare with the largest singular value σ_1 of A. It solves this problem:

$$\boldsymbol{\sigma_1} = \textbf{maximum ratio } \frac{||A\boldsymbol{x}||}{||\boldsymbol{x}||}. \quad \text{The winning vector is } \boldsymbol{x} = \boldsymbol{v}_1 \text{ with } A\boldsymbol{v}_1 = \sigma_1\boldsymbol{u}_1. \quad (9)$$

This "one at a time approach" applies also to λ_2 and σ_2. But not all $\boldsymbol{x}$'s are allowed:

$$\boldsymbol{\lambda_2} = \textbf{maximum ratio } \frac{\boldsymbol{x}^{\mathrm{T}}S\boldsymbol{x}}{\boldsymbol{x}^{\mathrm{T}}\boldsymbol{x}} \text{ among all } \boldsymbol{x}\text{'s with } \boldsymbol{q}_1^{\mathrm{T}}\boldsymbol{x} = 0. \quad \boldsymbol{x} = \boldsymbol{q_2} \text{ will win.} \quad (10)$$

$$\boldsymbol{\sigma_2} = \textbf{maximum ratio } \frac{||A\boldsymbol{x}||}{||\boldsymbol{x}||} \text{ among all } \boldsymbol{x}\text{'s with } \boldsymbol{v}_1^{\mathrm{T}}\boldsymbol{x} = 0. \quad \boldsymbol{x} = \boldsymbol{v_2} \text{ will win.} \quad (11)$$

When $S = A^{\mathrm{T}}A$ we find $\lambda_1 = \sigma_1^2$ and $\lambda_2 = \sigma_2^2$. Why does this approach succeed?

Start with the ratio $r(\boldsymbol{x}) = \boldsymbol{x}^{\mathrm{T}}S\boldsymbol{x}/\boldsymbol{x}^{\mathrm{T}}\boldsymbol{x}$. This is called the *Rayleigh quotient*. To maximize $r(\boldsymbol{x})$, set its partial derivatives to zero: $\partial r/\partial x_i = 0$ for $i = 1, \ldots, n$. Those derivatives are messy and here is the result: one vector equation for the winning $\boldsymbol{x}$:

$$\textbf{The derivatives of } r(\boldsymbol{x}) = \frac{\boldsymbol{x}^{\mathrm{T}}S\boldsymbol{x}}{\boldsymbol{x}^{\mathrm{T}}\boldsymbol{x}} \textbf{ are zero when } \qquad S\boldsymbol{x} = r(\boldsymbol{x})\boldsymbol{x}. \qquad (12)$$

So the winning $\boldsymbol{x}$ is an eigenvector of S. The maximum ratio $r(\boldsymbol{x})$ is the largest eigenvalue λ_1 of S. All good. Now turn to A—and notice the connection to $S = A^{\mathrm{T}}A$!

$$\text{Maximizing } \frac{||A\boldsymbol{x}||}{||\boldsymbol{x}||} \text{ also maximizes } \left(\frac{||A\boldsymbol{x}||}{||\boldsymbol{x}||}\right)^2 = \frac{\boldsymbol{x}^{\mathrm{T}}A^{\mathrm{T}}A\boldsymbol{x}}{\boldsymbol{x}^{\mathrm{T}}\boldsymbol{x}} = \frac{\boldsymbol{x}^{\mathrm{T}}S\boldsymbol{x}}{\boldsymbol{x}^{\mathrm{T}}\boldsymbol{x}}.$$

So the winning $\boldsymbol{x} = \boldsymbol{v}_1$ in (9) is the same as the top eigenvector $\boldsymbol{q}_1$ of $S = A^{\mathrm{T}}A$ in (8).

Now I have to explain why $\boldsymbol{q}_2$ and $\boldsymbol{v}_2$ are the winning vectors in (10) and (11). We know they are orthogonal to $\boldsymbol{q}_1$ and $\boldsymbol{v}_1$, so they are allowed in those competitions. These paragraphs can be optional for readers who aim to see the SVD in action (Section 7.3).

Start with any orthogonal matrix Q_1 that has $\boldsymbol{q}_1$ in its first column. The other $n-1$ orthonormal columns just have to be orthogonal to $\boldsymbol{q}_1$. Then use $S\boldsymbol{q}_1 = \lambda_1 \boldsymbol{q}_1$:

$$SQ_1 = S\begin{bmatrix} \boldsymbol{q}_1 & \boldsymbol{q}_2 \cdots \boldsymbol{q}_n \end{bmatrix} = \begin{bmatrix} \boldsymbol{q}_1 & \boldsymbol{q}_2 \cdots \boldsymbol{q}_n \end{bmatrix}\begin{bmatrix} \lambda_1 & \boldsymbol{w}^{\mathrm{T}} \\ \mathbf{0} & S_{n-1} \end{bmatrix} = Q_1\begin{bmatrix} \lambda_1 & \boldsymbol{w}^{\mathrm{T}} \\ \mathbf{0} & S_{n-1} \end{bmatrix}. \quad (13)$$

Multiply by Q_1^{T}, remember $Q_1^{\mathrm{T}}Q_1 = I$, and recognize that $Q_1^{\mathrm{T}}SQ_1$ is symmetric like S:

$$\text{The symmetry of } Q_1^{\mathrm{T}}SQ_1 = \begin{bmatrix} \lambda_1 & \boldsymbol{w}^{\mathrm{T}} \\ \mathbf{0} & S_{n-1} \end{bmatrix} \text{ forces } \boldsymbol{w} = \mathbf{0} \text{ and } S_{n-1}^{\mathrm{T}} = S_{n-1}.$$

The requirement $\boldsymbol{q}_1^{\mathrm{T}}\boldsymbol{x} = 0$ has reduced the maximum problem (10) to size $n-1$. The largest eigenvalue of S_{n-1} will be the *second largest* for S. **It is $\boldsymbol{\lambda_2}$.** The winning vector in (10) will be the eigenvector $\boldsymbol{q}_2$ with $S\boldsymbol{q}_2 = \lambda_2\boldsymbol{q}_2$.

We just keep going—or use the magic word *induction*—to produce all the eigenvectors $\boldsymbol{q}_1, \ldots, \boldsymbol{q}_n$ and their eigenvalues $\lambda_1, \ldots, \lambda_n$. The Spectral Theorem $S = Q\Lambda Q^{\mathrm{T}}$ is proved even with repeated eigenvalues. All symmetric matrices can be diagonalized.

Similarly the SVD is found one step at a time from (9) and (11) and onwards. Section 7.4 will show the geometry—we are finding the axes of an ellipse. Here I ask a different question: **How are the $\boldsymbol{\lambda}$'s and $\boldsymbol{\sigma}$'s actually computed**?

Computing the Eigenvalues of S and Singular Values of A

The singular values σ_i of A are the square roots of the eigenvalues λ_i of $S = A^{\mathrm{T}}A$. This connects the SVD to a *symmetric eigenvalue problem* (good). But in the end we don't want to multiply A^{T} times A (squaring is time-consuming: not good).

The first idea is ***to produce zeros in A and S without changing any σ's and λ's.*** Singular vectors and eigenvectors will change—no problem. The similar matrix $Q^{-1}SQ$ has the same λ's as S. If Q is orthogonal, this matrix is $Q^{\mathrm{T}}SQ$ and still symmetric.

Section 11.3 will show how to build Q from 2 by 2 rotations so that $Q^{\mathrm{T}}SQ$ is **symmetric and tridiagonal** (many zeros). But rotations can't get all the way to a diagonal matrix. To show all the eigenvalues of S needs a new idea and more work.

For the SVD, what is the parallel to $Q^{\mathrm{T}}SQ$? Now we don't want to change any singular values of A. Natural answer: You can multiply A by *two different orthogonal matrices Q_1 and Q_2*. Use them to produce zeros in $Q_1^{\mathrm{T}}AQ_2$. The σ's don't change:

$$(Q_1^{\mathrm{T}}AQ_2)^{\mathrm{T}}(Q_1^{\mathrm{T}}AQ_2) = Q_2^{\mathrm{T}}A^{\mathrm{T}}AQ_2 = Q_2^{\mathrm{T}}SQ_2 \text{ gives the same } \sigma(A) \text{ and } \lambda(S).$$

The freedom of two Q's allows us to reach $Q_1^{\mathrm{T}}AQ_2 =$ **bidiagonal matrix** (2 diagonals). This compares perfectly to $Q^{\mathrm{T}}SQ = 3$ diagonals. It is nice to notice the connection between them: *(bidiagonal)*^{T} *(bidiagonal)* $=$ *tridiagonal.*

The final steps to a *diagonal* Λ and a *diagonal* Σ need more ideas. This problem can't be easy, because underneath we are solving $\det(S - \lambda I) = 0$ for polynomials of degree $n = 100$ or 1000 or more. We certainly don't use those polynomials!

The favorite way to find λ's and σ's in **LAPACK** uses simple orthogonal matrices to approach $Q^{\mathrm{T}}SQ = \Lambda$ and $U^{\mathrm{T}}AV = \Sigma$. **We stop when very close to Λ and Σ.**

This 2-step approach (zeros first) is built into the commands **eig**(S) and **svd**(A).

■ REVIEW OF THE KEY IDEAS ■

1. The SVD factors A into $U\Sigma V^{\mathrm{T}}$, with r singular values $\sigma_1 \geq \ldots \geq \sigma_r > 0$.
2. The numbers $\sigma_1^2, \ldots, \sigma_r^2$ are the nonzero eigenvalues of AA^{T} and $A^{\mathrm{T}}A$.
3. The orthonormal columns of U and V are eigenvectors of AA^{T} and $A^{\mathrm{T}}A$.
4. Those columns hold orthonormal bases for the four fundamental subspaces of A.
5. Those bases diagonalize the matrix: $A\boldsymbol{v}_i = \sigma_i \boldsymbol{u}_i$ for $i \leq r$. This is $\boldsymbol{AV = U\Sigma}$.
6. $A = \sigma_1 \boldsymbol{u}_1 \boldsymbol{v}_1^{\mathrm{T}} + \cdots + \sigma_r \boldsymbol{u}_r \boldsymbol{v}_r^{\mathrm{T}}$ and σ_1 is the maximum of the ratio $||A\boldsymbol{x}|| \,/\, ||\boldsymbol{x}||$.

■ WORKED EXAMPLES ■

7.2 A Identify by name these decompositions of A into a sum of columns times rows:

1. *Orthogonal* columns $\boldsymbol{u}_1\sigma_1, \ldots, \boldsymbol{u}_r\sigma_r$ times *orthonormal* rows $\boldsymbol{v}_1^{\mathrm{T}}, \ldots, \boldsymbol{v}_r^{\mathrm{T}}$.
2. *Orthonormal* columns $\boldsymbol{q}_1, \ldots, \boldsymbol{q}_r$ times *triangular* rows $\boldsymbol{r}_1^{\mathrm{T}}, \ldots, \boldsymbol{r}_r^{\mathrm{T}}$.
3. *Triangular* columns $\boldsymbol{l}_1, \ldots, \boldsymbol{l}_r$ times *triangular* rows $\boldsymbol{u}_1^{\mathrm{T}}, \ldots, \boldsymbol{u}_r^{\mathrm{T}}$.

Where do the rank and the pivots and the singular values of A come into this picture?

Solution These three factorizations are basic to linear algebra, pure or applied:

1. **Singular Value Decomposition $\boldsymbol{A = U\Sigma V^{\mathrm{T}}}$**
2. **Gram-Schmidt Orthogonalization $\boldsymbol{A = QR}$**
3. **Gaussian Elimination $\boldsymbol{A = LU}$**

You might prefer to separate out singular values $\boldsymbol{\sigma_i}$ and heights $\boldsymbol{h_i}$ and pivots $\boldsymbol{d_i}$:

1. $A = U\Sigma V^{\mathrm{T}}$ with unit vectors in U and V. ***The r singular values $\boldsymbol{\sigma_i}$ are in $\boldsymbol{\Sigma}$.***
2. $A = QHR$ with unit vectors in Q and diagonal 1's in R. ***The r heights $\boldsymbol{h_i}$ are in $\boldsymbol{H}$.***
3. $A = LDU$ with diagonal 1's in L and U. ***The r pivots $\boldsymbol{d_i}$ are in $\boldsymbol{D}$.***

Each h_i tells the height of column i above the plane of columns 1 to $i-1$. The volume of the full n-dimensional box ($r = m = n$) comes from $A = U\Sigma V^{\mathrm{T}} = LDU = QHR$:

$$|\,\mathbf{det}\,\boldsymbol{A}\,| = |\,\textit{product of } \boldsymbol{\sigma}\text{'s}\,| = |\,\textit{product of } \boldsymbol{d}\text{'s}\,| = |\,\textit{product of } \boldsymbol{h}\text{'s}\,|.$$

7.2 B Show that $\sigma_1 \geq |\lambda|_{\max}$. The largest singular value dominates all eigenvalues.

Solution Start from $A = U\Sigma V^{\mathrm{T}}$. Remember that multiplying by an orthogonal matrix *does not change length*: $\|Q\boldsymbol{x}\| = \|\boldsymbol{x}\|$ because $\|Q\boldsymbol{x}\|^2 = \boldsymbol{x}^{\mathrm{T}}Q^{\mathrm{T}}Q\boldsymbol{x} = \boldsymbol{x}^{\mathrm{T}}\boldsymbol{x} = \|\boldsymbol{x}\|^2$. This applies to $Q = U$ and $Q = V^{\mathrm{T}}$. In between is the diagonal matrix Σ.

$$\|A\boldsymbol{x}\| = \|U\Sigma V^{\mathrm{T}}\boldsymbol{x}\| = \|\Sigma V^{\mathrm{T}}\boldsymbol{x}\| \leq \sigma_1\|V^{\mathrm{T}}\boldsymbol{x}\| = \sigma_1\|\boldsymbol{x}\|. \tag{14}$$

An eigenvector has $\|A\boldsymbol{x}\| = |\lambda|\|\boldsymbol{x}\|$. So (14) says that $|\lambda|\|\boldsymbol{x}\| \leq \sigma_1\|\boldsymbol{x}\|$. Then $|\boldsymbol{\lambda}| \leq \boldsymbol{\sigma_1}$.

Apply also to the unit vector $\boldsymbol{x} = (1, 0, \ldots, 0)$. Now $A\boldsymbol{x}$ is the first column of A. Then by inequality (14), this column has length $\leq \sigma_1$. Every entry must have $|a_{ij}| \leq \sigma_1$.

Equation (14) shows again that ***the maximum value of*** $||A\boldsymbol{x}||/||\boldsymbol{x}||$ ***equals*** $\boldsymbol{\sigma_1}$.

Section 11.2 will explain how the ratio $\sigma_{\max}/\sigma_{\min}$ governs the roundoff error in solving $A\boldsymbol{x} = \boldsymbol{b}$. MATLAB warns you if this "*condition number*" is large. Then $\boldsymbol{x}$ is unreliable.

Problem Set 7.2

1 Find the eigenvalues of these matrices. Then find singular values from $A^{\mathrm{T}}A$:

$$A = \begin{bmatrix} 0 & 4 \\ 0 & 0 \end{bmatrix} \qquad\qquad A = \begin{bmatrix} 0 & 4 \\ 1 & 0 \end{bmatrix}$$

For each A, construct V from the eigenvectors of $A^{\mathrm{T}}A$ and U from the eigenvectors of AA^{T}. Check that $A = U\Sigma V^{\mathrm{T}}$.

2 Find $A^{\mathrm{T}}A$ and V and Σ and $\boldsymbol{u}_i = A\boldsymbol{v}_i/\sigma_i$ and the full SVD:

$$A = \begin{bmatrix} 2 & 2 \\ -1 & 1 \end{bmatrix} = U\Sigma V^{\mathrm{T}}.$$

3 In Problem 2, show that AA^{T} is diagonal. Its eigenvectors $\boldsymbol{u}_1, \boldsymbol{u}_2$ are _____. Its eigenvalues σ_1^2, σ_2^2 are _____. The rows of A are orthogonal but they are not _____. So the columns of A are not orthogonal.

4 Compute $A^{\mathrm{T}}A$ and AA^{T} and their eigenvalues and unit eigenvectors for V and U.

$$\textbf{Rectangular matrix} \qquad A = \begin{bmatrix} 1 & 1 & 0 \\ 0 & 1 & 1 \end{bmatrix}.$$

Check $AV = U\Sigma$ (this decides $\pm$ signs in U). Σ has the same shape as A: 2×3.

5 (a) The row space of $A = \begin{bmatrix} 1 & 1 \\ 3 & 3 \end{bmatrix}$ is 1-dimensional. Find $\boldsymbol{v}_1$ in the row space and $\boldsymbol{u}_1$ in the column space. What is σ_1? Why is there no σ_2?

(b) Choose $\boldsymbol{v}_2$ and $\boldsymbol{u}_2$ in U and V. Then $A = U\Sigma V^{\mathrm{T}} = \boldsymbol{u}_1\sigma_1\boldsymbol{v}_1^{\mathrm{T}}$ (one term only).

6 Substitute the SVD for A and A^{T} to show that $A^{\mathrm{T}}A$ has its eigenvalues in $\Sigma^{\mathrm{T}}\Sigma$ and AA^{T} has its eigenvalues in $\Sigma\Sigma^{\mathrm{T}}$. Since a diagonal $\Sigma^{\mathrm{T}}\Sigma$ has the same nonzeros as $\Sigma\Sigma^{\mathrm{T}}$, we see again that $A^{\mathrm{T}}A$ and AA^{T} have the same nonzero eigenvalues.

7 If $(A^{\mathrm{T}}A)\boldsymbol{v} = \sigma^2\boldsymbol{v}$, multiply by A. *Move the parentheses to get* $(AA^{\mathrm{T}})A\boldsymbol{v} = \sigma^2(A\boldsymbol{v})$. **If $\boldsymbol{v}$ is an eigenvector of $A^{\mathrm{T}}A$, then ______ is an eigenvector of AA^{T}.**

8 Find the eigenvalues and unit eigenvectors $\boldsymbol{v}_1, \boldsymbol{v}_2$ of $A^{\mathrm{T}}A$. Then find $\boldsymbol{u}_1 = A\boldsymbol{v}_1/\sigma_1$:

$$A = \begin{bmatrix} 1 & 2 \\ 3 & 6 \end{bmatrix} \text{ and } A^{\mathrm{T}}A = \begin{bmatrix} 10 & 20 \\ 20 & 40 \end{bmatrix} \text{ and } AA^{\mathrm{T}} = \begin{bmatrix} 5 & 15 \\ 15 & 45 \end{bmatrix}.$$

Verify that $\boldsymbol{u}_1$ is a unit eigenvector of AA^{T}. Complete the matrices U, Σ, V.

$$\textbf{SVD} \qquad \begin{bmatrix} 1 & 2 \\ 3 & 6 \end{bmatrix} = \begin{bmatrix} \boldsymbol{u}_1 & \boldsymbol{u}_2 \end{bmatrix} \begin{bmatrix} \sigma_1 & \\ & 0 \end{bmatrix} \begin{bmatrix} \boldsymbol{v}_1 & \boldsymbol{v}_2 \end{bmatrix}^{\mathrm{T}}.$$

9 Write down orthonormal bases for the four fundamental subspaces of this A.

10 (a) Why is the trace of $A^{\mathrm{T}}A$ equal to the sum of all a_{ij}^2 ? In Example 3 it is 50.

(b) For every rank-one matrix, why is $\sigma_1^2 =$ sum of all a_{ij}^2?

11 Find the eigenvalues and unit eigenvectors of $A^{\mathrm{T}}A$ and AA^{T}. Keep each $A\boldsymbol{v} = \sigma\boldsymbol{u}$. Then construct the singular value decomposition and verify that A equals $U\Sigma V^{\mathrm{T}}$.

$$\textbf{Fibonacci matrix} \qquad A = \begin{bmatrix} 1 & 1 \\ 1 & 0 \end{bmatrix}$$

12 Use the **svd** part of the MATLAB demo **eigshow** to find those $\boldsymbol{v}$'s graphically.

13 If $A = U\Sigma V^{\mathrm{T}}$ is a square invertible matrix then $A^{-1} =$ ______ ______ ______. Check $A^{-1}A$. This shows that *the singular values of* A^{-1} *are* $1/\sigma_i$.

Note: The largest singular value of A^{-1} is therefore $1/\sigma_{\min}(A)$. The largest eigenvalue $|\lambda(A^{-1})|_{\max}$ is $1/|\lambda(A)|_{\min}$. Then equation (14) says that $\sigma_{\min}(A) \le |\lambda(A)|_{\min}$.

14 Suppose $\boldsymbol{u}_1, \ldots, \boldsymbol{u}_n$ and $\boldsymbol{v}_1, \ldots, \boldsymbol{v}_n$ are orthonormal bases for $\mathbf{R}^n$. Construct the matrix $A = U\Sigma V^{\mathrm{T}}$ that transforms each $\boldsymbol{v}_j$ into $\boldsymbol{u}_j$ to give $A\boldsymbol{v}_1 = \boldsymbol{u}_1, \ldots, A\boldsymbol{v}_n = \boldsymbol{u}_n$.

15 Construct the matrix with rank one that has $A\boldsymbol{v} = 12\boldsymbol{u}$ for $\boldsymbol{v} = \frac{1}{2}(1,1,1,1)$ and $\boldsymbol{u} = \frac{1}{3}(2,2,1)$. Its only singular value is $\sigma_1 =$ ______.

16 Suppose A has orthogonal columns $\boldsymbol{w}_1, \boldsymbol{w}_2, \ldots, \boldsymbol{w}_n$ of lengths $\sigma_1, \sigma_2, \ldots, \sigma_n$. What are U, Σ, and V in the SVD?

17 Suppose A is a 2 by 2 symmetric matrix with unit eigenvectors $\boldsymbol{u}_1$ and $\boldsymbol{u}_2$. If its eigenvalues are $\lambda_1 = 3$ and $\lambda_2 = -2$, what are the matrices U, Σ, V^{T} in its SVD?

18 If $A = QR$ with an orthogonal matrix Q, the SVD of A is almost the same as the SVD of R. Which of the three matrices U, Σ, V is changed because of Q?

19 Suppose A is invertible (with $\sigma_1 > \sigma_2 > 0$). Change A by ***as small a matrix as possible*** to produce a singular matrix A_0. Hint: U and V do not change:

$$\text{From} \quad A = \begin{bmatrix} \boldsymbol{u}_1 & \boldsymbol{u}_2 \end{bmatrix} \begin{bmatrix} \sigma_1 & \\ & \sigma_2 \end{bmatrix} \begin{bmatrix} \boldsymbol{v}_1 & \boldsymbol{v}_2 \end{bmatrix}^{\mathrm{T}} \quad \text{find the nearest } A_0.$$

20 Find the singular values of A from the command **svd**(A) or by hand.

$$A = \begin{bmatrix} 1 & 0 \\ 100 & 1 \end{bmatrix}. \text{ Why is } \sigma_2 = \frac{1}{\sigma_1} \text{ for this matrix?}$$

21 Why doesn't the SVD for $A + I$ just use $\Sigma + I$?

22 If $A = U\Sigma V^{\mathrm{T}}$ then $Q_1 A Q_2^{\mathrm{T}} = (Q_1 U)\Sigma(Q_2 V)^{\mathrm{T}}$. Why will any orthogonal matrices Q_1 and Q_2 leave $Q_1 U$ = orthogonal matrix and $Q_2 V$ = orthogonal matrix? Then Σ sees **no change in the singular values: $Q_1 A Q_2^{\mathrm{T}}$ has the same σ's as A.**

23 If Q is an orthogonal matrix, why do all its singular values equal 1?

24 (a) Find the maximum of $\dfrac{\boldsymbol{x}^{\mathrm{T}} S \boldsymbol{x}}{\boldsymbol{x}^{\mathrm{T}} \boldsymbol{x}} = \dfrac{3x_1^2 + 2x_1x_2 + 3x_2^2}{x_1^2 + x_2^2}$. What matrix is S?

(b) Find the maximum of $\dfrac{(x_1 + 4x_2)^2}{x_1^2 + x_2^2}$. For what matrix A is this $\dfrac{||A\boldsymbol{x}||^2}{||\boldsymbol{x}||^2}$?

25 What are the **minimum values** of the ratios $\dfrac{\boldsymbol{x}^{\mathrm{T}} S \boldsymbol{x}}{\boldsymbol{x}^{\mathrm{T}} \boldsymbol{x}}$ and $\dfrac{||A\boldsymbol{x}||^2}{||\boldsymbol{x}||^2}$? We should take $\boldsymbol{x}$ to be which eigenvectors of S? Should $\boldsymbol{x}$ always be an eigenvector of A?

26 Every matrix $A = U\Sigma V^{\mathrm{T}}$ takes **circles to ellipses**. $AV = U\Sigma$ says that the radius vectors $\boldsymbol{v}_1$ and $\boldsymbol{v}_2$ of the circle go to the semi-axes $\sigma_1 \boldsymbol{u}_1$ and $\sigma_2 \boldsymbol{u}_2$ of the ellipse. Draw the circle and the ellipse for $\theta = 30°$:

$$V = \begin{bmatrix} 0 & 1 \\ 1 & 0 \end{bmatrix} \qquad U = \begin{bmatrix} \cos\theta & -\sin\theta \\ \sin\theta & \cos\theta \end{bmatrix} \qquad \Sigma = \begin{bmatrix} 2 & 0 \\ 0 & 1 \end{bmatrix}.$$

Section 7.4 will start with an important SVD picture for 2 by 2 matrices:

A = **(rotate) (stretch) (rotate)**. With symmetry S = **(rotate) (stretch) (rotate back)**.

27 This problem looks for all matrices A with a given column space in $\mathbf{R}^m$ and a given row space in $\mathbf{R}^n$. Suppose $\boldsymbol{c}_1, \ldots, \boldsymbol{c}_r$ and $\boldsymbol{b}_1, \ldots, \boldsymbol{b}_r$ are bases for those two spaces. Make them columns of C and B. The goal is to show that A has this form:

$\boldsymbol{A = CMB^{\mathrm{T}}}$ ***for an r by r invertible matrix M***. *Hint: Start from* $A = U\Sigma V^{\mathrm{T}}$.

The first r columns of U and V must be connected to C and B by invertible matrices, because they contain bases for the same column space (in U) and row space (in V).

7.3 Principal Component Analysis (PCA by the SVD)

1 Data often comes in a matrix : n samples and m measurements per sample.

2 Center each row of the matrix A by subtracting the mean from each measurement.

3 The SVD finds combinations of the data that contain the most information.

4 Largest singular value $\boldsymbol{\sigma_1}$ $\leftrightarrow$ greatest variance $\leftrightarrow$ most information in $\boldsymbol{u}_1$.

This section explains a major application of the SVD to statistics and data analysis. Our examples will come from human genetics and face recognition and finance. The problem is to understand a large matrix of data ($=$ measurements). For each of n samples we are measuring m variables. The data matrix A_0 has n columns and m rows.

Graphically, the columns of A_0 are n points in $\mathbf{R}^m$. After we subtract the average of each row to reach A, the n points are often clustered along a line or close to a plane (or other low-dimensional subspace of $\mathbf{R}^m$). What is that line or plane or subspace ?

Let me start with a picture instead of numbers. For $m = 2$ variables like age and height, the n points lie in the plane $\mathbf{R}^2$. Subtract the average age and height to center the data. If the n recentered points cluster along a line, *how will linear algebra find that line* ?

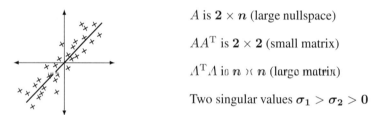

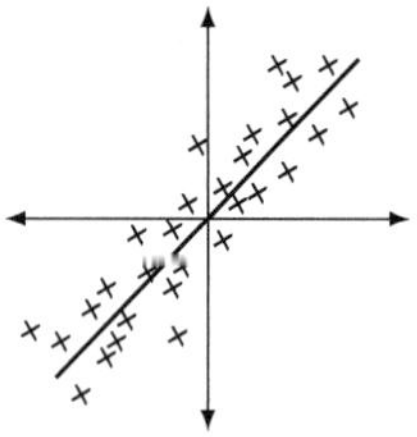

A is $\mathbf{2} \times \boldsymbol{n}$ (large nullspace)

AA^{T} is $\mathbf{2} \times \mathbf{2}$ (small matrix)

$A^{\mathrm{T}}A$ is $\boldsymbol{n} \times \boldsymbol{n}$ (large matrix)

Two singular values $\boldsymbol{\sigma_1} > \boldsymbol{\sigma_2} > \mathbf{0}$

Figure 7.2: Data points in A are often close to a line in $\mathbf{R}^2$ or a subspace in $\mathbf{R}^m$.

Let me go more carefully in constructing the data matrix. Start with the measurements in A_0 : the sample data. Find the average (the *mean*) $\mu_1, \mu_2, \ldots, \mu_m$ of each row. *Subtract each mean μ_i from row i to center the data.* The average along each row is now zero, for the centered matrix A. So the point $(0, 0)$ in Figure 7.2 is now the true center of the n points.

$$\textbf{The ``sample covariance matrix'' is defined by } S = \frac{\boldsymbol{AA^{\mathrm{T}}}}{\boldsymbol{n-1}}.$$

A shows the distance $a_{ij} - \mu_i$ from each measurement to the row average μ_i.

$(AA^{\mathrm{T}})_{11}$ and $(AA^{\mathrm{T}})_{22}$ show the **sum of squared distances (sample variances $\boldsymbol{s_1^2}$, $\boldsymbol{s_2^2}$)**.

$(AA^{\mathrm{T}})_{12}$ shows the **sample covariance $\boldsymbol{s_{12}}$** $=$ (row 1 of A)$\cdot$(row 2 of A).

The variance is a key number throughout statistics. An average exam score $\mu = 85$ tells you it was a decent exam. A variance of $s^2 = 25$ (standard deviation $s = 5$) means that most grades were in the 80's : closely packed. A sample variance $s^2 = 225$ ($s = 15$) means that grades were widely scattered. Chapter 12 explains variances.

The *covariance* of a math exam and a history exam is a dot product of those rows of A, with average grades subtracted out. Covariance below zero means: One subject strong when the other is weak. High covariance means: Both strong or both weak.

We divide by $n-1$ instead of n for reasons known best to statisticians. They tell me that one degree of freedom was used by the mean, leaving $n-1$. (I think the best plan is to agree with them.) In any case n should be a big number to count on reliable statistics. Since the rows of A have n entries, the numbers in AA^{T} have size growing like n and the division by $n-1$ keeps them steady.

Example 1 **Six math and history scores (notice the zero mean in each row)**

$$\boldsymbol{A}=\begin{bmatrix}3 & -4 & 7 & 1 & -4 & -3\\ 7 & -6 & 8 & -1 & -1 & -7\end{bmatrix} \text{ has sample covariance } \boldsymbol{S}=\frac{AA^{\mathrm{T}}}{5}=\begin{bmatrix}20 & 25\\ 25 & 40\end{bmatrix}.$$

The two rows of A are highly correlated: $s_{12}=25$. Above average math went with above average history. Changing all the signs in row 2 would produce *negative covariance* $s_{12}=-25$. Notice that S has positive trace and determinant; AA^{T} is positive definite.

The eigenvalues of S are near 57 and 3. So the first rank one piece $\sqrt{57}\,\boldsymbol{u}_1\boldsymbol{v}_1^{\mathrm{T}}$ is much larger than the second piece $\sqrt{3}\,\boldsymbol{u}_2\boldsymbol{v}_2^{\mathrm{T}}$. **The leading eigenvector $\boldsymbol{u}_1$ shows the direction that you see in the scatter graph of Figure 7.2.** That eigenvector is close to $\boldsymbol{u}_1=(.6,.8)$ and the direction in the graph nearly gives a $6-8-10$ or $3-4-5$ right triangle.

The SVD of A (centered data) shows the dominant direction in the scatter plot.

The second singular vector $\boldsymbol{u}_2$ is perpendicular to $\boldsymbol{u}_1$. The second singular value $\sigma_2\approx\sqrt{3}$ measures the spread across the dominant line. If the data points in A fell exactly on a line ($\boldsymbol{u}_1$ direction), then σ_2 would be zero. Actually there would only be σ_1.

The Essentials of Principal Component Analysis (PCA)

PCA gives a way to understand a data plot in dimension $m=$ the number of measured variables (here age and height). Subtract average age and height ($m=2$ for n samples) to center the m by n data matrix A. *The crucial connection to linear algebra* is in the singular values and singular vectors of A. Those come from the eigenvalues $\lambda=\sigma^2$ and the eigenvectors $\boldsymbol{u}$ of the sample covariance matrix $S=AA^{\mathrm{T}}/(n-1)$.

- The total variance in the data is the sum of all eigenvalues and of sample variances s^2:

  Total variance $\boldsymbol{T}=\boldsymbol{\sigma_1^2}+\cdots+\boldsymbol{\sigma_m^2}=\boldsymbol{s_1^2}+\cdots+\boldsymbol{s_m^2}=$ **trace** (*diagonal sum*).

- The first eigenvector $\boldsymbol{u}_1$ of S points in the most significant direction of the data. That direction accounts for (or *explains*) a fraction σ_1^2/T of the total variance.

- The next eigenvector $\boldsymbol{u}_2$ (orthogonal to $\boldsymbol{u}_1$) accounts for a smaller fraction σ_2^2/T.

- Stop when those fractions are small. You have the R directions that explain most of the data. The n data points are very near an R-dimensional subspace with basis $\boldsymbol{u}_1$ to $\boldsymbol{u}_R$. These $\boldsymbol{u}$'s are the **principal components** in m-dimensional space.

- R is the "effective rank" of A. The true rank r is probably m or n: full rank matrix.

Perpendicular Least Squares

It may not be widely recognized that the best line in Figure 7.2 (the line in the $\boldsymbol{u}_1$ direction) also solves a problem of *perpendicular least squares* (= orthogonal regression):

The sum of squared distances from the points to the line is a minimum.

Proof. Separate each column $\boldsymbol{a}_j$ into its components along the $\boldsymbol{u}_1$ line and $\boldsymbol{u}_2$ line:

$$\textbf{Right triangles} \quad \sum_{j=1}^{n} ||\boldsymbol{a}_j||^2 = \sum_{j=1}^{n} |\boldsymbol{a}_j^{\mathrm{T}}\boldsymbol{u}_1|^2 + \sum_{j=1}^{n} |\boldsymbol{a}_j^{\mathrm{T}}\boldsymbol{u}_2|^2 \qquad (1)$$

The sum on the left is fixed by the data points $\boldsymbol{a}_j$ (columns of A). The first sum on the right is $\boldsymbol{u}_1^{\mathrm{T}}AA^{\mathrm{T}}\boldsymbol{u}_1$. So when we maximize that sum in PCA by choosing the eigenvector $\boldsymbol{u}_1$, we minimize the second sum. That second sum (squared distances from the data points to the best line) is a minimum for perpendicular least squares.

Ordinary least squares in Chapter 4 reached a linear equation $A^{\mathrm{T}}A\widehat{\boldsymbol{x}} = A^{\mathrm{T}}\boldsymbol{b}$ by using *vertical distances* to the best line. PCA produces an eigenvalue problem for $\boldsymbol{u}_1$ by using *perpendicular distances*. "Total least squares" will allow for errors in A as well as $\boldsymbol{b}$.

The Sample Correlation Matrix

Data analysis works mostly with A (centered data). But the measurements in A might have different units like inches and pounds and years and dollars. Changing one set of units (inches to meters or years to seconds) would have a big effect on that row of A and S. If scaling is a problem, **we change from covariance matrix S to correlation matrix C**:

A diagonal matrix D rescales A. Each row of DA has length $\sqrt{n-1}$.
The sample correlation matrix $C = DAA^{\mathrm{T}}D/(n-1)$ has 1's on its diagonal.

Chapter 12 on Probability and Statistics will introduce the *expected* covariance matrix V and the *expected* correlation matrix (with diagonal 1's). Those use probabilities instead of actual measurements. The covariance matrix *predicts* the spread of future measurements around their mean, while A and the sample covariances S and the scaled correlation matrix $C = DSD$ use real data. All are highly important—a big connection between statistics and the linear algebra of positive definite matrices and the SVD.

Genetic Variation in Europe

We can follow changes in human populations by looking at genomes. To manage the huge amount of data, one good way to see genetic variation is from SNP's. The uncommon alleles (bases A/C/T/G in a pair from father and mother) are counted by the SNP:

SNP $= 0$ No change from the common base in that population: normal genotype
SNP $= 1$ The base pair shows one change from the usual pair
SNP $= 2$ Both bases are the less common allele

The uncentered matrix A_0 has a column for every person and a row for every base pair. The entries are mostly 0, quite a few 1, not so many 2. We don't test all 3 billion pairs. After subtracting row averages from A_0, the eigenvectors of AA^{T} are extremely revealing. **In Figure 7.4 the first singular vectors of A almost reproduce a map of Europe.**

This means: The SNP's from France and Germany and Italy are quite different. Even from the French and German and Italian parts of Switzerland those "snips" are different! Only Spain and Portugal are surprisingly confounded and harder to separate. More often than not, the DNA of an individual reveals his birthplace within 300 kilometers or 200 miles. A mixture of grandparents usually places the grandchild between their origins.

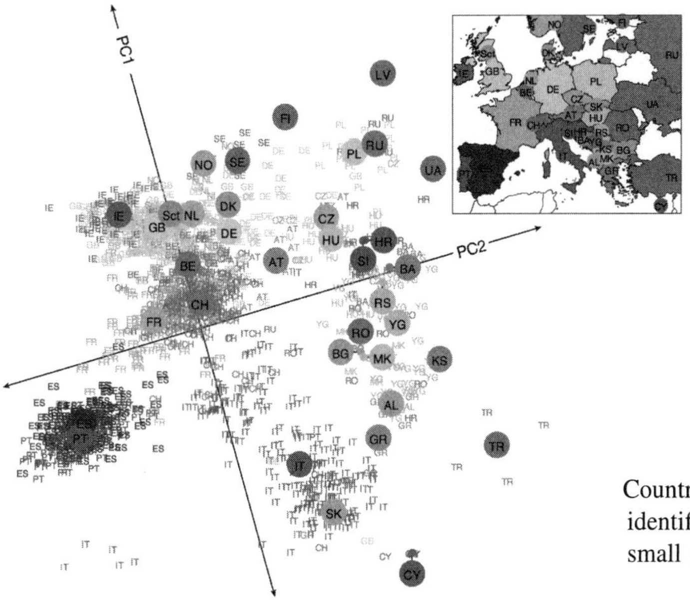

Countries are identified by small circles

Figure 7.3: *Nature* (2008) Novembre et al: vol. 456 pp.98-101/doc:10.1038/nature07331.

What is the significant message? If we test genomes to understand how they correlate with diseases, we must not forget their spatial variation. Without correcting for geography, what looks medically significant can be very misleading. *Confounding* is a serious problem in medical genetics that PCA and population genetics can help to solve—to remove effects due to geography that don't have medical importance.

In fact "spatial statistics" is a tricky world. *Example*: Every matrix with three diagonals of 1, C, 1 shows a not surprising influence of next door neighbors (from the 1's). But its singular vectors have sine and cosine oscillations going across the map, independent of C. You might think those are true wave-like variations but they can be meaningless.

Maybe statistics produces more arguments than mathematics does? Reducing big data to a single small "P-*value*" can be instructive or it can be extremely deceptive. The expression P-*value* appears in many articles. P stands for the probability that an observation is consistent with the *null hypothesis* (= pure chance). If you see 5 heads in a row, the probability is $P = 1/32$ that this came by chance from a fair coin (or $P = 2/32$ if your observation is taken to be 5 heads or 5 tails in a row). Often a P-value below 0.05 makes the null hypothesis doubtful—maybe a crook is flipping the coin. As here, P-values are not the most reliable guides in statistics—but they are extremely convenient.

Eigenfaces

Recognizing faces would not seem to depend—at first glance—on linear algebra. But an early and well publicized application of the SVD was to **face recognition**. We are not compressing an image, we are identifying it.

The plan is to start with a "training set" A_0 of n images of a wide variety of faces. Each image becomes a very long vector by stacking all pixel grayscales into a column. Then A_0 must be centered: subtract the average of every *column* of A_0 to reach A.

The singular vector $\boldsymbol{v}_1$ of this A tells us the combination of known faces that best identifies a new face. Then $\boldsymbol{v}_2$ tells us the next best combination.

Probably we will use the R best vectors $\boldsymbol{v}_1, \ldots, \boldsymbol{v}_R$ with largest singular values $\sigma_1 \geq \cdots \geq \sigma_R$ of A. Those identify new faces more accurately than any other R vectors. Perhaps $R = 100$ of those **eigenfaces** $A\boldsymbol{v}$ will capture nearly all the variance in the training set. Those R eigenfaces span "face space".

This plan of attack was suggested by Matthew Turk and Alex Pentland. It developed the suggestion by Sirovich and Kirby to use PCA in compressing images of faces. I learned a lot from Jeff Jauregui's description on the Web. His summary is this: **PCA provides a mechanism to recognize geometric/photometric similarity through algebraic means.** He assembled the first principal component (first singular vector) into the first eigenface. Of course the average of each column was added back or you wouldn't see a face!

Note PCA is compared to NMF in a fascinating letter to *Nature* (Lee and Seung, vol. 401, 21 Oct. 1999). Nonnegative Matrix Factorization does not allow the negative entries that always appear in the singular vectors $\boldsymbol{v}$. So everything *adds*—which needs more vectors but they are often more meaningful.

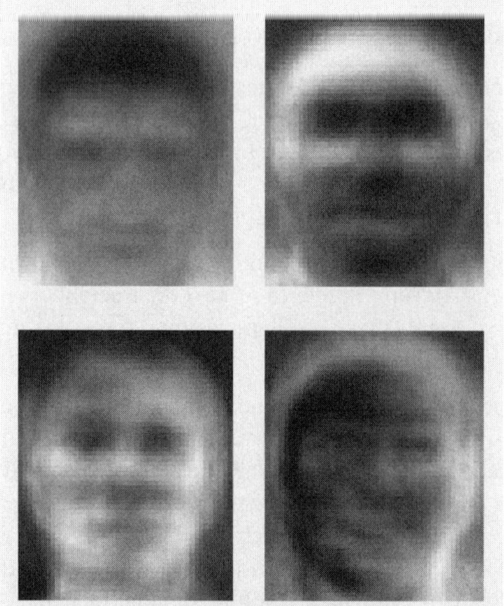

Figure 7.4: Eigenfaces pick out hairline and mouth and eyes and shape.

Applications of Eigenfaces

The first commercial use of PCA face recognition was for law enforcement and security. An early test at Super Bowl 35 in Tampa produced a very negative reaction from the crowd! The test was without the knowledge of the fans. Newspapers began calling it the "Snooper Bowl". I don't think the original eigenface idea is still used commercially (even in secret).

New applications of the SVD approach have come for other identification problems: Eigenvoices, Eigengaits, Eigeneyes, Eigenexpressions. I learned this from Matthew Turk (now in Santa Barbara, originally an MIT grad student. He told me he was in my class). The original eigenfaces in his thesis had problems accounting for rotation and scaling and lighting in the facial images. But the key ideas live on.

In the end, face space is nonlinear. So eventually we want nonlinear PCA.

Model Order Reduction

For a large-scale dynamic problem, the computational cost can become unmanageable. "Dynamic" means that the solution $\boldsymbol{u}(t)$ evolves as time goes forward. Fluid flow, chemical reactions, wave propagation, biological growth, electronic systems, these problems are everywhere. **A reduced model tries to identify important states of the system.** From a reduced problem we compute the needed information at much lower cost.

Model reduction is a truly important computational approach. Many good ideas have been proposed to reduce the original large problem. One simple and often useful idea is to take "snapshots" of the flow, put them in a matrix A, find the principal components (the left singular vectors of A), and work in their much smaller subspace:

A **snapshot** is a column vector that describes the state of the system
It can be an approximation to a typical true state $\boldsymbol{u}(t^*)$
From n snapshots, build a matrix A whose columns span a useful range of states

Now find the first R left singular vectors $\boldsymbol{u}_1$ to $\boldsymbol{u}_R$ of A. They are a basis for a Proper Orthogonal Decomposition (**POD** basis). In practice we choose R so that

$$\textbf{Variance} \approx \textbf{Energy} \qquad \sigma_1^2 + \cdots + \sigma_R^2 \text{ is } 99\% \text{ or } 99.9\% \text{ of } \sigma_1^2 + \cdots + \sigma_n^2.$$

These vectors are an optimal basis for reconstructing the snapshots in A. If those snapshots are well chosen, then combinations of $\boldsymbol{u}_1$ to $\boldsymbol{u}_R$ will be close to the exact solution $\boldsymbol{u}(t)$ for desired times t and parameters p.

So much depends on the snapshots! *SIAM Review* 2015 includes an excellent survey by Beiner, Gugercin, and Willcox. The SVD compresses data as well as images.

Searching the Web

We believe that Google creates rankings by a walk that follows web links. When this walk goes often to a site, the ranking is high. The frequency of visits gives the leading eigenvector $(\lambda = 1)$ of the "Web matrix"—the largest eigenvalue problem ever solved.

That Markov matrix has more than 3 billion rows and columns, from 3 billion web sites.

Many of the important techniques are well-kept secrets of Google. Probably they start with an earlier eigenvector as a first approximation, and they run the random walk very fast. To get a high ranking, you want a lot of links from important sites.

Here is an application of the SVD to web search engines. When you google a word, you get a list of web sites in order of importance. You could try typing "four subspaces".

The HITS algorithm was an early proposal to produce that ranked list. It begins with about 200 sites found from an index of key words. After that we look *only at links between pages*. Search engines are link-based more than content-based.

Start with the 200 sites and all sites that link to them and all sites they link to. That is our list, to be put in order. Importance can be measured by links out and links in.

1. The site may be an ***authority***: *Links come in* from many sites. Especially from hubs.

2. The site may be a ***hub***: *Links go out* to many sites in the list. Especially to authorities.

We want numbers $x_1, \ldots, x_N$ to rank the authorities and $y_1, \ldots, y_N$ to rank the hubs. Start with a simple count: x_i^0 and y_i^0 count the links into and out of site i.

Here is the point: *A good authority has links from important sites* (like hubs). Links from universities count more heavily than links from friends. *A good hub is linked to important sites* (like authorities). A link to **amazon.com** unfortunately means more than a link to **wellesleycambridge.com**. The raw counts $\boldsymbol{x}^0$ and $\boldsymbol{y}^0$ are updated to $\boldsymbol{x}^1$ and $\boldsymbol{y}^1$ by taking account of *good* links (measuring their quality by $\boldsymbol{x}^0$ and $\boldsymbol{y}^0$):

$$\textbf{Authority / Hub} \quad x_i^1 \;/\; y_i^1 = \text{Add up } y_j^0 \;/\; x_j^0 \text{ for all links } \textbf{into } i \;/\; \textbf{out from } i \qquad (2)$$

In matrix language those are $\boldsymbol{x}^1 = A^{\mathrm{T}}\boldsymbol{y}^0$ and $\boldsymbol{y}^1 = A\boldsymbol{x}^0$. The matrix A contains 1's and 0's, with $a_{ij} = 1$ when i links to j. In the language of graphs, A is an "adjacency matrix" for the Web (an enormous matrix). The new $\boldsymbol{x}^1$ and $\boldsymbol{y}^1$ give better rankings, but not the best. Take another step like (2), to reach $\boldsymbol{x}^2$ and $\boldsymbol{y}^2$ from $A^{\mathrm{T}}A\boldsymbol{x}^0$ and $AA^{\mathrm{T}}\boldsymbol{y}^0$:

$$\textbf{Authority} \quad \boldsymbol{x}^2 = A^{\mathrm{T}}\boldsymbol{y}^1 = A^{\mathrm{T}}A\boldsymbol{x}^0 \qquad\qquad \textbf{Hub} \quad \boldsymbol{y}^2 = A\boldsymbol{x}^1 = AA^{\mathrm{T}}\boldsymbol{y}^0. \qquad (3)$$

In two steps we are multiplying by $A^{\mathrm{T}}A$ and AA^{T}. Twenty steps will multiply by $(A^{\mathrm{T}}A)^{10}$ and $(AA^{\mathrm{T}})^{10}$. **When we take powers, the largest eigenvalue σ_1^2 begins to dominate.** The vectors $\boldsymbol{x}$ and $\boldsymbol{y}$ line up with the leading eigenvectors $\boldsymbol{v}_1$ and $\boldsymbol{u}_1$ of $A^{\mathrm{T}}A$ and AA^{T}. We are computing the top terms in the SVD, by the **power method** that is discussed in Section 11.3. It is wonderful that linear algebra helps to understand the Web.

This HITS algorithm is described in the 1999 *Scientific American* (June 16). But I don't think the SVD is mentioned there. . . The excellent book by Langville and Meyer, *Google's PageRank and Beyond*, explains in detail the science of search engines.

PCA in Finance: The Dynamics of Interest Rates

The mathematics of finance constantly applies linear algebra and PCA. We choose one application: the **yield curve for Treasury securities**. The "yield" is the interest rate paid on the bonds or notes or bills. That rate depends on time to maturity. For longer bonds (3 years to 20 years) the rate increases with length. The Federal Reserve adjusts short term yields to slow or stimulate the economy. This is the *yield curve*, used by risk managers and traders and investors.

Here is data for the first 6 business days of 2001—each column is a yield curve for investments on a particular day. The time to maturity is the "tenor". The six columns at the left are the interest rates, changing from day to day. The five columns at the right are interest rate *differences between days*, with the mean difference subtracted from each row. **This is the centered matrix A with its rows adding to zero**. A real world application might start with 252 business days instead of 5 or 6 (a year instead of a week).

Table 1. U.S. Treasury Yields : 6 Days and 5 Centered Daily Differences

	US Treasury Yields in 2001						Matrix A in Basis Points (0.01 %)				
Tenor	Jan 3	Jan 4	Jan 5	Jan 6	Jan 7	Jan 10	Jan 4	Jan 5	Jan 6	Jan 7	Jan 10
3 MO	5.87	5.69	5.37	5.12	5.19	5.24	−5.4	−19.4	−12.4	19.6	17.6
6 MO	5.58	5.44	5.20	4.98	5.03	5.11	−4.6	−14.6	−12.6	14.4	17.4
1 YR	5.11	5.04	4.82	4.60	4.61	4.71	1.0	−14.0	−14.0	9.0	18.0
2 YR	4.87	4.92	4.77	4.56	4.54	4.64	9.6	−10.4	−16.4	2.6	14.0
3 YR	4.82	4.92	4.78	4.57	4.55	4.65	13.4	−10.6	−17.6	1.4	13.4
5 YR	4.76	4.94	4.82	4.66	4.65	4.73	18.6	−11.4	−15.4	−0.4	8.6
7 YR	4.97	5.18	5.07	4.93	4.94	4.98	20.8	−11.2	−14.2	0.8	3.8
10 YR	4.92	5.14	5.03	4.93	4.94	4.98	20.8	−12.2	−11.2	−0.2	2.8
20 YR	5.46	5.62	5.56	5.50	5.52	5.53	14.6	−7.4	−7.4	0.6	−0.4

With five columns we might expect five singular values. But the five column vectors add to the zero vector (since every row of A adds to zero after centering). So $S = AA^{\mathrm{T}}/(5-1)$ has four nonzero eigenvalues $\sigma_1^2 > \sigma_2^2 > \sigma_3^2 > \sigma_4^2$. Here are the singular values σ_i and their squares σ_i^2 and the fractions of the total variance $T = \sigma_1^2 + \cdots + \sigma_4^2 =$ trace of S that are "explained" by each principal component (each eigenvector $\boldsymbol{u}_i$ of S).

	σ_i	σ_i^2	σ_i^2/T
Principal component $\boldsymbol{u}_1$	36.39	1323.9	.7536
Principal component $\boldsymbol{u}_2$	19.93	397.2	.2261
Principal component $\boldsymbol{u}_3$	5.85	34.2	.0195
Principal component $\boldsymbol{u}_4$	1.19	1.4	.0008
Principal component $\boldsymbol{u}_5$	0.00	0.0	.0000
		$T = 1756.7$	1.0000

A "scree plot" graphs those fractions σ_i^2/T dropping quickly to zero. In a larger problem you often see fast dropoff followed by a flatter part at the bottom (near $\sigma^2 = 0$). Locating the elbow between those two parts (significant and insignificant PC's) is important.

We also aim to understand each principal component. Those singular vectors $\boldsymbol{u}_i$ of A are eigenvectors of S. The entries in those vectors are the "*loadings*". Here are $\boldsymbol{u}_1$ to $\boldsymbol{u}_5$ for this yield curve example (with $S\boldsymbol{u}_5 = \mathbf{0}$).

	$\boldsymbol{u}_1$	$\boldsymbol{u}_2$	$\boldsymbol{u}_3$	$\boldsymbol{u}_4$	$\boldsymbol{u}_5$
3 MO	0.383	0.529	−0.478	0.060	0.084
6 MO	0.336	0.436	−0.046	0.210	−0.263
1 YR	0.358	0.263	0.225	−0.491	0.237
2 YR	0.352	−0.028	0.460	0.096	0.242
3 YR	0.371	−0.131	0.430	0.258	−0.555
5 YR	0.349	−0.293	0.117	−0.188	0.446
7 YR	0.323	−0.365	−0.228	0.459	0.081
10 YR	0.297	−0.378	−0.351	−0.579	−0.470
20 YR	0.184	−0.280	−0.361	0.227	0.268

Those five $\boldsymbol{u}$'s are orthonormal. They give bases for the four-dimensional column space of A and the one-dimensional nullspace of A^{T}. What financial meaning do they have?

$\boldsymbol{u}_1$ measures a weighted average of the daily changes in the 9 yields

$\boldsymbol{u}_2$ gauges the daily change in the yield spread between long and short bonds

$\boldsymbol{u}_3$ shows daily changes in the curvature (short and long bonds versus medium)

These graphs show the nine loadings on $\boldsymbol{u}_1, \boldsymbol{u}_2, \boldsymbol{u}_3$ above from 3 months to 20 years.

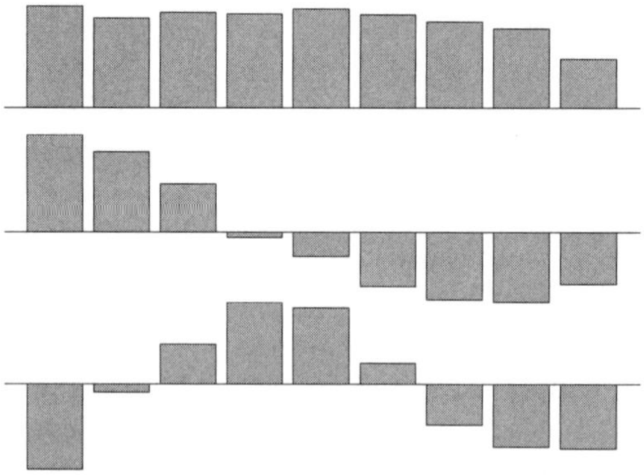

The output from a typical code (written in R) will include two more tables—which are going on the book's website. One will show the *right* singular vectors $\boldsymbol{v}_i$ of A. These are eigenvectors of $A^{\mathrm{T}}A$. They are proportional to the vectors $A^{\mathrm{T}}\boldsymbol{u}$. They have 5 components and they show the movement of yields and short-long spreads during the week.

The total variance $T = 1756.7$ (the trace $\sigma_1^2 + \sigma_2^2 + \sigma_3^2 + \sigma_4^2$ of S) is also the sum of the diagonal entries of S. Those are the sample variances of the rows of A. Here they are:

$$\boldsymbol{s_1^2}+\cdots+\boldsymbol{s_9^2} = 313.3+225.8+199.5+172.3+195.8+196.8+193.7+178.7+80.8 = \mathbf{1756.7}.$$

Every s^2 is below σ_1^2. And 1756.7 is also the trace of $A^{\mathrm{T}}A/(n-1)$: column variances.

Note that this PCA section 7.3 is working with centered *rows* in A. In some applications (like finance), the matrix is usually transposed and the *columns* are centered. Then the sample covariance matrix S uses $A^{\mathrm{T}}A$, and the $\boldsymbol{v}$'s are the more important principal components. Linear algebra with practical interpretations tells us so much.

Problem Set 7.3

1 Suppose A_0 holds these 2 measurements of 5 samples:

$$A_0 = \begin{bmatrix} 5 & 4 & 3 & 2 & 1 \\ -1 & 1 & 0 & 1 & -1 \end{bmatrix}$$

Find the average of each row and subtract it to produce the centered matrix A. Compute the sample covariance matrix $S = AA^{\mathrm{T}}/(n-1)$ and find its eigenvalues λ_1 and λ_2. What line through the origin is closest to the 5 samples in columns of A?

2 Take the steps of Problem 1 for this 2 by 6 matrix A_0 :

$$A_0 = \begin{bmatrix} 1 & 0 & 1 & 0 & 1 & 0 \\ 1 & 2 & 3 & 3 & 2 & 1 \end{bmatrix}$$

3 The sample variances s_1^2, s_2^2 and the sample covariance s_{12} are the entries of S. What is S (after subtracting means) when $A_0 = \begin{bmatrix} 1 & 2 & 3 \\ 5 & 2 & 2 \end{bmatrix}$? What is σ_1 ?

4 From the eigenvectors of $S = AA^{\mathrm{T}}$, find the line (the $\boldsymbol{u}_1$ direction through the center point) and then the plane ($\boldsymbol{u}_1, \boldsymbol{u}_2$ directions) closest to these four points in three-dimensional space :

$$A = \begin{bmatrix} 1 & -1 & 0 & 0 \\ 0 & 0 & 2 & -2 \\ 1 & 1 & -1 & -1 \end{bmatrix}.$$

5 From this sample covariance matrix S, find the correlation matrix DSD with 1's down its main diagonal. D is a positive diagonal matrix that produces those 1's.

$$S = \begin{bmatrix} 4 & 2 & 0 \\ 2 & 4 & 1 \\ 0 & 1 & 1 \end{bmatrix}.$$

6 Choose the diagonal matrix D that produces DSD and find the correlations c_{ij}:

$$S = \begin{bmatrix} s_1^2 & s_{12} & s_{13} \\ s_{12} & s_2^2 & s_{23} \\ s_{13} & s_{23} & s_3^2 \end{bmatrix} \qquad DSD = \begin{bmatrix} 1 & c_{12} & c_{13} \\ c_{12} & 1 & c_{23} \\ c_{13} & c_{23} & 1 \end{bmatrix}.$$

7 Suppose A_0 is a 5 by 10 matrix with average grades for 5 courses over 10 years. How would you create the centered matrix A and the sample covariance matrix S ? When you find the leading eigenvector of S, what does it tell you ?

7.4 The Geometry of the SVD

1 A typical square matrix $A = U\Sigma V^{\mathrm{T}}$ factors into (rotation) (stretching) (rotation).

2 The geometry shows how A transforms vectors $\boldsymbol{x}$ on a circle to vectors $A\boldsymbol{x}$ on an ellipse.

3 The **norm** of A is $||A|| = \sigma_1$. This singular value is its maximum growth factor $||A\boldsymbol{x}|| \,/\, ||\boldsymbol{x}||$.

4 **Polar decomposition** factors A into QS: rotation $Q = UV^{\mathrm{T}}$ times stretching $S = V\Sigma V^{\mathrm{T}}$.

5 The **pseudoinverse** $A^+ = V\Sigma^+ U^{\mathrm{T}}$ brings $A\boldsymbol{x}$ in the column space back to $\boldsymbol{x}$ in the row space.

The SVD separates a matrix into three steps: **(orthogonal)**×**(diagonal)**×**(orthogonal)**. Ordinary words can express the geometry behind it: **(rotation)**×**(stretching)**×**(rotation)**. $U\Sigma V^{\mathrm{T}}\boldsymbol{x}$ starts with the rotation to $V^{\mathrm{T}}\boldsymbol{x}$. Then Σ stretches that vector to $\Sigma V^{\mathrm{T}}\boldsymbol{x}$, and U rotates to $A\boldsymbol{x} = U\Sigma V^{\mathrm{T}}\boldsymbol{x}$. Here is the picture.

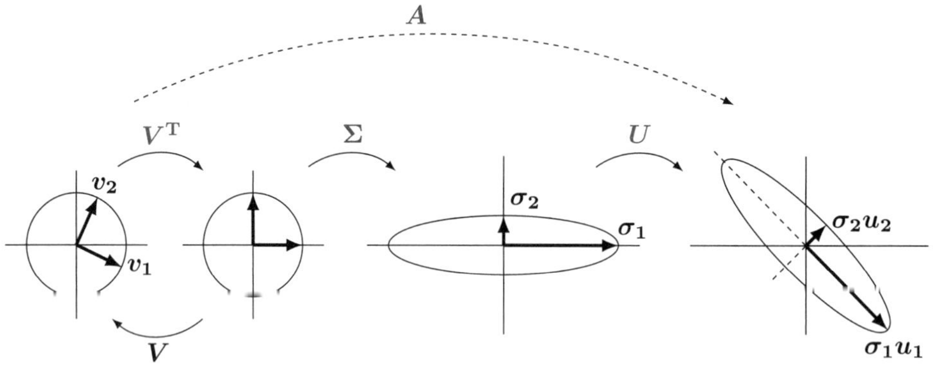

Figure 7.5: U and V are rotations and possible reflections. Σ stretches circle to ellipse.

Admittedly, this picture applies to a 2 by 2 matrix. And not every 2 by 2 matrix, because U and V didn't allow for a reflection—all three matrices have determinant > 0. This A would have to be invertible because the three steps are shown as invertible:

$$\begin{bmatrix} a & b \\ c & d \end{bmatrix} = \begin{bmatrix} \cos\theta & -\sin\theta \\ \sin\theta & \cos\theta \end{bmatrix} \begin{bmatrix} \sigma_1 & \\ & \sigma_2 \end{bmatrix} \begin{bmatrix} \cos\phi & \sin\phi \\ -\sin\phi & \cos\phi \end{bmatrix} = U\Sigma V^{\mathrm{T}}. \quad (1)$$

The four numbers a, b, c, d in the matrix A led to four numbers $\theta, \sigma_1, \sigma_2, \phi$ in its SVD.

This picture will guide us to three neat ideas in the algebra of matrices:

1 **The norm $||A||$ of a matrix**—its maximum growth factor.

2 **The polar decomposition $A = QS$**—orthogonal Q times positive definite S.

3 **The pseudoinverse A^+**—the best inverse when the matrix A is not invertible.

The Norm of a Matrix

If I choose one crucial number in the picture it is $\boldsymbol{\sigma_1}$. That number is the *largest growth factor of any vector $\boldsymbol{x}$*. If you follow the vector $\boldsymbol{v}_1$ on the left, you see it rotate to $(1, 0)$ and stretch to $(\sigma_1, 0)$ and finally rotate to $\sigma_1 \boldsymbol{u}_1$. The statement $A\boldsymbol{v}_1 = \sigma_1 \boldsymbol{u}_1$ is exactly the SVD equation. This largest singular value σ_1 is the "*norm*" of the matrix A.

The norm $||A||$ is the largest ratio $\dfrac{||A\boldsymbol{x}||}{||\boldsymbol{x}||}$ $\quad ||A|| = \max\limits_{\boldsymbol{x} \neq \boldsymbol{0}} \dfrac{||A\boldsymbol{x}||}{||\boldsymbol{x}||} = \boldsymbol{\sigma_1} \qquad (2)$

MATLAB uses norm $(\boldsymbol{x})$ for vector lengths and the same word norm (A) for matrix norms. The math symbols have double bars: $||\boldsymbol{x}||$ and $||A||$. Here $||\boldsymbol{x}||$ means the standard length of a vector with $||\boldsymbol{x}||^2 = |x_1|^2 + \cdots + |x_n|^2$. The matrix norm comes from this vector norm when $\boldsymbol{x} = \boldsymbol{v}_1$ and $A\boldsymbol{x} = \sigma_1 \boldsymbol{u}_1$ and $||A\boldsymbol{x}|| \,/\, ||\boldsymbol{x}|| = \sigma_1 =$ largest ratio $= ||A||$.

Two valuable properties of that number norm (A) come directly from its definition:

Triangle inequality $\quad ||A + B|| \leq ||A|| + ||B|| \qquad$ **Product inequality** $\quad ||AB|| \leq ||A|| \, ||B|| \qquad (3)$

The definition (2) says that $||\boldsymbol{Ax}|| \leq ||\boldsymbol{A}|| \, ||\boldsymbol{x}||$ for every vector $\boldsymbol{x}$. That is what we know! Then the triangle inequality for vectors leads to the triangle inequality for matrices:

For vectors $\quad ||(A + B)\boldsymbol{x}|| \leq ||A\boldsymbol{x}|| + ||B\boldsymbol{x}|| \leq ||A|| \, ||\boldsymbol{x}|| + ||B|| \, ||\boldsymbol{x}||.$

Divide this by $||\boldsymbol{x}||$. Takc thc maximum over all $\boldsymbol{x}$. Then $||\boldsymbol{A} + \boldsymbol{B}|| \leq ||\boldsymbol{A}|| + ||\boldsymbol{B}||$.

The product inequality comes quickly from $||AB\boldsymbol{x}|| \leq ||A|| \, ||B\boldsymbol{x}|| \leq ||A|| \, ||B|| \, ||\boldsymbol{x}||$. Again divide by $||\boldsymbol{x}||$. Take the maximum over all $\boldsymbol{x}$. The result is $||\boldsymbol{AB}|| \leq ||\boldsymbol{A}|| \, ||\boldsymbol{B}||$.

Example 1 A rank-one matrix $A = \boldsymbol{u}\boldsymbol{v}^{\mathrm{T}}$ is as basic as we can get. It has one nonzero eigenvalue λ_1 and one nonzero singular value σ_1. Neatly, its eigenvector is $\boldsymbol{u}$ and its singular vectors (left and right) are $\boldsymbol{u}$ and $\boldsymbol{v}$.

Eigenvector $\quad A\boldsymbol{u} = (\boldsymbol{u}\boldsymbol{v}^{\mathrm{T}})\boldsymbol{u} = \boldsymbol{u}(\boldsymbol{v}^{\mathrm{T}}\boldsymbol{u}) = \lambda_1 \boldsymbol{u} \qquad$ So $\lambda_1 = \boldsymbol{v}^{\mathrm{T}}\boldsymbol{u}$

Singular vector $\quad A^{\mathrm{T}}A\boldsymbol{v} = (\boldsymbol{v}\boldsymbol{u}^{\mathrm{T}})(\boldsymbol{u}\boldsymbol{v}^{\mathrm{T}})\boldsymbol{v} = \boldsymbol{v}(\boldsymbol{u}^{\mathrm{T}}\boldsymbol{u})(\boldsymbol{v}^{\mathrm{T}}\boldsymbol{v}) = \sigma_1^2 \boldsymbol{v} \qquad$ So $\sigma_1 = ||\boldsymbol{u}|| \, ||\boldsymbol{v}||$.

It makes you feel good that $|\lambda_1| \leq \sigma_1$ is exactly the Schwarz inequality $|\boldsymbol{v}^{\mathrm{T}}\boldsymbol{u}| \leq ||\boldsymbol{u}|| \, ||\boldsymbol{v}||$.

How do we know that $|\lambda_1| \leq \sigma_1$? The eigenvector for $A\boldsymbol{x} = \lambda_1 \boldsymbol{x}$ will give the ratio $||A\boldsymbol{x}|| \,/\, ||\boldsymbol{x}|| = ||\lambda_1 \boldsymbol{x}|| \,/\, ||\boldsymbol{x}||$ which is $|\lambda_1|$. The maximum ratio σ_1 can't be less than $|\lambda_1|$.

Is it also true that $|\lambda_2| \leq \sigma_2$? **No**. That is completely wrong. In fact a 2 by 2 matrix will have $|\det A| = |\lambda_1 \lambda_2| = \sigma_1 \sigma_2$. In this case $|\lambda_1| \leq \sigma_1$ will force $|\lambda_2| \geq \sigma_2$.

The closest rank k matrix to A is $A_k = \sigma_1 u_1 v_1^{\mathrm{T}} + \cdots + \sigma_k u_k v_k^{\mathrm{T}}$

This is the key fact in matrix approximation: The Eckart-Young-Mirsky Theorem says that

$$||A - B|| \geq ||A - A_k|| = \sigma_{k+1} \textbf{ for all matrices } B \textbf{ of rank } k.$$

To me this completes the Fundamental Theorem of Linear Algebra. The v's and u's give orthonormal bases for the four fundamental subspaces, and the *first k v's and u's and* σ's give the best matrix approximation to A.

Polar Decomposition $A = QS$

Every complex number $x + iy$ has the polar form $re^{i\theta}$. A number $r \geq 0$ multiplies a number $e^{i\theta}$ on the unit circle. We have $x + iy = r\cos\theta + ir\sin\theta = r(\cos\theta + i\sin\theta) = re^{i\theta}$. Think of these numbers as 1 by 1 matrices. Then $e^{i\theta}$ is an *orthogonal matrix* Q and $r \geq 0$ is a *positive semidefinite matrix* (call it S). The ***polar decomposition*** extends the same idea to n by n matrices: orthogonal times positive semidefinite, $A = QS$.

> Every real square matrix can be factored into $A = QS$, where Q is ***orthogonal*** and S is ***symmetric positive semidefinite***. If A is invertible, S is positive definite.

For the proof we just insert $V^{\mathrm{T}}V = I$ into the middle of the SVD:

$$\textbf{Polar decomposition} \qquad A = U\Sigma V^{\mathrm{T}} = (UV^{\mathrm{T}})(V\Sigma V^{\mathrm{T}}) = (Q)(S). \tag{4}$$

The first factor UV^{T} is Q. The product of orthogonal matrices is orthogonal. The second factor $V\Sigma V^{\mathrm{T}}$ is S. It is positive semidefinite because its eigenvalues are in Σ.

If A is invertible then Σ and S are also invertible. ***S is the symmetric positive definite square root of $A^{\mathrm{T}}A$***, because $S^2 = V\Sigma^2V^{\mathrm{T}} = A^{\mathrm{T}}A$. So the eigenvalues of S are the singular values of A. The eigenvectors of S are the singular vectors v of A.

There is also a polar decomposition $A = KQ$ in the reverse order. Q is the same but now $K = U\Sigma U^{\mathrm{T}}$. Then K is the symmetric positive definite square root of AA^{T}.

Example 2 The SVD example in Section 7.2 was $A = \begin{bmatrix} 3 & 0 \\ 4 & 5 \end{bmatrix} = U\Sigma V^{\mathrm{T}}$. Find the factors Q and S (rotation and stretch) in the polar decomposition $A = QS$.

Solution I will just copy the matrices U and Σ and V from Section 7.2:

$$Q = UV^{\mathrm{T}} = \frac{1}{\sqrt{20}}\begin{bmatrix} 1 & -3 \\ 3 & 1 \end{bmatrix}\begin{bmatrix} 1 & -1 \\ -1 & 1 \end{bmatrix} = \frac{1}{\sqrt{20}}\begin{bmatrix} 4 & -2 \\ 2 & 4 \end{bmatrix} = \frac{1}{\sqrt{5}}\begin{bmatrix} 2 & -1 \\ 1 & 2 \end{bmatrix}$$

$$S = V\Sigma V^{\mathrm{T}} = \frac{\sqrt{5}}{2}\begin{bmatrix} 1 & -1 \\ 1 & 1 \end{bmatrix}\begin{bmatrix} 3 & \\ & 1 \end{bmatrix}\begin{bmatrix} 1 & 1 \\ -1 & 1 \end{bmatrix} = \sqrt{5}\begin{bmatrix} 2 & 1 \\ 1 & 2 \end{bmatrix}. \text{ Then } \boldsymbol{A = QS}.$$

In mechanics, the polar decomposition separates the ***rotation*** (in Q) from the ***stretching*** (in S). The eigenvalues of S give the stretching factors as in Figure 7.5. The eigenvectors of S give the stretching directions (the principal axes of the ellipse). The orthogonal matrix Q includes both rotations U and V^{T}.

Here is a fact about rotations. $Q = UV^{\mathrm{T}}$ is the **nearest orthogonal matrix** to A. This Q makes the norm $||Q - A||$ as small as possible. That corresponds to the fact that $e^{i\theta}$ is the nearest number on the unit circle to $re^{i\theta}$.

The SVD tells us an even more important fact about nearest singular matrices:

The nearest singular matrix A_0 to A comes by changing the smallest $\sigma_{\min}$ to zero.

So $\sigma_{\min}$ is measuring the distance from A to singularity. For the matrix in Example 2 that distance is $\sigma_{\min} = \sqrt{5}$. If I change $\sigma_{\min}$ to zero, this knocks out the last (smallest) piece in $A = \sigma_1 \boldsymbol{u}_1 \boldsymbol{v}_1^{\mathrm{T}} + \sigma_2 \boldsymbol{u}_2 \boldsymbol{v}_2^{\mathrm{T}}$. Then only the rank-one (singular!) matrix $\sigma_1 \boldsymbol{u}_1 \boldsymbol{v}_1^{\mathrm{T}}$ will be left: the closest to A. The smallest change had norm $\sigma_2 = \sqrt{5}$ (*smaller than* 3).

In computational practice we often do knock out a very small σ. Working with singular matrices is better than coming too close to zero and not noticing.

The Pseudoinverse A^+

By choosing good bases, A multiplies $\boldsymbol{v}_i$ in the row space to give $\sigma_i \boldsymbol{u}_i$ in the column space. A^{-1} must do the opposite! If $A\boldsymbol{v} = \sigma \boldsymbol{u}$ then $A^{-1}\boldsymbol{u} = \boldsymbol{v}/\sigma$. The singular values of A^{-1} are $1/\sigma$, just as the eigenvalues of A^{-1} are $1/\lambda$. The bases are reversed. The $\boldsymbol{u}$'s are in the row space of A^{-1}, the $\boldsymbol{v}$'s are in the column space.

Until this moment we would have added "*if A^{-1} exists.*" Now we don't. A matrix that multiplies $\boldsymbol{u}_i$ to produce $\boldsymbol{v}_i/\sigma_i$ *does* exist. It is the pseudoinverse A^+:

Pseudoinverse of A

$$\boldsymbol{A^+ = V\Sigma^+ U^{\mathrm{T}}} = \underset{n \text{ by } n}{\begin{bmatrix} \boldsymbol{v}_1 \cdots \boldsymbol{v}_r \cdots \boldsymbol{v}_n \end{bmatrix}} \underset{n \text{ by } m}{\begin{bmatrix} \sigma_1^{-1} & & \\ & \ddots & \\ & & \sigma_r^{-1} \\ & & \end{bmatrix}} \underset{m \text{ by } m}{\begin{bmatrix} \boldsymbol{u}_1 \cdots \boldsymbol{u}_r \cdots \boldsymbol{u}_m \end{bmatrix}^{\mathrm{T}}}$$

The ***pseudoinverse*** $\boldsymbol{A^+}$ is an n by m matrix. If A^{-1} exists (we said it again), then A^+ is the same as A^{-1}. In that case $m = n = r$ and we are inverting $U\Sigma V^{\mathrm{T}}$ to get $V\Sigma^{-1}U^{\mathrm{T}}$. The new symbol A^+ is needed when $r < m$ or $r < n$. Then A has no two-sided inverse, but it has a *pseudo*inverse A^+ with that same rank r:

$$A^+\boldsymbol{u}_i = \frac{1}{\sigma_i}\boldsymbol{v}_i \quad \text{for } i \le r \qquad \text{and} \qquad A^+\boldsymbol{u}_i = \boldsymbol{0} \quad \text{for } i > r.$$

The vectors $\boldsymbol{u}_1, \ldots, \boldsymbol{u}_r$ in the column space of A go back to $\boldsymbol{v}_1, \ldots, \boldsymbol{v}_r$ in the row space. The other vectors $\boldsymbol{u}_{r+1}, \ldots, \boldsymbol{u}_m$ are in the left nullspace, and A^+ sends them to zero. When we know what happens to all those basis vectors, we know A^+.

Notice the pseudoinverse of the diagonal matrix Σ. Each σ in Σ is replaced by σ^{-1} in Σ^+. The product $\Sigma^+\Sigma$ is as near to the identity as we can get. It is a projection matrix, $\Sigma^+\Sigma$ is partly I and otherwise zero. We can invert the σ's, but we can't do anything about the zero rows and columns. This example has $\sigma_1 = 2$ and $\sigma_2 = 3$:

$$\Sigma^+\Sigma = \begin{bmatrix} \mathbf{1/2} & 0 & 0 \\ 0 & \mathbf{1/3} & 0 \\ 0 & 0 & \mathbf{0} \end{bmatrix} \begin{bmatrix} \mathbf{2} & 0 & 0 \\ 0 & \mathbf{3} & 0 \\ 0 & 0 & \mathbf{0} \end{bmatrix} = \begin{bmatrix} \mathbf{1} & 0 & 0 \\ 0 & \mathbf{1} & 0 \\ 0 & 0 & \mathbf{0} \end{bmatrix} = \begin{bmatrix} \boldsymbol{I} & 0 \\ 0 & \mathbf{0} \end{bmatrix}.$$

The pseudoinverse A^+ is the n by m matrix that makes AA^+ and A^+A into projections.

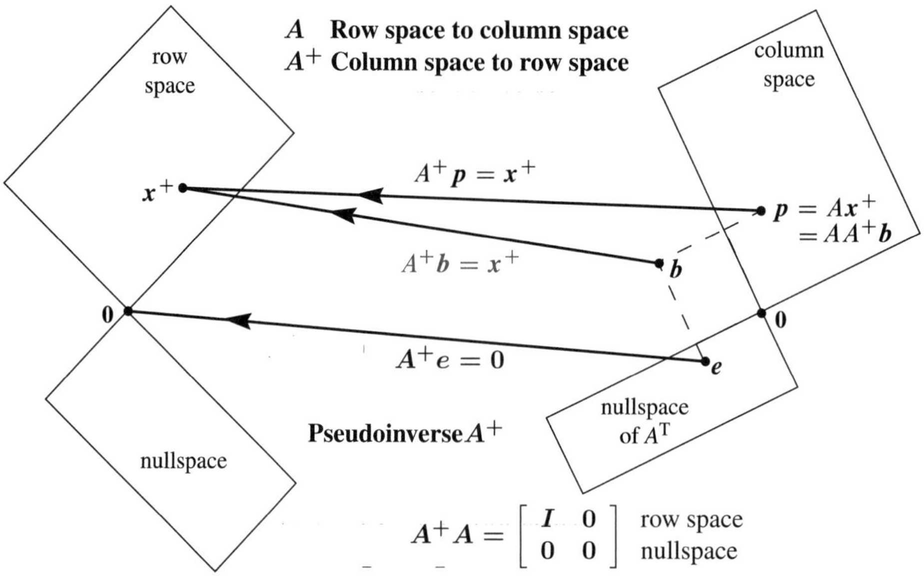

Figure 7.6: $A\boldsymbol{x}^+$ in the column space goes back to $A^+A\boldsymbol{x}^+ = \boldsymbol{x}^+$ in the row space.

Trying for $AA^{-1} = A^{-1}A = I$	$AA^+ =$ projection matrix onto the column space of A $A^+A =$ projection matrix onto the row space of A

Example 3 Every rank one matrix is a column times a row. With unit vectors $\boldsymbol{u}$ and $\boldsymbol{v}$, that is $A = \sigma \boldsymbol{u}\boldsymbol{v}^{\mathrm{T}}$. Its pseudoinverse is $A^+ = \boldsymbol{v}\boldsymbol{u}^{\mathrm{T}}/\sigma$. The product AA^+ is $\boldsymbol{u}\boldsymbol{u}^{\mathrm{T}}$, the projection onto the line through $\boldsymbol{u}$. The product A^+A is $\boldsymbol{v}\boldsymbol{v}^{\mathrm{T}}$.

Example 4 Find the pseudoinverse of $A = \begin{bmatrix} 1 & 1 \\ 1 & 1 \end{bmatrix}$. This matrix is not invertible. The rank is 1. The only singular value is $\sigma_1 = 2$. That is inverted to $1/2$ in Σ^+ (also rank 1).

$$A^+ = V\Sigma^+U^{\mathrm{T}} = \frac{1}{\sqrt{2}}\begin{bmatrix} 1 & 1 \\ 1 & -1 \end{bmatrix}\begin{bmatrix} 1/2 & 0 \\ 0 & 0 \end{bmatrix}\frac{1}{\sqrt{2}}\begin{bmatrix} 1 & 1 \\ 1 & -1 \end{bmatrix} = \frac{\mathbf{1}}{\mathbf{4}}\begin{bmatrix} \mathbf{1} & \mathbf{1} \\ \mathbf{1} & \mathbf{1} \end{bmatrix}.$$

A^+ also has rank 1. Its column space is always the row space of A.

Least Squares with Dependent Columns

That matrix A with four 1's appeared in Section 4.3 on least squares. It broke the requirement of independent columns. The matrix appeared when we made two measurements, both at time $t = 1$. The closest straight line went halfway between the measurements 3 and 1, but there was no way to decide on the slope of the best line.

In matrix language, $A^{\mathrm{T}}A$ was singular. The equation $A^{\mathrm{T}}A\boldsymbol{x} = $ **$A^{\mathrm{T}}\boldsymbol{b}$ had infinitely many solutions**. The pseudoinverse gives us a way to choose a "best solution" $\boldsymbol{x}^+ = A^+\boldsymbol{b}$.

Let me repeat the unsolvable $A\boldsymbol{x} = \boldsymbol{b}$ and the infinitely solvable $A^{\mathrm{T}}A\widehat{\boldsymbol{x}} = A^{\mathrm{T}}\boldsymbol{b}$:

$$A\boldsymbol{x} = \begin{bmatrix} 1 & 1 \\ 1 & 1 \end{bmatrix} \begin{bmatrix} x_1 \\ x_2 \end{bmatrix} = \begin{bmatrix} 3 \\ 1 \end{bmatrix} = \boldsymbol{b} \qquad A^{\mathrm{T}}A\widehat{\boldsymbol{x}} = \begin{bmatrix} 2 & 2 \\ 2 & 2 \end{bmatrix} \begin{bmatrix} \widehat{x}_1 \\ \widehat{x}_2 \end{bmatrix} = \begin{bmatrix} 4 \\ 4 \end{bmatrix} = A^{\mathrm{T}}\boldsymbol{b}$$

Any vector $\widehat{\boldsymbol{x}} = (1 + c,\ 1 - c)$ will solve those normal equations $A^{\mathrm{T}}A\widehat{\boldsymbol{x}} = A^{\mathrm{T}}\boldsymbol{b}$. The purpose of the pseudoinverse is to choose one solution $\widehat{\boldsymbol{x}} = \boldsymbol{x}^+$.

> $\boldsymbol{x}^+ = \boldsymbol{A}^+\boldsymbol{b} = (\boldsymbol{1}, \boldsymbol{1})$ **is the shortest solution to** $A^{\mathrm{T}}A\widehat{\boldsymbol{x}} = A^{\mathrm{T}}\boldsymbol{b}$ and $A\widehat{\boldsymbol{x}} = \boldsymbol{p}$.

You can see that $\boldsymbol{x}^+ = (1, 1)$ is shorter than any other solution $\widehat{\boldsymbol{x}} = (1 + c, 1 - c)$. The length squared of $\widehat{\boldsymbol{x}}$ is $(1 + c)^2 + (1 - c)^2 = 2 + 2c^2$. The shortest choice is $c = 0$. That gives the solution $\boldsymbol{x}^+ = (1, 1)$ in the row space of A.

The geometry tells us what A^+ should do: Take the column space of A back to the row space. Both spaces have dimension r. Kill off the error vector $\boldsymbol{e}$ in the left nullspace.

The pseudoinverse A^+ and this best solution $\boldsymbol{x}^+$ are essential in statistics, because experiments often have a matrix with dependent columns as well as dependent rows.

■ REVIEW OF THE KEY IDEAS ■

1. The ellipse of vectors $A\boldsymbol{x}$ has axes along the singular vectors $\boldsymbol{u}_i$.
2. The matrix norm $||A|| = \sigma_1$ comes from the vector length: Maximize $||A\boldsymbol{x}||/||\boldsymbol{x}||$.
3. Invertible matrix = (orthogonal matrix) (positive definite matrix): $A = QS$.
4. Every $A = U\Sigma V^{\mathrm{T}}$ has a pseudoinverse $A^+ = V\Sigma^+U^{\mathrm{T}}$ that sends $\boldsymbol{N}(A^{\mathrm{T}})$ to $\boldsymbol{Z}$.

■ WORKED EXAMPLES ■

7.4 A If A has rank n (full column rank) then it has a **left inverse** $L = (A^{\mathrm{T}}A)^{-1}A^{\mathrm{T}}$. This matrix L gives $\boldsymbol{LA} = \boldsymbol{I}$. Explain why the pseudoinverse is $\boldsymbol{A}^+ = \boldsymbol{L}$ in this case.

If A has rank m (full row rank) then it has a **right inverse** $R = A^{\mathrm{T}}(AA^{\mathrm{T}})^{-1}$. This matrix R gives $\boldsymbol{AR} = \boldsymbol{I}$. Explain why the pseudoinverse is $\boldsymbol{A}^+ = \boldsymbol{R}$ in this case.

Find L for A_1 and find R for A_2. Find A^+ for all three matrices A_1, A_2, A_3:

$$A_1 = \begin{bmatrix} \mathbf{2} \\ \mathbf{2} \end{bmatrix} \quad A_2 = \begin{bmatrix} \mathbf{2} & \mathbf{2} \end{bmatrix} \quad A_3 = \begin{bmatrix} \mathbf{2} & \mathbf{2} \\ \mathbf{1} & \mathbf{1} \end{bmatrix}.$$

Solution If A has independent columns then $A^{\mathrm{T}}A$ is invertible—this is a key point of Section 4.2. Certainly $L = (A^{\mathrm{T}}A)^{-1}A^{\mathrm{T}}$ multiplies A to give $LA = I$: a left inverse.

$AL = A(A^{\mathrm{T}}A)^{-1}A^{\mathrm{T}}$ is the projection matrix (Section 4.2) on the column space. So L meets the requirements on A^+: *LA and AL are projections* on $\boldsymbol{C}(A)$ and $\boldsymbol{C}(A^{\mathrm{T}})$.

If A has rank m (full row rank) then AA^{T} is invertible. Certainly A multiplies $R = A^{\mathrm{T}}(AA^{\mathrm{T}})^{-1}$ to give $AR = I$. In the opposite order, $RA = A^{\mathrm{T}}(AA^{\mathrm{T}})^{-1}A$ is the projection matrix onto the row space (column space of A^{T}). So R equals the pseudoinverse A^+.

The example A_1 has full column rank (for L) and A_2 has full row rank (for R):

$$\boldsymbol{A_1^+} = (A_1^{\mathrm{T}}A_1)^{-1}A_1^{\mathrm{T}} = \frac{1}{\sqrt{8}}\begin{bmatrix} 2 & 2 \end{bmatrix} \qquad \boldsymbol{A_2^+} = A_2^{\mathrm{T}}(A_2A_2^{\mathrm{T}})^{-1} = \frac{1}{\sqrt{8}}\begin{bmatrix} 2 \\ 2 \end{bmatrix}.$$

Notice $A_1^+A_1 = [1]$ and $A_2A_2^+ = [1]$. But A_3 has no left or right inverse. **Its rank is not full. Its pseudoinverse brings the column space of $\boldsymbol{A_3}$ to the row space.**

$$A_3^+ = \begin{bmatrix} 2 & 2 \\ 1 & 1 \end{bmatrix}^+ = \frac{\boldsymbol{v}_1\boldsymbol{u}_1^{\mathrm{T}}}{\sigma_1} = \frac{1}{10}\begin{bmatrix} \mathbf{2} & \mathbf{1} \\ \mathbf{2} & \mathbf{1} \end{bmatrix}.$$

Problem Set 7.4

Problems 1–4 compute and use the SVD of a particular matrix (not invertible).

1 (a) Compute $A^{\mathrm{T}}A$ and its eigenvalues and unit eigenvectors $\boldsymbol{v}_1$ and $\boldsymbol{v}_2$. Find σ_1.

$$\textbf{Rank one matrix} \quad A = \begin{bmatrix} 1 & 2 \\ 3 & 6 \end{bmatrix}$$

(b) Compute AA^{T} and its eigenvalues and unit eigenvectors $\boldsymbol{u}_1$ and $\boldsymbol{u}_2$.

(c) Verify that $A\boldsymbol{v}_1 = \sigma_1\boldsymbol{u}_1$. Put numbers into $\boldsymbol{A} = \boldsymbol{U\Sigma V}^{\mathbf{T}}$ (this is the SVD).

2 (a) From the $\boldsymbol{u}$'s and $\boldsymbol{v}$'s in Problem 1 write down orthonormal bases for the four fundamental subspaces of this matrix A.

(b) Describe all matrices that have those same four subspaces. Multiples of A?

3 From U, V, and Σ in Problem 1 find the orthogonal matrix $Q = UV^{\mathrm{T}}$ and the symmetric matrix $S = V\Sigma V^{\mathrm{T}}$. Verify the polar decomposition $A = QS$. This S is only semidefinite because ______. Test $S^2 = A$.

4 Compute the pseudoinverse $A^+ = V\Sigma^+U^{\mathrm{T}}$. The diagonal matrix Σ^+ contains $1/\sigma_1$. Rename the four subspaces (for A) in Figure 7.6 as four subspaces for A^+. Compute the projections A^+A and AA^+ on the row and column spaces of A.

Problems 5–9 are about the SVD of an invertible matrix.

5 Compute $A^{\mathrm{T}}A$ and its eigenvalues and unit eigenvectors $\boldsymbol{v}_1$ and $\boldsymbol{v}_2$. What are the singular values σ_1 and σ_2 for this matrix A?

$$A = \begin{bmatrix} 3 & 3 \\ -1 & 1 \end{bmatrix}.$$

6 AA^{T} has the same eigenvalues σ_1^2 and σ_2^2 as $A^{\mathrm{T}}A$. Find unit eigenvectors $\boldsymbol{u}_1$ and $\boldsymbol{u}_2$. Put numbers into the SVD:

$$A = \begin{bmatrix} 3 & 3 \\ -1 & 1 \end{bmatrix} = \begin{bmatrix} \boldsymbol{u}_1 & \boldsymbol{u}_2 \end{bmatrix} \begin{bmatrix} \sigma_1 & \\ & \sigma_2 \end{bmatrix} \begin{bmatrix} \boldsymbol{v}_1 & \boldsymbol{v}_2 \end{bmatrix}^{\mathrm{T}}.$$

7 In Problem 6, multiply columns times rows to show that $A = \sigma_1\boldsymbol{u}_1\boldsymbol{v}_1^{\mathrm{T}} + \sigma_2\boldsymbol{u}_2\boldsymbol{v}_2^{\mathrm{T}}$. Prove from $A = U\Sigma V^{\mathrm{T}}$ that every matrix of rank r is the sum of r matrices of rank one.

8 From U, V, and Σ find the orthogonal matrix $Q = UV^{\mathrm{T}}$ and the symmetric matrix $K = U\Sigma U^{\mathrm{T}}$. Verify the polar decomposition in reverse order $A = KQ$.

9 The pseudoinverse of this A is the same as _____ because _____.

Problems 10–11 compute and use the SVD of a 1 by 3 rectangular matrix.

10 Compute $A^{\mathrm{T}}A$ and AA^{T} and their eigenvalues and unit eigenvectors when the matrix is $A = \begin{bmatrix} 3 & 4 & 0 \end{bmatrix}$. What are the singular values of A?

11 Put numbers into the singular value decomposition of A:

$$A = \begin{bmatrix} 3 & 4 & 0 \end{bmatrix} = [\boldsymbol{u}_1] \begin{bmatrix} \sigma_1 & 0 & 0 \end{bmatrix} \begin{bmatrix} \boldsymbol{v}_1 & \boldsymbol{v}_2 & \boldsymbol{v}_3 \end{bmatrix}^{\mathrm{T}}.$$

Put numbers into the pseudoinverse $V\Sigma^+U^{\mathrm{T}}$ of A. *Compute AA^+ and A^+A*:

$$\textbf{Pseudoinverse} \quad A^+ = \begin{bmatrix} \\ \\ \\ \end{bmatrix} = \begin{bmatrix} \boldsymbol{v}_1 & \boldsymbol{v}_2 & \boldsymbol{v}_3 \end{bmatrix} \begin{bmatrix} 1/\sigma_1 \\ 0 \\ 0 \end{bmatrix} \begin{bmatrix} \boldsymbol{u}_1 \end{bmatrix}^{\mathrm{T}}.$$

12 What is the only 2 by 3 matrix that has no pivots and no singular values? What is Σ for that matrix? A^+ is the zero matrix, but what is its shape?

13 If $\det A = 0$ why is $\det A^+ = 0$? If A has rank r, why does A^+ have rank r?

14 For vectors in the unit circle $||\boldsymbol{x}|| = 1$, the vectors $\boldsymbol{y} = A\boldsymbol{x}$ in the ellipse will have $||A^{-1}\boldsymbol{y}|| = 1$. This ellipse has axes along the singular vectors with lengths $= \sigma_1, \ldots, \sigma_r$ (as in Figure 7.5). Expand $||A^{-1}\boldsymbol{y}||^2 = 1$ for $A = [2\ \ 1\,;1\ \ 2]$.

Problems 15–18 bring out the main properties of A^+ and $x^+ = A^+b$.

15 All matrices in this problem have rank one. The vector $\boldsymbol{b}$ is (b_1, b_2).

$$A = \begin{bmatrix} 2 & 2 \\ 1 & 1 \end{bmatrix} \quad AA^{\mathrm{T}} = \begin{bmatrix} 8 & 4 \\ 4 & 2 \end{bmatrix} \quad A^{\mathrm{T}}A = \begin{bmatrix} 5 & 5 \\ 5 & 5 \end{bmatrix} \quad A^+ = \begin{bmatrix} .2 & .1 \\ .2 & .1 \end{bmatrix}$$

(a) The equation $A^{\mathrm{T}}A\widehat{\boldsymbol{x}} = A^{\mathrm{T}}\boldsymbol{b}$ has many solutions because $A^{\mathrm{T}}A$ is ____ .

(b) Verify that $\boldsymbol{x}^+ = A^+\boldsymbol{b} = (.2b_1 + .1b_2, .2b_1 + .1b_2)$ solves $A^{\mathrm{T}}A\boldsymbol{x}^+ = A^{\mathrm{T}}\boldsymbol{b}$.

(c) Add $(1, -1)$ to that $\boldsymbol{x}^+$ to get another solution to $A^{\mathrm{T}}A\widehat{\boldsymbol{x}} = A^{\mathrm{T}}\boldsymbol{b}$. Show that $\|\widehat{\boldsymbol{x}}\|^2 = \|\boldsymbol{x}^+\|^2 + 2$, and $\boldsymbol{x}^+$ is shorter.

16 ***The vector $\boldsymbol{x}^+ = A^+\boldsymbol{b}$ is the shortest possible solution to $A^{\mathrm{T}}A\widehat{\boldsymbol{x}} = A^{\mathrm{T}}\boldsymbol{b}$.*** Reason: The difference $\widehat{\boldsymbol{x}} - \boldsymbol{x}^+$ is in the nullspace of $A^{\mathrm{T}}A$. This is also the nullspace of A, orthogonal to $\boldsymbol{x}^+$. Explain how it follows that $\|\widehat{\boldsymbol{x}}\|^2 = \|\boldsymbol{x}^+\|^2 + \|\widehat{\boldsymbol{x}} - \boldsymbol{x}^+\|^2$.

17 Every $\boldsymbol{b}$ in $\mathbf{R}^m$ is $\boldsymbol{p} + \boldsymbol{e}$. This is the column space part plus the left nullspace part. Every $\boldsymbol{x}$ in $\mathbf{R}^n$ is $\boldsymbol{x}^+ + \boldsymbol{x}_n$. This is the row space part plus the nullspace part. Then

$$AA^+\boldsymbol{p} = \underline{\quad\quad} \qquad AA^+\boldsymbol{e} = \underline{\quad\quad} \qquad A^+A\boldsymbol{x}^+ = \underline{\quad\quad} \qquad A^+A\boldsymbol{x}_n = \underline{\quad\quad}$$

18 Find A^+ and A^+A and AA^+ and $\boldsymbol{x}^+$ for this matrix $A = U\Sigma V^{\mathrm{T}}$ and these $\boldsymbol{b}$:

$$A = \begin{bmatrix} 3 \\ 4 \end{bmatrix} = \begin{bmatrix} .6 & -.8 \\ .8 & .6 \end{bmatrix} \begin{bmatrix} 5 \\ 0 \end{bmatrix} [1] \qquad \boldsymbol{b} = \begin{bmatrix} 3 \\ 4 \end{bmatrix} \text{ and } \boldsymbol{b} = \begin{bmatrix} -4 \\ 3 \end{bmatrix}.$$

19 A general 2 by 2 matrix A is determined by four numbers. If triangular, it is determined by three. If diagonal, by two. If a rotation, by one. If a unit eigenvector, also by one. Check that the total count is four for each factorization of A:

$$\textbf{Four numbers in} \quad LU \quad LDU \quad QR \quad U\Sigma V^{\mathrm{T}} \quad X\Lambda X^{-1}.$$

20 Following Problem 18, check that LDL^{T} and $Q\Lambda Q^{\mathrm{T}}$ are determined by *three* numbers. This is correct because the matrix is now ____ .

21 From A and A^+ show that A^+A is correct and $(A^+A)^2 = A^+A = $ projection.

$$A = \sum_1^r \sigma_i \boldsymbol{u}_i \boldsymbol{v}_i^{\mathrm{T}} \qquad A^+ = \sum_1^r \frac{\boldsymbol{v}_i \boldsymbol{u}_i^{\mathrm{T}}}{\sigma_i} \qquad A^+A = \sum_1^r \boldsymbol{v}_i \boldsymbol{v}_i^{\mathrm{T}} \qquad AA^+ = \sum_1^r \boldsymbol{u}_i \boldsymbol{u}_i^{\mathrm{T}}$$

22 Each pair of singular vectors $\boldsymbol{v}$ and $\boldsymbol{u}$ has $A\boldsymbol{v} = \sigma\boldsymbol{u}$ and $A^{\mathrm{T}}\boldsymbol{u} = \sigma\boldsymbol{v}$. Show that the double vector $\begin{bmatrix} \boldsymbol{v} \\ \boldsymbol{u} \end{bmatrix}$ is an eigenvector of the symmetric block matrix $M = \begin{bmatrix} 0 & A^{\mathrm{T}} \\ A & 0 \end{bmatrix}$. The SVD of A is equivalent to the diagonalization of that symmetric matrix M.

Chapter 8

Linear Transformations

8.1 The Idea of a Linear Transformation

1 A **linear transformation** T takes vectors $\boldsymbol{v}$ to vectors $T(\boldsymbol{v})$. Linearity requires

$\boxed{T(c\boldsymbol{v}+d\boldsymbol{w}) = cT(\boldsymbol{v})+dT(\boldsymbol{w})}$ Note $T(\mathbf{0}) = \mathbf{0}$ so $T(\boldsymbol{v}) = \boldsymbol{v}+\boldsymbol{u}_0$ is not linear.

2 The input vectors $\boldsymbol{v}$ and outputs $T(\boldsymbol{v})$ can be in $\mathbf{R}^n$ or matrix space or function space.

3 If A is m by $n, T(\boldsymbol{x}) = A\boldsymbol{x}$ is linear from the input space $\mathbf{R}^n$ to the output space $\mathbf{R}^m$.

4 The derivative $T(f) = \dfrac{df}{dx}$ is linear. The integral $T^+(f) = \displaystyle\int_0^x f(t)\,dt$ is its pseudoinverse.

5 The product ST of two linear transformations is still linear: $\boxed{(ST)(\boldsymbol{v}) = S(T(\boldsymbol{v})).}$

When a matrix A multiplies a vector $\boldsymbol{v}$, it "transforms" $\boldsymbol{v}$ into another vector $A\boldsymbol{v}$. ***In goes*** $\boldsymbol{v}$, ***out comes*** $T(\boldsymbol{v}) = A\boldsymbol{v}$. A transformation T follows the same idea as a function. In goes a number x, out comes $f(x)$. For one vector $\boldsymbol{v}$ or one number x, we multiply by the matrix or we evaluate the function. The deeper goal is to see all vectors $\boldsymbol{v}$ at once. We are transforming the whole space $\mathbf{V}$ when we multiply every $\boldsymbol{v}$ by A.

Start again with a matrix A. It transforms $\boldsymbol{v}$ to $A\boldsymbol{v}$. It transforms $\boldsymbol{w}$ to $A\boldsymbol{w}$. Then we *know* what happens to $\boldsymbol{u} = \boldsymbol{v}+\boldsymbol{w}$. There is no doubt about $A\boldsymbol{u}$, it has to equal $A\boldsymbol{v}+A\boldsymbol{w}$. Matrix multiplication $T(\boldsymbol{v}) = A\boldsymbol{v}$ gives a ***linear transformation***:

> A ***transformation*** T assigns an output $T(\boldsymbol{v})$ to each input vector $\boldsymbol{v}$ in $\mathbf{V}$.
> The transformation is ***linear*** if it meets these requirements for all $\boldsymbol{v}$ and $\boldsymbol{w}$:
>
> (a) $T(\boldsymbol{v}+\boldsymbol{w}) = T(\boldsymbol{v})+T(\boldsymbol{w})$ (b) $T(c\boldsymbol{v}) = cT(\boldsymbol{v})$ for all c.

If the input is $\boldsymbol{v} = \mathbf{0}$, the output must be $T(\boldsymbol{v}) = \mathbf{0}$. We combine rules (a) and (b) into one:

Linear transformation $\quad T(c\boldsymbol{v} + d\boldsymbol{w})$ ***must equal*** $cT(\boldsymbol{v}) + dT(\boldsymbol{w})$.

Again I can test matrix multiplication for linearity: $A(c\boldsymbol{v} + d\boldsymbol{w}) = cA\boldsymbol{v} + dA\boldsymbol{w}$ is *true*.

A linear transformation is highly restricted. Suppose T adds $\boldsymbol{u}_0$ to every vector. Then $T(\boldsymbol{v}) = \boldsymbol{v} + \boldsymbol{u}_0$ and $T(\boldsymbol{w}) = \boldsymbol{w} + \boldsymbol{u}_0$. This isn't good, or at least *it isn't linear*. Applying T to $\boldsymbol{v} + \boldsymbol{w}$ produces $\boldsymbol{v} + \boldsymbol{w} + \boldsymbol{u}_0$. That is not the same as $T(\boldsymbol{v}) + T(\boldsymbol{w})$:

Shift is not linear $\quad \boldsymbol{v} + \boldsymbol{w} + \boldsymbol{u}_0$ is not $T(\boldsymbol{v}) + T(\boldsymbol{w}) = (\boldsymbol{v} + \boldsymbol{u}_0) + (\boldsymbol{w} + \boldsymbol{u}_0)$.

The exception is when $\boldsymbol{u}_0 = \mathbf{0}$. The transformation reduces to $T(\boldsymbol{v}) = \boldsymbol{v}$. This is the ***identity transformation*** (nothing moves, as in multiplication by the identity matrix). That is certainly linear. In this case the input space $\mathbf{V}$ is the same as the output space $\mathbf{W}$.

The linear-plus-shift transformation $T(\boldsymbol{v}) = A\boldsymbol{v} + \boldsymbol{u}_0$ is called *"affine"*. Straight lines stay straight although T is not linear. Computer graphics works with affine transformations in Section 10.6, because we must be able to move images.

Example 1 Choose a fixed vector $\boldsymbol{a} = (1, 3, 4)$, and let $T(\boldsymbol{v})$ be the dot product $\boldsymbol{a} \cdot \boldsymbol{v}$:

The input is $\boldsymbol{v} = (v_1, v_2, v_3)$. $\quad$ The output is $T(\boldsymbol{v}) = \boldsymbol{a} \cdot \boldsymbol{v} = v_1 + 3v_2 + 4v_3$.

Dot products are linear. The inputs $\boldsymbol{v}$ come from three-dimensional space, so $\mathbf{V} = \mathbf{R}^3$. The outputs are just numbers, so the output space is $\mathbf{W} = \mathbf{R}^1$. We are multiplying by the row matrix $A = [1 \ \ 3 \ \ 4]$. Then $T(\boldsymbol{v}) = A\boldsymbol{v}$.

You will get good at recognizing which transformations are linear. If the output involves squares or products or lengths, v_1^2 or $v_1 v_2$ or $\|\boldsymbol{v}\|$, then T is not linear.

Example 2 The length $T(\boldsymbol{v}) = \|\boldsymbol{v}\|$ is not linear. Requirement (a) for linearity would be $\|\boldsymbol{v} + \boldsymbol{w}\| = \|\boldsymbol{v}\| + \|\boldsymbol{w}\|$. Requirement (b) would be $\|c\boldsymbol{v}\| = c\|\boldsymbol{v}\|$. Both are false!

Not (a): The sides of a triangle satisfy an *inequality* $\|\boldsymbol{v} + \boldsymbol{w}\| \le \|\boldsymbol{v}\| + \|\boldsymbol{w}\|$.
Not (b): The length $\|-\boldsymbol{v}\|$ is $\|\boldsymbol{v}\|$ and not $-\|\boldsymbol{v}\|$. For negative c, linearity fails.

Example 3 (Rotation) T is the transformation that *rotates every vector by* $30°$. The "*domain*" of T is the xy plane (all input vectors $\boldsymbol{v}$). The "*range*" of T is also the xy plane (all rotated vectors $T(\boldsymbol{v})$). We described T without a matrix: rotate the plane by $30°$.

Is rotation linear? *Yes it is.* We can rotate two vectors and add the results. The sum of rotations $T(\boldsymbol{v}) + T(\boldsymbol{w})$ is the same as the rotation $T(\boldsymbol{v} + \boldsymbol{w})$ of the sum. **The whole plane is turning together, in this linear transformation.**

Lines to Lines, Triangles to Triangles, Basis Tells All

Figure 8.1 shows the line from $\boldsymbol{v}$ to $\boldsymbol{w}$ in the input space. It also shows the line from $T(\boldsymbol{v})$ to $T(\boldsymbol{w})$ in the output space. Linearity tells us: Every point on the input line goes onto the output line. And more than that: ***Equally spaced points go to equally spaced points***. The middle point $\boldsymbol{u} = \frac{1}{2}\boldsymbol{v} + \frac{1}{2}\boldsymbol{w}$ goes to the middle point $T(\boldsymbol{u}) = \frac{1}{2}T(\boldsymbol{v}) + \frac{1}{2}T(\boldsymbol{w})$.

The second figure moves up a dimension. Now we have three corners $\boldsymbol{v}_1$, $\boldsymbol{v}_2$, $\boldsymbol{v}_3$. Those inputs have three outputs $T(\boldsymbol{v}_1)$, $T(\boldsymbol{v}_2)$, $T(\boldsymbol{v}_3)$. *The input triangle goes onto the output triangle.* Equally spaced points stay equally spaced (along the edges, and then between the edges). The middle point $\boldsymbol{u} = \frac{1}{3}(\boldsymbol{v}_1 + \boldsymbol{v}_2 + \boldsymbol{v}_3)$ goes to the middle point $T(\boldsymbol{u}) = \frac{1}{3}(T(\boldsymbol{v}_1) + T(\boldsymbol{v}_2) + T(\boldsymbol{v}_3))$.

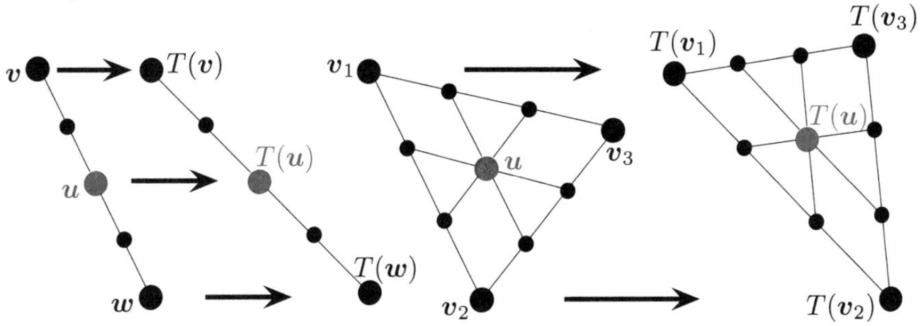

Figure 8.1: Lines to lines, equal spacing to equal spacing, $\boldsymbol{u} = \mathbf{0}$ to $T(\boldsymbol{u}) = \mathbf{0}$.

The rule of linearity extends to combinations of three vectors or n vectors:

Linearity $\quad \boldsymbol{u} = c_1\boldsymbol{v}_1 + c_2\boldsymbol{v}_2 + \cdots + c_n\boldsymbol{v}_n \quad$ **must transform to**

$$T(\boldsymbol{u}) = c_1T(\boldsymbol{v}_1) + c_2T(\boldsymbol{v}_2) + \cdots + c_nT(\boldsymbol{v}_n) \tag{1}$$

The 2-vector rule starts the 3-vector proof: $T(c\boldsymbol{u} + d\boldsymbol{v} + e\boldsymbol{w}) = T(c\boldsymbol{u}) + T(d\boldsymbol{v} + e\boldsymbol{w})$. Then linearity applies to both of those parts, to give $cT(\boldsymbol{u}) + dT(\boldsymbol{v}) + eT(\boldsymbol{w})$.

The n-vector rule (1) leads to the most important fact about linear transformations:

Suppose you know $T(\boldsymbol{v})$ for all vectors $\boldsymbol{v}_1, \ldots, \boldsymbol{v}_n$ in a basis
Then you know $T(\boldsymbol{u})$ for every vector $\boldsymbol{u}$ in the space.

You see the reason: Every $\boldsymbol{u}$ in the space is a combination of the basis vectors $\boldsymbol{v}_j$. Then linearity tells us that $T(\boldsymbol{u})$ is the same combination of the outputs $T(\boldsymbol{v}_j)$.

Example 4 **The transformation T takes the derivative of the input**: $T(\boldsymbol{u}) = d\boldsymbol{u}/d\boldsymbol{x}$.

How do you find the derivative of $u = 6 - 4x + 3x^2$? You start with the derivatives of 1, x, and x^2. Those are the basis vectors. Their derivatives are 0, 1, and $2x$. Then you use linearity for the derivative of any combination:

$$\frac{du}{dx} = 6\,(\text{derivative of } 1) - 4\,(\text{derivative of } x) + 3\,(\text{derivative of } x^2) = -4 + 6x.$$

All of calculus depends on linearity! Precalculus finds a few key derivatives, for x^n and $\sin x$ and $\cos x$ and e^x. Then linearity applies to all their combinations.

I would say that the only rule special to calculus is the *chain rule*. That produces the derivative of a chain of functions $f(g(x))$.

Nullspace of $T(u) = du/dx$. For the nullspace we solve $T(u) = 0$. The derivative is zero when u *is a constant function*. So the one-dimensional nullspace is a line in function space—all multiples of the special solution $u = 1$.

Column space of $T(u) = du/dx$. In our example the input space contains all quadratics $a + bx + cx^2$. The outputs (the column space) are all linear functions $b + 2cx$. Notice that the **Counting Theorem** is still true : $\boldsymbol{r} + (\boldsymbol{n} - \boldsymbol{r}) = \boldsymbol{n}$.

dimension (**column space**)+dimension (**nullspace**) = **2**+**1**=**3**=dimension (**input space**)

What is the matrix for d/dx? I can't leave derivatives without asking for a matrix. We have a linear transformation $T = d/dx$. We know what T does to the basis functions:

$$\boldsymbol{v}_1, \boldsymbol{v}_2, \boldsymbol{v}_3 = \mathbf{1}, \boldsymbol{x}, \boldsymbol{x}^2 \qquad \frac{dv_1}{dx} = 0 \qquad \frac{dv_2}{dx} = 1 = \boldsymbol{v}_1 \qquad \frac{dv_3}{dx} = 2x = \mathbf{2}\boldsymbol{v}_2. \qquad (2)$$

The 3-dimensional input space **V** (= quadratics) transforms to the 2-dimensional output space **W** (= linear functions). If $\boldsymbol{v}_1, \boldsymbol{v}_2, \boldsymbol{v}_3$ were vectors, I would know the matrix.

$$\boxed{\boldsymbol{A} = \begin{bmatrix} 0 & \mathbf{1} & 0 \\ 0 & 0 & \mathbf{2} \end{bmatrix} = \text{matrix form of the derivative } T = \frac{d}{dx}.} \qquad (3)$$

The linear transformation du/dx is perfectly copied by the matrix multiplication $A\boldsymbol{u}$.

$$\begin{array}{l} \textbf{Input } \boldsymbol{u} \\ a + bx + cx^2 \end{array} \qquad \textbf{Multiplication } A\boldsymbol{u} = \begin{bmatrix} 0 & 1 & 0 \\ 0 & 0 & 2 \end{bmatrix} \begin{bmatrix} a \\ b \\ c \end{bmatrix} = \begin{bmatrix} b \\ 2c \end{bmatrix} \qquad \textbf{Output } \frac{d\boldsymbol{u}}{d\boldsymbol{x}} = b + 2cx.$$

The connection from T to A (we will connect every transformation to a matrix) depended on choosing an input basis $1, x, x^2$ and an output basis $1, x$.

Next we look at integrals. They give the pseudoinverse T^+ of the derivative! I can't write T^{-1} and I can't say "*inverse of T*" when the derivative of 1 is 0.

Example 5 **Integration T^+ is also linear** : $\int_0^x (D + Ex)\,dx = Dx + \frac{1}{2}Ex^2$.

The input basis is now $1, x$. The output basis is $1, x, x^2$. The matrix A^+ for T^+ is 3 by 2:

$$\textbf{Input } \boldsymbol{v} \quad \textbf{Multiplication } A^+\boldsymbol{v} = \begin{bmatrix} 0 & 0 \\ 1 & 0 \\ 0 & \frac{1}{2} \end{bmatrix} \begin{bmatrix} D \\ E \end{bmatrix} = \begin{bmatrix} \mathbf{0} \\ \boldsymbol{D} \\ \frac{1}{2}\boldsymbol{E} \end{bmatrix} \quad \textbf{Output = Integral of } \boldsymbol{v}$$
$$D + Ex \qquad\qquad T^+(\boldsymbol{v}) = Dx + \tfrac{1}{2}Ex^2$$

The Fundamental Theorem of Calculus says that integration is the (pseudo)inverse of differentiation. For linear algebra, the matrix A^+ is the (pseudo)inverse of the matrix A:

$$A^+A = \begin{bmatrix} 0 & 0 \\ 1 & 0 \\ 0 & \frac{1}{2} \end{bmatrix} \begin{bmatrix} 0 & 1 & 0 \\ 0 & 0 & 2 \end{bmatrix} = \begin{bmatrix} \mathbf{0} & 0 & 0 \\ 0 & \mathbf{1} & 0 \\ 0 & 0 & \mathbf{1} \end{bmatrix} \quad \text{and} \quad AA^+ = \begin{bmatrix} \mathbf{1} & 0 \\ 0 & \mathbf{1} \end{bmatrix}. \tag{4}$$

The derivative of a constant function is zero. That zero is on the diagonal of A^+A. Calculus wouldn't be calculus without that 1-dimensional nullspace of $T = d/dx$.

Examples of Transformations (mostly linear)

Example 6 Project every 3-dimensional vector onto the horizontal plane $z = 1$. The vector $\boldsymbol{v} = (x, y, z)$ is transformed to $T(\boldsymbol{v}) = (x, y, 1)$. This transformation is not linear. Why not? It doesn't even transform $\boldsymbol{v} = \mathbf{0}$ into $T(\boldsymbol{v}) = \mathbf{0}$.

Example 7 Suppose A is an *invertible matrix.* Certainly $T(\boldsymbol{v} + \boldsymbol{w}) = A\boldsymbol{v} + A\boldsymbol{w} = T(\boldsymbol{v}) + T(\boldsymbol{w})$. Another linear transformation is multiplication by A^{-1}. This produces the ***inverse transformation*** T^{-1}, which brings every vector $T(\boldsymbol{v})$ back to $\boldsymbol{v}$:

$$T^{-1}(T(\boldsymbol{v})) = \boldsymbol{v} \quad \text{matches the matrix multiplication} \quad A^{-1}(A\boldsymbol{v}) = \boldsymbol{v}.$$

If $T(\boldsymbol{v}) = A\boldsymbol{v}$ and $S(\boldsymbol{u}) = B\boldsymbol{u}$, then the product $T(S(u))$ matches the product $AB\boldsymbol{u}$.

We are reaching an unavoidable question. ***Are all linear transformations from*** $\mathbf{V} = \mathbf{R}^n$ ***to*** $\mathbf{W} = \mathbf{R}^m$ ***produced by matrices?*** When a linear T is described as a "rotation" or "projection" or ". . .", is there always a matrix A hiding behind T? Is $T(\boldsymbol{v})$ always $A\boldsymbol{v}$?

The answer is *yes*! This is an approach to linear algebra that doesn't start with matrices. We still end up with matrices—*after we choose an input basis and output basis.*

Note Transformations have a language of their own. For a matrix, the column space contains all outputs $A\boldsymbol{v}$. The nullspace contains all inputs for which $A\boldsymbol{v} = \mathbf{0}$. Translate those words into "*range*" and "*kernel*":

Range of T = set of *all outputs $T(\boldsymbol{v})$*. Range corresponds to column space.

Kernel of T = set of *all inputs for which $T(\boldsymbol{v}) = \mathbf{0}$*. Kernel corresponds to nullspace.

The range is in the output space $\mathbf{W}$. The kernel is in the input space $\mathbf{V}$. When T is multiplication by a matrix, $T(\boldsymbol{v}) = A\boldsymbol{v}$, range is column space and kernel is nullspace.

Linear Transformations of the Plane

It is more interesting to *see* a transformation than to define it. When a 2 by 2 matrix A multiplies all vectors in $\mathbf{R}^2$, we can watch how it acts. Start with a "house" that has eleven endpoints. Those eleven vectors $\boldsymbol{v}$ are transformed into eleven vectors $A\boldsymbol{v}$. Straight lines between $\boldsymbol{v}$'s become straight lines between the transformed vectors $A\boldsymbol{v}$. (The transformation from house to house is linear!) Applying A to a standard house produces a new house—possibly stretched or rotated or otherwise unlivable.

This part of the book is visual, not theoretical. We will show four houses and the matrices that produce them. The columns of H are the eleven corners of the first house. (H is 2 by 12, so **plot2d** in Problem 25 will connect the 11th corner to the first.) A multiplies the 11 points in the house matrix H to produce the corners AH of the other houses.

House matrix $$H = \begin{bmatrix} -6 & -6 & -7 & 0 & 7 & 6 & 6 & -3 & -3 & 0 & 0 & -6 \\ -7 & 2 & 1 & 8 & 1 & 2 & -7 & -7 & -2 & -2 & -7 & -7 \end{bmatrix}.$$

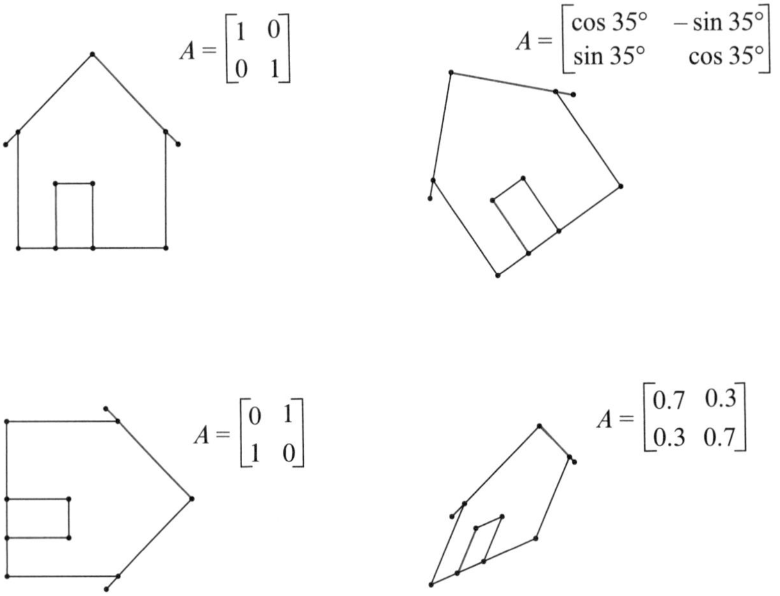

Figure 8.2: Linear transformations of a house drawn by **plot2d**$(A * H)$.

■ REVIEW OF THE KEY IDEAS ■

1. A transformation T takes each $\boldsymbol{v}$ in the input space to $T(\boldsymbol{v})$ in the output space.

2. T is **linear** if $T(\boldsymbol{v}+\boldsymbol{w}) = T(\boldsymbol{v}) + T(w)$ and $T(c\boldsymbol{v}) = cT(\boldsymbol{v})$: lines to lines.

3. Combinations to combinations: $T\,(c_1\boldsymbol{v}_1 + \cdots + c_n\boldsymbol{v}_n) = c_1\,T(\boldsymbol{v}_1) + \cdots + c_n\,T(\boldsymbol{v}_n)$.

4. $T =$ *derivative* and $T^+ =$ *integral* are linear. So is $T(\boldsymbol{v}) = A\boldsymbol{v}$ from $\mathbf{R}^n$ to $\mathbf{R}^m$.

■ WORKED EXAMPLES ■

8.1 A The elimination matrix $\left[\begin{smallmatrix}1 & 0\\ 1 & 1\end{smallmatrix}\right]$ gives a ***shearing transformation*** from (x, y) to $\boldsymbol{T(x, y) = (x, x + y)}$. If the inputs fill a square, draw the transformed square.

Solution The points $(1, 0)$ and $(2, 0)$ on the x axis transform by T to $(1, 1)$ and $(2, 2)$ on the $45°$ line. Points on the y axis are *not moved*: $T(0, y) = (0, y) =$ eigenvectors with $\lambda = 1$.

Vertical lines slide up
This is the shearing
Squares go to parallelograms

$$A = \begin{bmatrix} 1 & 0 \\ 1 & 1 \end{bmatrix}$$

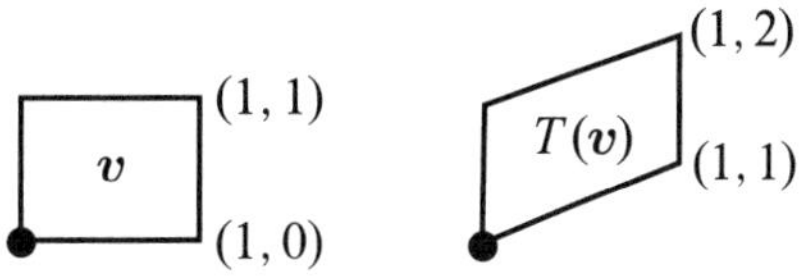

8.1 B A **nonlinear transformation** T is invertible if every $\boldsymbol{b}$ in the output space comes from exactly one $\boldsymbol{x}$ in the input space: $T(\boldsymbol{x}) = \boldsymbol{b}$ always has exactly one solution. Which of these transformations (on real numbers x) is invertible and what is T^{-1}? ***None are linear, not even T_3.*** When you solve $T(\boldsymbol{x}) = \boldsymbol{b}$, you are inverting T:

$$T_1(x) = x^2 \quad T_2(x) = x^3 \quad T_3(x) = x + 9 \quad T_4(x) = e^x \quad T_5(x) = \frac{1}{x} \text{ for nonzero } x\text{'s}$$

Solution T_1 is not invertible: $x^2 = 1$ has *two* solutions and $x^2 = -1$ has *no* solution.
T_4 is not invertible because $e^x = -1$ has no solution. (If the output space changes to *positive* b's then the inverse of $e^x = b$ is $x = \ln b$.)

Notice $T_5^2 =$ identity. But $T_3^2(x) = x + 18$. What are $T_2^2(x)$ and $T_4^2(x)$?

T_2, T_3, T_5 are invertible: $x^3 = b$ and $x + 9 = b$ and $\frac{1}{x} = b$ have one solution x.

$$x = T_2^{-1}(b) = b^{1/3} \qquad x = T_3^{-1}(b) = b - 9 \qquad x = T_5^{-1}(b) = 1/b$$

Problem Set 8.1

1 A linear transformation must leave the zero vector fixed: $T(\mathbf{0}) = \mathbf{0}$. Prove this from $T(\boldsymbol{v} + \boldsymbol{w}) = T(\boldsymbol{v}) + T(\boldsymbol{w})$ by choosing $\boldsymbol{w} =$ _____ (and finish the proof). Prove it also from $T(c\boldsymbol{v}) = cT(\boldsymbol{v})$ by choosing $c =$ _____.

2 Requirement (b) gives $T(c\boldsymbol{v}) = cT(\boldsymbol{v})$ and also $T(d\boldsymbol{w}) = dT(\boldsymbol{w})$. Then by addition, requirement (a) gives $T(\quad) = (\quad)$. What is $T(c\boldsymbol{v} + d\boldsymbol{w} + e\boldsymbol{u})$?

3 Which of these transformations are not linear? The input is $\boldsymbol{v} = (v_1, v_2)$:

(a) $T(\boldsymbol{v}) = (v_2, v_1)$ (b) $T(\boldsymbol{v}) = (v_1, v_1)$ (c) $T(\boldsymbol{v}) = (0, v_1)$
(d) $T(\boldsymbol{v}) = (0, 1)$ (e) $T(\boldsymbol{v}) = v_1 - v_2$ (f) $T(\boldsymbol{v}) = v_1 v_2$.

4 If S and T are linear transformations, is $T(S(\boldsymbol{v}))$ linear or quadratic?

(a) (Special case) If $S(\boldsymbol{v}) = \boldsymbol{v}$ and $T(\boldsymbol{v}) = \boldsymbol{v}$, then $T(S(\boldsymbol{v})) = \boldsymbol{v}$ or $\boldsymbol{v}^2$?

(b) (General case) $S(\boldsymbol{v}_1 + \boldsymbol{v}_2) = S(\boldsymbol{v}_1) + S(\boldsymbol{v}_2)$ and $T(\boldsymbol{v}_1 + \boldsymbol{v}_2) = T(\boldsymbol{v}_1) + T(\boldsymbol{v}_2)$ combine into

$$T(S(\boldsymbol{v}_1 + \boldsymbol{v}_2)) = T(\underline{\qquad}) = \underline{\qquad} + \underline{\qquad}.$$

5 Suppose $T(\boldsymbol{v}) = \boldsymbol{v}$ except that $T(0, v_2) = (0, 0)$. Show that this transformation satisfies $T(c\boldsymbol{v}) = cT(\boldsymbol{v})$ but does not satisfy $T(\boldsymbol{v} + \boldsymbol{w}) = T(\boldsymbol{v}) + T(\boldsymbol{w})$.

6 Which of these transformations satisfy $T(\boldsymbol{v} + \boldsymbol{w}) = T(\boldsymbol{v}) + T(\boldsymbol{w})$ and which satisfy $T(c\boldsymbol{v}) = cT(\boldsymbol{v})$?

(a) $T(\boldsymbol{v}) = \boldsymbol{v}/\|\boldsymbol{v}\|$ (b) $T(\boldsymbol{v}) = v_1 + v_2 + v_3$ (c) $T(\boldsymbol{v}) = (v_1, 2v_2, 3v_3)$

(d) $T(\boldsymbol{v}) =$ largest component of $\boldsymbol{v}$.

7 For these transformations of $\mathbf{V} = \mathbf{R}^2$ to $\mathbf{W} = \mathbf{R}^2$, find $T(T(\boldsymbol{v}))$. Show that when $T(\boldsymbol{v})$ is linear, then also $T(T(\boldsymbol{v}))$ is linear.

(a) $T(\boldsymbol{v}) = -\boldsymbol{v}$ (b) $T(\boldsymbol{v}) = \boldsymbol{v} + (1, 1)$

(c) $T(\boldsymbol{v}) = 90^\circ$ rotation $= (-v_2, v_1)$

(d) $T(\boldsymbol{v}) =$ projection $= \frac{1}{2}(v_1 + v_2,\ v_1 + v_2)$.

8 Find the range and kernel (like the column space and nullspace) of T:

(a) $T(v_1, v_2) = (v_1 - v_2, 0)$ (b) $T(v_1, v_2, v_3) = (v_1, v_2)$

(c) $T(v_1, v_2) = (0, 0)$ (d) $T(v_1, v_2) = (v_1, v_1)$.

9 The "cyclic" transformation T is defined by $T(v_1, v_2, v_3) = (v_2, v_3, v_1)$. What is $T(T(\boldsymbol{v}))$? What is $T^3(\boldsymbol{v})$? What is $T^{100}(\boldsymbol{v})$? Apply T a hundred times to $\boldsymbol{v}$.

10 A linear transformation from $\mathbf{V}$ to $\mathbf{W}$ has an *inverse* from $\mathbf{W}$ to $\mathbf{V}$ when the range is all of $\mathbf{W}$ and the kernel contains only $\boldsymbol{v} = \mathbf{0}$. Then $T(\boldsymbol{v}) = \boldsymbol{w}$ has one solution $\boldsymbol{v}$ for each $\boldsymbol{w}$ in $\mathbf{W}$. Why are these T's not invertible?

(a) $T(v_1, v_2) = (v_2, v_2)$ $\mathbf{W} = \mathbf{R}^2$

(b) $T(v_1, v_2) = (v_1, v_2, v_1 + v_2)$ $\mathbf{W} = \mathbf{R}^3$

(c) $T(v_1, v_2) = v_1$ $\mathbf{W} = \mathbf{R}^1$

11 If $T(\boldsymbol{v}) = A\boldsymbol{v}$ and A is m by n, then T is "multiplication by A."

(a) What are the input and output spaces $\mathbf{V}$ and $\mathbf{W}$?

(b) Why is range of T = column space of A?

(c) Why is kernel of T = nullspace of A?

12 Suppose a linear T transforms $(1,1)$ to $(2,2)$ and $(2,0)$ to $(0,0)$. Find $T(\boldsymbol{v})$:

(a) $\boldsymbol{v} = (2,2)$ (b) $\boldsymbol{v} = (3,1)$ (c) $\boldsymbol{v} = (-1,1)$ (d) $\boldsymbol{v} = (a,b)$.

Problems 13-19 may be harder. The input space V contains all 2 by 2 matrices M.

13 M is any 2 by 2 matrix and $A = \left[\begin{smallmatrix}1 & 2\\ 3 & 4\end{smallmatrix}\right]$. The transformation T is defined by $T(M) = AM$. What rules of matrix multiplication show that T is linear?

14 Suppose $A = \left[\begin{smallmatrix}1 & 2\\ 3 & 5\end{smallmatrix}\right]$. Show that the range of T is the whole matrix space **V** and the kernel is the zero matrix:

(1) If $AM = 0$ prove that M must be the zero matrix.

(2) Find a solution to $AM = B$ for any 2 by 2 matrix B.

15 Suppose $A = \left[\begin{smallmatrix}1 & 2\\ 3 & 6\end{smallmatrix}\right]$. Show that the identity matrix I is not in the range of T. Find a nonzero matrix M such that $T(M) = AM$ is zero.

16 Suppose T transposes every 2 by 2 matrix M. Try to find a matrix A which gives $AM = M^{\mathrm{T}}$. *Show that no matrix A will do it. To professors:* Is this a linear transformation that doesn't come from a matrix? The matrix should be 4 by 4!

17 The transformation T that transposes every 2 by 2 matrix is definitely linear. Which of these extra properties are true?

(a) $T^2 =$ identity transformation.

(b) The kernel of T is the zero matrix.

(c) Every 2 by 2 matrix is in the range of T.

(d) $T(M) = -M$ is impossible.

18 Suppose $T(M) = \left[\begin{smallmatrix}1 & 0\\ 0 & 0\end{smallmatrix}\right]\left[\,\mathrm{M}\,\right]\left[\begin{smallmatrix}0 & 0\\ 0 & 1\end{smallmatrix}\right]$. Find a matrix with $T(M) \neq 0$. Describe all matrices with $T(M) = 0$ (the kernel) and all output matrices $T(M)$ (the range).

19 If A and B are invertible and $T(M) = AMB$, find $T^{-1}(M)$ in the form $(\quad)M(\quad)$.

Questions 20–26 are about house transformations. The output is $T(H) = AH$.

20 How can you tell from the picture of T (house) that A is

(a) a diagonal matrix?

(b) a rank-one matrix?

(c) a lower triangular matrix?

21 Draw a picture of T (house) for these matrices:

$$D = \begin{bmatrix}2 & 0\\ 0 & 1\end{bmatrix} \quad \text{and} \quad A = \begin{bmatrix}.7 & .7\\ .3 & .3\end{bmatrix} \quad \text{and} \quad U = \begin{bmatrix}1 & 1\\ 0 & 1\end{bmatrix}.$$

22 What are the conditions on $A = \left[\begin{smallmatrix} \mathrm{a} & \mathrm{b} \\ \mathrm{c} & \mathrm{d} \end{smallmatrix}\right]$ to ensure that T (house) will

(a) sit straight up?

(b) expand the house by 3 in all directions?

(c) rotate the house with no change in its shape?

23 Describe T (house) when $T(\boldsymbol{v}) = -\boldsymbol{v} + (1, 0)$. This T is "affine".

24 Change the house matrix H to add a chimney.

25 The standard house is drawn by **plot2d**(H). Circles from o and lines from –:

$$x = H(1,:)'; y = H(2,:)';$$
$$\text{axis}([-1010\text{–}1010]), \text{axis('square')}$$
$$\text{plot}(x, y, 'o', x, y, '-');$$

Test **plot2d**(A$'*$ H) and **plot2d**(A$'*$ A $*$ H) with the matrices in Figure 8.1.

26 Without a computer sketch the houses $A * H$ for these matrices A:

$$\begin{bmatrix} 1 & 0 \\ 0 & .1 \end{bmatrix} \quad \text{and} \quad \begin{bmatrix} .5 & .5 \\ .5 & .5 \end{bmatrix} \quad \text{and} \quad \begin{bmatrix} .5 & .5 \\ -.5 & .5 \end{bmatrix} \quad \text{and} \quad \begin{bmatrix} 1 & 1 \\ 1 & 0 \end{bmatrix}.$$

27 This code creates a vector theta of 50 angles. It draws the unit circle and then it draws T (circle) = ellipse. $T(\boldsymbol{v}) = A\boldsymbol{v}$ **takes circles to ellipses.**

```
A = [2 1;1 2]    % You can change A
theta = [0:2 * pi/50:2 * pi];
oirolo = [ooo(thota); oin(thota)];
ellipse = A * circle;
axis([–4 4 –4 4]); axis('square')
plot(circle(1,:), circle(2,:), ellipse(1,:), ellipse(2,:))
```

28 Add two eyes and a smile to the circle in Problem 27. (If one eye is dark and the other is light, you can tell when the face is reflected across the y axis.) Multiply by matrices A to get new faces.

29 What conditions on $\det A = ad - bc$ ensure that the output house AH will

(a) be squashed onto a line?

(b) keep its endpoints in clockwise order (not reflected)?

(c) have the same area as the original house?

30 Why does every linear transformation T from $\mathbf{R}^2$ to $\mathbf{R}^2$ take squares to parallelograms? Rectangles also go to parallelograms (squashed if T is not invertible).

8.2 The Matrix of a Linear Transformation

1 We know all $\boldsymbol{T(v)}$ if we know $T(\boldsymbol{v}_1),\ldots,T(\boldsymbol{v}_n)$ for an input basis $\boldsymbol{v}_1,\ldots,\boldsymbol{v}_n$: use **linearity**.

2 Column j in the "matrix for T" comes from applying T to the input basis vector $\boldsymbol{v}_j$.

3 Write $T(\boldsymbol{v_j}) = a_{1j}\boldsymbol{w}_1 + \cdots + a_{mj}\boldsymbol{w}_m$ in the output basis of $\boldsymbol{w}$'s. Those a_{ij} go into column j.

4 The matrix for $T(\boldsymbol{x}) = A\boldsymbol{x}$ is A, *if* the input and output bases = columns of $I_{n\times n}$ and $I_{m\times m}$.

5 When the bases change to $\boldsymbol{v}$'s and $\boldsymbol{w}$'s, the matrix for the same T changes from A to $W^{-1}AV$.

6 Best bases: $V = W =$ eigenvectors and $V, W =$ singular vectors give diagonal Λ and Σ.

The next pages assign a matrix A to every linear transformation T. For ordinary column vectors, the input $\boldsymbol{v}$ is in $\mathbf{V} = \mathbf{R}^n$ and the output $T(\boldsymbol{v})$ is in $\mathbf{W} = \mathbf{R}^m$. The matrix A for this transformation will be m by n. Our choice of bases in $\mathbf{V}$ and $\mathbf{W}$ will decide A.

The standard basis vectors for $\mathbf{R}^n$ and $\mathbf{R}^m$ are the columns of I. That choice leads to a standard matrix. Then $T(\boldsymbol{v}) = A\boldsymbol{v}$ in the normal way. But these spaces also have other bases, so *the same transformation T is represented by other matrices.* A main theme of linear algebra is to choose the bases that give the best matrix (a diagonal matrix) for T.

All vector spaces $\mathbf{V}$ and $\mathbf{W}$ have bases. Each choice of those bases leads to a matrix for T. When the input basis is different from the output basis, the matrix for $T(\boldsymbol{v}) = \boldsymbol{v}$ will not be the identity I. It will be the "change of basis matrix". Here is the key idea:

> **Suppose we know $\boldsymbol{T(v)}$ for the input basis vectors $\boldsymbol{v}_1$ to $\boldsymbol{v}_n$.**
> **Columns 1 to n of the matrix will contain those outputs $\boldsymbol{T(v_1)}$ to $\boldsymbol{T(v_n)}$.**
> **$\boldsymbol{A}$ times $\boldsymbol{c}$ = matrix times vector = combination of those n columns.**
> **$\boldsymbol{Ac}$ is the correct combination $\boldsymbol{c_1T(v_1) + \cdots + c_nT(v_n) = T(v)}$.**

Reason Every $\boldsymbol{v}$ is a unique combination $c_1\boldsymbol{v}_1 + \cdots + c_n\boldsymbol{v}_n$ of the basis vectors $\boldsymbol{v}_j$. Since T is a linear transformation (here is the moment for linearity), $T(\boldsymbol{v})$ must be **the same combination** $c_1T(\boldsymbol{v}_1) + \cdots + c_nT(\boldsymbol{v}_n)$ **of the outputs** $T(\boldsymbol{v}_j)$ **in the columns**.

Our first example gives the matrix A for the standard basis vectors in $\mathbf{R}^2$ and $\mathbf{R}^3$.

Example 1 Suppose T transforms $\boldsymbol{v}_1 = (1,0)$ to $T(\boldsymbol{v}_1) = (2,3,4)$. Suppose the second basis vector $\boldsymbol{v}_2 = (0,1)$ goes to $T(\boldsymbol{v}_2) = (5,5,5)$. If T is linear from $\mathbf{R}^2$ to $\mathbf{R}^3$ then its "standard matrix" is 3 by 2. Those outputs $T(\boldsymbol{v}_1)$ and $T(\boldsymbol{v}_2)$ go into the columns of A:

$$A = \begin{bmatrix} 2 & 5 \\ 3 & 5 \\ 4 & 5 \end{bmatrix} \qquad c_1 = 1 \text{ and } c_2 = 1 \text{ give } T(\boldsymbol{v}_1 + \boldsymbol{v}_2) = \begin{bmatrix} 2 & 5 \\ 3 & 5 \\ 4 & 5 \end{bmatrix} \begin{bmatrix} 1 \\ 1 \end{bmatrix} = \begin{bmatrix} 7 \\ 8 \\ 9 \end{bmatrix}.$$

Change of Basis

Example 2 Suppose the input space $\mathbf{V} = \mathbf{R}^2$ is also the output space $\mathbf{W} = \mathbf{R}^2$. *Suppose that $T(\boldsymbol{v}) = \boldsymbol{v}$ is the identity transformation.* You might expect its matrix to be I, but that only happens when the input basis is the same as the output basis. I will choose different bases to see how the matrix is constructed.

For this special case $T(\boldsymbol{v}) = \boldsymbol{v}$, I will call the matrix B instead of A. We are just changing basis from the $\boldsymbol{v}$'s to the $\boldsymbol{w}$'s. Each $\boldsymbol{v}$ is a combination of $\boldsymbol{w}_1$ and $\boldsymbol{w}_2$.

Input basis $\begin{bmatrix} \boldsymbol{v}_1 & \boldsymbol{v}_2 \end{bmatrix} = \begin{bmatrix} 3 & 6 \\ 3 & 8 \end{bmatrix}$ **Output basis** $\begin{bmatrix} \boldsymbol{w}_1 & \boldsymbol{w}_2 \end{bmatrix} = \begin{bmatrix} 3 & 0 \\ 1 & 2 \end{bmatrix}$ **Change of basis** $\boldsymbol{v}_1 = \mathbf{1}\boldsymbol{w}_1 + \mathbf{1}\boldsymbol{w}_2$, $\boldsymbol{v}_2 = \mathbf{2}\boldsymbol{w}_1 + \mathbf{3}\boldsymbol{w}_2$

Please notice! I wrote the input basis $\boldsymbol{v}_1, \boldsymbol{v}_2$ in terms of the output basis $\boldsymbol{w}_1, \boldsymbol{w}_2$. That is because of our key rule. We apply the identity transformation T to each input basis vector: $T(\boldsymbol{v}_1) = \boldsymbol{v}_1$ and $T(\boldsymbol{v}_2) = \boldsymbol{v}_2$. **Then we write those outputs $\boldsymbol{v}_1$ and $\boldsymbol{v}_2$ in the output basis $\boldsymbol{w}_1$ and $\boldsymbol{w}_2$.** Those bold numbers $\mathbf{1}, \mathbf{1}$ and $\mathbf{2}, \mathbf{3}$ tell us column 1 and column 2 of the matrix B (the change of basis matrix): $WB = V$ so $\boldsymbol{B} = \boldsymbol{W}^{-1}\boldsymbol{V}$.

Matrix B for change of basis
$$\begin{bmatrix} \boldsymbol{w}_1 & \boldsymbol{w}_2 \end{bmatrix} \begin{bmatrix} \boldsymbol{B} \end{bmatrix} = \begin{bmatrix} \boldsymbol{v}_1 & \boldsymbol{v}_2 \end{bmatrix} \text{ is } \begin{bmatrix} 3 & 0 \\ 1 & 2 \end{bmatrix} \begin{bmatrix} \mathbf{1} & \mathbf{2} \\ \mathbf{1} & \mathbf{3} \end{bmatrix} = \begin{bmatrix} 3 & 6 \\ 3 & 8 \end{bmatrix}. \quad (1)$$

> When the input basis is in the columns of a matrix $\boldsymbol{V}$, and the output basis is in the columns of $\boldsymbol{W}$, the change of basis matrix for $T = I$ is $\boldsymbol{B} = \boldsymbol{W}^{-1}\boldsymbol{V}$.

The key I see a clear way to understand that rule $B = W^{-1}V$. Suppose the same vector $\boldsymbol{u}$ is written in the input basis of $\boldsymbol{v}$'s and the output basis of $\boldsymbol{w}$'s. I will do that three ways:

$$\begin{aligned} \boldsymbol{u} &= c_1\boldsymbol{v}_1 + \cdots + c_n\boldsymbol{v}_n \\ \boldsymbol{u} &= d_1\boldsymbol{w}_1 + \cdots + d_n\boldsymbol{w}_n \end{aligned} \quad \text{is} \quad \begin{bmatrix} \boldsymbol{v}_1 & \cdots & \boldsymbol{v}_n \end{bmatrix} \begin{bmatrix} c_1 \\ \vdots \\ c_n \end{bmatrix} = \begin{bmatrix} \boldsymbol{w}_1 & \cdots & \boldsymbol{w}_n \end{bmatrix} \begin{bmatrix} d_1 \\ \vdots \\ d_n \end{bmatrix} \quad \text{and } \boldsymbol{Vc} = \boldsymbol{Wd}.$$

$$\text{The coefficients } \boldsymbol{d} \text{ in the new basis of } \boldsymbol{w}\text{'s are } \boldsymbol{d} = \boldsymbol{W}^{-1}\boldsymbol{Vc}. \text{ Then } \boldsymbol{B} \text{ is } \boldsymbol{W}^{-1}\boldsymbol{V}. \quad (2)$$

This formula $\boldsymbol{B} = \boldsymbol{W}^{-1}\boldsymbol{V}$ produces one of the world's greatest mysteries: When the *standard basis* $\boldsymbol{V} = \boldsymbol{I}$ is changed to a different basis $\boldsymbol{W}$, **the change of basis matrix is not $\boldsymbol{W}$ but $\boldsymbol{B} = \boldsymbol{W}^{-1}$.** Larger basis vectors have smaller coefficients!

$$\begin{bmatrix} x \\ y \end{bmatrix} \text{ in the standard basis has coefficients } \begin{bmatrix} \boldsymbol{w}_1 & \boldsymbol{w}_2 \end{bmatrix}^{-1} \begin{bmatrix} x \\ y \end{bmatrix} \text{ in the } \boldsymbol{w}_1, \boldsymbol{w}_2 \text{ basis.}$$

Construction of the Matrix

Now we construct a matrix for any linear transformation. Suppose T transforms the space $\mathbf{V}$ (n-dimensional) to the space $\mathbf{W}$ (m-dimensional). We choose a basis $\boldsymbol{v}_1, \ldots, \boldsymbol{v}_n$ for $\mathbf{V}$ and we choose a basis $\boldsymbol{w}_1, \ldots, \boldsymbol{w}_m$ for $\mathbf{W}$. The matrix A will be m by n. To find the first column of A, apply T to the first basis vector $\boldsymbol{v}_1$. The output $T(\boldsymbol{v}_1)$ is in $\mathbf{W}$.

> $T(\boldsymbol{v}_1)$ ***is a combination*** $a_{11}\boldsymbol{w}_1 + \cdots + a_{m1}\boldsymbol{w}_m$ ***of the output basis for*** $\mathbf{W}$.

These numbers $a_{11}, \ldots, a_{m1}$ *go into the first column of* A. Transforming $\boldsymbol{v}_1$ to $T(\boldsymbol{v}_1)$ matches multiplying $(1, 0, \ldots, 0)$ by A. It yields that first column of the matrix. When T is the derivative and the first basis vector is 1, its derivative is $T(\boldsymbol{v}_1) = \mathbf{0}$. So for the derivative matrix below, the first column of A is all zero.

Example 3 The input basis of $\boldsymbol{v}$'s is $1, x, x^2, x^3$. The output basis of $\boldsymbol{w}$'s is $1, x, x^2$. Then T **takes the derivative**: $\boldsymbol{T(v) = \dfrac{dv}{dx}}$ and A = "derivative matrix".

If $\boldsymbol{v} = c_1 + c_2x + c_3x^2 + c_4x^3$
then $\dfrac{d\boldsymbol{v}}{dx} = \mathbf{1}c_2 + \mathbf{2}c_3x + \mathbf{3}c_4x^2$

$$A\boldsymbol{c} = \begin{bmatrix} 0 & \mathbf{1} & 0 & 0 \\ 0 & 0 & \mathbf{2} & 0 \\ 0 & 0 & 0 & \mathbf{3} \end{bmatrix} \begin{bmatrix} c_1 \\ c_2 \\ c_3 \\ c_4 \end{bmatrix} = \begin{bmatrix} c_2 \\ 2c_3 \\ 3c_4 \end{bmatrix}$$

> **Key rule:** The jth column of A is found by applying T to the jth basis vector $\boldsymbol{v}_j$
>
> $$T(\boldsymbol{v}_j) = \text{combination of output basis vectors} = a_{1j}\boldsymbol{w}_1 + \cdots + a_{mj}\boldsymbol{w}_m. \qquad (3)$$

These numbers a_{ij} go into A. *The matrix is constructed to get the basis vectors right.* ***Then linearity gets all other vectors right.*** Every $\boldsymbol{v}$ is a combination $c_1\boldsymbol{v}_1 + \cdots + c_n\boldsymbol{v}_n$, and $T(\boldsymbol{v})$ is a combination of the $\boldsymbol{w}$'s. When A multiplies the vector $\boldsymbol{c} = (c_1, \ldots, c_n)$ in the $\boldsymbol{v}$ combination, $A\boldsymbol{c}$ produces the coefficients in the $T(\boldsymbol{v})$ combination. This is because matrix multiplication (combining columns) is linear like T.

The matrix A tells us what T does. Every linear transformation from $\mathbf{V}$ to $\mathbf{W}$ can be converted to a matrix. This matrix depends on the bases.

Example 4 For the integral $T^+(\boldsymbol{v})$, the first basis function is again 1. Its integral is the second basis function x. So the first column of the "integral matrix" A^+ is $(0, 1, 0, 0)$.

The integral of $\boldsymbol{d_1 + d_2x + d_3x^2}$

is $\boldsymbol{d_1x + \frac{1}{2}d_2x^2 + \frac{1}{3}d_3x^3}$

$$A^+\boldsymbol{d} = \begin{bmatrix} 0 & 0 & 0 \\ \mathbf{1} & 0 & 0 \\ 0 & \frac{1}{2} & 0 \\ 0 & 0 & \frac{1}{3} \end{bmatrix} \begin{bmatrix} d_1 \\ d_2 \\ d_3 \end{bmatrix} = \begin{bmatrix} 0 \\ \boldsymbol{d_1} \\ \frac{1}{2}\boldsymbol{d_2} \\ \frac{1}{3}\boldsymbol{d_3} \end{bmatrix}$$

If you integrate a function and then differentiate, you get back to the start. So $AA^+ = I$. But if you differentiate before integrating, the constant term is lost. So A^+A is not I. ***The integral of the derivative of* 1 *is zero:***

$$T^+T(1) = \text{integral of zero function } = 0.$$

This matches A^+A, whose first column is all zero. The derivative T has a kernel (the constant functions). Its matrix A has a nullspace. Main idea again: $A\boldsymbol{v}$ copies $T(\boldsymbol{v})$.

The examples of the derivative and integral made three points. First, linear transformations T are everywhere—in calculus and differential equations and linear algebra. Second, spaces other than $\mathbf{R}^n$ are important—we had functions in $\mathbf{V}$ and $\mathbf{W}$. Third, **if we differentiate and then integrate, we can multiply their matrices A^+A.**

Matrix Products AB Match Transformations TS

We have come to something important—the real reason for the rule to multiply matrices. *At last we discover why*! Two linear transformations T and S are represented by two matrices A and B. Now compare TS with the multiplication AB:

When we apply the transformation T to the output from S, we get TS by this rule: $(TS)(\boldsymbol{u})$ *is defined to be* $\boldsymbol{T(S(u))}$. The output $S(\boldsymbol{u})$ becomes the input to T.

When we apply the matrix A to the output from B, we multiply AB by this rule: $(AB)(\boldsymbol{x})$ is defined to be $\boldsymbol{A(Bx)}$. The output $B\boldsymbol{x}$ becomes the input to A.
Matrix multiplication gives the correct matrix AB to represent TS.

The transformation S is from a space $\mathbf{U}$ to $\mathbf{V}$. Its matrix B uses a basis $\boldsymbol{u}_1, \ldots, \boldsymbol{u}_p$ for $\mathbf{U}$ and a basis $\boldsymbol{v}_1, \ldots, \boldsymbol{v}_n$ for $\mathbf{V}$. That matrix is n by p. The transformation T is from $\mathbf{V}$ to $\mathbf{W}$ as before. *Its matrix A must use the same basis $\boldsymbol{v}_1, \ldots, \boldsymbol{v}_n$ for* $\mathbf{V}$—this is the output space for S and the input space for T. ***Then the matrix AB matches TS.***

Multiplication The linear transformation TS starts with any vector $\boldsymbol{u}$ in $\mathbf{U}$, goes to $S(\boldsymbol{u})$ in $\mathbf{V}$ and then to $T(S(\boldsymbol{u}))$ in $\mathbf{W}$. The matrix AB starts with any $\boldsymbol{x}$ in $\mathbf{R}^p$, goes to $B\boldsymbol{x}$ in $\mathbf{R}^n$ and then to $AB\boldsymbol{x}$ in $\mathbf{R}^m$. **The matrix AB correctly represents TS:**

$$TS: \quad \mathbf{U} \to \mathbf{V} \to \mathbf{W} \qquad\qquad AB: \quad (m \text{ by } n)(n \text{ by } p) = (m \text{ by } p).$$

The input is $\boldsymbol{u} = x_1\boldsymbol{u}_1 + \cdots + x_p\boldsymbol{u}_p$. The output $T(S(\boldsymbol{u}))$ matches the output $AB\boldsymbol{x}$. ***Product of transformations TS matches product of matrices AB.***

The most important cases are when the spaces $\mathbf{U}, \mathbf{V}, \mathbf{W}$ are the same and their bases are the same. With $m = n = p$ we have square matrices that we can multiply.

Example 5 S rotates the plane by θ and T also rotates by θ. Then TS rotates by 2θ. This transformation T^2 corresponds to the rotation matrix A^2 through 2θ:

$$T = S \qquad A = B \qquad T^2 = \text{rotation by } 2\theta \qquad A^2 = \begin{bmatrix} \cos 2\theta & -\sin 2\theta \\ \sin 2\theta & \cos 2\theta \end{bmatrix}. \tag{4}$$

By matching (transformation)^2 with (matrix)^2, we pick up the formulas for $\cos 2\theta$ and $\sin 2\theta$. Multiply A times A:

$$\begin{bmatrix}\cos\theta & -\sin\theta\\ \sin\theta & \cos\theta\end{bmatrix}\begin{bmatrix}\cos\theta & -\sin\theta\\ \sin\theta & \cos\theta\end{bmatrix}=\begin{bmatrix}\cos^2\theta-\sin^2\theta & -2\sin\theta\cos\theta\\ 2\sin\theta\cos\theta & \cos^2\theta-\sin^2\theta\end{bmatrix}. \quad (5)$$

Comparing (4) with (5) produces $\cos 2\theta = \cos^2\theta - \sin^2\theta$ and $\sin 2\theta = 2\sin\theta\cos\theta$. Trigonometry (the double angle rule) comes from linear algebra.

Example 6 S rotates by the angle θ and T rotates by $-\theta$. Then $TS = I$ leads to $AB = I$.

In this case $T(S(\boldsymbol{u}))$ is $\boldsymbol{u}$. We rotate forward and back. For the matrices to match, $AB\boldsymbol{x}$ must be $\boldsymbol{x}$. *The two matrices are inverses.* Check this by putting $\cos(-\theta) = \cos\theta$ and $\sin(-\theta) = -\sin\theta$ into the backward rotation matrix A:

$$AB=\begin{bmatrix}\cos\theta & \sin\theta\\ -\sin\theta & \cos\theta\end{bmatrix}\begin{bmatrix}\cos\theta & -\sin\theta\\ \sin\theta & \cos\theta\end{bmatrix}=\begin{bmatrix}\cos^2\theta+\sin^2\theta & 0\\ 0 & \cos^2\theta+\sin^2\theta\end{bmatrix}=I.$$

Choosing the Best Bases

Now comes the final step in this section of the book. **Choose bases that diagonalize the matrix.** With the standard basis (the columns of I) our transformation T produces some matrix A—probably not diagonal. That same T is represented by different matrices when we choose different bases. The two great choices are eigenvectors and singular vectors:

> **Eigenvectors** If T transforms $\mathbf{R}^n$ to $\mathbf{R}^n$, its matrix A is square. But using the standard basis, that matrix A is probably not diagonal. If there are n independent eigenvectors, *choose those as the input and output basis.* In this good basis, **the matrix for T is the diagonal eigenvalue matrix Λ.**

Example 7 **The projection matrix** T projects every $\boldsymbol{v} = (x, y)$ in $\mathbf{R}^2$ onto the line $y = -x$. Using the standard basis, $\boldsymbol{v}_1 = (1, 0)$ projects to $T(\boldsymbol{v}_1) = \left(\frac{1}{2}, -\frac{1}{2}\right)$. For $\boldsymbol{v}_2 = (0, 1)$ the projection is $T(\boldsymbol{v}_2) = \left(-\frac{1}{2}, \frac{1}{2}\right)$. Those are the columns of A:

Projection matrix
Standard bases
Not diagonal

$$A=\begin{bmatrix}\frac{1}{2} & -\frac{1}{2}\\ -\frac{1}{2} & \frac{1}{2}\end{bmatrix} \quad \text{has } A^{\mathrm{T}} = A \text{ and } A^2 = A.$$

Now comes the main point of eigenvectors. Make them the basis vectors ! Diagonalize !

When the basis vectors are eigenvectors, the matrix becomes diagonal.

$$\begin{aligned}\boldsymbol{v}_1 = \boldsymbol{w}_1 = (1,-1) \text{ projects to itself} : T(\boldsymbol{v}_1) = \boldsymbol{v}_1 \text{ and } \lambda_1 = 1\\ \boldsymbol{v}_2 = \boldsymbol{w}_2 = (1,\ \ 1) \text{ projects to zero} : T(\boldsymbol{v}_2) = \mathbf{0} \text{ and } \lambda_2 = 0\end{aligned}$$

Eigenvector bases
Diagonal matrix

$$\text{The new matrix is } \begin{bmatrix}1 & 0\\ 0 & 0\end{bmatrix} = \begin{bmatrix}\lambda_1 & 0\\ 0 & \lambda_2\end{bmatrix} = \Lambda. \tag{6}$$

Eigenvectors are the perfect basis vectors. They produce the eigenvalue matrix Λ.

What about other choices of *input basis* $=$ *output basis*? Put those basis vectors into the columns of B. We saw above that the change of basis matrices (between standard basis and new basis) are $\boldsymbol{B}_{\text{in}} = \boldsymbol{B}$ and $\boldsymbol{B}_{\text{out}} = \boldsymbol{B}^{-1}$. The new matrix for T is **similar** to A:

$\boldsymbol{A}_{\text{new}} = \boldsymbol{B}^{-1}\boldsymbol{A}\boldsymbol{B}$ in the new basis of $\boldsymbol{b}$'s is similar to $\boldsymbol{A}$ in the standard basis:

$$\boldsymbol{A}_{\boldsymbol{b}\text{'s to }\boldsymbol{b}\text{'s}} = \boldsymbol{B}^{-1}{}_{\text{standard to }\boldsymbol{b}\text{'s}} \quad \boldsymbol{A}_{\text{standard}} \quad \boldsymbol{B}_{\boldsymbol{b}\text{'s to standard}} \tag{7}$$

I used the multiplication rule for the transformation ITI. *The matrices for* I, T, I *were* B^{-1}, A, B. The matrix B contains the input vectors $\boldsymbol{b}$ in the standard basis.

Finally we allow *different spaces* $\boldsymbol{V}$ *and* $\boldsymbol{W}$, *and different bases* $\boldsymbol{v}$'s and $\boldsymbol{w}$'s. When we know T and we choose bases, we get a matrix A. Probably A is not symmetric or even square. But we can always choose $\boldsymbol{v}$'s and $\boldsymbol{w}$'s that produce a diagonal matrix. This will be the *singular value matrix* $\boldsymbol{\Sigma} = \textbf{diag}\,(\boldsymbol{\sigma}_1, \ldots, \boldsymbol{\sigma}_r)$ in the decomposition $A = U\Sigma V^{\mathrm{T}}$.

> **Singular vectors** The SVD says that $U^{-1}AV = \Sigma$. The right singular vectors $\boldsymbol{v}_1, \ldots, \boldsymbol{v}_n$ will be the input basis. The left singular vectors $\boldsymbol{u}_1, \ldots, \boldsymbol{u}_m$ will be the output basis. By the rule for matrix multiplication, the matrix for the same transformation in these new bases is $B_{\text{out}}^{-1}AB_{\text{in}} = U^{-1}AV = \Sigma$.

I can't say that Σ is "similar" to A. We are working now with two bases, input and output. But those are *orthonormal bases* and they preserve the lengths of vectors. Following a good suggestion by David Vogan, I propose that we say: **$\boldsymbol{\Sigma}$ is "isometric" to $\boldsymbol{A}$.**

Definition $\quad C = Q_1^{-1}AQ_2$ *is isometric to* A *if* Q_1 *and* Q_2 *are orthogonal.*

Example 8 To construct the matrix A for the transformation $T = \frac{d}{dx}$, we chose the input basis $1, x, x^2, x^3$ and the output basis $1, x, x^2$. The matrix A was simple but unfortunately it wasn't diagonal. But we can take each basis *in the opposite order*.

Now the input basis is $x^3, x^2, x, 1$ and the output basis is $x^2, x, 1$. The change of basis matrices B_{in} and B_{out} are permutations. The matrix for $T(\boldsymbol{u}) = d\boldsymbol{u}/dx$ with the new bases is **the diagonal singular value matrix $\boldsymbol{B}_{\text{out}}^{-1}\boldsymbol{A}\boldsymbol{B}_{\text{in}} = \boldsymbol{\Sigma}$** with σ's $= 3, 2, 1$:

$$\boldsymbol{B}_{\text{out}}^{-1}\boldsymbol{A}\boldsymbol{B}_{\text{in}} = \begin{bmatrix} & & 1\\ & 1 & \\ 1 & & \end{bmatrix}\begin{bmatrix}0 & \mathbf{1} & 0 & 0\\ 0 & 0 & \mathbf{2} & 0\\ 0 & 0 & 0 & \mathbf{3}\end{bmatrix}\begin{bmatrix} & & & 1\\ & & 1 & \\ & 1 & & \\ 1 & & & \end{bmatrix} = \begin{bmatrix}\mathbf{3} & 0 & 0 & 0\\ 0 & \mathbf{2} & 0 & 0\\ 0 & 0 & \mathbf{1} & 0\end{bmatrix}. \tag{8}$$

Well, this was a tough section. We found that $x^3,\ x^2,\ x,\ 1$ have derivatives $3x^2,\ 2x,\ 1,\ 0$.

■ REVIEW OF THE KEY IDEAS ■

1. If we know $T(\boldsymbol{v}_1),\ldots,T(\boldsymbol{v}_n)$ for a basis, linearity will determine all other $T(\boldsymbol{v})$.

2. $\left\{\begin{array}{l}\text{Linear transformation } T\\ \text{Input basis } \boldsymbol{v}_1,\ldots,\boldsymbol{v}_n\\ \text{Output basis } \boldsymbol{w}_1,\ldots,\boldsymbol{w}_m\end{array}\right\} \rightarrow \begin{array}{c}\text{Matrix } A \ (m \text{ by } n)\\ \text{represents } T\\ \text{in these bases}\end{array}$

3. The change of basis matrix $B = W^{-1}V = B_{\text{out}}^{-1}B_{\text{in}}$ represents the identity $T(\boldsymbol{v}) = \boldsymbol{v}$.

4. If A and B represent T and S, and the output basis for S is the input basis for T, then the matrix AB represents the transformation $T(S(\boldsymbol{u}))$.

5. The best input-output bases are eigenvectors and/or singular vectors of A. Then

$$B^{-1}AB = \Lambda = \text{eigenvalues} \qquad B_{\text{out}}^{-1}AB_{\text{in}} = \Sigma = \text{singular values.}$$

■ WORKED EXAMPLES ■

8.2 A The space of 2 by 2 matrices has these four "vectors" as a basis:

$$\boldsymbol{v}_1 = \begin{bmatrix}1&0\\0&0\end{bmatrix} \qquad \boldsymbol{v}_2 = \begin{bmatrix}0&1\\0&0\end{bmatrix} \qquad \boldsymbol{v}_3 = \begin{bmatrix}0&0\\1&0\end{bmatrix} \qquad \boldsymbol{v}_4 = \begin{bmatrix}0&0\\0&1\end{bmatrix}.$$

T is the linear transformation that *transposes* every 2 by 2 matrix. What is the matrix A that represents T in this basis (output basis = input basis)? What is the inverse matrix A^{-1}? What is the transformation T^{-1} that inverts the transpose operation?

Solution Transposing those four "basis matrices" just reverses $\boldsymbol{v}_2$ and $\boldsymbol{v}_3$:

$$\begin{array}{l}T(\boldsymbol{v}_1) = \boldsymbol{v}_1\\ T(\boldsymbol{v}_2) = \boldsymbol{v}_3\\ T(\boldsymbol{v}_3) = \boldsymbol{v}_2\\ T(\boldsymbol{v}_4) = \boldsymbol{v}_4\end{array} \quad \text{gives the four columns of} \quad A = \begin{bmatrix}1&0&0&0\\0&0&1&0\\0&1&0&0\\0&0&0&1\end{bmatrix}$$

The inverse matrix A^{-1} is the same as A. The inverse transformation T^{-1} is the same as T. If we transpose and transpose again, the final matrix equals the original matrix.

Notice that the space of 2 by 2 matrices is 4-dimensional. So the matrix A (for the transpose T) is 4 by 4. The nullspace of A is $\boldsymbol{Z}$ and the kernel of T is the zero matrix—the only matrix that transposes to zero. The eigenvalues of A are $1, 1, 1, -1$.

Which line of matrices has $T(A) = A^{\mathrm{T}} = -A$ with that eigenvalue $\lambda = -1$?

Problem Set 8.2

Questions 1–4 extend the first derivative example to higher derivatives.

1 The transformation S takes the ***second derivative***. Keep $1, x, x^2, x^3$ as the input basis $\boldsymbol{v}_1, \boldsymbol{v}_2, \boldsymbol{v}_3, \boldsymbol{v}_4$ and also as output basis $\boldsymbol{w}_1, \boldsymbol{w}_2, \boldsymbol{w}_3, \boldsymbol{w}_4$. Write $S(\boldsymbol{v}_1), S(\boldsymbol{v}_2), S(\boldsymbol{v}_3), S(\boldsymbol{v}_4)$ in terms of the $\boldsymbol{w}$'s. Find the 4 by 4 matrix A_2 for S.

2 What functions have $S(\boldsymbol{v}) = \mathbf{0}$? They are in the kernel of the second derivative S. What vectors are in the nullspace of its matrix A_2 in Problem 1?

3 The second derivative A_2 is not the square of a rectangular first derivative matrix A_1:

$$A_1 = \begin{bmatrix} 0 & 1 & 0 & 0 \\ 0 & 0 & 2 & 0 \\ 0 & 0 & 0 & 3 \end{bmatrix} \quad \text{does not allow } A_1^2 = A_2.$$

Add a zero row 4 to A_1 so that output space = input space. Compare A_1^2 with A_2. Conclusion: We want output basis = _____ basis. Then $m = n$.

4 (a) The product TS of first and second derivatives produces the *third* derivative. Add zeros to make 4 by 4 matrices, then compute $A_1 A_2 = A_3$.

(b) The matrix A_2^2 corresponds to $S^2 =$ *fourth* derivative. Why is this zero?

Questions 5–9 are about a particular transformation T and its matrix A.

5 With bases $\boldsymbol{v}_1, \boldsymbol{v}_2, \boldsymbol{v}_3$ and $\boldsymbol{w}_1, \boldsymbol{w}_2, \boldsymbol{w}_3$, suppose $T(\boldsymbol{v}_1) = \boldsymbol{w}_2$ and $T(\boldsymbol{v}_2) = T(\boldsymbol{v}_3) = \boldsymbol{w}_1 + \boldsymbol{w}_3$. T is a linear transformation. Find the matrix A and multiply by the vector $(1, 1, 1)$. What is the output from T when the input is $\boldsymbol{v}_1 + \boldsymbol{v}_2 + \boldsymbol{v}_3$?

6 Since $T(\boldsymbol{v}_2) = T(\boldsymbol{v}_3)$, the solutions to $T(\boldsymbol{v}) = \mathbf{0}$ are $\boldsymbol{v} =$ _____. What vectors are in the nullspace of A? Find all solutions to $T(\boldsymbol{v}) = \boldsymbol{w}_2$.

7 Find a vector that is not in the column space of A. Find a combination of $\boldsymbol{w}$'s that is not in the range of the transformation T.

8 You don't have enough information to determine T^2. Why is its matrix not necessarily A^2? What more information do you need?

9 Find the *rank* of A. The rank is not the dimension of the whole output space $\mathbf{W}$. It is the dimension of the _____ of T.

Questions 10–13 are about invertible linear transformations.

10 Suppose $T(\boldsymbol{v}_1) = \boldsymbol{w}_1 + \boldsymbol{w}_2 + \boldsymbol{w}_3$ and $T(\boldsymbol{v}_2) = \boldsymbol{w}_2 + \boldsymbol{w}_3$ and $T(\boldsymbol{v}_3) = \boldsymbol{w}_3$. Find the matrix A for T using these basis vectors. What input vector $\boldsymbol{v}$ gives $T(\boldsymbol{v}) = \boldsymbol{w}_1$?

11 Invert the matrix A in Problem 10. Also invert the transformation T—what are $T^{-1}(\boldsymbol{w}_1)$ and $T^{-1}(\boldsymbol{w}_2)$ and $T^{-1}(\boldsymbol{w}_3)$?

12 Which of these are true and why is the other one ridiculous?

(a) $T^{-1}T = I$ (b) $T^{-1}(T(\boldsymbol{v}_1)) = \boldsymbol{v}_1$ (c) $T^{-1}(T(\boldsymbol{w}_1)) = \boldsymbol{w}_1$.

13 Suppose the spaces $\mathbf{V}$ and $\mathbf{W}$ have the same basis $\boldsymbol{v}_1, \boldsymbol{v}_2$.

(a) Describe a transformation T (not I) that is its own inverse.

(b) Describe a transformation T (not I) that equals T^2.

(c) Why can't the same T be used for both (a) and (b)?

Questions 14–19 are about changing the basis.

14 (a) What matrix B transforms $(1,0)$ into $(2,5)$ and transforms $(0,1)$ to $(1,3)$?

(b) What matrix C transforms $(2,5)$ to $(1,0)$ and $(1,3)$ to $(0,1)$?

(c) Why does no matrix transform $(2,6)$ to $(1,0)$ and $(1,3)$ to $(0,1)$?

15 (a) What matrix M transforms $(1,0)$ and $(0,1)$ to (r,t) and (s,u)?

(b) What matrix N transforms (a,c) and (b,d) to $(1,0)$ and $(0,1)$?

(c) What condition on a, b, c, d will make part (b) impossible?

16 (a) How do M and N in Problem 15 yield the matrix that transforms (a,c) to (r,t) and (b,d) to (s,u)?

(b) What matrix transforms $(2,5)$ to $(1,1)$ and $(1,3)$ to $(0,2)$?

17 If you keep the same basis vectors but put them in a different order, the change of basis matrix B is a ______ matrix. If you keep the basis vectors in order but change their lengths, B is a ______ matrix.

18 The matrix that rotates the axis vectors $(1,0)$ and $(0,1)$ through an angle θ is Q. What are the coordinates (a,b) of the original $(1,0)$ using the new (rotated) axes? This *inverse* can be tricky. Draw a figure or solve for a and b:

$$Q = \begin{bmatrix} \cos\theta & -\sin\theta \\ \sin\theta & \cos\theta \end{bmatrix} \qquad \begin{bmatrix} 1 \\ 0 \end{bmatrix} = a \begin{bmatrix} \cos\theta \\ \sin\theta \end{bmatrix} + b \begin{bmatrix} -\sin\theta \\ \cos\theta \end{bmatrix}.$$

19 The matrix that transforms $(1,0)$ and $(0,1)$ to $(1,4)$ and $(1,5)$ is $B =$ ______. The combination $a(1,4) + b(1,5)$ that equals $(1,0)$ has $(a,b) = (\ \ ,\ \ )$. How are those new coordinates of $(1,0)$ related to B or B^{-1}?

Questions 20–23 are about the space of quadratic polynomials $y = A + Bx + Cx^2$.

20 The parabola $\boldsymbol{w}_1 = \frac{1}{2}(x^2 + x)$ equals one at $x = 1$, and zero at $x = 0$ and $x = -1$. Find the parabolas $\boldsymbol{w}_2, \boldsymbol{w}_3$, and then find $\boldsymbol{y}(x)$ by linearity.

(a) $\boldsymbol{w}_2$ equals one at $x = 0$ and zero at $x = 1$ and $x = -1$.

(b) $\boldsymbol{w}_3$ equals one at $x = -1$ and zero at $x = 0$ and $x = 1$.

(c) $\boldsymbol{y}(x)$ equals 4 at $x = 1$ and 5 at $x = 0$ and 6 at $x = -1$. Use $\boldsymbol{w}_1, \boldsymbol{w}_2, \boldsymbol{w}_3$.

21 One basis for second-degree polynomials is $\boldsymbol{v}_1 = 1$ and $\boldsymbol{v}_2 = x$ and $\boldsymbol{v}_3 = x^2$. Another basis is $\boldsymbol{w}_1, \boldsymbol{w}_2, \boldsymbol{w}_3$ from Problem 20. Find two change of basis matrices, from the $\boldsymbol{w}$'s to the $\boldsymbol{v}$'s and from the $\boldsymbol{v}$'s to the $\boldsymbol{w}$'s.

22 What are the three equations for A, B, C if the parabola $y = A + Bx + Cx^2$ equals 4 at $x = a$ and 5 at $x = b$ and 6 at $x = c$? Find the determinant of the 3 by 3 matrix. That matrix transforms values like $4, 5, 6$ to parabolas y—or is it the other way?

23 Under what condition on the numbers $m_1, m_2, \ldots, m_9$ do these three parabolas give a basis for the space of all parabolas $a + bx + cx^2$?

$$\boldsymbol{v}_1 = m_1 + m_2x + m_3x^2, \quad \boldsymbol{v}_2 = m_4 + m_5x + m_6x^2, \quad \boldsymbol{v}_3 = m_7 + m_8x + m_9x^2.$$

24 The Gram-Schmidt process changes a basis $\boldsymbol{a}_1, \boldsymbol{a}_2, \boldsymbol{a}_3$ to an orthonormal basis $\boldsymbol{q}_1, \boldsymbol{q}_2, \boldsymbol{q}_3$. These are columns in $A = QR$. Show that R is the change of basis matrix from the $\boldsymbol{a}$'s to the $\boldsymbol{q}$'s ($\boldsymbol{a}_2$ is what combination of $\boldsymbol{q}$'s when $A = QR$?).

25 Elimination changes the rows of A to the rows of U with $A = LU$. Row 2 of A is what combination of the rows of U? Writing $A^{\mathrm{T}} = U^{\mathrm{T}}L^{\mathrm{T}}$ to work with columns, the change of basis matrix is $B = L^{\mathrm{T}}$. We have *bases* if the matrices are ______ .

26 Suppose $\boldsymbol{v}_1, \boldsymbol{v}_2, \boldsymbol{v}_3$ are **eigenvectors** for T. This means $T(\boldsymbol{v}_i) = \lambda_i \boldsymbol{v}_i$ for $i = 1, 2, 3$. What is the matrix for T when the input and output bases are the $\boldsymbol{v}$'s?

27 Every invertible linear transformation can have I as its matrix! Choose any input basis $\boldsymbol{v}_1, \ldots, \boldsymbol{v}_n$. For output basis choose $\boldsymbol{w}_i = T(\boldsymbol{v}_i)$. Why must T be invertible?

28 Using $\boldsymbol{v}_1 = \boldsymbol{w}_1$ and $\boldsymbol{v}_2 = \boldsymbol{w}_2$ find the standard matrix for these T's:

(a) $T(\boldsymbol{v}_1) = \boldsymbol{0}$ and $T(\boldsymbol{v}_2) = 3\boldsymbol{v}_1$ (b) $T(\boldsymbol{v}_1) = \boldsymbol{v}_1$ and $T(\boldsymbol{v}_1 + \boldsymbol{v}_2) = \boldsymbol{v}_1$.

29 Suppose T reflects the xy plane across the x axis and S is reflection across the y axis. If $\boldsymbol{v} = (x, y)$ what is $S(T(\boldsymbol{v}))$? Find a simpler description of the product ST.

30 Suppose T is reflection across the $45°$ line, and S is reflection across the y axis. If $\boldsymbol{v} = (2, 1)$ then $T(\boldsymbol{v}) = (1, 2)$. Find $S(T(\boldsymbol{v}))$ and $T(S(\boldsymbol{v}))$. Usually $ST \neq TS$.

31 **The product of two reflections is a rotation**. Multiply these reflection matrices to find the rotation angle:

$$\begin{bmatrix} \cos 2\theta & \sin 2\theta \\ \sin 2\theta & -\cos 2\theta \end{bmatrix} \begin{bmatrix} \cos 2\alpha & \sin 2\alpha \\ \sin 2\alpha & -\cos 2\alpha \end{bmatrix}.$$

32 Suppose A is a 3 by 4 matrix of rank $r = 2$, and $T(\boldsymbol{v}) = A\boldsymbol{v}$. Choose input basis vectors $\boldsymbol{v}_1, \boldsymbol{v}_2$ from the row space of A and $\boldsymbol{v}_3, \boldsymbol{v}_4$ from the nullspace. Choose output basis vectors $\boldsymbol{w}_1 = A\boldsymbol{v}_1$, $\boldsymbol{w}_2 = A\boldsymbol{v}_2$ in the column space and $\boldsymbol{w}_3$ from the nullspace of A^{T}. What specially simple matrix represents T in these special bases?

33 The space $\mathbf{M}$ of 2 by 2 matrices has the basis $\boldsymbol{v}_1, \boldsymbol{v}_2, \boldsymbol{v}_3, \boldsymbol{v}_4$ in Worked Example **8.2 A**. Suppose T multiplies each matrix by $\left[\begin{smallmatrix} a & b \\ c & d \end{smallmatrix}\right]$. With $\boldsymbol{w}$'s equal to $\boldsymbol{v}$'s, what 4 by 4 matrix A represents this transformation T on matrix space?

34 True or False: If we know $T(\boldsymbol{v})$ for n different nonzero vectors in $\mathbf{R}^n$, then we know $T(\boldsymbol{v})$ for every vector $\boldsymbol{v}$ in $\mathbf{R}^n$.

8.3 The Search for a Good Basis

1 With a new input basis B_{in} and output basis B_{out}, every matrix A becomes $\boldsymbol{B}_{\textbf{out}}^{-1}\boldsymbol{A}\boldsymbol{B}_{\textbf{in}}$.

2 $B_{\text{in}} = B_{\text{out}} =$ **"generalized eigenvectors of** $\boldsymbol{A}$**"** produces the **Jordan form** $J = B^{-1}AB$.

3 The **Fourier matrix** $\boldsymbol{F} = B_{\text{in}} = B_{\text{out}}$ diagonalizes every circulant matrix (use the **FFT**).

4 Sines and cosines, Legendre and Chebyshev: those are great bases for **function space**.

This is an important section of the book. I am afraid that most readers will skip it—or won't get this far. The first chapters prepared the way by explaining the idea of a **basis**. Chapter 6 introduced the eigenvectors $\boldsymbol{x}$ and Chapter 7 found singular vectors $\boldsymbol{v}$ and $\boldsymbol{u}$. Those are two winners but many other choices are very valuable.

First comes the pure algebra from Section 8.2 and then come good bases. The input basis vectors will be the columns of $B_{\textbf{in}}$. The output basis vectors will be the columns of $B_{\textbf{out}}$. Always $B_{\textbf{in}}$ and $B_{\textbf{out}}$ are *invertible*—basis vectors are independent!

Pure algebra If A is the matrix for a transformation T in the standard basis, then

$$\boldsymbol{B}_{\textbf{out}}^{-1}\boldsymbol{A}\boldsymbol{B}_{\textbf{in}} \text{ is the matrix in the new bases.} \tag{1}$$

The standard basis vectors are the *columns of the identity*: $B_{\textbf{in}} = I_{\mathbf{n}\times\mathbf{n}}$ and $B_{\textbf{out}} = I_{\mathbf{m}\times\mathbf{m}}$. Now we are choosing special bases to make the matrix clearer and simpler than A. When $B_{\text{in}} = B_{\text{out}} = B$, the square matrix $B^{-1}AB$ is *similar* to A.

Applied algebra Applications are all about choosing good bases. Here are four important choices for vectors and three choices for functions. Eigenvectors and singular vectors led to Λ and Σ in Section 8.2. The Jordan form is new.

1 $\boldsymbol{B}_{\textbf{in}} = \boldsymbol{B}_{\textbf{out}} =$ **eigenvector matrix** $\boldsymbol{X}$. Then $\boldsymbol{X}^{-1}\boldsymbol{A}\boldsymbol{X} =$ **eigenvalues in** $\boldsymbol{\Lambda}$.

This choice requires A to be a square matrix with n independent eigenvectors. "A must be diagonalizable." We get Λ when $B_{\textbf{in}} = B_{\textbf{out}}$ is the eigenvector matrix X.

2 $\boldsymbol{B}_{\textbf{in}} = \boldsymbol{V}$ **and** $\boldsymbol{B}_{\textbf{out}} = \boldsymbol{U}$: **singular vectors of** $\boldsymbol{A}$. Then $\boldsymbol{U}^{-1}\boldsymbol{A}\boldsymbol{V} =$ **diagonal** $\boldsymbol{\Sigma}$.

Σ is the singular value matrix (with $\sigma_1, \ldots, \sigma_r$ on its diagonal) when B_{in} and B_{out} are the singular vector matrices V and U. Recall that those columns of B_{in} and B_{out} are orthonormal eigenvectors of $A^{\mathrm{T}}A$ and AA^{T}. Then $A = U\Sigma V^{\mathrm{T}}$ gives $\Sigma = U^{-1}AV$.

3 $\boldsymbol{B}_{\textbf{in}} = \boldsymbol{B}_{\textbf{out}} =$ **generalized eigenvectors of** $\boldsymbol{A}$. Then $\boldsymbol{B}^{-1}\boldsymbol{A}\boldsymbol{B} =$ **Jordan form** $\boldsymbol{J}$.

A is a square matrix but it may only have s independent eigenvectors. (If $s = n$ then B is X and J is Λ.) In all cases Jordan constructed $n - s$ additional "generalized" eigenvectors, aiming to make the Jordan form J *as diagonal as possible*:

i) There are s square blocks along the diagonal of J.

ii) Each block has one eigenvalue λ, one eigenvector, and 1's above the diagonal.

The good case has n 1×1 blocks, each containing an eigenvalue. Then J is Λ (diagonal).

Example 1 This Jordan matrix J has eigenvalues $\lambda = 2, 2, 3, 3$ (two double eigenvalues). Those eigenvalues lie along the diagonal because J is triangular. There are two independent eigenvectors for $\lambda = 2$, but there is only *one line of eigenvectors for* $\lambda = 3$. This will be true for every matrix $C = BJB^{-1}$ that is similar to J.

$$\textbf{Jordan matrix} \quad J = \begin{bmatrix} 2 & & & \\ & 2 & & \\ & & 3 & 1 \\ & & 0 & 3 \end{bmatrix} \quad \begin{matrix} \text{Two 1 by 1 blocks} \\ \text{One 2 by 2 block} \\ \text{Three eigenvectors} \\ \text{Eigenvalues } 2, 2, 3, 3 \end{matrix}$$

Two eigenvectors for $\lambda = 2$ are $\boldsymbol{x}_1 = (1, 0, 0, 0)$ and $\boldsymbol{x}_2 = (0, 1, 0, 0)$. One eigenvector for $\lambda = 3$ is $\boldsymbol{x}_3 = (0, 0, 1, 0)$. The "generalized eigenvector" for this Jordan matrix is the fourth standard basis vector $\boldsymbol{x}_4 = (0, 0, 0, 1)$. The eigenvectors for J (normal and generalized) are just the columns $\boldsymbol{x}_1, \boldsymbol{x}_2, \boldsymbol{x}_3, \boldsymbol{x}_4$ of the identity matrix I.

Notice $(\boldsymbol{J} - \boldsymbol{3I})\boldsymbol{x}_4 = \boldsymbol{x}_3$. **The generalized eigenvector x_4 connects to the true eigenvector x_3.** A true $\boldsymbol{x}_4$ would have $(J - 3I)\boldsymbol{x}_4 = \mathbf{0}$, but that doesn't happen here.

Every matrix $C = BJB^{-1}$ that is similar to this J will have true eigenvectors $\boldsymbol{b}_1, \boldsymbol{b}_2, \boldsymbol{b}_3$ in the first three columns of B. The fourth column of B will be a generalized eigenvector $\boldsymbol{b}_4$ of C, tied to the true $\boldsymbol{b}_3$. Here is a quick proof that uses $B\boldsymbol{x}_3 = \boldsymbol{b}_3$ and $B\boldsymbol{x}_4 = \boldsymbol{b}_4$ to show: The fourth column $\boldsymbol{b}_4$ is tied to $\boldsymbol{b}_3$ by $(\boldsymbol{C} - \boldsymbol{3I})\boldsymbol{b}_4 = \boldsymbol{b}_3$.

$$(BJB^{-1} - 3I)\,\boldsymbol{b}_4 = BJ\,\boldsymbol{x}_4 - 3B\,\boldsymbol{x}_4 = B(J - 3I)\,\boldsymbol{x}_4 = B\,\boldsymbol{x}_3 = \boldsymbol{b}_3. \qquad (2)$$

The point of Jordan's theorem is that every square matrix A has a complete set of eigenvectors and generalized eigenvectors. When those go into the columns of B, the matrix $B^{-1}AB = J$ is in Jordan form. Based on Example 1, here is a description of J.

The Jordan Form

For every A, we want to choose B so that $B^{-1}AB$ is as ***nearly diagonal as possible***. When A has a full set of n eigenvectors, they go into the columns of B. Then $B = X$. The matrix $X^{-1}AX$ is diagonal, period. This is the Jordan form of A—when A can be diagonalized. In the general case, eigenvectors are missing and Λ can't be reached.

Suppose A has s independent eigenvectors. Then it is similar to a Jordan matrix with s blocks. Each block has an *eigenvalue on the diagonal with* 1's *just above it.* This block accounts for exactly one eigenvector of A. Then B contains generalized eigenvectors as well as ordinary eigenvectors.

When there are n eigenvectors, all n blocks will be 1 by 1. In that case $J = \Lambda$.

The Jordan form solves the differential equation $du/dt = Au$ for **any square matrix** $\boldsymbol{A} = \boldsymbol{BJB^{-1}}$. The solution $e^{At}\boldsymbol{u}(0)$ becomes $\boldsymbol{u}(t) = Be^{Jt}B^{-1}\boldsymbol{u}(0)$. J is triangular and its matrix exponential e^{Jt} involves $e^{\lambda t}$ times powers $1, t, \ldots, t^{s-1}$.

(Jordan form) If A has s independent eigenvectors, it is similar to a matrix J that has s Jordan blocks $J_1 \ldots, J_s$ on its diagonal. Some matrix B puts A into Jordan form:

Jordan form
$$B^{-1}AB = \begin{bmatrix} J_1 & & \\ & \ddots & \\ & & J_s \end{bmatrix} = J. \quad (3)$$

Each block J_i has one eigenvalue λ_i, one eigenvector, and 1's just above the diagonal:

Jordan block
$$J_i = \begin{bmatrix} \lambda_i & 1 & & \\ & \cdot & \cdot & \\ & & \cdot & 1 \\ & & & \lambda_i \end{bmatrix}. \quad (4)$$

Matrices are similar if they share the same Jordan form J—not otherwise.

The Jordan form J has an off-diagonal 1 for each missing eigenvector (and the 1's are next to the eigenvalues). In every family of similar matrices, we are picking one outstanding member called J. It is nearly diagonal (or if possible completely diagonal). We can quickly solve $d\boldsymbol{u}/dt = J\boldsymbol{u}$ and take powers J^k. Every other matrix in the family has the form BJB^{-1}.

Jordan's Theorem is proved in my textbook *Linear Algebra and Its Applications*. Please refer to that book (or more advanced books) for the proof. The reasoning is rather intricate and in actual computations the Jordan form is not at all popular—its calculation is not stable. A slight change in A will separate the repeated eigenvalues and remove the off-diagonal 1's—switching Jordan to a diagonal Λ.

Proved or not, you have caught the central idea of similarity—to make A as simple as possible while preserving its essential properties. The best basis B gives $B^{-1}AB = J$.

Question Find the eigenvalues and all possible Jordan forms if $\boldsymbol{A^2} =$ zero matrix.

Answer The eigenvalues must all be zero, because $A\boldsymbol{x} = \lambda\boldsymbol{x}$ leads to $A^2\boldsymbol{x} = \lambda^2\boldsymbol{x} = 0\boldsymbol{x}$. The Jordan form of A has $J^2 = 0$ because $J^2 = (B^{-1}AB)(B^{-1}AB) = B^{-1}A^2B = 0$. Every block in J has $\lambda = 0$ on the diagonal. Look at J_k^2 for block sizes $1, 2, 3$:

$$\begin{bmatrix} 0 \end{bmatrix}^2 = \begin{bmatrix} 0 \end{bmatrix} \qquad \begin{bmatrix} 0 & 1 \\ 0 & 0 \end{bmatrix}^2 = \begin{bmatrix} 0 & 0 \\ 0 & 0 \end{bmatrix} \qquad \begin{bmatrix} 0 & 1 & 0 \\ 0 & 0 & 1 \\ 0 & 0 & 0 \end{bmatrix}^2 = \begin{bmatrix} 0 & 0 & \mathbf{1} \\ 0 & 0 & 0 \\ 0 & 0 & 0 \end{bmatrix}$$

Conclusion: If $J^2 = 0$ then all block sizes must be 1 or 2. J^2 is not zero for 3 by 3.

The rank of J (and A) will be the total number of 1's. **The maximum rank is $\boldsymbol{n/2}$.** This happens when there are $n/2$ blocks, each of size 2 and rank 1.

Now come the great bases of applied mathematics. Their discrete forms are vectors in $\mathbf{R}^n$. Their continuous forms are functions in a function space. Since they are chosen once and for all, *without knowing the matrix* A, these bases $B_{\text{in}} = B_{\text{out}}$ probably don't diagonalize A. But for many important matrices A in applied mathematics, the matrices $B^{-1}AB$ are *close to diagonal.*

4 $B_{\text{in}} = B_{\text{out}} =$ **Fourier matrix** F **Then** $F\boldsymbol{x}$ **is a Discrete Fourier Transform of** $\boldsymbol{x}$.

Those words are telling us: The Fourier matrix with columns $(1, \lambda, \lambda^2, \lambda^3)$ in equation (6) is important. Those are good basis vectors to work with.

We ask: Which matrices are diagonalized by F? This time we are starting with the eigenvectors $(1, \lambda, \lambda^2, \lambda^3)$ and finding the matrices that have those eigenvectors:

$$\text{If } \boldsymbol{\lambda^4 = 1} \text{ then} \quad P\boldsymbol{x} = \begin{bmatrix} 0 & 1 & 0 & 0 \\ 0 & 0 & 1 & 0 \\ 0 & 0 & 0 & 1 \\ 1 & 0 & 0 & 0 \end{bmatrix} \begin{bmatrix} 1 \\ \lambda \\ \lambda^2 \\ \lambda^3 \end{bmatrix} = \lambda \begin{bmatrix} 1 \\ \lambda \\ \lambda^2 \\ \lambda^3 \end{bmatrix} = \lambda \boldsymbol{x}. \tag{5}$$

P is a permutation matrix. The equation $P\boldsymbol{x} = \lambda\boldsymbol{x}$ says that $\boldsymbol{x}$ is an eigenvector and λ is an eigenvalue of P. Notice how the fourth row of this vector equation is $1 = \lambda^4$. That rule for λ makes everything work.

Does this give four different eigenvalues λ? *Yes.* The four numbers $\lambda = \boldsymbol{1, i, -1, -i}$ all satisfy $\lambda^4 = 1$. (You know $i^2 = -1$. Squaring both sides gives $i^4 = 1$.) So those four numbers are the eigenvalues of P, each with its eigenvector $\boldsymbol{x} = (1, \lambda, \lambda^2, \lambda^3)$. **The eigenvector matrix F diagonalizes the permutation matrix P**:

$$\textbf{Eigenvalue matrix } \boldsymbol{\Lambda} \quad \begin{bmatrix} 1 & & & \\ & i & & \\ & & -1 & \\ & & & -i \end{bmatrix} \qquad \textbf{Eigenvector matrix is Fourier matrix } \boldsymbol{F} \quad \begin{bmatrix} 1 & 1 & 1 & 1 \\ 1 & i & -1 & -i \\ 1 & i^2 & 1 & (-i)^2 \\ 1 & i^3 & -1 & (-i)^3 \end{bmatrix} \tag{6}$$

Those columns of F are orthogonal because they are eigenvectors of P (an orthogonal matrix). Unfortunately this Fourier matrix F is complex (it is the most important complex matrix in the world). Multiplications $F\boldsymbol{x}$ are done millions of times very quickly, by the Fast Fourier Transform. The FFT comes in Section 9.3.

Key question: What other matrices beyond P have this same eigenvector matrix F? We know that P^2 and P^3 and P^4 have the same eigenvectors as P. The same matrix F diagonalizes all powers of P. And the eigenvalues of P^2 and P^3 and P^4 are the numbers λ^2 and λ^3 and λ^4. For example $P^2\boldsymbol{x} = \lambda^2\boldsymbol{x}$:

$$P^2\boldsymbol{x} = \begin{bmatrix} 0 & 0 & 1 & 0 \\ 0 & 0 & 0 & 1 \\ 1 & 0 & 0 & 0 \\ 0 & 1 & 0 & 0 \end{bmatrix} \begin{bmatrix} 1 \\ \lambda \\ \lambda^2 \\ \lambda^3 \end{bmatrix} = \lambda^2 \begin{bmatrix} 1 \\ \lambda \\ \lambda^2 \\ \lambda^3 \end{bmatrix} = \lambda^2 \boldsymbol{x} \text{ when } \boldsymbol{\lambda^4 = 1}.$$

The fourth power is special because $\boldsymbol{P^4 = I}$. When we do the "cyclic permutation" four times, $P^4\boldsymbol{x}$ is the same vector $\boldsymbol{x}$ that we started with. The eigenvalues of $P^4 = I$ are just $1, 1, 1, 1$. And that number 1 agrees with the fourth power of all the eigenvalues of P: $1^4 = 1$ and $i^4 = 1$ and $(-1)^4 = 1$ and $(-i)^4 = 1$.

One more step brings in many more matrices. If P and P^2 and P^3 and $P^4 = I$ have the same eigenvector matrix F, so does any combination $C = c_1P + c_2P^2 + c_3P^3 + c_0I$:

$$\textbf{Circulant matrix } \boldsymbol{C} = \begin{bmatrix} c_0 & \boldsymbol{c_1} & c_2 & c_3 \\ c_3 & c_0 & \boldsymbol{c_1} & c_2 \\ c_2 & c_3 & c_0 & \boldsymbol{c_1} \\ \boldsymbol{c_1} & c_2 & c_3 & c_0 \end{bmatrix} \quad \begin{array}{l} \text{has eigenvectors in the Fourier matrix } F \\ \text{has four eigenvalues } c_0 + c_1\lambda + c_2\lambda^2 + c_3\lambda^3 \\ \quad\text{from the four numbers } \lambda = 1, i, -1, -i \\ \text{The eigenvalue from } \lambda = 1 \text{ is } c_0 + c_1 + c_2 + c_3 \end{array}$$

That was a big step. We have found all the matrices (circulant matrices C) whose eigenvectors are the Fourier vectors in F. We also know the four eigenvalues of C, but we haven't given them a good formula or a name until now:

The four eigenvalues of C are given by the Fourier transform Fc

$$\boldsymbol{Fc} = \begin{bmatrix} 1 & 1 & 1 & 1 \\ 1 & i & -1 & -i \\ 1 & -1 & 1 & -1 \\ 1 & -i & -1 & i \end{bmatrix} \begin{bmatrix} c_0 \\ c_1 \\ c_2 \\ c_3 \end{bmatrix} = \begin{matrix} c_0 + c_1 + c_2 + c_3 \\ c_0 + ic_1 - c_2 - ic_3 \\ c_0 - c_1 + c_2 - c_3 \\ c_0 - ic_1 - c_2 + ic_3 \end{matrix}$$

Example 2 The same ideas work for a Fourier matrix F and a circulant matrix C of any size. Two by two matrices look trivial but they are very useful. Now eigenvalues of P have $\lambda^2 = 1$ instead of $\lambda^4 = 1$ and the complex number i is not needed: $\boldsymbol{\lambda = \pm 1}$.

Fourier matrix F from eigenvectors of P and C $\quad F = \begin{bmatrix} \mathbf{1} & \mathbf{1} \\ \mathbf{1} & \mathbf{-1} \end{bmatrix} \quad P = \begin{bmatrix} 0 & 1 \\ 1 & 0 \end{bmatrix}$ Circulant $c_0I + c_1P$ $\quad C = \begin{bmatrix} \boldsymbol{c_0} & \boldsymbol{c_1} \\ \boldsymbol{c_1} & \boldsymbol{c_0} \end{bmatrix}.$

The eigenvalues of C are $c_0 + c_1$ and $c_0 - c_1$. Those are given by the Fourier transform $F\boldsymbol{c}$ when the vector $\boldsymbol{c}$ is (c_0, c_1). This transform $F\boldsymbol{c}$ gives the eigenvalues of C for any size n.

Notice that **circulant matrices have constant diagonals**. The same number c_0 goes down the main diagonal. The number c_1 is on the diagonal above, and that diagonal "wraps around" or "circles around" to the southwest corner of C. This explains the name *circulant* and it indicates that these matrices are *periodic* or *cyclic*. Even the powers of λ cycle around because $\lambda^4 = 1$ leads to $\lambda^5, \lambda^6, \lambda^7, \lambda^8 = \lambda, \lambda^2, \lambda^3, \lambda^4$.

Constancy down the diagonals is a crucial property of C. It corresponds to *constant coefficients* in a differential equation. This is exactly when Fourier works perfectly!

$$\text{The equation} \quad \frac{d^2u}{dt^2} = -u \quad \text{is solved by} \quad u = c_0 \cos t + c_1 \sin t.$$

$$\text{The equation} \quad \frac{d^2u}{dt^2} = tu \quad \text{cannot be solved by elementary functions.}$$

These equations are linear. The first is the oscillation equation for a simple spring. It is Newton's Law $f = ma$ with mass $m = 1$, $a = d^2u/dt^2$, and force $f = -u$. Constant coefficients produce the differential equations that you can really solve.

The equation $u'' = tu$ has a variable coefficient t. This is Airy's equation in physics and optics (it was derived to explain a rainbow). The solutions change completely when t passes through zero, and those solutions require infinite series. *We won't go there.*

The point is that equations with constant coefficients have simple solutions like $e^{\lambda t}$. You discover λ by substituting $e^{\lambda t}$ into the differential equation. That number λ is like an eigenvalue. For $u = \cos t$ and $u = \sin t$ the number is $\lambda = i$. Euler's great formula $e^{it} = \cos t + i \sin t$ introduces complex numbers as we saw in the eigenvalues of P and C.

Bases for Function Space

For functions of x, the first basis I would think of contains the powers $1, x, x^2, x^3, \ldots$ Unfortunately this is a terrible basis. Those functions x^n are just barely independent. x^{10} is *almost* a combination of other basis vectors $1, x, \ldots, x^9$. It is virtually impossible to compute with this poor "ill-conditioned" basis.

If we had vectors instead of functions, the test for a good basis would look at $B^{\mathrm{T}}B$. This matrix contains all inner products between the basis vectors (columns of B). *The basis is orthonormal when* $B^{\mathrm{T}}B = I$. That is best possible. But the basis $1, x, x^2, \ldots$ produces the evil **Hilbert matrix**: $B^{\mathrm{T}}B$ has an enormous ratio between its largest and smallest eigenvalues. A large condition number signals an unhappy choice of basis.

Note Now the columns of B are functions instead of vectors. We still use $B^{\mathrm{T}}B$ to test for independence. So we need to know the dot product (inner product is a better name) of two functions—those are the numbers in $B^{\mathrm{T}}B$.

The dot product of vectors is just $\boldsymbol{x}^{\mathrm{T}}\boldsymbol{y} = x_1y_1 + \cdots + x_ny_n$. The inner product of functions will integrate instead of adding, but the idea is completely parallel:

$$\text{Inner product } (\boldsymbol{f}, \boldsymbol{g}) = \int f(x)g(x)\,dx$$

$$\text{Complex inner product } (\boldsymbol{f}, \boldsymbol{g}) = \int \overline{f(x)}\,g(x)\,dx, \quad \overline{f} = \text{ complex conjugate}$$

$$\text{Weighted inner product } (\boldsymbol{f}, \boldsymbol{g})_w = \int w(x)\,\overline{f(x)}\,g(x)\,dx, \quad w = \text{ weight function}$$

When the integrals go from $x = 0$ to $x = 1$, the inner product of x^i with x^j is

$$\int_0^1 x^i\,x^j\,dx = \left.\frac{x^{i+j+1}}{i+j+1}\right]_{x=0}^{x=1} = \frac{1}{i+j+1} = \text{entries of Hilbert matrix } B^{\mathrm{T}}B$$

By changing to the symmetric interval from $x = -1$ to $x = 1$, we immediately have *orthogonality between all even functions and all odd functions*:

$$\textbf{Interval } [-1,\ 1] \qquad \int_{-1}^{1} x^2\,x^5\,dx = 0 \qquad \int_{-1}^{1} \textbf{even}(x)\,\textbf{odd}(x)\,dx = 0.$$

This change makes half of the basis functions orthogonal to the other half. It is so simple that we continue using the symmetric interval -1 to 1 (or $-\pi$ to π). But we want a better basis than the powers x^n—hopefully an orthogonal basis.

Orthogonal Bases for Function Space

Here are the three leading even-odd bases for theoretical and numerical computations:

5. The **Fourier basis**	$1, \sin x,\ \cos x,\ \sin 2x,\ \cos 2x, \dots$
6. The **Legendre basis**	$1,\ x,\ x^2 - \frac{1}{3},\ x^3 - \frac{3}{5}x, \dots$
7. The **Chebyshev basis**	$1,\ x,\ 2x^2 - 1,\ 4x^3 - 3x, \dots$

The Fourier basis functions (sines and cosines) are all *periodic*. They repeat over every 2π interval because $\cos(x+2\pi) = \cos x$ and $\sin(x+2\pi) = \sin x$. So this basis is especially good for functions $f(x)$ that are themselves periodic: $f(x + 2\pi) = f(x)$.

This basis is also *orthogonal*. Every sine and cosine is orthogonal to every other sine and cosine. Of course we don't expect the basis function $\cos nx$ to be orthogonal to itself.

Most important, the sine-cosine basis is also *excellent for approximation*. If we have a smooth periodic function $f(x)$, then a few sines and cosines (low frequencies) are all we need. Jumps in $f(x)$ and noise in the signal are seen in higher frequencies (larger n). We hope and expect that the signal is not drowned by the noise.

The *Fourier transform* connects $f(x)$ to the coefficients a_k and b_k in its Fourier series:

Fourier series $\quad f(x) = a_0 + b_1 \sin x + a_1 \cos x + b_2 \sin 2x + a_2 \cos 2x + \cdots$

We see that **function space is infinite-dimensional**. It takes infinitely many basis functions to capture perfectly a typical $f(x)$. But the formula for each coefficient (for example a_3) is just like the formula $\boldsymbol{b}^{\mathrm{T}}\boldsymbol{a}/\boldsymbol{a}^{\mathrm{T}}\boldsymbol{a}$ for projecting a vector $\boldsymbol{b}$ onto the line through $\boldsymbol{a}$.

Here we are projecting the function $f(x)$ onto the line in function space through $\cos 3x$:

$$\textbf{Fourier coefficient} \quad a_3 = \frac{(f(x), \cos 3x)}{(\cos 3x, \cos 3x)} = \frac{\int f(x) \cos 3x\, dx}{\int \cos 3x \cos 3x\, dx}. \qquad (7)$$

Example 3 The double angle formula in trigonometry is $\cos 2x = 2\cos^2 x - 1$. This tells us that $\cos^2 x = \frac{1}{2} + \frac{1}{2} \cos 2x$. A very short Fourier series. So is $\sin^2 x = \frac{1}{2} - \frac{1}{2} \cos 2x$.

Fourier series is just linear algebra in function space. Let me explain that properly as a highlight of Chapter 10 about applications.

Legendre Polynomials and Chebyshev Polynomials

The Legendre polynomials are the result of applying the Gram-Schmidt idea (Section 4.4). The plan is to orthogonalize the powers $1, x, x^2, \ldots$ To start, the odd function $\boldsymbol{x}$ is already orthogonal to the even function 1 over the interval from -1 to 1. Their product $(x)(1) = x$ integrates to zero. But the inner product between x^2 and 1 is $\int x^2\,dx = 2/3$:

$$\frac{(x^2,1)}{(1,1)} = \frac{\int x^2\,dx}{\int 1\,dx} = \frac{2/3}{2} = \frac{1}{3} \qquad \text{Gram-Schmidt gives } \boldsymbol{x^2 - \frac{1}{3}} = \textbf{Legendre}$$

Similarly the odd power x^3 has a component $3x/5$ in the direction of the odd function x:

$$\frac{(x^3,x)}{(x,x)} = \frac{\int x^4\,dx}{\int x^2\,dx} = \frac{2/5}{2/3} = \frac{3}{5} \qquad \text{Gram-Schmidt gives } \boldsymbol{x^3 - \frac{3}{5}x} = \textbf{Legendre}$$

Continuing Gram-Schmidt for $x^4, x^5, \ldots$ produces every Legendre function—a good basis.

Finally we turn to the Chebyshev polynomials $1, x, 2x^2 - 1, 4x^3 - 3x$. They don't come from Gram-Schmidt. Instead they are connected to $1, \cos\theta, \cos 2\theta, \cos 3\theta$. This gives a giant computational advantage—we can use the Fast Fourier Transform. The connection of Chebyshev to Fourier appears when we set $\boldsymbol{x = \cos\theta}$:

Chebyshev to Fourier	$\boldsymbol{2x^2 - 1} = 2(\cos\theta)^2 - 1 = \boldsymbol{\cos 2\theta}$ $\boldsymbol{4x^3 - 3x} = 4(\cos\theta)^3 - 3(\cos\theta) = \boldsymbol{\cos 3\theta}$

The nth degree Chebyshev polynomial $\boldsymbol{T_n(x)}$ converts to Fourier's $\boldsymbol{\cos\, n\theta = T_n(\cos\theta)}$.

Note These polynomials are the basis for a big software project called **"chebfun"**. Every function $f(x)$ is replaced by a super-accurate Chebyshev approximation. Then you can integrate $f(x)$, and solve $f(x) = 0$, and find its maximum or minimum. More than that, you can solve differential equations involving $f(x)$—fast and to high accuracy.

When **chebfun** replaces $f(x)$ by a polynomial, you are ready to solve problems.

■ REVIEW OF THE KEY IDEAS ■

1. A basis is good if its matrix B is well-conditioned. Orthogonal bases are best.
2. Also good if $\Lambda = B^{-1}AB$ is diagonal. But the Jordan form J can be very unstable.
3. The Fourier matrix diagonalizes constant-coefficient periodic equations: perfection.
4. The basis $1, x, x^2, \ldots$ leads to $B^{\mathrm{T}}B =$ Hillbert matrix: Terrible for computations.
5. Legendre and Chebyshev polynomials are excellent bases for function space.

Problem Set 8.3

1 In Example 1, what is the rank of $J - 3I$? What is the dimension of its nullspace? This dimension gives the number of independent eigenvectors for $\lambda = 3$.

The algebraic multiplicity is 2, because $\det(J - \lambda I)$ has the repeated factor $(\lambda - 3)^2$. The geometric multiplicity is 1, because there is only 1 independent eigenvector.

2 These matrices A_1 and A_2 are similar to J. Solve $A_1 B_1 = B_1 J$ and $A_2 B_2 = B_2 J$ to find the basis matrices B_1 and B_2 with $J = B_1^{-1} A_1 B_1$ and $J = B_2^{-1} A_2 B_2$.

$$J = \begin{bmatrix} 0 & 1 \\ 0 & 0 \end{bmatrix} \quad A_1 = \begin{bmatrix} 0 & 4 \\ 0 & 0 \end{bmatrix} \quad A_2 = \begin{bmatrix} 4 & -8 \\ 2 & -4 \end{bmatrix}$$

3 This transpose block J^{T} has the same triple eigenvalue 2 (with only one eigenvector) as J. Find the basis change B so that $J = B^{-1} J^{\mathrm{T}} B$ (which means $BJ = J^{\mathrm{T}} B$):

$$J = \begin{bmatrix} 2 & 1 & 0 \\ 0 & 2 & 1 \\ 0 & 0 & 2 \end{bmatrix} \qquad J^{\mathrm{T}} = \begin{bmatrix} 2 & 0 & 0 \\ 1 & 2 & 0 \\ 0 & 1 & 2 \end{bmatrix}$$

4 J and K are Jordan forms with the same zero eigenvalues and the same rank 2. But show that no invertible B solves $BK = JB$, so K *is not similar to* J:

$$J = \begin{bmatrix} 0 & 1 & 0 & 0 \\ & 0 & 0 & 0 \\ & & 0 & 1 \\ & & & 0 \end{bmatrix} \qquad K = \begin{bmatrix} 0 & 1 & 0 & 0 \\ & 0 & 1 & 0 \\ & & 0 & 0 \\ & & & 0 \end{bmatrix}$$

5 If $A^3 = 0$ show that all $\lambda = 0$, and all Jordan blocks with $J^3 = 0$ have size $1, 2,$ or 3. It follows that rank $(A) \leq 2n/3$. If $A^n = 0$ why is rank $(A) < n$?

6 Show that $\boldsymbol{u}(t) = \begin{bmatrix} te^{\lambda t} \\ e^{\lambda t} \end{bmatrix}$ solves $\dfrac{d\boldsymbol{u}}{dt} = J\boldsymbol{u}$ with $J = \begin{bmatrix} \lambda & 1 \\ 0 & \lambda \end{bmatrix}$ and $\boldsymbol{u}(0) = \begin{bmatrix} 0 \\ 1 \end{bmatrix}$.

J is not diagonalizable so $te^{\lambda t}$ enters the solution.

7 Show that the difference equation $v_{k+2} - 2\lambda v_{k+1} + \lambda^2 v_k = 0$ is solved by $v_k = \lambda^k$ and also by $v_k = k\lambda^k$. Those correspond to $e^{\lambda t}$ and $te^{\lambda t}$ in Problem 6.

8 What are the 3 solutions to $\lambda^3 = 1$? They are complex numbers $\lambda = \cos\theta + i\sin\theta = e^{i\theta}$. Then $\lambda^3 = e^{3i\theta} = 1$ when the angle 3θ is 0 or 2π or 4π. Write the 3 by 3 Fourier matrix F with columns $(1, \lambda, \lambda^2)$.

9 Check that any 3 by 3 circulant C has eigenvectors $(1, \lambda, \lambda^2)$ from Problem 8. If the diagonals of your matrix C contain c_0, c_1, c_2 then its eigenvalues are in $F\boldsymbol{c}$.

10 Using formula (7) find $a_3 \cos 3x$ in the Fourier series of $f(x) = \begin{cases} 1 \text{ for } -L \leq x \leq L \\ 0 \text{ for } \quad L \leq |x| \leq 2\pi \end{cases}$

Chapter 9

Complex Vectors and Matrices

Real versus Complex

R = line of all real numbers $-\infty < x < \infty$ $\leftrightarrow$ **C** = plane of all complex numbers $z = x + iy$

$|x|$ = absolute value of x $\leftrightarrow$ $|z| = \sqrt{x^2 + y^2} = r$ = absolute value (or modulus) of z

1 and -1 solve $x^2 = 1$ $\leftrightarrow$ $z = 1, w, \ldots, w^{n-1}$ solve $z^n = 1$ where $w = e^{2\pi i/n}$

The **complex conjugate** of $z = x + iy$ is $\overline{z} = x - iy$. $|z|^2 = x^2 + y^2 = z\overline{z}$ and $\dfrac{1}{z} = \dfrac{\overline{z}}{|z|^2}$.

The **polar form** of $z = x + iy$ is $|z|e^{i\theta} = re^{i\theta} = r\cos\theta + ir\sin\theta$. The angle has $\tan\theta = \dfrac{y}{x}$.

$\mathbf{R}^n$: vectors with n real components	$\leftrightarrow$	$\mathbf{C}^n$: vectors with n complex components
length: $\|\boldsymbol{x}\|^2 = x_1^2 + \cdots + x_n^2$	$\leftrightarrow$	length: $\|\boldsymbol{z}\|^2 = |z_1|^2 + \cdots + |z_n|^2$
transpose: $(A^{\mathrm{T}})_{ij} = A_{ji}$	$\leftrightarrow$	conjugate transpose: $(A^{\mathrm{H}})_{ij} = \overline{A_{ji}}$
dot product: $\boldsymbol{x}^{\mathrm{T}}\boldsymbol{y} = x_1y_1 + \cdots + x_ny_n$	$\leftrightarrow$	inner product: $\boldsymbol{u}^{\mathrm{H}}\boldsymbol{v} = \overline{u}_1v_1 + \cdots + \overline{u}_nv_n$
reason for A^{T}: $(A\boldsymbol{x})^{\mathrm{T}}\boldsymbol{y} = \boldsymbol{x}^{\mathrm{T}}(A^{\mathrm{T}}\boldsymbol{y})$	$\leftrightarrow$	reason for A^{H}: $(A\boldsymbol{u})^{\mathrm{H}}\boldsymbol{v} = \boldsymbol{u}^{\mathrm{H}}(A^{\mathrm{H}}\boldsymbol{v})$
orthogonality: $\boldsymbol{x}^{\mathrm{T}}\boldsymbol{y} = 0$	$\leftrightarrow$	orthogonality: $\boldsymbol{u}^{\mathrm{H}}\boldsymbol{v} = 0$
symmetric matrices: $S = S^{\mathrm{T}}$	$\leftrightarrow$	Hermitian matrices: $S = S^{\mathrm{H}}$
$\boldsymbol{S = Q\Lambda Q^{-1} = Q\Lambda Q^{\mathrm{T}}}$ (real Λ)	$\leftrightarrow$	$\boldsymbol{S = U\Lambda U^{-1} = U\Lambda U^{\mathrm{H}}}$ (real Λ)
skew-symmetric matrices: $K^{\mathrm{T}} = -K$	$\leftrightarrow$	skew-Hermitian matrices $K^{\mathrm{H}} = -K$
orthogonal matrices: $Q^{\mathrm{T}} = Q^{-1}$	$\leftrightarrow$	unitary matrices: $U^{\mathrm{H}} = U^{-1}$
orthonormal columns: $Q^{\mathrm{T}}Q = I$	$\leftrightarrow$	orthonormal columns: $U^{\mathrm{H}}U = I$
$(Q\boldsymbol{x})^{\mathrm{T}}(Q\boldsymbol{y}) = \boldsymbol{x}^{\mathrm{T}}\boldsymbol{y}$ and $\|Q\boldsymbol{x}\| = \|\boldsymbol{x}\|$	$\leftrightarrow$	$(U\boldsymbol{x})^{\mathrm{H}}(U\boldsymbol{y}) = \boldsymbol{x}^{\mathrm{H}}\boldsymbol{y}$ and $\|U\boldsymbol{z}\| = \|\boldsymbol{z}\|$

A complete presentation of linear algebra must include complex numbers $z = x + iy$. Even when the matrix is real, ***the eigenvalues and eigenvectors are often complex***. Example: A 2 by 2 rotation matrix has complex eigenvectors $\boldsymbol{x} = (1, i)$ and $\overline{\boldsymbol{x}} = (1, -i)$. I will summarize Sections 9.1 and 9.2 in these few unforgettable words: When you transpose a vector $\boldsymbol{v}$ or a matrix A, *take the conjugate of every entry* ($\boldsymbol{i}$ **changes to** $-\boldsymbol{i}$). Section 9.3 is about the most important complex matrix of all—*the Fourier matrix* F.

9.1 Complex Numbers

Start with the imaginary number i. Everybody knows that $x^2 = -1$ has no real solution. When you square a real number, the answer is never negative. So the world has agreed on a solution called i. (Except that electrical engineers call it j.) Imaginary numbers follow the normal rules of addition and multiplication, with one difference. ***Replace*** i^2 ***by*** -1.

This section gives the main facts about complex numbers. It is a review for some students and a reference for everyone. Everything comes from $i^2 = -1$ and $e^{2\pi i} = 1$.

> ***A complex number*** (say $3+2i$) ***is a real number*** (3) ***plus an imaginary number*** $(2i)$. Addition keeps the real and imaginary parts separate. Multiplication uses $i^2 = -1$:
>
> **Add:** $(3+2i)+(3+2i) = 6+4i$
>
> **Multiply:** $(3+2i)(1-i) = 3+2i-3i-2i^2 = 5-i.$

If I add $3+i$ to $1-i$, the answer is 4. The real numbers $3+1$ stay separate from the imaginary numbers $i-i$. We are adding the vectors $(3,1)$ and $(1,-1)$ to get $(4,0)$.

The number $(1+i)^2$ is $1+i$ times $1+i$. The rules give the surprising answer $2i$:

$$(1+i)(1+i) = 1+i+i+i^2 = 2i.$$

In the complex plane, $1+i$ is at an angle of 45°. It is like the vector $(1,1)$. When we square $1+i$ to get $2i$, the angle doubles to 90°. If we square again, the answer is $(2i)^2 = -4$. The 90° angle doubled to 180°, the direction of a negative real number.

A real number is just a complex number $z = a+bi$, with zero imaginary part: $b = 0$.

The ***real part*** is $a = \text{Re}\,(a+bi)$. The ***imaginary part*** is $b = \text{Im}\,(a+bi)$.

The Complex Plane

Complex numbers correspond to points in a plane. Real numbers go along the x axis. Pure imaginary numbers are on the y axis. ***The complex number*** $\mathbf{3+2}\boldsymbol{i}$ ***is at the point with coordinates*** $(\mathbf{3},\mathbf{2})$. The number zero, which is $0+0i$, is at the origin.

Adding and subtracting complex numbers is like adding and subtracting vectors in the plane. The real component stays separate from the imaginary component. The vectors go head-to-tail as usual. The complex plane $\mathbf{C}^1$ is like the ordinary two-dimensional plane $\mathbf{R}^2$, except that we multiply complex numbers and we didn't multiply vectors.

Now comes an important idea. ***The complex conjugate of*** $\mathbf{3+2}\boldsymbol{i}$ ***is*** $\mathbf{3-2}\boldsymbol{i}$. The complex conjugate of $z = 1-i$ is $\overline{z} = 1+i$. In general the conjugate of $z = a+bi$ is $\overline{z} = a-bi$. (**Some writers use a "*bar*" on the number and others use a "*star*":** $\overline{\boldsymbol{z}} = \boldsymbol{z}^*$.) The imaginary parts of z and "z bar" have opposite signs. In the complex plane, $\overline{z}$ is the image of z on the other side of the real axis.

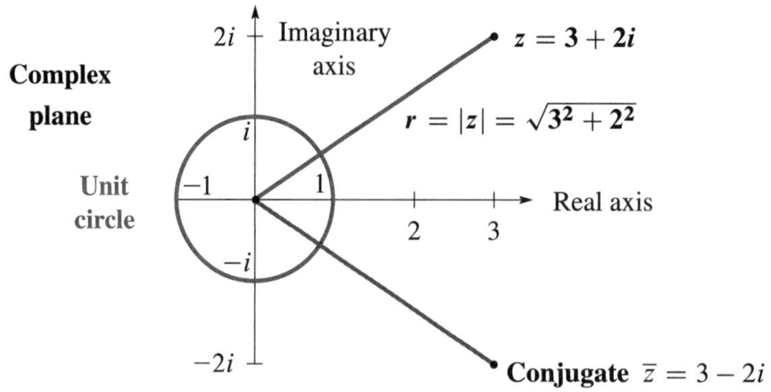

Figure 9.1: The number $z = a + bi$ corresponds to the point (a, b) and the vector $\begin{bmatrix} a \\ b \end{bmatrix}$.

Two useful facts. ***When we multiply conjugates $\overline{z}_1$ and $\overline{z}_2$, we get the conjugate of $z_1 z_2$.*** And when we add $\overline{z}_1$ and $\overline{z}_2$, we get the conjugate of $z_1 + z_2$:

$$\overline{z}_1 + \overline{z}_2 = (3 - 2i) + (1 + i) = \mathbf{4 - i}. \text{ This is the conjugate of } z_1 + z_2 = \mathbf{4 + i}.$$
$$\overline{z}_1 \times \overline{z}_2 = (3 - 2i) \times (1 + i) = \mathbf{5 + i}. \text{ This is the conjugate of } z_1 \times z_2 = \mathbf{5 - i}.$$

Adding and multiplying is exactly what linear algebra needs. By taking conjugates of $A\boldsymbol{x} = \lambda\boldsymbol{x}$, when A is real, we have another eigenvalue $\overline{\lambda}$ and its eigenvector $\overline{\boldsymbol{x}}$:

$$\textit{Eigenvalues } \boldsymbol{\lambda} \textit{ and } \overline{\boldsymbol{\lambda}} \qquad \textit{If } Ax = \boldsymbol{\lambda x} \textit{ and } A \textit{ is real then } \boldsymbol{A\overline{x}} = \overline{\boldsymbol{\lambda}}\overline{\boldsymbol{x}}. \tag{1}$$

Something special happens when $z = 3 + 2i$ combines with its own complex conjugate $\overline{z} = 3 - 2i$. The result from adding $z + \overline{z}$ or multiplying $z\overline{z}$ is always real:

$$\begin{aligned} \boldsymbol{z + \overline{z} = \text{real}} \qquad & (3 + 2i) + (3 - 2i) = 6 \quad \text{(real)} \\ \boldsymbol{z\overline{z} = \text{real}} \qquad & (3 + 2i) \times (3 - 2i) = 9 + 6i - 6i - 4i^2 = 13 \quad \text{(real)}. \end{aligned}$$

The sum of $z = a + bi$ and its conjugate $\overline{z} = a - bi$ is the real number $2a$. The product of z times $\overline{z}$ is the real number $a^2 + b^2$:

$$\textbf{Multiply } \boldsymbol{z} \textbf{ times } \boldsymbol{\overline{z}} \textbf{ to get } \boldsymbol{|z|^2 = r^2} \qquad (a + bi)(a - bi) = a^2 + b^2. \tag{2}$$

The next step with complex numbers is $1/z$. How to divide by $a + ib$? The best idea is to multiply first by $\overline{z}/\overline{z} = 1$. That produces $z\overline{z}$ in the denominator, which is $a^2 + b^2$:

$$\frac{1}{a + ib} = \frac{1}{a + ib}\,\frac{a - ib}{a - ib} = \frac{a - ib}{a^2 + b^2} \qquad \frac{1}{3 + 2i} = \frac{1}{3 + 2i}\,\frac{3 - 2i}{3 - 2i} = \frac{3 - 2i}{13}.$$

In case $a^2 + b^2 = 1$, this says that $(a + ib)^{-1}$ is $a - ib$. ***On the unit circle, $1/z$ equals $\overline{z}$.*** Later we will say: $1/e^{i\theta}$ is $e^{-i\theta}$. Use distance r and angle θ to multiply and divide.

The Polar Form $re^{i\theta}$

The square root of $a^2 + b^2$ is $|z|$. This is the ***absolute value*** (or ***modulus***) of the number $z = a + ib$. The square root $|z|$ is also written r, because it is the distance from 0 to z. ***The real number r in the polar form gives the size of the complex number z***:

$$\text{The absolute value of} \quad z = a + ib \quad \text{is} \quad |z| = \sqrt{a^2 + b^2}. \quad \textbf{This is called } r.$$

$$\text{The absolute value of} \quad z = 3 + 2i \quad \text{is} \quad |z| = \sqrt{3^2 + 2^2}. \quad \text{This is } r = \sqrt{13}.$$

The other part of the polar form is the angle θ. The angle for $z = 5$ is $\theta = 0$ (because this z is real and positive). The angle for $z = 3i$ is $\pi/2$ radians. The angle for a negative $z = -9$ is π radians. ***The angle doubles when the number is squared***. The polar form is excellent for multiplying complex numbers (not good for addition).

When the distance is r and the angle is θ, trigonometry gives the other two sides of the triangle. The real part (along the bottom) is $a = r\cos\theta$. The imaginary part (up or down) is $b = r\sin\theta$. Put those together, and the rectangular form becomes the polar form $re^{i\theta}$.

$$\textit{The number} \quad z = a + ib \quad \textit{is also} \quad z = r\cos\theta + ir\sin\theta. \quad \textit{This is } re^{i\theta}$$

Note: $\cos\theta + i\sin\theta$ ***has absolute value*** $r = 1$ ***because*** $\cos^2\theta + \sin^2\theta = 1$. Thus $\cos\theta + i\sin\theta$ lies on the circle of radius 1—*the unit circle*.

Example 1 Find r and θ for $z = 1 + i$ and also for the conjugate $\overline{z} = 1 - i$.

Solution The absolute value is the same for z and $\overline{z}$. It is $r = \sqrt{1+1} = \sqrt{2}$:

$$|z|^2 = 1^2 + 1^2 = 2 \qquad \text{and also} \qquad |\overline{z}|^2 = 1^2 + (-1)^2 = 2.$$

The distance from the center is $r = \sqrt{2}$. What about the angle θ? The number $1 + i$ is at the point $(1, 1)$ in the complex plane. The angle to that point is $\pi/4$ radians or $45°$. The cosine is $1/\sqrt{2}$ and the sine is $1/\sqrt{2}$. Combining r and θ brings back $z = 1 + i$:

$$r\cos\theta + ir\sin\theta = \sqrt{2}\left(\frac{1}{\sqrt{2}}\right) + i\sqrt{2}\left(\frac{1}{\sqrt{2}}\right) = 1 + i.$$

The angle to the conjugate $1 - i$ can be positive or negative. We can go to $7\pi/4$ radians which is $315°$. Or we can go *backwards through a negative angle*, to $-\pi/4$ radians or $-45°$. ***If z is at angle θ, its conjugate $\overline{z}$ is at $2\pi - \theta$ and also at $-\theta$.***

We can freely add 2π or 4π or -2π to any angle! Those go full circles so the final point is the same. This explains why there are infinitely many choices of θ. Often we select the angle between 0 and 2π. But $-\theta$ is very useful for the conjugate $\overline{z}$. And $1 = e^0 = e^{2\pi i}$.

Powers and Products: Polar Form

Computing $(1+i)^2$ and $(1+i)^8$ is quickest in polar form. That form has $r = \sqrt{2}$ and $\theta = \pi/4$ (or $45°$). If we square the absolute value to get $r^2 = 2$, and double the angle to get $2\theta = \pi/2$ (or $90°$), we have $(1+i)^2$. For the eighth power we need r^8 and 8θ:

$$\boldsymbol{(1+i)^8} \qquad r^8 = 2 \cdot 2 \cdot 2 \cdot 2 = 16 \text{ and } 8\theta = 8 \cdot \frac{\pi}{4} = 2\pi.$$

This means: $(1+i)^8$ has absolute value 16 and angle 2π. *So* $(1+i)^8 = 16$.

Powers are easy in polar form. So is multiplication of complex numbers.

$$\textit{The nth power of} \quad \boldsymbol{z = r(\cos\theta + i\sin\theta)} \quad \textit{is} \quad \boldsymbol{z^n = r^n(\cos n\theta + i\sin n\theta)}. \qquad (3)$$

In that case z multiplies itself. To multiply z times z', ***multiply r's and add angles***:

$$r(\cos\theta + i\sin\theta) \text{ times } r'(\cos\theta' + i\sin\theta') = \boldsymbol{rr'}\big(\cos(\boldsymbol{\theta + \theta'}) + i\sin(\boldsymbol{\theta + \theta'})\big). \qquad (4)$$

One way to understand this is by trigonometry. Why do we get the double angle 2θ for z^2?

$$(\cos\theta + i\sin\theta) \times (\cos\theta + i\sin\theta) = \cos^2\theta + i^2\sin^2\theta + 2i\sin\theta\cos\theta.$$

The real part $\cos^2\theta - \sin^2\theta$ is $\cos 2\theta$. The imaginary part $2\sin\theta\cos\theta$ is $\sin 2\theta$. Those are the "double angle" formulas. They show that θ in z becomes 2θ in z^2.

There is a second way to understand the rule for z^n. It uses the only amazing formula in this section. Remember that $\cos\theta + i\sin\theta$ has absolute value 1. The cosine is made up of even powers, starting with $1 - \frac{1}{2}\theta^2$. The sine is made up of odd powers, starting with $\theta - \frac{1}{6}\theta^3$. The beautiful fact is that $e^{i\theta}$ combines both of those series into $\cos\theta + i\sin\theta$:

$$e^x = 1 + x + \frac{1}{2}x^2 + \frac{1}{6}x^3 + \cdots \quad \text{becomes} \quad e^{i\theta} = 1 + i\theta + \frac{1}{2}i^2\theta^2 + \frac{1}{6}i^3\theta^3 + \cdots$$

Write -1 for i^2 to see $1 - \frac{1}{2}\theta^2$. ***The complex number*** $\boldsymbol{e^{i\theta}}$ ***is*** $\boldsymbol{\cos\theta + i\sin\theta}$:

$$\textit{Euler's Formula} \quad \boldsymbol{e^{i\theta} = \cos\theta + i\sin\theta} \text{ gives } \boldsymbol{z = r\cos\theta + ir\sin\theta = re^{i\theta}} \qquad (5)$$

The special choice $\theta = 2\pi$ gives $\cos 2\pi + i\sin 2\pi$ which is 1. Somehow the infinite series $e^{2\pi i} = 1 + 2\pi i + \frac{1}{2}(2\pi i)^2 + \cdots$ adds up to 1.

Now multiply $e^{i\theta}$ times $e^{i\theta'}$. Angles add for the same reason that exponents add:

$$e^2 \text{ times } e^3 \text{ is } e^5 \qquad e^{i\theta} \text{ times } e^{i\theta} \text{ is } e^{2i\theta} \qquad e^{i\theta} \text{ times } e^{i\theta'} \text{ is } e^{i(\theta+\theta')}$$

The powers $(re^{i\theta})^n$ are equal to $r^n e^{in\theta}$. They stay on the unit circle when $r = 1$ and $r^n = 1$. Then we find n different numbers whose nth powers equal 1:

Set $w = e^{2\pi i/n}$. ***The nth powers of*** $1, w, w^2, \ldots, w^{n-1}$ ***all equal 1.***

Those are the "nth roots of 1." They solve the equation $z^n = 1$. They are equally spaced around the unit circle in Figure 9.2b, where the full 2π is divided by n. Multiply their angles by n to take nth powers. That gives $w^n = e^{2\pi i}$ which is 1. Also $(w^2)^n = e^{4\pi i} = 1$. Each of those numbers, to the nth power, comes around the unit circle to 1.

These n roots of 1 are the key numbers for signal processing. The Discrete Fourier Transform uses $w = e^{2\pi i/n}$ and its powers. Section 9.3 shows how to decompose a vector (a signal) into n frequencies by the Fast Fourier Transform.

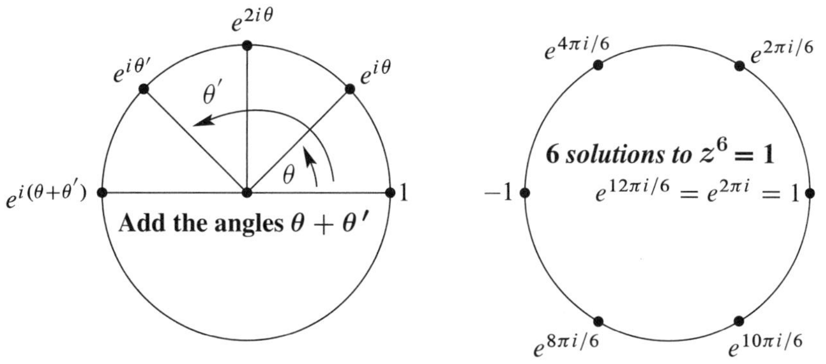

Figure 9.2: (a) $e^{i\theta}$ times $e^{i\theta'}$ is $e^{i(\theta+\theta')}$. (b) The nth power of $e^{2\pi i/n}$ is $e^{2\pi i} = 1$.

■ REVIEW OF THE KEY IDEAS ■

1. Adding $a + ib$ to $c + id$ is like adding $(a, b) + (c, d)$. Use $i^2 = -1$ to multiply.

2. The conjugate of $z = a + bi = re^{i\theta}$ is $\overline{z} = z^* = a - bi = re^{-i\theta}$.

3. z times $\overline{z}$ is $re^{i\theta}$ times $re^{-i\theta}$. This is $r^2 = |z|^2 = a^2 + b^2$ (real).

4. Powers and products are easy in polar form $z = re^{i\theta}$. *Multiply r's and add θ's.*

Problem Set 9.1

Questions 1–8 are about operations on complex numbers.

1 Add and multiply each pair of complex numbers:

(a) $2+i, 2-i$ (b) $-1+i, -1+i$ (c) $\cos\theta + i\sin\theta, \cos\theta - i\sin\theta$

2 Locate these points on the complex plane. Simplify them if necessary:

(a) $2+i$ (b) $(2+i)^2$ (c) $\frac{1}{2+i}$ (d) $|2+i|$

3 Find the absolute value $r = |z|$ of these four numbers. If θ is the angle for $6-8i$, what are the angles for the other three numbers?

(a) $6-8i$ (b) $(6-8i)^2$ (c) $\frac{1}{6-8i}$ (d) $(6+8i)^2$

4 If $|z| = 2$ and $|w| = 3$ then $|z \times w| =$ _____ and $|z+w| \leq$ _____ and $|z/w| =$ _____ and $|z-w| \leq$ _____.

5 Find $a+ib$ for the numbers at angles $30°, 60°, 90°, 120°$ on the unit circle. If w is the number at $30°$, check that w^2 is at $60°$. What power of w equals 1?

6 If $z = r\cos\theta + ir\sin\theta$ then $1/z$ has absolute value _____ and angle _____. Its polar form is _____. Multiply $z \times 1/z$ to get 1.

7 The complex multiplication $M = (a+bi)(c+di)$ is a 2 by 2 real multiplication

$$\begin{bmatrix} a & -b \\ b & a \end{bmatrix} \begin{bmatrix} c \\ d \end{bmatrix} = \begin{bmatrix} \; \\ \; \end{bmatrix}.$$

The right side contains the real and imaginary parts of M. Test $M = (1+3i)(1-3i)$.

8 $A = A_1 + iA_2$ is a complex n by n matrix and $\boldsymbol{b} = \boldsymbol{b}_1 + i\boldsymbol{b}_2$ is a complex vector. The solution to $A\boldsymbol{x} = \boldsymbol{b}$ is $\boldsymbol{x}_1 + i\boldsymbol{x}_2$. Write $A\boldsymbol{x} = \boldsymbol{b}$ as a real system of size $2n$:

Complex n by n
Real $2n$ by $2n$
$$\begin{bmatrix} \quad & \quad \\ \quad & \quad \end{bmatrix} \begin{bmatrix} \boldsymbol{x}_1 \\ \boldsymbol{x}_2 \end{bmatrix} = \begin{bmatrix} \boldsymbol{b}_1 \\ \boldsymbol{b}_2 \end{bmatrix}.$$

Questions 9–16 are about the conjugate $\overline{z} = a - ib = re^{-i\theta} = z^*$.

9 Write down the complex conjugate of each number by changing i to $-i$:

(a) $2-i$ (b) $(2-i)(1-i)$ (c) $e^{i\pi/2}$ (which is i)

(d) $e^{i\pi} = -1$ (e) $\frac{1+i}{1-i}$ (which is also i) (f) $i^{103} =$ _____.

10 The sum $z + \overline{z}$ is always _____. The difference $z - \overline{z}$ is always _____. Assume $z \neq 0$. The product $z \times \overline{z}$ is always _____. The ratio $z/\overline{z}$ has absolute value _____.

11 For a real matrix, the conjugate of $A\boldsymbol{x} = \lambda \boldsymbol{x}$ is $A\overline{\boldsymbol{x}} = \overline{\lambda}\overline{\boldsymbol{x}}$. This proves two things: $\overline{\lambda}$ is another eigenvalue and $\overline{\boldsymbol{x}}$ is its eigenvector. Find the eigenvalues $\lambda, \overline{\lambda}$ and eigenvectors $\boldsymbol{x}, \overline{\boldsymbol{x}}$ of $A = [\,a \;\; b; \;\; -b \;\; a\,]$.

12 The eigenvalues of a real 2 by 2 matrix come from the quadratic formula:

$$\det \begin{bmatrix} a-\lambda & b \\ c & d-\lambda \end{bmatrix} = \lambda^2 - (a+d)\lambda + (ad-bc) = 0$$

gives the two eigenvalues $\lambda = \left[a + d \pm \sqrt{(a+d)^2 - 4(ad-bc)}\right]/2$.

(a) If $a = b = d = 1$, the eigenvalues are complex when c is ____.

(b) What are the eigenvalues when $ad = bc$?

13 In Problem 12 the eigenvalues are not real when $(\text{trace})^2 = (a+d)^2$ is smaller than ____. Show that the λ's *are* real when $bc > 0$.

14 A real skew-symmetric matrix $(A^{\mathrm{T}} = -A)$ has pure imaginary eigenvalues. First proof: If $A\boldsymbol{x} = \lambda\boldsymbol{x}$ then block multiplication gives

$$\begin{bmatrix} 0 & A \\ -A & 0 \end{bmatrix} \begin{bmatrix} \boldsymbol{x} \\ i\boldsymbol{x} \end{bmatrix} = i\lambda \begin{bmatrix} \boldsymbol{x} \\ i\boldsymbol{x} \end{bmatrix}.$$

This block matrix is symmetric. Its eigenvalues must be ____! So λ is ____.

Questions 15–22 are about the form $re^{i\theta}$ of the complex number $r\cos\theta + ir\sin\theta$.

15 Write these numbers in Euler's form $re^{i\theta}$. Then square each number:

(a) $1 + \sqrt{3}i$ (b) $\cos 2\theta + i \sin 2\theta$ (c) $-7i$ (d) $5 - 5i$.

16 (A favorite) Find the absolute value and the angle for $z = \sin\theta + i\cos\theta$ (careful). Locate this z in the complex plane. Multiply z by $\cos\theta + i\sin\theta$ to get ____.

17 Draw all eight solutions of $z^8 = 1$ in the complex plane. What is the rectangular form $a + ib$ of the root $z = \overline{w} = \exp(-2\pi i/8)$?

18 Locate the cube roots of 1 in the complex plane. Locate the cube roots of -1. Together these are the sixth roots of ____.

19 By comparing $e^{3i\theta} = \cos 3\theta + i\sin 3\theta$ with $(e^{i\theta})^3 = (\cos\theta + i\sin\theta)^3$, find the "triple angle" formulas for $\cos 3\theta$ and $\sin 3\theta$ in terms of $\cos\theta$ and $\sin\theta$.

20 Suppose the conjugate $\overline{z}$ is equal to the reciprocal $1/z$. What are all possible z's?

21 (a) Why do e^i and i^e both have absolute value 1?

(b) In the complex plane put stars near the points e^i and i^e.

(c) The number i^e could be $(e^{i\pi/2})^e$ or $(e^{5i\pi/2})^e$. Are those equal?

22 Draw the paths of these numbers from $t = 0$ to $t = 2\pi$ in the complex plane:

(a) e^{it} (b) $e^{(-1+i)t} = e^{-t}e^{it}$ (c) $(-1)^t = e^{t\pi i}$.

9.2 Hermitian and Unitary Matrices

The main message of this section can be presented in one sentence: ***When you transpose a complex vector z or matrix A, take the complex conjugate too***. Don't stop at z^{T} or A^{T}. Reverse the signs of all imaginary parts. From a column vector with $z_j = a_j + ib_j$, the good row vector $\overline{z}^{\mathrm{T}}$ is the *conjugate transpose* with components $a_j - ib_j$:

$$\textbf{Conjugate transpose} \qquad \overline{\boldsymbol{z}}^{\mathrm{T}} = \begin{bmatrix} \overline{z}_1 & \cdots & \overline{z}_n \end{bmatrix} = \begin{bmatrix} a_1 - ib_1 & \cdots & a_n - ib_n \end{bmatrix}. \tag{1}$$

Here is one reason to go to $\overline{z}$. The length squared of a real vector is $x_1^2 + \cdots + x_n^2$. The length squared of a complex vector is *not* $z_1^2 + \cdots + z_n^2$. With that wrong definition, the length of $(1, i)$ would be $1^2 + i^2 = 0$. A nonzero vector would have zero length—not good. Other vectors would have complex lengths. Instead of $(a + bi)^2$ we want $a^2 + b^2$, the *absolute value squared*. This is $(a + bi)$ times $(a - bi)$.

For each component we want z_j times $\overline{z}_j$, which is $|z_j|^2 = a_j^2 + b_j^2$. That comes when the components of $\boldsymbol{z}$ multiply the components of $\overline{\boldsymbol{z}}$:

$$\begin{matrix}\textbf{Length}\\ \textbf{squared}\end{matrix} \quad \begin{bmatrix} \overline{z}_1 & \cdots & \overline{z}_n \end{bmatrix} \begin{bmatrix} z_1 \\ \vdots \\ z_n \end{bmatrix} = |z_1|^2 + \cdots + |z_n|^2. \quad \textbf{This is} \quad \overline{\boldsymbol{z}}^{\mathrm{T}} \boldsymbol{z} = \|\boldsymbol{z}\|^2. \tag{2}$$

Now the squared length of $(1, i)$ is $1^2 + |i|^2 = 2$. The length is $\sqrt{2}$. The squared length of $(1 + i, 1 - i)$ is 4. The only vectors with zero length are zero vectors.

> ***The length $\|\boldsymbol{z}\|$ is the square root of*** $\overline{\boldsymbol{z}}^{\mathrm{T}}\boldsymbol{z} = \boldsymbol{z}^{\mathrm{H}}\boldsymbol{z} = |z_1|^2 + \cdots + |z_n|^2$

Before going further we replace two symbols by one symbol. Instead of a bar for the conjugate and T for the transpose, we just use a superscript H. Thus $\overline{\boldsymbol{z}}^{\mathbf{T}} = \boldsymbol{z}^{\mathbf{H}}$. This is "$\boldsymbol{z}$ Hermitian," the *conjugate transpose* of $\boldsymbol{z}$. The new word is pronounced "Hermeeshan." The new symbol applies also to matrices: The conjugate transpose of a matrix A is A^{H}.

Another popular notation is A^*. The MATLAB transpose command $'$ automatically takes complex conjugates ($\boldsymbol{z}'$ is $\boldsymbol{z}^{\mathrm{H}} = \overline{\boldsymbol{z}}^{\mathrm{T}}$ and A' is $A^{\mathrm{H}} = \overline{A}^{\mathrm{T}}$).

$$\boldsymbol{A}^{\mathrm{H}} \textbf{ is "}\boldsymbol{A}\textbf{ Hermitian"} \quad \text{If} \quad A = \begin{bmatrix} 1 & i \\ 0 & 1+i \end{bmatrix} \quad \text{then} \quad A^{\mathrm{H}} = \begin{bmatrix} 1 & 0 \\ -i & 1-i \end{bmatrix}$$

Complex Inner Products

For real vectors, the length squared is $\boldsymbol{x}^{\mathrm{T}}\boldsymbol{x}$—*the inner product of $\boldsymbol{x}$ with itself.* For complex vectors, the length squared is $\boldsymbol{z}^{\mathrm{H}}\boldsymbol{z}$. It will be very desirable if $\boldsymbol{z}^{\mathrm{H}}\boldsymbol{z}$ is the inner product of $\boldsymbol{z}$ with itself. To make that happen, the complex inner product should use the conjugate transpose (not just the transpose). This has no effect on real vectors.

DEFINITION The inner product of real or complex vectors $\boldsymbol{u}$ and $\boldsymbol{v}$ is $\boldsymbol{u}^{\mathrm{H}}\boldsymbol{v}$:

$$\boldsymbol{u}^{\mathrm{H}}\boldsymbol{v} = \begin{bmatrix} \overline{u}_1 & \cdots & \overline{u}_n \end{bmatrix} \begin{bmatrix} v_1 \\ \vdots \\ v_n \end{bmatrix} = \overline{u}_1 v_1 + \cdots + \overline{u}_n v_n. \qquad (3)$$

With complex vectors, $\boldsymbol{u}^{\mathrm{H}}\boldsymbol{v}$ is different from $\boldsymbol{v}^{\mathrm{H}}\boldsymbol{u}$. *The order of the vectors is now important.* In fact $\boldsymbol{v}^{\mathrm{H}}\boldsymbol{u} = \overline{v}_1 u_1 + \cdots + \overline{v}_n u_n$ is the complex conjugate of $\boldsymbol{u}^{\mathrm{H}}\boldsymbol{v}$. We have to put up with a few inconveniences for the greater good.

Example 1 The inner product of $\boldsymbol{u} = \begin{bmatrix} 1 \\ i \end{bmatrix}$ with $\boldsymbol{v} = \begin{bmatrix} i \\ 1 \end{bmatrix}$ is $\begin{bmatrix} 1 & -i \end{bmatrix} \begin{bmatrix} i \\ 1 \end{bmatrix} = 0$.

Example 1 is surprising. Those vectors $(1, i)$ and $(i, 1)$ don't look perpendicular. But they are. ***A zero inner product still means that the*** (complex) ***vectors are orthogonal.*** Similarly the vector $(1, i)$ is orthogonal to the vector $(1, -i)$. Their inner product is $1 - 1$. We are correctly getting zero for the inner product—where we would be incorrectly getting zero for the length of $(1, i)$ if we forgot to take the conjugate.

Note We have chosen to conjugate the first vector $\boldsymbol{u}$. Some authors choose the second vector $\boldsymbol{v}$. Their complex inner product would be $\boldsymbol{u}^{\mathrm{T}}\overline{\boldsymbol{v}}$. I think it is a free choice.

The inner product of Au ***with*** v ***equals the inner product of*** u ***with*** $A^{\mathrm{H}}\boldsymbol{v}$:

$$\boldsymbol{A}^{\mathrm{H}} \textbf{ is also called the “adjoint” of } \boldsymbol{A} \qquad (A\boldsymbol{u})^{\mathrm{H}}\boldsymbol{v} = \boldsymbol{u}^{\mathrm{H}}(A^{\mathrm{H}}\boldsymbol{v}). \qquad (4)$$

The conjugate of $A\boldsymbol{u}$ is $\overline{A\boldsymbol{u}}$. Transposing $\overline{A\boldsymbol{u}}$ gives $\overline{\boldsymbol{u}}^{\mathrm{T}}\overline{A}^{\mathrm{T}}$ as usual. This is $\boldsymbol{u}^{\mathrm{H}}A^{\mathrm{H}}$. Everything that should work, does work. The rule for ^{H} comes from the rule for ^{T}.
We constantly use the fact that $(a - ib)(c - id)$ is the conjugate of $(a + ib)(c + id)$.

The conjugate transpose of AB is $\boldsymbol{(AB)^{\mathrm{H}} = B^{\mathrm{H}}A^{\mathrm{H}}}$.

Hermitian Matrices $S = S^{\mathrm{H}}$

Among real matrices, *symmetric matrices* form the most important special class: $S = S^{\mathrm{T}}$. They have real eigenvalues and the orthogonal eigenvectors in an orthogonal matrix Q. Every real symmetric matrix can be written as $S = Q\Lambda Q^{-1}$ and also as $S = Q\Lambda Q^{\mathrm{T}}$ (because $Q^{-1} = Q^{\mathrm{T}}$). All this follows from $S^{\mathrm{T}} = S$, when S is real.

Among complex matrices, the special class contains the ***Hermitian matrices***: $S = S^{\mathrm{H}}$. The condition on the entries is $s_{ij} = \overline{s_{ji}}$. In this case we say that "S ***is*** Hermitian." *Every real symmetric matrix is Hermitian*, because taking its conjugate has no effect. The next matrix is also Hermitian, $S = S^{\mathrm{H}}$:

Example 2 $S = \begin{bmatrix} 2 & 3-3i \\ 3+3i & 5 \end{bmatrix}$ The main diagonal must be real since $s_{ii} = \overline{s_{ii}}$. Across it are conjugates $3+3i$ and $3-3i$.

This example will illustrate the three crucial properties of all Hermitian matrices.

If $S = S^{\mathrm{H}}$ and z is any real or complex column vector, the number $z^{\mathrm{H}}Sz$ is real.

Quick proof: $\boldsymbol{z}^{\mathrm{H}}S\boldsymbol{z}$ is certainly 1 by 1. Take its conjugate transpose:

$$(\boldsymbol{z}^{\mathrm{H}}S\boldsymbol{z})^{\mathrm{H}} = \boldsymbol{z}^{\mathrm{H}}S^{\mathrm{H}}(\boldsymbol{z}^{\mathrm{H}})^{\mathrm{H}} \quad \text{which is } \boldsymbol{z}^{\mathrm{H}}S\boldsymbol{z} \text{ again.}$$

So the number $\boldsymbol{z}^{\mathrm{H}}S\boldsymbol{z}$ equals its conjugate and must be real. Here is that "energy" $\boldsymbol{z}^{\mathrm{H}}S\boldsymbol{z}$:

$$\begin{bmatrix} \overline{z}_1 & \overline{z}_2 \end{bmatrix} \begin{bmatrix} 2 & 3-3i \\ 3+3i & 5 \end{bmatrix} \begin{bmatrix} z_1 \\ z_2 \end{bmatrix} = \underset{\textit{diagonal}}{2\overline{z}_1 z_1 + 5\overline{z}_2 z_2} + \underset{\textit{off-diagonal}}{(3-3i)\overline{z}_1 z_2 + (3+3i) z_1 \overline{z}_2}.$$

The terms $2|z_1|^2$ and $5|z_2|^2$ from the diagonal are both real. The off-diagonal terms are conjugates of each other—so their sum is real. (The imaginary parts cancel when we add.) The whole expression $\boldsymbol{z}^{\mathrm{H}}S\boldsymbol{z}$ is real, and this will make λ real.

Every eigenvalue of a Hermitian matrix is real.

Proof Suppose $S\boldsymbol{z} = \lambda\boldsymbol{z}$. *Multiply both sides by $\boldsymbol{z}^{\mathrm{H}}$ to get $\boldsymbol{z}^{\mathrm{H}}S\boldsymbol{z} = \lambda\boldsymbol{z}^{\mathrm{H}}\boldsymbol{z}$.* On the left side, $\boldsymbol{z}^{\mathrm{H}}S\boldsymbol{z}$ is real. On the right side, $\boldsymbol{z}^{\mathrm{H}}\boldsymbol{z}$ is the length squared, real and positive. So the ratio $\lambda = \boldsymbol{z}^{\mathrm{H}}S\boldsymbol{z}/\boldsymbol{z}^{\mathrm{H}}\boldsymbol{z}$ is a real number. Q.E.D.

The example above has eigenvalues $\lambda = 8$ and $\lambda = -1$, real because $S = S^{\mathrm{H}}$:

$$\begin{vmatrix} 2-\lambda & 3-3i \\ 3+3i & 5-\lambda \end{vmatrix} = \lambda^2 - 7\lambda + 10 - |3+3i|^2$$
$$= \lambda^2 - 7\lambda + 10 - 18 = (\lambda - 8)(\lambda + 1).$$

The eigenvectors of a Hermitian matrix are orthogonal (when they correspond to different eigenvalues). If $S\boldsymbol{z} = \lambda\boldsymbol{z}$ and $S\boldsymbol{y} = \beta\boldsymbol{y}$ then $\boldsymbol{y}^{\mathrm{H}}\boldsymbol{z} = 0$.

Proof Multiply $S\boldsymbol{z} = \lambda\boldsymbol{z}$ on the left by $\boldsymbol{y}^{\mathrm{H}}$. Multiply $\boldsymbol{y}^{\mathrm{H}}S^{\mathrm{H}} = \beta\boldsymbol{y}^{\mathrm{H}}$ on the right by $\boldsymbol{z}$:

$$\boldsymbol{y}^{\mathrm{H}}S\boldsymbol{z} = \lambda\boldsymbol{y}^{\mathrm{H}}\boldsymbol{z} \quad \text{and} \quad \boldsymbol{y}^{\mathrm{H}}S^{\mathrm{H}}\boldsymbol{z} = \beta\boldsymbol{y}^{\mathrm{H}}\boldsymbol{z}. \tag{5}$$

The left sides are equal so $\lambda\boldsymbol{y}^{\mathrm{H}}\boldsymbol{z} = \beta\boldsymbol{y}^{\mathrm{H}}\boldsymbol{z}$. Then $\boldsymbol{y}^{\mathrm{H}}\boldsymbol{z}$ **must be zero.**

The eigenvectors are orthogonal in our example with $\lambda = 8$ and $\beta = -1$:

$$(S - 8I)\boldsymbol{z} = \begin{bmatrix} -6 & 3-3i \\ 3+3i & -3 \end{bmatrix} \begin{bmatrix} z_1 \\ z_2 \end{bmatrix} = \begin{bmatrix} 0 \\ 0 \end{bmatrix} \quad \text{and} \quad \boldsymbol{z} = \begin{bmatrix} 1 \\ 1+i \end{bmatrix}$$

$$(S + I)\boldsymbol{y} = \begin{bmatrix} 3 & 3-3i \\ 3+3i & 6 \end{bmatrix} \begin{bmatrix} y_1 \\ y_2 \end{bmatrix} = \begin{bmatrix} 0 \\ 0 \end{bmatrix} \quad \text{and} \quad \boldsymbol{y} = \begin{bmatrix} 1-i \\ -1 \end{bmatrix}.$$

Orthogonal eigenvectors $\quad \boldsymbol{y}^{\mathrm{H}}\boldsymbol{z} = \begin{bmatrix} 1+i & -1 \end{bmatrix} \begin{bmatrix} 1 \\ 1+i \end{bmatrix} = 0.$

These eigenvectors have squared length $1^2 + 1^2 + 1^2 = 3$. After division by $\sqrt{3}$ they are unit vectors. They were orthogonal, now they are ***orthonormal***. They go into the columns of the *eigenvector matrix* X, which diagonalizes S.

When S is real and symmetric, X is Q—an orthogonal matrix. Now S is complex and Hermitian. Its eigenvectors are complex and orthonormal. ***The eigenvector matrix X is like Q, but complex***: $\boldsymbol{Q^{\mathrm{H}}Q = I}$. We assign Q a new name "*unitary*" but still call it Q.

Unitary Matrices

A ***unitary matrix*** Q is a (complex) square matrix that has ***orthonormal columns***.

Unitary matrix that diagonalizes S: $\quad Q = \dfrac{1}{\sqrt{3}} \begin{bmatrix} 1 & 1-i \\ 1+i & -1 \end{bmatrix}$

This Q is also a Hermitian matrix. I didn't expect that! The example is almost too perfect. We will see that the eigenvalues of this Q must be 1 and -1.

The matrix test for real orthonormal columns was $Q^{\mathrm{T}}Q = I$. The zero inner products appear off the diagonal. In the complex case, Q^{T} becomes Q^{H}. The columns show themselves as orthonormal when Q^{H} multiplies Q. The inner products fill up $\boldsymbol{Q^{\mathrm{H}}Q = I}$:

> ***Every matrix Q with orthonormal columns has*** $\boldsymbol{Q^{\mathrm{H}}Q = I}$.
>
> ***If Q is square, it is a unitary matrix. Then*** $\boldsymbol{Q^{\mathrm{H}} = Q^{-1}}$.

Suppose Q (with orthonormal columns) multiplies any $\boldsymbol{z}$. The vector length stays the same, because $\boldsymbol{z}^{\mathrm{H}}Q^{\mathrm{H}}Q\boldsymbol{z} = \boldsymbol{z}^{\mathrm{H}}\boldsymbol{z}$. If $\boldsymbol{z}$ is an eigenvector of Q we learn something more: ***The eigenvalues of unitary (and orthogonal) matrices Q all have absolute value*** $|\lambda| = 1$.

> ***If Q is unitary then*** $\|Q\boldsymbol{z}\| = \|\boldsymbol{z}\|$. ***Therefore*** $Q\boldsymbol{z} = \lambda\boldsymbol{z}$ ***leads to*** $\boldsymbol{|\lambda| = 1}$.

Our 2 by 2 example is both Hermitian ($Q = Q^{\mathrm{H}}$) and unitary ($Q^{-1} = Q^{\mathrm{H}}$). That means real eigenvalues and it means $|\lambda| = 1$. A real number with $|\lambda| = 1$ has only two possibilities: *The eigenvalues are* 1 *or* -1. The trace of Q is zero so $\lambda = 1$ and $\lambda = -1$.

Example 3 The 3 by 3 ***Fourier matrix*** is in Figure 9.3. Is it Hermitian? Is it unitary? F_3 is certainly symmetric. It equals its transpose. But it doesn't equal its conjugate transpose—*it is not Hermitian.* If you change i to $-i$, you get a different matrix.

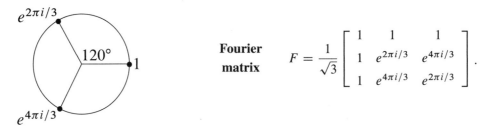

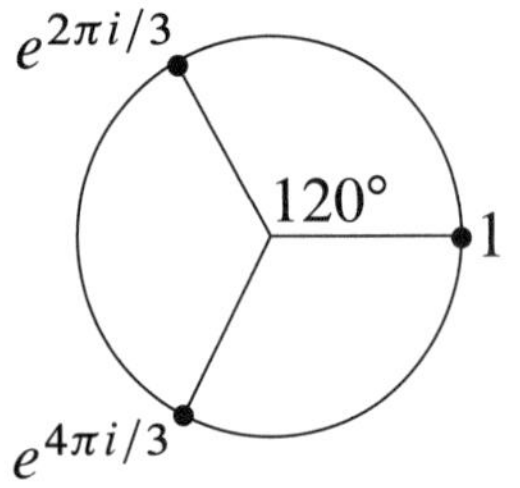

Fourier matrix $$F = \frac{1}{\sqrt{3}}\begin{bmatrix} 1 & 1 & 1 \\ 1 & e^{2\pi i/3} & e^{4\pi i/3} \\ 1 & e^{4\pi i/3} & e^{2\pi i/3} \end{bmatrix}.$$

Figure 9.3: The cube roots of 1 go into the Fourier matrix $F = F_3$.

Is F unitary? *Yes.* The squared length of every column is $\frac{1}{3}(1+1+1)$ (unit vector). The first column is orthogonal to the second column because $1 + e^{2\pi i/3} + e^{4\pi i/3} = 0$. This is the sum of the three numbers marked in Figure 9.3.

Notice the symmetry of the figure. If you rotate it by $120°$, the three points are in the same position. Therefore their sum S also stays in the same position! The only possible sum in the same position after $120°$ rotation is $S = 0$.

Is column 2 of F orthogonal to column 3? Their dot product looks like

$$\tfrac{1}{3}(1 + e^{6\pi i/3} + e^{6\pi i/3}) = \tfrac{1}{3}(1+1+1).$$

This is not zero. The answer is wrong because we forgot to take complex conjugates. The complex inner product uses ^{H} not ^{T}.

$$(\text{column } 2)^{\mathrm{H}}(\text{column } 3) = \tfrac{1}{3}(1\cdot 1 + e^{-2\pi i/3}e^{4\pi i/3} + e^{-4\pi i/3}e^{2\pi i/3})$$
$$= \tfrac{1}{3}(1 + e^{2\pi i/3} + e^{-2\pi i/3}) = 0.$$

So we do have orthogonality. ***Conclusion:*** F ***is a unitary matrix.***

The next section will study the n by n Fourier matrices. Among all complex unitary matrices, these are the most important. When we multiply a vector by F, we are computing its ***Discrete Fourier Transform***. When we multiply by F^{-1}, we are computing the *inverse transform*. The special property of unitary matrices is that $F^{-1} = F^{\mathrm{H}}$. The inverse transform only differs by changing i to $-i$:

Change i to $-i$ $$F^{-1} = F^{\mathrm{H}} = \frac{1}{\sqrt{3}}\begin{bmatrix} 1 & 1 & 1 \\ 1 & e^{-2\pi i/3} & e^{-4\pi i/3} \\ 1 & e^{-4\pi i/3} & e^{-2\pi i/3} \end{bmatrix}.$$

Everyone who works with F recognizes its value. The last section of this chapter will bring together Fourier analysis and complex numbers and linear algebra.

Problem Set 9.2

1 Find the lengths of $\boldsymbol{u} = (1+i, 1-i, 1+2i)$ and $\boldsymbol{v} = (i, i, i)$. Find $\boldsymbol{u}^{\mathrm{H}}\boldsymbol{v}$ and $\boldsymbol{v}^{\mathrm{H}}\boldsymbol{u}$.

2 Compute $A^{\mathrm{H}}A$ and AA^{H}. Those are both _____ matrices:

$$A = \begin{bmatrix} i & 1 & i \\ 1 & i & i \end{bmatrix}.$$

3 Solve $A\boldsymbol{z} = \mathbf{0}$ to find a vector $\boldsymbol{z}$ in the nullspace of A in Problem 2. Show that $\boldsymbol{z}$ is orthogonal to the columns of A^{H}. Show that $\boldsymbol{z}$ is *not* orthogonal to the columns of A^{T}. ***The good row space is no longer $\boldsymbol{C}(A^{\mathrm{T}})$. Now it is $\boldsymbol{C}(A^{\mathrm{H}})$.***

4 Problem 3 indicates that the four fundamental subspaces are $\boldsymbol{C}(A)$ and $\boldsymbol{N}(A)$ and _____ and _____. Their dimensions are still r and $n-r$ and r and $m-r$. They are still orthogonal subspaces. *The symbol* ^{H} *takes the place of* ^{T}.

5 (a) Prove that $A^{\mathrm{H}}A$ is always a Hermitian matrix.

(b) If $A\boldsymbol{z} = \mathbf{0}$ then $A^{\mathrm{H}}A\boldsymbol{z} = \mathbf{0}$. If $A^{\mathrm{H}}A\boldsymbol{z} = \mathbf{0}$, multiply by $\boldsymbol{z}^{\mathrm{H}}$ to prove that $A\boldsymbol{z} = \mathbf{0}$. The nullspaces of A and $A^{\mathrm{H}}A$ are _____. Therefore $A^{\mathrm{H}}A$ is an invertible Hermitian matrix when the nullspace of A contains only $\boldsymbol{z} = 0$.

6 True or false (give a reason if true or a counterexample if false):

(a) If A is a real matrix then $A + iI$ is invertible.

(b) If S is a Hermitian matrix then $S + iI$ is invertible.

(c) If Q is a unitary matrix then $Q + iI$ is invertible.

7 When you multiply a Hermitian matrix by a real number c, is cS still Hermitian? Show that iS is skew-Hermitian when S is Hermitian. The 3 by 3 Hermitian matrices are a subspace provided the "scalars" are real numbers.

8 Which classes of matrices does P belong to: invertible, Hermitian, unitary?

$$P = \begin{bmatrix} 0 & i & 0 \\ 0 & 0 & i \\ i & 0 & 0 \end{bmatrix}.$$

Compute P^2, P^3, and P^{100}. What are the eigenvalues of P?

9 Find the unit eigenvectors of P in Problem 8, and put them into the columns of a unitary matrix Q. What property of P makes these eigenvectors orthogonal?

10 Write down the 3 by 3 circulant matrix $C = 2I + 5P$. It has the same eigenvectors as P in Problem 8. Find its eigenvalues.

11 If Q and U are unitary matrices, show that Q^{-1} is unitary and also QU is unitary. Start from $Q^{\mathrm{H}}Q = I$ and $U^{\mathrm{H}}U = I$.

12 How do you know that the determinant of every Hermitian matrix is real?

13 The matrix $A^{\mathrm{H}}A$ is not only Hermitian but also positive definite, when the columns of A are independent. Proof: $\boldsymbol{z}^{\mathrm{H}}A^{\mathrm{H}}A\boldsymbol{z}$ is positive if $\boldsymbol{z}$ is nonzero because ______.

14 Diagonalize these Hermitian matrices to reach $S = Q\Lambda Q^{\mathrm{H}}$:

$$S = \begin{bmatrix} 0 & 1-i \\ i+1 & 1 \end{bmatrix} \quad \text{and} \quad S = \begin{bmatrix} 2 & 1+i \\ i-1 & 3 \end{bmatrix}.$$

15 Diagonalize this skew-Hermitian matrix to reach $K = Q\Lambda Q^{\mathrm{H}}$. All λ's are ______:

$$K = \begin{bmatrix} 0 & -1+i \\ 1+i & i \end{bmatrix}.$$

16 Diagonalize this orthogonal matrix to reach $U = Q\Lambda Q^{\mathrm{H}}$. Now all λ's are ______:

$$U = \begin{bmatrix} \cos\theta & -\sin\theta \\ \sin\theta & \cos\theta \end{bmatrix}.$$

17 Diagonalize this unitary matrix to reach $U = Q\Lambda Q^{\mathrm{H}}$. Again all λ's are ______:

$$U = \frac{1}{\sqrt{3}} \begin{bmatrix} 1 & 1-i \\ 1+i & -1 \end{bmatrix}.$$

18 If $\boldsymbol{v}_1, \ldots, \boldsymbol{v}_n$ is an orthonormal basis for $\mathbf{C}^n$, the matrix with those columns is a ______ matrix. Show that any vector $\boldsymbol{z}$ equals $(\boldsymbol{v}_1^{\mathrm{H}}\boldsymbol{z})\boldsymbol{v}_1 + \cdots + (\boldsymbol{v}_n^{\mathrm{H}}\boldsymbol{z})\boldsymbol{v}_n$.

19 $\boldsymbol{v} = (1, i, 1)$, $\boldsymbol{w} = (i, 1, 0)$ and $\boldsymbol{z} =$ ______ are an orthogonal basis for ______.

20 If $S = A + iB$ is a Hermitian matrix, are its real and imaginary parts symmetric?

21 The (complex) dimension of $\mathbf{C}^n$ is ______. Find a non-real basis for $\mathbf{C}^n$.

22 Describe all 1 by 1 and 2 by 2 Hermitian matrices and unitary matrices.

23 How are the eigenvalues of A^{H} related to the eigenvalues of the square matrix A?

24 If $\boldsymbol{u}^{\mathrm{H}}\boldsymbol{u} = 1$ show that $I - 2\boldsymbol{u}\boldsymbol{u}^{\mathrm{H}}$ is Hermitian and also unitary. The rank-one matrix $\boldsymbol{u}\boldsymbol{u}^{\mathrm{H}}$ is the projection onto what line in $\mathbf{C}^n$?

25 If $A + iB$ is a unitary matrix (A and B are real) show that $Q = \left[\begin{smallmatrix} \mathbf{A} & -\mathbf{B} \\ \mathbf{B} & \mathbf{A} \end{smallmatrix}\right]$ is an orthogonal matrix.

26 If $A + iB$ is Hermitian (A and B are real) show that $\left[\begin{smallmatrix} \mathbf{A} & -\mathbf{B} \\ \mathbf{B} & \mathbf{A} \end{smallmatrix}\right]$ is symmetric.

27 Prove that the inverse of a Hermitian matrix is also Hermitian (transpose $S^{-1}S = I$).

28 A matrix with orthonormal eigenvectors has the form $N = Q\Lambda Q^{-1} = Q\Lambda Q^{\mathrm{H}}$. *Prove that* $NN^{\mathrm{H}} = N^{\mathrm{H}}N$. These N are exactly the **normal matrices**. Examples are Hermitian, skew-Hermitian, and unitary matrices. Construct a 2 by 2 normal matrix from $Q\Lambda Q^{\mathrm{H}}$ by choosing complex eigenvalues in Λ.

9.3 The Fast Fourier Transform

Many applications of linear algebra take time to develop. It is not easy to explain them in an hour. The teacher and the author must choose between completing the theory and adding new applications. Often the theory wins, but this section is an exception. It explains the most valuable numerical algorithm in the last century.

We want to multiply quickly by F and F^{-1}, the Fourier matrix and its inverse. This is achieved by the Fast Fourier Transform. An ordinary product $F\boldsymbol{c}$ uses n^2 multiplications (F has n^2 entries). The FFT needs only n times $\frac{1}{2}\log_2 n$. We will see how.

The FFT has revolutionized signal processing. Whole industries are speeded up by this one idea. Electrical engineers are the first to know the difference—they take your Fourier transform as they meet you (if you are a function). Fourier's idea is to represent f as a sum of harmonics $c_k e^{ikx}$. The function is seen in *frequency space* through the coefficients c_k, instead of *physical space* through its values $f(x)$. The passage backward and forward between c's and f's is by the Fourier transform. Fast passage is by the FFT.

Roots of Unity and the Fourier Matrix

Quadratic equations have two roots (or one repeated root). Equations of degree n have n roots (counting repetitions). This is the Fundamental Theorem of Algebra, and to make it true we must allow complex roots. This section is about the very special equation $z^n = 1$. The solutions z are the "nth roots of unity." They are n evenly spaced points around the unit circle in the complex plane.

Figure 9.4 shows the eight solutions to $z^8 = 1$. Their spacing is $\frac{1}{8}(360°) = 45°$. The first root is at $45°$ or $\theta = 2\pi/8$ radians. ***It is the complex number*** $w = e^{i\theta} = \boldsymbol{e^{i2\pi/8}}$. We call this number w_8 to emphasize that it is an 8th root. You could write it in terms of $\cos\frac{2\pi}{8}$ and $\sin\frac{2\pi}{8}$, but don't do it. The seven other 8th roots are $w^2, w^3, \ldots, w^8$, going around the circle. Powers of w are best in polar form, because we work only with the angles $\frac{2\pi}{8}, \frac{4\pi}{8}, \ldots, \frac{16\pi}{8} = 2\pi$. Those 8 angles in degrees are $45°, 90°, 135°, \ldots, 360°$.

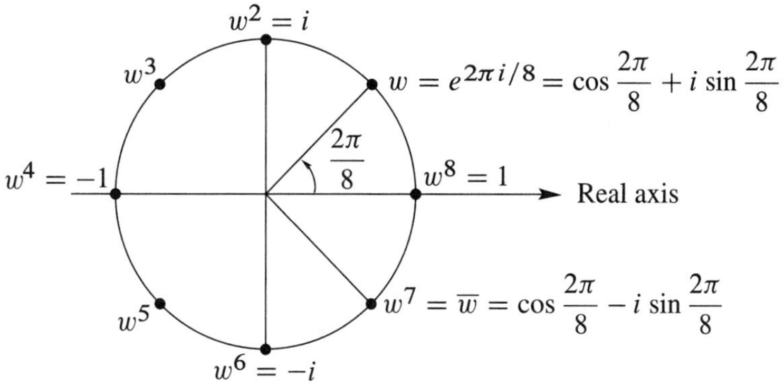

Figure 9.4: The eight solutions to $z^8 = 1$ are $1, w, w^2, \ldots, w^7$ with $w = (1+i)/\sqrt{2}$.

The fourth roots of 1 are also in the figure. They are $i, -1, -i, 1$. The angle is now $2\pi/4$ or $90°$. The first root $w_4 = e^{2\pi i/4}$ is nothing but i. Even the square roots of 1 are seen, with $w_2 = e^{i2\pi/2} = -1$. Do not despise those square roots 1 and -1. The idea behind the FFT is to go from an **8 by 8** Fourier matrix (containing powers of w_8) to the **4 by 4** matrix below (with powers of $w_4 = i$). The same idea goes from 4 to 2. By exploiting the connections of F_8 down to F_4 and up to F_{16} (and beyond), the FFT makes multiplication by F_{1024} very quick.

We describe the *Fourier matrix*, first for $n = 4$. Its rows contain powers of 1 and w and w^2 and w^3. These are the fourth roots of 1, and their powers come in a special order.

Fourier matrix $n = 4$ $\boldsymbol{w = i}$

$$F = \begin{bmatrix} 1 & 1 & 1 & 1 \\ 1 & w & w^2 & w^3 \\ 1 & w^2 & w^4 & w^6 \\ 1 & w^3 & w^6 & w^9 \end{bmatrix} = \begin{bmatrix} 1 & 1 & 1 & 1 \\ 1 & i & i^2 & i^3 \\ 1 & i^2 & i^4 & i^6 \\ 1 & i^3 & i^6 & i^9 \end{bmatrix}.$$

The matrix is symmetric ($F = F^{\mathrm{T}}$). It is *not* Hermitian. Its main diagonal is not real. But $\frac{1}{2}F$ is a ***unitary matrix***, which means that $(\frac{1}{2}F^{\mathrm{H}})(\frac{1}{2}F) = I$:

The columns of F give $\boldsymbol{F^{\mathrm{H}}F = 4I}$. ***Its inverse is*** $\boldsymbol{\frac{1}{4}F^{\mathrm{H}}}$ ***which is*** $\boldsymbol{F^{-1} = \frac{1}{4}\overline{F}}$.

The inverse changes from $w = i$ to $\overline{w} = -i$. That takes us from F to $\overline{F}$. When the Fast Fourier Transform gives a quick way to multiply by F, it does the same for $\overline{F}$ and F^{-1}.

Every column has length $\sqrt{n}$. So the unitary matrices are $Q = F/\sqrt{n}$ and $Q^{-1} = \overline{F}/\sqrt{n}$. We avoid $\sqrt{n}$ and just use F and $\boldsymbol{F^{-1} = \overline{F}/n}$. The main point is to multiply F times c_0, c_1, c_2, c_3:

4-point Fourier series

$$\begin{bmatrix} y_0 \\ y_1 \\ y_2 \\ y_3 \end{bmatrix} = F\boldsymbol{c} = \begin{bmatrix} 1 & 1 & 1 & 1 \\ 1 & w & w^2 & w^3 \\ 1 & w^2 & w^4 & w^6 \\ 1 & w^3 & w^6 & w^9 \end{bmatrix} \begin{bmatrix} c_0 \\ c_1 \\ c_2 \\ c_3 \end{bmatrix}. \tag{1}$$

The input is four complex coefficients c_0, c_1, c_2, c_3. The output is four function values y_0, y_1, y_2, y_3. The first output $y_0 = c_0 + c_1 + c_2 + c_3$ is the value of the Fourier series $\sum c_k e^{ikx}$ at $x = 0$. *The second output is the value of that series* $\sum c_k e^{ikx}$ *at* $x = 2\pi/4$:

$$y_1 = c_0 + c_1 e^{i2\pi/4} + c_2 e^{i4\pi/4} + c_3 e^{i6\pi/4} = c_0 + c_1 w + c_2 w^2 + c_3 w^3.$$

The third and fourth outputs y_2 and y_3 are the values of $\sum c_k e^{ikx}$ at $x = 4\pi/4$ and $x = 6\pi/4$. These are *finite* Fourier series! *They contain* $n = 4$ *terms and they are evaluated at* $n = 4$ *points*. Those points $x = 0, 2\pi/4, 4\pi/4, 6\pi/4$ are equally spaced.

The next point would be $x = 8\pi/4$ which is 2π. Then the series is back to y_0, because $e^{2\pi i}$ is the same as $e^0 = 1$. Everything cycles around with period 4. In this world $2 + 2$ is 0 because $(w^2)(w^2) = w^0 = 1$. We follow the convention that ***j and k go from*** $\boldsymbol{0}$ ***to*** $\boldsymbol{n - 1}$ (instead of 1 to n). The "zeroth row" and "zeroth column" of F contain all ones.

The n by n Fourier matrix contains powers of $w = e^{2\pi i/n}$:

$$F_n\boldsymbol{c} = \begin{bmatrix} 1 & 1 & 1 & \cdot & 1 \\ 1 & w & w^2 & \cdot & w^{n-1} \\ 1 & w^2 & w^4 & \cdot & w^{2(n-1)} \\ \cdot & \cdot & \cdot & \cdot & \cdot \\ 1 & w^{n-1} & w^{2(n-1)} & \cdot & w^{(n-1)^2} \end{bmatrix} \begin{bmatrix} c_0 \\ c_1 \\ c_2 \\ \cdot \\ c_{n-1} \end{bmatrix} = \begin{bmatrix} y_0 \\ y_1 \\ y_2 \\ \cdot \\ y_{n-1} \end{bmatrix} = \boldsymbol{y}. \qquad (2)$$

F_n is symmetric but not Hermitian. ***Its columns are orthogonal***, and $F_n\overline{F}_n = nI$. *Then* F_n^{-1} *is* $\overline{F}_n/n$. The inverse contains powers of $\overline{w}_n = e^{-2\pi i/n}$. Look at the pattern in F:

The entry in row j, column k ***is*** w^{jk}. ***Row zero and column zero contain*** $w^0 = 1$.

When we multiply $\boldsymbol{c}$ by F_n, we sum the series at n points. *When we multiply* $\boldsymbol{y}$ *by* F_n^{-1}, *we find the coefficients* $\boldsymbol{c}$ *from the function values* $\boldsymbol{y}$. In MATLAB that command is $\boldsymbol{c} = \mathrm{fft}(\boldsymbol{y})$. The matrix F passes from "frequency space" to "physical space."

Important note. Many authors prefer to work with $\omega = e^{-2\pi i/N}$, which is the *complex conjugate* of our w. (They often use the Greek omega, and I will do that to keep the two options separate.) With this choice, their DFT matrix contains powers of ω not w. It is $\overline{F}$, the conjugate of our F. $\overline{F}$ goes from physical space to frequency space.

$\overline{F}$ is a completely reasonable choice! MATLAB uses $\omega = e^{-2\pi i/N}$. The DFT matrix fft(eye(N)) contains powers of this number $\omega = \overline{w}$. **The Fourier matrix F with w's reconstructs $\boldsymbol{y}$ from $\boldsymbol{c}$. The matrix $\overline{F}$ with ω's computes Fourier coefficients as fft($\boldsymbol{y}$).**

Also important. When a function $f(x)$ has period 2π, and we change x to $e^{i\theta}$, the function is defined around the unit circle (where $z = e^{i\theta}$). The Discrete Fourier Transform is the same as interpolation. Find the polynomial $p(z) = c_0 + c_1 z + \cdots + c_{n-1}z^{n-1}$ that matches n values $f_0, \ldots, f_{n-1}$:

Interpolation Find $c_0, \ldots, c_{n-1}$ so that $p(z) = f$ at n points $z = 1, \ldots, w^{n-1}$

The Fourier matrix is the Vandermonde matrix for interpolation at those n special points.

One Step of the Fast Fourier Transform

We want to multiply F times $\boldsymbol{c}$ as quickly as possible. Normally a matrix times a vector takes n^2 separate multiplications—the matrix has n^2 entries. You might think it is impossible to do better. (If the matrix has zero entries then multiplications can be skipped. But the Fourier matrix has no zeros!) By using the special pattern w^{jk} for its entries, F can be factored in a way that produces many zeros. This is the **FFT**.

The key idea is to connect F_n with the half-size Fourier matrix $F_{n/2}$. Assume that n is a power of 2 (say $n = 2^{10} = 1024$). We will connect F_{1024} to ***two copies of F_{512}***.

When $n = 4$, the key is in the relation between F_4 and two copies of F_2:

$$\boldsymbol{F_4} = \begin{bmatrix} 1 & 1 & 1 & 1 \\ 1 & i & i^2 & i^3 \\ 1 & i^2 & i^4 & i^6 \\ 1 & i^3 & i^6 & i^9 \end{bmatrix} \quad \text{and} \quad \begin{bmatrix} \boldsymbol{F_2} & \\ & \boldsymbol{F_2} \end{bmatrix} = \begin{bmatrix} 1 & 1 & & \\ 1 & i^2 & & \\ & & 1 & 1 \\ & & 1 & i^2 \end{bmatrix}.$$

On the left is F_4, with no zeros. On the right is a matrix that is half zero. The work is cut in half. But wait, those matrices are not the same. We need two sparse and simple matrices to complete the FFT factorization:

Factors for FFT
$$F_4 = \begin{bmatrix} 1 & & 1 & \\ & 1 & & i \\ 1 & & -1 & \\ & 1 & & -i \end{bmatrix} \begin{bmatrix} 1 & 1 & & \\ 1 & i^2 & & \\ & & 1 & 1 \\ & & 1 & i^2 \end{bmatrix} \begin{bmatrix} 1 & & & \\ & & 1 & \\ & 1 & & \\ & & & 1 \end{bmatrix}. \tag{3}$$

The last matrix is a permutation. It puts the even c's (c_0 and c_2) ahead of the odd c's (c_1 and c_3). The middle matrix performs half-size transforms F_2 and F_2 on the even c's and odd c's. The matrix at the left combines the two half-size outputs—in a way that produces the correct full-size output $\boldsymbol{y} = F_4\boldsymbol{c}$.

The same idea applies when $n = 1024$ and $m = \frac{1}{2}n = 512$. The number w is $e^{2\pi i/1024}$. It is at the angle $\theta = 2\pi/1024$ on the unit circle. The Fourier matrix F_{1024} is full of powers of w. The first stage of the FFT is the great factorization discovered by Cooley and Tukey (and foreshadowed in 1805 by Gauss):

$$\boldsymbol{F_{1024}} = \begin{bmatrix} I_{512} & D_{512} \\ I_{512} & -D_{512} \end{bmatrix} \begin{bmatrix} \boldsymbol{F_{512}} & \\ & \boldsymbol{F_{512}} \end{bmatrix} \begin{bmatrix} \text{even-odd} \\ \text{permutation} \end{bmatrix}. \tag{4}$$

I_{512} is the identity matrix. D_{512} is the diagonal matrix with entries $(1, w, \ldots, w^{511})$. The two copies of F_{512} are what we expected. Don't forget that they use the 512th root of unity (which is nothing but w^2!!) The permutation matrix separates the incoming vector $\boldsymbol{c}$ into its even and odd parts $\boldsymbol{c}' = (c_0, c_2, \ldots, c_{1022})$ and $\boldsymbol{c}'' = (c_1, c_3, \ldots, c_{1023})$.

Here are the algebra formulas which say the same thing as that factorization of F_{1024}:

(One step of the FFT) Set $\boldsymbol{m = \frac{1}{2}n}$. The first m and last m components of $\boldsymbol{y} = F_n\boldsymbol{c}$ combine the half-size transforms $\boldsymbol{y}' = F_m\boldsymbol{c}'$ and $\boldsymbol{y}'' = F_m\boldsymbol{c}''$. Equation (4) shows this step from n to $m = n/2$ as $I\boldsymbol{y}' + D\boldsymbol{y}''$ and $I\boldsymbol{y}' - D\boldsymbol{y}''$:

$$\begin{aligned} y_j &= y'_j + (w_n)^j y''_j, \quad j = 0, \ldots, m-1 \\ y_{j+m} &= y'_j - (w_n)^j y''_j, \quad j = 0, \ldots, m-1. \end{aligned} \tag{5}$$

Split $\boldsymbol{c}$ into $\boldsymbol{c}'$ and $\boldsymbol{c}''$, transform them by F_m into $\boldsymbol{y}'$ and $\boldsymbol{y}''$, then (5) reconstructs $\boldsymbol{y}$.

Those formulas come from separating $c_0 \ldots, c_{n-1}$ into even c_{2k} and odd c_{2k+1} : w is w_n.

$$\boldsymbol{y} = \boldsymbol{F}\boldsymbol{c} \qquad y_j = \sum_0^{n-1} w^{jk} c_k = \sum_0^{m-1} w^{2jk} c_{2k} + \sum_0^{m-1} w^{j(2k+1)} c_{2k+1} \text{ with } m = \tfrac{1}{2}n. \tag{6}$$

The even c's go into $\boldsymbol{c}' = (c_0, c_2, \ldots)$ and the odd c's go into $\boldsymbol{c}'' = (c_1, c_3, \ldots)$. Then come the transforms $F_m c'$ and $F_m c''$. **The key is $\boldsymbol{w_n^2 = w_m}$.** This gives $w_n^{2jk} = w_m^{jk}$.

$$\textbf{Rewrite (6)} \quad y_j = \sum (w_m)^{jk} c_k' + (w_n)^j \sum (w_m)^{jk} c_k'' = y_j' + (w_n)^j y_j''. \tag{7}$$

For $j \geq m$, the minus sign in (5) comes from factoring out $(w_n)^m = -1$ from $(w_n)^j$.

MATLAB easily separates even c's from odd c's and multiplies by w_n^j. We use conj(F) or equivalently MATLAB's inverse transform ifft, because fft is based on $\omega = \overline{w} = e^{-2\pi i/n}$. Problem 16 shows that F and conj(F) are linked by permuting rows.

FFT step from n to $n/2$ in MATLAB

$$\begin{aligned} y' &= \text{ifft}\,(c(0:2:n-2)) * n/2; \\ y'' &= \text{ifft}\,(c(1:2:n-1)) * n/2; \\ d &= w.\hat{}\,(0:n/2-1)'; \\ y &= [y' + d.*y''\,;\ y' - d.*y'']; \end{aligned}$$

The flow graph shows c' and c'' going through the half-size F_2. Those steps are called "*butterflies*," from their shape. Then the outputs y' and y'' are combined (multiplying y'' by $1, i$ from D and also by $-1, -i$ from $-D$) to produce $y = F_4 c$.

This reduction from F_n to two F_m's almost cuts the work in half—you see the zeros in the matrix factorization. That reduction is good but not great. The full idea of the **FFT** is much more powerful. It saves much more than half the time.

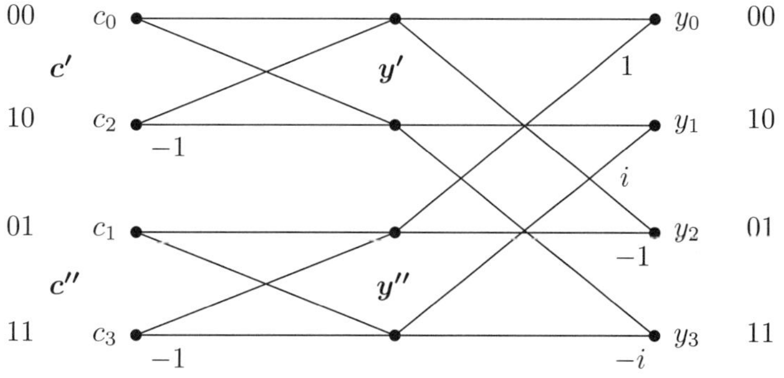

The Full FFT by Recursion

If you have read this far, you probably guessed what comes next. We reduced F_n to $F_{n/2}$. ***Keep going to $\boldsymbol{F_{n/4}}$.*** Every F_{512} leads to F_{256}. Then 256 leads to 128. *That is recursion.*

Recursion is a basic principle of many fast algorithms. Here is step 2 with four copies of F_{256} and D (256 powers of ω_{512}). Evens of evens $c_0, c_4, c_8, \ldots$ come first:

$$\begin{bmatrix} F_{512} & \\ & F_{512} \end{bmatrix} = \begin{bmatrix} I & D & & \\ I & -D & & \\ & & I & D \\ & & I & -D \end{bmatrix} \begin{bmatrix} F & & & \\ & F & & \\ & & F & \\ & & & F \end{bmatrix} \begin{bmatrix} \text{pick } 0,4,8,\ldots \\ \text{pick } 2,6,10,\ldots \\ \text{pick } 1,5,9,\ldots \\ \text{pick } 3,7,11,\ldots \end{bmatrix}.$$

We will count the individual multiplications, to see how much is saved. Before the **FFT** was invented, the count was the usual $n^2 = (1024)^2$. This is about a million multiplications. I am not saying that they take a long time. The cost becomes large when we have many, many transforms to do—which is typical. Then the saving by the FFT is also large:

The final count for size $n = 2^\ell$ is reduced from n^2 to $\frac{1}{2}n\ell$.

The number 1024 is 2^{10}, so $\ell = 10$. The original count of $(1024)^2$ is reduced to (5)(1024). The saving is a factor of 200. A million is reduced to five thousand. That is why the FFT has revolutionized signal processing.

Here is the reasoning behind $\frac{1}{2}n\ell$. There are ℓ levels, going from $n = 2^\ell$ down to $n = 1$. Each level has $n/2$ multiplications from the diagonal D's, to reassemble the half-size outputs from the lower level. This yields the final count $\frac{1}{2}n\ell$, which is $\frac{1}{2}n \log_2 n$.

One last note about this remarkable algorithm. There is an amazing rule for the order that the c's enter the FFT, after all the even-odd permutations. Write the numbers 0 to $n-1$ in binary (like $00, 01, 10, 11$ for $n = 4$). Reverse the order of those digits: $00, 10, 01, 11$. That gives the **bit-reversed order $\mathbf{0, 2, 1, 3}$** with evens before odds (See Problem 17.) The complete picture shows the c's in bit-reversed order, the $\ell = \log_2 n$ steps of the recursion, and the final output $y_0, \ldots, y_{n-1}$ which is F_n times $\boldsymbol{c}$.

The chapter ends with that very fundamental idea, a matrix multiplying a vector.

Problem Set 9.3

1 Multiply the three matrices in equation (3) and compare with F. In which six entries do you need to know that $i^2 = -1$?

2 Invert the three factors in equation (3) to find a fast factorization of F^{-1}.

3 F is symmetric. So transpose equation (3) to find a new Fast Fourier Transform!

4 All entries in the factorization of F_6 involve powers of $w_6 =$ sixth root of 1:

$$F_6 = \begin{bmatrix} I & D \\ I & -D \end{bmatrix} \begin{bmatrix} F_3 & \\ & F_3 \end{bmatrix} \begin{bmatrix} P \end{bmatrix}.$$

Write down these matrices with $1, w_6, w_6^2$ in D and $w_3 = w_6^2$ in F_3. Multiply!

5 If $\boldsymbol{v} = (1,0,0,0)$ and $\boldsymbol{w} = (1,1,1,1)$, show that $F\boldsymbol{v} = \boldsymbol{w}$ and $F\boldsymbol{w} = 4\boldsymbol{v}$. Therefore $F^{-1}\boldsymbol{w} = \boldsymbol{v}$ and $F^{-1}\boldsymbol{v} =$ ______.

6 What is F^2 and what is F^4 for the 4 by 4 Fourier matrix?

7 Put the vector $\boldsymbol{c} = (1,0,1,0)$ through the three steps of the FFT to find $\boldsymbol{y} = F\boldsymbol{c}$. Do the same for $\boldsymbol{c} = (0,1,0,1)$.

8 Compute $\boldsymbol{y} = F_8\boldsymbol{c}$ by the three FFT steps for $\boldsymbol{c} = (1,0,1,0,1,0,1,0)$. Repeat the computation for $\boldsymbol{c} = (0,1,0,1,0,1,0,1)$.

9 If $w = e^{2\pi i/64}$ then w^2 and $\sqrt{w}$ are among the _____ and _____ roots of 1.

10 (a) Draw all the sixth roots of 1 on the unit circle. Prove they add to zero.

(b) What are the three cube roots of 1? Do they also add to zero?

11 The columns of the Fourier matrix F are the *eigenvectors* of the cyclic permutation P (see Section 8.3). Multiply PF to find the eigenvalues $\lambda_1, \lambda_2, \lambda_3, \lambda_4$:

$$\begin{bmatrix} 0 & 1 & 0 & 0 \\ 0 & 0 & 1 & 0 \\ 0 & 0 & 0 & 1 \\ 1 & 0 & 0 & 0 \end{bmatrix} \begin{bmatrix} 1 & 1 & 1 & 1 \\ 1 & i & i^2 & i^3 \\ 1 & i^2 & i^4 & i^6 \\ 1 & i^3 & i^6 & i^9 \end{bmatrix} = \begin{bmatrix} 1 & 1 & 1 & 1 \\ 1 & i & i^2 & i^3 \\ 1 & i^2 & i^4 & i^6 \\ 1 & i^3 & i^6 & i^9 \end{bmatrix} \begin{bmatrix} \lambda_1 & & & \\ & \lambda_2 & & \\ & & \lambda_3 & \\ & & & \lambda_4 \end{bmatrix}.$$

This is $PF = F\Lambda$ or $P = F\Lambda F^{-1}$. The eigenvector matrix (usually X) is F.

12 The equation $\det(P - \lambda I) = 0$ is $\lambda^4 = 1$. This shows again that the eigenvalues are $\lambda =$ _____. Which permutation P has eigenvalues = cube roots of 1?

13 (a) Two eigenvectors of C are $(1, 1, 1, 1)$ and $(1, i, i^2, i^3)$. Find the eigenvalues e.

$$\begin{bmatrix} c_0 & c_1 & c_2 & c_3 \\ c_3 & c_0 & c_1 & c_2 \\ c_2 & c_3 & c_0 & c_1 \\ c_1 & c_2 & c_3 & c_0 \end{bmatrix} \begin{bmatrix} 1 \\ 1 \\ 1 \\ 1 \end{bmatrix} = e_1 \begin{bmatrix} 1 \\ 1 \\ 1 \\ 1 \end{bmatrix} \quad \text{and} \quad C \begin{bmatrix} 1 \\ i \\ i^2 \\ i^3 \end{bmatrix} = e_2 \begin{bmatrix} 1 \\ i \\ i^2 \\ i^3 \end{bmatrix}.$$

(b) $P = F\Lambda F^{-1}$ immediately gives $P^2 = F\Lambda^2 F^{-1}$ and $P^3 = F\Lambda^3 F^{-1}$. Then $\boldsymbol{C} = c_0 I + c_1 P + c_2 P^2 + c_3 P^3 = F(c_0 I + c_1 \Lambda + c_2 \Lambda^2 + c_3 \Lambda^3) F^{-1} = \boldsymbol{FEF^{-1}}$. That matrix E in parentheses is diagonal. It contains the _____ of C.

14 Find the eigenvalues of the "periodic" $-1, 2, -1$ matrix from $E = 2I - \Lambda - \Lambda^3$, with the eigenvalues of P in Λ. The -1's in the corners make this matrix periodic:

$$C = \begin{bmatrix} 2 & -1 & 0 & -1 \\ -1 & 2 & -1 & 0 \\ 0 & -1 & 2 & -1 \\ -1 & 0 & -1 & 2 \end{bmatrix} \quad \text{has } c_0 = 2, c_1 = -1, c_2 = 0, c_3 = -1.$$

15 ***Fast convolution = Fast multiplication by C:*** To multiply C times a vector $\boldsymbol{x}$, we can multiply $F(E(F^{-1}\boldsymbol{x}))$ instead. The direct way uses n^2 separate multiplications. Knowing E and F, the second way uses only $n \log_2 n + n$ multiplications. How many of those come from E, how many from F, and how many from F^{-1}?

16 **Notice.** Why is row i of $\overline{F}$ the same as row $N - i$ of F (numbered 0 to $N - 1$)?

17 What is the *bit-reversed order* of the numbers $0, 1, \ldots, 7$? Write them all in binary (base 2) as $000, 001, \ldots, 111$ and reverse each order. The 8 numbers are now _____.

Chapter 10

Applications

10.1 Graphs and Networks

Over the years I have seen one model so often, and I found it so basic and useful, that I always put it first. The model consists of ***nodes connected by edges***. This is called a ***graph***.

Graphs of the usual kind display functions $f(x)$. Graphs of this node-edge kind lead to matrices. This section is about the ***incidence matrix*** of a graph—which tells how the n nodes are connected by the m edges. Normally $m > n$, there are more edges than nodes.

For any m by n matrix there are two fundamental subspaces in $\mathbf{R}^n$ and two in $\mathbf{R}^m$. They are the row spaces and nullspaces of A and A^{T}. Their *dimensions* $r, n-r$ and $r, m-r$ come from the most important theorem in linear algebra. The second part of that theorem is the *orthogonality* of the row space and nullspace. Our goal is to show how examples from graphs illuminate this Fundamental Theorem of Linear Algebra.

When I construct a ***graph*** and its ***incidence matrix***, the subspace dimensions will be easy to discover. But we want the subspaces themselves—and orthogonality helps. It is essential to connect the subspaces to the graph they come from. By specializing to incidence matrices, **the laws of linear algebra become Kirchhoff's laws**. Please don't be put off by the words "current" and "voltage." These rectangular matrices are the best.

Every entry of an incidence matrix is 0 or 1 or -1. This continues to hold during elimination. All pivots and multipliers are ± 1. Therefore both factors in $A = LU$ also contain $0, 1, -1$. So do the nullspace matrices! All four subspaces have basis vectors with these exceptionally simple components. The matrices are not concocted for a textbook, they come from a model that is absolutely essential in pure and applied mathematics.

The Incidence Matrix

Figure 10.1 displays a graph with $m = 6$ edges and $n = 4$ nodes. The 6 by 4 matrix A tells which nodes are connected by which edges. The first row $-1, 1, 0, 0$ shows that the first edge goes *from node* 1 *to node* 2 (-1 for node 1 because the arrow goes out, $+1$ for node 2 with arrow in).

Row numbers in A are edge numbers, column numbers $1, 2, 3, 4$ are node numbers!

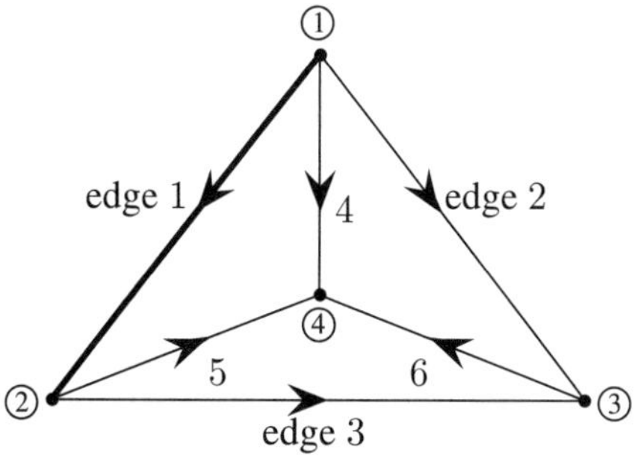

node

① ② ③ ④

$$A = \begin{bmatrix} -1 & 1 & 0 & 0 \\ -1 & 0 & 1 & 0 \\ 0 & -1 & 1 & 0 \\ -1 & 0 & 0 & 1 \\ 0 & -1 & 0 & 1 \\ 0 & 0 & -1 & 1 \end{bmatrix} \begin{matrix} 1 \\ 2 \\ 3 \\ 4 \\ 5 \\ 6 \end{matrix} \quad \textbf{edge}$$

Figure 10.1: Complete graph with $m=6$ edges and $n=4$ nodes: 6 by 4 incidence matrix A.

You can write down the matrix by looking at the graph. The second graph has the same four nodes but only three edges. Its incidence matrix B is 3 by 4.

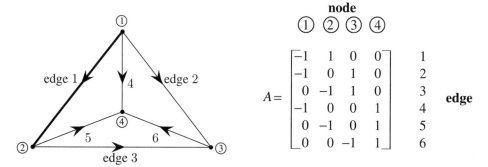

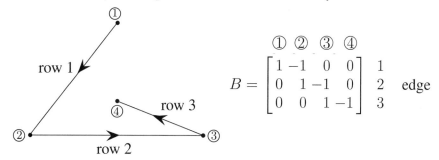

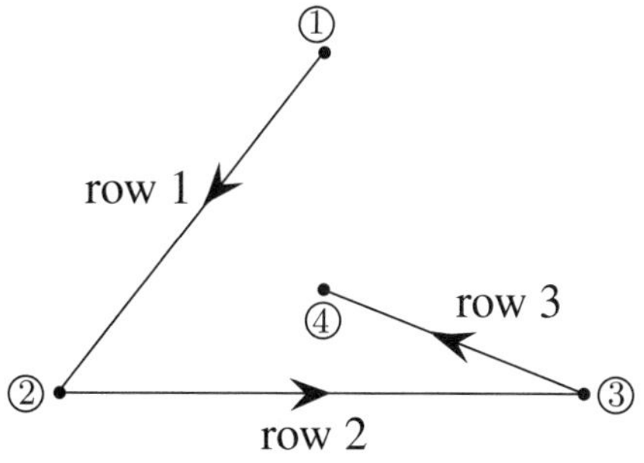

① ② ③ ④

$$B = \begin{bmatrix} 1 & -1 & 0 & 0 \\ 0 & 1 & -1 & 0 \\ 0 & 0 & 1 & -1 \end{bmatrix} \begin{matrix} 1 \\ 2 \\ 3 \end{matrix} \quad \text{edge}$$

Figure 10.1*: Tree with 3 edges and 4 nodes and no loops. Then B has independent rows.

The first graph is *complete*—every pair of nodes is connected by an edge. The second graph is a ***tree***—the graph has ***no closed loops***. Those are the two extremes. The maximum number of edges is $\frac{1}{2}n(n-1) = 6$ and the minimum to stay connected is $n - 1 = 3$.

Elimination reduces every graph to a tree. Loops produce dependent rows in A and zero rows in the echelon forms U and R. Look at the large loop from edges 1, 2, 3 in the first graph, which leads to a zero row in U:

$$\begin{bmatrix} -1 & 1 & 0 & 0 \\ -1 & 0 & 1 & 0 \\ 0 & -1 & 1 & 0 \end{bmatrix} \longrightarrow \begin{bmatrix} -1 & 1 & 0 & 0 \\ 0 & -1 & 1 & 0 \\ 0 & -1 & 1 & 0 \end{bmatrix} \longrightarrow \begin{bmatrix} -1 & 1 & 0 & 0 \\ 0 & -1 & 1 & 0 \\ \mathbf{0} & \mathbf{0} & \mathbf{0} & \mathbf{0} \end{bmatrix}$$

Those steps are typical. When edges 1 and 2 share node 1, elimination produces the "shortcut edge" without node 1. If the graph already has this shortcut edge making a loop, then elimination gives a row of zeros. When the dust clears we have a tree.

An idea suggests itself: ***Rows are dependent when edges form a loop***. Independent rows come from trees. This is the key to the row space. We are assuming that the graph is connected, and the arrows could go either way. On each edge, *flow with the arrow is "positive."* Flow in the opposite direction counts as negative. The flow might be a current or a signal or a force—or even oil or gas or water.

When x_1, x_2, x_3, x_4 are voltages at the nodes, Ax gives voltage differences:

$$Ax = \begin{bmatrix} -1 & 1 & 0 & 0 \\ -1 & 0 & 1 & 0 \\ 0 & -1 & 1 & 0 \\ -1 & 0 & 0 & 1 \\ 0 & -1 & 0 & 1 \\ 0 & 0 & -1 & 1 \end{bmatrix} \begin{bmatrix} x_1 \\ x_2 \\ x_3 \\ x_4 \end{bmatrix} = \begin{bmatrix} x_2 - x_1 \\ x_3 - x_1 \\ x_3 - x_2 \\ x_4 - x_1 \\ x_4 - x_2 \\ x_4 - x_3 \end{bmatrix}. \qquad (1)$$

Let me say that again. The incidence matrix A is a difference matrix. The input vector $\boldsymbol{x}$ gives voltages, the output vector $A\boldsymbol{x}$ gives voltage differences (along edges 1 to 6). If the voltages are equal, the differences are zero. This tells us the nullspace of A.

1 The ***nullspace*** contains the solutions to $A\boldsymbol{x} = \mathbf{0}$. All six voltage differences are zero. This means: *All four voltages are equal.* Every $\boldsymbol{x}$ in the nullspace is a **constant vector**: $\boldsymbol{x} = (c, c, c, c)$. The nullspace of A is a line in $\mathbf{R}^n$—its dimension is $n - r = 1$.

The second incidence matrix B has the same nullspace. It contains $(\mathbf{1}, \mathbf{1}, \mathbf{1}, \mathbf{1})$:

1–dimensional nullspace: same for the tree

$$B\boldsymbol{x} = \begin{bmatrix} -1 & 1 & 0 & 0 \\ 0 & -1 & 1 & 0 \\ 0 & 0 & -1 & 1 \end{bmatrix} \begin{bmatrix} 1 \\ 1 \\ 1 \\ 1 \end{bmatrix} = \begin{bmatrix} 0 \\ 0 \\ 0 \end{bmatrix}.$$

We can raise or lower all voltages by the same amount c, without changing the differences. There is an "arbitrary constant" in the voltages. Compare this with the same statement for functions. We can raise or lower a function by C, without changing its derivative.

Calculus adds "$+C$" to indefinite integrals. Graph theory adds (c, c, c, c) to the vector $\boldsymbol{x}$. Linear algebra adds any vector $\boldsymbol{x}_n$ in the nullspace to one particular solution of $A\boldsymbol{x} = \boldsymbol{b}$.

The "$+C$" disappears in calculus when a definite integral starts at a known point. Similarly the nullspace disappears when we fix $\boldsymbol{x_4} = \mathbf{0}$. The unknown x_4 is removed and so are the fourth columns of A and B (those columns multiplied x_4). Electrical engineers would say that node 4 has been "grounded."

2 The ***row space*** contains all combinations of the six rows. Its dimension is certainly not 6. The equation $r + (n - r) = n$ must be $3 + 1 = 4$. The rank is $r = 3$, as we saw from elimination. After 3 edges, we start forming loops! The new rows are not independent.

How can we tell if $\boldsymbol{v} = (v_1, v_2, v_3, v_4)$ is in the row space? The slow way is to combine rows. The quick way is by orthogonality:

v is in the row space if and only if it is perpendicular to $(1, 1, 1, 1)$ in the nullspace.

The vector $\boldsymbol{v} = (0, 1, 2, 3)$ fails this test—its components add to 6. The vector $(-6, 1, 2, 3)$ is in the row space: $-6+1+2+3 = 0$. That vector equals $6(\text{row } 1) + 5(\text{row } 3) + 3(\text{row } 6)$.

Each row of A adds to zero. This must be true for every vector in the row space.

3 The ***column space*** contains all combinations of the four columns. We expect three independent columns, since there were three independent rows. The first three columns of A are independent (so are any three). But the four columns add to the zero vector, which says again that $(1,1,1,1)$ is in the nullspace. ***How can we tell if a particular vector b is in the column space of an incidence matrix?***

First answer Try to solve $A\boldsymbol{x} = \boldsymbol{b}$. That misses all the insight. As before, orthogonality gives a better answer. We are now coming to Kirchhoff's two famous laws of circuit theory—the voltage law and current law (**KVL** and **KCL**). Those are natural expressions of "laws" of linear algebra. It is especially pleasant to see the key role of the left nullspace.

Second answer $A\boldsymbol{x}$ is the vector of voltage differences $x_i - x_j$. If we add differences around a closed loop in the graph, they cancel to leave zero. Around the big triangle formed by edges $1, 3, -2$ (*the arrow goes backward on edge* 2) the differences cancel:

Sum of differences is 0 $$(x_2 - x_1) + (x_3 - x_2) - (x_3 - x_1) = \mathbf{0}.$$

Kirchhoff's Voltage Law: The components of $A\boldsymbol{x} = \boldsymbol{b}$ add to zero around every loop.

Around the big triangle: $b_1 + b_3 - b_2 = 0.$

By testing each loop, the Voltage Law decides whether $\boldsymbol{b}$ is in the column space. $A\boldsymbol{x} = \boldsymbol{b}$ can be solved exactly when the components of $\boldsymbol{b}$ satisfy all the same dependencies as the rows of A. Then elimination leads to $0 = 0$, and $A\boldsymbol{x} = \boldsymbol{b}$ is consistent.

4 The ***left nullspace*** contains the solutions to $A^{\mathrm{T}}\boldsymbol{y} = \mathbf{0}$. Its dimension is $m - r = 6 - 3$:

Current Law
$$A^{\mathrm{T}}\boldsymbol{y} = \begin{bmatrix} -1 & -1 & 0 & -1 & 0 & 0 \\ 1 & 0 & -1 & 0 & -1 & 0 \\ 0 & 1 & 1 & 0 & 0 & -1 \\ 0 & 0 & 0 & 1 & 1 & 1 \end{bmatrix} \begin{bmatrix} y_1 \\ y_2 \\ y_3 \\ y_4 \\ y_5 \\ y_6 \end{bmatrix} = \begin{bmatrix} 0 \\ 0 \\ 0 \\ 0 \end{bmatrix}. \quad (2)$$

The true number of equations is $r = 3$ and not $n = 4$. Reason: The four equations add to $0 = 0$. The fourth equation follows automatically from the first three.

What do the equations mean? The first equation says that $-y_1 - y_2 - y_4 = 0$. ***The net flow into node* 1 *is zero***. The fourth equation says that $y_4 + y_5 + y_6 = 0$. ***Flow into node 4 minus flow out is zero***. The equations $A^{\mathrm{T}}\boldsymbol{y} = \mathbf{0}$ are famous and fundamental:

Kirchhoff's Current Law: $A^{\mathrm{T}}y = 0$ *Flow in equals flow out at each node.*

This law deserves first place among the equations of applied mathematics. It expresses "*conservation*" and "*continuity*" and "*balance*." Nothing is lost, nothing is gained. When currents or forces are balanced, the equation to solve is $A^{\mathrm{T}}\boldsymbol{y} = \mathbf{0}$. Notice the beautiful fact that the matrix in this balance equation is the transpose of the incidence matrix A.

What are the actual solutions to $A^{\mathrm{T}}\boldsymbol{y} = \mathbf{0}$? The currents must balance themselves. The easiest way is to **flow around a loop**. If a unit of current goes around the big triangle (forward on edge 1 and 3, backward on 2), the six currents are $\boldsymbol{y} = (1, -1, 1, 0, 0, 0)$. This satisfies $A^{\mathrm{T}}\boldsymbol{y} = \mathbf{0}$. *Every loop current is a solution to the Current Law.* Flow in equals flow out at every node. A smaller loop goes forward on edge 1, forward on 5, back on 4. Then $\boldsymbol{y} = (\mathbf{1}, 0, 0, -\mathbf{1}, \mathbf{1}, 0)$ is also in the left nullspace.

We expect three independent $\boldsymbol{y}$'s: $m - r = 6 - 3 = 3$. The three small loops in the graph are independent. The big triangle seems to give a fourth $\boldsymbol{y}$, but that flow is the sum of flows around the small loops. *Flows around the* 3 *small loops are a basis for the left nullspace.*

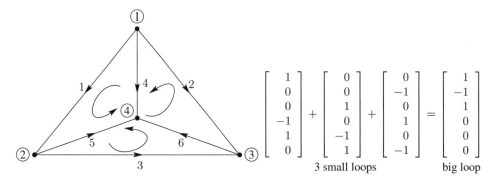

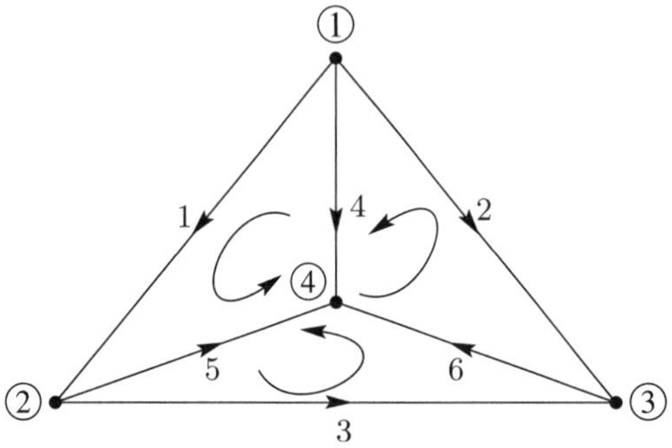

$$\begin{bmatrix} 1 \\ 0 \\ 0 \\ -1 \\ 1 \\ 0 \end{bmatrix} + \begin{bmatrix} 0 \\ 0 \\ 1 \\ 0 \\ -1 \\ 1 \end{bmatrix} + \begin{bmatrix} 0 \\ -1 \\ 0 \\ 1 \\ 0 \\ -1 \end{bmatrix} = \begin{bmatrix} 1 \\ -1 \\ 1 \\ 0 \\ 0 \\ 0 \end{bmatrix}$$

3 small loops big loop

The incidence matrix A comes from a connected graph with n nodes and m edges. The row space and column space have dimensions $r = n - 1$. The nullspaces of A and A^{T} have dimensions 1 and $m - n + 1$:

$\boldsymbol{N}(A)$ The constant vectors $(c, c, \ldots, c)$ make up the nullspace of A : dim $= 1$.

$\boldsymbol{C}(A^{\mathrm{T}})$ The edges of any tree give r independent rows of A : $r = n - 1$.

$\boldsymbol{C}(A)$ ***Voltage Law***: The components of $A\boldsymbol{x}$ add to zero around all loops: dim $= n - 1$.

$\boldsymbol{N}(A^{\mathrm{T}})$ ***Current Law***: $A^{\mathrm{T}}\boldsymbol{y} =$ (**flow in**) $-$ (**flow out**) $= \mathbf{0}$ is solved by loop currents.

There are $m - r = m - n + 1$ independent small loops in the graph.

For every graph in a plane, linear algebra yields ***Euler's formula***: Theorem 1 in topology!

$$(\textit{number of nodes}) - (\textit{number of edges}) + (\textit{number of small loops}) = \mathbf{1}.$$

This is $(\boldsymbol{n}) - (\boldsymbol{m}) + (\boldsymbol{m} - \boldsymbol{n} + \mathbf{1}) = \mathbf{1}$. The graph in our example has $4 - 6 + 3 = 1$.

A single triangle has (3 nodes) $-$ (3 edges) $+$ (1 loop). On a 10-node tree with 9 edges and no loops, Euler's count is $10 - 9 + 0$. All planar graphs lead to the answer 1.

The next figure shows a network with a current source. Kirchhoff's Current Law changes from $A^{\mathrm{T}}\boldsymbol{y} = \mathbf{0}$ to $A^{\mathrm{T}}\boldsymbol{y} = \boldsymbol{f}$, to balance the source $\boldsymbol{f}$ from outside. *Flow into each node still equals flow out.* The six edges would have conductances $c_1, \ldots, c_6$, and the current source goes into node 1. The source comes out from node 4 to keep the overall balance (**in** $=$ **out**). The problem is: ***Find the currents $y_1, \ldots, y_6$ on the six edges.***
Flows in networks now lead us from the incidence matrix A to the Laplacian matrix $A^{\mathrm{T}}A$.

Voltages and Currents and $A^{\mathrm{T}}Ax = f$

We started with voltages $\boldsymbol{x} = (x_1, \ldots, x_n)$ at the nodes. So far we have $A\boldsymbol{x}$ to find voltage differences $x_i - x_j$ along edges. And we have the Current Law $A^{\mathrm{T}}\boldsymbol{y} = \mathbf{0}$ to find edge currents $\boldsymbol{y} = (y_1, \ldots y_m)$. If all resistances in the network are 1, Ohm's Law will match $\boldsymbol{y} = A\boldsymbol{x}$. Then $A^{\mathrm{T}}\boldsymbol{y} = A^{\mathrm{T}}A\boldsymbol{x} = \mathbf{0}$. We are close but not quite there.

Without any sources, the solution to $A^{\mathrm{T}}A\boldsymbol{x} = \mathbf{0}$ will just be no flow: $\boldsymbol{x} = \mathbf{0}$ and $\boldsymbol{y} = \mathbf{0}$. I can see three ways to produce $\boldsymbol{x} \neq \mathbf{0}$ and $\boldsymbol{y} \neq \mathbf{0}$.

1 Assign fixed voltages x_i to one or more nodes.

2 Add batteries (voltage sources) in one or more edges.

3 Add current sources going into one or more nodes. See Figure 10.2

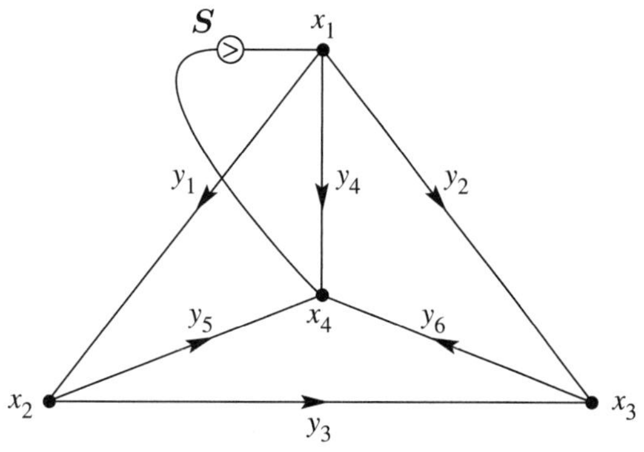

Figure 10.2: The currents y_1 to y_6 in a network with a source S from node 4 to node 1.

Example Figure 10.2 includes a current source S from node 4 to node 1. That current will trickle back through the network to node 4. Some current y_4 will go directly on edge 4. Other current will go the long way from node 1 to 2 to 4, or 1 to 3 to 4. By symmetry I expect no current ($y_3 = 0$) from node 2 to node 3. Solving the network equations will confirm this. **The matrix in those equations is $A^{\mathrm{T}}A$, *the graph Laplacian matrix*:**

$$\begin{bmatrix} -1 & -1 & 0 & -1 & 0 & 0 \\ 1 & 0 & -1 & 0 & -1 & 0 \\ 0 & 1 & 1 & 0 & 0 & -1 \\ 0 & 0 & 0 & 1 & 1 & 1 \end{bmatrix} \begin{bmatrix} -1 & 1 & 0 & 0 \\ -1 & 0 & 1 & 0 \\ 0 & -1 & 1 & 0 \\ -1 & 0 & 0 & 1 \\ 0 & -1 & 0 & 1 \\ 0 & 0 & -1 & 1 \end{bmatrix} = \underset{\boldsymbol{A^{\mathrm{T}}A}}{\begin{bmatrix} 3 & -1 & -1 & -1 \\ -1 & 3 & -1 & -1 \\ -1 & -1 & 3 & -1 \\ -1 & -1 & -1 & 3 \end{bmatrix}}$$

That Laplacian matrix is not invertible! We cannot solve for all four potentials because $(1,1,1,1)$ is in the nullspace of A and $A^{\mathrm{T}}A$. *One node has to be grounded*. Setting $x_4 = 0$ removes the fourth row and column, and this leaves a 3 by 3 invertible matrix. Now we solve $A^{\mathrm{T}}A\boldsymbol{x} = \boldsymbol{f}$ for the unknown potentials x_1, x_2, x_3, with source S into node 1:

$$\textbf{Voltages} \quad A^{\mathrm{T}}Ax = f \qquad \begin{bmatrix} 3 & -1 & -1 \\ -1 & 3 & -1 \\ -1 & -1 & 3 \end{bmatrix} \begin{bmatrix} x_1 \\ x_2 \\ x_3 \end{bmatrix} = \begin{bmatrix} S \\ 0 \\ 0 \end{bmatrix} \quad \text{gives} \quad \begin{bmatrix} x_1 \\ x_2 \\ x_3 \end{bmatrix} = \begin{bmatrix} S/2 \\ S/4 \\ S/4 \end{bmatrix}.$$

$$\textbf{Currents} \quad y = -Ax \qquad \begin{bmatrix} y_1 \\ y_2 \\ y_3 \\ y_4 \\ y_5 \\ y_6 \end{bmatrix} = - \begin{bmatrix} -1 & 1 & 0 & 0 \\ -1 & 0 & 1 & 0 \\ 0 & -1 & 1 & 0 \\ -1 & 0 & 0 & 1 \\ 0 & -1 & 0 & 1 \\ 0 & 0 & -1 & 1 \end{bmatrix} \begin{bmatrix} S/2 \\ S/4 \\ S/4 \\ 0 \end{bmatrix} = \begin{bmatrix} S/4 \\ S/4 \\ 0 \\ S/2 \\ S/4 \\ S/4 \end{bmatrix}.$$

Half the current goes directly on edge 4. That is $y_4 = S/2$. No current crosses from node 2 to node 3. Symmetry indicated $y_3 = 0$ and now the solution proves it.

Admission of error I remembered that current flows from high voltage to low voltage. That produces the minus sign in $\boldsymbol{y} = -A\boldsymbol{x}$. And the correct form of Ohm's Law will be $R\boldsymbol{y} = -A\boldsymbol{x}$ when the resistances on the edges are not all 1. *Conductances* are neater than resistances: $C = R^{-1} =$ diagonal matrix. **We now present Ohm's Law $\boldsymbol{y} = -\boldsymbol{C}\boldsymbol{A}\boldsymbol{x}$.**

Networks and $A^{\mathrm{T}}CA$

In a real network, the current $\boldsymbol{y}$ along an edge is the product of two numbers. One number is the difference between the potentials $\boldsymbol{x}$ at the ends of the edge. This voltage difference is $A\boldsymbol{x}$ and it drives the flow. The other number c is the "***conductance***"—which measures how easily flow gets through.

In physics and engineering, c is decided by the material. For electrical currents, c is high for metal and low for plastics. For a superconductor, c is nearly infinite. If we consider elastic stretching, c might be low for metal and higher for plastics. In economics, c measures the capacity of an edge or its cost.

To summarize, the graph is known from its incidence matrix A. This tells the node-edge connections. A ***network*** goes further, and assigns a conductance c to each edge. *These numbers $c_1, \ldots, c_m$ go into the "conductance matrix" C—which is diagonal.*

For a network of resistors, the conductance is $c = 1/(\text{resistance})$. In addition to Kirchhoff's Laws for the whole system of currents, we have Ohm's Law for each current. Ohm's Law connects the current y_1 on edge 1 to the voltage difference $x_2 - x_1$:

Ohm's Law: Current along edge = conductance times voltage difference.

Ohm's Law for all m currents is $\boldsymbol{y} = -CA\boldsymbol{x}$. The vector $A\boldsymbol{x}$ gives the potential differences, and C multiplies by the conductances. Combining Ohm's Law with Kirchhoff's

Current Law $A^{\mathrm{T}}\boldsymbol{y}=\mathbf{0}$, we get $A^{\mathrm{T}}CA\boldsymbol{x}=\mathbf{0}$. This is *almost* the central equation for network flows. The only thing wrong is the zero on the right side! The network needs power from outside—a voltage source or a current source—to make something happen.

Note about signs In circuit theory we change from $A\boldsymbol{x}$ to $-A\boldsymbol{x}$. The flow is from higher potential to lower potential. There is (positive) current from node 1 to node 2 when x_1-x_2 is positive—whereas $A\boldsymbol{x}$ was constructed to yield x_2-x_1. The minus sign in physics and electrical engineering is a plus sign in mechanical engineering and economics. $A\boldsymbol{x}$ versus $-A\boldsymbol{x}$ is a general headache but unavoidable.

Note about applied mathematics Every new application has its own form of Ohm's Law. For springs it is Hooke's Law. The stress $\boldsymbol{y}$ is (elasticity C) times (stretching $A\boldsymbol{x}$). For heat conduction, $A\boldsymbol{x}$ is a temperature gradient. For oil flows it is a pressure gradient. For least squares regression in statistics (Chapter 12) C^{-1} is the covariance matrix.

My textbooks *Introduction to Applied Mathematics* and *Computational Science and Engineering* (Wellesley-Cambridge Press) are practically built on $A^{\mathrm{T}}CA$. This is the key to equilibrium in matrix equations and also in differential equations. Applied mathematics is more organized than it looks! *In new problems I have learned to watch for $A^{\mathrm{T}}CA$.*

Problem Set 10.1

Problems 1–7 and 8–14 are about the incidence matrices for these graphs.

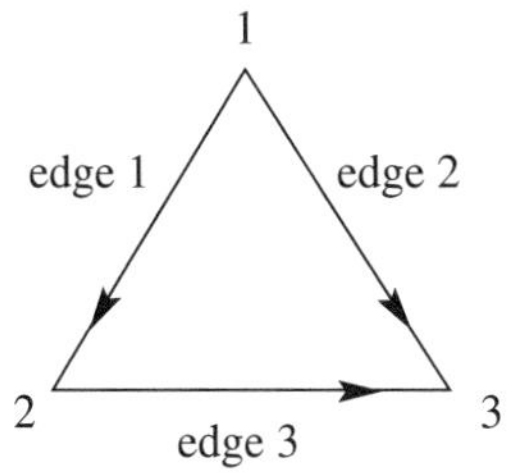

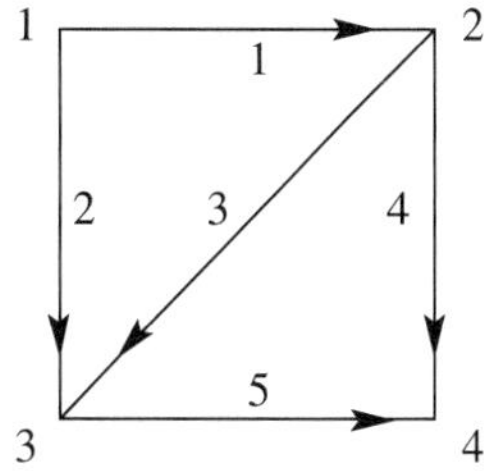

1 Write down the 3 by 3 incidence matrix A for the triangle graph. The first row has -1 in column 1 and $+1$ in column 2. What vectors (x_1, x_2, x_3) are in its nullspace? How do you know that $(1, 0, 0)$ is not in its row space?

2 Write down A^{T} for the triangle graph. Find a vector $\boldsymbol{y}$ in its nullspace. The components of $\boldsymbol{y}$ are currents on the edges—how much current is going around the triangle?

3 Eliminate x_1 and x_2 from the third equation to find the echelon matrix U. What tree corresponds to the two nonzero rows of U?

$$\begin{aligned}-x_1+x_2&=b_1\\-x_1+x_3&=b_2\\-x_2+x_3&=b_3.\end{aligned}$$

4 Choose a vector (b_1, b_2, b_3) for which $A\boldsymbol{x} = \boldsymbol{b}$ can be solved, and another vector $\boldsymbol{b}$ that allows no solution. How are those $\boldsymbol{b}$'s related to $\boldsymbol{y} = (1, -1, 1)$?

5 Choose a vector (f_1, f_2, f_3) for which $A^{\mathrm{T}}\boldsymbol{y} = \boldsymbol{f}$ can be solved, and a vector $\boldsymbol{f}$ that allows no solution. How are those $\boldsymbol{f}$'s related to $\boldsymbol{x} = (1, 1, 1)$? The equation $A^{\mathrm{T}}\boldsymbol{y} = \boldsymbol{f}$ is Kirchhoff's _____ law.

6 Multiply matrices to find $A^{\mathrm{T}}A$. Choose a vector $\boldsymbol{f}$ for which $A^{\mathrm{T}}A\boldsymbol{x} = \boldsymbol{f}$ can be solved, and solve for $\boldsymbol{x}$. Put those potentials $\boldsymbol{x}$ and the currents $\boldsymbol{y} = -A\boldsymbol{x}$ and current sources $\boldsymbol{f}$ onto the triangle graph. Conductances are 1 because $C = I$.

7 With conductances $c_1 = 1$ and $c_2 = c_3 = 2$, multiply matrices to find $A^{\mathrm{T}}CA$. For $\boldsymbol{f} = (1, 0, -1)$ find a solution to $A^{\mathrm{T}}CA\boldsymbol{x} = \boldsymbol{f}$. Write the potentials $\boldsymbol{x}$ and currents $\boldsymbol{y} = -CA\boldsymbol{x}$ on the triangle graph, when the current source $\boldsymbol{f}$ goes into node 1 and out from node 3.

8 Write down the 5 by 4 incidence matrix A for the square graph with two loops. Find one solution to $A\boldsymbol{x} = \boldsymbol{0}$ and two solutions to $A^{\mathrm{T}}\boldsymbol{y} = \boldsymbol{0}$.

9 Find two requirements on the b's for the five differences $x_2 - x_1, x_3 - x_1, x_3 - x_2, x_4 - x_2, x_4 - x_3$ to equal b_1, b_2, b_3, b_4, b_5. You have found Kirchhoff's _____ law around the two _____ in the graph.

10 Reduce A to its echelon form U. The three nonzero rows give the incidence matrix for what graph? You found one tree in the square graph—find the other seven trees.

11 Multiply matrices to find $A^{\mathrm{T}}A$ and guess how its entries come from the graph:

(a) The diagonal of $A^{\mathrm{T}}A$ tells how many _____ into each node.

(b) The off-diagonals -1 or 0 tell which pairs of nodes are _____.

12 Why is each statement true about $A^{\mathrm{T}}A$? *Answer for $A^{\mathrm{T}}A$ not A.*

(a) Its nullspace contains $(1, 1, 1, 1)$. Its rank is $n - 1$.

(b) It is positive semidefinite but not positive definite.

(c) Its four eigenvalues are real and their signs are _____.

13 With conductances $c_1 = c_2 = 2$ and $c_3 = c_4 = c_5 = 3$, multiply the matrices $A^{\mathrm{T}}CA$. Find a solution to $A^{\mathrm{T}}CA\boldsymbol{x} = \boldsymbol{f} = (1, 0, 0, -1)$. Write these potentials $\boldsymbol{x}$ and currents $\boldsymbol{y} = -CA\boldsymbol{x}$ on the nodes and edges of the square graph.

14 The matrix $A^{\mathrm{T}}CA$ is not invertible. What vectors $\boldsymbol{x}$ are in its nullspace? Why does $A^{\mathrm{T}}CA\boldsymbol{x} = \boldsymbol{f}$ have a solution if and only if $f_1 + f_2 + f_3 + f_4 = 0$?

15 A connected graph with 7 nodes and 7 edges has how many loops?

16 For the graph with 4 nodes, 6 edges, and 3 loops, add a new node. If you connect it to one old node, Euler's formula becomes $(\quad) - (\quad) + (\quad) = 1$. If you connect it to two old nodes, Euler's formula becomes $(\quad) - (\quad) + (\quad) = 1$.

17 Suppose A is a 12 by 9 incidence matrix from a connected (but unknown) graph.

(a) How many columns of A are independent?

(b) What condition on $\boldsymbol{f}$ makes it possible to solve $A^{\mathrm{T}}\boldsymbol{y} = \boldsymbol{f}$?

(c) The diagonal entries of $A^{\mathrm{T}}A$ give the number of edges into each node. What is the sum of those diagonal entries?

18 Why does a complete graph with $n = 6$ nodes have $m = 15$ edges? A tree connecting 6 nodes has _____ edges.

Note The ***stoichiometric matrix*** in chemistry is an important "generalized" incidence matrix. Its entries show how much of each chemical species (each column) goes into each reaction (each row).

10.2 Matrices in Engineering

This section will show how engineering problems produce symmetric matrices K (often K is positive definite). The "linear algebra reason" for symmetry and positive definiteness is their form $K = A^{\mathrm{T}}A$ and $K = A^{\mathrm{T}}CA$. The "physical reason" is that the expression $\frac{1}{2}\boldsymbol{u}^{\mathrm{T}}K\boldsymbol{u}$ represents *energy*—and energy is never negative. The matrix C, often diagonal, contains positive physical constants like conductance or stiffness or diffusivity.

Our best examples come from mechanical and civil and aeronautical engineering. K is the ***stiffness matrix***, and $K^{-1}\boldsymbol{f}$ is the structure's response to forces $\boldsymbol{f}$ from outside. Section 10.1 turned to electrical engineering—the matrices came from networks and circuits. The exercises involve chemical engineering and I could go on! Economics and management and engineering design come later in this chapter (the key is optimization).

Engineering leads to linear algebra in two ways, directly and indirectly:

Direct way The physical problem has only a finite number of pieces. The laws connecting their position or velocity are *linear* (movement is not too big or too fast). The laws are expressed by *matrix equations.*

Indirect way The physical system is "continuous". Instead of individual masses, the mass density and the forces and the velocities are functions of x or x, y or x, y, z. The laws are expressed by *differential equations.* ***To find accurate solutions we approximate by finite difference equations or finite element equations***.

Both ways produce matrix equations and linear algebra. I really believe that you cannot do modern engineering without matrices.

Here we present equilibrium equations $K\boldsymbol{u} = \boldsymbol{f}$. With motion, $Md^2\boldsymbol{u}/dt^2 + K\boldsymbol{u} = \boldsymbol{f}$ becomes dynamic. Then we would use eigenvalues from $K\boldsymbol{x} = \lambda M\boldsymbol{x}$, or finite differences.

Differential Equation to Matrix Equation

Differential equations are continuous. Our basic example will be $-d^2u/dx^2 = f(x)$. Matrix equations are discrete. Our basic example will be $K_0\boldsymbol{u} = \boldsymbol{f}$. By taking the step from second derivatives to second differences, you will see the big picture in a very short space. *Start with fixed boundary conditions at both ends $x = 0$ and $x = 1$:*

Fixed-fixed boundary value problem
$$-\frac{d^2u}{dx^2} = 1 \text{ with } u(0) = 0 \text{ and } u(1) = 0. \qquad (1)$$

That differential equation is linear. A particular solution is $u_p = -\frac{1}{2}x^2$ (then $d^2u/dx^2 = -1$). We can add any function "in the nullspace". Instead of solving $A\boldsymbol{x} = \boldsymbol{0}$ for a vector $\boldsymbol{x}$, we solve $-d^2u/dx^2 = 0$ for a function $u_n(x)$. (Main point: The right side is zero.)

The nullspace solutions are $u_n(x) = C + Dx$ (a 2-dimensional nullspace for a second order differential equation). The complete solution is $u_p + u_n$:

Complete solution to	$-\dfrac{d^2u}{dx^2} = 1$	$u(x) = -\dfrac{1}{2}x^2 + C + Dx.$	(2)

Now find C and D from the two boundary conditions: Set $x = 0$ and then $x = 1$. At $x = 0, u(0) = 0$ forces $\boldsymbol{C = 0}$. At $x = 1, u(1) = 0$ forces $-\frac{1}{2} + D = 0$. Then $\boldsymbol{D = \frac{1}{2}}$:

$$u(x) = -\frac{1}{2}x^2 + \frac{1}{2}x = \frac{\mathbf{1}}{\mathbf{2}}(\boldsymbol{x - x^2}) \text{ solves the fixed-fixed boundary value problem.} \quad (3)$$

Differences Replace Derivatives

To get matrices instead of derivatives, we have three basic choices—*forward or backward or centered differences.* Start with first derivatives and first differences:

$$\frac{\boldsymbol{du}}{\boldsymbol{dx}} \approx \frac{u(x + \Delta x) - u(x)}{\Delta x} \text{ or } \frac{u(x) - u(x - \Delta x)}{\Delta x} \text{ or } \frac{u(x + \Delta x) - u(x - \Delta x)}{2\Delta x}.$$

Between $x = 0$ and $x = 1$, we divide the interval into $n + 1$ equal pieces. The pieces have width $\Delta x = 1/(n + 1)$. The values of u at the n breakpoints $\Delta x, 2\Delta x, \ldots$ will be the unknowns u_1 to u_n in our matrix equation $K\boldsymbol{u} = \boldsymbol{f}$:

Solution to compute: $\boldsymbol{u} = (u_1, u_2, \ldots, u_n) \approx (u(\Delta x), u(2\Delta x), \ldots, u(n\Delta x))$.
Zero values $u_0 = u_{n+1} = 0$ come from the boundary conditions $u(0) = u(1) = 0$.

Replace the derivatives in $-\dfrac{d}{dx}\left(\dfrac{du}{dx}\right) = 1$ *by forward and backward differences*:

$$\frac{1}{(\Delta x)^2}\begin{bmatrix} 1 & -1 & 0 & 0 \\ 0 & 1 & -1 & 0 \\ 0 & 0 & 1 & -1 \end{bmatrix}\begin{bmatrix} 1 & 0 & 0 \\ -1 & 1 & 0 \\ 0 & -1 & -1 \\ 0 & 0 & -1 \end{bmatrix}\begin{bmatrix} u_1 \\ u_2 \\ u_3 \end{bmatrix} = \begin{bmatrix} 1 \\ 1 \\ 1 \end{bmatrix} \quad (4)$$

This is our matrix equation when $n = 3$ and $\Delta x = \frac{1}{4}$. The two first differences are transposes of each other! The equation is $A^{\mathrm{T}}A\boldsymbol{u} = (\Delta x)^2\boldsymbol{f}$. When we multiply $A^{\mathrm{T}}A$, we get the positive definite second difference matrix K_0:

$\boldsymbol{K_0\, u} =$ $\boldsymbol{(\Delta x)^2 f}$	$\begin{bmatrix} 2 & -1 & 0 \\ -1 & 2 & -1 \\ 0 & -1 & 2 \end{bmatrix}\begin{bmatrix} u_1 \\ u_2 \\ u_3 \end{bmatrix} = \dfrac{1}{16}\begin{bmatrix} 1 \\ 1 \\ 1 \end{bmatrix}$	gives	$\begin{bmatrix} u_1 \\ u_2 \\ u_3 \end{bmatrix} = \dfrac{1}{32}\begin{bmatrix} 3 \\ 4 \\ 3 \end{bmatrix}.$	(5)

The wonderful fact in this example is that those numbers u_1, u_2, u_3 are exactly correct! They agree with the true solution $u = \frac{1}{2}(x - x^2)$ at the three meshpoints $x = \frac{1}{4}, \frac{2}{4}, \frac{3}{4}$. Figure 10.3 shows the true solution (continuous curve) and the approximations u_1, u_2, u_3 (lying exactly on the curve). This curve is a parabola.

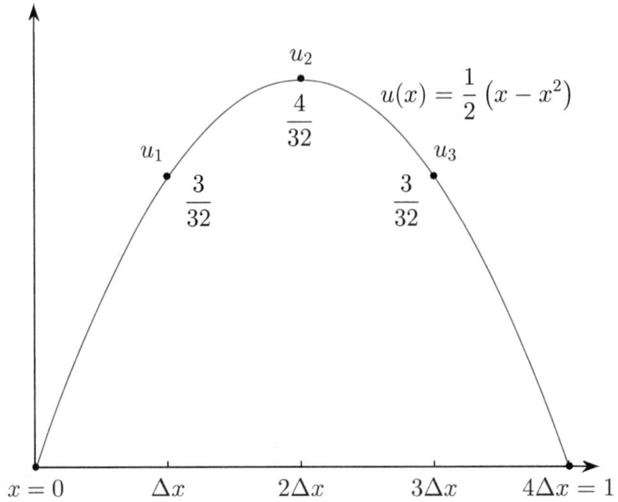

Figure 10.3: Solutions to $-\dfrac{d^2u}{dx^2} = 1$ and $K_0\boldsymbol{u} = (\boldsymbol{\Delta x})^2\boldsymbol{f}$ with fixed-fixed boundaries.

How to explain this perfect answer, lying right on the graph of $u(x)$? In the matrix equation, $K_0 = A^{\mathrm{T}}A$ is a "second difference matrix." It gives a centered approximation to $-d^2u/dx^2$. I included the minus sign because the first derivative is *antisymmetric*. The second derivative by itself is *negative*:

$$\text{The "transpose" of } \frac{d}{dx} \text{ is } -\frac{d}{dx}. \text{ Then } \left(-\frac{d}{dx}\right)\left(\frac{d}{dx}\right) \text{ is positive definite.}$$

You can see that in the matrices A and A^{T}. The transpose of A = *forward difference* is $A^{\mathrm{T}} = -$ *backward difference*. I don't want to choose a centered $u(x+\Delta x) - u(x-\Delta x)$. Centered is the best for a first difference, but then the second difference $A^{\mathrm{T}}A$ would stretch from $u(x + \mathbf{2}\Delta x)$ to $u(x - \mathbf{2}\Delta x)$: not good.

Now we can explain the perfect answers, exactly on the true curve $u(x) = \frac{1}{2}(x - x^2)$. Second differences $-1, 2, -1$ are exactly correct for straight lines $y = x$ and parabolas!

$$\boldsymbol{y = x} \qquad -\frac{d^2y}{dx^2} = \mathbf{0} \qquad -(x+\Delta x) \;\; +2x \;\; -(x-\Delta x) \;=\; \mathbf{0}(\Delta x)^2$$

$$\boldsymbol{y = x^2} \qquad -\frac{d^2y}{dx^2} = \mathbf{-2} \qquad -(x+\Delta x)^2 \; +2x^2 \; -(x-\Delta x)^2 = \; \mathbf{-2}(\Delta x)^2$$

The miracle continues to $y = x^3$. The correct $-d^2y/dx^2 = -6x$ is produced by second differences. But for $y = x^4$ we return to earth. Second differences don't exactly match $-y'' = -12x^2$. The approximations u_1, u_2, u_3 won't fall on the graph of $u(x)$.

Fixed End and Free End and Variable Coefficient $c(x)$

To see two new possibilities, I will change the equation and also one boundary condition:

$$\boxed{-\frac{d}{dx}\left((1+x)\frac{du}{dx}\right) = f(x) \text{ with } u(0)=0 \text{ and } \frac{\boldsymbol{du}}{\boldsymbol{dx}}(\mathbf{1}) = \mathbf{0}.} \qquad (6)$$

The end $x = 1$ is now **free**. There is no support at that end. "A hanging bar is fixed only at the top." There is no force at the free end $x = 1$. That translates to $du/dx = 0$ instead of the fixed condition $u = 0$ at $x = 1$.

The other change is in the coefficient $c(x) = 1 + x$. The stiffness of the bar is varying as you go from $x = 0$ to $x = 1$. Maybe its width is changing, or the material changes. This coefficient $1 + x$ will bring a new matrix C into the difference equation.

Since u_4 is no longer fixed at 0, it becomes a new unknown. The backward difference A is 4 by 4. And the multiplication by $c(x) = 1 + x$ becomes a diagonal matrix C—which multiplies by $1 + \Delta x, \ldots, 1 + 4\Delta x$ at the meshpoints. Here are A^{T}, C, and A:

$$A^{\mathrm{T}}CA = \begin{bmatrix} 1 & -1 & 0 & 0 \\ 0 & 1 & -1 & 0 \\ 0 & 0 & 1 & -1 \\ 0 & 0 & 0 & 1 \end{bmatrix} \begin{bmatrix} 1.25 & & & \\ & 1.5 & & \\ & & 1.75 & \\ & & & 2.0 \end{bmatrix} \begin{bmatrix} 1 & 0 & 0 & 0 \\ -1 & 1 & 0 & 0 \\ 0 & -1 & 1 & 0 \\ 0 & 0 & -1 & 1 \end{bmatrix}. \qquad (7)$$

This matrix $K = A^{\mathrm{T}}CA$ will be symmetric and positive definite! Symmetric because $(A^{\mathrm{T}}CA)^{\mathrm{T}} = A^{\mathrm{T}}C^{\mathrm{T}}A^{\mathrm{TT}} = A^{\mathrm{T}}CA$. Positive definite because it passes the energy test: A has independent columns, so $A\boldsymbol{x} \neq \mathbf{0}$ when $\boldsymbol{x} \neq \mathbf{0}$.

$$\text{Energy } = \boldsymbol{x}^{\mathrm{T}}A^{\mathrm{T}}CA\boldsymbol{x} = (A\boldsymbol{x})^{\mathrm{T}}C(A\boldsymbol{x}) > 0 \text{ for every } \boldsymbol{x} \neq \mathbf{0}, \text{ because } A\boldsymbol{x} \neq \mathbf{0}.$$

When you multiply the matrices $A^{\mathrm{T}}A$ and $A^{\mathrm{T}}CA$ for this fixed-free combination, watch how 1 replaces 2 in the last corner of $A^{\mathrm{T}}A$. That fourth equation has $u_4 - u_3$, a first (not second) difference coming from the free boundary condition $du/dx = 0$.

Notice in $A^{\mathrm{T}}CA$ how c_1, c_2, c_3, c_4 come from $c(x) = 1 + x$ in equation (7). Previously the c's were simply $1, 1, 1, 1$. Here are the **fixed-free** matrices:

$$A^{\mathrm{T}}A = \begin{bmatrix} 2 & -1 & & \\ -1 & 2 & -1 & \\ & -1 & 2 & -1 \\ & & -1 & \mathbf{1} \end{bmatrix} \quad A^{\mathrm{T}}CA = \begin{bmatrix} c_1 + c_2 & -c_2 & & \\ -c_2 & c_2 + c_3 & -c_3 & \\ & -c_3 & c_3 + c_4 & -c_4 \\ & & -c_4 & \boldsymbol{c_4} \end{bmatrix}. \qquad (8)$$

Free-free Boundary Conditions

Suppose both ends of the bar are free. Now $du/dx = 0$ at both $x = 0$ and $x = 1$. Nothing is holding the bar in place! Physically it is unstable—it can move with no force. Mathematically all constant functions like $u = 1$ satisfy these free conditions. **Algebraically our matrices $A^T A$ and $A^T C A$ will not be invertible**:

Free-free examples
Unknown u_0, u_1, u_2
$\Delta x = 0.5$

$$A^T A = \begin{bmatrix} \mathbf{1} & -1 & 0 \\ -1 & 2 & -1 \\ 0 & -1 & \mathbf{1} \end{bmatrix} \qquad A^T C A = \begin{bmatrix} \boldsymbol{c_0} & -c_0 & \\ -c_0 & c_0 + c_1 & -c_1 \\ & -c_1 & \boldsymbol{c_1} \end{bmatrix}.$$

The vector $(1, 1, 1)$ is in both nullspaces. This matches $u(x) = 1$ in the continuous problem. Free-free $A^T A \boldsymbol{u} = \boldsymbol{f}$ and $A^T C A \boldsymbol{u} = \boldsymbol{f}$ are generally unsolvable.

Before explaining more physical examples, may I write down six of the matrices? The tridiagonal K_0 appears many times in this textbook. Now we are seeing its applications. These matrices are all symmetric, and the first four are positive definite:

$$K_0 = A_0^T A_0 = \begin{bmatrix} \mathbf{2} & -1 & \\ -1 & 2 & -1 \\ & -1 & \mathbf{2} \end{bmatrix} \qquad A_0^T C_0 A_0 = \begin{bmatrix} c_1 + c_2 & -c_2 & \\ -c_2 & c_2 + c_3 & -c_3 \\ & -c_3 & c_3 + c_4 \end{bmatrix}$$

Fixed-fixed **Spring constants included**

$$K_1 = A_1^T A_1 = \begin{bmatrix} \mathbf{2} & -1 & \\ -1 & 2 & -1 \\ & -1 & \mathbf{1} \end{bmatrix} \qquad A_1^T C_1 A_1 = \begin{bmatrix} c_1 + c_2 & -c_2 & \\ -c_2 & c_2 + c_3 & -c_3 \\ & -c_3 & c_3 \end{bmatrix}$$

Fixed-free **Spring constants included**

$$K_{\textbf{singular}} = \begin{bmatrix} \mathbf{1} & -1 & \\ -1 & 2 & -1 \\ & -1 & \mathbf{1} \end{bmatrix} \qquad K_{\textbf{circular}} = \begin{bmatrix} 2 & -1 & \mathbf{-1} \\ -1 & 2 & -1 \\ \mathbf{-1} & -1 & 2 \end{bmatrix}$$

Free-free **Periodic $u(0) = u(1)$**

The matrices $K_0, K_1, K_{\textbf{singular}}$, and $K_{\textbf{circular}}$ have $C = I$ for simplicity. This means that all the "spring constants" are $c_i = 1$. We included $A_0^T C_0 A_0$ and $A_1^T C_1 A_1$ to show how the spring constants enter the matrix (without changing its positive definiteness). Our next goal is to see these same stiffness matrices in other engineering problems.

A Line of Springs and Masses

Figure 10.4 shows three masses m_1, m_2, m_3 connected by a line of springs. The fixed-fixed case has four springs, with top and bottom fixed. That leads to K_0 and $A_0^{\mathrm{T}} C_0 A_0$. The fixed-free case has only three springs; the lowest mass hangs freely. That will lead to K_1 and $A_1^{\mathrm{T}} C_1 A_1$. A **free-free** problem produces $K_{\textbf{singular}}$.

We want equations for the mass movements $\boldsymbol{u}$ and the spring tensions $\boldsymbol{y}$:

$$\boldsymbol{u} = (u_1, u_2, u_3) = \textbf{movements of the masses} \text{ (down is positive)}$$
$$\boldsymbol{y} = (y_1, y_2, y_3, y_4) \text{ or } (y_1, y_2, y_3) = \textbf{tensions in the springs}$$

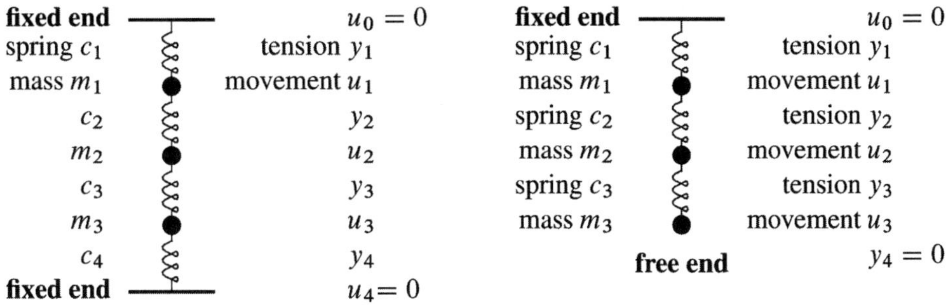

Figure 10.4: Lines of springs and masses: **fixed-fixed** and **fixed-free** ends.

When a mass moves downward, its displacement is positive ($u_j > 0$). For the springs, tension is positive and compression is negative ($y_i < 0$). In tension, the spring is stretched so it pulls the masses inward. Each spring is controlled by its own Hooke's Law $y = c\,e$: (*stretching force* $\boldsymbol{y}$) = (*spring constant* $\mathbf{c}$) times (*stretching distance* $\mathbf{e}$).

Our job is to link these one-spring equations $y = c\,e$ into a vector equation $K\boldsymbol{u} = \boldsymbol{f}$ for the whole system. The force vector $\boldsymbol{f}$ comes from gravity. The gravitational constant g will multiply each mass to produce downward forces $\boldsymbol{f} = (m_1 g, m_2 g, m_3 g)$.

The real problem is to find the stiffness matrix (**fixed-fixed** and **fixed-free**). The best way to create K is in three steps, not one. Instead of connecting the movements $\boldsymbol{u}_j$ directly to the forces f_i, it is much better to connect each vector to the next in this list:

$$\begin{aligned}
u &= \mathit{Movements} \text{ of } n \text{ masses} &&= (u_1, \ldots, u_n)\\
e &= \mathit{Elongations} \text{ of } m \text{ springs} &&= (e_1, \ldots, e_m)\\
y &= \mathit{Internal\ forces} \text{ in } m \text{ springs} &&= (y_1, \ldots, y_m)\\
f &= \mathit{External\ forces} \text{ on } n \text{ masses} &&= (f_1, \ldots, f_n)
\end{aligned}$$

A great framework for applied mathematics connects $\boldsymbol{u}$ to $\boldsymbol{e}$ to $\boldsymbol{y}$ to $\boldsymbol{f}$. Then $\boldsymbol{A}^{\mathrm{T}}\boldsymbol{C}\boldsymbol{A}\boldsymbol{u} = \boldsymbol{f}$:

$$\begin{array}{ccc}
\boxed{u} & & \boxed{f}\\
A\downarrow & & \uparrow A^{\mathrm{T}}\\
\boxed{e} & \xrightarrow{C} & \boxed{y}
\end{array}
\qquad
\begin{array}{ll}
\boldsymbol{e} = A\boldsymbol{u} & A \text{ is } m \text{ by } n\\
\boldsymbol{y} = C\boldsymbol{e} & C \text{ is } m \text{ by } m\\
\boldsymbol{f} = A^{\mathrm{T}}\boldsymbol{y} & A^{\mathrm{T}} \text{ is } n \text{ by } m
\end{array}$$

We will write down the matrices A and C and A^{T} for the two examples, first with fixed ends and then with the lower end free. Forgive the simplicity of these matrices, it is their form that is so important. Especially the appearance of A together with A^{T}.

The elongation $\boldsymbol{e}$ *is the stretching distance*—how far the springs are extended. Originally there is no stretching—the system is lying on a table. When it becomes vertical and upright, gravity acts. The masses move down by distances u_1, u_2, u_3. Each spring is stretched or compressed by $e_i = u_i - u_{i-1}$, *the difference in displacements of its ends*:

	First spring:	$e_1 = \boldsymbol{u}_1$	(the top is fixed so $u_0 = 0$)
Stretching of	Second spring:	$e_2 = \boldsymbol{u}_2 - \boldsymbol{u}_1$	
each spring	Third spring:	$e_3 = \boldsymbol{u}_3 - \boldsymbol{u}_2$	
	Fourth spring:	$e_4 = \quad\; - \boldsymbol{u}_3$	(the bottom is fixed so $u_4 = 0$)

If both ends move the same distance, that spring is not stretched: $u_j = u_{j-1}$ and $e_j = 0$. The matrix in those four equations is a 4 by 3 *difference matrix* A, and $\boldsymbol{e} = A\boldsymbol{u}$:

Stretching distances (elongations)

$$\boldsymbol{e} = A\boldsymbol{u} \quad \text{is} \quad \begin{bmatrix} e_1 \\ e_2 \\ e_3 \\ e_4 \end{bmatrix} = \begin{bmatrix} 1 & 0 & 0 \\ -1 & 1 & 0 \\ 0 & -1 & 1 \\ 0 & 0 & -1 \end{bmatrix} \begin{bmatrix} u_1 \\ u_2 \\ u_3 \end{bmatrix}. \tag{9}$$

The next equation $\boldsymbol{y} = C\boldsymbol{e}$ connects spring elongation $\boldsymbol{e}$ with spring tension $\boldsymbol{y}$. *This is Hooke's Law* $y_i = c_i e_i$ *for each separate spring.* It is the "constitutive law" that depends on the material in the spring. A soft spring has small c, so a moderate force y can produce a large stretching e. Hooke's linear law is nearly exact for real springs, before they are overstretched and the material becomes plastic.

Since each spring has its own law, the matrix in $\boldsymbol{y} = C\boldsymbol{e}$ is a diagonal matrix C:

Hooke's Law $\boldsymbol{y} = C\boldsymbol{e}$

$$\begin{aligned} y_1 &= c_1 e_1 \\ y_2 &= c_2 e_2 \\ y_3 &= c_3 e_3 \\ y_4 &= c_4 e_4 \end{aligned} \quad \text{is} \quad \begin{bmatrix} y_1 \\ y_2 \\ y_3 \\ y_4 \end{bmatrix} = \begin{bmatrix} c_1 & & & \\ & c_2 & & \\ & & c_3 & \\ & & & c_4 \end{bmatrix} \begin{bmatrix} e_1 \\ e_2 \\ e_3 \\ e_4 \end{bmatrix} \tag{10}$$

Combining $\boldsymbol{e} = A\boldsymbol{u}$ with $\boldsymbol{y} = C\boldsymbol{e}$, the spring forces (tension forces) are $\boldsymbol{y} = CA\boldsymbol{u}$.

Finally comes the balance equation, the most fundamental law of applied mathematics. The internal forces from the springs balance the external forces on the masses. Each mass is pulled or pushed by the spring force y_j above it. From below it feels the spring force y_{j+1} plus f_j from gravity. Thus $y_j = y_{j+1} + f_j$ or $f_j = y_j - y_{j+1}$:

Force balance $\boldsymbol{f} = A^{\mathrm{T}}\boldsymbol{y}$

$$\begin{aligned} f_1 &= y_1 - y_2 \\ f_2 &= y_2 - y_3 \\ f_3 &= y_3 - y_4 \end{aligned} \quad \text{is} \quad \begin{bmatrix} f_1 \\ f_2 \\ f_3 \end{bmatrix} = \begin{bmatrix} 1 & -1 & 0 & 0 \\ 0 & 1 & -1 & 0 \\ 0 & 0 & 1 & -1 \end{bmatrix} \begin{bmatrix} y_1 \\ y_2 \\ y_3 \\ y_4 \end{bmatrix} \tag{11}$$

That matrix is A^{T} ! *The equation for balance of forces is* $\boldsymbol{f} = A^{\mathrm{T}}\boldsymbol{y}$. Nature transposes the rows and columns of the $\boldsymbol{e} - \boldsymbol{u}$ matrix to produce the $\boldsymbol{f} - \boldsymbol{y}$ matrix. This is the beauty of the framework, that A^{T} appears along with A. The three equations combine into $K\boldsymbol{u} = \boldsymbol{f}$.

$$\left\{\begin{array}{lll} \boldsymbol{e} & = & A\boldsymbol{u} \\ \boldsymbol{y} & = & C\boldsymbol{e} \\ \boldsymbol{f} & = & A^{\mathrm{T}}\boldsymbol{y} \end{array}\right\}$$

combine into $A^{\mathrm{T}}CA\boldsymbol{u} = \boldsymbol{f}$ or $K\boldsymbol{u} = \boldsymbol{f}$
$K = A^{\mathrm{T}}CA$ is the **stiffness matrix** (mechanics)
$K = A^{\mathrm{T}}CA$ is the **conductance matrix** (networks)

Finite element programs spend major effort on assembling $K = A^{\mathrm{T}}CA$ from thousands of smaller pieces. We find K for four springs (**fixed-fixed**) by multiplying A^{T} times CA:

$$\begin{bmatrix} 1 & -1 & 0 & 0 \\ 0 & 1 & -1 & 0 \\ 0 & 0 & 1 & -1 \end{bmatrix} \begin{bmatrix} c_1 & 0 & 0 \\ -c_2 & c_2 & 0 \\ 0 & -c_3 & c_3 \\ 0 & 0 & -c_4 \end{bmatrix} = \begin{bmatrix} c_1+c_2 & -c_2 & 0 \\ -c_2 & c_2+c_3 & -c_3 \\ 0 & -c_3 & c_3+c_4 \end{bmatrix}$$

If all springs are identical, with $c_1 = c_2 = c_3 = c_4 = 1$, then $C = I$. The stiffness matrix reduces to $A^{\mathrm{T}}A$. It becomes the special $-1, 2, -1$ matrix K_0.

Note the difference between $A^{\mathrm{T}}A$ from engineering and LU from linear algebra. The matrix A from four springs is 4 by 3. The triangular matrices from elimination are square. The stiffness matrix K is assembled from $A^{\mathrm{T}}A$, and then broken up into LU. One step is applied mathematics, the other is computational mathematics. Each K is built from rectangular matrices and factored into square matrices.

May I list some properties of $K = A^{\mathrm{T}}CA$? You know almost all of them:

1. K is **tridiagonal**, because mass 3 is not connected to mass 1.
2. K is **symmetric**, because C is symmetric and A^{T} comes with A.
3. K is **positive definite**, because $c_i > 0$ and A has **independent columns**.
4. K^{-1} is a **full matrix** (not sparse) with **all positive entries**.

Property 4 leads to an important fact about $\boldsymbol{u} = K^{-1}\boldsymbol{f}$: *If all forces act downwards* ($f_j > 0$) *then all movements are downwards* ($u_j > 0$). Notice that "positive" is different from "positive definite". K^{-1} is positive (K is not). Both are positive definite.

Example 1 Suppose all $c_i = c$ and $m_j = m$. Find the movements $\boldsymbol{u}$ and tensions $\boldsymbol{y}$.

All springs are the same and all masses are the same. But all movements and elongations and tensions will *not* be the same. K^{-1} includes $\frac{1}{c}$ because $A^{\mathrm{T}}CA$ includes c:

Movements $$\boldsymbol{u} = K^{-1}\boldsymbol{f} = \frac{1}{4c}\begin{bmatrix} 3 & 2 & 1 \\ 2 & 4 & 2 \\ 1 & 2 & 3 \end{bmatrix}\begin{bmatrix} mg \\ mg \\ mg \end{bmatrix} = \frac{mg}{c}\begin{bmatrix} \mathbf{3/2} \\ \mathbf{2} \\ \mathbf{3/2} \end{bmatrix}$$

The displacement u_2, for the mass in the middle, is greater than u_1 and u_3. The units are correct: the force mg divided by force per unit length c gives a length u. Then

Elongations $$\boldsymbol{e} = A\boldsymbol{u} = \begin{bmatrix} 1 & 0 & 0 \\ -1 & 1 & 0 \\ 0 & -1 & 1 \\ 0 & 0 & -1 \end{bmatrix} \frac{mg}{c} \begin{bmatrix} \frac{3}{2} \\ 2 \\ \frac{3}{2} \end{bmatrix} = \frac{mg}{c}\begin{bmatrix} \mathbf{3/2} \\ \mathbf{1/2} \\ \mathbf{-1/2} \\ \mathbf{-3/2} \end{bmatrix}.$$

Warning: *Normally you cannot write* $K^{-1} = A^{-1}C^{-1}(A^{\mathrm{T}})^{-1}$.

The three matrices are mixed together by $A^{\mathrm{T}}CA$, and they cannot easily be untangled. In general, $A^{\mathrm{T}}\boldsymbol{y} = \boldsymbol{f}$ has many solutions. And four equations $A\boldsymbol{u} = \boldsymbol{e}$ would usually have no solution with three unknowns. But $A^{\mathrm{T}}CA$ gives the correct solution to all three equations in the framework. Only when $m = n$ and the matrices are square can we go from $\boldsymbol{y} = (A^{\mathrm{T}})^{-1}\boldsymbol{f}$ to $\boldsymbol{e} = C^{-1}\boldsymbol{y}$ to $\boldsymbol{u} = A^{-1}\boldsymbol{e}$. We will see that now.

Fixed End and Free End

Remove the fourth spring. All matrices become 3 by 3. The pattern does not change! The matrix A loses its fourth row and (of course) A^{T} loses its fourth column. The new stiffness matrix K_1 becomes a product of square matrices:

$$A_1^{\mathrm{T}}C_1A_1 = \begin{bmatrix} 1 & -1 & 0 \\ 0 & 1 & -1 \\ 0 & 0 & 1 \end{bmatrix} \begin{bmatrix} c_1 & & \\ & c_2 & \\ & & c_3 \end{bmatrix} \begin{bmatrix} 1 & 0 & 0 \\ -1 & 1 & 0 \\ 0 & -1 & 1 \end{bmatrix}.$$

The missing column of A^{T} and row of A multiplied the missing c_4. So the quickest way to find the new $A^{\mathrm{T}}CA$ is to set $c_4 = 0$ in the old one:

FIXED
FREE
$$A_1^{\mathrm{T}}C_1A_1 = \begin{bmatrix} c_1+c_2 & -c_2 & 0 \\ -c_2 & c_2+c_3 & -c_3 \\ 0 & -c_3 & c_3 \end{bmatrix}. \qquad (12)$$

Example 2 If $c_1 = c_2 = c_3 = 1$ and $C = I$, this is the $-1, 2, -1$ tridiagonal matrix K_1. The last entry of K_1 is 1 instead of 2 because the spring at the bottom is free. Suppose all $m_j = m$:

Fixed-free
$$\boldsymbol{u} = K_1^{-1}\boldsymbol{f} = \frac{1}{c}\begin{bmatrix} 1 & 1 & 1 \\ 1 & 2 & 2 \\ 1 & 2 & 3 \end{bmatrix} \begin{bmatrix} mg \\ mg \\ mg \end{bmatrix} = \frac{mg}{c}\begin{bmatrix} 3 \\ 5 \\ 6 \end{bmatrix}.$$

Those movements are greater than the free-free case. The number 3 appears in u_1 because all three masses are pulling the first spring down. The next mass moves by that 3 plus an additional 2 from the masses below it. The third mass drops even more $(3 + 2 + 1 = 6)$. The elongations $\boldsymbol{e} = A\boldsymbol{u}$ in the springs display those numbers $3, 2, 1$:

$$\boldsymbol{e} = \begin{bmatrix} 1 & 0 & 0 \\ -1 & 1 & 0 \\ 0 & -1 & 1 \end{bmatrix} \frac{mg}{c}\begin{bmatrix} 3 \\ 5 \\ 6 \end{bmatrix} = \frac{mg}{c}\begin{bmatrix} 3 \\ 2 \\ 1 \end{bmatrix}.$$

Two Free Ends: K is Singular

Freedom at *both ends* means trouble. The whole line can move. A is 2 by 3:

$$\textbf{FREE-FREE} \quad \boldsymbol{e} = A\boldsymbol{u} \qquad \begin{bmatrix} e_1 \\ e_2 \end{bmatrix} = \begin{bmatrix} u_2 - u_1 \\ u_3 - u_2 \end{bmatrix} = \begin{bmatrix} -1 & 1 & 0 \\ 0 & -1 & 1 \end{bmatrix} \begin{bmatrix} u_1 \\ u_2 \\ u_3 \end{bmatrix}. \tag{13}$$

Now there is a nonzero solution to $A\boldsymbol{u} = \boldsymbol{0}$. ***The masses can move with no stretching of the springs***. The whole line can shift by $\boldsymbol{u} = (1, 1, 1)$ and this leaves $\boldsymbol{e} = (0, 0)$:

$$A\boldsymbol{u} = \begin{bmatrix} -1 & 1 & 0 \\ 0 & -1 & 1 \end{bmatrix} \begin{bmatrix} 1 \\ 1 \\ 1 \end{bmatrix} = \begin{bmatrix} 0 \\ 0 \end{bmatrix} = \textbf{no stretching}. \tag{14}$$

$A\boldsymbol{u} = \boldsymbol{0}$ certainly leads to $A^{\mathrm{T}}CA\boldsymbol{u} = \boldsymbol{0}$. Then $A^{\mathrm{T}}CA$ is only *positive semidefinite*, without c_1 and c_4. The pivots will be c_2 and c_3 and *no third pivot*. The rank is only 2:

$$\begin{bmatrix} -1 & 0 \\ 1 & -1 \\ 0 & 1 \end{bmatrix} \begin{bmatrix} c_2 & \\ & c_3 \end{bmatrix} \begin{bmatrix} -1 & 1 & 0 \\ 0 & -1 & 1 \end{bmatrix} = \begin{bmatrix} c_2 & -c_2 & 0 \\ -c_2 & c_2 + c_3 & -c_3 \\ 0 & -c_3 & c_3 \end{bmatrix} \tag{15}$$

Two eigenvalues will be positive but $\boldsymbol{x} = (1, 1, 1)$ is an eigenvector for $\lambda = 0$. We can solve $A^{\mathrm{T}}CA\boldsymbol{u} = \boldsymbol{f}$ only for special vectors $\boldsymbol{f}$. The forces have to add to $f_1 + f_2 + f_3 = 0$, or the whole line of springs (with both ends free) will take off like a rocket.

Circle of Springs

A third spring will complete the circle from mass 3 back to mass 1. This doesn't make K invertible—the stiffness matrix $K_{circular}$ matrix is still singular:

$$A^{\mathrm{T}}_{\textbf{circular}} A_{\textbf{circular}} = \begin{bmatrix} 1 & -1 & 0 \\ 0 & 1 & -1 \\ -1 & 0 & 1 \end{bmatrix} \begin{bmatrix} 1 & 0 & -1 \\ -1 & 1 & 0 \\ 0 & -1 & 1 \end{bmatrix} = \begin{bmatrix} 2 & -1 & -1 \\ -1 & 2 & -1 \\ -1 & -1 & 2 \end{bmatrix}. \tag{16}$$

The only pivots are 2 and $\frac{3}{2}$. The eigenvalues are 3 and 3 and 0. The determinant is zero. The nullspace still contains $\boldsymbol{x} = (1, 1, 1)$, when all the masses move together. This movement vector $(1, 1, 1)$ is in the nullspace of $A_{\textbf{circular}}$ and $K_{\textbf{circular}} = A^{\mathrm{T}}CA$.

May I summarize this section? I hope the example will help you connect calculus with linear algebra, replacing differential equations by difference equations. If your step Δx is small enough, you will have a totally satisfactory solution.

$$\textbf{The equation is} \quad -\frac{d}{dx}\left(c(x)\frac{du}{dx}\right) = f(x) \ \ \textbf{with} \ \ u(0) = 0 \ \ \textbf{and} \ \ \left[u(1) \text{ or } \frac{du}{dx}(1)\right] = 0$$

Divide the bar into N pieces of length Δx. Replace du/dx by $A\boldsymbol{u}$ and $-dy/dx$ by $A^{\mathrm{T}}\boldsymbol{y}$. Now A and A^{T} include $1/\Delta x$. The end conditions are $u_0 = 0$ and $[u_N = 0 \text{ or } y_N = 0]$.

The three steps $-d/dx$ and $c(x)$ and d/dx correspond to A^{T} and C and A:

$$\boldsymbol{f} = A^{\mathrm{T}}\boldsymbol{y} \text{ and } \boldsymbol{y} = C\boldsymbol{e} \text{ and } \boldsymbol{e} = A\boldsymbol{u} \text{ give } A^{\mathrm{T}}CA\boldsymbol{u} = \boldsymbol{f}.$$

This is a fundamental example in computational science and engineering.

1. Model the problem by a differential equation
2. Discretize the differential equation to a difference equation
3. Understand and solve the difference equation (and boundary conditions!)
4. Interpret the solution; visualize it; redesign if needed.

Numerical simulation has become a third branch of science, beside experiment and deduction. Computer design of the Boeing 777 was much less expensive than a wind tunnel.

The two texts *Introduction to Applied Mathematics* and *Computational Science and Engineering* (Wellesley-Cambridge Press) develop this whole subject further—see the course page **math.mit.edu/18085** with video lectures (The lectures are also on **ocw.mit.edu** and **YouTube**). I hope this book helps you to see the framework behind the computations.

Problem Set 10.2

1 Show that $\det A_0^{\mathrm{T}}C_0A_0 = c_1c_2c_3+c_1c_3c_4+c_1c_2c_4+c_2c_3c_4$. Find also $\det A_1^{\mathrm{T}}C_1A_1$ in the fixed-free example.

2 Invert $A_1^{\mathrm{T}}C_1A_1$ in the fixed-free example by multiplying $A_1^{-1}C_1^{-1}(A_1^{\mathrm{T}})^{-1}$.

3 In the free-free case when $A^{\mathrm{T}}CA$ in equation (15) is singular, add the three equations $A^{\mathrm{T}}CA\boldsymbol{u} = \boldsymbol{f}$ to show that we need $f_1+f_2+f_3 = 0$. Find a solution to $A^{\mathrm{T}}CA\boldsymbol{u} = \boldsymbol{f}$ when the forces $\boldsymbol{f} = (-1, 0, 1)$ balance themselves. Find all solutions!

4 Both end conditions for the free-free differential equation are $du/dx = 0$:

$$-\frac{d}{dx}\left(c(x)\frac{du}{dx}\right) = f(x) \quad \text{with} \quad \frac{du}{dx} = 0 \text{ at both ends.}$$

Integrate both sides to show that the force $f(x)$ must balance itself, $\int f(x)\,dx = 0$, or there is no solution. The complete solution is one particular solution $u(x)$ plus any constant. The constant corresponds to $\boldsymbol{u} = (1, 1, 1)$ in the nullspace of $A^{\mathrm{T}}CA$.

5 In the fixed-free problem, the matrix A is square and invertible. We can solve $A^{\mathrm{T}}\boldsymbol{y} = \boldsymbol{f}$ separately from $A\boldsymbol{u} = \boldsymbol{e}$. Do the same for the differential equation:

$$\text{Solve } -\frac{dy}{dx} = f(x) \text{ with } y(1) = 0. \text{ Graph } y(x) \text{ if } f(x) = 1.$$

6 The 3 by 3 matrix $K_1 = A_1^{\mathrm{T}} C_1 A_1$ in equation (6) splits into three "element matrices" $c_1 E_1 + c_2 E_2 + c_3 E_3$. Write down those pieces, one for each c. Show how they come from *column times row* multiplication of $A_1^{\mathrm{T}} C_1 A_1$. This is how finite element stiffness matrices are actually assembled.

7 For five springs and four masses with both ends fixed, what are the matrices A and C and K? With $C = I$ solve $K\boldsymbol{u} = \mathsf{ones}(4)$.

8 Compare the solution $\boldsymbol{u} = (u_1, u_2, u_3, u_4)$ in Problem 7 to the solution of the continuous problem $-u'' = 1$ with $u(0) = 0$ and $u(1) = 0$. The parabola $u(x)$ should correspond at $x = \frac{1}{5}, \frac{2}{5}, \frac{3}{5}, \frac{4}{5}$ to $\boldsymbol{u}$—is there a $(\Delta x)^2$ factor to account for?

9 Solve the fixed-free problem $-u'' = mg$ with $u(0) = 0$ and $u'(1) = 0$. Compare $u(x)$ at $x = \frac{1}{3}, \frac{2}{3}, \frac{3}{3}$ with the vector $\boldsymbol{u} = (3mg, 5mg, 6mg)$ in Example 2.

10 Suppose $c_1 = c_2 = c_3 = c_4 = 1$, $m_1 = 2$ and $m_2 = m_3 = 1$. Solve $A^{\mathrm{T}} C A \boldsymbol{u} = (2, 1, 1)$ for this fixed-fixed line of springs. Which mass moves the most (largest u)?

11 (MATLAB) Find the displacements $u(1), \ldots, u(100)$ of 100 masses connected by springs all with $c = 1$. Each force is $f(i) = .01$. Print graphs of $\boldsymbol{u}$ with **fixed-fixed** and **fixed-free** ends. Note that $\mathsf{diag}(\mathsf{ones}(n, 1), d)$ is a matrix with n ones along diagonal d. This print command will graph a vector u:

```
plot(u, '+');    xlabel('mass number');    ylabel('movement');    print
```

12 (MATLAB) Chemical engineering has a first derivative du/dx from fluid velocity as well as d^2u/dx^2 from diffusion. Replace du/dx by a *forward* difference, then a *centered* difference, then a *backward* difference, with $\Delta x = \frac{1}{8}$. Graph your three numerical solutions of

$$-\frac{d^2u}{dx^2} + 10\,\frac{du}{dx} = 1 \quad \text{with} \quad u(0) = u(1) = 0.$$

This ***convection-diffusion equation*** appears everywhere. It transforms to the Black-Scholes equation for option prices in mathematical finance.

Problem 12 is developed into the first MATLAB homework in my 18.085 course on Computational Science and Engineering at MIT. Videos on *ocw.mit.edu*.

10.3 Markov Matrices, Population, and Economics

This section is about ***positive matrices***: every $a_{ij} > 0$. The key fact is quick to state: ***The largest eigenvalue is real and positive and so is its eigenvector.*** In economics and ecology and population dynamics and random walks, that fact leads a long way:

Markov $\lambda_{\text{max}} = 1$ **Population** $\lambda_{\text{max}} > 1$ **Consumption** $\lambda_{\text{max}} < 1$

λ_{max} controls the powers of A. We will see this first for $\lambda_{\text{max}} = 1$.

Markov Matrices

Multiply a positive vector $\boldsymbol{u}_0$ again and again by this matrix A:

Markov matrix $$A = \begin{bmatrix} .8 & .3 \\ .2 & .7 \end{bmatrix} \qquad \boldsymbol{u}_1 = A\boldsymbol{u}_0 \qquad \boldsymbol{u}_2 = A\boldsymbol{u}_1 = A^2\boldsymbol{u}_0$$

After k steps we have $A^k\boldsymbol{u}_0$. The vectors $\boldsymbol{u}_1, \boldsymbol{u}_2, \boldsymbol{u}_3, \ldots$ will approach a "*steady state*" $\boldsymbol{u}_\infty = (.6, .4)$. This final outcome does not depend on the starting vector $\boldsymbol{u}_0$. ***For every*** $\boldsymbol{u}_0 = (a, 1-a)$ ***we converge to the same*** $\boldsymbol{u}_{\infty(.6,.4)}$. The question is why.

The steady state equation $A\boldsymbol{u}_\infty = \boldsymbol{u}_\infty$ makes $\boldsymbol{u}_\infty$ ***an eigenvector with eigenvalue* 1**:

Steady state $$\begin{bmatrix} .8 & .3 \\ .2 & .7 \end{bmatrix} \begin{bmatrix} .6 \\ .4 \end{bmatrix} = \begin{bmatrix} .6 \\ .4 \end{bmatrix} = \boldsymbol{u}_\infty.$$

Multiplying by A does not change $\boldsymbol{u}_\infty$. But this does not explain why so many vectors $\boldsymbol{u}_0$ lead to $\boldsymbol{u}_\infty$. Other examples might have a steady state, but it is not necessarily attractive:

Not Markov $B = \begin{bmatrix} 1 & 0 \\ 0 & 2 \end{bmatrix}$ has the unattractive steady state $B\begin{bmatrix} 1 \\ 0 \end{bmatrix} = \begin{bmatrix} 1 \\ 0 \end{bmatrix}$.

In this case, the starting vector $\boldsymbol{u}_0 = (0, 1)$ will give $\boldsymbol{u}_1 = (0, 2)$ and $\boldsymbol{u}_2 = (0, 4)$. The second components are doubled. In the language of eigenvalues, B has $\lambda = 1$ but also $\lambda = 2$— this produces instability. The component of $\boldsymbol{u}$ along that unstable eigenvector is multiplied by λ, and $|\lambda| > 1$ means blowup.

This section is about two special properties of A that guarantee a *stable steady state*. These properties define a positive ***Markov matrix***, and A above is one particular example:

Markov matrix	**1.** ***Every entry of*** A ***is positive***: $a_{ij} > 0$. **2.** ***Every column of*** A ***adds to* 1.**

Column 2 of B adds to 2, not 1. When A is a Markov matrix, two facts are immediate:
Because of **1**: Multiplying $\boldsymbol{u}_0 \geq 0$ by A produces a nonnegative $\boldsymbol{u}_1 = A\boldsymbol{u}_0 \geq 0$.
Because of **2**: If the components of $\boldsymbol{u}_0$ add to 1, so do the components of $\boldsymbol{u}_1 = A\boldsymbol{u}_0$.

Reason: The components of $\boldsymbol{u}_0$ add to 1 when $\begin{bmatrix}1 & \dots & 1\end{bmatrix}\boldsymbol{u}_0 = 1$. This is true for each column of A by Property 2. Then by matrix multiplication $[1 \ \dots \ 1]A = [1 \ \dots \ 1]$:

$$\textbf{Components of } A\boldsymbol{u}_0 \textbf{ add to 1} \qquad [1 \ \dots \ 1]A\boldsymbol{u}_0 = [1 \ \dots \ 1]\boldsymbol{u}_0 = 1.$$

The same facts apply to $\boldsymbol{u}_2 = A\boldsymbol{u}_1$ and $\boldsymbol{u}_3 = A\boldsymbol{u}_2$. ***Every vector $A^k\boldsymbol{u}_0$ is nonnegative with components adding to*** 1. These are "***probability vectors***." The limit $\boldsymbol{u}_\infty$ is also a probability vector—but we have to prove that there is a limit. We will show that $\lambda_{\max} = 1$ for a positive Markov matrix.

Example 1 The fraction of rental cars in Denver starts at $\frac{1}{50} = .02$. The fraction outside Denver is .98. Every month, 80% of the Denver cars stay in Denver (and 20% leave). Also 5% of the outside cars come in (95% stay outside). This means that the fractions $\boldsymbol{u}_0 = (.02, .98)$ are multiplied by A:

$$\textbf{First month} \qquad A = \begin{bmatrix} \mathbf{.80} & \mathbf{.05} \\ \mathbf{.20} & \mathbf{.95} \end{bmatrix} \quad \text{leads to} \quad \boldsymbol{u}_1 = A\boldsymbol{u}_0 = A\begin{bmatrix} .02 \\ .98 \end{bmatrix} = \begin{bmatrix} .065 \\ .935 \end{bmatrix}.$$

Notice that $.065 + .935 = 1$. All cars are accounted for. Each step multiplies by A:

$$\textbf{Next month} \qquad \boldsymbol{u}_2 = A\boldsymbol{u}_1 = (.09875,\ .90125). \ \text{This is } A^2\boldsymbol{u}_0.$$

All these vectors are positive because A is positive. Each vector $\boldsymbol{u}_k$ will have its components adding to 1. The first component has grown from .02 and cars are moving toward Denver. What happens in the long run?

This section involves powers of matrices. The understanding of A^k was our first and best application of diagonalization. Where A^k can be complicated, the diagonal matrix Λ^k is simple. The eigenvector matrix X connects them: A^k equals $X\Lambda^kX^{-1}$. The new application to Markov matrices uses the eigenvalues (in Λ) and the eigenvectors (in X). We will show that $\boldsymbol{u}_\infty$ **is an eigenvector of A corresponding to** $\lambda = 1$.

Since every column of A adds to 1, nothing is lost or gained. We are moving rental cars or populations, and no cars or people suddenly appear (or disappear). The fractions add to 1 and the matrix A keeps them that way. The question is how they are distributed after k time periods—which leads us to A^k.

Solution $A^k\boldsymbol{u}_0$ gives the fractions in and out of Denver after k steps. We diagonalize A to understand A^k. The eigenvalues are $\lambda = \mathbf{1}$ and $\mathbf{.75}$ (the trace is 1.75).

$$Ax = \lambda x \qquad A\begin{bmatrix} .2 \\ .8 \end{bmatrix} = 1\begin{bmatrix} .2 \\ .8 \end{bmatrix} \quad \textbf{and} \quad A\begin{bmatrix} -1 \\ 1 \end{bmatrix} = .75\begin{bmatrix} -1 \\ 1 \end{bmatrix}.$$

The starting vector $\boldsymbol{u}_0$ combines $\boldsymbol{x}_1$ and $\boldsymbol{x}_2$, in this case with coefficients 1 and .18:

$$\textbf{Combination of eigenvectors} \qquad \boldsymbol{u}_0 = \begin{bmatrix} .02 \\ .98 \end{bmatrix} = \begin{bmatrix} .2 \\ .8 \end{bmatrix} + .18\begin{bmatrix} -1 \\ 1 \end{bmatrix}.$$

Now multiply by A to find $\boldsymbol{u}_1$. The eigenvectors are multiplied by $\lambda_1 = 1$ and $\lambda_2 = .75$:

$$\textbf{Each } \boldsymbol{x} \textbf{ is multiplied by } \boldsymbol{\lambda} \qquad \boldsymbol{u}_1 = 1\begin{bmatrix} .2 \\ .8 \end{bmatrix} + (.75)(.18)\begin{bmatrix} -1 \\ 1 \end{bmatrix}.$$

Every month, another $\lambda = .75$ multiplies the vector $\boldsymbol{x}_2$. The eigenvector $\boldsymbol{x}_1$ is unchanged:

After k steps $$\boldsymbol{u}_k = A^k\boldsymbol{u}_0 = 1^k \begin{bmatrix} .2 \\ .8 \end{bmatrix} + (.75)^k(.18)\begin{bmatrix} -1 \\ 1 \end{bmatrix}.$$

This equation reveals what happens. ***The eigenvector $\boldsymbol{x}_1$ with $\lambda = 1$ is the steady state.*** The other eigenvector $\boldsymbol{x}_2$ disappears because $|\lambda| < 1$. The more steps we take, the closer we come to $\boldsymbol{u}_\infty = (.2, .8)$. In the limit, $\frac{2}{10}$ of the cars are in Denver and $\frac{8}{10}$ are outside. This is the pattern for Markov chains, even starting from $\boldsymbol{u}_0 = (0, 1)$:

> If A is a *positive* Markov matrix (entries $a_{ij} > 0$, each column adds to $\mathbf{1}$), then $\lambda_1 = 1$ is larger than any other eigenvalue. The eigenvector $\boldsymbol{x}_1$ is the ***steady state***:
>
> $$\boldsymbol{u}_k = \boldsymbol{x}_1 + c_2(\lambda_2)^k\boldsymbol{x}_2 + \cdots + c_n(\lambda_n)^k\boldsymbol{x}_n \quad \textit{always approaches} \quad \boldsymbol{u}_\infty = \boldsymbol{x}_1.$$

The first point is to see that $\lambda = 1$ is an eigenvalue of A. *Reason*: Every column of $A - I$ adds to $1 - 1 = 0$. The rows of $A - I$ add up to the zero row. Those rows are linearly dependent, so $A - I$ is singular. Its determinant is zero and $\lambda = 1$ is an eigenvalue.

The second point is that no eigenvalue can have $|\lambda| > 1$. With such an eigenvalue, the powers A^k would grow. But A^k is also a Markov matrix! A^k has positive entries still adding to 1—and that leaves no room to get large.

A lot of attention is paid to the possibility that another eigenvalue has $|\lambda| = 1$.

Example 2 $A = \begin{bmatrix} 0 & 1 \\ 1 & 0 \end{bmatrix}$ has no steady state because $\lambda_2 = -1$.

This matrix sends all cars from inside Denver to outside, and vice versa. The powers A^k alternate between A and I. The second eigenvector $\boldsymbol{x}_2 = (-1, 1)$ will be multiplied by $\lambda_2 = -1$ at every step—and does not become smaller: No steady state.

Suppose the entries of A or any power of A are all *positive*—zero is not allowed. In this "regular" or "primitive" case, $\lambda = 1$ is strictly larger than any other eigenvalue. The powers A^k approach the rank one matrix that has the steady state in every column.

Example 3 (**"Everybody moves"**) Start with three groups. At each time step, half of group 1 goes to group 2 and the other half goes to group 3. The other groups also *split in half and move*. Take one step from the starting populations p_1, p_2, p_3:

New populations $$\boldsymbol{u}_1 = A\boldsymbol{u}_0 = \begin{bmatrix} 0 & \frac{1}{2} & \frac{1}{2} \\ \frac{1}{2} & 0 & \frac{1}{2} \\ \frac{1}{2} & \frac{1}{2} & 0 \end{bmatrix}\begin{bmatrix} p_1 \\ p_2 \\ p_3 \end{bmatrix} = \begin{bmatrix} \frac{1}{2}p_2 + \frac{1}{2}p_3 \\ \frac{1}{2}p_1 + \frac{1}{2}p_3 \\ \frac{1}{2}p_1 + \frac{1}{2}p_2 \end{bmatrix}.$$

A is a Markov matrix. Nobody is born or lost. A contains zeros, which gave trouble in Example 2. But after two steps in this new example, the zeros disappear from A^2:

Two-step matrix $$\boldsymbol{u}_2 = A^2\boldsymbol{u}_0 = \begin{bmatrix} \frac{1}{2} & \frac{1}{4} & \frac{1}{4} \\ \frac{1}{4} & \frac{1}{2} & \frac{1}{4} \\ \frac{1}{4} & \frac{1}{4} & \frac{1}{2} \end{bmatrix}\begin{bmatrix} p_1 \\ p_2 \\ p_3 \end{bmatrix}.$$

The eigenvalues of A are $\lambda_1 = 1$ (because A is Markov) and $\lambda_2 = \lambda_3 = -\frac{1}{2}$. For $\lambda = 1$, ***the eigenvector*** $\boldsymbol{x}_1 = (\frac{1}{3}, \frac{1}{3}, \frac{1}{3})$ ***will be the steady state***. When three equal populations split in half and move, the populations are again equal. Starting from $\boldsymbol{u}_0 = (8, 16, 32)$, the Markov chain approaches its steady state:

$$\boldsymbol{u}_0 = \begin{bmatrix} 8 \\ 16 \\ 32 \end{bmatrix} \qquad \boldsymbol{u}_1 = \begin{bmatrix} 24 \\ 20 \\ 12 \end{bmatrix} \qquad \boldsymbol{u}_2 = \begin{bmatrix} 16 \\ 18 \\ 22 \end{bmatrix} \qquad \boldsymbol{u}_3 = \begin{bmatrix} 20 \\ 19 \\ 17 \end{bmatrix}.$$

The step to $\boldsymbol{u}_4$ will split some people in half. This cannot be helped. The total population is $8 + 16 + 32 = 56$ at every step. The steady state is 56 times $(\frac{1}{3}, \frac{1}{3}, \frac{1}{3})$. You can see the three populations approaching, but never reaching, their final limits $56/3$.

Challenge Problem 6.7.16 created a Markov matrix A from the number of links between websites. The steady state $\boldsymbol{u}$ will give the Google rankings. ***Google finds $\boldsymbol{u}_\infty$ by a random walk that follows links*** (*random surfing*). That eigenvector comes from counting the fraction of visits to each website—a quick way to compute the steady state.

The size $|\lambda_2|$ of the second eigenvalue controls the speed of convergence to steady state.

Perron-Frobenius Theorem

One matrix theorem dominates this subject. The Perron-Frobenius Theorem applies when all $a_{ij} \geq 0$. There is no requirement that columns add to 1. We prove the neatest form, when all $a_{ij} > 0$: any positive matrix A (not necessarily positive definite!).

Perron-Frobenius for $A > 0$ ***All numbers in $A\boldsymbol{x} = \lambda_{\max}\boldsymbol{x}$ are strictly positive.***

Proof The key idea is to look at all numbers t such that $A\boldsymbol{x} \geq t\boldsymbol{x}$ for some nonnegative vector $\boldsymbol{x}$ (other than $\boldsymbol{x} = \boldsymbol{0}$). We are allowing inequality in $A\boldsymbol{x} \geq t\boldsymbol{x}$ in order to have many small positive candidates t. For the largest value $t_{\max}$ (which is attained), we will show that ***equality holds***: $A\boldsymbol{x} = t_{\max}\boldsymbol{x}$.

Otherwise, if $A\boldsymbol{x} \geq t_{\max}\boldsymbol{x}$ is not an equality, multiply by A. Because A is positive that produces a strict inequality $A^2\boldsymbol{x} > t_{\max}A\boldsymbol{x}$. Therefore the positive vector $\boldsymbol{y} = A\boldsymbol{x}$ satisfies $A\boldsymbol{y} > t_{\max}\boldsymbol{y}$, and $t_{\max}$ could be increased. This contradiction forces the equality $A\boldsymbol{x} = t_{\max}\boldsymbol{x}$, and *we have an eigenvalue*. Its eigenvector $\boldsymbol{x}$ is positive because on the left side of that equality, $A\boldsymbol{x}$ is sure to be positive.

To see that no eigenvalue can be larger than $t_{\max}$, suppose $A\boldsymbol{z} = \lambda\boldsymbol{z}$. Since λ and $\boldsymbol{z}$ may involve negative or complex numbers, we take absolute values: $|\lambda||\boldsymbol{z}| = |A\boldsymbol{z}| \leq A|\boldsymbol{z}|$ by the "triangle inequality." This $|\boldsymbol{z}|$ is a nonnegative vector, so this $|\lambda|$ is one of the possible candidates t. Therefore $|\lambda|$ cannot exceed $t_{\max}$—which must be $\lambda_{\max}$.

Population Growth

Divide the population into three age groups: age < 20, age 20 to 39, and age 40 to 59. At year T the sizes of those groups are n_1, n_2, n_3. Twenty years later, the sizes have changed for three reasons: births, deaths, and getting older.

1. **Reproduction** $n_1^{\text{new}} = F_1\,n_1 + F_2\,n_2 + F_3\,n_3$ gives a new generation

2. **Survival** $n_2^{\text{new}} = P_1 n_1$ and $n_3^{\text{new}} = P_2 n_2$ gives the older generations

The fertility rates are F_1, F_2, F_3 (F_2 largest). The *Leslie matrix* A might look like this:

$$\begin{bmatrix} n_1 \\ n_2 \\ n_3 \end{bmatrix}^{\textbf{new}} = \begin{bmatrix} F_1 & F_2 & F_3 \\ P_1 & 0 & 0 \\ 0 & P_2 & 0 \end{bmatrix} \begin{bmatrix} n_1 \\ n_2 \\ n_3 \end{bmatrix} = \begin{bmatrix} .04 & \mathbf{1.1} & .01 \\ .98 & 0 & 0 \\ 0 & \mathbf{.92} & 0 \end{bmatrix} \begin{bmatrix} n_1 \\ \boldsymbol{n_2} \\ n_3 \end{bmatrix}.$$

This is population projection in its simplest form, the same matrix A at every step. In a realistic model, A will change with time (from the environment or internal factors). Professors may want to include a fourth group, age ≥ 60, but we don't allow it.

The matrix has $A \geq 0$ but not $A > 0$. The Perron-Frobenius theorem still applies because $A^3 > 0$. The largest eigenvalue is $\lambda_{\max} \approx 1.06$. You can watch the generations move, starting from $n_2 = 1$ in the middle generation:

$$\textbf{eig}(A) = \begin{matrix} \mathbf{1.06} \\ -1.01 \\ -0.01 \end{matrix} \quad A^2 = \begin{bmatrix} 1.08 & \mathbf{0.05} & .00 \\ 0.04 & \mathbf{1.08} & .01 \\ 0.90 & 0 & 0 \end{bmatrix} \quad A^3 = \begin{bmatrix} 0.10 & \mathbf{1.19} & .01 \\ 0.06 & \mathbf{0.05} & .00 \\ 0.04 & \mathbf{0.99} & .01 \end{bmatrix}.$$

A fast start would come from $\boldsymbol{u}_0 = (0, 1, 0)$. That middle group will reproduce 1.1 and also survive .92. The newest and oldest generations are in $\boldsymbol{u}_1 = (1.1, 0, .92) =$ column 2 of A. Then $\boldsymbol{u}_2 = A\boldsymbol{u}_1 = A^2\boldsymbol{u}_0$ is the second column of A^2. The early numbers (transients) depend a lot on $\boldsymbol{u}_0$, but ***the asymptotic growth rate*** $\lambda_{\max}$ ***is the same from every start.*** Its eigenvector $\boldsymbol{x} = (.63, .58, .51)$ shows all three groups growing steadily together.

Caswell's book on *Matrix Population Models* emphasizes sensitivity analysis. The model is never exactly right. If the F's or P's in the matrix change by 10%, does $\lambda_{\max}$ go below 1 (which means extinction)? Problem 19 will show that a matrix change ΔA produces an eigenvalue change $\Delta\lambda = \boldsymbol{y}^{\mathrm{T}}(\Delta A)\boldsymbol{x}$. Here $\boldsymbol{x}$ and $\boldsymbol{y}^{\mathrm{T}}$ are the right and left eigenvectors of A, with $A\boldsymbol{x} = d\boldsymbol{x}$ and $A^{\mathrm{T}}\boldsymbol{y} = \lambda\boldsymbol{y}$.

Linear Algebra in Economics: The Consumption Matrix

A long essay about linear algebra in economics would be out of place here. A short note about one matrix seems reasonable. The ***consumption matrix*** tells how much of each input goes into a unit of output. This describes the manufacturing side of the economy.

Consumption matrix We have n industries like chemicals, food, and oil. To produce a unit of chemicals may require .2 units of chemicals, .3 units of food, and .4 units of oil. Those numbers go into row 1 of the consumption matrix A:

$$\begin{bmatrix} \text{chemical output} \\ \text{food output} \\ \text{oil output} \end{bmatrix} = \begin{bmatrix} .2 & .3 & .4 \\ .4 & .4 & .1 \\ .5 & .1 & .3 \end{bmatrix} \begin{bmatrix} \text{chemical input} \\ \text{food input} \\ \text{oil input} \end{bmatrix}.$$

Row 2 shows the inputs to produce food—a heavy use of chemicals and food, not so much oil. Row 3 of A shows the inputs consumed to refine a unit of oil. The real consumption matrix for the United States in 1958 contained 83 industries. The models in the 1990's are much larger and more precise. We chose a consumption matrix that has a convenient eigenvector.

Now comes the question: Can this economy meet demands y_1, y_2, y_3 for chemicals, food, and oil? To do that, the inputs p_1, p_2, p_3 will have to be higher—because part of $\boldsymbol{p}$ is consumed in producing $\boldsymbol{y}$. The input is $\boldsymbol{p}$ and the consumption is $A\boldsymbol{p}$, which leaves the output $\boldsymbol{p} - A\boldsymbol{p}$. This net production is what meets the demand $\boldsymbol{y}$:

Problem Find a vector $\boldsymbol{p}$ such that $\boldsymbol{p} - A\boldsymbol{p} = \boldsymbol{y}$ or $\boldsymbol{p} = (I - A)^{-1}\boldsymbol{y}$.

Apparently the linear algebra question is whether $I - A$ is invertible. But there is more to the problem. The vector $\boldsymbol{y}$ of required outputs is nonnegative, and so is A. *The production levels in $\boldsymbol{p} = (I - A)^{-1}\boldsymbol{y}$ must also be nonnegative.* The real question is:

When is $(I - A)^{-1}$ a nonnegative matrix?

This is the test on $(I - A)^{-1}$ for a productive economy, which can meet any demand. If A is small compared to I, then $A\boldsymbol{p}$ is small compared to $\boldsymbol{p}$. There is plenty of output. If A is too large, then production consumes too much and the demand $\boldsymbol{y}$ cannot be met.

"Small" or "large" is decided by the largest eigenvalue λ_1 of A (which is positive):

If $\lambda_1 > 1$ then $(I - A)^{-1}$ has negative entries
If $\lambda_1 = 1$ then $(I - A)^{-1}$ fails to exist
If $\lambda_1 < 1$ then $(I - A)^{-1}$ is nonnegative as desired.

The main point is that last one. The reasoning uses a nice formula for $(I - A)^{-1}$, which we give now. The most important infinite series in mathematics is the ***geometric series*** $1 + x + x^2 + \cdots$. This series adds up to $1/(1 - x)$ provided x lies between -1 and 1. When $x = 1$ the series is $1 + 1 + 1 + \cdots = \infty$. When $|x| \geq 1$ the terms x^n don't go to zero and the series has no chance to converge.

The nice formula for $(I - A)^{-1}$ is the ***geometric series of matrices***:

Geometric series $(I - A)^{-1} = I + A + A^2 + A^3 + \cdots.$

If you multiply the series $S = I + A + A^2 + \cdots$ by A, you get the same series except for I. Therefore $S - AS = I$, which is $(I - A)S = I$. The series adds to $S = (I - A)^{-1}$ if it converges. ***And it converges if all eigenvalues of*** A ***have*** $|\lambda| < 1$.

In our case $A \geq 0$. All terms of the series are nonnegative. Its sum is $(I - A)^{-1} \geq 0$.

Example 4 $A = \begin{bmatrix} .2 & .3 & .4 \\ .4 & .4 & .1 \\ .5 & .1 & .3 \end{bmatrix}$ has $\lambda_{\text{max}} = \mathbf{.9}$ and $(I - A)^{-1} = \frac{1}{93}\begin{bmatrix} 41 & 25 & 27 \\ 33 & 36 & 24 \\ 34 & 23 & 36 \end{bmatrix}$.

This economy is productive. A is small compared to I, because λ_{max} is .9. To meet the demand $\boldsymbol{y}$, start from $\boldsymbol{p} = (I - A)^{-1}\boldsymbol{y}$. Then $A\boldsymbol{p}$ is consumed in production, leaving $\boldsymbol{p} - A\boldsymbol{p}$. This is $(I - A)\boldsymbol{p} = \boldsymbol{y}$, and the demand is met.

Example 5 $A = \begin{bmatrix} 0 & 4 \\ 1 & 0 \end{bmatrix}$ has $\lambda_{\text{max}} = \mathbf{2}$ and $(I - A)^{-1} = -\frac{1}{3}\begin{bmatrix} 1 & 4 \\ 1 & 1 \end{bmatrix}$.

This consumption matrix A is too large. Demands can't be met, because production consumes more than it yields. The series $I + A + A^2 + \ldots$ does not converge to $(I - A)^{-1}$ because $\lambda_{\text{max}} > 1$. The series is growing while $(I - A)^{-1}$ is actually negative.

In the same way $1 + 2 + 4 + \cdots$ is not really $1/(1 - 2) = -1$. But not entirely false !

Problem Set 10.3

Questions 1–12 are about Markov matrices and their eigenvalues and powers.

1 Find the eigenvalues of this Markov matrix (their sum is the trace).

$$A = \begin{bmatrix} .90 & .15 \\ .10 & .85 \end{bmatrix}.$$

What is the steady state eigenvector for the eigenvalue $\lambda_1 = 1$?

2 Diagonalize the Markov matrix in Problem 1 to $A = X\Lambda X^{-1}$ by finding its other eigenvector:

$$A = \begin{bmatrix} & \\ & \end{bmatrix}\begin{bmatrix} 1 & \\ & .75 \end{bmatrix}\begin{bmatrix} & \\ & \end{bmatrix}.$$

What is the limit of $A^k = X\Lambda^k X^{-1}$ when $\Lambda^k = \begin{bmatrix} 1 & 0 \\ 0 & .75^k \end{bmatrix}$ approaches $\begin{bmatrix} 1 & 0 \\ 0 & 0 \end{bmatrix}$?

3 What are the eigenvalues and steady state eigenvectors for these Markov matrices?

$$A = \begin{bmatrix} 1 & .2 \\ 0 & .8 \end{bmatrix} \quad A = \begin{bmatrix} .2 & 1 \\ .8 & 0 \end{bmatrix} \quad A = \begin{bmatrix} \frac{1}{2} & \frac{1}{4} & \frac{1}{4} \\ \frac{1}{4} & \frac{1}{2} & \frac{1}{4} \\ \frac{1}{4} & \frac{1}{4} & \frac{1}{2} \end{bmatrix}.$$

4 For every 4 by 4 Markov matrix, what eigenvector of A^{T} corresponds to the (known) eigenvalue $\lambda = 1$?

5 Every year 2% of young people become old and 3% of old people become dead. (No births.) Find the steady state for

$$\begin{bmatrix} \text{young} \\ \text{old} \\ \text{dead} \end{bmatrix}_{k+1} = \begin{bmatrix} .98 & .00 & 0 \\ .02 & .97 & 0 \\ .00 & .03 & 1 \end{bmatrix} \begin{bmatrix} \text{young} \\ \text{old} \\ \text{dead} \end{bmatrix}_{k}.$$

6 For a Markov matrix, the sum of the components of $\boldsymbol{x}$ equals the sum of the components of $A\boldsymbol{x}$. If $A\boldsymbol{x} = \lambda\boldsymbol{x}$ with $\lambda \neq 1$, prove that the components of this non-steady eigenvector $\boldsymbol{x}$ add to zero.

7 Find the eigenvalues and eigenvectors of A. Explain why A^k approaches A^∞:

$$A = \begin{bmatrix} .8 & .3 \\ .2 & .7 \end{bmatrix} \qquad A^\infty = \begin{bmatrix} .6 & .6 \\ .4 & .4 \end{bmatrix}.$$

Challenge problem: Which Markov matrices produce that steady state $(.6, .4)$?

8 The steady state eigenvector of a permutation matrix is $(\frac{1}{4}, \frac{1}{4}, \frac{1}{4}, \frac{1}{4})$. This is *not* approached when $\boldsymbol{u}_0 = (0, 0, 0, 1)$. What are $\boldsymbol{u}_1$ and $\boldsymbol{u}_2$ and $\boldsymbol{u}_3$ and $\boldsymbol{u}_4$? What are the four eigenvalues of P, which solve $\lambda^4 = 1$?

Permutation matrix = Markov matrix $\quad P = \begin{bmatrix} 0 & 1 & 0 & 0 \\ 0 & 0 & 1 & 0 \\ 0 & 0 & 0 & 1 \\ 1 & 0 & 0 & 0 \end{bmatrix}.$

9 Prove that the square of a Markov matrix is also a Markov matrix.

10 If $A = \left[\begin{smallmatrix} a & b \\ c & d \end{smallmatrix}\right]$ is a Markov matrix, its eigenvalues are 1 and _____. The steady state eigenvector is $\boldsymbol{x}_1 =$ _____.

11 Complete A to a Markov matrix and find the steady state eigenvector. When A is a symmetric Markov matrix, why is $\boldsymbol{x}_1 = (1, \ldots, 1)$ its steady state?

$$A = \begin{bmatrix} .7 & .1 & .2 \\ .1 & .6 & .3 \\ \_ & \_ & \_ \end{bmatrix}.$$

12 A Markov differential equation is not $d\boldsymbol{u}/dt = A\boldsymbol{u}$ but $d\boldsymbol{u}/dt = (A - I)\boldsymbol{u}$. The diagonal is negative, the rest of $A - I$ is positive. The columns add to zero, not 1.

Find λ_1 and λ_2 for $B = A - I = \begin{bmatrix} -.2 & .3 \\ .2 & -.3 \end{bmatrix}$. Why does $A - I$ have $\lambda_1 = 0$?

When $e^{\lambda_1 t}$ and $e^{\lambda_2 t}$ multiply $\boldsymbol{x}_1$ and $\boldsymbol{x}_2$, what is the steady state as $t \to \infty$?

Questions 13–15 are about linear algebra in economics.

13 Each row of the consumption matrix in Example 4 adds to .9. Why does that make $\lambda = .9$ an eigenvalue, and what is the eigenvector?

14 Multiply $I + A + A^2 + A^3 + \cdots$ by $I - A$ to get I. The series adds to $(I - A)^{-1}$. For $A = \begin{bmatrix} 0 & \frac{1}{2} \\ 1 & 0 \end{bmatrix}$, find A^2 and A^3 and use the pattern to add up the series.

15 For which of these matrices does $I + A + A^2 + \cdots$ yield a nonnegative matrix $(I - A)^{-1}$? Then the economy can meet any demand:

$$A = \begin{bmatrix} 0 & 1 \\ 0 & 0 \end{bmatrix} \qquad A = \begin{bmatrix} 0 & 4 \\ .2 & 0 \end{bmatrix} \qquad A = \begin{bmatrix} .5 & 1 \\ .5 & 0 \end{bmatrix}.$$

If the demands are $\boldsymbol{y} = (2, 6)$, what are the vectors $\boldsymbol{p} = (I - A)^{-1}\boldsymbol{y}$?

16 (Markov again) This matrix has zero determinant. What are its eigenvalues?

$$A = \begin{bmatrix} .4 & .2 & .3 \\ .2 & .4 & .3 \\ .4 & .4 & .4 \end{bmatrix}.$$

Find the limits of $A^k \boldsymbol{u}_0$ starting from $\boldsymbol{u}_0 = (1, 0, 0)$ and then $\boldsymbol{u}_0 = (100, 0, 0)$.

17 If A is a Markov matrix, why doesn't $I + A + A^2 + \cdots$ add up to $(I - A)^{-1}$?

18 For the Leslie matrix show that $\det(A - \lambda I) = 0$ gives $F_1\lambda^2 + F_2 P_1 \lambda + F_3 P_1 P_2 = \lambda^3$. The right side λ^3 is larger as $\lambda \longrightarrow \infty$. The left side is larger at $\lambda = 1$ if $F_1 + F_2 P_1 + F_3 P_1 P_2 > 1$. In that case the two sides are equal at an eigenvalue $\lambda_{\max} > 1$: *growth.*

19 **Sensitivity of eigenvalues**: A matrix change ΔA produces eigenvalue changes $\Delta\Lambda$. *Those changes* $\Delta\lambda_1, \ldots, \Delta\lambda_n$ *are on the diagonal of* $(X^{-1}\,\Delta A\, X)$. **Challenge**: Start from $AX = X\Lambda$. The eigenvectors and eigenvalues change by ΔX and $\Delta\Lambda$:

$$(A{+}\Delta A)(X{+}\Delta X) = (X{+}\Delta X)(\Lambda{+}\Delta\Lambda) \text{ becomes } A(\Delta X){+}(\Delta A)X = X(\Delta\Lambda){+}(\Delta X)\Lambda.$$

Small terms $(\Delta A)(\Delta X)$ and $(\Delta X)(\Delta\Lambda)$ are ignored. *Multiply the last equation by* X^{-1}. From the inner terms, the diagonal part of $X^{-1}(\Delta A)X$ gives $\Delta\Lambda$ as we want. *Why do the outer terms* $X^{-1}\, A\, \Delta X$ *and* $X^{-1}\, \Delta X\, \Lambda$ *cancel on the diagonal*?

$$\text{Explain } X^{-1}A = \Lambda X^{-1} \text{ and then } \quad \textbf{diag}(\Lambda\, X^{-1}\, \Delta X) = \textbf{diag}(X^{-1}\, \Delta X\, \Lambda).$$

20 Suppose $B > A > 0$, meaning that each $b_{ij} > a_{ij} > 0$. How does the Perron-Frobenius discussion show that $\lambda_{\max}(B) > \lambda_{\max}(A)$?

10.4 Linear Programming

Linear programming is linear algebra plus two new ideas: ***inequalities*** and ***minimization***. The starting point is still a matrix equation $A\boldsymbol{x} = \boldsymbol{b}$. But the only acceptable solutions are *nonnegative*. We require $\boldsymbol{x} \geq 0$ (meaning that no component of $\boldsymbol{x}$ can be negative). The matrix has $n > m$, more unknowns than equations. If there are any solutions $\boldsymbol{x} \geq \mathbf{0}$ to $A\boldsymbol{x} = \boldsymbol{b}$, there are probably a lot. Linear programming picks the solution $\boldsymbol{x}^* \geq \mathbf{0}$ that minimizes the cost:

> ***The cost is*** $c_1x_1 + \cdots + c_nx_n$. ***The winning vector*** x^* ***is the nonnegative solution of*** $Ax = b$ ***that has smallest cost.***

Thus a linear programming problem starts with a matrix A and two vectors $\boldsymbol{b}$ and $\boldsymbol{c}$:

i) A has $n > m$: for example $A = \begin{bmatrix} 1 & 1 & 2 \end{bmatrix}$ (one equation, three unknowns)

ii) $\boldsymbol{b}$ has m components for m equations $A\boldsymbol{x} = \boldsymbol{b}$: for example $\boldsymbol{b} = \begin{bmatrix} 4 \end{bmatrix}$

iii) The ***cost vector c*** has n components: for example $\boldsymbol{c} = \begin{bmatrix} 5 & 3 & 8 \end{bmatrix}$.

Then the problem is to minimize $\boldsymbol{c} \cdot \boldsymbol{x}$ subject to the requirements $A\boldsymbol{x} = \boldsymbol{b}$ and $\boldsymbol{x} \geq 0$:

$$\textit{Minimize}\quad 5x_1 + 3x_2 + 8x_3 \quad \textit{subject to} \quad x_1 + x_2 + 2x_3 = 4 \quad \textit{and} \quad x_1, x_2, x_3 \geq 0.$$

We jumped right into the problem, without explaining where it comes from. Linear programming is actually the most important application of mathematics to management. Development of the fastest algorithm and fastest code is highly competitive. You will see that finding $\boldsymbol{x}^*$ is harder than solving $A\boldsymbol{x} = \boldsymbol{b}$, because of the extra requirements: $\boldsymbol{x}^* \geq 0$ and minimum cost $\boldsymbol{c}^{\mathrm{T}}\boldsymbol{x}^*$. We will explain the background, and the famous *simplex method*, and *interior point methods*, after solving the example.

Look first at the "constraints": $A\boldsymbol{x} = \boldsymbol{b}$ and $\boldsymbol{x} \geq 0$. The equation $x_1 + x_2 + 2x_3 = 4$ gives a plane in three dimensions. The nonnegativity $x_1 \geq 0, x_2 \geq 0, x_3 \geq 0$ chops the plane down to a triangle. The solution $\boldsymbol{x}^*$ must lie in the triangle PQR in Figure 8.6.

Inside that triangle, all components of $\boldsymbol{x}$ are positive. On the edges of PQR, one component is zero. At the corners P and Q and R, two components are zero. ***The optimal solution*** x^* ***will be one of those corners!*** We will now show why.

The triangle contains all vectors $\boldsymbol{x}$ that satisfy $A\boldsymbol{x} = \boldsymbol{b}$ and $\boldsymbol{x} \geq 0$. Those $\boldsymbol{x}$'s are called *feasible points*, and the triangle is the ***feasible set***. These points are the allowed candidates in the minimization of $\boldsymbol{c} \cdot \boldsymbol{x}$, which is the final step:

> ***Find*** x^* ***in the triangle*** PQR ***to minimize the cost*** $5x_1 + 3x_2 + 8x_3$.

The vectors that have *zero* cost lie on the plane $5x_1 + 3x_2 + 8x_3 = 0$. That plane does not meet the triangle. We cannot achieve zero cost, while meeting the requirements on $\boldsymbol{x}$. So increase the cost C until the plane $5x_1 + 3x_2 + 8x_3 = C$ does meet the triangle. As C increases, we have *parallel planes moving toward the triangle*.

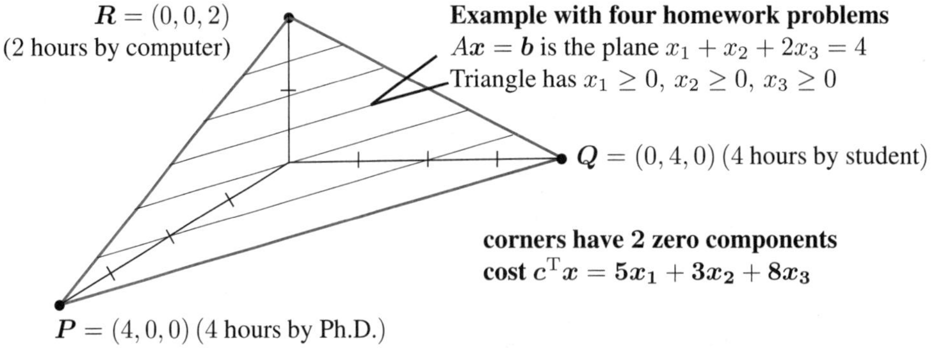

Figure 10.5: The triangle contains all nonnegative solutions: $Ax = b$ and $x \geq 0$. The lowest cost solution x^* is a corner P, Q, or R of this feasible set.

The first plane $5x_1 + 3x_2 + 8x_3 = C$ to touch the triangle has minimum cost C. ***The point where it touches is the solution x^*.*** This touching point must be one of the corners P or Q or R. A moving plane could not reach the inside of the triangle before it touches a corner! So check the cost $5x_1 + 3x_2 + 8x_3$ at each corner:

$P = (4,0,0)$ costs 20	$Q = (0,4,0)$ costs 12	$R = (0,0,2)$ costs 16.

The winner is Q. Then $x^* = (0,4,0)$ solves the linear programming problem.

If the cost vector c is changed, the parallel planes are tilted. For small changes, Q is still the winner. For the cost $c \cdot x = 5x_1 + 4x_2 + 7x_3$, the optimum x^* moves to $R = (0,0,2)$. The minimum cost is now $7 \cdot 2 = 14$.

Note 1 Some linear programs *maximize profit* instead of minimizing cost. The mathematics is almost the same. The parallel planes start with a large value of C, instead of a small value. They move toward the origin (instead of away), as C gets smaller. *The first touching point is still a corner.*

Note 2 The requirements $Ax = b$ and $x \geq 0$ could be impossible to satisfy. The equation $x_1 + x_2 + x_3 = -1$ cannot be solved with $x \geq 0$. *That feasible set is empty.*

Note 3 It could also happen that the feasible set is *unbounded.* If the requirement is $x_1 + x_2 - 2x_3 = 4$, the large positive vector $(100, 100, 98)$ is now a candidate. So is the larger vector $(1000, 1000, 998)$. The plane $Ax = b$ is no longer chopped off to a triangle. The two corners P and Q are still candidates for x^*, but R moved to infinity.

Note 4 With an unbounded feasible set, the minimum cost could be $-\infty$ (*minus infinity*). Suppose the cost is $-x_1 - x_2 + x_3$. Then the vector $(100, 100, 98)$ costs $C = -102$. The vector $(1000, 1000, 998)$ costs $C = -1002$. We are being paid to include x_1 and x_2, instead of paying a cost. In realistic applications this will not happen. But it is theoretically possible that A, b, and c can produce unexpected triangles and costs.

The Primal and Dual Problems

This first problem will fit $\boldsymbol{A}, \boldsymbol{b}, \boldsymbol{c}$ in that example. The unknowns x_1, x_2, x_3 represent hours of work by a Ph.D. and a student and a machine. The costs per hour are \$5, \$3, and \$8. (*I apologize for such low pay.*) The number of hours cannot be negative: $x_1 \geq 0, x_2 \geq 0, x_3 \geq 0$. The Ph.D. and the student get through one homework problem per hour. *The machine solves two problems in one hour.* In principle they can share out the homework, which has four problems to be solved: $x_1 + x_2 + 2x_3 = 4$.

The problem is to finish the four problems at minimum cost $\boldsymbol{c}^{\mathrm{T}}\boldsymbol{x}$.

If all three are working, the job takes one hour: $x_1 = x_2 = x_3 = 1$. The cost is $5 + 3 + 8 = 16$. But certainly the Ph.D. should be put out of work by the student (who is just as fast and costs less—this problem is getting realistic). When the student works two hours and the machine works one, the cost is $6 + 8$ and all four problems get solved. We are on the edge $\boldsymbol{QR}$ of the triangle because the Ph.D. is not working: $x_1 = 0$. But the best point is all work by student (at $\boldsymbol{Q}$) or all work by machine (at $\boldsymbol{R}$). In this example the student solves four problems in four hours for \$12—the minimum cost.

With only one equation in $A\boldsymbol{x} = \boldsymbol{b}$, the corner $(0, 4, 0)$ has only one nonzero component. ***When $A\boldsymbol{x} = \boldsymbol{b}$ has m equations, corners have m nonzeros.*** We solve $A\boldsymbol{x} = \boldsymbol{b}$ for those m variables, with $n - m$ free variables set to zero. But unlike Chapter 3, ***we don't know which m variables to choose.***

The number of possible corners is the number of ways to choose m components out of n. This number "n choose m" is heavily involved in gambling and probability. With $n = 20$ unknowns and $m = 8$ equations (still small numbers), the "feasible set" can have $20!/8!12!$ corners. That number is $(20)(19)\cdots(13) = 5{,}079{,}110{,}400$.

Checking three corners for the minimum cost was fine. Checking five billion corners is not the way to go. The simplex method described below is much faster.

The Dual Problem In linear programming, problems come in pairs. There is a minimum problem and a maximum problem—the original and its "dual." The original problem was specified by a matrix A and two vectors $\boldsymbol{b}$ and $\boldsymbol{c}$. The dual problem transposes A and switches $\boldsymbol{b}$ and $\boldsymbol{c}$: ***Maximize $\boldsymbol{b} \cdot \boldsymbol{y}$***. Here is the dual to our example:

> **A cheater offers to solve homework problems by selling the answers.** The charge is y dollars per problem, or $4y$ altogether. (Note how $\boldsymbol{b} = 4$ has gone into the cost.) The cheater must be as cheap as the Ph.D. or student or machine: $y \leq 5$ and $y \leq 3$ and $2y \leq 8$. (Note how $\boldsymbol{c} = (5, 3, 8)$ has gone into inequality constraints). The cheater maximizes the income $4y$.

Dual Problem	***Maximize $\boldsymbol{b} \cdot \boldsymbol{y}$ subject to $A^{\mathrm{T}}\boldsymbol{y} \leq \boldsymbol{c}$*** .

The maximum occurs when $y = 3$. The income is $4y = 12$. The maximum in the dual problem (\$12) equals the minimum in the original (\$12). *Max = min* is duality.

If either problem has a best vector (x^* or y^*) then so does the other.
Minimum cost $c \cdot x^*$ equals maximum income $b \cdot y^*$

This book started with a row picture and a column picture. The first "duality theorem" was about rank: The number of independent rows equals the number of independent columns. That theorem, like this one, was easy for small matrices. Minimum cost = maximum income is proved in our text *Linear Algebra and Its Applications*. One line will establish the easy half of the theorem: ***The cheater's income $b^{\mathrm{T}}y$ cannot exceed the honest cost***:

$$\textit{If}\ \ Ax = b, x \geq 0, A^{\mathrm{T}}y \leq c \quad \textit{then} \quad b^{\mathrm{T}}y = (Ax)^{\mathrm{T}}y = x^{\mathrm{T}}(A^{\mathrm{T}}y) \leq x^{\mathrm{T}}c. \quad (1)$$

The full duality theorem says that when $\boldsymbol{b}^{\mathrm{T}}\boldsymbol{y}$ reaches its maximum and $\boldsymbol{x}^{\mathrm{T}}\boldsymbol{c}$ reaches its minimum, they are equal: $\boldsymbol{b} \cdot \boldsymbol{y}^* = \boldsymbol{c} \cdot \boldsymbol{x}^*$. Look at the last step in (1), with $\leq$ sign:

The dot product of $\boldsymbol{x} \geq 0$ and $\boldsymbol{s} = \boldsymbol{c} - A^{\mathrm{T}}\boldsymbol{y} \geq 0$ gave $\boldsymbol{x}^{\mathrm{T}}\boldsymbol{s} \geq 0$. This is $\boldsymbol{x}^{\mathrm{T}}A^{\mathrm{T}}\boldsymbol{y} \leq \boldsymbol{x}^{\mathrm{T}}\boldsymbol{c}$.

Equality needs $x^{\mathrm{T}}s = 0$ So the optimal solution has $x_j^* = 0$ or $s_j^* = 0$ for each j.

The Simplex Method

Elimination is the workhorse for linear equations. The simplex method is the workhorse for linear inequalities. We cannot give the simplex method as much space as elimination, but the idea can be clear. *The simplex method goes from one corner to a neighboring corner of lower cost.* Eventually (and quite soon in practice) it reaches the corner of minimum cost.

A ***corner*** is a vector $\boldsymbol{x} \geq 0$ that satisfies the m equations $A\boldsymbol{x} = \boldsymbol{b}$ with at most m positive components. *The other $n - m$ components are zero.* (Those are the free variables. Back substitution gives the m basic variables. All variables must be nonnegative or $\boldsymbol{x}$ is a false corner.) For a *neighboring corner*, one zero component of x becomes positive and one positive component becomes zero.

The simplex method must decide which component "enters" by becoming positive, and which component "leaves" by becoming zero. That exchange is chosen so as to lower the total cost. This is one step of the simplex method, moving toward x^*.

Here is the overall plan. Look at each zero component at the current corner. If it changes from 0 to 1, the other nonzeros have to adjust to keep $A\boldsymbol{x} = \boldsymbol{b}$. Find the new $\boldsymbol{x}$ by back substitution and compute the change in the total cost $\boldsymbol{c} \cdot \boldsymbol{x}$. This change is the "reduced cost" r of the new component. The ***entering variable*** is the one that gives the ***most negative*** r. This is the greatest cost reduction for a single unit of a new variable.

Example 1 Suppose the current corner is $\boldsymbol{P} = (4, 0, 0)$, with the Ph.D. doing all the work (the cost is \$20). If the student works one hour, the cost of $\boldsymbol{x} = (3, 1, 0)$ is down to \$18. The reduced cost is $r = -2$. If the machine works one hour, then $\boldsymbol{x} = (2, 0, 1)$ also costs \$18. The reduced cost is also $r = -2$. In this case the simplex method can choose either the student or the machine as the entering variable.

Even in this small example, the first step may not go immediately to the best $\boldsymbol{x}^*$. The method chooses the entering variable before it knows how much of that variable to include. We computed r when the entering variable changes from 0 to 1, but one unit may be too much or too little. The method now chooses the leaving variable (the Ph.D.). It moves to corner $\boldsymbol{Q}$ or $\boldsymbol{R}$ in the figure.

The more of the entering variable we include, the lower the cost. This has to stop when one of the positive components (which are adjusting to keep $A\boldsymbol{x} = \boldsymbol{b}$) hits zero. *The **leaving variable** is the first positive x_i to reach zero.* When that happens, a neighboring corner has been found. Then start again (from the new corner) to find the next variables to enter and leave.

When all reduced costs are positive, the current corner is the optimal $\boldsymbol{x}^*$. No zero component can become positive without increasing $\boldsymbol{c} \cdot \boldsymbol{x}$. No new variable should enter. The problem is solved (and we can show that $\boldsymbol{y}^*$ is found too).

Note Generally $\boldsymbol{x}^*$ is reached in αn steps, where α is not large. But examples have been invented which use an exponential number of simplex steps. Eventually a different approach was developed, which is guaranteed to reach $\boldsymbol{x}^*$ in fewer (but more difficult) steps. The new methods travel through the *interior* of the feasible set.

Example 2 Minimize the cost $\boldsymbol{c} \cdot \boldsymbol{x} = 3x_1 + x_2 + 9x_3 + x_4$. The constraints are $\boldsymbol{x} \geq 0$ and two equations $A\boldsymbol{x} = \boldsymbol{b}$:

$$\begin{aligned} x_1 + 2x_3 + x_4 &= 4 \qquad m = 2 \quad \text{equations} \\ x_2 + x_3 - x_4 &= 2 \qquad n = 4 \quad \text{unknowns.} \end{aligned}$$

A starting corner is $\boldsymbol{x} = (4, 2, 0, 0)$ which costs $\boldsymbol{c} \cdot \boldsymbol{x} = 14$. It has $m = 2$ nonzeros and $n - m = 2$ zeros. The zeros are x_3 and x_4. The question is whether x_3 or x_4 should enter (become nonzero). Try one unit of each of them:

$$\begin{aligned} &\text{If } x_3 = 1 \text{ and } x_4 = 0, \quad \text{then } \boldsymbol{x} = (2, 1, 1, 0) \text{ costs } 16. \\ &\text{If } x_4 = 1 \text{ and } x_3 = 0, \quad \text{then } \boldsymbol{x} = (3, 3, 0, 1) \text{ costs } 13. \end{aligned}$$

Compare those costs with 14. The reduced cost of x_3 is $r = 2$, positive and useless. The reduced cost of x_4 is $r = -1$, negative and helpful. *The entering variable is x_4.*

How much of x_4 can enter? One unit of x_4 made x_1 drop from 4 to 3. Four units will make x_1 drop from 4 to zero (while x_2 increases all the way to 6). *The leaving variable is x_1.* The new corner is $\boldsymbol{x} = (0, 6, 0, 4)$, which costs only $\boldsymbol{c} \cdot \boldsymbol{x} = 10$. This is the optimal $\boldsymbol{x}^*$, but to know that we have to try another simplex step from $(0, 6, 0, 4)$. Suppose x_1 or x_3 tries to enter:

Start from the corner $(0, 6, 0, 4)$

$$\begin{aligned} &\text{If } x_1 = 1 \text{ and } x_3 = 0, \quad \text{then } \boldsymbol{x} = (1, 5, 0, 3) \text{ costs } 11. \\ &\text{If } x_3 = 1 \text{ and } x_1 = 0, \quad \text{then } \boldsymbol{x} = (0, 3, 1, 2) \text{ costs } 14. \end{aligned}$$

Those costs are higher than 10. Both r's are positive—it does not pay to move. The current corner $(0,6,0,4)$ is the solution $\boldsymbol{x}^*$.

These calculations can be streamlined. Each simplex step solves three linear systems with the same matrix B. (This is the m by m matrix that keeps the m basic columns of A.) When a column enters and an old column leaves, there is a quick way to update B^{-1}. That is how most codes organize the simplex method.

Our text on *Computational Science and Engineering* includes a short code with comments. (The code is also on **math.mit.edu/cse**) The best $\boldsymbol{y}^*$ solves m equations $A^{\mathrm{T}}\boldsymbol{y}^* = \boldsymbol{c}$ in the m components that are nonzero in $\boldsymbol{x}^*$. Then we have optimality $\boldsymbol{x}^{\mathrm{T}}\boldsymbol{s} = 0$ and this is duality: *Either* $x_j^* = 0$ *or the "slack" in* $\boldsymbol{s}^* = \boldsymbol{c} - A^{\mathrm{T}}y^*$ *has* $s_j^* = 0$.

When $\boldsymbol{x}^* = (0,4,0)$ was the optimal corner $\boldsymbol{Q}$, the cheater's price was set by $y^* = 3$.

Interior Point Methods

The simplex method moves along the edges of the feasible set, eventually reaching the optimal corner $\boldsymbol{x}^*$. ***Interior point methods move inside the feasible set*** (where $\boldsymbol{x} > \boldsymbol{0}$). These methods hope to go more directly to $\boldsymbol{x}^*$. They work well.

One way to stay inside is to put a barrier at the boundary. Add extra cost as a *logarithm that blows up* when any variable x_j touches zero. The best vector has $\boldsymbol{x} > \boldsymbol{0}$. The number θ is a small parameter that we move toward zero.

Barrier problem **Minimize** $\boldsymbol{c}^{\mathrm{T}}\boldsymbol{x} - \theta\,(\log x_1 + \cdots + \log x_n)$ **with** $A\boldsymbol{x} = \boldsymbol{b}$ (2)

This cost is nonlinear (but linear programming is already nonlinear from inequalities). The constraints $x_j \geq 0$ are not needed because $\log x_j$ becomes infinite at $x_j = 0$.

The barrier gives an *approximate problem* for each θ. The m constraints $A\boldsymbol{x} = \boldsymbol{b}$ have Lagrange multipliers $y_1, \ldots, y_m$. This is the good way to deal with constraints.

$\boldsymbol{y}$ from Lagrange $$L(x,y,\theta) = \boldsymbol{c}^{\mathrm{T}}\boldsymbol{x} - \theta\,(\textstyle\sum \log x_i) - \boldsymbol{y}^{\mathrm{T}}(A\boldsymbol{x} - \boldsymbol{b}) \qquad (3)$$

$\partial L/\partial y = 0$ brings back $A\boldsymbol{x} = \boldsymbol{b}$. The derivatives $\partial L/\partial x_j$ are interesting !

Optimality in barrier pbm $$\frac{\partial L}{\partial x_j} = c_j - \frac{\theta}{x_j} - (A^{\mathrm{T}}y)_j = 0 \quad \text{which is} \quad \boldsymbol{x_j s_j = \theta}. \qquad (4)$$

The true problem has $x_j s_j = 0$. The barrier problem has $x_j s_j = \theta$. The solutions $\boldsymbol{x}^*(\theta)$ lie on the ***central path*** to $\boldsymbol{x}^*(0)$. Those n optimality equations $x_j s_j = \theta$ are nonlinear, and we solve them iteratively by Newton's method.

The current $\boldsymbol{x}, \boldsymbol{y}, \boldsymbol{s}$ will satisfy $A\boldsymbol{x} = \boldsymbol{b}, \boldsymbol{x} \geq \boldsymbol{0}$ and $A^{\mathrm{T}}\boldsymbol{y} + \boldsymbol{s} = \boldsymbol{c}$, *but not* $x_j s_j = \theta$. Newton's method takes a step $\Delta\boldsymbol{x}, \Delta\boldsymbol{y}, \Delta\boldsymbol{s}$. By ignoring the second-order term $\Delta\boldsymbol{x}\Delta\boldsymbol{s}$

in $(\boldsymbol{x}+\Delta\boldsymbol{x})(\boldsymbol{s}+\Delta\boldsymbol{s})=\boldsymbol{\theta}$, the corrections in $\boldsymbol{x},\boldsymbol{y},\boldsymbol{s}$ come from linear equations:

Newton step

$$\begin{aligned} A\,\Delta\boldsymbol{x} &= 0 \\ A^{\mathrm{T}}\Delta\boldsymbol{y}+\Delta\boldsymbol{s} &= 0 \\ s_j\Delta x_j + x_j\Delta s_j &= \theta - x_j s_j \end{aligned} \tag{5}$$

Newton iteration has quadratic convergence for each θ, and then θ approaches zero. The duality gap $\boldsymbol{x}^{\mathrm{T}}\boldsymbol{s}$ generally goes below 10^{-8} after 20 to 60 steps. The explanation in my *Computational Science and Engineering* textbook takes one Newton step in detail, for the example with four homework problems. I didn't intend that the student should end up doing all the work, but $\boldsymbol{x}^*$ turned out that way.

This interior point method is used almost "as is" in commercial software, for a large class of linear and nonlinear optimization problems.

Problem Set 10.4

1 Draw the region in the xy plane where $x+2y=6$ and $x\geq 0$ and $y\geq 0$. Which point in this "feasible set" minimizes the cost $c=x+3y$? Which point gives maximum cost? Those points are at corners.

2 Draw the region in the xy plane where $x+2y\leq 6,\ 2x+y\leq 6,\ x\geq 0,\ y\geq 0$. It has four corners. Which corner minimizes the cost $c=2x-y$?

3 What are the corners of the set $x_1+2x_2-x_3=4$ with x_1,x_2,x_3 all ≥ 0? Show that the cost x_1+2x_3 can be very negative in this feasible set. This is an example of unbounded cost: no minimum.

4 Start at $\boldsymbol{x}=(0,0,2)$ where the machine solves all four problems for \$16. Move to $\boldsymbol{x}=(0,1,\quad)$ to find the reduced cost r (the savings per hour) for work by the student. Find r for the Ph.D. by moving to $\boldsymbol{x}=(1,0,\quad)$ with 1 hour of Ph.D. work.

5 Start Example 1 from the Ph.D. corner $(4,0,0)$ with $\boldsymbol{c}$ changed to $[\,5\ \ 3\ \ 7\,]$. Show that r is better for the machine even when the total cost is lower for the student. The simplex method takes two steps, first to the machine and then to the student for $\boldsymbol{x}^*$.

6 Choose a different cost vector $\boldsymbol{c}$ so the Ph.D. gets the job. Rewrite the dual problem (maximum income to the cheater).

7 A six-problem homework on which the Ph.D. is fastest gives a second constraint $2x_1+x_2+x_3=6$. Then $\boldsymbol{x}=(2,2,0)$ shows two hours of work by Ph.D. and student on each homework. Does this $\boldsymbol{x}$ minimize the cost $\boldsymbol{c}^{\mathrm{T}}\boldsymbol{x}$ with $\boldsymbol{c}=(5,3,8)$?

8 These two problems are also dual. Prove weak duality, that always $\boldsymbol{y}^{\mathrm{T}}\boldsymbol{b}\leq\boldsymbol{c}^{\mathrm{T}}\boldsymbol{x}$:

Primal problem Minimize $\boldsymbol{c}^{\mathrm{T}}\boldsymbol{x}$ with $A\boldsymbol{x}\geq\boldsymbol{b}$ and $\boldsymbol{x}\geq\boldsymbol{0}$.
Dual problem Maximize $\boldsymbol{y}^{\mathrm{T}}\boldsymbol{b}$ with $A^{\mathrm{T}}\boldsymbol{y}\leq\boldsymbol{c}$ and $\boldsymbol{y}\geq\boldsymbol{0}$.

10.5 Fourier Series: Linear Algebra for Functions

This section goes from finite dimensions to *infinite* dimensions. I want to explain linear algebra in infinite-dimensional space, and to show that it still works. First step: look back. This book began with vectors and dot products and linear combinations. We begin by converting those basic ideas to the infinite case—then the rest will follow.

What does it mean for a vector to have infinitely many components? There are two different answers, both good:

1. The vector is infinitely long: $\boldsymbol{v} = (v_1, v_2, v_3, \ldots)$. It could be $(1, \frac{1}{2}, \frac{1}{4}, \ldots)$.

2. The vector is a function $f(x)$. It could be $\boldsymbol{v} = \sin x$.

We will go both ways. Then the idea of a Fourier series will connect them.

After vectors come *dot products*. The natural dot product of two infinite vectors $(v_1, v_2, \ldots)$ and $(w_1, w_2, \ldots)$ is an infinite series:

$$\textbf{Dot product} \qquad \boldsymbol{v} \cdot \boldsymbol{w} = v_1 w_1 + v_2 w_2 + \cdots . \tag{1}$$

This brings a new question, which never occurred to us for vectors in $\mathbf{R}^n$. Does this infinite sum add up to a finite number? Does the series converge? Here is the first and biggest difference between finite and infinite.

When $\boldsymbol{v} = \boldsymbol{w} = (1, 1, 1, \ldots)$, the sum certainly does not converge. In that case $\boldsymbol{v} \cdot \boldsymbol{w} = 1 + 1 + 1 + \cdots$ is infinite. Since $\boldsymbol{v}$ equals $\boldsymbol{w}$, we are really computing $\boldsymbol{v} \cdot \boldsymbol{v} = \|\boldsymbol{v}\|^2$, the length squared. The vector $(1, 1, 1, \ldots)$ has infinite length. *We don't want that vector.* Since we are making the rules, we don't have to include it. The only vectors to be allowed are those with finite length:

DEFINITION The vector $\boldsymbol{v} = (v_1, v_2, \ldots)$ and the function $f(x)$ are in our infinite-dimensional "***Hilbert spaces***" if and only if their lengths $\|\boldsymbol{v}\|$ and $\|f\|$ are finite:

$$\|\boldsymbol{v}\|^2 = \boldsymbol{v} \cdot \boldsymbol{v} = v_1^2 + v_2^2 + v_3^2 + \cdots \qquad \text{must add to a finite number.}$$

$$\|f\|^2 = (f, f) = \int_0^{2\pi} |f(x)|^2 \, dx \qquad \text{must be a finite integral.}$$

Example 1 The vector $\boldsymbol{v} = (1, \frac{1}{2}, \frac{1}{4}, \ldots)$ is included in Hilbert space, because its length is $2/\sqrt{3}$. We have a geometric series that adds to $4/3$. The length of $\boldsymbol{v}$ is the square root:

$$\textbf{Length squared} \qquad \boldsymbol{v} \cdot \boldsymbol{v} = 1 + \tfrac{1}{4} + \tfrac{1}{16} + \cdots = \frac{1}{1 - \frac{1}{4}} = \tfrac{4}{3}.$$

Question If $\boldsymbol{v}$ and $\boldsymbol{w}$ have finite length, how large can their dot product be?

Answer The sum $\boldsymbol{v} \cdot \boldsymbol{w} = v_1 w_1 + v_2 w_2 + \cdots$ also adds to a finite number. We can safely take dot products. The Schwarz inequality is still true:

$$\textbf{Schwarz inequality} \qquad |\boldsymbol{v} \cdot \boldsymbol{w}| \leq \|\boldsymbol{v}\| \, \|\boldsymbol{w}\|. \tag{2}$$

The ratio of $\boldsymbol{v} \cdot \boldsymbol{w}$ to $\|\boldsymbol{v}\| \, \|\boldsymbol{w}\|$ is still the cosine of θ (the angle between $\boldsymbol{v}$ and $\boldsymbol{w}$). Even in infinite-dimensional space, $|\cos\theta|$ is not greater than 1.

Now change over to functions. Those are the "vectors." The space of functions $f(x)$, $g(x), h(x), \ldots$ defined for $0 \le x \le 2\pi$ must be somehow bigger than $\mathbf{R}^n$. ***What is the dot product of*** $f(x)$ ***and*** $g(x)$***? What is the length of*** $f(x)$?

Key point in the continuous case: *Sums are replaced by integrals.* Instead of a sum of v_j times w_j, the dot product is an integral of $f(x)$ times $g(x)$. Change the "dot" to parentheses with a comma, and change the words "dot product" to *inner product*:

DEFINITION The ***inner product*** of $f(x)$ and $g(x)$, and the ***length squared*** of $f(x)$, are

$$(f,g)=\int_0^{2\pi} f(x)g(x)\,dx \quad \text{and} \quad \|f\|^2=\int_0^{2\pi}\left(f(x)\right)^2 dx. \tag{3}$$

The interval $[0,2\pi]$ where the functions are defined could change to a different interval like $[0,1]$ or $(-\infty,\infty)$. We chose 2π because our first examples are $\sin x$ and $\cos x$.

Example 2 The length of $f(x)=\sin x$ comes from its inner product with itself:

$$(f,f)=\int_0^{2\pi}(\sin\ x)^2\,dx=\pi. \quad \text{The length of } \sin x \text{ is } \sqrt{\pi}.$$

That is a standard integral in calculus—not part of linear algebra. By writing $\sin^2 x$ as $\frac{1}{2}-\frac{1}{2}\cos 2x$, we see it go above and below its average value $\frac{1}{2}$. Multiply that average by the interval length 2π to get the answer π.

More important: $\sin x$ ***and*** $\cos x$ ***are orthogonal in function space***: $(f,g)=0$

Inner product is zero

$$\int_0^{2\pi}\sin x\cos x\,dx=\int_0^{2\pi}\tfrac{1}{2}\sin 2x\,dx=\left[-\tfrac{1}{4}\cos 2x\right]_0^{2\pi}=0. \tag{4}$$

This zero is no accident. It is highly important to science. The orthogonality goes beyond the two functions $\sin x$ and $\cos x$, to an infinite list of sines and cosines. The list contains $\cos 0x$ (which is 1), $\sin x, \cos x, \sin 2x, \cos 2x, \sin 3x, \cos 3x, \ldots$.

Every function in that list is orthogonal to every other function in the list.

Fourier Series

The Fourier series of a function $f(x)$ is its expansion into sines and cosines:

$$f(x)=\boldsymbol{a_0+a_1\cos x+b_1\sin x+a_2\cos 2x+b_2\sin 2x}+\cdots. \tag{5}$$

We have an orthogonal basis! The vectors in "function space" are combinations of the sines and cosines. On the interval from $x=2\pi$ to $x=4\pi$, all our functions repeat what they did from 0 to 2π. They are "*periodic*." The distance between repetitions is the period 2π.

Remember: The list is infinite. The Fourier series is an infinite series. We avoided the vector $\boldsymbol{v} = (1, 1, 1, \ldots)$ because its length is infinite, now we avoid a function like $\frac{1}{2} + \cos x + \cos 2x + \cos 3x + \cdots$. (*Note*: This is π times the famous **delta function** $\delta(x)$. It is an infinite "spike" above a single point. At $x = 0$ its height $\frac{1}{2} + 1 + 1 + \cdots$ is infinite. At all points inside $0 < x < 2\pi$ the series adds in some average way to zero.) The integral of $\delta(x)$ is 1. But $\int \delta^2(x) = \infty$, so delta functions are not allowed into Hilbert space.

Compute the length of a typical sum $f(x)$:

$$(f, f) = \int_0^{2\pi} (a_0 + a_1 \cos x + b_1 \sin x + a_2 \cos 2x + \cdots)^2 \, dx$$

$$= \int_0^{2\pi} (a_0^2 + a_1^2 \cos^2 x + b_1^2 \sin^2 x + a_2^2 \cos^2 2x + \cdots) \, dx$$

$$\|\boldsymbol{f}\|^2 = 2\pi a_0^2 + \pi(a_1^2 + b_1^2 + a_2^2 + \cdots). \tag{6}$$

The step from line 1 to line 2 used orthogonality. All products like $\cos x \cos 2x$ integrate to give zero. Line 2 contains what is left—the integrals of each sine and cosine squared. Line 3 evaluates those integrals. (The integral of 1^2 is 2π, when all other integrals give π.) If we divide by their lengths, our functions become *orthonormal*:

$$\frac{1}{\sqrt{2\pi}}, \frac{\cos x}{\sqrt{\pi}}, \frac{\sin x}{\sqrt{\pi}}, \frac{\cos 2x}{\sqrt{\pi}}, \ldots$$ ***is an orthonormal basis for our function space.***

These are unit vectors. We could combine them with coefficients $A_0, A_1, B_1, A_2, \ldots$ to yield a function $F(x)$. Then the 2π and the π's drop out of the formula for length.

Function length = vector length $\qquad \|F\|^2 = (F, F) = A_0^2 + A_1^2 + B_1^2 + A_2^2 + \cdots. \quad (7)$

Here is the important point, for $f(x)$ as well as $F(x)$. *The function has finite length exactly when the vector of coefficients has finite length.* Fourier series gives us a perfect match between the Hilbert spaces for functions and for vectors. The function is in L^2, its Fourier coefficients are in ℓ^2.

The function space contains $f(x)$ exactly when the Hilbert space contains the vector $\boldsymbol{v} = (a_0, a_1, b_1, \ldots)$ of Fourier coefficients of $f(x)$. Both must have finite length.

Example 3 Suppose $f(x)$ is a "square wave," equal to 1 for $0 \le x < \pi$. Then $f(x)$ drops to -1 for $\pi \le x < 2\pi$. The $+1$ and -1 repeat forever. This $f(x)$ is an odd function like the sines, and all its cosine coefficients are zero. We will find its Fourier series, containing only sines:

$$\textbf{Square wave} \qquad f(x) = \frac{4}{\pi}\left[\frac{\sin x}{1} + \frac{\sin 3x}{3} + \frac{\sin 5x}{5} + \cdots\right]. \tag{8}$$

The length of this function is $\sqrt{2\pi}$, because at every point $\big(f(x)\big)^2$ is $(-1)^2$ or $(+1)^2$:

$$\|f\|^2 = \int_0^{2\pi} \big(f(x)\big)^2 \, dx = \int_0^{2\pi} 1 \, dx = 2\pi.$$

At $x = 0$ the sines are zero and the Fourier series gives zero. This is half way up the jump from -1 to $+1$. The Fourier series is also interesting when $x = \frac{\pi}{2}$. At this point the square wave equals 1, and the sines in (8) alternate between $+1$ and -1:

$$\textbf{Formula for } \pi \qquad 1 = \frac{4}{\pi}\left(1 - \frac{1}{3} + \frac{1}{5} - \frac{1}{7} + \cdots\right). \tag{9}$$

Multiply by π to find a magical formula $4(1 - \frac{1}{3} + \frac{1}{5} - \frac{1}{7} + \cdots)$ for that famous number.

The Fourier Coefficients

How do we find the a's and b's which multiply the cosines and sines? For a given function $f(x)$, we are asking for its Fourier coefficients a_k and b_k:

$$\textbf{Fourier series} \qquad f(x) = a_0 + a_1 \cos x + b_1 \sin x + a_2 \cos 2x + \cdots .$$

Here is the way to find a_1. *Multiply both sides by* $\cos x$. *Then integrate from* 0 *to* 2π. The key is orthogonality! All integrals on the right side are zero, except for $\cos^2 x$:

$$\textbf{For coefficient } a_1 \qquad \int_0^{2\pi} f(x) \cos x \, dx = \int_0^{2\pi} a_1 \cos^2 x \, dx = \pi a_1. \tag{10}$$

Divide by π and you have a_1. To find any other a_k, multiply the Fourier series by $\cos kx$. Integrate from 0 to 2π. Use orthogonality, so only the integral of $a_k \cos^2 kx$ is left. That integral is πa_k, and divide by π:

$$\boldsymbol{a_k} = \frac{1}{\pi}\int_0^{2\pi} f(x) \cos kx \, dx \quad \text{and similarly} \quad \boldsymbol{b_k} = \frac{1}{\pi}\int_0^{2\pi} f(x) \sin kx \, dx. \tag{11}$$

The exception is a_0. This time we multiply by $\cos 0x = 1$. The integral of 1 is 2π:

$$\textbf{Constant term} \qquad \boldsymbol{a_0} = \frac{1}{2\pi}\int_0^{2\pi} f(x) \cdot 1 \, dx = \textbf{\textit{average value}} \text{ of } f(x). \tag{12}$$

I used those formulas to find the Fourier coefficients for the square wave in equation (8). The integral of $f(x) \cos kx$ was zero. The integral of $f(x) \sin kx$ was $4/k$ for odd k.

Compare Linear Algebra in $\mathbf{R}^n$

Infinite-dimensional Hilbert space is very much like the n-dimensional space $\mathbf{R}^n$. Suppose the nonzero vectors $\boldsymbol{v}_1, \ldots, \boldsymbol{v}_n$ are orthogonal in $\mathbf{R}^n$. We want to write the vector $\boldsymbol{b}$ (instead of the function $f(x)$) as a combination of those $\boldsymbol{v}$'s:

$$\textbf{Finite orthogonal series} \qquad \boldsymbol{b} = c_1\boldsymbol{v}_1 + c_2\boldsymbol{v}_2 + \cdots + c_n\boldsymbol{v}_n. \tag{13}$$

Multiply both sides by $\boldsymbol{v}_1^{\mathrm{T}}$. Use orthogonality, so $\boldsymbol{v}_1^{\mathrm{T}}\boldsymbol{v}_2 = 0$. Only the c_1 term is left:

$$\textbf{Coefficient } c_1 \qquad \boldsymbol{v}_1^{\mathrm{T}}\boldsymbol{b} = c_1\boldsymbol{v}_1^{\mathrm{T}}\boldsymbol{v}_1 + 0 + \cdots + 0. \quad \text{Therefore } c_1 = \boldsymbol{v}_1^{\mathrm{T}}\boldsymbol{b}/\boldsymbol{v}_1^{\mathrm{T}}\boldsymbol{v}_1. \tag{14}$$

The denominator $\boldsymbol{v}_1^{\mathrm{T}}\boldsymbol{v}_1$ is the length squared, like π in equation (11). The numerator $\boldsymbol{v}_1^{\mathrm{T}}\boldsymbol{b}$ is the inner product like $\int f(x) \cos kx \, dx$. ***Coefficients are easy to find when the***

basis vectors are orthogonal. We are just doing one-dimensional projections, to find the components along each basis vector.

The formulas are even better when the vectors are orthonormal. Then we have unit vectors in Q. The denominators $\boldsymbol{v}_k^{\mathrm{T}}\boldsymbol{v}_k$ are all 1. You know $c_k = \boldsymbol{v}_k^{\mathrm{T}}\boldsymbol{b}$ in another form:

$$\textbf{Equation for c's} \quad c_1\boldsymbol{v}_1 + \cdots + c_n\boldsymbol{v}_n = \boldsymbol{b} \quad \text{or} \quad \begin{bmatrix} \boldsymbol{v}_1 & \cdots & \boldsymbol{v}_n \end{bmatrix} \begin{bmatrix} c_1 \\ \vdots \\ c_n \end{bmatrix} = \boldsymbol{b}.$$

$$Q\boldsymbol{c} = \boldsymbol{b} \quad \text{yields} \quad \boldsymbol{c} = Q^{\mathrm{T}}\boldsymbol{b}. \quad \text{Row by row this is } c_k = \boldsymbol{q}_k^{\mathrm{T}}\boldsymbol{b}.$$

Fourier series is like having a matrix with infinitely many orthogonal columns. Those columns are the basis functions $1, \cos x, \sin x, \ldots$. After dividing by their lengths we have an "infinite orthogonal matrix." Its inverse is its transpose, Q^{T}. Orthogonality is what reduces a series of terms to one single term, when we integrate.

Problem Set 10.5

1 Integrate the trig identity $2\cos jx \cos kx = \cos(j+k)x + \cos(j-k)x$ to show that $\cos jx$ is orthogonal to $\cos kx$, provided $j \neq k$. What is the result when $j = k$?

2 Show that $1, x$, and $x^2 - \frac{1}{3}$ are orthogonal, when the integration is from $x = -1$ to $x = 1$. Write $f(x) = 2x^2$ as a combination of those orthogonal functions.

3 Find a vector $(w_1, w_2, w_3, \ldots)$ that is orthogonal to $\boldsymbol{v} = (1, \frac{1}{2}, \frac{1}{4}, \ldots)$. Compute its length $\|\boldsymbol{w}\|$.

4 The first three *Legendre polynomials* are $1, x$, and $x^2 - \frac{1}{3}$. Choose c so that the fourth polynomial $x^3 - cx$ is orthogonal to the first three. All integrals go from -1 to 1.

5 For the square wave $f(x)$ in Example 3 jumping from 1 to -1, show that

$$\int_0^{2\pi} f(x)\cos x\,dx = 0 \qquad \int_0^{2\pi} f(x)\sin x\,dx = 4 \qquad \int_0^{2\pi} f(x)\sin 2x\,dx = 0.$$

Which three Fourier coefficients come from those integrals?

6 The square wave has $\|f\|^2 = 2\pi$. Then (6) gives what remarkable sum for π^2?

7 Graph the square wave. Then graph by hand the sum of two sine terms in its series, or graph by machine the sum of 2, 3, and 10 terms. The famous ***Gibbs phenomenon*** is the oscillation that overshoots the jump (this doesn't die down with more terms).

8 Find the lengths of these vectors in Hilbert space:

(a) $\boldsymbol{v} = \left(\frac{1}{\sqrt{1}}, \frac{1}{\sqrt{2}}, \frac{1}{\sqrt{4}}, \frac{1}{\sqrt{8}}, \ldots\right)$

(b) $\boldsymbol{v} = (1, a, a^2, \ldots)$

(c) $f(x) = 1 + \sin x$.

9 Compute the Fourier coefficients a_k and b_k for $f(x)$ defined from 0 to 2π:

(a) $f(x) = 1$ for $0 \leq x \leq \pi$, $f(x) = 0$ for $\pi < x < 2\pi$

(b) $f(x) = x$.

10 When $f(x)$ has period 2π, why is its integral from $-\pi$ to π the same as from 0 to 2π? If $f(x)$ is an *odd* function, $f(-x) = -f(x)$, show that $\int_{-\pi}^{\pi} f(x)\, dx$ is zero. Odd functions only have sine terms, even functions only have cosines.

11 Using trigonometric identities find the two terms in the Fourier series for $f(x)$:

(a) $f(x) = \cos^2 x$ (b) $f(x) = \cos\left(x + \frac{\pi}{3}\right)$ (c) $f(x) = \sin^3 x$

12 The functions $1, \cos x, \sin x, \cos 2x, \sin 2x, \ldots$ are a basis for Hilbert space. Write the derivatives of those first five functions as combinations of the same five functions. What is the 5 by 5 "differentiation matrix" for these functions?

13 Find the Fourier coefficients a_k and b_k of the square pulse $F(x)$ centered at $x = 0$: $F(x) = 1/h$ for $|x| \leq h/2$ and $F(x) = 0$ for $h/2 < |x| \leq \pi$.

As $h \to 0$, this $F(x)$ approaches a delta function. Find the limits of a_k and b_k.

Section 4.1 of *Computational Science and Engineering* explains the sine series, cosine series, complete series, and complex series $\Sigma\, c_k e^{ikx}$ on **math.mit.edu/cse**.

Section 9.3 of this book explains the *Discrete Fourier Transform*. This is "Fourier series for vectors" and it is computed by the **Fast Fourier Transform**. That fast algorithm comes quickly from special complex numbers $z = e^{i\theta} = \cos\theta + i\sin\theta$ when the angle is $\theta = 2\pi k/n$.

10.6 Computer Graphics

Computer graphics deals with images. The images are moved around. Their scale is changed. Three dimensions are projected onto two dimensions. All the main operations are done by matrices—but the shape of these matrices is surprising.

The transformations of three-dimensional space are done with 4 *by* 4 *matrices.* You would expect 3 by 3. The reason for the change is that one of the four key operations cannot be done with a 3 by 3 matrix multiplication. Here are the four operations:

Translation	**(shift the origin to another point** $P_0 = (x_0, y_0, z_0)$**)**
Rescaling	**(by** c **in all directions or by different factors** c_1, c_2, c_3**)**
Rotation	**(around an axis through the origin or an axis through** P_0**)**
Projection	**(onto a plane through the origin or a plane through** P_0**).**

Translation is the easiest—just add (x_0, y_0, z_0) to every point. But this is not linear! No 3 by 3 matrix can move the origin. So we change the coordinates of the origin to $(0, 0, 0, 1)$. This is why the matrices are 4 by 4. The "*homogeneous coordinates*" of the point (x, y, z) are $(x, y, z, 1)$ and we now show how they work.

1. Translation Shift the whole three-dimensional space along the vector $\boldsymbol{v}_0$. The origin moves to (x_0, y_0, z_0). This vector $\boldsymbol{v}_0$ is added to every point $\boldsymbol{v}$ in $\mathbf{R}^3$. Using homogeneous coordinates, the 4 by 4 matrix T shifts the whole space by $\boldsymbol{v}_0$:

$$\textbf{\textit{Translation matrix}} \qquad T = \begin{bmatrix} 1 & 0 & 0 & 0 \\ 0 & 1 & 0 & 0 \\ 0 & 0 & 1 & 0 \\ x_0 & y_0 & z_0 & 1 \end{bmatrix}.$$

Important: *Computer graphics works with row vectors.* We have row times matrix instead of matrix times column. You can quickly check that $[0\ \ 0\ \ 0\ \ 1]\,T = [x_0\ \ y_0\ \ z_0\ \ 1]$.

To move the points $(0, 0, 0)$ and (x, y, z) by $\boldsymbol{v}_0$, change to homogeneous coordinates $(0, 0, 0, 1)$ and $(x, y, z, 1)$. Then multiply by T. A row vector times T gives a row vector. ***Every*** $\boldsymbol{v}$ ***moves to*** $\boldsymbol{v} + \boldsymbol{v}_0$: $[x\ \ y\ \ z\ \ 1]\,T = [x + x_0\ \ y + y_0\ \ z + z_0\ \ 1]$.

The output tells where any $\boldsymbol{v}$ will move. (It goes to $\boldsymbol{v} + \boldsymbol{v}_0$.) Translation is now achieved by a matrix, which was impossible in $\mathbf{R}^3$.

2. Scaling To make a picture fit a page, we change its width and height. A copier will rescale a figure by 90%. In linear algebra, we multiply by .9 times the identity matrix. That matrix is normally 2 by 2 for a plane and 3 by 3 for a solid. In computer graphics, with homogeneous coordinates, the matrix is *one size larger:*

$$\textbf{\textit{Rescale the plane:}} \quad S = \begin{bmatrix} .9 & & \\ & .9 & \\ & & 1 \end{bmatrix} \qquad \textbf{\textit{Rescale a solid:}} \quad S = \begin{bmatrix} c & 0 & 0 & 0 \\ 0 & c & 0 & 0 \\ 0 & 0 & c & 0 \\ 0 & 0 & 0 & 1 \end{bmatrix}.$$

Important: S is not cI. We keep the "1" in the lower corner. Then $[x, y, 1]$ times S is the correct answer in homogeneous coordinates. The origin stays in its normal position because $[0\ 0\ 1]S = [0\ 0\ 1]$.

If we change that 1 to c, the result is strange. ***The point*** (cx, cy, cz, c) ***is the same as*** $(x, y, z, 1)$. The special property of homogeneous coordinates is that *multiplying by cI does not move the point.* The origin in $\mathbf{R}^3$ has homogeneous coordinates $(0, 0, 0, 1)$ and $(0, 0, 0, c)$ for every nonzero c. This is the idea behind the word "homogeneous."

Scaling can be different in different directions. To fit a full-page picture onto a half-page, scale the y direction by $\frac{1}{2}$. To create a margin, scale the x direction by $\frac{3}{4}$. The graphics matrix is diagonal but not 2 by 2. It is 3 by 3 to rescale a plane and 4 by 4 to rescale a space:

$$\textbf{\textit{Scaling matrices}} \quad S = \begin{bmatrix} \frac{3}{4} & & \\ & \frac{1}{2} & \\ & & 1 \end{bmatrix} \quad \text{and} \quad S = \begin{bmatrix} c_1 & & & \\ & c_2 & & \\ & & c_3 & \\ & & & 1 \end{bmatrix}.$$

That last matrix S rescales the x, y, z directions by positive numbers c_1, c_2, c_3. The extra column in all these matrices leaves the extra 1 at the end of every vector.

Summary The scaling matrix S is the same size as the translation matrix T. They can be multiplied. To translate and then rescale, multiply $\boldsymbol{v}TS$. To rescale and then translate, multiply $\boldsymbol{v}ST$. Are those different? *Yes.*

The point (x, y, z) in $\mathbf{R}^3$ has homogeneous coordinates $(x, y, z, 1)$ in $\mathbf{P}^3$. This "projective space" is not the same as $\mathbf{R}^4$. It is still three-dimensional. To achieve such a thing, (cx, cy, cz, c) is the same point as $(x, y, z, 1)$. Those points of projective space $\mathbf{P}^3$ are really lines through the origin in $\mathbf{R}^4$.

Computer graphics uses ***affine*** transformations, *linear plus shift.* An affine transformation T is executed on $\mathbf{P}^3$ by a 4 by 4 matrix with a special fourth column:

$$A = \begin{bmatrix} a_{11} & a_{12} & a_{13} & 0 \\ a_{21} & a_{22} & a_{23} & 0 \\ a_{31} & a_{32} & a_{33} & 0 \\ a_{41} & a_{42} & a_{43} & 1 \end{bmatrix} = \begin{bmatrix} T(1,0,0) & 0 \\ T(0,1,0) & 0 \\ T(0,0,1) & 0 \\ T(0,0,0) & 1 \end{bmatrix}.$$

The usual 3 by 3 matrix tells us three outputs, this tells four. The usual outputs come from the inputs $(1, 0, 0)$ and $(0, 1, 0)$ and $(0, 0, 1)$. When the transformation is linear, three outputs reveal everything. When the transformation is affine, the matrix also contains the output from $(0, 0, 0)$. Then we know the shift.

3. Rotation A rotation in $\mathbf{R}^2$ or $\mathbf{R}^3$ is achieved by an orthogonal matrix Q. The determinant is $+1$. (With determinant -1 we get an extra reflection through a mirror.) Include the extra column when you use homogeneous coordinates!

$$\textbf{\textit{Plane rotation}} \quad Q = \begin{bmatrix} \cos\theta & -\sin\theta \\ \sin\theta & \cos\theta \end{bmatrix} \quad \text{becomes} \quad R = \begin{bmatrix} \cos\theta & -\sin\theta & 0 \\ \sin\theta & \cos\theta & 0 \\ 0 & 0 & 1 \end{bmatrix}.$$

This matrix rotates the plane around the origin. ***How would we rotate around a different point*** $(4,5)$? The answer brings out the beauty of homogeneous coordinates. ***Translate*** $(4,5)$ ***to*** $(0,0)$***, then rotate by*** θ***, then translate*** $(0,0)$ ***back to*** $(4,5)$:

$$\boldsymbol{v}\,T_-RT_+ = \begin{bmatrix} x & y & 1 \end{bmatrix} \begin{bmatrix} 1 & 0 & 0 \\ 0 & 1 & 0 \\ -4 & -5 & 1 \end{bmatrix} \begin{bmatrix} \cos\theta & -\sin\theta & 0 \\ \sin\theta & \cos\theta & 0 \\ 0 & 0 & 1 \end{bmatrix} \begin{bmatrix} 1 & 0 & 0 \\ 0 & 1 & 0 \\ 4 & 5 & 1 \end{bmatrix}.$$

I won't multiply. The point is to apply the matrices one at a time: $\boldsymbol{v}$ translates to $\boldsymbol{v}T_-$, then rotates to $\boldsymbol{v}T_-R$, and translates back to $\boldsymbol{v}T_-RT_+$. Because each point $\begin{bmatrix} x & y & 1 \end{bmatrix}$ is a row vector, T_- acts first. The center of rotation $(4,5)$—otherwise known as $(4,5,1)$—moves first to $(0,0,1)$. Rotation doesn't change it. Then T_+ moves it back to $(4,5,1)$. All as it should be. The point $(4,6,1)$ moves to $(0,1,1)$, then turns by θ and moves back.

In three dimensions, every rotation Q turns around an axis. The axis doesn't move—it is a line of eigenvectors with $\lambda = 1$. Suppose the axis is in the z direction. The 1 in Q is to leave the z axis alone, the extra 1 in R is to leave the origin alone:

$$Q = \begin{bmatrix} \cos\theta & -\sin\theta & 0 \\ \sin\theta & \cos\theta & 0 \\ 0 & 0 & 1 \end{bmatrix} \quad \text{and} \quad R = \begin{bmatrix} & & & 0 \\ & Q & & 0 \\ & & & 0 \\ 0 & 0 & 0 & 1 \end{bmatrix}.$$

Now suppose the rotation is around the unit vector $\boldsymbol{a} = (a_1, a_2, a_3)$. With this axis $\boldsymbol{a}$, the rotation matrix Q which fits into R has three parts:

$$Q = (\cos\theta)I + (1-\cos\theta)\begin{bmatrix} a_1^2 & a_1a_2 & a_1a_3 \\ a_1a_2 & a_2^2 & a_2a_3 \\ a_1a_3 & a_2a_3 & a_3^2 \end{bmatrix} - \sin\theta \begin{bmatrix} 0 & a_3 & -a_2 \\ -a_3 & 0 & a_1 \\ a_2 & -a_1 & 0 \end{bmatrix}. \qquad (1)$$

The axis doesn't move because $\boldsymbol{a}Q = \boldsymbol{a}$. When $\boldsymbol{a} = (0,0,1)$ is in the z direction, this Q becomes the previous Q—for rotation around the z axis.

The linear transformation Q always goes in the upper left block of R. Below it we see zeros, because rotation leaves the origin in place. When those are not zeros, the transformation is affine and the origin moves.

4. Projection In a linear algebra course, most planes go through the origin. In real life, most don't. A plane through the origin is a vector space. The other planes are affine spaces, sometimes called "flats." An affine space is what comes from translating a vector space.

We want to project three-dimensional vectors onto planes. Start with a plane through the origin, whose unit normal vector is $\boldsymbol{n}$. (We will keep $\boldsymbol{n}$ as a column vector.) The vectors in the plane satisfy $\boldsymbol{n}^{\mathrm{T}}\boldsymbol{v} = 0$. ***The usual projection onto the plane is the matrix*** $\boldsymbol{I - nn^{\mathrm{T}}}$. To project a vector, multiply by this matrix. The vector $\boldsymbol{n}$ is projected to zero, and the in-plane vectors $\boldsymbol{v}$ are projected onto themselves:

$$(I - \boldsymbol{n}\boldsymbol{n}^{\mathrm{T}})\boldsymbol{n} = \boldsymbol{n} - \boldsymbol{n}(\boldsymbol{n}^{\mathrm{T}}\boldsymbol{n}) = \boldsymbol{0} \quad \text{and} \quad (I - \boldsymbol{n}\boldsymbol{n}^{\mathrm{T}})\boldsymbol{v} = \boldsymbol{v} - \boldsymbol{n}(\boldsymbol{n}^{\mathrm{T}}\boldsymbol{v}) = \boldsymbol{v}.$$

In homogeneous coordinates the projection matrix becomes 4 by 4 (but the origin doesn't move):

$$\textbf{\textit{Projection onto the plane}} \quad \boldsymbol{n}^{\mathrm{T}}\boldsymbol{v} = 0 \qquad P = \begin{bmatrix} & & & 0 \\ & I - \boldsymbol{n}\boldsymbol{n}^{\mathrm{T}} & & 0 \\ & & & 0 \\ 0 & 0 & 0 & 1 \end{bmatrix}.$$

Now project onto a plane $\boldsymbol{n}^{\mathrm{T}}(\boldsymbol{v} - \boldsymbol{v}_0) = 0$ that does *not* go through the origin. One point on the plane is $\boldsymbol{v}_0$. This is an affine space (or a ***flat***). It is like the solutions to $A\boldsymbol{v} = \boldsymbol{b}$ when the right side is not zero. One particular solution $\boldsymbol{v}_0$ is added to the nullspace—to produce a flat.

The projection onto the flat has three steps. Translate $\boldsymbol{v}_0$ to the origin by T_-. Project along the $\boldsymbol{n}$ direction, and translate back along the row vector $\boldsymbol{v}_0$:

$$\textbf{\textit{Projection onto a flat}} \qquad T_-PT_+ = \begin{bmatrix} I & 0 \\ -\boldsymbol{v}_0 & 1 \end{bmatrix} \begin{bmatrix} I - \boldsymbol{n}\boldsymbol{n}^{\mathrm{T}} & 0 \\ 0 & 1 \end{bmatrix} \begin{bmatrix} I & 0 \\ \boldsymbol{v}_0 & 1 \end{bmatrix}.$$

I can't help noticing that T_- and T_+ are inverse matrices: translate and translate back. They are like the elementary matrices of Chapter 2.

The exercises will include reflection matrices, also known as *mirror matrices*. These are the fifth type needed in computer graphics. A reflection moves each point twice as far as a projection—***the reflection goes through the plane and out the other side***. So change the projection $I - \boldsymbol{n}\boldsymbol{n}^{\mathrm{T}}$ to $I - 2\boldsymbol{n}\boldsymbol{n}^{\mathrm{T}}$ for a mirror matrix.

The matrix P gave a "*parallel*" projection. All points move parallel to $\boldsymbol{n}$, until they reach the plane. The other choice in computer graphics is a "*perspective*" projection. This is more popular because it includes foreshortening. With perspective, an object looks larger as it moves closer. Instead of staying parallel to $\boldsymbol{n}$ (and parallel to each other), the lines of projection come *toward the eye*—the center of projection. This is how we perceive depth in a two-dimensional photograph.

The basic problem of computer graphics starts with a scene and a viewing position. Ideally, the image on the screen is what the viewer would see. The simplest image assigns just one bit to every small picture element—called a ***pixel***. It is light or dark. This gives a black and white picture with no shading. You would not approve. In practice, we assign shading levels between 0 and 2^8 for three colors like red, green, and blue. That means $8 \times 3 = 24$ bits for each pixel. Multiply by the number of pixels, and a lot of memory is needed!

Physically, a *raster frame buffer* directs the electron beam. It scans like a television set. The quality is controlled by the number of pixels and the number of bits per pixel. In this area, the standard text is *Computer Graphics*: *Principles and Practice* by Hughes, Van Dam, McGuire, Skylar, Foley, Feiner, and Akeley (3rd edition, Addison-Wesley, 2014). Notes by Ronald Goldman and by Tony DeRose were excellent references.

■ REVIEW OF THE KEY IDEAS ■

1. Computer graphics needs shift operations $T(\boldsymbol{v}) = \boldsymbol{v} + \boldsymbol{v}_0$ as well as linear operations $T(\boldsymbol{v}) = A\boldsymbol{v}$.

2. A shift in $\mathbf{R}^n$ can be executed by a matrix of order $n+1$, using homogeneous coordinates.

3. The extra component 1 in $[x\ y\ z\ 1]$ is preserved when all matrices have the numbers $0, 0, 0, 1$ as last column.

Problem Set 10.6

1 A typical point in $\mathbf{R}^3$ is $x\boldsymbol{i} + y\boldsymbol{j} + z\boldsymbol{k}$. The coordinate vectors $\boldsymbol{i}$, $\boldsymbol{j}$, and $\boldsymbol{k}$ are $(1, 0, 0)$, $(0, 1, 0)$, $(0, 0, 1)$. The coordinates of the point are (x, y, z).

This point in computer graphics is $x\boldsymbol{i} + y\boldsymbol{j} + z\boldsymbol{k} +$ **origin**. Its homogeneous coordinates are (, , ,). Other coordinates for the same point are (, , ,).

2 A linear transformation T is determined when we know $T(\boldsymbol{i}), T(\boldsymbol{j}), T(\boldsymbol{k})$. For an affine transformation we also need $T($____$)$. The input point $(x, y, z, 1)$ is transformed to $xT(\boldsymbol{i}) + yT(\boldsymbol{j}) + zT(\boldsymbol{k}) +$ ____.

3 Multiply the 4 by 4 matrix T for translation along $(1, 4, 3)$ and the matrix T_1 for translation along $(0, 2, 5)$. The product TT_1 is translation along ____.

4 Write down the 4 by 4 matrix S that scales by a constant c. Multiply ST and also TS, where T is translation by $(1, 4, 3)$. To blow up the picture around the center point $(1, 4, 3)$, would you use $\boldsymbol{v}ST$ or $\boldsymbol{v}TS$?

5 What scaling matrix S (in homogeneous coordinates, so 3 by 3) would produce a 1 by 1 square page from a standard 8.5 by 11 page?

6 What 4 by 4 matrix would move a corner of a cube to the origin and then multiply all lengths by 2? The corner of the cube is originally at $(1, 1, 2)$.

7 When the three matrices in equation 1 multiply the unit vector $\boldsymbol{a}$, show that they give $(\cos\theta)\boldsymbol{a}$ and $(1 - \cos\theta)\boldsymbol{a}$ and $\boldsymbol{0}$. Addition gives $\boldsymbol{a}Q = \boldsymbol{a}$ and the rotation axis is not moved.

8 If $\boldsymbol{b}$ is perpendicular to $\boldsymbol{a}$, multiply by the three matrices in 1 to get $(\cos\theta)\boldsymbol{b}$ and $\boldsymbol{0}$ and a vector perpendicular to $\boldsymbol{b}$. So $Q\boldsymbol{b}$ makes an angle θ with $\boldsymbol{b}$. ***This is rotation***.

9 What is the 3 by 3 projection matrix $I - \boldsymbol{n}\boldsymbol{n}^{\mathrm{T}}$ onto the plane $\frac{2}{3}x + \frac{2}{3}y + \frac{1}{3}z = 0$? In homogeneous coordinates add $0, 0, 0, 1$ as an extra row and column in P.

10 With the same 4 by 4 matrix P, multiply T_-PT_+ to find the projection matrix onto the plane $\frac{2}{3}x+\frac{2}{3}y+\frac{1}{3}z=1$. The translation T_- moves a point on that plane (choose one) to $(0,0,0,1)$. The inverse matrix T_+ moves it back.

11 Project $(3,3,3)$ onto those planes. Use P in Problem 9 and T_-PT_+ in Problem 10.

12 If you project a square onto a plane, what shape do you get?

13 If you project a cube onto a plane, what is the outline of the projection? Make the projection plane perpendicular to a diagonal of the cube.

14 The 3 by 3 mirror matrix that reflects through the plane $\boldsymbol{n}^{\mathrm{T}}\boldsymbol{v}=0$ is $M=I-2\boldsymbol{n}\boldsymbol{n}^{\mathrm{T}}$. Find the reflection of the point $(3,3,3)$ in the plane $\frac{2}{3}x+\frac{2}{3}y+\frac{1}{3}z=0$.

15 Find the reflection of $(3,3,3)$ in the plane $\frac{2}{3}x+\frac{2}{3}y+\frac{1}{3}z=1$. Take three steps T_-MT_+ using 4 by 4 matrices: translate by T_- so the plane goes through the origin, reflect the translated point $(3,3,3,1)T_-$ in that plane, then translate back by T_+.

16 The vector between the origin $(0,0,0,1)$ and the point $(x,y,z,1)$ is the difference $\boldsymbol{v}=$ _____. In homogeneous coordinates, vectors end in _____. So we add a _____ to a point, not a point to a point.

17 If you multiply only the *last* coordinate of each point to get (x,y,z,c), you rescale the whole space by the number _____. This is because the point (x,y,z,c) is the same as $(\;,\;,\;,1)$.

10.7 Linear Algebra for Cryptography

1 Codes can use finite fields as alphabets: letters in the message become numbers $0, 1, \ldots, p-1$.

2 The numbers are added and multiplied $(mod\ p)$. Divide by p, keep the remainder.

3 A Hill Cipher multiplies blocks of the message by a secret matrix $E\ (mod\ p)$.

4 To decode, multiply each block by the inverse matrix $D\ (mod\ p)$. Not a very secure cipher!

Cryptography is about encoding and decoding messages. Banks do this all the time with financial information. Amazingly, modern algorithms can involve extremely deep mathematics. "Elliptic curves" play a part in cryptography, as they did in the sensational proof by Andrew Wiles of Fermat's Last Theorem.

This section will not go that far! But it will be our first experience with *finite fields* and *finite vector spaces*. The field for $\mathbf{R}^n$ contains all real numbers. The field for "modular arithmetic" contains only p integers $0, 1, \ldots, p-1$. There were infinitely many vectors in $\mathbf{R}^n$—now there will only be p^n messages of length n in message space. The alphabet from A to Z is finite (as in $p = 26$).

The codes in this section will be easily breakable—they are much too simple for practical security. The power of computers demands more complex cryptography, because that power would quickly detect a small encoding matrix. But a matrix code (the Hill Cipher) will allow us to see linear algebra at work in a new way.

All our calculations in encoding and decoding will be "**mod** p". But the central concepts of linear independence and bases and inverse matrices and determinants survive this change. We will be doing "linear algebra with finite fields". Here is the meaning of *mod* p:

$\mathbf{27 \equiv 2\ (mod\ 5)}$	means that **27 − 2 is divisible by 5**
$\boldsymbol{y \equiv x\ (mod\ p)}$	means that $\boldsymbol{y} -$ $\boldsymbol{x}$ **is divisible by** $\boldsymbol{p}$

Dividing y by 5 produces one of the five possible remainders $x = 0, 1, 2, 3, 4$. All the numbers $5, -5, 10, -10, \ldots$ with no remainder are congruent to zero $(mod\ 5)$. The numbers $y = 6, -4, 11, -9, \ldots$ are all congruent to $x = 1 (mod\ 5)$.

We use the word **congruent** for the symbol $\equiv$ and we call this "modular arithmetic". Every integer y produces one of the values $x = 0, 1, 2, \ldots, p-1$.

The theory is best if p is a prime number. With $p = 26$ letters from A to Z, we unfortunately don't start with a prime p. Cryptography can deal with this problem.

Modular Arithmetic

Linear algebra is based on linear combinations of vectors. Now our vectors $(x_1, \ldots, x_n)$ are strings of integers limited to $x = 0, 1, \ldots, p-1$. All calculations produce these integers when we work "$mod\ p$". This means: *Every integer y outside that range is divided by p and x is the remainder*:

$y = qp + x$	$y \equiv x\ (mod\ p)$	**y divided by p has remainder x**

Addition mod 3 $10 \equiv 1\ (mod\ 3)$ and $16 \equiv 1\ (mod\ 3)$ and $10 + 16 \equiv 1 + 1\ (mod\ 3)$

I could add $10 + 16$ and divide 26 by 3 to get the remainder 2.
Or I can just add remainders $1 + 1$ to reach the same answer 2.

Addition mod 2 $11 \equiv 1\ (mod\ 2)$ and $17 \equiv 1\ (mod\ 2)$ and $11 + 17 = 28 \equiv 0\ (mod\ 2)$

The remainders added to $1 + 1$ *but this is not* 2. The final step was $2 \equiv 0\ (mod\ 2)$.

Addition $mod\ p$ is completely reasonable. So is **multiplication $mod\ p$**. Here $p = 3$:

$$10 \equiv 1\ (mod\ 3) \text{ times } 16 \equiv 1\ (mod\ 3) \text{ gives } 1 \text{ times } 1 \equiv 1 \qquad 160 \equiv 1\ (mod\ 3)$$

$$5 \equiv 2\ (mod\ 3) \text{ times } 8 \equiv 2\ (mod\ 3) \text{ gives } 2 \text{ times } 2 \equiv 1 \qquad 40 \equiv 1\ (mod\ 3)$$

Conclusion: We can safely add and multiply modulo p. So we can take linear combinations. This is the key operation in linear algebra. **But can we divide ?**

In the real number field, the inverse is $1/y$ (for any number except $y = 0$). This means: We found another real number z so that $yz = 1$. Invertibility is a requirement for a field. **Is inversion always possible $mod\ p$?** For every number $y = 1, \ldots, p-1$ can we find another number $z = 1, \ldots, p-1$ so that $yz \equiv 1\ mod\ p$?

The examples $3^{-1} \equiv 4\ (mod\ 11)$ and $2^{-1} \equiv 6\ (mod\ 11)$ and $5^{-1} \equiv 9\ (mod\ 11)$ all succeed. Can you solve $7z \equiv 1\ (mod\ 11)$? Inverting numbers will be the key to inverting matrices.

Let me show that inversion $mod\ p$ has a problem when p is not a prime number. The example $p = 26$ factors into 2 times 13. **Then $y = 2$ cannot have an inverse z $(mod\ 26)$.** The requirement $2z \equiv 1\ (mod\ 26)$ is impossible to satisfy because $2z$ and 26 are even.

Similarly 5 has no inverse z when p is 25. We can't solve $5z \equiv 1\ (mod\ 25)$. The number $5z - 1$ is never going to be a multiple of 5, so it can't be a multiple of 25.

Inversion of every y $(0 < y < p)$ will be possible if and only if p is prime.

Inversion needs $y, 2y, 3y, \ldots, py$ to have different remainders when divided by p.

If my and ny had the same remainder x then $(m-n)y$ would be divisible by p.

The prime number p would have to divide either $m - n$ or y. Both are impossible.

So $y, \ldots, py$ have different remainders: **One of those remainders must be $x = 1$.**

The Enigma Machine and the Hill Cipher

Lester Hill published his cipher (his system for encoding and decoding) in the American Mathematical Monthly (1929). The idea was simple, but in some way it started the transition of cryptography from linguistics to mathematics. Codes up to that time mainly mixed up alphabets and rearranged messages. The **Enigma code** used by the German Navy in World War II was a giant advance—using machines that look to us like primitive computers. The English set up Bletchley Park to break Enigma. They hired puzzle solvers and language majors. And by good luck they also happened to get Alan Turing.

I don't know if you have seen the movie about him: *The Imitation Game*. A lot of it is unrealistic (like *Good Will Hunting* and *A Beautiful Mind* at MIT). But the core idea of breaking the Enigma code was correct, using human weaknesses in the encoding and broadcasting. The German naval command openly sent out their coded orders—knowing that the codes were too complicated to break (if it hadn't been for those weaknesses). The codebreaking required English electronics to undo the German electronics. It also required genius.

Alan Turing was surely a genius—England's most exceptional mathematician. His life was ultimately tragic and he ended it in 1954. The biography by Andrew Hodges is excellent. Turing arrived at Bletchley Park the day after Poland was invaded. It is to Winston Churchill's credit that he gave fast and full support when his support was needed.

The Enigma Machine had gears and wheels. The Hill Cipher only needs a matrix. That is the code to be explained now, using linear algebra. You will see how decoding involved inverse matrices. All steps use modular arithmetic, multiplying and inverting *mod p*.

I will follow the neat exposition of Professor Spickler of Salisbury State University, which he made available on the Web: **facultyfp.salisbury.edu/despickler/personal/index.asp**

Modular Arithmetic with Matrices

Addition, subtraction, and multiplication are all we need for $A\boldsymbol{x}$ (matrix times vector). To multiply $mod\ p$ we can multiply the integers in A times the integers in $\boldsymbol{x}$ as usual—and then replace every entry of $A\boldsymbol{x}$ by its value $mod\ p$.

Key questions: When can we solve $A\boldsymbol{x} \equiv \boldsymbol{b}\,(mod\ p)$? Do we still have the four subspaces $\boldsymbol{C}(A), \boldsymbol{N}(A), \boldsymbol{C}(A^{\mathrm{T}}), \boldsymbol{N}(A^{\mathrm{T}})$? Are they still orthogonal in pairs? Is there still an inverse matrix $mod\ p$ whenever the determinant of A is nonzero $mod\ p$? I am happy to say that the last three answers are *yes* (but the inverse question requires p to be a prime number).

We can find $A^{-1}\,(mod\ p)$ by Gauss-Jordan elimination, reducing $[A\ \ I]$ to $[I\ \ A^{-1}]$ as in Section 2.5. Or we can use determinants and the cofactor matrix C in the formula $A^{-1} = (\det A)^{-1}\,C^{\mathrm{T}}$. I will work mod 3 with a 2 by 2 integer matrix A :

$$[\,A\ \ I\,] = \begin{bmatrix} \mathbf{2} & \mathbf{0} & 1 & 0 \\ \mathbf{2} & \mathbf{1} & 0 & 1 \end{bmatrix} \to \begin{bmatrix} 2 & 0 & 1 & 0 \\ 0 & 1 & 2 & 1 \end{bmatrix} \xrightarrow{\substack{\text{multiply row 1} \\ \text{by } 2^{-1} \equiv 2}} \begin{bmatrix} 1 & 0 & \mathbf{2} & \mathbf{0} \\ 0 & 1 & \mathbf{2} & \mathbf{1} \end{bmatrix} = [\,I\ \ A^{-1}\,]$$

By pure chance $A^{-1} \equiv A$! Multiplying A times A *mod* 3 does give the identity matrix:

$$A^2 = AA^{-1} = \begin{bmatrix} 2 & 0 \\ 2 & 1 \end{bmatrix}\begin{bmatrix} 2 & 0 \\ 2 & 1 \end{bmatrix} = \begin{bmatrix} 4 & 0 \\ 6 & 1 \end{bmatrix} \equiv \begin{bmatrix} 1 & 0 \\ 0 & 1 \end{bmatrix} \ (mod\ 3).$$

The determinant of A is 2, and the cofactor formula from Section 5.3 also gives $A^{-1} \equiv A$:

$$\begin{bmatrix} 2 & 0 \\ 2 & 1 \end{bmatrix}^{-1} = 2^{-1}\begin{bmatrix} 1 & -0 \\ -2 & 2 \end{bmatrix} \equiv 2\begin{bmatrix} 1 & -0 \\ -2 & 2 \end{bmatrix} \equiv \begin{bmatrix} 2 & 0 \\ 2 & 1 \end{bmatrix} \ (mod\ 3).$$

Theorem. A^{-1} exists *mod* p if and only if $(\det A)^{-1}$ exists *mod* p.
The requirement is: $\det A$ and p have no common factors.

Encryption with the Hill Cipher

The original cipher used the letters A to Z with $p = 26$. Hill chose an n by n encryption matrix E so that $\det E$ is not divisible by 2 or 13. Then the number $\det E$ has an inverse *mod* 26 and so does the matrix E. The inverse matrix $E^{-1} \equiv D\,(mod\ 26)$ will be the decryption matrix that decodes the message.

Now convert each letter of the message into a number from 0 to 25. The obvious choice from $A = 0$ to $Z = 25$ is acceptable because the matrix will make this cipher stronger.

> Ignore spaces and divide the message into blocks $\boldsymbol{v}_1, \boldsymbol{v}_2, \ldots$ of size n. Then multiply each message block ($mod\ p$) by the encryption matrix E. The coded message is $E\boldsymbol{v}_1, E\boldsymbol{v}_2, \ldots$ and you know what the decoder will do.

Spikler's example has $D = E^{-1} =$
$\det E = 583 \equiv 11\,(mod\ 26)$

$$\begin{bmatrix} 2 & 3 & 15 \\ 5 & 8 & 12 \\ 1 & 13 & 4 \end{bmatrix}^{-1} \equiv \begin{bmatrix} 10 & 19 & 16 \\ 4 & 23 & 7 \\ 17 & 5 & 19 \end{bmatrix} \ (mod\ 26).$$

Of course a codebreaker will not know E or D. And the block size n is generally unknown too. For the matrices Hill had in mind n would not be very large and a computer could quickly discover E and D.

I am not sure if Hill's Cipher could become seriously difficulty to break by choosing very large matrices and a large prime number p. And by encoding the coded message a second time, using a different block size n_2 and large matrix E_2 and large prime p_2.

Finite Fields and Finite Vector Spaces

In algebra, a field $\mathbf{F}$ is a set of scalars that can be added and multiplied and inverted (except 0 can't be inverted). Familiar examples are the real numbers $\mathbf{R}$ and the complex numbers $\mathbf{C}$ and the rational numbers $\mathbf{Q}$ (containing every ratio p/q of integers). From a field you build vectors $\boldsymbol{v} = (f_1, f_2, \ldots, f_n)$. From linear combinations of vectors you build vector spaces. *So linear algebra begins with a field* $\mathbf{F}$.

I taught for ten years from a textbook that started with fields. On the way to $\mathbf{R}^n$, we lost a lot of students. That was a signal—the emphasis was misplaced if we wanted the course to be useful. I believe the right way is to understand $\mathbf{R}^n$ and its subspaces first, as you do. Then you can look at other fields and vector spaces with a natural question in mind: *What is new when the field is not* $\mathbf{R}$?

These pages are asking that question for **finite fields**. The possibilities become more limited but also highly interesting. The starting point (and not quite the ending point) is the finite field $\mathbf{F}_p$. It contains only the numbers $0, 1, \ldots, p-1$ and p is a prime number. I will focus first on the field $\mathbf{F}_2$ with only 2 members "0" and "1". You could think of $\mathbf{0}$ and $\mathbf{1}$ as "even" and "odd" because the rules to add and multiply are obeyed by the even numbers and odd numbers: even $+$ odd $=$ *odd* and even $\times$ odd $=$ *even*.

Addition table	0	1
0	0	1
1	1	0

Multiplication table	0	1
0	0	0
1	0	1

This is addition and multiplication "mod 2".

From this field $\mathbf{F}_2$ we can build vectors like $\boldsymbol{v} = (0,0,1)$ and $\boldsymbol{w} = (1,0,1)$. There are three components with two choices each: a total of $2^3 = 8$ different vectors in the vector space $(\mathbf{F}_2)^3$. You know the requirements on a subspace and the possibilities it opens up:

a) The zero-dimensional subspace containing only $\mathbf{0} = (0,0,0)$.

b) One-dimensional subspaces containing $\mathbf{0}$ and a vector like $\boldsymbol{v}$. Notice $\boldsymbol{v} + \boldsymbol{v} = \mathbf{0}$!

c) Two-dimensional subspaces with a basis like $\boldsymbol{v}$ and $\boldsymbol{w}$ and 4 vectors $\mathbf{0}, \boldsymbol{v}, \boldsymbol{w}, \boldsymbol{v} + \boldsymbol{w}$.

d) The full three-dimensional subspace $(\mathbf{F}_2)^3$ with 8 vectors.

What are the possible bases for $(\mathbf{F}_2)^3$? The standard basis contains $(1,0,0)$ and $(0,1,0)$ and $(0,0,1)$. Those vectors are linearly independent and they span $(\mathbf{F}_2)^3$. Their eight combinations with coefficients 0 and 1 fill all of $(\mathbf{F}_2)^3$.

What about matrices that multiply those vectors? The matrices will be 1 by 3, or 2 by 3, or 3 by 3. When they are 3 by 3 we can ask if they are invertible. Their determinants can only be 0 (singular matrix) or 1 (invertible matrix). Let me leave you the pleasure of deciding whether these matrices are invertible. *And how would you find the inverse* ?

$$A = \begin{bmatrix} 1 & 0 & 0 \\ 1 & 1 & 0 \\ 1 & 1 & 1 \end{bmatrix} \qquad B = \begin{bmatrix} 1 & 1 & 0 \\ 0 & 1 & 1 \\ 1 & 0 & 1 \end{bmatrix} \qquad C = \begin{bmatrix} 1 & 1 & 1 \\ 0 & 0 & 1 \\ 1 & 0 & 0 \end{bmatrix}$$

Out of 2^9 possible matrices over $\mathbf{F}_2$, I will guess that most are singular.

To conclude this discussion of $\mathbf{F}_2$, I mention a field with $2^2 = 4$ members. It will not come from multiplication (mod 4), because 4 is not prime. The multiplication 2 times 2 will give 0 (and 2 has no inverse): *not a field.* But we can start with the numbers 0 and 1 in $\mathbf{F}_2$ and invent two more numbers a and $1 + a$—provided they follow these two rules: $(\boldsymbol{a} + \boldsymbol{a} = \mathbf{0})$ and $(\boldsymbol{a} \times \boldsymbol{a} = \mathbf{1} + \boldsymbol{a})$. Then a and $1 + a$ are inverses. Not obvious !

Add	0	1	a	$1+a$
0	0	1	a	$1+a$
1	1	0	$1+a$	a
a	a	$1+a$	0	1
$1+a$	$1+a$	a	1	0

Multiply	0	1	a	$1+a$
0	0	0	0	0
1	0	1	a	$1+a$
a	0	a	$1+a$	1
$1+a$	0	$1+a$	1	a

Beyond $p = 2$, we have the fields $\mathbf{F}_p$ for all prime numbers p. They use addition and multiplication $mod\ p$. They are alphabets for codes. They provide the components for vectors $\boldsymbol{v} = (f_1, \ldots, f_n)$ in the space $(\mathbf{F}_p)^n$. They provide the entries for matrices that multiply those vectors. These fields $\mathbf{F}_p$ are the most frequently used finite fields.

The only other finite fields have p^k members. The example above of $0, 1, a, 1 + a$ had $2^2 = 4$ members. We will leave it there and get back safely to $\mathbf{R}$.

Problem Set 10.7

1 If you multiply n whole numbers (even or odd) when is the answer odd? Translate into multiplication $(mod\ 2)$: If you multiply 0's and 1's when is the answer 1 ?

2 If you add n whole numbers (even or odd) when is the sum of the numbers odd? Translate into adding 0's and 1's $(mod\ 2)$. When do they add to 1 ?

3 (a) If $y_1 \equiv x_1$ and $y_2 \equiv x_2$, why is $y_1 + y_2 \equiv x_1 + x_2$? All are $mod\ p$. *Suggestion*: $y_1 = p\,q_1 + x_1$ and $y_2 = p\,q_2 + x_2$. Now add $y_1 + y_2$.

(b) Can you be sure that $x_1 + x_2$ is smaller than p ? *No.* Give an example where there is a smaller x with $(y_1 + y_2) = x\ (mod\ p)$.

4 $p = 39$ is not prime. Find a number a that has no inverse $z\ (mod\ 39)$. This means that $az \equiv 1\ (mod\ 39)$ has no solution. Then find a 2 by 2 matrix A that has no inverse matrix $Z\ (mod\ 39)$. This means that $AZ \equiv I\ (mod\ 39)$ has no solution.

5 Show that $y \equiv x\ (mod\ p)$ *leads to* $-y \equiv -x\ (mod\ p)$.

6 Find a matrix that has independent columns in $\mathbf{R}^2$ but dependent columns $(mod\ 5)$.

7 What are all the 2 by 2 matrices of 0's and 1's that are invertible $(mod\ 2)$?

8 Is the row space of A still orthogonal to the nullspace in modular arithmetic $(mod\ 11)$? Are bases for those subspaces still bases $(mod\ 11)$?

9 (Hill's Cipher) Separate the message THISWHOLEBOOKISINCODE into blocks of 3 letters. Replace each letter by a number from 1 to 26 (normal order). Multiply each block by the 3 by 3 matrix L with 1's on and below the diagonal. What is the coded message (in numbers) and how would you decode it?

10 Suppose you know the original message (the plaintext). Suppose you also see the coded message. How would you start to discover the matrix in Hill's Cipher ? For a very long message do you expect success ?

Chapter 11

Numerical Linear Algebra

1 The goals of numerical linear algebra are **speed** and **accuracy** and **stability** : $n > 10^3$ or 10^6.

2 Matrices can be full or sparse or banded or structured : special algorithms for each.

3 Accuracy of elimination is controlled by the **condition number** $||\boldsymbol{A}||\ ||\boldsymbol{A}^{-1}||$.

4 Gram-Schmidt is often computed by using **Householder reflections** $H = I - 2\boldsymbol{u}\boldsymbol{u}^{\mathrm{T}}$ to find Q.

5 Eigenvalues use $\boldsymbol{QR}$ **iterations** $\boldsymbol{A_0} = Q_0R_0 \to R_0Q_0 = \boldsymbol{A_1} = Q_1R_1 \to \to \boldsymbol{A_n}$.

6 **Shifted** $\boldsymbol{QR}$ is even better : Shift to $A_k - c_kI = Q_kR_k$, shift back $A_{k+1} = R_kQ_k + c_kI$.

7 Iteration $S\boldsymbol{x}_{k+1} = \boldsymbol{b} - T\boldsymbol{x}_k$ solves $(S+T)\,\boldsymbol{x} = \boldsymbol{b}$ if all eigenvalues of $S^{-1}T$ have $|\lambda| < 1$.

8 Iterative methods often use **preconditioners** P. Change $A\boldsymbol{x} = \boldsymbol{b}$ to $PA\boldsymbol{x} = P\boldsymbol{b}$ with $PA \approx I$.

9 **Conjugate gradients** and **GMRES** are Krylov methods; see Trefethen-Bau (and other texts).

11.1 Gaussian Elimination in Practice

Numerical linear algebra is a struggle for *quick* solutions and also *accurate* solutions. We need efficiency but we have to avoid instability. In Gaussian elimination, the main freedom (always available) is to ***exchange equations***. This section explains when to exchange rows for the sake of speed, and when to do it for the sake of accuracy.

The key to accuracy is to avoid unnecessarily large numbers. Often that requires us to avoid small numbers! A small pivot generally means large multipliers (since we divide by the pivot). A good plan is "***partial pivoting***", to choose the *largest available pivot* in each new column. We will see why this pivoting strategy is built into computer programs.

Other row exchanges are done to save elimination steps. In practice, most large matrices are ***sparse***—almost all entries are zeros. Elimination is fastest when the equations are ordered *to produce a narrow band of nonzeros*. Zeros inside the band "fill in" during elimination—those zeros are destroyed and don't save computing time.

Section 11.2 is about instability that can't be avoided. It is built into the problem, and this sensitivity is measured by the "***condition number***". Then Section 11.3 describes how to solve $A\boldsymbol{x} = \boldsymbol{b}$ by ***iterations***. Instead of direct elimination, the computer solves an easier equation many times. Each answer $\boldsymbol{x}_k$ leads to the next guess $\boldsymbol{x}_{k+1}$. For good iterations (the **conjugate gradient method** is extremely good), the $\boldsymbol{x}_k$ converge quickly to $\boldsymbol{x} = A^{-1}\boldsymbol{b}$.

The Fastest Supercomputer

A new supercomputing record was announced by IBM and Los Alamos on May 20, 2008. The Roadrunner was the first to achieve a quadrillion (10^{15}) floating-point operations per second: *a petaflop machine*. The benchmark for this world record was a large dense linear system $A\boldsymbol{x} = \boldsymbol{b}$: computer speed is tested by linear algebra.

That machine was shut down in 2013! The TOP500 project ranks the 500 most powerful computer systems in the world. As I write this page in October 2015, the first four are from NUDT in China, Cray and IBM in the US, and Fujitsu in Japan. They all use a LINUX-based system. And all vector processors have fallen out of the top 500.

Looking ahead, the Summit is expected to take first place with 150-300 petaflops. President Obama has just ordered the development of an exascale system (1000 petaflops). Up to now we are following Moore's Law of doubling every 14 months.

The LAPACK software does elimination with partial pivoting. The biggest difference from this book is to organize the steps to use large submatrices and never single numbers. And graphics processing units (GPU's) are now almost required for success. The market for video games dwarfs scientific computing and led to astonishing acceleration in the chips.

Before IBM's BlueGene, a key issue was to count the standard quad-core processors that a petaflop machine would need: 32,000. The new architecture uses much less power, but its hybrid design has a price: a code needs three separate compilers and explicit instructions to move all the data. Please see the excellent article in *SIAM News* (**siam.org**, July 2008) and the update on **www.lanl.gov/roadrunner**.

Our thinking about matrix calculations is reflected in the highly optimized **BLAS** (*Basic Linear Algebra Subroutines*). They come at levels 1, 2, and 3:

Level 1 Linear combinations of vectors $a\boldsymbol{u} + \boldsymbol{v}$: $O(n)$ work

Level 2 Matrix-vector multiplications $A\boldsymbol{u} + \boldsymbol{v}$: $O(n^2)$ work

Level 3 Matrix-matrix multiplications $AB + C$: $O(n^3)$ work

Level 1 is an elimination step (multiply row j by ℓ_{ij} and subtract from row i). Level 2 can eliminate a whole column at once. A high performance solver is rich in Level 3 BLAS (AB has $2n^3$ flops and $2n^2$ data, a good ratio of work to talk).

It is *data passing* and *storage retrieval* that limit the speed of parallel processing. The high-velocity cache between main memory and floating-point computation has to be fully used! Top speed demands a ***block matrix approach*** to elimination.

The big change, coming now, is parallel processing at the chip level.

Roundoff Error and Partial Pivoting

Up to now, any pivot (nonzero of course) was accepted. In practice a small pivot is dangerous. A catastrophe can occur when numbers of different sizes are added. Computers keep a fixed number of significant digits (say three decimals, for a very weak machine). The sum $10{,}000 + 1$ is rounded off to $10{,}000$. The "1" is completely lost. Watch how that changes the solution to this problem:

$$\begin{aligned} .0001u + v &= 1 \\ -u + v &= 0 \end{aligned} \qquad \text{starts with coefficient matrix} \qquad A = \begin{bmatrix} .0001 & 1 \\ -1 & 1 \end{bmatrix}.$$

If we accept .0001 as the pivot, elimination adds 10,000 times row 1 to row 2. Roundoff leaves

$$10{,}000v = 10{,}000 \qquad \text{instead of} \qquad 10{,}001v = 10{,}000.$$

The computed answer $v = 1$ is near the true $v = .9999$. But then back substitution puts the wrong $v = 1$ into the equation for u:

$$.0001\,u + 1 = 1 \qquad \text{instead of} \qquad .0001\,u + .9999 = 1.$$

The first equation gives $u = 0$. The correct answer (look at the second equation) is $u = 1.000$. By losing the "1" in the matrix, we have lost the solution. ***The small change from*** $\mathbf{10{,}001}$ ***to*** $\mathbf{10{,}000}$ ***has changed the answer from*** $\boldsymbol{u = 1}$ ***to*** $\boldsymbol{u = 0}$ (100% error!).

If we exchange rows, even this weak computer finds an answer that is correct to 3 places:

$$\begin{aligned} -u + v &= 0 \\ .0001u + v &= 1 \end{aligned} \quad \longrightarrow \quad \begin{aligned} -u + v &= 0 \\ v &= 1 \end{aligned} \quad \longrightarrow \quad \begin{aligned} u &= 1 \\ v &= 1. \end{aligned}$$

The original pivots were .0001 and 10,000—badly scaled. After a row exchange the exact pivots are -1 and 1.0001—well scaled. The computed pivots -1 and 1 come close to the exact values. Small pivots bring numerical instability, and the remedy is ***partial pivoting***. Here is our strategy when we reach and search column k for the best available pivot:

Choose the largest number in row k or below. Exchange its row with row k.

The strategy of *complete pivoting* looks also in later columns for the largest pivot. It exchanges columns as well as rows. This expense is seldom justified, and all major codes use partial pivoting. Multiplying a row or column by a scaling constant can also be very worthwhile. *If the first equation above is* $u + 10{,}000v = 10{,}000$ *and we don't rescale, then* 1 *looks like a good pivot and we would miss the essential row exchange.*

For positive definite matrices, row exchanges are *not* required. It is safe to accept the pivots as they appear. Small pivots can occur, but the matrix is not improved by row exchanges. When its condition number is high, the problem is in the matrix and not in the code. In this case the output is unavoidably sensitive to the input.

The reader now understands how a computer actually solves $Ax = b$—***by elimination with partial pivoting***. Compared with the theoretical description—***find*** A^{-1} ***and multiply*** $A^{-1}b$—the details took time. But in computer time, elimination is much faster. I believe that elimination is also the best approach to the algebra of row spaces and nullspaces.

Operation Counts: Full Matrices

Here is a practical question about cost. *How many separate operations are needed to solve $Ax = b$ by elimination?* This decides how large a problem we can afford.

Look first at A, which changes gradually into U. When a multiple of row 1 is subtracted from row 2, we do n operations. The first is a division by the pivot, to find the multiplier ℓ. For the other $n-1$ entries along the row, the operation is a "multiply-subtract". For convenience, we count this as a single operation. If you regard multiplying by ℓ and subtracting from the existing entry as two separate operations, *multiply all our counts by* 2.

The matrix A is n by n. The operation count applies to all $n-1$ rows below the first. Thus it requires n times $n-1$ operations, or $n^2 - n$, to produce zeros below the first pivot. *Check: All n^2 entries are changed, except the n entries in the first row.*

When elimination is down to k equations, the rows are shorter. We need only $k^2 - k$ operations (instead of $n^2 - n$) to clear out the column below the pivot. This is true for $1 \le k \le n$. The last step requires no operations ($1^2 - 1 = 0$); forward elimination is complete. The total count to reach U is the sum of $k^2 - k$ over all values of k from 1 to n:

$$(1^2 + \cdots + n^2) - (1 + \cdots + n) = \frac{n(n+1)(2n+1)}{6} - \frac{n(n+1)}{2} = \boxed{\frac{n^3 - n}{3}}\,.$$

Those are known formulas for the sum of the first n numbers and their squares. Substituting $n = 100$ gives a million minus a hundred—then divide by 3. (That translates into one second on a workstation.) We will ignore n in comparison with n^3, to reach our main conclusion:

The multiply-subtract count is $\frac{1}{3}n^3$ for forward elimination (A to U, producing L).

That means $\frac{1}{3}n^3$ multiplications and subtractions. Doubling n increases this cost by eight (because n is cubed). 100 equations are easy, 1000 are more expensive,10000 densc equations are close to impossible. We need a faster computer or a lot of zeros or a new idea.

On the right side of the equations, the steps go much faster. We operate on single numbers, not whole rows. ***Each right side needs exactly n^2 operations***. Down and back up we are solving two triangular systems, $L\boldsymbol{c} = \boldsymbol{b}$ forward and $U\boldsymbol{x} = \boldsymbol{c}$ backward. In back substitution, the last unknown needs only division by the last pivot. The equation above it needs two operations—substituting x_n and dividing by *its* pivot. The kth step needs k multiply-subtract operations, and the total for back substitution is

$$1 + 2 + \cdots + n = \frac{n(n+1)}{2} \approx \tfrac{1}{2}n^2 \quad \text{operations.}$$

The forward part is similar. *The n^2 total exactly equals the count for multiplying $A^{-1}\boldsymbol{b}$!* This leaves Gaussian elimination with two big advantages over $A^{-1}\boldsymbol{b}$:

1 Elimination requires $\frac{1}{3}n^3$ multiply-subtracts, compared to n^3 for A^{-1}.

2 If A is *banded* so are L and U: by comparison A^{-1} is full of nonzeros.

Band Matrices

These counts are improved when A has "*good zeros*". A good zero is an entry that remains zero in L and U. ***The best zeros are at the beginning of a row.*** They require no elimination steps (the multipliers are zero). So we also find those same good zeros in L. That is especially clear for this *tridiagonal matrix* A (and for band matrices in Figure 11.1):

Tridiagonal Bidiagonal times bidiagonal
$$\begin{bmatrix} 1 & -1 & & \\ -1 & 2 & -1 & \\ & -1 & 2 & -1 \\ & & -1 & 2 \end{bmatrix} = \begin{bmatrix} 1 & & & \\ -1 & 1 & & \\ & -1 & 1 & \\ & & -1 & 1 \end{bmatrix} \begin{bmatrix} 1 & -1 & & \\ & 1 & -1 & \\ & & 1 & -1 \\ & & & 1 \end{bmatrix}.$$

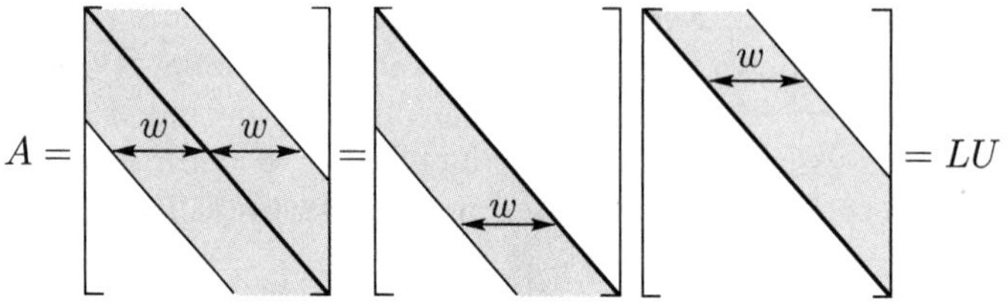

Figure 11.1: $A = LU$ for a band matrix. Good zeros in A *stay zero* in L and U.

These zeros lead to a complete change in the operation count, for "half-bandwidth" w:

A band matrix has $a_{ij} = 0$ *when* $|i - j| > w$.

Thus $w = 1$ for a diagonal matrix, $w = 2$ for tridiagonal, $w = n$ for dense. The length of the pivot row is at most w. There are no more than $w - 1$ nonzeros below any pivot. Each stage of elimination is complete after $w(w-1)$ operations, and *the band structure survives*. There are n columns to clear out. Therefore:

Elimination on a band matrix (A to L and U) needs less than $w^2 n$ operations.

For a band matrix, the count is proportional to n instead of n^3. It is also proportional to w^2. A full matrix has $w = n$ and we are back to n^3. For an exact count, remember that the bandwidth drops below w in the lower right corner (not enough space):

$$\textbf{Band} \quad \frac{w(w-1)(3n-2w+1)}{3} \qquad \textbf{Dense} \quad \frac{n(n-1)(n+1)}{3} = \frac{n^3-n}{3}$$

On the right side of $A\boldsymbol{x} = \boldsymbol{b}$, to find $\boldsymbol{x}$ from $\boldsymbol{b}$, the cost is about $2wn$ (compared to the usual n^2). *Main point*: *For a band matrix the operation counts are* ***proportional to*** $\boldsymbol{n}$. This is extremely fast. A tridiagonal matrix of order 10,000 is very cheap, provided *we don't compute* A^{-1}. That inverse matrix has no zeros at all:

$$A = \begin{bmatrix} 1 & -1 & 0 & 0 \\ -1 & 2 & -1 & 0 \\ 0 & -1 & 2 & -1 \\ 0 & 0 & -1 & 2 \end{bmatrix} \quad \text{has} \quad A^{-1} = U^{-1}L^{-1} = \begin{bmatrix} 4 & 3 & 2 & 1 \\ 3 & 3 & 2 & 1 \\ 2 & 2 & 2 & 1 \\ 1 & 1 & 1 & 1 \end{bmatrix}.$$

We are actually worse off knowing A^{-1} than knowing L and U. Multiplication by A^{-1} needs the full n^2 steps. Solving $L\boldsymbol{c} = \boldsymbol{b}$ and $U\boldsymbol{x} = \boldsymbol{c}$ needs only $2wn$.

A band structure is very common in practice, when the matrix reflects connections between near neighbors: $a_{13} = 0$ and $a_{14} = 0$ because 1 is not a neighbor of 3 and 4.

We close with counts for Gauss-Jordan and Gram-Schmidt-Householder:

A^{-1} costs n^3 multiply-subtract steps.	***QR costs $\frac{2}{3}n^3$ steps.***

In $AA^{-1} = I$, the jth column of A^{-1} solves $A\boldsymbol{x}_j = j$th column of I. The left side costs $\frac{1}{3}n^3$ as usual. (This is a one-time cost! L and U are not repeated.) The special saving for the jth column of I comes from its first $j-1$ zeros. No work is required on the right side until elimination reaches row j. The forward cost is $\frac{1}{2}(n-j)^2$ instead of $\frac{1}{2}n^2$. Summing over j, the total for forward elimination on the n right sides is $\frac{1}{6}n^3$. The final multiply-subtract count for A^{-1} is n^3 if we actually want the inverse:

$$\textbf{For}\,\boldsymbol{A}^{-1} \qquad \frac{n^3}{3}\,(L \text{ and } U) + \frac{n^3}{6}\,(\text{forward}) + n\left(\frac{n^2}{2}\right)(\text{back substitutions}) = \boldsymbol{n}^3. \quad (1)$$

Orthogonalization *(A to Q)*: The key difference from elimination is that *each multiplier is decided by a dot product.* That takes n operations, where elimination just divides by the pivot. Then there are n "multiply-subtract" operations to remove from column k its projection along column $j < k$ (see Section 4.4). The combined cost is $2n$ where for elimination it is n. This factor 2 is the price of orthogonality. We are changing a dot product to zero where elimination changes an entry to zero.

Caution To judge a numerical algorithm, it is **not enough** to count the operations. Beyond "flop counting" is a study of stability (Householder wins) and the flow of data.

Reordering Sparse Matrices

For band matrices with constant width w, the row ordering is optimal. But for most sparse matrices in real computations, the width of the band is *not constant* and there are many zeros inside the band. Those zeros can fill in as elimination proceeds—they are lost. We need to ***renumber the equations to reduce fill-in***, and thereby speed up elimination.

Generally speaking, we want to move zeros to early rows and columns. Later rows and columns are shorter anyway. The "**a**pproximate **m**inimum **d**egree" algorithm in sparse MATLAB is *greedy*—it chooses the row to eliminate without counting all the consequences. We may reach a nearly full matrix near the end, but the total operation count to reach LU is still much smaller. To find the absolute minimum of nonzeros in L and U is an NP-hard problem, much too expensive, and **amd** is a good compromise.

Fill-in is famous when each point on a square grid is connected to its four nearest neighbors. It is impossible to number all the gridpoints so that neighbors stay together! If we number by rows of the grid, there is a long wait to come around to the gridpoint above.

$$\begin{matrix} j \\ i \\ k \end{matrix} \begin{bmatrix} 1 & 1 & 1 \\ -2 & 1 & \mathbf{0} \\ -2 & \mathbf{0} & 2 \end{bmatrix} \longrightarrow \begin{bmatrix} 1 & 1 & 1 \\ \mathbf{0} & 3 & \mathbf{2} \\ \mathbf{0} & \mathbf{2} & 4 \end{bmatrix}$$

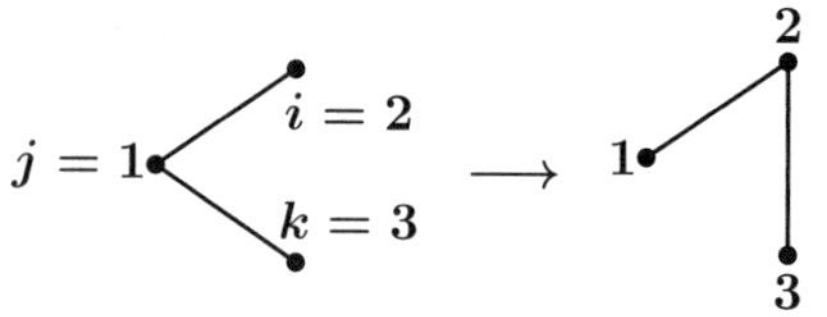

$a_{32} = 0$ $\quad$ $a_{32} = \mathbf{2}$ $\quad$ $a_{32} = \mathbf{0}$ **before** $\quad$ $a_{32} \neq \mathbf{0}$ **after**

We only need the *positions* of the nonzeros, not their exact values. Think of the graph of nonzeros: *Node i is connected to node j if $a_{ij} \neq 0$.* Watch to see how elimination can create nonzeros (new edges), which we are trying to avoid.

The command **nnz**(L) counts the nonzero multipliers in the lower triangular L, **find** (L) will list them, and **spy**(L) shows them all.

The goal of **colamd** and **symamd** is a better ordering (permutation P) that reduces fill-in for AP and $P^{\mathrm{T}}AP$—*by choosing the pivot with the fewest nonzeros below it.*

Fast Orthogonalization

There are three ways to reach the important factorization $A = QR$. Gram-Schmidt works to find the orthonormal vectors in Q. Then R is upper triangular because of the order of Gram-Schmidt steps. Now we look at better methods (Householder and Givens), which use a product of specially simple Q's that we *know* are orthogonal.

Elimination gives $A = LU$, orthogonalization gives $A = QR$. We don't want a triangular L, we want an orthogonal Q. L is a product of E's from elimination, with 1's on the diagonal and the multiplier ℓ_{ij} below. *Q will be a product of orthogonal matrices.*

There are two simple orthogonal matrices to take the place of the E's. The ***reflection matrices*** $I - 2\boldsymbol{u}\boldsymbol{u}^{\mathrm{T}}$ are named after Householder. The ***plane rotation matrices*** are named after Givens. The simple matrix that rotates the xy plane by θ is Q_{21}:

Givens rotation in the 1-2 plane
$$Q_{21} = \begin{bmatrix} \cos\theta & -\sin\theta & 0 \\ \sin\theta & \cos\theta & 0 \\ 0 & 0 & 1 \end{bmatrix}.$$

Use Q_{21} the way you used E_{21}, to produce a zero in the $(2,1)$ position. That determines the angle θ. Bill Hager gives this example in *Applied Numerical Linear Algebra*:

$$Q_{21}A = \begin{bmatrix} .6 & .8 & 0 \\ -.8 & .6 & 0 \\ 0 & 0 & 1 \end{bmatrix} \begin{bmatrix} 90 & -153 & 114 \\ 120 & -79 & -223 \\ 200 & -40 & 395 \end{bmatrix} = \begin{bmatrix} 150 & -155 & -110 \\ 0 & 75 & -225 \\ 200 & -40 & 395 \end{bmatrix}.$$

The zero came from $-.8(90) + .6(120)$. No need to find θ, what we needed was $\cos\theta$:

$$\cos\theta = \frac{90}{\sqrt{90^2 + 120^2}} \quad \text{and} \quad \sin\theta = \frac{-120}{\sqrt{90^2 + 120^2}}. \tag{2}$$

Now we attack the $(3,1)$ entry. The rotation will be in rows and columns 3 and 1. The numbers $\cos\theta$ and $\sin\theta$ are determined from 150 and 200, instead of 90 and 120.

$$Q_{31}Q_{21}A = \begin{bmatrix} .6 & 0 & .8 \\ 0 & 1 & 0 \\ -.8 & 0 & .6 \end{bmatrix} \begin{bmatrix} 150 & \cdot & \cdot \\ 0 & \cdot & \cdot \\ 200 & \cdot & \cdot \end{bmatrix} = \begin{bmatrix} 250 & -125 & 250 \\ 0 & 75 & -225 \\ 0 & 100 & 325 \end{bmatrix}.$$

One more step to R. The $(3,2)$ entry has to go. The numbers $\cos\theta$ and $\sin\theta$ now come from 75 and 100. The rotation is now in rows and columns 2 and 3:

$$Q_{32}Q_{31}Q_{21}A = \begin{bmatrix} 1 & 0 & 0 \\ 0 & .6 & .8 \\ 0 & -.8 & .6 \end{bmatrix} \begin{bmatrix} 250 & -125 & \cdot \\ 0 & 75 & \cdot \\ 0 & 100 & \cdot \end{bmatrix} = \begin{bmatrix} 250 & -125 & 250 \\ 0 & 125 & 125 \\ 0 & 0 & 375 \end{bmatrix}.$$

We have reached the upper triangular R. What is Q? Move the plane rotations Q_{ij} to the other side to find $A = QR$—just as you moved the elimination matrices E_{ij} to the other side to find $A = LU$:

$$Q_{32}Q_{31}Q_{21}A = R \qquad \text{means} \qquad A = (Q_{21}^{-1}Q_{31}^{-1}Q_{32}^{-1})R = QR. \tag{3}$$

The inverse of each Q_{ij} is Q_{ij}^{T} (rotation through $-\theta$). The inverse of E_{ij} was not an orthogonal matrix! LU and QR are similar but L and Q are not the same.

Householder reflections are faster than rotations because each one clears out a whole column below the diagonal. Watch how the first column $\boldsymbol{a}_1$ of A becomes column $\boldsymbol{r}_1$ of R:

Reflection by H_1

$H_1 = I - 2\boldsymbol{u}_1\boldsymbol{u}_1^{\mathrm{T}}$

$$H_1\,\boldsymbol{a}_1 = \begin{bmatrix} \|\boldsymbol{a}_1\| \\ 0 \\ \cdot \\ 0 \end{bmatrix} \quad \text{or} \quad \begin{bmatrix} -\|\boldsymbol{a}_1\| \\ 0 \\ \cdot \\ 0 \end{bmatrix} = \boldsymbol{r}_1\,. \tag{4}$$

The length was not changed, and $\boldsymbol{u}_1$ is in the direction of $\boldsymbol{a}_1 - \boldsymbol{r}_1$. We have $n-1$ entries in the unit vector $\boldsymbol{u}_1$ to get $n-1$ zeros in $\boldsymbol{r}_1$. (Rotations had one angle θ to get one zero.) When we reach column k, we have $n-k$ available choices in the unit vector $\boldsymbol{u}_k$. This leads to $n-k$ zeros in $\boldsymbol{r}_k$. *We just store the $\boldsymbol{u}$'s and $\boldsymbol{r}$'s to know the final Q and R:*

Inverse of H_i is H_i $\quad (H_{n-1}\ldots H_1)A = R$ means $A = (H_1 \ldots H_{n-1})R = QR.$ (5)

This is how LAPACK improves on 19th century Gram-Schmidt. Q is *exactly* orthogonal.

Section 11.3 explains how $A = QR$ is used in the other big computation of linear algebra—***the eigenvalue problem***. The factors QR are reversed to give $A_1 = RQ$ which is $Q^{-1}AQ$. Since A_1 is similar to A, the eigenvalues are unchanged. Then A_1 is factored into Q_1R_1, and reversing the factors gives A_2. Amazingly, the entries below the diagonal get smaller in $A_1, A_2, A_3, \ldots$ and we can identify the eigenvalues. This is the "QR method" for $A\boldsymbol{x} = \lambda\boldsymbol{x}$, a big success of numerical linear algebra.

Problem Set 11.1

1 Find the two pivots with and without row exchange to maximize the pivot:

$$A = \begin{bmatrix} .001 & 0 \\ 1 & 1000 \end{bmatrix}.$$

With row exchanges to maximize pivots, why are no entries of L larger than 1? Find a 3 by 3 matrix A with all $|a_{ij}| \leq 1$ and $|\ell_{ij}| \leq 1$ but third pivot $= 4$.

2 Compute the exact inverse of the Hilbert matrix A by elimination. Then compute A^{-1} again by rounding all numbers to three figures:

Ill-conditioned matrix $$A = \mathsf{hilb(3)} = \begin{bmatrix} 1 & \frac{1}{2} & \frac{1}{3} \\ \frac{1}{2} & \frac{1}{3} & \frac{1}{4} \\ \frac{1}{3} & \frac{1}{4} & \frac{1}{5} \end{bmatrix}.$$

3 For the same A compute $\boldsymbol{b} = A\boldsymbol{x}$ for $\boldsymbol{x} = (1,1,1)$ and $\boldsymbol{x} = (0,6,-3.6)$. A small change $\Delta\boldsymbol{b}$ produces a large change $\Delta\boldsymbol{x}$.

4 Find the eigenvalues (by computer) of the 8 by 8 Hilbert matrix $a_{ij} = 1/(i+j-1)$. In the equation $A\boldsymbol{x} = \boldsymbol{b}$ with $\|\boldsymbol{b}\| = 1$, how large can $\|\boldsymbol{x}\|$ be? If $\boldsymbol{b}$ has roundoff error less than 10^{-16}, how large an error can this cause in $\boldsymbol{x}$? See Section 9.2.

5 For back substitution with a band matrix (width w), show that the number of multiplications to solve $U\boldsymbol{x} = \boldsymbol{c}$ is approximately wn.

6 If you know L and U and Q and R, is it faster to solve $LU\boldsymbol{x} = \boldsymbol{b}$ or $QR\boldsymbol{x} = \boldsymbol{b}$?

7 Show that the number of multiplications to invert an upper triangular n by n matrix is about $\frac{1}{6}n^3$. Use back substitution on the columns of I, upward from 1's.

8 Choosing the largest available pivot in each column (partial pivoting), factor each A into $PA = LU$:

$$A = \begin{bmatrix} 1 & 0 \\ 2 & 2 \end{bmatrix} \quad \text{and} \quad A = \begin{bmatrix} 1 & 0 & 1 \\ 2 & 2 & 0 \\ 0 & 2 & 0 \end{bmatrix}.$$

9 Put 1's on the three central diagonals of a 4 by 4 tridiagonal matrix. Find the cofactors of the six zero entries. Those entries are nonzero in A^{-1}.

10 (Suggested by C. Van Loan.) Find the LU factorization and solve by elimination when $\varepsilon = 10^{-3}, 10^{-6}, 10^{-9}, 10^{-12}, 10^{-15}$:

$$\begin{bmatrix} \varepsilon & 1 \\ 1 & 1 \end{bmatrix} \begin{bmatrix} x_1 \\ x_2 \end{bmatrix} = \begin{bmatrix} 1+\varepsilon \\ 2 \end{bmatrix}.$$

The true $\boldsymbol{x}$ is $(1,1)$. Make a table to show the error for each ε. Exchange the two equations and solve again—the errors should almost disappear.

11 (a) Choose $\sin\theta$ and $\cos\theta$ to triangularize A, and find R:

$$\textbf{Givens rotation} \qquad Q_{21}A = \begin{bmatrix}\cos\theta & -\sin\theta\\ \sin\theta & \cos\theta\end{bmatrix}\begin{bmatrix}1 & -1\\ 3 & 5\end{bmatrix} = \begin{bmatrix}* & *\\ 0 & *\end{bmatrix} = R.$$

(b) Choose $\sin\theta$ and $\cos\theta$ to make QAQ^{-1} triangular. What are the eigenvalues?

12 When A is multiplied by a plane rotation Q_{ij}, which entries of A are changed? When $Q_{ij}A$ is multiplied on the right by Q_{ij}^{-1}, which entries are changed now?

13 How many multiplications and how many additions are used to compute $Q_{ij}A$? Careful organization of the whole sequence of rotations gives $\frac{2}{3}n^3$ multiplications and $\frac{2}{3}n^3$ additions—the same as for QR by reflectors and twice as many as for LU.

Challenge Problems

14 (**Turning a robot hand**) The robot produces any 3 by 3 rotation A from plane rotations around the x, y, z axes. Then $Q_{32}Q_{31}Q_{21}A = R$, where A is orthogonal so R is I! The three robot turns are in $A = Q_{21}^{-1}Q_{31}^{-1}Q_{32}^{-1}$. The three angles are "Euler angles" and $\det Q = 1$ to avoid reflection. Start by choosing $\cos\theta$ and $\sin\theta$ so that

$$Q_{21}A = \begin{bmatrix}\cos\theta & -\sin\theta & 0\\ \sin\theta & \cos\theta & 0\\ 0 & 0 & 1\end{bmatrix}\frac{1}{3}\begin{bmatrix}-1 & 2 & 2\\ 2 & -1 & 2\\ 2 & 2 & -1\end{bmatrix} \quad \text{is zero in the } (2,1) \text{ position.}$$

15 Create the 10 by 10 second difference matrix $K = \textbf{toeplitz}([2\ -1\ \textbf{zeros}(1,8)])$. Permute rows and columns randomly by $KK = K(\mathbf{randperm}(10), \mathbf{randperm}(10))$. Factor by $[L, U] = \textbf{lu}(K)$ and $[LL, UU] = \textbf{lu}(KK)$, and count nonzeros by $\textbf{nnz}(L)$ and $\textbf{nnz}(LL)$. In this case L is in perfect tridiagonal order, but not LL.

16 Another ordering for this matrix K colors the meshpoints alternately red and black. This permutation P changes the normal $1,\ldots,10$ to $1,3,5,7,9,2,4,6,8,10$:

$$\textbf{Red-black ordering} \qquad PKP^{\mathrm{T}} = \begin{bmatrix}2I & D\\ D^{\mathrm{T}} & 2I\end{bmatrix}. \qquad \textit{Find the matrix } D.$$

So many interesting experiments are possible. If you send good ideas they can go on the linear algebra website math.mit.edu/linearalgebra. I also recommend learning the command $B = \mathbf{sparse}(A)$, after which $\textbf{find}(B)$ will list the nonzero entries and $\boldsymbol{\ell}\mathbf{u}(B)$ will factor B using that sparse format for L and U. Only the nonzeros are computed, where ordinary (dense) MATLAB computes all the zeros too.

17 Jeff Stuart has created a student activity that brilliantly demonstrates ill-conditioning:

$$\begin{bmatrix}1 & 1.0001\\ 1 & 1.0000\end{bmatrix}\begin{bmatrix}x\\ y\end{bmatrix} = \begin{bmatrix}3.0001 + \boldsymbol{e}\\ 3.0000 + \boldsymbol{E}\end{bmatrix} \quad \textbf{With errors } \boldsymbol{e} \textbf{ and } \boldsymbol{E} \quad \begin{array}{l} x = 2 - 10000(\boldsymbol{e} - \boldsymbol{E})\\ y = 1 + 10000(\boldsymbol{e} - \boldsymbol{E})\end{array}$$

When those equations are shown by nearly parallel long sticks, a small shake gives a big jump in the crossing point (x, y). Errors $\boldsymbol{e}$ and $\boldsymbol{E}$ are amplified by 10000.

11.2 Norms and Condition Numbers

How do we measure the size of a matrix? For a vector, the length is $\|\boldsymbol{x}\|$. For a matrix, ***the norm is*** $\|A\|$. This word "*norm*" is sometimes used for vectors, instead of length. It is always used for matrices, and there are many ways to measure $\|A\|$. We look at the requirements on all "matrix norms" and then choose one.

Frobenius squared all the $|a_{ij}|^2$ and added; his norm $\|A\|_{\mathrm{F}}$ is the square root. This treats A like a long vector with n^2 components: sometimes useful, but not the choice here.

I prefer to start with a vector norm. The triangle inequality says that $\|\boldsymbol{x}+\boldsymbol{y}\|$ is not greater than $\|\boldsymbol{x}\|+\|\boldsymbol{y}\|$. The length of $2\boldsymbol{x}$ or $-2\boldsymbol{x}$ is doubled to $2\|\boldsymbol{x}\|$. The same rules will apply to matrix norms:

$$\|A+B\| \le \|A\| + \|B\| \quad \text{and} \quad \|cA\| = |c|\,\|A\|. \tag{1}$$

The second requirements for a matrix norm are new, because matrices multiply. The norm $\|A\|$ controls the growth from $\boldsymbol{x}$ to $A\boldsymbol{x}$, and from B to AB:

Growth factor $\|A\|$
$$\|A\boldsymbol{x}\| \le \|A\|\,\|\boldsymbol{x}\| \quad \text{and} \quad \|AB\| \le \|A\|\,\|B\|. \tag{2}$$

This leads to a natural way to define $\|A\|$, the norm of a matrix:

The norm of A is the largest ratio $\|A\boldsymbol{x}\|/\|\boldsymbol{x}\|$:
$$\|A\| = \max_{\boldsymbol{x}\neq \boldsymbol{0}} \frac{\|A\boldsymbol{x}\|}{\|\boldsymbol{x}\|}. \tag{3}$$

$\|A\boldsymbol{x}\|/\|\boldsymbol{x}\|$ is never larger than $\|A\|$ (its maximum). This says that $\|A\boldsymbol{x}\| \le \|A\|\,\|\boldsymbol{x}\|$.

Example 1 If A is the identity matrix I, the ratios are $\|\boldsymbol{x}\|/\|\boldsymbol{x}\|$. Therefore $\|I\| = 1$. If A is an orthogonal matrix Q, lengths are again preserved: $\|Q\boldsymbol{x}\| = \|\boldsymbol{x}\|$. The ratios still give $\|Q\| = 1$. An orthogonal Q is good to compute with: errors don't grow.

Example 2 The norm of a diagonal matrix is its largest entry (using absolute values):

$$A = \begin{bmatrix} 2 & 0 \\ 0 & 3 \end{bmatrix} \quad \text{has norm} \quad \|A\| = 3. \quad \text{The eigenvector} \quad \boldsymbol{x} = \begin{bmatrix} 0 \\ 1 \end{bmatrix} \quad \text{has} \quad A\boldsymbol{x} = 3\boldsymbol{x}.$$

The eigenvalue is 3. For this A (but not all A), the largest eigenvalue equals the norm.

For a positive definite symmetric matrix the norm is $\|A\| = \lambda_{\max}(A)$.

Choose $\boldsymbol{x}$ to be the eigenvector with maximum eigenvalue. Then $\|A\boldsymbol{x}\|/\|\boldsymbol{x}\|$ equals $\lambda_{\max}$. The point is that no other $\boldsymbol{x}$ can make the ratio larger. The matrix is $A = Q\Lambda Q^{\mathrm{T}}$, and the orthogonal matrices Q and Q^{T} leave lengths unchanged. So the ratio to maximize is really $\|\Lambda\boldsymbol{x}\|/\|\boldsymbol{x}\|$. The norm is the largest eigenvalue in the diagonal Λ.

Symmetric matrices Suppose A is symmetric but not positive definite. $A = Q\Lambda Q^{\mathrm{T}}$ is still true. Then the norm is the largest of $|\lambda_1|$, $|\lambda_2|$, ..., $|\lambda_n|$. We take absolute values, because the norm is only concerned with length. For an eigenvector $\|A\boldsymbol{x}\| = \|\lambda \boldsymbol{x}\| = |\lambda|$ times $\|\boldsymbol{x}\|$. The $\boldsymbol{x}$ that gives the maximum ratio is the eigenvector for the maximum $|\lambda|$.

Unsymmetric matrices If A is not symmetric, its eigenvalues may not measure its true size. *The norm can be larger than any eigenvalue.* A very unsymmetric example has $\lambda_1 = \lambda_2 = 0$ but its norm is not zero:

$$\|\boldsymbol{A}\| > \boldsymbol{\lambda}_{\max} \quad A = \begin{bmatrix} 0 & 2 \\ 0 & 0 \end{bmatrix} \quad \text{has norm} \quad \|A\| = \max_{\boldsymbol{x} \neq \mathbf{0}} \frac{\|A\boldsymbol{x}\|}{\|\boldsymbol{x}\|} = 2.$$

The vector $\boldsymbol{x} = (0, 1)$ gives $A\boldsymbol{x} = (2, 0)$. The ratio of lengths is $2/1$. This is the maximum ratio $\|A\|$, even though $\boldsymbol{x}$ is not an eigenvector.

It is the ***symmetric matrix*** $A^{\mathrm{T}}A$, not the unsymmetric A, that has eigenvector $\boldsymbol{x} = (0, 1)$. The norm is really decided by ***the largest eigenvalue of*** $A^{\mathrm{T}}A$:

The norm of A (symmetric or not) ***is the square root of*** $\boldsymbol{\lambda}_{\max}(A^{\mathrm{T}}A)$***:***

$$\boxed{\|A\|^2} = \max_{\boldsymbol{x} \neq \mathbf{0}} \frac{\|A\boldsymbol{x}\|^2}{\|\boldsymbol{x}\|^2} = \max_{\boldsymbol{x} \neq \mathbf{0}} \frac{\boldsymbol{x}^{\mathrm{T}}A^{\mathrm{T}}A\boldsymbol{x}}{\boldsymbol{x}^{\mathrm{T}}\boldsymbol{x}} = \boxed{\lambda_{\max}(A^{\mathrm{T}}A)}. \tag{4}$$

The unsymmetric example with $\lambda_{\max}(A) = 0$ has $\lambda_{\max}(A^{\mathrm{T}}A) = 4$:

$$A = \begin{bmatrix} 0 & 2 \\ 0 & 0 \end{bmatrix} \text{ leads to } A^{\mathrm{T}}A = \begin{bmatrix} 0 & 0 \\ 0 & 4 \end{bmatrix} \text{ with } \lambda_{\max} = 4. \text{ So the norm is } \|A\| = \sqrt{4}.$$

For any A Choose $\boldsymbol{x}$ to be the eigenvector of $A^{\mathrm{T}}A$ with largest eigenvalue $\lambda_{\max}$. The ratio in equation (4) is $\boldsymbol{x}^{\mathrm{T}}A^{\mathrm{T}}A\boldsymbol{x} = \boldsymbol{x}^{\mathrm{T}}(\lambda_{\max})\boldsymbol{x}$ divided by $\boldsymbol{x}^{\mathrm{T}}\boldsymbol{x}$. This is $\lambda_{\max}$.

No $\boldsymbol{x}$ can give a larger ratio. The symmetric matrix $A^{\mathrm{T}}A$ has eigenvalues $\lambda_1, \ldots, \lambda_n$ and orthonormal eigenvectors $\boldsymbol{q}_1, \boldsymbol{q}_2, \ldots, \boldsymbol{q}_n$. Every $\boldsymbol{x}$ is a combination of those vectors. Try this combination in the ratio and remember that $\boldsymbol{q}_i^{\mathrm{T}}\boldsymbol{q}_j = 0$:

$$\frac{\boldsymbol{x}^{\mathrm{T}}A^{\mathrm{T}}A\boldsymbol{x}}{\boldsymbol{x}^{\mathrm{T}}\boldsymbol{x}} = \frac{(c_1\boldsymbol{q}_1 + \cdots + c_n\boldsymbol{q}_n)^{\mathrm{T}}(c_1\lambda_1\boldsymbol{q}_1 + \cdots + c_n\lambda_n\boldsymbol{q}_n)}{(c_1\boldsymbol{q}_1 + \cdots + c_n\boldsymbol{q}_n)^{\mathrm{T}}(c_1\boldsymbol{q}_1 + \cdots + c_n\boldsymbol{q}_n)} = \frac{c_1^2\lambda_1 + \cdots + c_n^2\lambda_n}{c_1^2 + \cdots + c_n^2}.$$

The maximum ratio $\lambda_{\max}$ is when all c's are zero, except the one that multiplies $\lambda_{\max}$.

Note 1 The ratio in equation (4) is the ***Rayleigh quotient*** for the symmetric matrix $A^{\mathrm{T}}A$. Its maximum is the largest eigenvalue $\lambda_{\max}(A^{\mathrm{T}}A)$. The minimum ratio is $\lambda_{\min}(A^{\mathrm{T}}A)$. If you substitute any vector $\boldsymbol{x}$ into the Rayleigh quotient $\boldsymbol{x}^{\mathrm{T}}A^{\mathrm{T}}A\boldsymbol{x}/\boldsymbol{x}^{\mathrm{T}}\boldsymbol{x}$, you are guaranteed to get a number between $\lambda_{\min}(A^{\mathrm{T}}A)$ and $\lambda_{\max}(A^{\mathrm{T}}A)$.

Note 2 The norm $\|A\|$ equals the ***largest singular value*** $\boldsymbol{\sigma_{\max}}$ of A. The singular values $\sigma_1, \ldots, \sigma_r$ are the square roots of the positive eigenvalues of $A^{\mathrm{T}}A$. So certainly $\sigma_{\max} = (\lambda_{\max})^{1/2}$. Since U and V are orthogonal in $A = U\Sigma V^{\mathrm{T}}$, the norm is $\|\boldsymbol{A}\| = \boldsymbol{\sigma_{\max}}$.

The Condition Number of A

Section 9.1 showed that roundoff error can be serious. Some systems are sensitive, others are not so sensitive. The sensitivity to error is measured by the ***condition number***. This is the first chapter in the book which intentionally introduces errors. We want to estimate how much they change $\boldsymbol{x}$.

The original equation is $A\boldsymbol{x} = \boldsymbol{b}$. Suppose the right side is changed to $\boldsymbol{b} + \Delta\boldsymbol{b}$ because of roundoff or measurement error. The solution is then changed to $\boldsymbol{x} + \Delta\boldsymbol{x}$. Our goal is to estimate the change $\Delta\boldsymbol{x}$ in the solution from the change $\Delta\boldsymbol{b}$ in the equation. Subtraction gives the *error equation* $A(\Delta\boldsymbol{x}) = \Delta\boldsymbol{b}$:

$$\text{Subtract } A\boldsymbol{x} = \boldsymbol{b} \text{ from } A(\boldsymbol{x} + \Delta\boldsymbol{x}) = \boldsymbol{b} + \Delta\boldsymbol{b} \quad \text{to find} \quad \boxed{A(\Delta\boldsymbol{x}) = \Delta\boldsymbol{b}.} \tag{5}$$

The error is $\Delta\boldsymbol{x} = A^{-1}\Delta\boldsymbol{b}$. It is large when A^{-1} is large (then A is nearly singular). The error $\Delta\boldsymbol{x}$ is especially large when $\Delta\boldsymbol{b}$ points in the worst direction—which is amplified most by A^{-1}. ***The worst error has*** $\|\Delta\boldsymbol{x}\| = \|A^{-1}\| \, \|\Delta\boldsymbol{b}\|$.

This error bound $\|A^{-1}\|$ has one serious drawback. If we multiply A by 1000, then A^{-1} is divided by 1000. The matrix looks a thousand times better. But a simple rescaling cannot change the reality of the problem. It is true that $\Delta\boldsymbol{x}$ will be divided by 1000, but so will the exact solution $\boldsymbol{x} = A^{-1}\boldsymbol{b}$. The ***relative error*** $\|\Delta\boldsymbol{x}\|/\|\boldsymbol{x}\|$ will stay the same. It is this relative change in $\boldsymbol{x}$ that should be compared to the relative change in $\boldsymbol{b}$.

Comparing relative errors will now lead to the "condition number" $c = \|A\| \, \|A^{-1}\|$. Multiplying A by 1000 does not change this number, because A^{-1} is divided by 1000 and the condition number c stays the same. It measures the sensitivity of $A\boldsymbol{x} = \boldsymbol{b}$.

The solution error is less than $c = \|A\| \, \|A^{-1}\|$ ***times the problem error:***

Condition number c
$$\frac{\|\Delta\boldsymbol{x}\|}{\|\boldsymbol{x}\|} \le c\frac{\|\Delta\boldsymbol{b}\|}{\|\boldsymbol{b}\|}. \tag{6}$$

If the problem error is ΔA (error in A instead of $\boldsymbol{b}$), still c controls $\Delta\boldsymbol{x}$:

Error ΔA in A
$$\frac{\|\Delta\boldsymbol{x}\|}{\|\boldsymbol{x} + \Delta\boldsymbol{x}\|} \le c\frac{\|\Delta A\|}{\|A\|}. \tag{7}$$

Proof The original equation is $\boldsymbol{b} = A\boldsymbol{x}$. The error equation (5) is $\Delta\boldsymbol{x} = A^{-1}\Delta\boldsymbol{b}$. Apply the key property $\|A\boldsymbol{x}\| \le \|A\|\|\boldsymbol{x}\|$ of matrix norms:

$$\|\boldsymbol{b}\| \le \|A\|\,\|\boldsymbol{x}\| \qquad \text{and} \qquad \|\Delta\boldsymbol{x}\| \le \|A^{-1}\|\,\|\Delta\boldsymbol{b}\|.$$

Multiply the left sides to get $\|\boldsymbol{b}\|\,\|\Delta\boldsymbol{x}\|$, and multiply the right sides to get $c\|\boldsymbol{x}\|\,\|\Delta\boldsymbol{b}\|$. Divide both sides by $\|\boldsymbol{b}\|\,\|\boldsymbol{x}\|$. The left side is now the relative error $\|\Delta\boldsymbol{x}\|/\|\boldsymbol{x}\|$. The right side is now the upper bound in equation (6).

The same condition number $c = \|A\|\,\|A^{-1}\|$ appears when the error is in the matrix. We have ΔA instead of $\Delta\boldsymbol{b}$ in the error equation:

$$\text{Subtract } A\boldsymbol{x} = \boldsymbol{b} \text{ from } (A + \Delta A)(\boldsymbol{x} + \Delta\boldsymbol{x}) = \boldsymbol{b} \text{ to find } A(\Delta\boldsymbol{x}) = -(\Delta A)(\boldsymbol{x} + \Delta\boldsymbol{x}).$$

Multiply the last equation by A^{-1} and take norms to reach equation (7):

$$\|\Delta\boldsymbol{x}\| \le \|A^{-1}\|\,\|\Delta A\|\,\|\boldsymbol{x} + \Delta\boldsymbol{x}\| \quad \text{or} \quad \frac{\|\Delta\boldsymbol{x}\|}{\|\boldsymbol{x} + \Delta\boldsymbol{x}\|} \le \|A\|\,\|A^{-1}\|\frac{\|\Delta A\|}{\|A\|}.$$

Conclusion Errors enter in two ways. They begin with an error ΔA or $\Delta\boldsymbol{b}$—a wrong matrix or a wrong $\boldsymbol{b}$. This problem error is amplified (a lot or a little) into the solution error $\Delta\boldsymbol{x}$. That error is bounded, relative to $\boldsymbol{x}$ itself, by the condition number c.

The error $\Delta\boldsymbol{b}$ depends on computer roundoff and on the original measurements of $\boldsymbol{b}$. The error ΔA also depends on the elimination steps. Small pivots tend to produce large errors in L and U. Then $L+\Delta L$ times $U+\Delta U$ equals $A+\Delta A$. When ΔA or the condition number is very large, the error $\Delta\boldsymbol{x}$ can be unacceptable.

Example 3 When A is symmetric, $c = \|A\|\,\|A^{-1}\|$ comes from the eigenvalues:

$$A = \begin{bmatrix} 6 & 0 \\ 0 & 2 \end{bmatrix} \text{ has norm } 6. \qquad A^{-1} = \begin{bmatrix} \frac{1}{6} & 0 \\ 0 & \frac{1}{2} \end{bmatrix} \text{ has norm } \tfrac{1}{2}.$$

This A is symmetric positive definite. Its norm is $\lambda_{\max} = 6$. The norm of A^{-1} is $1/\lambda_{\min} = \frac{1}{2}$. Multiplying norms gives the *condition number* $\|A\|\,\|A^{-1}\| = \lambda_{\max}/\lambda_{\min}$:

$$\textbf{Condition number for positive definite } \boldsymbol{A} \qquad c = \frac{\lambda_{\max}}{\lambda_{\min}} = \frac{6}{2} = 3.$$

Example 4 Keep the same A, with eigenvalues 6 and 2. To make $\boldsymbol{x}$ small, choose $\boldsymbol{b}$ along the first eigenvector $(1, 0)$. To make $\Delta\boldsymbol{x}$ large, choose $\Delta\boldsymbol{b}$ along the second eigenvector $(0, 1)$. Then $\boldsymbol{x} = \frac{1}{6}\boldsymbol{b}$ and $\Delta\boldsymbol{x} = \frac{1}{2}\Delta\boldsymbol{b}$. The ratio $\|\Delta\boldsymbol{x}\|/\|\boldsymbol{x}\|$ is exactly $c = 3$ times the ratio $\|\Delta\boldsymbol{b}\|/\|\boldsymbol{b}\|$.

This shows that the worst error allowed by the condition number $\|A\|\,\|A^{-1}\|$ can actually happen. Here is a useful rule of thumb, experimentally verified for Gaussian elimination: *The computer can lose* $\log c$ *decimal places to roundoff error.*

Problem Set 11.2

1 Find the norms $||A|| = \lambda_{\max}$ and condition numbers $c = \lambda_{\max}/\lambda_{\min}$ of these positive definite matrices:

$$\begin{bmatrix} .5 & 0 \\ 0 & 2 \end{bmatrix} \quad \begin{bmatrix} 2 & 1 \\ 1 & 2 \end{bmatrix} \quad \begin{bmatrix} 3 & 1 \\ 1 & 1 \end{bmatrix}.$$

2 Find the norms and condition numbers from the square roots of $\lambda_{\max}(A^{\mathrm{T}}A)$ and $\lambda_{\min}(A^{\mathrm{T}}A)$. Without positive definiteness in A, we go to $A^{\mathrm{T}}A$!

$$\begin{bmatrix} -2 & 0 \\ 0 & 2 \end{bmatrix} \quad \begin{bmatrix} 1 & 1 \\ 0 & 0 \end{bmatrix} \quad \begin{bmatrix} 1 & 1 \\ -1 & 1 \end{bmatrix}.$$

3 Explain these two inequalities from the definitions (3) of $||A||$ and $||B||$:

$$||AB\boldsymbol{x}|| \le ||A||\,||B\boldsymbol{x}|| \le ||A||\,||B||\,||\boldsymbol{x}||.$$

From the ratio of $||AB\boldsymbol{x}||$ to $||\boldsymbol{x}||$, deduce that $||AB|| \le ||A||\,||B||$. This is the key to using matrix norms. The norm of A^n is never larger than $||A||^n$.

4 Use $||AA^{-1}|| \le ||A||\,||A^{-1}||$ to prove that the condition number is at least 1.

5 Why is I the only symmetric positive definite matrix that has $\lambda_{\max} = \lambda_{\min} = 1$? Then the only other matrices with $||A|| = 1$ and $||A^{-1}|| = 1$ must have $A^{\mathrm{T}}A = I$. Those are _____ matrices: perfectly conditioned.

6 Orthogonal matrices have norm $||Q|| = 1$. If $A = QR$ show that $||A|| \le ||R||$ and also $||R|| \le ||A||$. Then $||A|| = ||Q||\,||R||$. Find an example of $A = LU$ with $||A|| < ||L||||U||$.

7 (a) Which famous inequality gives $||(A+B)\boldsymbol{x}|| \le ||A\boldsymbol{x}|| + ||B\boldsymbol{x}||$ for every $\boldsymbol{x}$?

(b) Why does the definition (3) of matrix norms lead to $||A+B|| \le ||A|| + ||B||$?

8 Show that if λ is any eigenvalue of A, then $|\lambda| \le ||A||$. Start from $A\boldsymbol{x} = \lambda\boldsymbol{x}$.

9 The "*spectral radius*" $\rho(A) = |\lambda_{\max}|$ is the largest absolute value of the eigenvalues. Show with 2 by 2 examples that $\rho(A+B) \le \rho(A) + \rho(B)$ and $\rho(AB) \le \rho(A)\rho(B)$ can both be *false*. The spectral radius is not acceptable as a norm.

10 (a) Explain why A and A^{-1} have the same condition number.

(b) Explain why A and A^{T} have the same norm, based on $\lambda(A^{\mathrm{T}}A)$ and $\lambda(AA^{\mathrm{T}})$.

11 Estimate the condition number of the ill-conditioned matrix $A = \begin{bmatrix} 1 & 1 \\ 1 & 1.0001 \end{bmatrix}$.

12 Why is the determinant of A no good as a norm? Why is it no good as a condition number?

13 (Suggested by C. Moler and C. Van Loan.) Compute $\boldsymbol{b} - A\boldsymbol{y}$ and $\boldsymbol{b} - A\boldsymbol{z}$ when

$$\boldsymbol{b} = \begin{bmatrix} .217 \\ .254 \end{bmatrix} \quad A = \begin{bmatrix} .780 & .563 \\ .913 & .659 \end{bmatrix} \quad \boldsymbol{y} = \begin{bmatrix} .341 \\ -.087 \end{bmatrix} \quad \boldsymbol{z} = \begin{bmatrix} .999 \\ -1.0 \end{bmatrix}.$$

Is $\boldsymbol{y}$ closer than $\boldsymbol{z}$ to solving $A\boldsymbol{x} = \boldsymbol{b}$? Answer in two ways: Compare the *residual* $\boldsymbol{b} - A\boldsymbol{y}$ to $\boldsymbol{b} - A\boldsymbol{z}$. Then compare $\boldsymbol{y}$ and $\boldsymbol{z}$ to the true $\boldsymbol{x} = (1, -1)$. Both answers can be right. Sometimes we want a small residual, sometimes a small $\Delta\boldsymbol{x}$.

14 (a) Compute the determinant of A in Problem 13. Compute A^{-1}.

(b) If possible compute $\|A\|$ and $\|A^{-1}\|$ and show that $c > 10^6$.

Problems 15–19 are about vector norms other than the usual $\|\boldsymbol{x}\| = \sqrt{\boldsymbol{x} \cdot \boldsymbol{x}}$.

15 The "$\boldsymbol{\ell}^1$ norm" and the "$\boldsymbol{\ell}^\infty$ norm" of $\boldsymbol{x} = (x_1, \ldots, x_n)$ are

$$\|\boldsymbol{x}\|_1 = |x_1| + \cdots + |x_n| \quad \text{and} \quad \|\boldsymbol{x}\|_\infty = \max_{1 \le i \le n} |x_i|.$$

Compute the norms $\|\boldsymbol{x}\|$ and $\|\boldsymbol{x}\|_1$ and $\|\boldsymbol{x}\|_\infty$ of these two vectors in $\mathbf{R}^5$:

$$\boldsymbol{x} = (1, 1, 1, 1, 1) \qquad \boldsymbol{x} = (.1, .7, .3, .4, .5).$$

16 Prove that $\|\boldsymbol{x}\|_\infty \le \|\boldsymbol{x}\| \le \|\boldsymbol{x}\|_1$. Show from the Schwarz inequality that the ratios $\|\boldsymbol{x}\|/\|\boldsymbol{x}\|_\infty$ and $\|\boldsymbol{x}\|_1/\|\boldsymbol{x}\|$ are never larger than $\sqrt{n}$. Which vector $(x_1, \ldots, x_n)$ gives ratios equal to $\sqrt{n}$?

17 All vector norms must satisfy the *triangle inequality*. Prove that

$$\|\boldsymbol{x} + \boldsymbol{y}\|_\infty \le \|\boldsymbol{x}\|_\infty + \|\boldsymbol{y}\|_\infty \qquad \text{and} \qquad \|\boldsymbol{x} + \boldsymbol{y}\|_1 \le \|\boldsymbol{x}\|_1 + \|\boldsymbol{y}\|_1.$$

18 Vector norms must also satisfy $\|c\boldsymbol{x}\| = |c|\,\|\boldsymbol{x}\|$. The norm must be positive except when $\boldsymbol{x} = \mathbf{0}$. Which of these are norms for vectors (x_1, x_2) in $\mathbf{R}^2$?

$$\|\boldsymbol{x}\|_A = |x_1| + 2|x_2| \qquad \|\boldsymbol{x}\|_B = \min(|x_1|, |x_2|)$$

$$\|\boldsymbol{x}\|_C = \|\boldsymbol{x}\| + \|\boldsymbol{x}\|_\infty \qquad \|\boldsymbol{x}\|_D = \|A\boldsymbol{x}\| \quad \text{(this answer depends on } A\text{)}.$$

Challenge Problems

19 Show that $\boldsymbol{x}^{\mathrm{T}}\boldsymbol{y} \le \|\boldsymbol{x}\|_1\,\|\boldsymbol{y}\|_\infty$ by choosing components $y_i = \pm 1$ to make $\boldsymbol{x}^{\mathrm{T}}\boldsymbol{y}$ as large as possible.

20 The eigenvalues of the $-1, 2, -1$ difference matrix K are $\lambda = 2 - 2\cos(j\pi/n{+}1)$. Estimate $\lambda_{\min}$ and $\lambda_{\max}$ and $c = \mathbf{cond}(K) = \lambda_{\max}/\lambda_{\min}$ as n increases: $c \approx Cn^2$ with what constant C?

Test this estimate with **eig**(K) and **cond**(K) for $n = 10, 100, 1000$.

11.3 Iterative Methods and Preconditioners

Up to now, our approach to $A\boldsymbol{x} = \boldsymbol{b}$ has been direct. We accepted $\boldsymbol{A}$ as it came. We attacked it by elimination with row exchanges. We now look at **iterative methods, which replace A by a simpler matrix S**. The difference $T = S - A$ is moved over to the right side of the equation. The problem becomes easier to solve, with S instead of A. But there is a price—*the simpler system has to be solved over and over.*

An iterative method is easy to invent. Just split A (carefully) into $S - T$.

$$\textbf{Rewrite } A\boldsymbol{x} = \boldsymbol{b} \qquad S\boldsymbol{x} = T\boldsymbol{x} + \boldsymbol{b}. \tag{1}$$

The novelty is to solve (1) iteratively. Each guess $\boldsymbol{x}_k$ leads to the next $\boldsymbol{x}_{k+1}$:

$$\textbf{Pure iteration} \qquad S\boldsymbol{x}_{k+1} = T\boldsymbol{x}_k + \boldsymbol{b}. \tag{2}$$

Start with any $\boldsymbol{x}_0$. Then solve $S\boldsymbol{x}_1 = T\boldsymbol{x}_0 + \boldsymbol{b}$. Continue to $S\boldsymbol{x}_2 = T\boldsymbol{x}_1 + \boldsymbol{b}$. A hundred iterations are very common—often more. Stop when (and if!) $\boldsymbol{x}_{k+1}$ is sufficiently close to $\boldsymbol{x}_k$—or when the **residual** $\boldsymbol{r}_k = \boldsymbol{b} - A\boldsymbol{x}_k$ is near zero. Our hope is to get near the true solution, more quickly than by elimination. When the $\boldsymbol{x}_k$ converge, their limit x_∞ does solve equation (1): $S\boldsymbol{x}_\infty = T\boldsymbol{x}_\infty + \boldsymbol{b}$ means $A\boldsymbol{x}_\infty = \boldsymbol{b}$.

The two goals of the splitting $A = S - T$ are ***speed per step*** and ***fast convergence***. The speed of each step depends on S and the speed of convergence depends on $S^{-1}T$:

1 Equation (2) should be easy to solve for $\boldsymbol{x}_{k+1}$. The "***preconditioner***" S could be the diagonal or triangular part of A. A fast way uses $S = L_0U_0$, where those factors have many zeros compared to the exact $A = LU$. This is "*incomplete LU*".

2 The difference $\boldsymbol{x} - \boldsymbol{x}_k$ (this is the error $\boldsymbol{e}_k$) should go quickly to zero. Subtracting equation (2) from (1) cancels $\boldsymbol{b}$, and it leaves the ***equation for the error*** $\boldsymbol{e}_k$:

$$\textbf{Error equation} \qquad S\boldsymbol{e}_{k+1} = T\boldsymbol{e}_k \ \text{ which means } \ \boldsymbol{e}_{k+1} = S^{-1}T\boldsymbol{e}_k. \tag{3}$$

At every step the error is multiplied by $S^{-1}T$. If $S^{-1}T$ is small, its powers go quickly to zero. But what is "small"?

The extreme splitting is $S = A$ and $T = 0$. Then the first step of the iteration is the original $A\boldsymbol{x} = \boldsymbol{b}$. Convergence is perfect and $S^{-1}T$ is zero. But the cost of that step is what we wanted to avoid. The choice of S is a battle between speed per step (a simple S) and fast convergence (S close to A). Here are some choices of S:

J $S =$ diagonal part of A (the iteration is called *Jacobi's method*)

GS $S =$ lower triangular part of A including the diagonal (*Gauss-Seidel method*)

ILU $S =$ approximate L times approximate U (*incomplete $L\,U$ method*).

Our first question is pure linear algebra: ***When do the x_k's converge to x?*** The answer uncovers the number $|\lambda|_{\max}$ that controls convergence. In examples of Jacobi and Gauss-Seidel, we will compute this "*spectral radius*" $|\lambda|_{\max}$. It is the largest eigenvalue of the ***iteration matrix*** $B = S^{-1}T$.

The Spectral Radius $\rho(B)$ Controls Convergence

Equation (3) is $\boldsymbol{e}_{k+1} = S^{-1}T\boldsymbol{e}_k$. Every iteration step multiplies the error by the same matrix $B = S^{-1}T$. The error after k steps is $\boldsymbol{e}_k = B^k\boldsymbol{e}_0$. ***The error approaches zero if the powers of $B = S^{-1}T$ approach zero.*** It is beautiful to see how the eigenvalues of B—the largest eigenvalue in particular—control the matrix powers B^k.

> The powers B^k approach zero if and only if every eigenvalue of B has $|\lambda| < 1$.
> ***The rate of convergence is controlled by the spectral radius of B:*** $\boldsymbol{\rho = \max|\lambda(B)|}$.

The test for convergence is $|\lambda|_{\max} < 1$. Real eigenvalues must lie between -1 and 1. Complex eigenvalues $\lambda = a + ib$ must have $|\lambda|^2 = a^2 + b^2 < 1$. The spectral radius "*rho*" is the largest distance from 0 to the eigenvalues of $B = S^{-1}T$. This is $\rho = |\lambda|_{\max}$.

To see why $|\lambda|_{\max} < 1$ is necessary, suppose the starting error $\boldsymbol{e}_0$ happens to be an eigenvector of B. After one step the error is $B\boldsymbol{e}_0 = \lambda\boldsymbol{e}_0$. After k steps the error is $B^k\boldsymbol{e}_0 = \lambda^k\boldsymbol{e}_0$. If we start with an eigenvector, we continue with that eigenvector—and the *factor* λ^k *only goes to zero when* $|\lambda| < 1$. This condition is required of every eigenvalue.

To see why $|\lambda|_{\max} < 1$ is sufficient for the error to approach zero, suppose $\boldsymbol{e}_0$ is a combination of eigenvectors:

$$\boldsymbol{e}_0 = c_1\boldsymbol{x}_1 + \cdots + c_n\boldsymbol{x}_n \quad \text{leads to} \quad \boldsymbol{e}_k = c_1(\lambda_1)^k\boldsymbol{x}_1 + \cdots + c_n(\lambda_n)^k\boldsymbol{x}_n. \tag{4}$$

This is the point of eigenvectors! When we multiply by B, each eigenvector $\boldsymbol{x}_i$ is multiplied by λ_i. If all $|\lambda_i| < 1$ then equation (4) ensures that $\boldsymbol{e}_k$ goes to zero.

Example 1 $B = \begin{bmatrix} .6 & .5 \\ .6 & .5 \end{bmatrix}$ has $\lambda_{\max} = 1.1$ $\quad B' = \begin{bmatrix} .6 & 1.1 \\ 0 & .5 \end{bmatrix}$ has $\lambda_{\max} = .6$

B^2 is 1.1 times B. Then B^3 is $(1.1)^2$ times B. The powers of B will blow up. Contrast with the powers of B'. The matrix $(B')^k$ has $(.6)^k$ and $(.5)^k$ on its diagonal. The off-diagonal entries also involve $\rho^k = (.6)^k$, which sets the speed of convergence.

Note When there are too few eigenvectors, equation (4) is not correct. We turn to the *Jordan form* when eigenvectors are missing and the matrix B can't be diagonalized:

$$\textbf{Jordan form } J \qquad B = MJM^{-1} \quad \text{and} \quad B^k = MJ^kM^{-1}. \tag{5}$$

Section 8.3 shows how J and J^k are made of "blocks" with one repeated eigenvalue:

$$\text{The powers of a 2 by 2 block in } J \text{ are} \quad \begin{bmatrix} \lambda & 1 \\ 0 & \lambda \end{bmatrix}^k = \begin{bmatrix} \lambda^k & k\lambda^{k-1} \\ 0 & \lambda^k \end{bmatrix}.$$

If $|\lambda| < 1$ then these powers approach zero. The extra factor k from a double eigenvalue is overwhelmed by the decreasing factor λ^{k-1}. This applies to every block:

Diagonalizable or not: Convergence $B^k \to 0$ ***and its speed depend on*** $\rho = |\lambda|_{\max} < 1$.

Jacobi versus Gauss-Seidel

We now solve a specific 2 by 2 problem by splitting A. Watch for that number $|\lambda|_{\max}$.

$$A\boldsymbol{x} = \boldsymbol{b} \qquad \begin{aligned} 2u - \ \ v &= \ \ 4 \\ -u + 2v &= -2 \end{aligned} \qquad \text{has the solution} \qquad \begin{bmatrix} u \\ v \end{bmatrix} = \begin{bmatrix} 2 \\ 0 \end{bmatrix}. \tag{6}$$

The first splitting is ***Jacobi's method***. Keep the *diagonal* of A on the left side (this is S). Move the off-diagonal part of A to the right side (this is T). Then iterate:

Jacobi iteration $\quad S\boldsymbol{x}_{k+1} = T\boldsymbol{x}_k + \boldsymbol{b} \quad$ $\begin{aligned} 2u_{k+1} &= v_k + 4 \\ 2v_{k+1} &= u_k - 2. \end{aligned}$

Start from $u_0 = v_0 = 0$. The first step finds $u_1 = 2$ and $v_1 = -1$. Keep going:

$$\begin{bmatrix} 0 \\ 0 \end{bmatrix} \quad \begin{bmatrix} 2 \\ -1 \end{bmatrix} \quad \begin{bmatrix} 3/2 \\ 0 \end{bmatrix} \quad \begin{bmatrix} 2 \\ -1/4 \end{bmatrix} \quad \begin{bmatrix} 15/8 \\ 0 \end{bmatrix} \quad \begin{bmatrix} 2 \\ -1/16 \end{bmatrix} \quad \text{approaches} \quad \begin{bmatrix} 2 \\ 0 \end{bmatrix}.$$

This shows convergence. At steps 1, 3, 5 the second component is -1, $-1/4$, $-1/16$. Those drop by 4 in each two steps. *The error equation is* $S\boldsymbol{e}_{k+1} = T\boldsymbol{e}_k$:

$$\textbf{Error equation} \qquad \begin{bmatrix} 2 & 0 \\ 0 & 2 \end{bmatrix} \boldsymbol{e}_{k+1} = \begin{bmatrix} 0 & 1 \\ 1 & 0 \end{bmatrix} \boldsymbol{e}_k \quad \text{or} \quad \boldsymbol{e}_{k+1} = \begin{bmatrix} 0 & \frac{1}{2} \\ \frac{1}{2} & 0 \end{bmatrix} \boldsymbol{e}_k. \tag{7}$$

That last matrix $S^{-1}T$ has eigenvalues $\frac{1}{2}$ and $-\frac{1}{2}$. So its spectral radius is $\rho(B) = \frac{1}{2}$:

$$B = S^{-1}T = \begin{bmatrix} 0 & \frac{1}{2} \\ \frac{1}{2} & 0 \end{bmatrix} \quad \text{has } |\lambda|_{\max} = \tfrac{1}{2} \quad \text{and} \quad \begin{bmatrix} 0 & \frac{1}{2} \\ \frac{1}{2} & 0 \end{bmatrix}^2 = \begin{bmatrix} \frac{1}{4} & 0 \\ 0 & \frac{1}{4} \end{bmatrix}.$$

Two steps multiply the error by $\frac{1}{4}$ exactly, in this special example. The important message is this: Jacobi's method works well when the main diagonal of A is large compared to the off-diagonal part. The diagonal part is S, the rest is $-T$. We want the diagonal to dominate.

The eigenvalue $\lambda = \frac{1}{2}$ is unusually small. Ten iterations reduce the error by $2^{10} = 1024$. More typical and more expensive is $|\lambda|_{\max} = .99$ or $.999$.

The ***Gauss-Seidel method*** keeps the whole lower triangular part of A as S:

$$\textbf{Gauss-Seidel} \qquad \begin{aligned} 2u_{k+1} \qquad\quad &= v_k + 4 \\ -u_{k+1} + 2v_{k+1} &= \quad\ -2 \end{aligned} \qquad \text{or} \qquad \begin{aligned} u_{k+1} &= \tfrac{1}{2}v_k \quad\ \ + 2 \\ v_{k+1} &= \tfrac{1}{2}u_{k+1} - 1. \end{aligned} \tag{8}$$

Notice the change. The new u_{k+1} from the first equation is used *immediately* in the second equation. With Jacobi, we saved the old u_k until the whole step was complete. With Gauss-Seidel, the new values enter right away and the old u_k is destroyed. This cuts the storage in half. It also speeds up the iteration (usually). And it costs no more than the Jacobi method.

Test the iteration starting from another start $u_0 = 0$ and $v_0 = -1$:

$$\begin{bmatrix} 0 \\ -1 \end{bmatrix} \quad \begin{bmatrix} 3/2 \\ -1/4 \end{bmatrix} \quad \begin{bmatrix} 15/8 \\ -1/16 \end{bmatrix} \quad \begin{bmatrix} 63/32 \\ -1/64 \end{bmatrix} \quad \text{approaches} \quad \begin{bmatrix} 2 \\ 0 \end{bmatrix}.$$

The errors in the first component are 2, 1/2, 1/8, 1/32. The errors in the second component are $-1, -1/4, -1/16, -1/32$. We divide by 4 in *one step* not two steps. ***Gauss-Seidel is twice as fast as Jacobi***. We have $\rho_{\mathbf{GS}} = (\rho_{\mathbf{J}})^2$ when A is positive definite tridiagonal:

$$S = \begin{bmatrix} 2 & 0 \\ -1 & 2 \end{bmatrix} \quad \text{and} \quad T = \begin{bmatrix} 0 & 1 \\ 0 & 0 \end{bmatrix} \quad \text{and} \quad S^{-1}T = \begin{bmatrix} \mathbf{0} & \frac{1}{2} \\ 0 & \mathbf{\frac{1}{4}} \end{bmatrix}.$$

The Gauss-Seidel eigenvalues are 0 and $\frac{1}{4}$. Compare with $\frac{1}{2}$ and $-\frac{1}{2}$ for Jacobi.

With a small push we can describe the ***successive overrelaxation method*** (**SOR**). The new idea is to introduce a parameter ω (omega) into the iteration. Then choose this number ω to make the spectral radius of $S^{-1}T$ as small as possible.

Rewrite $A\boldsymbol{x} = \boldsymbol{b}$ as $\omega A\boldsymbol{x} = \omega\boldsymbol{b}$. The matrix S in **SOR** has the diagonal of the original A, but below the diagonal we use ωA. On the right side T is $S - \omega A$:

$$\textbf{SOR} \qquad \begin{aligned} 2u_{k+1} \qquad\qquad &= (2-2\omega)u_k + \omega v_k + 4\omega \\ -\omega u_{k+1} + 2v_{k+1} &= \qquad\qquad (2-2\omega)v_k - 2\omega. \end{aligned} \tag{9}$$

This looks more complicated to us, but the computer goes as fast as ever. SOR is like Gauss-Seidel, with an adjustable number ω. The best ω makes it faster.

I will put on record the most valuable test matrix of order n. It is our favorite -1, 2, -1 tridiagonal matrix K. The diagonal is $2I$. Below and above are -1's. Our example had $n = 2$, which leads to $\cos\frac{\pi}{3} = \frac{1}{2}$ as the Jacobi eigenvalue found above. Notice especially that this $|\lambda|_{\max}$ is squared for Gauss-Seidel:

The splittings of the -1, 2, -1 matrix K of order n yield these eigenvalues of B:

Jacobi ($S = 0, 2, 0$ matrix): $\quad S^{-1}T$ has $|\lambda|_{\max} = \cos\dfrac{\pi}{n+1}$

Gauss-Seidel ($S = -1, 2, 0$ matrix): $\quad S^{-1}T$ has $|\lambda|_{\max} = \left(\cos\dfrac{\pi}{n+1}\right)^2$

SOR (with the best ω): $\quad S^{-1}T$ has $|\lambda|_{\max} = \left(\cos\dfrac{\pi}{n+1}\right)^2 \Big/ \left(1 + \sin\dfrac{\pi}{n+1}\right)^2$.

Let me be clear: For the -1, 2, -1 matrix you should not use any of these iterations! Elimination on a tridiagonal matrix is very fast (exact LU). Iterations are intended for a large sparse matrix that has nonzeros far from the central diagonal. Those create many more nonzeros in the exact L and U. This **fill-in** is why elimination becomes expensive.

We mention one more splitting. The idea of "***incomplete*** LU" is to set the small nonzeros in L and U *back to zero*. This leaves triangular matrices L_0 and U_0 which are again sparse. The splitting has $S = L_0U_0$ on the left side. Each step is quick:

Incomplete LU $$L_0U_0\boldsymbol{x}_{k+1} = (L_0U_0 - A)\boldsymbol{x}_k + \boldsymbol{b}.$$

On the right side we do sparse matrix-vector multiplications. Don't multiply L_0 times U_0, those are matrices. Multiply $\boldsymbol{x}_k$ by U_0 and then multiply that vector by L_0. On the left side we do forward and back substitutions. If L_0U_0 is close to A, then $|\lambda|_{\max}$ is small. A few iterations will give a close answer.

Multigrid and Conjugate Gradients

I cannot leave the impression that Jacobi and Gauss-Seidel are great methods. Generally the "low-frequency" part of the error decays very slowly, and many iterations are needed. Here are two important ideas that bring tremendous improvement. **Multigrid** can solve problems of size n in $O(n)$ steps. With a good preconditioner, **conjugate gradients** becomes one of the most popular and powerful algorithms in numerical linear algebra.

Multigrid Solve smaller problems with coarser grids. Each iteration will be cheaper and faster. Then interpolate between the coarse grid values to get a quick headstart on the full-size problem. Multigrid might go 4 levels down and back.

Conjugate gradients An ordinary iteration like $\boldsymbol{x}_{k+1} = \boldsymbol{x}_k - A\boldsymbol{x}_k + \boldsymbol{b}$ involves multiplication by A at each step. If A is sparse, this is not too expensive: $A\boldsymbol{x}_k$ is what we are willing to do. It adds one more basis vector to the growing "Krylov spaces" that contain our approximations. But $\boldsymbol{x}_{k+1}$ is **not the best combination** of $\boldsymbol{x}_0, A\boldsymbol{x}_0, \ldots, A^k\boldsymbol{x}_0$. The ordinary iterations are simple but far from optimal.

The conjugate gradient method chooses **the best combination** $\boldsymbol{x}_k$ at every step. The extra cost (beyond one multiplication by A) is not great. We will give the CG iteration, emphasizing that this method was created for a *symmetric positive definite matrix*. When A is not symmetric, one good choice is GMRES. When $A = A^{\mathrm{T}}$ is not positive definite, there is MINRES. A world of high-powered iterative methods has been created around the idea of making optimal choices of each successive $\boldsymbol{x}_k$.

My textbook *Computational Science and Engineering* describes multigrid and CG in much more detail. Among books on numerical linear algebra, Trefethen-Bau is deservedly popular (others are terrific too). Golub-Van Loan is a level up.

The Problem Set reproduces the five steps in each conjugate gradient cycle from $\boldsymbol{x}_{k-1}$ to $\boldsymbol{x}_k$. We compute that new approximation $\boldsymbol{x}_k$, the new residual $\boldsymbol{r}_k = \boldsymbol{b} - A\boldsymbol{x}_k$, and the new search direction $\boldsymbol{d}_k$ to look for the next $\boldsymbol{x}_{k+1}$.

I wrote those steps for the original matrix A. But a **preconditioner** S can make convergence much faster. Our original equation is $A\boldsymbol{x} = \boldsymbol{b}$. The preconditioned equation is $S^{-1}A\boldsymbol{x} = S^{-1}\boldsymbol{b}$. Small changes in the code give the *preconditioned conjugate gradient method*—the leading iterative method to solve positive definite systems.

The biggest competition is direct elimination, with the equations reordered to take maximum advantage of the zeros in A. It is not easy to outperform Gauss.

Iterative Methods for Eigenvalues

We move from $A\boldsymbol{x} = \boldsymbol{b}$ to $A\boldsymbol{x} = \lambda\boldsymbol{x}$. Iterations are an option for linear equations. They are a necessity for eigenvalue problems. The eigenvalues of an n by n matrix are the roots of an nth degree polynomial. The determinant of $A - \lambda I$ starts with $(-\lambda)^n$. This book must not leave the impression that eigenvalues should be computed that way! Working from $\det(A - \lambda I) = 0$ is a *very poor approach*—except when n is small.

For $n > 4$ there is no formula to solve $\det(A - \lambda I) = 0$. Worse than that, the λ's can be very unstable and sensitive. It is much better to work with A itself, gradually making it diagonal or triangular. (Then the eigenvalues appear on the diagonal.) Good computer codes are available in the LAPACK library—individual routines are free on **www.netlib.org/lapack**. This library combines the earlier LINPACK and EISPACK, with many improvements (to use matrix-matrix operations in the Level 3 BLAS). It is a collection of Fortran 77 programs for linear algebra on high-performance computers. For your computer and mine, a high quality matrix package is all we need. For supercomputers with parallel processing, move to ScaLAPACK and block elimination.

We will briefly discuss the power method and the QR method (chosen by LAPACK) for computing eigenvalues. It makes no sense to give full details of the codes.

1 Power methods and inverse power methods. Start with any vector $\boldsymbol{u}_0$. Multiply by A to find $\boldsymbol{u}_1$. Multiply by A again to find $\boldsymbol{u}_2$. If $\boldsymbol{u}_0$ is a combination of the eigenvectors, then A multiplies each eigenvector $\boldsymbol{x}_i$ by λ_i. After k steps we have $(\lambda_i)^k$:

$$\boldsymbol{u}_k = A^k\boldsymbol{u}_0 = c_1(\lambda_1)^k\boldsymbol{x}_1 + \cdots + c_n(\lambda_n)^k\boldsymbol{x}_n. \tag{10}$$

As the power method continues, ***the largest eigenvalue begins to dominate***. The vectors $\boldsymbol{u}_k$ point toward that dominant eigenvector $\boldsymbol{x}_1$. We saw this for Markov matrices:

$$A = \begin{bmatrix} .9 & .3 \\ .1 & .7 \end{bmatrix} \quad \text{has} \quad \lambda_{\max} = 1 \quad \text{with eigenvector} \quad \begin{bmatrix} .75 \\ .25 \end{bmatrix}.$$

Start with $\boldsymbol{u}_0$ and multiply at every step by A:

$$\boldsymbol{u}_0 = \begin{bmatrix} 1 \\ 0 \end{bmatrix}, \; \boldsymbol{u}_1 = \begin{bmatrix} .9 \\ .1 \end{bmatrix}, \; \boldsymbol{u}_2 = \begin{bmatrix} .84 \\ .16 \end{bmatrix} \quad \text{is approaching} \quad \boldsymbol{u}_\infty = \begin{bmatrix} .75 \\ .25 \end{bmatrix}.$$

The speed of convergence depends on the *ratio* of the second largest eigenvalue λ_2 to the largest λ_1. We don't want λ_1 to be small, we want λ_2/λ_1 to be small. Here $\lambda_2 = .6$ and $\lambda_1 = 1$, giving good speed. For large matrices it often happens that $|\lambda_2/\lambda_1|$ is very close to 1. Then the power method is too slow.

Is there a way to find the *smallest* eigenvalue—which is often the most important in applications? Yes, by the *inverse* power method: Multiply $\boldsymbol{u}_0$ by A^{-1} instead of A. Since we never want to compute A^{-1}, we actually solve $A\boldsymbol{u}_1 = \boldsymbol{u}_0$. By saving the LU factors, the next step $A\boldsymbol{u}_2 = \boldsymbol{u}_1$ is fast. Step k has $A\boldsymbol{u}_k = \boldsymbol{u}_{k-1}$:

Inverse power method $$\boldsymbol{u}_k = A^{-k}\boldsymbol{u}_0 = \frac{c_1\boldsymbol{x}_1}{(\lambda_1)^k} + \cdots + \frac{c_n\boldsymbol{x}_n}{(\lambda_n)^k}. \tag{11}$$

Now the *smallest* eigenvalue $\lambda_{\min}$ is in control. When it is very small, the factor $1/\lambda_{\min}^k$ is large. For high speed, we make $\boldsymbol{\lambda}_{\min}$ even smaller by shifting the matrix to $A - \lambda^* I$.

That shift doesn't change the eigenvectors. (λ^* might come from the diagonal of A, even better is a Rayleigh quotient $x^{\mathrm{T}}Ax/x^{\mathrm{T}}x$). If λ^* is close to $\lambda_{\min}$ then $(A - \lambda^* I)^{-1}$ has the very large eigenvalue $(\lambda_{\min} - \lambda^*)^{-1}$. Each ***shifted inverse power step*** multiplies the eigenvector by this big number, and that eigenvector quickly dominates.

2 The QR Method This is a major achievement in numerical linear algebra. Sixty years ago, eigenvalue computations were slow and inaccurate. We didn't even realize that solving $\det(A - \lambda I) = 0$ was a terrible method. Jacobi had suggested earlier that A should gradually be made triangular—then the eigenvalues appear automatically on the diagonal. He used 2 by 2 rotations to produce off-diagonal zeros. (Unfortunately the previous zeros can become nonzero again. But Jacobi's method made a partial comeback with parallel computers.) The ***QR method*** is now a leader in eigenvalue computations.

The basic step is to factor A, whose eigenvalues we want, into QR. Remember from Gram-Schmidt (Section 4.4) that Q has orthonormal columns and R is triangular. For eigenvalues the key idea is: ***Reverse*** Q ***and*** R. The new matrix (same λ's) is $A_1 = RQ$. *The eigenvalues are not changed* in RQ *because* $A = QR$ *is similar to* $A_1 = Q^{-1}AQ$:

$$\boldsymbol{A_1 = RQ} \textbf{ has the same } \boldsymbol{\lambda} \qquad QR\boldsymbol{x} = \lambda\boldsymbol{x} \quad \text{gives} \quad RQ(Q^{-1}\boldsymbol{x}) = \lambda(Q^{-1}\boldsymbol{x}). \tag{12}$$

This process continues. Factor the new matrix A_1 into Q_1R_1. Then reverse the factors to R_1Q_1. This is the similar matrix A_2 and again no change in the eigenvalues. Amazingly, those eigenvalues begin to show up on the diagonal. Soon the last entry of A_4 holds an accurate eigenvalue. In that case we remove the last row and column and continue with a smaller matrix to find the next eigenvalue.

Two extra ideas make this method a success. One is to shift the matrix by a multiple of I, before factoring into QR. Then RQ is shifted back to give A_{k+1}:

$$\text{Factor } A_k - c_k I \text{ into } Q_k R_k. \text{ The next matrix is } A_{k+1} = R_k Q_k + c_k I.$$

A_{k+1} has the same eigenvalues as A_k, and the same as the original $A_0 = A$. A good shift chooses c near an (unknown) eigenvalue. That eigenvalue appears more accurately on the diagonal of A_{k+1}—which tells us a better c for the next step to A_{k+2}.

The second idea is to obtain off-diagonal zeros before the QR method starts. An elimination step E will do it, or a Givens rotation, but don't forget E^{-1} (or λ will change):

$$EAE^{-1} = \begin{bmatrix} 1 & & \\ & 1 & \\ & -1 & 1 \end{bmatrix} \begin{bmatrix} 1 & 2 & 3 \\ \mathbf{1} & 4 & 5 \\ \mathbf{1} & \mathbf{6} & 7 \end{bmatrix} \begin{bmatrix} 1 & & \\ & 1 & \\ & 1 & 1 \end{bmatrix} = \begin{bmatrix} 1 & 5 & 3 \\ \mathbf{1} & 9 & 5 \\ \mathbf{0} & \mathbf{4} & 2 \end{bmatrix}. \textbf{ Same } \boldsymbol{\lambda}\textbf{'s.}$$

We must leave those nonzeros 1 and 4 along *one subdiagonal*. More E's could remove them, but E^{-1} would fill them in again. This is a "***Hessenberg matrix***" (one nonzero subdiagonal). The zeros in the lower left corner will stay zero through the QR method. The operation count for each QR factorization drops from $O(n^3)$ to $O(n^2)$.

Golub and Van Loan give this example of one shifted QR step on a Hessenberg matrix. The shift is $7I$, taking 7 from all diagonal entries of A (then shifting back for A_1):

$$A = \begin{bmatrix} 1 & 2 & 3 \\ 4 & 5 & 6 \\ 0 & .001 & 7 \end{bmatrix} \quad \text{leads to} \quad A_1 = \begin{bmatrix} -.54 & 1.69 & 0.835 \\ .31 & 6.53 & -6.656 \\ 0 & .00002 & 7.012 \end{bmatrix}.$$

Factoring $A - 7I$ into QR produced $A_1 = RQ + 7I$. Notice the very small number .00002. The diagonal entry 7.012 is almost an exact eigenvalue of A_1, and therefore of A. Another QR step on A_1 with shift by $7.012I$ would give terrific accuracy.

For a few eigenvalues of a large sparse matrix I would look to **ARPACK**. Problems 25–27 describe the Arnoldi iteration that orthogonalizes the basis—each step has only three terms when A is symmetric. The matrix becomes *tridiagonal*: a wonderful start for computing eigenvalues.

Problem Set 11.3

Problems 1–12 are about iterative methods for $Ax = b$.

1 Change $A\boldsymbol{x} = \boldsymbol{b}$ to $\boldsymbol{x} = (I - A)\boldsymbol{x} + \boldsymbol{b}$. What are S and T for this splitting? What matrix $S^{-1}T$ controls the convergence of $\boldsymbol{x}_{k+1} = (I - A)\boldsymbol{x}_k + \boldsymbol{b}$?

2 If λ is an eigenvalue of A, then _____ is an eigenvalue of $B = I - A$. The real eigenvalues of B have absolute value less than 1 if the real eigenvalues of A lie between _____ and _____.

3 Show why the iteration $\boldsymbol{x}_{k+1} = (I - A)\boldsymbol{x}_k + \boldsymbol{b}$ does not converge for $A = \begin{bmatrix} 2 & -1 \\ -1 & 2 \end{bmatrix}$.

4 Why is the norm of B^k never larger than $\|B\|^k$? Then $\|B\| < 1$ guarantees that the powers B^k approach zero (convergence). No surprise since $|\lambda|_{\max}$ is below $\|B\|$.

5 If A is singular then all splittings $A = S - T$ must fail. From $A\boldsymbol{x} = \boldsymbol{0}$ show that $S^{-1}T\boldsymbol{x} = \boldsymbol{x}$. So this matrix $B = S^{-1}T$ has $\lambda = 1$ and fails.

6 Change the 2's to 3's and find the eigenvalues of $S^{-1}T$ for Jacobi's method:

$$S\boldsymbol{x}_{k+1} = T\boldsymbol{x}_k + \boldsymbol{b} \quad \text{is} \quad \begin{bmatrix} 3 & 0 \\ 0 & 3 \end{bmatrix} \boldsymbol{x}_{k+1} = \begin{bmatrix} 0 & 1 \\ 1 & 0 \end{bmatrix} \boldsymbol{x}_k + \boldsymbol{b}.$$

7 Find the eigenvalues of $S^{-1}T$ for the Gauss-Seidel method applied to Problem 6:

$$\begin{bmatrix} 3 & 0 \\ -1 & 3 \end{bmatrix} \boldsymbol{x}_{k+1} = \begin{bmatrix} 0 & 1 \\ 0 & 0 \end{bmatrix} \boldsymbol{x}_k + \boldsymbol{b}.$$

Does $|\lambda|_{\max}$ for Gauss-Seidel equal $|\lambda|^2_{\max}$ for Jacobi?

8 For any 2 by 2 matrix $\left[\begin{smallmatrix} a & b \\ c & d \end{smallmatrix}\right]$ show that $|\lambda|_{\max}$ equals $|bc/ad|$ for Gauss-Seidel and $|bc/ad|^{1/2}$ for Jacobi. We need $ad \neq 0$ for the matrix S to be invertible.

9 Write a computer code (MATLAB or other) for the Gauss-Seidel method. You can define S and T from A, or set up the iteration loop directly from the entries a_{ij}. Test it on the $-1, 2, -1$ matrices A of order 10, 20, 50 with $\boldsymbol{b} = (1, 0, \ldots, 0)$.

10 The Gauss-Seidel iteration at component i uses earlier parts of $\boldsymbol{x}^{\text{new}}$:

$$\textbf{Gauss-Seidel} \qquad x_i^{\text{new}} = x_i^{\text{old}} + \frac{1}{a_{ii}}\Big(b_i - \sum_{j=1}^{i-1} a_{ij} x_j^{\text{new}} - \sum_{j=i}^{n} a_{ij} x_j^{\text{old}}\Big).$$

If every $x_i^{\text{new}} = x_i^{\text{old}}$ how does this show that the solution $\boldsymbol{x}$ is correct? How does the formula change for Jacobi's method? For **SOR** insert ω outside the parentheses.

11 Divide equation (10) by λ_1^k and explain why $|\lambda_2/\lambda_1|$ controls the convergence of the power method. Construct a matrix A for which this method *does not converge*.

12 The Markov matrix $A = \left[\begin{smallmatrix} .9 & .3 \\ .1 & .7 \end{smallmatrix}\right]$ has $\lambda = 1$ and $.6$, and the power method $\boldsymbol{u}_k = A^k \boldsymbol{u}_0$ converges to $\left[\begin{smallmatrix} .75 \\ .25 \end{smallmatrix}\right]$. Find the eigenvectors of A^{-1}. What does the inverse power method $\boldsymbol{u}_{-k} = A^{-k}\boldsymbol{u}_0$ converge to (after you multiply by $.6^k$)?

13 The tridiagonal matrix of size $n-1$ with diagonals $-1, 2, -1$ has eigenvalues $\lambda_j = 2 - 2\cos(j\pi/n)$. Why are the smallest eigenvalues approximately $(j\pi/n)^2$? The inverse power method converges at the speed $\lambda_1/\lambda_2 \approx 1/4$.

14 For $A = \left[\begin{smallmatrix} 2 & -1 \\ -1 & 2 \end{smallmatrix}\right]$ apply the power method $\boldsymbol{u}_{k+1} = A\boldsymbol{u}_k$ three times starting with $\boldsymbol{u}_0 = \left[\begin{smallmatrix} 1 \\ 0 \end{smallmatrix}\right]$. What eigenvector is the power method converging to?

15 For $A = -1,\ 2, -1$ matrix, apply the *inverse* power method $\boldsymbol{u}_{k+1} = A^{-1}\boldsymbol{u}_k$ three times with the same $\boldsymbol{u}_0$. What eigenvector are the $\boldsymbol{u}_k$'s approaching?

16 In the QR method for eigenvalues when A is shifted to make $A_{22} = 0$, show that the 2, 1 entry drops from $\sin\boldsymbol{\theta}$ in $A = QR$ to $-\sin^3\boldsymbol{\theta}$ in RQ. (*Compute R and RQ.*) This "cubic convergence" makes the method a success:

$$A = \begin{bmatrix} \cos\theta & \sin\theta \\ \sin\theta & 0 \end{bmatrix} = QR = \begin{bmatrix} \cos\theta & -\sin\theta \\ \sin\theta & \cos\theta \end{bmatrix} \begin{bmatrix} 1 & ? \\ 0 & ? \end{bmatrix}.$$

17 If A is an orthogonal matrix, its QR factorization has $Q =$ _____ and $R =$ _____. Therefore $RQ =$ _____. These are among the rare examples when the QR method goes nowhere.

18 The shifted QR method factors $A - cI$ into QR. Show that the next matrix $A_1 = RQ + cI$ equals $Q^{-1}AQ$. Therefore A_1 has the _____ eigenvalues as A (but A_1 is closer to triangular).

19 When $A = A^{\mathrm{T}}$, the "*Lanczos method*" finds a's and b's and orthonormal $\boldsymbol{q}$'s so that $A\boldsymbol{q}_j = b_{j-1}\boldsymbol{q}_{j-1} + a_j\boldsymbol{q}_j + b_j\boldsymbol{q}_{j+1}$ (with $\boldsymbol{q}_0 = \boldsymbol{0}$). Multiply by $\boldsymbol{q}_j^{\mathrm{T}}$ to find a formula for a_j. The equation says that $AQ = QT$ where T is a tridiagonal matrix.

20 The equation in Problem 19 develops from this loop with $b_0 = 1$ and $\boldsymbol{r}_0 = $ any $\boldsymbol{q}_1$:

$$\boldsymbol{q}_{j+1} = \boldsymbol{r}_j/b_j;\ j = j+1;\ a_j = \boldsymbol{q}_j^{\mathrm{T}} A\boldsymbol{q}_j;\ \boldsymbol{r}_j = A\boldsymbol{q}_j - b_{j-1}\boldsymbol{q}_{j-1} - a_j\boldsymbol{q}_j;\ b_j = \|\boldsymbol{r}_j\|.$$

Write a code and test it on the -1, 2, -1 matrix A. $Q^{\mathrm{T}}Q$ should be I.

21 Suppose A is *tridiagonal and symmetric in the QR method.* From $A_1 = Q^{-1}AQ$ show that A_1 is symmetric. Write $A_1 = RAR^{-1}$ to show that A_1 is also tridiagonal. (If the lower part of A_1 is proved tridiagonal then by symmetry the upper part is too.) Symmetric tridiagonal matrices are the best way to start in the QR method.

Problems 22–25 present two fundamental iterations. Each step involves Aq or Ad.

The key point for large matrices is that matrix-vector multiplication is much faster than matrix-matrix multiplication. A crucial construction starts with a vector $\boldsymbol{b}$. Repeated multiplication will produce $A\boldsymbol{b}$, $A^2\boldsymbol{b}, \ldots$ but those vectors are far from orthogonal. The "**Arnoldi iteration**" creates an orthonormal basis $\boldsymbol{q}_1, \boldsymbol{q}_2, \ldots$ for the same space by the Gram-Schmidt idea: *orthogonalize each new $A\boldsymbol{q}_n$ against the previous $\boldsymbol{q}_1, \ldots, \boldsymbol{q}_{n-1}$*. The "Krylov space" spanned by $\boldsymbol{b}, A\boldsymbol{b}, \ldots, A^{n-1}\boldsymbol{b}$ then has a much better basis $\boldsymbol{q}_1, \ldots, \boldsymbol{q}_n$.

Here in pseudocode are two of the most important algorithms in numerical linear algebra: Arnoldi gives a good basis and CG gives a good approximation to $\boldsymbol{x} = A^{-1}\boldsymbol{b}$.

Arnoldi Iteration	**Conjugate Gradient Iteration for Positive Definite A**	
$\boldsymbol{q}_1 = \boldsymbol{b}/\|\boldsymbol{b}\|$	$\boldsymbol{x}_0 = 0, \boldsymbol{r}_0 = \boldsymbol{b}, \boldsymbol{d}_0 = \boldsymbol{r}_0$	
for $n = 1$ **to** $N-1$	**for** $n = 1$ **to** N	
$\boldsymbol{v} = A\boldsymbol{q}_n$	$\alpha_n = (\boldsymbol{r}_{n-1}^{\mathrm{T}}\boldsymbol{r}_{n-1})/(\boldsymbol{d}_{n-1}^{\mathrm{T}}\boldsymbol{A}\boldsymbol{d}_{n-1})$	step length $\boldsymbol{x}_{n-1}$ to $\boldsymbol{x}_n$
for $j = 1$ **to** n	$\boldsymbol{x}_n = \boldsymbol{x}_{n-1} + \alpha_n\boldsymbol{d}_{n-1}$	approximate solution
$h_{jn} = \boldsymbol{q}_j^{\mathrm{T}}\boldsymbol{v}$	$\boldsymbol{r}_n = \boldsymbol{r}_{n-1} - \alpha_n A\boldsymbol{d}_{n-1}$	new residual $\boldsymbol{b} - A\boldsymbol{x}_n$
$\boldsymbol{v} = \boldsymbol{v} - h_{jn}\boldsymbol{q}_j$	$\beta_n = (\boldsymbol{r}_n^{\mathrm{T}}\boldsymbol{r}_n)/(\boldsymbol{r}_{n-1}^{\mathrm{T}}\boldsymbol{r}_{n-1})$	improvement this step
$h_{n+1,n} = \|\boldsymbol{v}\|$	$\boldsymbol{d}_n = \boldsymbol{r}_n + \beta_n\boldsymbol{d}_{n-1}$	next search direction
$\boldsymbol{q}_{n+1} = \boldsymbol{v}/h_{n+1,n}$	% *Notice: only* 1 *matrix-vector multiplication $A\boldsymbol{q}$ and $A\boldsymbol{d}$*	

For conjugate gradients, the residuals $\boldsymbol{r}_n$ are orthogonal and the search directions are A-orthogonal: all $\boldsymbol{d}_j^{\mathrm{T}} A\boldsymbol{d}_k = 0$. The iteration solves $A\boldsymbol{x} = \boldsymbol{b}$ by minimizing the error $\boldsymbol{e}^{\mathrm{T}}A\boldsymbol{e}$ over all vectors in the *Krylov space* = span of $\boldsymbol{b}, A\boldsymbol{b}, \ldots, A^{n-1}\boldsymbol{b}$. It is a fantastic algorithm.

22 For the diagonal matrix $A = \mathtt{diag}([1\ \ 2\ \ 3\ \ 4])$ and the vector $\boldsymbol{b} = (1,1,1,1)$, go through one Arnoldi step to find the orthonormal vectors $\boldsymbol{q}_1$ and $\boldsymbol{q}_2$.

23 Arnoldi's method is finding Q so that $AQ = QH$ (column by column):

$$AQ = \begin{bmatrix} A\boldsymbol{q}_1 & \cdots & A\boldsymbol{q}_N \end{bmatrix} = \begin{bmatrix} \boldsymbol{q}_1 & \cdots & \boldsymbol{q}_N \end{bmatrix} \begin{bmatrix} h_{11} & h_{12} & \cdot & h_{1N} \\ h_{21} & h_{22} & \cdot & h_{2N} \\ 0 & h_{32} & \cdot & \cdot \\ 0 & 0 & \cdot & h_{NN} \end{bmatrix} = QH$$

H is a "Hessenberg matrix" with one nonzero subdiagonal. Here is the crucial fact when A is symmetric: ***The Hessenberg matrix*** $H = Q^{-1}AQ = Q^{\mathrm{T}}AQ$ ***is symmetric and therefore it is tridiagonal***. Explain that sentence.

24 This tridiagonal H (when A is symmetric) gives the **Lanczos iteration**:

$$\textbf{Three terms only} \qquad \boldsymbol{q}_{j+1} = (A\boldsymbol{q}_j - h_{j,j}\boldsymbol{q}_j - h_{j-1,j}\boldsymbol{q}_{j-1})/h_{j+1,j}$$

From $H = Q^{-1}AQ$, why are the eigenvalues of H the same as the eigenvalues of A? For large matrices, the "Lanczos method" computes the leading eigenvalues by stopping at a smaller tridiagonal matrix H_k. The QR method in the text is applied to compute the eigenvalues of H_k.

25 Apply the conjugate gradient method to solve $A\boldsymbol{x} = \boldsymbol{b} = \textbf{ones}(100, 1)$, where A is the $-1, 2, -1$ second difference matrix $A = \textbf{toeplitz}([2 \;\; -1 \;\; \textbf{zeros}(1, 98)])$. Graph $\boldsymbol{x}_{10}$ and $\boldsymbol{x}_{20}$ from CG, along with the exact solution $\boldsymbol{x}$. (Its 100 components are $x_i = (ih - i^2h^2)/2$ with $h = 1/101$. "**plot**$(i, x(i))$" should produce a parabola.)

26 For unsymmetric matrices, the spectral radius $\rho = \max|\lambda_i|$ is not a norm. But still $\|A^n\|$ grows or decays like ρ^n for large n. Compare those numbers for $A = [1 \;\; 1; \;\; 0 \;\; 1.1]$ using the command **norm**.

$\boldsymbol{A^n \to 0}$ **if and only if** $\boldsymbol{\rho < 1}$. When $A = S^{-1}T$, this is the key to convergence.

Chapter 12

Linear Algebra in Probability & Statistics

12.1 Mean, Variance, and Probability

We are starting with the three fundamental words of this chapter: *mean, variance, and probability*. Let me give a rough explanation of their meaning before I write any formulas:

The **mean** is the *average value* or expected value

The **variance** $\boldsymbol{\sigma^2}$ measures the average *squared distance* from the mean m

The **probabilities** of n different outcomes are positive numbers $p_1, \ldots, p_n$ adding to 1.

Certainly the mean is easy to understand. We will start there. But right away we have two different situations that you have to keep straight. On the one hand, we may have the results (*sample values*) from a completed trial. On the other hand, we may have the expected results (*expected values*) from future trials. Let me give examples:

Sample values Five random freshmen have ages $\mathbf{18, 17, 18, 19, 17}$

Sample mean $\frac{\mathbf{1}}{\mathbf{5}}(18 + 17 + 18 + 19 + 17) = \mathbf{17.8}$

Probabilities The ages in a freshmen class are $17\,(\mathbf{20\%}), 18\,(\mathbf{50\%}), 19\,(\mathbf{30\%})$

A random freshman has **expected age** $\mathbf{E}\,[\mathrm{x}] = (0.2)\,17 + (0.5)\,18 + (0.3)\,19 = \mathbf{18.1}$

Both numbers 17.8 and 18.1 are correct averages. The sample mean starts with N samples $x_1, \ldots, x_N$ from a completed trial. Their mean is the *average* of the N observed samples:

$$\textbf{Sample mean} \qquad m = \mu = \frac{\mathbf{1}}{\boldsymbol{N}}(x_1 + x_2 + \cdots + x_N) \tag{1}$$

The **expected value of x** starts with the probabilities $p_1, \ldots, p_n$ of the ages $x_1, \ldots, x_n$:

$$\textbf{Expected value} \quad m = \mathrm{E}[x] = p_1x_1 + p_2x_2 + \cdots + p_nx_n. \tag{2}$$

This is $\boldsymbol{p} \cdot \boldsymbol{x}$. Notice that $m = \mathrm{E}[x]$ tells us what to expect, $m = \mu$ tells us what we got.

By taking many samples (large N), the sample results will come close to the probabilities. The "Law of Large Numbers" says that with probability 1, the sample mean will converge to its expected value $\mathrm{E}[x]$ as the sample size N increases. A fair coin has probability $p_0 = \frac{1}{2}$ of tails and $p_1 = \frac{1}{2}$ of heads. Then $\mathrm{E}[x] = \left(\frac{1}{2}\right) 0 + \frac{1}{2}(1)$. The fraction of heads in N flips of the coin is the sample mean, expected to approach $\mathrm{E}[x] = \frac{1}{2}$.

This does *not* mean that if we have seen more tails than heads, the next sample is likely to be heads. The odds remain 50-50. The first 100 or 1000 flips do affect the sample mean. *But* 1000 *flips will not affect its limit*—because you are dividing by $N \to \infty$.

Variance (around the mean)

The **variance σ^2** measures expected distance (squared) from the expected mean $\mathrm{E}[x]$. The **sample variance S^2** measures actual distance (squared) from the sample mean. The square root is the **standard deviation σ or S**. After an exam, I email μ and S to the class. I don't know the expected mean and variance because I don't know the probabilities p_1 to p_{100} for each score. (After teaching for 50 years, I still have no idea what to expect.)

The deviation is always deviation *from the mean*—sample or expected. We are looking for the size of the "spread" around the mean value $x = m$. Start with N samples.

$$\textbf{Sample variance} \quad S^2 = \frac{1}{N-1}\left[(x_1 - m)^2 + \cdots + (x_N - m)^2\right] \tag{3}$$

The sample ages $x = 18, 17, 18, 19, 17$ have mean $m = 17.8$. That sample has variance 0.7 :

$$S^2 = \frac{1}{4}\left[(.2)^2 + (-.8)^2 + (.2)^2 + (1.2)^2 + (-.8)^2\right] = \frac{1}{4}(2.8) = \mathbf{0.7}$$

The minus signs disappear when we compute squares. Please notice ! Statisticians divide by $N - 1 = 4$ (and not $N = 5$) so that S^2 is an unbiased estimate of σ^2. One degree of freedom is already accounted for in the sample mean.

An important identity comes from splitting each $(x - m)^2$ into $x^2 - 2mx + m^2$:

$$\begin{aligned} \text{sum of } (x_i - m)^2 &= (\text{sum of } x_i^2) - 2m(\text{sum of } x_i) + (\text{sum of } m^2) \\ &= (\text{sum of } x_i^2) - 2m(Nm) + Nm^2 \\ \textbf{sum of } (\boldsymbol{x_i} - \boldsymbol{m})^2 &= (\textbf{sum of } \boldsymbol{x_i^2}) - \boldsymbol{Nm^2}. \end{aligned} \tag{4}$$

This is an equivalent way to find $(x_1 - m)^2 + \cdots + (x_N - m^2)$ by adding $x_1^2 + \cdots + x_N^2$.

Now start with probabilities p_i (never negative !) instead of samples. We find expected values instead of sample values. The variance σ^2 is the crucial number in statistics.

$$\textbf{Variance} \quad \boldsymbol{\sigma^2} = \mathbf{E}\left[(\boldsymbol{x}-\boldsymbol{m})^2\right] = \boldsymbol{p_1(x_1-m)^2} + \cdots + \boldsymbol{p_n(x_n-m)^2}. \qquad (5)$$

We are squaring the distance from the expected value $m = \mathrm{E}[x]$. We don't have samples, only expectations. We know probabilities but we don't know experimental outcomes.

Example 1 Find the variance σ^2 of the ages of college freshmen.

Solution The probabilities of ages $x_i = 17, 18, 19$ were $p_i = 0.2$ and 0.5 and 0.3. The expected value was $\boldsymbol{m} = \sum \boldsymbol{p_i x_i} = \mathbf{18.1}$. The variance uses those same probabilities :

$$\begin{aligned}\boldsymbol{\sigma^2} &= (0.2)(\mathbf{17}-\mathbf{18.1})^2 + (0.5)(\mathbf{18}-\mathbf{18.1})^2 + (0.3)(\mathbf{19}-\mathbf{18.1})^2 \\ &= (0.2)(\mathbf{1.21}) + (0.5)(\mathbf{0.01}) + (0.3)(\mathbf{0.81}) = \mathbf{0.49}.\end{aligned}$$

The **standard deviation** is the square root $\boldsymbol{\sigma} = \mathbf{0.7}$.
This measures the spread of $17, 18, 19$ around $\mathrm{E}[x]$, weighted by probabilities $.2, .5, .3$.

Continuous Probability Distributions

Up to now we have allowed for n possible outcomes $x_1, \ldots, x_n$. With ages 17, 18, 19, we only had $n = 3$. If we measure age in days instead of years, there will be a thousand possible ages (too many). Better to allow *every number between* 17 *and* 20—a continuum of possible ages. Then the probabilities p_1, p_2, p_3 for ages x_1, x_2, x_3 have to move to a **probability distribution** $\boldsymbol{p(x)}$ for a whole continuous range of ages $17 \leq x \leq 20$.

The best way to explain probability distributions is to give you two examples. They will be the **uniform distribution** and the **normal distribution**. The first (uniform) is easy. The normal distribution is all-important.

Uniform distribution Suppose ages are uniformly distributed between 17.0 and 20.0. All ages between those numbers are "equally likely". Of course any one exact age has no chance at all. There is zero probability that you will hit the exact number $x = 17.1$ or $x = 17 + \sqrt{2}$. What you can truthfully provide (assuming our uniform distribution) is **the chance $\boldsymbol{F(x)}$ that a random freshman has age less than $\boldsymbol{x}$** :

The chance of age less than $x = 17$ is $F(17) = \mathbf{0}$	$x \leq 17$ won't happen
The chance of age less than $x = 20$ is $F(20) = \mathbf{1}$	$x \leq 20$ will happen
The chance of age less than x is $\boldsymbol{F(x)} = \frac{1}{3}(\boldsymbol{x} - \mathbf{17})$	***F* goes from 0 to 1**

That formula $F(x) = \frac{1}{3}(x - 17)$ gives $F = 0$ at $x = 17$; then $x \leq 17$ won't happen. It gives $F(x) = 1$ at $x = 20$; then $x \leq 20$ is sure. Between 17 and 20, the graph of the **cumulative distribution $\boldsymbol{F(x)}$** increases linearly for this uniform model.

Let me draw the graphs of $F(x)$ and its derivative $p(x) =$ "probability density function".

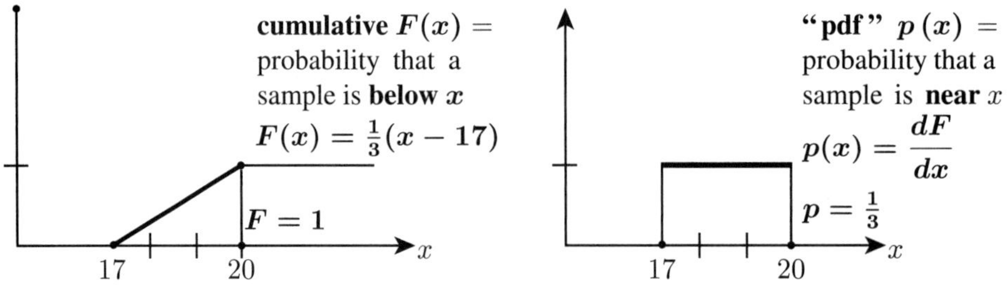

Figure 12.1: $F(x)$ is the cumulative distribution and its derivative $p(x) = dF/dx$ is the **probability density function (pdf)**. For this uniform distribution, $p(x)$ is constant between 17 and 20. The total area under the graph of $p(x)$ is the total probability $F = 1$.

You could say that $p(x)\,dx$ is the probability of a sample falling in between x and $x + dx$. This is "infinitesimally true": $p(x)\,dx$ is $F(x + dx) - F(x)$. Here is the full truth:

$$\boldsymbol{F = \textbf{integral of } p \qquad \textbf{Probability of } a \le x \le b = \int_a^b p(x)\,dx = F(b) - F(a)} \tag{6}$$

$F(b)$ is the probability of $x \le b$. I subtract $F(a)$ to keep $x \ge a$. That leaves $a \le x \le b$.

Mean and Variance of $p(x)$

What are the mean m and variance σ^2 for a probability distribution? Previously we added $\boldsymbol{p_i x_i}$ to get the mean (expected value). With a continuous distribution we **integrate** $\boldsymbol{x p(x)}$:

$$\textbf{Mean} \qquad m = \mathrm{E}[x] = \int x\,p(x)\,dx = \int_{x=17}^{20} (x)\left(\frac{1}{3}\right) dx = 18.5$$

For this uniform distribution, the mean m is halfway between 17 and 20. Then the probability of a random value x below this halfway point $m = 18.5$ is $F(m) = \frac{1}{2}$.

In MATLAB, $x =$ rand(1) chooses a random number uniformly between 0 and 1. Then the expected mean is $m = \frac{1}{2}$. The interval from 0 to x has probability $F(x) = x$. The interval below the mean m always has probability $F(m) = \frac{1}{2}$.

The variance is the average squared distance to the mean. With N outcomes, σ^2 is the sum of $p_i(x_i - m)^2$. For a continuous random variable x, the sum changes to an **integral**.

$$\textbf{Variance} \qquad \sigma^2 = \mathrm{E}\left[(x - m)^2\right] = \int p(x)\,(x - m)^2\,dx \tag{7}$$

When ages are uniform between $17 \le x \le 20$, the integral can shift to $0 \le x \le 3$:

$$\sigma^2 = \int_{17}^{20} \frac{1}{3}(x-18.5)^2\,dx = \int_0^3 \frac{1}{3}(x-1.5)^2\,dx = \frac{1}{9}(x-1.5)^3\Big]_{x=0}^{x=3} = \frac{2}{9}(1.5)^3 = \frac{3}{4}.$$

That is a typical example, and here is the complete picture for a uniform $p(x)$, 0 to a.

Uniform distribution for $0 \le x \le a$ **Density** $p(x) = \dfrac{1}{a}$ **Cumulative** $F(x) = \dfrac{x}{a}$

$$\textbf{Mean } m = \frac{a}{2} \text{ halfway} \qquad \textbf{Variance } \sigma^2 = \int_0^a \frac{1}{a}\left(x - \frac{a}{2}\right)^2 dx = \frac{a^2}{12} \qquad (8)$$

The mean is a multiple of a, the variance is a multiple of a^2. For $a = 3$, $\sigma^2 = \frac{9}{12} = \frac{3}{4}$. For one random number between 0 and 1 $\left(\text{mean } \frac{1}{2}\right)$ the variance is $\sigma^2 = \frac{1}{12}$.

Normal Distribution : Bell-shaped Curve

The normal distribution is also called the "Gaussian" distribution. It is the most important of all probability density functions $p(x)$. The reason for its overwhelming importance comes from repeating an experiment and averaging the outcomes. The experiments have their own distribution (like heads and tails). *The average approaches a normal distribution.*

Central Limit Theorem (informal) The average of N samples of "any" probability distribution approaches a normal distribution as $N \to \infty$.

Start with the "standard normal distribution". It is symmetric around $x = 0$, so its mean value is $m = 0$. It is chosen to have a standard variance $\sigma^2 = 1$. It is called $\mathbf{N}(0, 1)$.

$$\textbf{Standard normal distribution} \qquad p(x) = \frac{1}{\sqrt{2\pi}} e^{-x^2/2}. \qquad (9)$$

The graph of $p(x)$ is the **bell-shaped curve** in Figure 12.2. The standard facts are

Total probability = 1 $\qquad \displaystyle\int_{-\infty}^{\infty} p(x)\,dx = \frac{1}{\sqrt{2\pi}} \int_{-\infty}^{\infty} e^{-x^2/2}\,dx = \mathbf{1}$

Mean E $[x]$ = 0 $\qquad \displaystyle m = \frac{1}{\sqrt{2\pi}} \int_{-\infty}^{\infty} x e^{-x^2/2}\,dx = \mathbf{0}$

Variance E $[x^2]$ = 1 $\qquad \displaystyle \sigma^2 = \frac{1}{\sqrt{2\pi}} \int_{-\infty}^{\infty} (x-0)^2 e^{-x^2/2}\,dx = \mathbf{1}$

The zero mean was easy because we are integrating an odd function. Changing x to $-x$ shows that "integral $= -$ integral". So that integral must be $m = 0$.

The other two integrals apply the idea in Problem 12 to reach 1. Figure 12.2 shows a graph of $p(x)$ for the normal distribution $\mathbf{N}\,(0, \sigma)$ and also its cumulative distribution $F(x) =$ integral of $p(x)$. From the symmetry of $p(x)$ you see *mean = zero*. From $F(x)$ you see a very important practical approximation for opinion polling:

The probability that a random sample falls between $-\sigma$ and σ is $\boldsymbol{F(\sigma) - F(-\sigma) \approx \frac{2}{3}}$.

This is because $\int_{-\sigma}^{\sigma} p(x)\,dx$ equals $\int_{-\infty}^{\sigma} p(x)\,dx - \int_{-\infty}^{-\sigma} p(x)\,dx = F(\sigma) - F(-\sigma)$.

Similarly, the probability that a random x lies between -2σ and 2σ ("*less than two standard deviations from the mean*") is $F(2\sigma) - F(-2\sigma) \approx 0.95$. If you have an experimental result further than 2σ from the mean, it is fairly sure to be not accidental: chance $= 0.05$. Drug tests may look for a tighter confirmation, like probability 0.001. Searching for the Higgs boson used a hyper-strict test of 5σ deviation from pure accident.

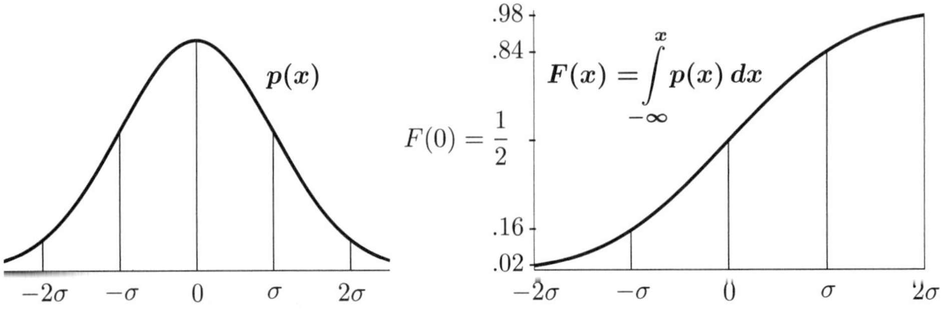

Figure 12.2: The standard normal distribution $p\,(x)$ has mean $m = 0$ and $\sigma = 1$.

The normal distribution with any mean m and standard deviation σ comes by shifting and stretching the standard $\mathbf{N}\,(0, 1)$. **Shift x to $x - m$. Stretch $x - m$ to $(x - m)/\sigma$.**

Gaussian density $p(x)$
Normal distribution $\mathbf{N}(m,\,\sigma)$

$$\boldsymbol{p(x) = \frac{1}{\sigma\sqrt{2\pi}}\,e^{-(x-m)^2/2\sigma^2}} \qquad (10)$$

The integral of $p(x)$ is $F(x)$—the probability that a random sample will fall below x. The differential $p(x)\,dx = F(x + dx) - F(x)$ is the probability that a random sample will fall between x and $x + dx$. There is no simple formula to integrate $e^{-x^2/2}$, so this cumulative distribution $F(x)$ is computed and tabulated very carefully.

N Coin Flips and $N \to \infty$

Example 2 **Suppose x is 1 or -1 with equal probabilities $p_1 = p_{-1} = \frac{1}{2}$.**

The mean value is $m = \frac{1}{2}(1) + \frac{1}{2}(-1) = \mathbf{0}$. The variance is $\sigma^2 = \frac{1}{2}(1)^2 + \frac{1}{2}(-1)^2 = \mathbf{1}$.

The key question is the *average* $A_N = (x_1 + \cdots + x_N)/N$. The independent x_i are ± 1 and we are dividing their sum by N. The expected mean of A_N is still zero. The law of large numbers says that this sample average approaches zero with probability 1. How fast does A_N approach zero? **What is its variance σ_N^2 ?**

$$\text{By linearity} \quad \boldsymbol{\sigma_N^2} = \frac{\sigma^2}{N^2} + \frac{\sigma^2}{N^2} + \cdots + \frac{\sigma^2}{N^2} = N\frac{\sigma^2}{N^2} = \frac{\mathbf{1}}{\boldsymbol{N}} \quad \text{since } \sigma^2 = 1. \tag{11}$$

Example 3 **Change outputs from 1 or -1 to $x = 1$ or $x = 0$. Keep $p_1 = p_0 = \frac{1}{2}$.**

The new mean value $m = \frac{1}{2}$ falls halfway between 0 and 1. The variance moves to $\sigma^2 = \frac{1}{4}$:

$$m = \frac{1}{2}(1) + \frac{1}{2}(0) = \frac{\mathbf{1}}{\mathbf{2}} \quad \text{and} \quad \boldsymbol{\sigma^2} = \frac{1}{2}\left(1 - \frac{1}{2}\right)^2 + \frac{1}{2}\left(0 - \frac{1}{2}\right)^2 = \frac{\mathbf{1}}{\mathbf{4}}.$$

$$\text{The average } A_N \text{ now has mean } \frac{1}{2} \text{ and variance } \frac{1}{4N^2} + \cdots + \frac{1}{4N^2} = \frac{\mathbf{1}}{\mathbf{4}\boldsymbol{N}} = \boldsymbol{\sigma_N^2}. \tag{12}$$

This σ_N is half the size of σ_N in Example 2. This must be correct because the new range 0 to 1 is half as long as -1 to 1. Examples 2-3 are showing a law of linearity.

The new $0 - 1$ variable x_{new} is $\frac{1}{2}\, x_{\text{old}} + \frac{1}{2}$. So the mean m is increased to $\frac{1}{2}$ and the variance is *multiplied* by $\left(\frac{1}{2}\right)^2$. A shift changes m and the rescaling changes σ^2.

$$\textbf{Linearity} \quad x_{\text{new}} = a x_{\text{old}} + b \text{ has } m_{\text{new}} = a m_{\text{old}} + b \text{ and } \sigma^2_{\text{new}} = a^2 \sigma^2_{\text{old}}. \tag{13}$$

Here are the results from three numerical tests: random 0 or 1 averaged over N trials.

[**48** 1's from $N = \mathbf{100}$] [**5035** 1's from $N = \mathbf{10000}$] [**19967** 1's from $N = \mathbf{40000}$].

The standardized X $= (x - m)/\sigma = \left(A_N - \frac{1}{2}\right) / 2\sqrt{N}$ was [$-$**.40**] [**.70**] [$-$**.33**].

The Central Limit Theorem says that the average of many coin flips will approach a normal distribution. Let us begin to see how that happens: **binomial approaches normal.**

For each flip, the probability of heads is $\frac{1}{2}$. For $N = 3$ flips, the probability of heads all three times is $\left(\frac{1}{2}\right)^3 = \frac{1}{8}$. The probability of heads twice and tails once is $\frac{3}{8}$, from three sequences HHT and HTH and THH. These numbers $\frac{1}{8}$ and $\frac{3}{8}$ are pieces of $\left(\frac{1}{2} + \frac{1}{2}\right)^3 = \frac{\mathbf{1}}{\mathbf{8}} + \frac{\mathbf{3}}{\mathbf{8}} + \frac{\mathbf{3}}{\mathbf{8}} + \frac{\mathbf{1}}{\mathbf{8}} = 1$. *The average number of heads in 3 flips is* 1.5.

$$\textbf{Mean} \;\; m = (3 \text{ heads})\frac{1}{8} + (2 \text{ heads})\frac{3}{8} + (1 \text{ head})\frac{3}{8} + 0 = \frac{3}{8} + \frac{6}{8} + \frac{3}{8} = \mathbf{1.5 \text{ heads}}$$

With N flips, Example 3 (or common sense) gives a mean of $\boldsymbol{m = \Sigma\, x_i p_i = \frac{1}{2}N}$ heads.

The variance σ^2 is based on the *squared distance* from this mean $N/2$. With $N = 3$ the variance is $\sigma^2 = \frac{\mathbf{3}}{\mathbf{4}}$ (*which is* $N/4$). To find σ^2 we add $(x_i - m)^2\, p_i$ with $m = 1.5$:

$$\boldsymbol{\sigma^2} = (3-1.5)^2\,\frac{1}{8} + (2-1.5)^2\,\frac{3}{8} + (1-1.5)^2\,\frac{3}{8} + (0-1.5)^2\,\frac{1}{8} = \frac{9+3+3+9}{32} = \frac{\mathbf{3}}{\mathbf{4}}.$$

For any N, the variance is $\boldsymbol{\sigma_N^2 = N/4}$. Then $\sigma_N = \sqrt{N}/2$.

Figure 12.3 shows how the probabilities of 0, 1, 2, 3, 4 heads in $N = 4$ flips come close to a bell-shaped Gaussian. That Gaussian is centered at the mean value $N/2 = 2$. To reach the standard Gaussian (mean 0 and variance 1) we shift and rescale that graph. If x is the number of heads in N flips—the average of N zero-one outcomes—then x is shifted by its mean $m = N/2$ and rescaled by $\sigma = \sqrt{N}/2$ to produce the standard X :

Shifted and scaled $$\boldsymbol{X = \frac{x-m}{\sigma} = \frac{x - \frac{1}{2}N}{\sqrt{N}/2}} \qquad (N = 4 \ \text{has}\ X = x - 2)$$

Subtracting m is "centering" or "detrending". The mean of X is zero.

Dividing by σ is "normalizing" or "standardizing". The variance of X is 1.

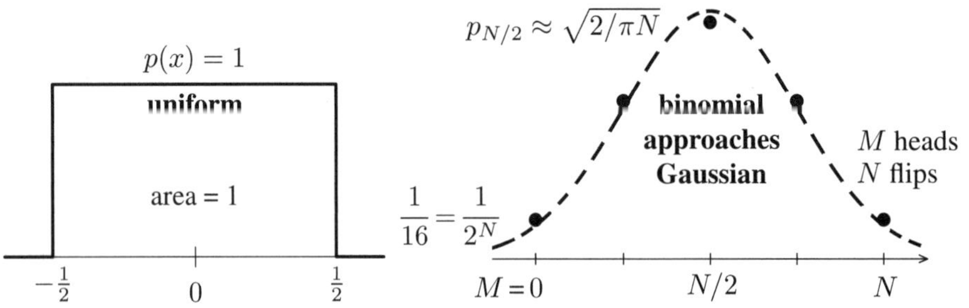

Figure 12.3: The probabilities $p = (1, 4, 6, 4, 1)/16$ for the number of heads in 4 flips. These p_i approach a Gaussian distribution with variance $\sigma^2 = N/4$ centered at $m = N/2$. For X, the Central Limit Theorem gives convergence to the normal distribution $\mathbf{N}(0, 1)$.

It is fun to see the Central Limit Theorem giving the right answer at the center point $X = 0$. At that point, the factor $e^{-X^2/2}$ equals 1. We know that the variance for N coin flips is $\sigma^2 = N/4$. The center of the bell-shaped curve has height $1/\sqrt{2\pi\sigma^2} = \boldsymbol{\sqrt{2/N\pi}}$.

What is the height at the center of the coin-flip distribution p_0 to p_N (the binomial distribution)? For $N = 4$, the probabilities for $0, 1, 2, 3, 4$ heads come from $\left(\frac{1}{2} + \frac{1}{2}\right)^4$.

Center probability $\boldsymbol{\frac{6}{16}}$ $$\left(\frac{1}{2} + \frac{1}{2}\right)^4 = \frac{1}{16} + \frac{4}{16} + \boldsymbol{\frac{6}{16}} + \frac{4}{16} + \frac{1}{16} = 1.$$

The binomial theorem in Problem 8 tells us the center probability $p_{N/2}$ for any even N :

$$\text{The center probability } \left(\frac{N}{2} \text{ heads}, \frac{N}{2} \text{ tails}\right) \text{ is } \quad \frac{1}{2^N}\frac{N!}{(N/2)!\,(N/2)!}$$

For $N = 4$, those factorials produce $4!/2!\,2! = 24/4 = 6$. For large N, Stirling's formula $\sqrt{2\pi N}(N/e)^N$ is a close approximation to $N!$. Use Stirling for N and twice for $N/2$:

Limit of coin-flip Center probability

$$p_{N/2} \approx \frac{1}{2^N}\frac{\sqrt{2\pi N}(N/e)^N}{\pi N(N/2e)^N} = \frac{\sqrt{2}}{\sqrt{\pi N}} = \frac{1}{\sqrt{2\pi}\sigma}. \tag{14}$$

At that last step we used the variance $\sigma^2 = N/4$ for the coin-tossing problem. The result $1/\sqrt{2\pi}\sigma$ matches the center value (above) for the Gaussian. The Central Limit Theorem is true: The "binomial distribution" approaches the normal distribution as $N \to \infty$.

Monte Carlo Estimation Methods

Scientific computing has to work with errors in the data. Financial computing has to work with unsure numbers and uncertain predictions. All of applied mathematics has moved to **accepting uncertainty in the inputs and estimating the variance in the outputs**.

How to estimate that variance? Often probability distributions $p(x)$ are not known. What we can do is to try different inputs $\boldsymbol{b}$ and compute the outputs $\boldsymbol{x}$ and take an average. This is the simplest form of a **Monte Carlo method** (named after the gambling palace on the Riviera, where I once saw a fight about whether the bet was placed in time). Monte Carlo approximates an expected value $\mathrm{E}[x]$ by a sample average $(x_1 + \cdots + x_N)/N$.

Please understand that every x_k can be expensive to compute. We are not just flipping coins. Each sample comes from a set of data b_k. *Monte Carlo randomly chooses this data b_k, it computes the outputs x_k, and then it averages those x's.* Decent accuracy for $\mathrm{E}[x]$ often requires many samples b and huge computing cost. The error in approximating $\mathrm{E}[x]$ by $(x_1 + \cdots + x_N)/N$ is normally of order $1/\sqrt{N}$. *Slow improvement as N increases.*

That $1/\sqrt{N}$ estimate came for coin flips in equation (11). Averaging N independent samples x_k of variance σ^2 reduces the variance to σ^2/N.

"Quasi-Monte Carlo" can sometimes reduce this variance to σ^2/N^2 : a big difference! The inputs b_k are selected very carefully—not just randomly. This QMC approach is surveyed in the journal *Acta Numerica* 2013. The newer idea of "Multilevel Monte Carlo" is outlined by Michael Giles in *Acta Numerica* 2015. Here is how it works.

Suppose it is much simpler to simulate another variable $y(b)$ close to $x(b)$. Then use N computations of $y(b_k)$ and only $N^* < N$ computations of $x(b_k)$ to estimate $\mathrm{E}[x]$.

2-level Monte Carlo

$$\mathrm{E}[x] \approx \frac{1}{N}\sum_{1}^{N} y(b_k) + \frac{1}{N^*}\sum_{1}^{N^*}\left[x(b_k) - y(b_k)\right].$$

The idea is that $x-y$ has a smaller variance σ^* than the original x. Therefore N^* can be smaller than N, with the same accuracy for $\mathrm{E}[x]$. We do N cheap simulations to find the y's. Those cost C each. We only do N^* expensive simulations involving x's. Those cost C^* each. The total computing cost is $NC + N^*C^*$.

Calculus minimizes the overall variance for a fixed total cost. The optimal ratio N^*/N is $\sqrt{C/C^*}\ \sigma^*/\sigma$. Three-level Monte Carlo would simulate x, y, and z:

$$\mathrm{E}[x] \approx \frac{1}{N}\sum_{1}^{N} \boldsymbol{z}(\boldsymbol{b_k}) + \frac{1}{N^*}\sum_{1}^{N^*}[\boldsymbol{y}(\boldsymbol{b_k}) - \boldsymbol{z}(\boldsymbol{b_k})] + \frac{1}{N^{**}}\sum_{1}^{N^{**}}[\boldsymbol{x}(\boldsymbol{b_k}) - \boldsymbol{y}(\boldsymbol{b_k})].$$

Giles optimizes $N, N^*, N^{**}, \ldots$ to keep $\mathrm{E}[x] \le$ fixed E_0, and provides a MATLAB code.

Review : Three Formulas for the Mean and the Variance

The formulas for m and σ^2 are the starting point for all of probability and statistics. There are three different cases to keep straight: **sample** values X_i, **expected** values (discrete p_i), and a range of **expected** values (continuous $p(x)$). Here are the mean and the variance:

Samples X_1 to X_N	$m = \dfrac{X_1 + \cdots + X_N}{N}$	$S^2 = \dfrac{(X_1-m)^2 + \cdots + (X_N-m)^2}{N-1}$
n possible outputs with probabilities p_i	$m = \sum_1^n p_i x_i$	$\sigma^2 = \sum_1^n p_i(x_i - m)^2$
Range of outputs with probability density	$m = \int x\, p(x)dx$	$\sigma^2 = \int (x-m)^2 p(x)dx$

A natural question: Why are there no probabilities p on the first line? How can these formulas be parallel? Answer: *We expect a fraction p_i of the samples to be $X = x_i$.* If this is exactly true, $X = x_i$ is repeated p_iN times. Then lines 1 and 2 give the same m.

When we work with samples, we don't know the p_i. We just include each output X as often as it comes. We get the "empirical" mean instead of the expected mean.

Problem Set 12.1

1 Add 7 to every output x. What happens to the mean and the variance? What are the new sample mean, the new expected mean, and the new variance?

2 We know: $\frac{1}{3}$ of all integers are divisible by 3 and $\frac{1}{7}$ of integers are divisible by 7. What fraction of integers will be divisible by 3 or 7 or both?

3 Suppose you sample from the numbers 1 to 1000 with equal probabilities 1/1000. What are the probabilities p_0 to p_9 that the last digit of your sample is $0, \ldots, 9$? What is the expected mean m of that last digit? What is its variance σ^2?

4 Sample again from 1 to 1000 but look at the last digit of the sample *squared*. That square could end with $x = 0, 1, 4, 5, 6,$ or 9. What are the probabilities $p_0, p_1, p_4, p_5, p_6, p_9$? What are the (expected) mean m and variance σ^2 of that number x?

5 (a little tricky) Sample again from 1 to 1000 with equal probabilities and let x be the *first* digit ($x = 1$ if the number is 15). What are the probabilities p_1 to p_9 (adding to 1) of $x = 1, \ldots, 9$? What are the mean and variance of x?

6 Suppose you have $N = 4$ samples $157, 312, 696, 602$ in Problem 5. What are the first digits x_1 to x_4 of the squares? What is the sample mean μ? What is the sample variance S^2? Remember to divide by $N - 1 = 3$ and not $N = 4$.

7 Equation (4) gave a second equivalent form for S^2 (the variance using samples):

$$\boldsymbol{S^2} = \frac{1}{N-1} \text{ sum of } (x_i - m)^2 = \frac{1}{N-1}\left[\left(\text{sum of } x_i^2\right) - Nm^2\right].$$

Verify the matching identity for the expected variance σ^2 (using $m = \Sigma\, p_i\, x_i$):

$$\boldsymbol{\sigma^2 = \textbf{ sum of } p_i\,(x_i - m)^2 = \left(\textbf{sum of } p_i\, x_i^2\right) - m^2}.$$

8 If all 24 samples from a population produce the same age $x = 20$, what are the sample mean μ and the sample variance S^2? What if $x = 20$ or 21, 12 times each?

9 Computer experiment as on page 541: Find the average $A_{1000000}$ of a million random 0-1 samples! What is $X = \left(A_N - \frac{1}{2}\right)/2\sqrt{N}$?

10 The probability p_i to get i heads in N coin flips is the *binomial number* $b_i = \binom{N}{i}$ divided by 2^N. The b_i add to $(1+1)^N = 2^N$ so the probabilities p_i add to 1.

$$p_0 + \cdots + p_N = \left(\frac{1}{2} + \frac{1}{2}\right)^N = \frac{1}{2^N}(b_0 + \cdots + b_N) \text{ with } b_i = \frac{N!}{i!\,(N-i)!}$$

$$N{=}4 \text{ leads to } b_0 = \frac{24}{24},\ b_1 = \frac{24}{(1)(6)} = 4,\ b_2 = \frac{24}{(2)(2)} = 6,\ p_i = \frac{1}{16}(1, 4, 6, 4, 1).$$

Notice $b_i = b_{N-i}$. *Problem*: Confirm that the mean $m = 0p_0 + \cdots + Np_N$ equals $\frac{N}{2}$.

11 For any function $f(x)$ the expected value is $\mathrm{E}[f] = \sum p_i\, f(x_i)$ or $\int p(x)\, f(x)\, dx$ (discrete probability or continuous probability). Suppose the mean is $\mathrm{E}[x] = m$ and the variance is $\mathrm{E}[(x-m)^2] = \sigma^2$. **What is $\mathbf{E}[x^2]$?**

12 Show that the standard normal distribution $p(x)$ has total probability $\int p(x)\, dx = 1$ as required. A famous trick multiplies $\int p(x)\, dx$ by $\int p(y)\, dy$ and computes the integral over all x and all y ($-\infty$ to ∞). The trick is to replace $dx\, dy$ in that double integral by $r\, dr\, d\theta$ (polar coordinates with $x^2 + y^2 = r^2$). Explain each step:

$$2\pi \int_{-\infty}^{\infty} p(x)\, dx \int_{-\infty}^{\infty} p(y)\, dy = \iint_{-\infty}^{\infty} e^{-(x^2+y^2)/2} dx\, dy = \int_{\theta=0}^{2\pi} \int_{r=0}^{\infty} e^{-r^2/2}\, r\, dr\, d\theta = \boldsymbol{2\pi}.$$

12.2 Covariance Matrices and Joint Probabilities

Linear algebra enters when we run M different experiments at once. We might measure age and height and weight ($M = 3$ measurements of N people). Each experiment has its own mean value. So we have a vector $\boldsymbol{m} = (m_1, m_2, m_3)$ containing the M mean values. Those could be *sample means* of age and height and weight. Or m_1, m_2, m_3 could be *expected values* of age, height, weight based on known probabilities.

A matrix becomes involved when we look at variances. Each experiment will have a sample variance S_i^2 or an expected $\sigma_i^2 = \mathrm{E}\left[(x_i - m_i)^2\right]$ based on the squared distance from its mean. Those M numbers $\sigma_1^2, \ldots, \sigma_M^2$ will go on the main diagonal of the matrix. So far we have made no connection between the M parallel experiments. They measure M different random variables, but the experiments are not necessarily independent!

If we measure age and height and weight $(\boldsymbol{a}, \boldsymbol{h}, \boldsymbol{w})$ for children, the results will be strongly correlated. Older children are generally taller and heavier. Suppose the means m_a, m_h, m_w are known. Then $\sigma_a^2, \sigma_h^2, \sigma_w^2$ are the separate variances in age, height, weight. **The new numbers are the covariances like $\boldsymbol{\sigma_{ah}}$, where age multiplies height.**

$$\textbf{Covariance} \qquad \boldsymbol{\sigma_{ah}} = \mathbf{E}\left[(\textbf{age} - \textbf{mean age})\ (\textbf{height} - \textbf{mean height})\right]. \tag{1}$$

This definition needs a close look. To compute σ_{ah}, it is not enough to know the probability of each age and the probability of each height. We have to know the **joint probability of each pair (age and height)**. This is because age is connected to height.

$\boldsymbol{p_{ah}}$ = probability that a random child has age = $\boldsymbol{a}$ **and** height = $\boldsymbol{h}$: both at once

$\boldsymbol{p_{ij}}$ = **probability that experiment 1 produces $\boldsymbol{x_i}$ and experiment 2 produces $\boldsymbol{y_j}$**

Suppose experiment 1 (age) has mean m_1. Experiment 2 (height) has mean m_2. The covariance in (1) between experiments 1 and 2 looks at **all pairs** of ages x_i, heights y_j :

$$\textbf{Covariance} \qquad \boldsymbol{\sigma_{12}} = \sum_{\textbf{all}\, \boldsymbol{i, j}} \sum \boldsymbol{p_{ij}(x_i - m_1)(y_j - m_2)} \tag{2}$$

To capture this idea of "joint probability p_{ij}" we begin with two small examples.

Example 1 Flip two coins separately. With 1 for heads and 0 for tails, the results can be $(1,1)$ or $(1,0)$ or $(0,1)$ or $(0,0)$. Those four outcomes all have probability $p_{11} = p_{10} = p_{01} = p_{00} = \frac{1}{4}$. **Independent experiments have Prob of $(\boldsymbol{i}, \boldsymbol{j})$ = (Prob of $\boldsymbol{i}$) (Prob of $\boldsymbol{j}$).**

Example 2 *Glue the coins together*, facing the same way. The only possibilities are $(1,1)$ and $(0,0)$. Those have probabilities $\frac{1}{2}$ and $\frac{1}{2}$. The probabilities p_{10} and p_{01} are zero. $(1,0)$ and $(0,1)$ won't happen because the coins stick together: both heads or both tails.

Probability matrices for Examples 1 and 2 $\qquad \boldsymbol{P} = \begin{bmatrix} p_{11} & p_{12} \\ p_{21} & p_{22} \end{bmatrix} = \begin{bmatrix} \frac{1}{4} & \frac{1}{4} \\ \frac{1}{4} & \frac{1}{4} \end{bmatrix} \qquad \boldsymbol{P} = \begin{bmatrix} \frac{1}{2} & 0 \\ 0 & \frac{1}{2} \end{bmatrix}.$

Let me stay longer with P, to show it in good matrix notation. The matrix shows the probability p_{ij} of each pair (x_i, y_j)—starting with $(x_1, y_1) =$ (heads, heads) and $(x_1, y_2) =$ (heads, tails). Notice the row sums p_i and column sums P_j and the total sum $= 1$.

$$\textbf{Probability matrix} \quad P = \begin{bmatrix} p_{11} & p_{12} \\ p_{21} & p_{22} \end{bmatrix} \qquad \begin{matrix} p_{11} + p_{12} = \boldsymbol{p_1} \\ p_{21} + p_{22} = \boldsymbol{p_2} \end{matrix} \quad \begin{pmatrix} \text{first} \\ \text{coin} \end{pmatrix}$$

$$\text{(second coin) column sums} \quad \boldsymbol{P_1} \quad \boldsymbol{P_2} \qquad\qquad 4 \text{ entries add to } 1$$

Those numbers p_1, p_2 and P_1, P_2 are called the **marginals** of the matrix P:

$p_1 = p_{11} + p_{12} =$ chance of heads from **coin 1** (coin 2 can be heads or tails)
$P_1 = p_{11} + p_{21} =$ chance of heads from **coin 2** (coin 1 can be heads or tails)

Example 1 showed *independent* variables. Every probability p_{ij} equals p_i times p_j $\left(\frac{1}{2}\right.$ times $\frac{1}{2}$ gave $p_{ij} = \frac{1}{4}$ in that example$\left.\right)$. In this case **the covariance $\boldsymbol{\sigma_{12}}$ will be zero.** Heads or tails from the first coin gave no information about the second coin.

$$\begin{matrix}\textbf{Zero covariance } \boldsymbol{\sigma_{12}} \\ \textbf{for independent trials}\end{matrix} \qquad V = \begin{bmatrix} \sigma_1^2 & 0 \\ 0 & \sigma_2^2 \end{bmatrix} = \textbf{diagonal covariance matrix}.$$

Independent experiments have $\sigma_{12} = 0$ because every $p_{ij} = (p_i)(p_j)$ in equation (2):

$$\sigma_{12} = \sum_i \sum_j (p_i)(p_j)(x_i - m_1)(y_j - m_2) = \left[\sum_i (p_i)(x_i - m_1)\right]\left[\sum_j (p_j)(y_j - m_2)\right] = [\mathbf{0}][\mathbf{0}].$$

The glued coins show perfect correlation. Heads on one means heads on the other. The covariance σ_{12} moves from 0 to $\sigma_1 \sigma_2 = \frac{1}{4}$—this is the largest possible value of σ_{12}:

$$\textbf{Means} = \frac{\mathbf{1}}{\mathbf{2}} \qquad \boldsymbol{\sigma_{12}} = \frac{\mathbf{1}}{\mathbf{2}}\left(1 - \frac{1}{2}\right)\left(1 - \frac{1}{2}\right) + \mathbf{0} + \mathbf{0} + \frac{\mathbf{1}}{\mathbf{2}}\left(0 - \frac{1}{2}\right)\left(0 - \frac{1}{2}\right) = \frac{\mathbf{1}}{\mathbf{4}}$$

Heads or tails from coin 1 gives complete information about heads or tails from coin 2:

$$\begin{matrix}\textbf{Glued coins give largest possible covariances} \\ \textbf{Singular covariance matrix: determinant} = 0\end{matrix} \qquad V_{\text{glue}} = \begin{bmatrix} \sigma_1^2 & \sigma_1\sigma_2 \\ \sigma_1\sigma_2 & \sigma_2^2 \end{bmatrix}$$

Always $\boldsymbol{\sigma_1^2 \sigma_2^2 \geq \sigma_{12}^2}$. Thus σ_{12} is *between* $-\sigma_1\sigma_2$ *and* $\sigma_1\sigma_2$. The covariance matrix V is **positive definite** (or in this singular case of glued coins, V is **positive semidefinite**). That is an important fact about M by M covariance matrices for M experiments.

Note that the **sample covariance matrix $\boldsymbol{S}$** from N trials is certainly semidefinite. Every new sample $X =$ (age, height, weight) contributes to the **sample mean $\overline{\boldsymbol{X}}$** and to S. Each term $(X_i - \overline{X})(X_i - \overline{X})^{\mathrm{T}}$ is positive semidefinite and we just add to reach S:

$$\overline{\boldsymbol{X}} = \frac{X_1 + \cdots + X_N}{N} \qquad \boldsymbol{S} = \frac{(X_1 - \overline{X})(X_1 - \overline{X})^{\mathrm{T}} + \cdots + (X_N - \overline{X})(X_N - \overline{X})^{\mathrm{T}}}{N - 1} \tag{3}$$

The Covariance Matrix V is Positive Semidefinite

Come back to the *expected* covariance σ_{12} between two experiments 1 and 2 (two coins):

$$\boxed{\begin{aligned}\sigma_{12} &= \text{expected value of } [(\textit{output}\,1 - \textit{mean}\,1) \text{ times } (\textit{output}\,2 - \textit{mean}\,2)] \\ &= \sum_{\text{all } i,j}\sum \boldsymbol{p_{ij}\,(x_i - m_1)\,(y_j - m_2)}.\end{aligned}} \tag{4}$$

$p_{ij} \geq 0$ is the probability of seeing output x_i in experiment 1 **and** y_j in experiment 2. Some pair of outputs must appear. Therefore the N^2 probabilities p_{ij} add to 1.

$$\textbf{Total probability (all pairs) is 1} \qquad \sum_{\text{all } i,j}\sum p_{ij} = 1. \tag{5}$$

Here is another fact we need. *Fix on one particular output* x_i in experiment 1. Allow *all outputs* y_j in experiment 2. Add the probabilities of $(x_i, y_1), (x_i, y_2), \ldots, (x_i, y_n)$:

$$\textbf{Row sum } \boldsymbol{p_i} \textbf{ of } \boldsymbol{P} \qquad \sum_{j=1}^{n} \boldsymbol{p_{ij}} = \textbf{probability } \boldsymbol{p_i} \textbf{ of } \boldsymbol{x_i} \textbf{ in experiment 1.} \tag{6}$$

Some y_j must happen in experiment 2 ! Whether the two coins are completely separate or glued together, we still get $\frac{1}{2}$ for the probability $p_H = p_{HH} + p_{HT}$ that coin 1 is heads:

$$\text{(separate) } P_{HH} + P_{HT} = \frac{1}{4} + \frac{1}{4} = \frac{\mathbf{1}}{\mathbf{2}} \qquad \text{(glued) } P_{HH} + P_{HT} = \frac{1}{2} + 0 = \frac{\mathbf{1}}{\mathbf{2}}.$$

That basic reasoning allows us to write one matrix formula that includes the covariance σ_{12} along with the separate variances σ_1^2 and σ_2^2 for experiment 1 and experiment 2. We get the whole covariance matrix V by adding the matrices V_{ij} for each pair (i, j):

$$\boxed{\begin{matrix}\textbf{Covariance matrix} \\ \boldsymbol{V = \Sigma\Sigma V_{ij}}\end{matrix} \qquad \boldsymbol{V} = \sum_{\text{all } i,j}\sum p_{ij} \begin{bmatrix} (x_i - m_1)^2 & (x_i - m_1)(y_j - m_2) \\ (x_i - m_1)(y_j - m_2) & (y_j - m_2)^2 \end{bmatrix}} \tag{7}$$

Off the diagonal, this is equation (2) for the covariance σ_{12}. On the diagonal, we are getting the ordinary variances σ_1^2 and σ_2^2. I will show in detail how we get $V_{11} = \sigma_1^2$ by using equation (6). Allowing all j just leaves the probability p_i of x_i in experiment 1:

$$\boldsymbol{V_{11}} = \sum_{\text{all } i,j}\sum p_{ij}(x_i - m_1)^2 = \sum_{\text{all } i} \text{(probability of } x_i\text{)}\ (x_i - m_1)^2 = \boldsymbol{\sigma_1^2}. \tag{8}$$

Please look at that twice. It is the key to producing the whole covariance matrix by one formula (7). The beauty of that formula is that it combines 2 by 2 matrices V_{ij}. And the matrix V_{ij} in (7) for each pair of outcomes i, j is **positive semidefinite**:

V_{ij} has diagonal entries $p_{ij}(x_i - m_1)^2 \geq 0$ and $p_{ij}(y_j - m_2)^2 \geq 0$ and $\det(V_{ij}) = 0$.

That matrix V_{ij} has rank 1. Equation (7) multiplies p_{ij} *times column* $\boldsymbol{U}$ *times row* $\boldsymbol{U}^{\mathrm{T}}$:

$$\begin{bmatrix} (x_i - m_1)^2 & (x_i - m_1)(y_j - m_2) \\ (x_i - m_1)(y_j - m_2) & (y_j - m_2)^2 \end{bmatrix} = \begin{bmatrix} x_i - m_1 \\ y_j - m_2 \end{bmatrix} \begin{bmatrix} x_i - m_1 & y_j - m_2 \end{bmatrix} \qquad (9)$$

Every matrix $\boldsymbol{U}\boldsymbol{U}^{\mathrm{T}}$ *is positive semidefinite.* So the whole matrix V (combining these matrices $\boldsymbol{U}\boldsymbol{U}^{\mathrm{T}}$ with weights $p_{ij} \geq 0$) is **at least semidefinite**—and probably V is definite.

The covariance matrix V is positive definite unless the experiments are dependent.

Now we move from two variables x and y to M variables like age-height-weight. The output from each trial is a vector X with M components. (Each child has an age-height-weight vector with 3 components.) The covariance matrix V is now M by M. V is created from the output vectors $\boldsymbol{X}$ and their average $\overline{\boldsymbol{X}} = \mathbf{E}[\boldsymbol{X}]$:

Covariance matrix $\quad \boldsymbol{V} = \mathbf{E}\left[\left(\boldsymbol{X} - \overline{\boldsymbol{X}}\right)\left(\boldsymbol{X} - \overline{\boldsymbol{X}}\right)^{\mathrm{T}}\right] \qquad (10)$

Remember that $\boldsymbol{X}\boldsymbol{X}^{\mathrm{T}}$ and $\overline{\boldsymbol{X}}\,\overline{\boldsymbol{X}}^{\mathrm{T}}$ = (column) (row) are M by M matrices.

For $M = 1$ (one variable) you see that $\overline{X}$ is the mean m and V is σ^2 (Section 12.1). For $M = 2$ (two coins) you see that $\overline{X}$ is (m_1, m_2) and V matches equation (10). The expectation E always adds up outputs times their probabilities. For age-height-weight the output could be $X = (5$ years, 31 inches, 48 pounds$)$ and its probability is $\boldsymbol{p_{5,31,48}}$.

Now comes a new idea. *Take any linear combination* $\boldsymbol{c}^{\mathrm{T}}\boldsymbol{X} = c_1 X_1 + \cdots + c_M X_M$. With $c = (6, 2, 5)$ this would be $\boldsymbol{c}^{\mathrm{T}}\boldsymbol{X} = 6\,(\text{age}) + 2\,(\text{height}) + 5\,(\text{weight})$. By linearity we know that its expected value $\mathrm{E}\,[\boldsymbol{c}^{\mathrm{T}}\boldsymbol{X}]$ is $\boldsymbol{c}^{\mathrm{T}}\mathrm{E}\,[\boldsymbol{X}] = \boldsymbol{c}^{\mathrm{T}}\overline{\boldsymbol{X}}$:

$$\mathbf{E}\,[\boldsymbol{c}^{\mathrm{T}}\boldsymbol{X}] = \boldsymbol{c}^{\mathrm{T}}\mathbf{E}\,[\boldsymbol{X}] = 6\,(\text{expected age}) + 2\,(\text{expected height}) + 5\,(\text{expected weight}).$$

More than that, we also know the *variance* σ^2 of that number $\boldsymbol{c}^{\mathrm{T}}\boldsymbol{X}$:

$$\begin{aligned} \text{Variance of } \boldsymbol{c}^{\mathrm{T}}\boldsymbol{X} &= \mathrm{E}\left[\left(\boldsymbol{c}^{\mathrm{T}}\boldsymbol{X} - \boldsymbol{c}^{\mathrm{T}}\overline{\boldsymbol{X}}\right)\left(\boldsymbol{c}^{\mathrm{T}}\boldsymbol{X} - \boldsymbol{c}^{\mathrm{T}}\overline{\boldsymbol{X}}\right)^{\mathrm{T}}\right] \\ &= \boldsymbol{c}^{\mathrm{T}}\mathrm{E}\left[\left(\boldsymbol{X} - \overline{\boldsymbol{X}}\right)\left(\boldsymbol{X} - \overline{\boldsymbol{X}}\right)^{\mathrm{T}}\right]\boldsymbol{c} = \boldsymbol{c}^{\mathrm{T}}\boldsymbol{V}\boldsymbol{c}! \end{aligned} \qquad (11)$$

Now the key point: *The variance of* $\boldsymbol{c}^{\mathrm{T}}\boldsymbol{X}$ *can never be negative.* So $\boldsymbol{c}^{\mathrm{T}}\boldsymbol{V}\boldsymbol{c} \geq \mathbf{0}$. *The covariance matrix V is therefore positive semidefinite by the energy test* $\boldsymbol{c}^{\mathrm{T}}\boldsymbol{V}\boldsymbol{c} \geq \mathbf{0}$.

Covariance matrices V open up the link between probability and linear algebra: V equals $Q\Lambda Q^{\mathrm{T}}$ with eigenvalues $\lambda_i \geq 0$ and orthonormal eigenvectors $\boldsymbol{q}_1$ to $\boldsymbol{q}_M$.

Diagonalizing the covariance matrix means finding M *independent* experiments as combinations of the original M experiments.

Confession I am not entirely happy with that proof based on $\boldsymbol{c}^{\mathrm{T}} V \boldsymbol{c} \geq \mathbf{0}$. The expectation symbol **E** is burying the key idea of **joint probability**. Allow me to show directly that V is positive semidefinite (at least for the age-height-weight example). The proof is simply that V **is the sum of the joint probability** p_{ahw} **of each combination (age, height, weight) times the positive semidefinite matrix** UU^{T}. Here U is $X - \overline{X}$:

$$\boldsymbol{V} = \sum_{\text{all } a,h,w} p_{ahw} \, \boldsymbol{U} \, \boldsymbol{U}^{\mathrm{T}} \quad \text{with} \quad \boldsymbol{U} = \begin{bmatrix} \text{age} \\ \text{height} \\ \text{weight} \end{bmatrix} - \begin{bmatrix} \text{mean age} \\ \text{mean height} \\ \text{mean weight} \end{bmatrix}. \qquad (12)$$

This is exactly like the 2 by 2 coin flip matrix V in equation (7). Now $M = 3$.

The value of the expectation symbol E is that it also allows *pdf*'s (probability density functions like $p(x, y, z)$ for continuous random variables x and y and z). If we allow all numbers as ages and heights and weights, instead of age $i = 0, 1, 2, 3 \ldots$, then we need $p(x, y, z)$ instead of p_{ijk}. The sums in this section of the book would all change to integrals. But we still have $V = \mathrm{E}\left[\boldsymbol{U}\boldsymbol{U}^{\mathrm{T}}\right]$:

$$\textbf{Covariance matrix} \quad V = \iiint p(x, y, z) \, \boldsymbol{U}\boldsymbol{U}^{\mathrm{T}} \, dx \, dy \, dz \quad \text{with} \quad \boldsymbol{U} = \begin{bmatrix} x - \overline{x} \\ y - \overline{y} \\ z - \overline{z} \end{bmatrix}. \qquad (13)$$

Always $\iiint p = 1$. Examples 1–2 emphasized how p can give diagonal V or singular V:

Independent variables $\boldsymbol{x}, \boldsymbol{y}, \boldsymbol{z}$ $\quad p(x, y, z) = \boldsymbol{p_1(x)\, p_2(y)\, p_3(z)}$.

Dependent variables $\boldsymbol{x}, \boldsymbol{y}, \boldsymbol{z}$ $\quad p(x, y, z) = \mathbf{0}$ except when $cx + dy + ez = 0$.

The Mean and Variance of $z = x + y$

Start with the sample mean. We have N samples of x. Their mean (= average) is m_x. We also have N samples of y and their mean is m_y. **The sample mean of** $\boldsymbol{z = x + y}$ **is clearly** $\boldsymbol{m_z = m_x + m_y}$:

$$\textbf{Mean of sum = Sum of means} \quad \frac{1}{N} \sum_{1}^{N} (x_i + y_i) = \frac{1}{N} \sum_{1}^{N} x_i + \frac{1}{N} \sum_{1}^{N} y_i. \qquad (14)$$

Nice to see something that simple. The *expected* mean of $z = x + y$ doesn't look so simple, but it must come out as $\mathbf{E}[z] = \mathbf{E}[x] + \mathbf{E}[y]$. Here is one way to see this.

The joint probability of the pair (x_i, y_j) is p_{ij}. Its value depends on whether the experiments are independent, which we don't know. But for the mean of the sum $z = x + y$, dependence or independence of x and y doesn't matter. *Expected values still add*:

$$\mathbf{E}[\boldsymbol{x}+\boldsymbol{y}]=\sum_i\sum_j p_{ij}(x_i+y_j)=\sum_i\sum_j p_{ij}x_i+\sum_i\sum_j p_{ij}y_j. \tag{15}$$

All the sums go from 1 to N. We can add in any order. For the first term on the right side, add the p_{ij} along row i of the probability matrix P to get p_i. That double sum gives $\mathrm{E}[x]$:

$$\sum_i\sum_j p_{ij}x_i=\sum_i(p_{i1}+\cdots+p_{iN})x_i=\sum_i p_ix_i=\mathrm{E}[x].$$

For the last term, add p_{ij} down column j of the matrix to get the probability P_j of y_j. Those pairs (x_1,y_j) and (x_2,y_j) and ... and (x_N,y_j) are all the ways to produce y_j:

$$\sum_i\sum_j p_{ij}y_j=\sum_j(p_{1j}+\cdots+p_{Nj})y_j=\sum_j P_jy_j=\mathrm{E}[y].$$

Now equation (15) says that $\mathrm{E}[\boldsymbol{x}+\boldsymbol{y}]=\mathrm{E}[\boldsymbol{x}]+\mathrm{E}[\boldsymbol{y}]$.

What about the variance of $z=x+y$? The joint probabilities p_{ij} and the covariance σ_{xy} will be involved. Let me separate the variance of $x+y$ into three simple pieces:

$$\begin{aligned}\boldsymbol{\sigma_z^2}&=\sum\sum p_{ij}(x_i+y_j-m_x-m_y)^2\\&=\sum\sum p_{ij}(x_i-m_x)^2+\sum\sum p_{ij}(y_j-m_y)^2+2\sum\sum p_{ij}(x_i-m_x)(y_j-m_y)\end{aligned}$$

The first piece is $\boldsymbol{\sigma_x^2}$. The second piece is $\boldsymbol{\sigma_y^2}$. The last piece is $\boldsymbol{2\sigma_{xy}}$.

$$\textbf{The variance of } \boldsymbol{z=x+y} \quad \textbf{is} \quad \boldsymbol{\sigma_z^2=\sigma_x^2+\sigma_y^2+2\sigma_{xy}}. \tag{16}$$

The Covariance Matrix for $Z = AX$

Here is a good way to see σ_z^2 when $z=x+y$. Think of (x,y) as a column vector $\boldsymbol{X}$. Think of the 1 by 2 matrix $A=\begin{bmatrix}1 & 1\end{bmatrix}$ multiplying that vector $\boldsymbol{X}$. Then $A\boldsymbol{X}$ is the sum $z=x+y$. The variance $\boldsymbol{\sigma_z^2}$ in equation (16) goes into matrix notation as

$$\boldsymbol{\sigma_z^2}=\begin{bmatrix}1 & 1\end{bmatrix}\begin{bmatrix}\sigma_x^2 & \sigma_{xy}\\ \sigma_{xy} & \sigma_y^2\end{bmatrix}\begin{bmatrix}1\\1\end{bmatrix} \quad \text{which is} \quad \boldsymbol{\sigma_z^2=AVA^{\mathrm{T}}}. \tag{17}$$

You can see that $\sigma_z^2=AVA^{\mathrm{T}}$ in (17) agrees with $\sigma_x^2+\sigma_y^2+2\sigma_{xy}$ in (16).

Now for the main point. The vector $\boldsymbol{X}$ could have M components coming from M experiments (instead of only 2). Those experiments will have an M by M covariance matrix $\boldsymbol{V_X}$. The matrix A could be K by M. Then $A\boldsymbol{X}$ is a vector with K combinations of the M outputs (instead of 1 combination $x+y$ of 2 outputs).

That vector $\boldsymbol{Z}=A\boldsymbol{X}$ of length K has a K by K covariance matrix $V_{\boldsymbol{Z}}$. Then the great rule for covariance matrices—of which equation (17) was only a 1 by 2 example—is this beautiful formula: Covariance matrix of $A\boldsymbol{X}$ is A (covariance matrix of $\boldsymbol{X}$) A^{T}:

$$\boxed{\textbf{The covariance matrix of } \boldsymbol{Z=AX} \textbf{ is } \boldsymbol{V_Z=AV_XA^{\mathrm{T}}}} \tag{18}$$

To me, this neat formula shows the beauty of matrix multiplication. I won't prove this formula, just admire it. It is constantly used in applications—coming in Section 12.3.

The Correlation ρ

Correlation ρ_{xy} is closely related to covariance σ_{xy}. They both measure dependence or independence. Start by rescaling or "standardizing" the random variables x and y **The new $X = x/\sigma_x$ and $Y = y/\sigma_y$ have variance $\sigma_X^2 = \sigma_Y^2 = 1$.** This is just like dividing a vector $\boldsymbol{v}$ by its length to produce a unit vector $\boldsymbol{v}/||\boldsymbol{v}||$ of length 1.

The correlation of x and y is the covariance of X and Y. If the original covariance of x and y was σ_{xy}, then rescaling to X and Y will divide by σ_x and σ_y :

$$\textbf{Correlation } \rho_{xy} = \frac{\sigma_{xy}}{\sigma_x \sigma_y} = \textbf{covariance of } \frac{x}{\sigma_x} \textbf{ and } \frac{y}{\sigma_y} \qquad \textbf{Always } -1 \le \rho_{xy} \le 1$$

Zero covariance gives zero correlation. *Independent random variables* produce $\rho_{xy} = 0$.

We know that always $\sigma_{xy}^2 \le \sigma_x^2 \sigma_y^2$ (the covariance matrix V is at least positive semidefinite). Then $\boldsymbol{\rho_{xy}^2 \le 1}$. Correlation near $\rho = +1$ means strong dependence in the same direction: often voting the same. Negative correlation means that y tends to be below its mean when x is above its mean: Voting in opposite directions.

Example 3 *Suppose that y is just $-x$.* A coin flip has outputs $x = 0$ or 1. The same flip has outputs $y = 0$ or -1. The mean m_x is $\frac{1}{2}$ for a fair coin, and m_y is $-\frac{1}{2}$. The covariance is $\sigma_{xy} = -\sigma_x \sigma_y$. The correlation divides by $\sigma_x \sigma_y$ to get $\boldsymbol{\rho_{xy} = -1}$. In this case the correlation matrix R has determinant zero (singular and only semidefinite):

$$\textbf{Correlation matrix } R = \begin{bmatrix} 1 & \rho_{xy} \\ \rho_{xy} & 1 \end{bmatrix} \qquad R = \begin{bmatrix} 1 & -1 \\ -1 & 1 \end{bmatrix} \text{ when } y = -x$$

R always has 1's on the diagonal because we normalized to $\sigma_X = \sigma_Y = 1$. R is the correlation matrix for x and y, and the covariance matrix for $X = x/\sigma_x$ and $Y = y/\sigma_y$.

That number ρ_{xy} is also called the Pearson coefficient.

Example 4 Suppose the random variables x, y, z are *independent. What matrix is R?*

Answer *R is the identity matrix.* All three correlations $\rho_{xx}, \rho_{yy}, \rho_{zz}$ are 1 by definition. All three cross-correlations $\rho_{xy}, \rho_{xz}, \rho_{yz}$ are zero by independence.

The correlation matrix R comes from the covariance matrix V, when we rescale every row and every column. Divide each row i and column i by the ith standard deviation σ_i.

(a) $R = DVD$ for the diagonal matrix $D = \text{diag}\,[1/\sigma_1, \ldots, 1/\sigma_M]$.

(b) If covariance V is positive definite, correlation $R = DVD$ is also positive definite.

■ WORKED EXAMPLES ■

12.2 A Suppose x and y are independent random variables with mean 0 and variance 1. Then the covariance matrix $\boldsymbol{V_X}$ for $\boldsymbol{X} = (x, y)$ is the 2 by 2 identity matrix. What are the mean $\boldsymbol{m_Z}$ and the covariance matrix $\boldsymbol{V_Z}$ for the 3-component vector $\boldsymbol{Z} = (x, y, ax+by)$?

Solution

$$\textbf{\textit{Z} is connected to \textit{X} by \textit{A}} \quad \boldsymbol{Z} = \begin{bmatrix} x \\ y \\ ax+by \end{bmatrix} = \begin{bmatrix} 1 & 0 \\ 0 & 1 \\ a & b \end{bmatrix} \begin{bmatrix} x \\ y \end{bmatrix} = A\boldsymbol{X}.$$

The vector $\boldsymbol{m_X}$ contains the means of the M components of $\boldsymbol{X}$. The vector $\boldsymbol{m_Z}$ contains the means of the K components of $\boldsymbol{Z} = A\boldsymbol{X}$. The matrix connection between the means of $\boldsymbol{X}$ and $\boldsymbol{Z}$ has to be linear: $\boldsymbol{m_Z} = A\boldsymbol{m_X}$. The mean of $ax+by$ is $am_x + bm_y$.

The covariance matrix for $\boldsymbol{Z}$ is $\boldsymbol{V_Z} = AA^{\mathrm{T}}$, when $\boldsymbol{V_X}$ is the 2 by 2 identity matrix:

$$\boldsymbol{V_Z} = \begin{matrix}\textbf{covariance matrix for} \\ \boldsymbol{Z = (x, y, ax+by)}\end{matrix} = \begin{bmatrix} 1 & 0 \\ 0 & 1 \\ a & b \end{bmatrix} \begin{bmatrix} 1 & 0 & a \\ 0 & 1 & b \end{bmatrix} = \begin{bmatrix} \mathbf{1} & \mathbf{0} & \boldsymbol{a} \\ \mathbf{0} & \mathbf{1} & \boldsymbol{b} \\ \boldsymbol{a} & \boldsymbol{b} & \boldsymbol{a^2+b^2} \end{bmatrix}.$$

Interpretation: x and y are independent so $\sigma_{xy} = 0$. Then the covariance of x with $ax+by$ is a and the covariance of y with $ax+by$ is b. Those just come from the two independent parts of $ax+by$. Finally, equation (18) gives the variance of $ax+by$:

$$\textbf{Use } \boldsymbol{V_Z = AV_XA^{\mathrm{T}}} \qquad \sigma^2_{ax+by} = \sigma^2_{ax} + \sigma^2_{by} + 2\sigma_{ax,by} = a^2 + b^2 + 0.$$

The 3 by 3 matrix $\boldsymbol{V_Z}$ is *singular*. Its determinant is $a^2 + b^2 - a^2 - b^2 = 0$. The third component $z = ax+by$ is completely dependent on x and y. The rank of $\boldsymbol{V_Z}$ is only 2.

GPS Example The signal from a GPS satellite includes its departure time. The receiver clock gives the arrival time. The receiver multiplies the travel time by the speed of light. Then it knows the distance from that satellite. Distances from four or more satellites pinpoint the receiver position (using least squares !).

One problem: The speed of light changes in the ionosphere. But the correction will be almost the same for all nearby receivers. If one receiver stays in a known position, we can take differences from that position. **Differential GPS** reduces the error variance:

$$\begin{matrix}\textbf{Difference matrix} \\ \boldsymbol{A} = [\,\mathbf{1} \ \ -\mathbf{1}\,]\end{matrix} \quad \begin{matrix}\textbf{Covariance matrix} \\ \boldsymbol{V_Z = AV_XA^{\mathrm{T}}}\end{matrix} \quad \begin{aligned} \boldsymbol{V_Z} &= \begin{bmatrix} 1 & -1 \end{bmatrix} \begin{bmatrix} \sigma_1^2 & \sigma_{12} \\ \sigma_{12} & \sigma_2^2 \end{bmatrix} \begin{bmatrix} 1 \\ -1 \end{bmatrix} \\ &= \boldsymbol{\sigma_1^2 - 2\sigma_{12} + \sigma_2^2} \end{aligned}$$

Errors in the speed of light are gone. Then centimeter positioning accuracy is achievable. (The key ideas are on page 320 of *Algorithms for Global Positioning* by Borre and Strang.) The GPS world is all about time and space and amazing accuracy.

Problem Set 12.2

1 (a) Compute the variance σ^2 when the coin flip probabilities are p and $1 - p$ (tails $= 0$, heads $= 1$).

(b) The sum of N independent flips (0 or 1) is the count of heads after N tries. The rule (16-17-18) for the variance of a sum gives $\sigma^2 =$ _____.

2 What is the covariance σ_{kl} between the results $x_1, \ldots, x_n$ of Experiment 3 and the results $y_1, \ldots, y_n$ of Experiment 5 ? Your formula will look like σ_{12} in equation (2). Then the $(3,5)$ and $(5,3)$ entries of the covariance matrix V are $\sigma_{35} = \sigma_{53}$.

3 For $M = 3$ experiments, the variance-covariance matrix V will be 3 by 3. There will be a probability p_{ijk} that the three outputs are x_i and y_j and z_k. Write down a formula like equation (7) for the matrix V.

4 What is the covariance matrix V for $M = 3$ independent experiments with means m_1, m_2, m_3 and variances $\sigma_1^2, \sigma_2^2, \sigma_3^2$?

Problems 5–9 are about the conditional probability that $Y = y_j$ when we know $X = x_i$.
Notation: **Prob $(Y = y_j | X = x_i)$** $=$ probability of the outcome y_j given that $X = x_i$.

Example 1 *Coin* 1 *is glued to coin* 2. Then Prob $(Y =$ heads when $X =$ heads$)$ is **1**.
Example 2 *Independent coin flips*: X gives no information about Y. Useless to know X. Then Prob $(Y =$ heads $|X =$ heads$)$ is the same as Prob $(Y =$ heads$)$.

5 Explain the **sum rule** of conditional probability:

$$\text{Prob}\,(Y = y_j) = \text{ sum over all outputs } x_i \text{ of Prob}\,(Y = y_j | X = x_i).$$

6 The n by n matrix P contains **joint probabilities** $p_{ij} =$ Prob $(X = x_i$ **and** $Y = y_j)$.

Explain why the conditional Prob $(Y = y_j | X = x_i)$ equals $\dfrac{p_{ij}}{p_{i1} + \cdots + p_{in}} = \dfrac{p_{ij}}{p_i}$.

7 For this joint probability matrix with Prob $(x_1, y_2) = 0.3$, find Prob $(y_2 | x_1)$ and Prob (x_1).

$$P = \begin{bmatrix} p_{11} & p_{12} \\ p_{21} & p_{22} \end{bmatrix} = \begin{bmatrix} 0.1 & 0.3 \\ 0.2 & 0.4 \end{bmatrix}$$ The entries p_{ij} add to 1. Some i, j must happen.

8 Explain the **product rule** of conditional probability:

$p_{ij} =$ Prob $(X = x_i$ **and** $Y = y_j)$ equals Prob $(Y = y_j | X = x_i)$ times Prob $(X = x_i)$.

9 Derive this **Bayes Theorem** for p_{ij} from the product rule in Problem 8:

$$\text{Prob}\,(Y = y_j \textbf{ and } X = x_i) = \frac{\text{Prob}\,(X = x_i | Y = y_j)\ \text{Prob}\,(Y = y_j)}{\text{Prob}\,(X = x_i)}$$

"Bayesians" use prior information. "Frequentists" only use sampling information.

12.3 Multivariate Gaussian and Weighted Least Squares

The normal probability density $p(x)$ (the Gaussian) depends on only two numbers:

$$\textbf{Mean } m \textbf{ and variance } \sigma^2 \qquad p(x) = \frac{1}{\sqrt{2\pi}\,\sigma}\, e^{-(x-m)^2/2\sigma^2}. \tag{1}$$

The graph of $p(x)$ is a bell-shaped curve centered at $x = m$. The continuous variable x can be anywhere between $-\infty$ and ∞. With probability close to $\frac{2}{3}$, that random x will lie between $m - \sigma$ and $m + \sigma$ (less than one standard deviation σ from its mean value m).

$$\int_{-\infty}^{\infty} p(x)\,dx = 1 \quad \text{and} \quad \int_{m-\sigma}^{m+\sigma} p(x)\,dx = \frac{1}{\sqrt{2\pi}} \int_{-1}^{1} e^{-X^2/2}\,dX \approx \frac{2}{3}. \tag{2}$$

That integral has a change of variables from x to $X = (x - m)/\sigma$. This simplifies the exponent to $-X^2/2$ and it simplifies the limits of integration to -1 and 1. Even the $1/\sigma$ from p disappears outside the integral because dX equals dx/σ. Every Gaussian turns into a **standard Gaussian** $p(X)$ with mean $m = 0$ and variance $\sigma^2 = 1$. Just call it $p(x)$:

$$\textbf{The standard normal distribution } N(0,1) \quad \textbf{has} \quad p(x) = \frac{1}{\sqrt{2\pi}}\, e^{-x^2/2}. \tag{3}$$

Integrating $p(x)$ from $-\infty$ to x gives the cumulative distribution $F(x)$: the probability that a random sample is below x. That probability will be $F = \frac{1}{2}$ at $x = 0$ (the mean).

Two-dimensional Gaussians

Now we have $M = 2$ Gaussian random variables x and y. They have means m_1 and m_2. They have variances σ_1^2 and σ_2^2. If they are *independent*, then their probability density $p(x, y)$ is just $\boldsymbol{p_1(x)}$ **times** $\boldsymbol{p_2(y)}$. Multiply probabilities when variables are independent:

$$\textbf{Independent } x \textbf{ and } y \quad p(x,y) = \frac{1}{2\pi\sigma_1\sigma_2}\, e^{-(x-m_1)^2/2\sigma_1^2}\, e^{-(y-m_2)^2/2\sigma_2^2} \tag{4}$$

The covariance of x and y will be $\boldsymbol{\sigma_{12} = 0}$. The covariance matrix V will be *diagonal*. The variances σ_1^2 and σ_2^2 are always on the main diagonal of V. The exponent in $p(x, y)$ is just the sum of the x-exponent and the y-exponent. Good to notice that the two exponents can be combined into $-\frac{1}{2}\,(\boldsymbol{x} - \boldsymbol{m})^{\mathrm{T}}\, V^{-1}\,(\boldsymbol{x} - \boldsymbol{m})$ with V^{-1} in the middle:

$$-\frac{(x-m_1)^2}{2\sigma_1^2} - \frac{(y-m_2)^2}{2\sigma_2^2} = -\frac{1}{2}\begin{bmatrix} x-m_1 & y-m_2 \end{bmatrix} \begin{bmatrix} \sigma_1^2 & 0 \\ 0 & \sigma_2^2 \end{bmatrix}^{-1} \begin{bmatrix} x-m_1 \\ y-m_2 \end{bmatrix} \tag{5}$$

Non-independent x and y

We are ready to give up independence. The exponent (5) with V^{-1} is still correct when V is no longer a diagonal matrix. **Now the Gaussian depends on a vector $\boldsymbol{m}$ and a matrix $\boldsymbol{V}$.**

When $M = 2$, the first variable x may give partial information about the second variable y (and vice versa). Maybe part of y is decided by x and part is truly independent. It is the M by M covariance matrix V that accounts for dependencies between the M variables $\boldsymbol{x} = x_1, \ldots, x_M$. Its inverse V^{-1} goes into $p(\boldsymbol{x})$:

Multivariate Gaussian probability distribution

$$\boxed{p(\boldsymbol{x}) = \frac{1}{(\sqrt{2\pi})^M \sqrt{\det V}}\, e^{-(\boldsymbol{x}-\boldsymbol{m})^{\mathrm{T}} \boldsymbol{V}^{-1} (\boldsymbol{x}-\boldsymbol{m})/2}} \tag{6}$$

The vectors $\boldsymbol{x} = (x_1, \ldots, x_M)$ and $\boldsymbol{m} = (m_1, \ldots, m_M)$ contain the random variables and their means. The M square roots of 2π and the determinant of V are included to make the total probability equal to 1. Let me check that by linear algebra. I use the eigenvalues λ and orthonormal eigenvectors $\boldsymbol{q}$ of the symmetric matrix $V = Q\Lambda Q^{\mathrm{T}}$. So $\boldsymbol{V}^{-1} = \boldsymbol{Q}\boldsymbol{\Lambda}^{-1}\boldsymbol{Q}^{\mathrm{T}}$:

$$\boldsymbol{X} = \boldsymbol{x} - \boldsymbol{m} \qquad (\boldsymbol{x} - \boldsymbol{m})^{\mathrm{T}} \boldsymbol{V}^{-1} (\boldsymbol{x} - \boldsymbol{m}) = \boldsymbol{X}^{\mathrm{T}} \boldsymbol{Q} \boldsymbol{\Lambda}^{-1} \boldsymbol{Q}^{\mathrm{T}} \boldsymbol{X} = \boldsymbol{Y}^{\mathrm{T}} \boldsymbol{\Lambda}^{-1} \boldsymbol{Y}$$

Notice! The combinations $\boldsymbol{Y} = \boldsymbol{Q}^{\mathrm{T}}\boldsymbol{X} = \boldsymbol{Q}^{\mathrm{T}}(\boldsymbol{x} - \boldsymbol{m})$ are statistically independent. *Their covariance matrix Λ is diagonal.*

This step of diagonalizing V by its eigenvector matrix Q is the same as "uncorrelating" the random variables. Covariances are zero for the new variables $X_1, \ldots X_m$. This is the point where linear algebra helps calculus to compute multidimensional integrals.

The integral of $p(\boldsymbol{x})$ is not changed when we center the variable $\boldsymbol{x}$ by subtracting $\boldsymbol{m}$ to reach $\boldsymbol{X}$, and rotate that variable to reach $\boldsymbol{Y} = \boldsymbol{Q}^{\mathrm{T}}\boldsymbol{X}$. The matrix $\boldsymbol{\Lambda}$ is diagonal! So the integral we want splits into M separate one-dimensional integrals that we know:

$$\int \cdots \int e^{-\boldsymbol{Y}^{\mathrm{T}} \boldsymbol{\Lambda}^{-1} \boldsymbol{Y}/2}\, d\boldsymbol{Y} = \left(\int_{-\infty}^{\infty} e^{-y_1^2/2\lambda_1}\, dy_1 \right) \cdots \left(\int_{-\infty}^{\infty} e^{-y_M^2/2\lambda_M}\, dy_M \right)$$

$$= \left(\sqrt{2\pi\lambda_1}\right) \cdots \left(\sqrt{2\pi\lambda_M}\right) = \left(\sqrt{2\pi}\right)^M \sqrt{\det V}. \tag{7}$$

The determinant of V (also the determinant of Λ) is the product $(\lambda_1) \ldots (\lambda_M)$ of the eigenvalues. Then (7) gives the correct number to divide by so that $p(x_1, \ldots, x_M)$ in equation (6) has integral $= 1$ as desired.

The mean and variance of $p(\boldsymbol{x})$ are also M-dimensional integrals. The same idea of diagonalizing V by its eigenvectors and introducing $\boldsymbol{Y} = \boldsymbol{Q}^{\mathrm{T}}\boldsymbol{X}$ will find those integrals:

Vector $\boldsymbol{m}$ of means $$\int \cdots \int \boldsymbol{x}\, p(\boldsymbol{x})\, d\boldsymbol{x} = (m_1, m_2, \ldots) = \boldsymbol{m} \tag{8}$$

Covariance matrix $\boldsymbol{V}$ $$\int \cdots \int (\boldsymbol{x} - \boldsymbol{m})\, p(\boldsymbol{x}) (\boldsymbol{x} - \boldsymbol{m})^{\mathrm{T}}\, d\boldsymbol{x} = \boldsymbol{V}. \tag{9}$$

Conclusion: Formula (6) for the probability density $p(\boldsymbol{x})$ has all the properties we want.

Weighted Least Squares

In Chapter 4, least squares started from an unsolvable system $Ax = b$. We chose $\widehat{x}$ to minimize the error $||b - Ax||^2$. That led us to the least squares equation $A^T A\widehat{x} = A^T b$. The best $A\widehat{x}$ is the projection of b onto the column space of A. But is this squared distance $E = ||b - Ax||^2$ the right error measure to minimize?

If the measurement errors in b are independent random variables, with mean $m = 0$ and variance $\sigma^2 = 1$ and a normal distribution, Gauss would say **yes**: *Use least squares.* If the errors are not independent or their variances are not equal. Gauss would say **no**: *Use **weighted** least squares.* This section will show that the good measure of error is $E = (b - Ax)^T V^{-1}(b - Ax)$. The equation for the best $\widehat{x}$ uses the covariance matrix V:

Weighted least squares $\quad A^T V^{-1} A\widehat{x} = A^T V^{-1} b. \quad (10)$

The most important examples have m *independent* errors in b. Those errors have variances $\sigma_1^2, \ldots, \sigma_m^2$. By independence, V is a diagonal matrix. The good weights $1/\sigma_1^2, \ldots, 1/\sigma_m^2$ come from V^{-1}. *We are weighting the errors in b to have* **variance = 1**:

Weighted least squares
Independent errors in b $\quad$ Minimize $\quad E = \sum_{i=1}^{m} \frac{(b - Ax)_i^2}{\sigma_i^2} \quad (11)$

By weighting the errors, we are "whitening" the noise. **White noise** is a quick description of independent errors based on the standard Gaussian $\mathbf{N}(0, 1)$ with mean zero and $\sigma^2 = 1$.

Let me write down the steps to equations (10) and (11) for the best $\widehat{x}$:

Start with $Ax = b$ $\quad$ (m equations, n unknowns, $m > n$, no solution)

Each right side b_i has mean zero and variance σ_i^2. The b_i are independent.

Divide the ith equation by σ_i to have variance $= 1$ for every b_i/σ_i

That division turns $Ax = b$ into $V^{-1/2}Ax = V^{-1/2}b$ with $V^{-1/2} = \text{diag}\,(1/\sigma_1, \ldots, 1/\sigma_m)$

Ordinary least squares on those weighted equations has $A \to V^{-1/2}A$ and $b \to V^{-1/2}b$

$$(V^{-1/2}A)^T(V^{-1/2}A)\widehat{x} = (V^{-1/2}A)^T V^{-1/2}\, b \;\text{ is }\; A^T V^{-1} A\widehat{x} = A^T V^{-1} b. \quad (12)$$

Because of $1/\sigma^2$ in V^{-1}, more reliable equations (*smaller* σ) get heavier weights. This is the main point of weighted least squares.

Those diagonal weightings (uncoupled equations) are the most frequent and the simplest. They apply to *independent errors in the* b_i. When these measurement errors are not independent, V is no longer diagonal—but (12) is still the correct weighted equation.

In practice, finding all the covariances can be serious work. Diagonal V is simpler.

The Variance in the Estimated $\widehat{x}$

One more point: Often the important question is not the best $\widehat{x}$ for one particular set of measurements $\boldsymbol{b}$. This is only one sample! The real goal is to know the reliability of the whole experiment. That is measured (as reliability always is) by the **variance in the estimate** $\widehat{\boldsymbol{x}}$. First, zero mean in $\boldsymbol{b}$ gives zero mean in $\widehat{x}$. Then the formula connecting variance $\boldsymbol{V}$ in the inputs $\boldsymbol{b}$ to variance $\boldsymbol{W}$ in the outputs $\widehat{x}$ turns out to be beautiful:

Variance-covariance matrix $\boldsymbol{W}$ for $\widehat{x}$ $\quad \mathrm{E}[(\widehat{x}-x)(\widehat{x}-x)^{\mathrm{T}}] = (\boldsymbol{A}^{\mathrm{T}}\boldsymbol{V}^{-1}\boldsymbol{A})^{-1}. \quad (13)$

That smallest possible variance comes from the best possible weighting, which is V^{-1}.

This key formula is a perfect application of Section 12.2. **If $\boldsymbol{b}$ has covariance matrix V, then $\widehat{x} = \boldsymbol{L}\boldsymbol{b}$ has covariance matrix $\boldsymbol{L}\boldsymbol{V}\boldsymbol{L}^{\mathrm{T}}$.** Equation (12) above tells us that L is $(A^{\mathrm{T}}V^{-1}A)^{-1}A^{\mathrm{T}}V^{-1}$. Now substitute this into LVL^{T} and watch equation (13) appear:

$$\boldsymbol{L}\boldsymbol{V}\boldsymbol{L}^{\mathrm{T}} = (A^{\mathrm{T}}V^{-1}A)^{-1}A^{\mathrm{T}}V^{-1} \quad V \quad V^{-1}A\,(A^{\mathrm{T}}V^{-1}A)^{-1} = (\boldsymbol{A}^{\mathrm{T}}\boldsymbol{V}^{-1}\boldsymbol{A})^{-1}.$$

This is the covariance W of the output, our best estimate $\widehat{x}$. It is time for examples.

Example 1 Suppose a doctor measures your heart rate x three times $(m = 3, n = 1)$:

$$\begin{array}{l} x = b_1 \\ x = b_2 \\ x = b_3 \end{array} \quad \text{is} \quad A\boldsymbol{x} = \boldsymbol{b} \quad \text{with} \quad A = \begin{bmatrix} 1 \\ 1 \\ 1 \end{bmatrix} \quad \text{and} \quad V = \begin{bmatrix} \sigma_1^2 & 0 & 0 \\ 0 & \sigma_2^2 & 0 \\ 0 & 0 & \sigma_3^2 \end{bmatrix}$$

The variances could be σ_1^2 [illegible]/9 and $\sigma_2^2 = 1/4$ and $\sigma_3^2 = \mathbf{1}$. You are getting more nervous as measurements are taken: [illegible] is less reliable than b_2 and b_1. All three measurements contain some information, so they all go into the best (weighted) estimate $\widehat{x}$:

$$V^{-1/2}A\widehat{x} = V^{-1/2}\boldsymbol{b} \quad \text{is} \quad \begin{array}{l} 3x = 3b_1 \\ 2x = 2b_2 \\ 1x = 1b_3 \end{array} \quad \text{leading to} \quad \boldsymbol{A}^{\mathrm{T}}\boldsymbol{V}^{-1}\boldsymbol{A}\widehat{x} = \boldsymbol{A}^{\mathrm{T}}\boldsymbol{V}^{-1}\boldsymbol{b}$$

$$\begin{bmatrix} 1 & 1 & 1 \end{bmatrix} \begin{bmatrix} 9 & & \\ & 4 & \\ & & 1 \end{bmatrix} \begin{bmatrix} 1 \\ 1 \\ 1 \end{bmatrix} \widehat{x} = \begin{bmatrix} 1 & 1 & 1 \end{bmatrix} \begin{bmatrix} 9 & & \\ & 4 & \\ & & 1 \end{bmatrix} \begin{bmatrix} b_1 \\ b_2 \\ b_3 \end{bmatrix}$$

$$\boxed{\widehat{x} = \frac{\mathbf{9}b_1 + \mathbf{4}b_2 + b_3}{\mathbf{14}} \quad \textbf{is a weighted average of } b_1, b_2, b_3}$$

Most weight is on b_1 since its variance σ_1 is smallest. The variance of $\widehat{\boldsymbol{x}}$ has the beautiful formula $W = (A^{\mathrm{T}}V^{-1}A)^{-1} = 1/14$:

$$\textbf{Variance of } \widehat{\boldsymbol{x}} \qquad \left(\begin{bmatrix} 1 & 1 & 1 \end{bmatrix} \begin{bmatrix} 9 & & \\ & 4 & \\ & & 1 \end{bmatrix} \begin{bmatrix} 1 \\ 1 \\ 1 \end{bmatrix} \right)^{-1} = \frac{1}{14} \quad \text{is smaller than} \quad \frac{1}{9}$$

The BLUE theorem of Gauss (proved on the website) says that our $\widehat{\boldsymbol{x}} = L\boldsymbol{b}$ is the best linear unbiased estimate of the solution to $A\boldsymbol{x} = \boldsymbol{b}$. Any other unbiased choice $\boldsymbol{x}^* = L^*\boldsymbol{b}$ has greater variance than $\widehat{\boldsymbol{x}}$. All unbiased choices have $L^*A = I$ so that an exact $A\boldsymbol{x} = \boldsymbol{b}$ will produce the right answer $\boldsymbol{x} = L^*\boldsymbol{b} = L^*A\boldsymbol{x}$.

Note. I must add that there are reasons not to minimize squared errors in the first place. One reason: This $\widehat{\boldsymbol{x}}$ often has many small components. The squares of small numbers are very small, and they appear when we minimize. It is easier to make sense of *sparse* vectors—only a few nonzeros. Statisticians often prefer to minimize **unsquared errors**: **the sum of** $|(\boldsymbol{b} - A\boldsymbol{x})_i|$. *This error measure is* L^1 *instead of* L^2. Because of the absolute values, the equation for $\widehat{\boldsymbol{x}}$ becomes nonlinear (it is actually piecewise linear).

Fast new algorithms are computing a sparse $\widehat{\boldsymbol{x}}$ quickly and the future may belong to L^1.

The Kalman Filter

The "Kalman filter" is the great algorithm in dynamic least squares. That word *dynamic* means that new measurements $\boldsymbol{b}_k$ keep coming. So the best estimate $\widehat{\boldsymbol{x}}_k$ keeps changing (based on all of $\boldsymbol{b}_0, \ldots, \boldsymbol{b}_k$). More than that, the matrix A is also changing. So $\widehat{\boldsymbol{x}}_2$ will be our best least squares estimate of the latest solution $\boldsymbol{x}_k$ to the **whole history of observation equations and update equations (state equations) up to time 2**:

$$A_0\boldsymbol{x}_0 = \boldsymbol{b}_0 \qquad \boldsymbol{x}_1 = F_0\boldsymbol{x}_0 \qquad A_1\boldsymbol{x}_1 = \boldsymbol{b}_1 \qquad \boldsymbol{x}_2 = F_1\boldsymbol{x}_1 \qquad A_2\boldsymbol{x}_2 = \boldsymbol{b}_2 \qquad (14)$$

The Kalman idea is to introduce one equation at a time. There will be errors in each equation. With every new equation, we update the best estimate $\widehat{\boldsymbol{x}}_k$ for the current $\boldsymbol{x}_k$. But history is not forgotten! This new estimate $\widehat{\boldsymbol{x}}_k$ uses all the past observations $\boldsymbol{b}_0$ to $\boldsymbol{b}_{k-1}$ and all the state equations $\boldsymbol{x}_{\text{new}} = F_{\text{old}}\,\boldsymbol{x}_{\text{old}}$. A large and growing least squares problem.

One more important point. Each least squares equation is **weighted** using the covariance matrix V_k for the error in b_k. There is even a covariance matrix C_k for errors in the update equations $\boldsymbol{x}_{k+1} = F_k x_k$. The best $\widehat{\boldsymbol{x}}_2$ then depends on $\boldsymbol{b}_0, \boldsymbol{b}_1, \boldsymbol{b}_2$ and V_0, V_1, V_2 and C_1, C_2. The good way to write $\widehat{\boldsymbol{x}}_k$ is as an update to the previous $\widehat{\boldsymbol{x}}_{k-1}$.

Let me concentrate on a simplified problem, without the matrices F_k and the covariances C_k. We are estimating the same true $\boldsymbol{x}$ at every step. How do we get $\widehat{\boldsymbol{x}}_1$ from $\widehat{\boldsymbol{x}}_0$?

$$\textbf{OLD} \quad A_0\,\boldsymbol{x}_0 = \boldsymbol{b}_0 \text{ leads to the weighted equation } A_0^{\mathrm{T}}\,V_0^{-1}\,A_0\,\widehat{\boldsymbol{x}}_0 = A_0^{\mathrm{T}}\,V_0^{-1}\,\boldsymbol{b}_0. \qquad (15)$$

$$\textbf{NEW} \quad \begin{bmatrix} A_0 \\ A_1 \end{bmatrix} \widehat{\boldsymbol{x}}_1 = \begin{bmatrix} \boldsymbol{b}_0 \\ \boldsymbol{b}_1 \end{bmatrix} \quad \text{leads to the following weighted equation for } \widehat{\boldsymbol{x}}_1:$$

$$\begin{bmatrix} A_0^{\mathrm{T}} & A_1^{\mathrm{T}} \end{bmatrix} \begin{bmatrix} V_0^{-1} & \\ & V_1^{-1} \end{bmatrix} \begin{bmatrix} A_0 \\ A_1 \end{bmatrix} \widehat{\boldsymbol{x}}_1 = \begin{bmatrix} A_0^{\mathrm{T}} & A_1^{\mathrm{T}} \end{bmatrix} \begin{bmatrix} V_0^{-1} & \\ & V_1^{-1} \end{bmatrix} \begin{bmatrix} \boldsymbol{b}_0 \\ \boldsymbol{b}_1 \end{bmatrix}. \quad (16)$$

Yes, we could just solve that new problem and forget the old one. But the old solution $\widehat{\boldsymbol{x}}_0$ needed work that we hope to reuse in $\widehat{\boldsymbol{x}}_1$. What we look for is **an update to** $\widehat{\boldsymbol{x}}_0$:

Kalman update gives $\widehat{\boldsymbol{x}}_1$ from $\widehat{\boldsymbol{x}}_0$ $\qquad \widehat{\boldsymbol{x}}_1 = \widehat{\boldsymbol{x}}_0 + K_1(\boldsymbol{b}_1 - A_1\,\widehat{\boldsymbol{x}}_0). \quad (17)$

The update correction is the mismatch $\boldsymbol{b}_1 - A_1\widehat{\boldsymbol{x}}_0$ between the old state $\widehat{\boldsymbol{x}}_0$ and the new measurements $\boldsymbol{b}_1$—multiplied by the *Kalman gain matrix* K_1. The formula for K_1 comes from comparing the solutions $\widehat{\boldsymbol{x}}_1$ and $\widehat{\boldsymbol{x}}_0$ to (15) and (16). And when we update $\widehat{\boldsymbol{x}}_0$ to $\widehat{\boldsymbol{x}}_1$ based on new data $\boldsymbol{b}_1$, **we also update the covariance matrix $\boldsymbol{W_0}$ to $\boldsymbol{W_1}$**. Remember $W_0 = (A_0^{\mathrm{T}}\,V_0^{-1}\,A_0)^{-1}$ from equation (13). Update its inverse to W_1^{-1}:

Covariance $\boldsymbol{W_1}$ of errors in $\widehat{\boldsymbol{x}}_1$ $\qquad \boldsymbol{W_1^{-1} = W_0^{-1} + A_1^{\mathrm{T}}\,V_1^{-1}\,A_1} \quad (18)$

Kalman gain matrix $\boldsymbol{K_1}$ $\qquad \boldsymbol{K_1 = W_1\,A_1^{\mathrm{T}}\,V_1^{-1}} \quad (19)$

This is the heart of the Kalman filter. Notice the importance of the W_k. Those matrices measure the reliability of the whole process, where the vector $\widehat{\boldsymbol{x}}_k$ estimates the current state based on the particular measurements $\boldsymbol{b}_0$ to $\boldsymbol{b}_k$.

Whole chapters and whole books are written to explain the dynamic Kalman filter, when the states $\boldsymbol{x}_k$ are also changing (based on the matrices F_k). There is a *prediction* of $\boldsymbol{x}_k$ using F, followed by a *correction* using the new data $\boldsymbol{b}$. Perhaps best to stop here.

This page was about **recursive least squares**: adding new data b_k and updating both $\widehat{\boldsymbol{x}}$ and W : the best current estimate based on all the data, and its covariance matrix.

Problem Set 12.3

1 Two measurements of the same variable x give two equations $x = b_1$ and $x = b_2$. Suppose the means are zero and the variances are σ_1^2 and σ_2^2, with independent errors: V is diagonal with entries σ_1^2 and σ_2^2. Write the two equations as $A\boldsymbol{x} = \boldsymbol{b}$ (A is 2 by 1). As in the text Example 1, find this best estimate $\widehat{\boldsymbol{x}}$ based on b_1 and b_2 :

$$\widehat{\boldsymbol{x}} = \frac{b_1/\sigma_1^2 + b_2/\sigma_2^2}{1/\sigma_1^2 + 1/\sigma_2^2} \qquad \mathrm{E}\left[\widehat{\boldsymbol{x}}\,\widehat{\boldsymbol{x}}^{\mathrm{T}}\right] = \left(\frac{1}{\sigma_1^2} + \frac{1}{\sigma_2^2}\right)^{-1}.$$

2 (a) In Problem 1, suppose the second measurement b_2 becomes super-exact and its variance $\sigma_2 \to 0$. What is the best estimate $\widehat{\boldsymbol{x}}$ when σ_2 reaches zero?

(b) The opposite case has $\sigma_2 \to \infty$ and no information in b_2. What is now the best estimate $\widehat{\boldsymbol{x}}$ based on b_1 and b_2 ?

3 If x and y are independent with probabilities $p_1(x)$ and $p_2(y)$, then $p(x,y) = p_1(x)\,p_2(y)$. By separating double integrals into products of single integrals ($-\infty$ to ∞) show that

$$\iint p(x,y)\,dx\,dy = \mathbf{1} \qquad \text{and} \qquad \iint (x+y)\,p(x,y)\,dx\,dy = \boldsymbol{m_1} + \boldsymbol{m_2}.$$

4 Continue Problem 3 for independent x, y to show that $p(x,y) = p_1(x)\,p_2(y)$ has

$$\iint (x-m_1)^2\,p(x,y)\,dx\,dy = \boldsymbol{\sigma_1^2} \qquad \iint (x-m_1)(y-m_2)\,p(x,y)\,dx\,dy = \mathbf{0}.$$

So the 2 by 2 covariance matrix V is diagonal and its entries are _____ .

5 Show that the inverse of a 2 by 2 covariance matrix V is

$$V^{-1} = \begin{bmatrix} \sigma_1^2 & \sigma_{12} \\ \sigma_{12} & \sigma_2^2 \end{bmatrix}^{-1} = \frac{1}{1-\rho^2}\begin{bmatrix} 1/\sigma_1^2 & -\rho/\sigma_1\sigma_2 \\ -\rho/\sigma_1\sigma_2 & 1/\sigma_2^2 \end{bmatrix} \quad \begin{matrix}\text{with correlation} \\ \rho = \sigma_{12}/\sigma_1\sigma_2.\end{matrix}$$

This produces the exponent $-(\boldsymbol{x}-\boldsymbol{m})^{\mathrm{T}}\,\boldsymbol{V}^{-1}(\boldsymbol{x}-\boldsymbol{m})$ in a 2-variable Gaussian.

6 Suppose $\widehat{x}_k$ is the average of $b_1, \ldots, b_k$. A new measurement b_{k+1} arrives and we want the new average $\widehat{x}_{k+1}$. The Kalman update equation (17) is

$$\textbf{New average} \qquad \widehat{x}_{k+1} = \widehat{x}_k + \frac{1}{k+1}\left(b_{k+1} - \widehat{x}_k\right).$$

Verify that $\widehat{x}_{k+1}$ is the correct average of $b_1 \ldots, b_{k+1}$.

7 Also check the update equation (18) for the variance $W_{k+1} = \sigma^2/(k+1)$ of this average $\widehat{x}$ assuming that $W_k = \sigma^2/k$ and b_{k+1} has variance $V = \sigma^2$.

8 (**Steady model**) Problems 6–7 were *static* least squares. All the sample averages $\widehat{x}_k$ were estimates of the same x. To make the Kalman filter *dynamic*, include also a *state equation $x_{k+1} = Fx_k$ with its own error variance s^2*. The dynamic least squares problem allows x to "drift" as k increases:

$$\begin{bmatrix} 1 & \\ -F & 1 \\ & 1 \end{bmatrix}\begin{bmatrix} x_0 \\ x_1 \end{bmatrix} = \begin{bmatrix} b_0 \\ 0 \\ b_1 \end{bmatrix} \quad \text{with variances} \quad \begin{bmatrix} \sigma^2 \\ s^2 \\ \sigma^2 \end{bmatrix}.$$

With $F = 1$, divide both sides of those three equations by σ, s, and σ. Find $\widehat{x_0}$ and $\widehat{x_1}$ by least squares, which gives more weight to the recent b_1. The Kalman filter is developed in *Algorithms for Global Positioning* (Borre and Strang, Wellesley-Cambridge Press).

Change in A^{-1} from a Change in A

This final page connects the beginning of the book (inverses and rank one matrices) with the end of the book (dynamic least squares and filters). Begin with this basic formula:

$$\textbf{The inverse of } M = \boldsymbol{I} - \boldsymbol{u}\boldsymbol{v}^{\mathrm{T}} \textbf{ is } M^{-1} = \boldsymbol{I} + \frac{\boldsymbol{u}\boldsymbol{v}^{\mathrm{T}}}{\boldsymbol{1} - \boldsymbol{v}^{\mathrm{T}}\boldsymbol{u}}$$

The quickest proof is $MM^{-1} = I - \boldsymbol{u}\boldsymbol{v}^{\mathrm{T}} + (1 - \boldsymbol{u}\boldsymbol{v}^{\mathrm{T}})\dfrac{\boldsymbol{u}\boldsymbol{v}^{\mathrm{T}}}{1 - \boldsymbol{v}^{\mathrm{T}}\boldsymbol{u}} = I - \boldsymbol{u}\boldsymbol{v}^{\mathrm{T}} + \boldsymbol{u}\boldsymbol{v}^{\mathrm{T}} = I.$

M is not invertible if $\boldsymbol{v}^{\mathrm{T}}\boldsymbol{u} = 1$ (then $M\boldsymbol{u} = \boldsymbol{0}$). Here $\boldsymbol{v}^{\mathrm{T}} = \boldsymbol{u}^{\mathrm{T}} = \begin{bmatrix} 1 & 1 & 1 \end{bmatrix}$:

Example The inverse of $M = I - \begin{bmatrix} 1\,1\,1 \\ 1\,1\,1 \\ 1\,1\,1 \end{bmatrix}$ is $M^{-1} = I + \dfrac{1}{\mathbf{1 - 3}} \begin{bmatrix} 1\,1\,1 \\ 1\,1\,1 \\ 1\,1\,1 \end{bmatrix}$

But we don't always start from the identity matrix. Many applications need to invert $\boldsymbol{M} = \boldsymbol{A} - \boldsymbol{u}\boldsymbol{v}^{\mathrm{T}}$. After we solve $A\boldsymbol{x} = \boldsymbol{b}$ we expect a rank one change to give $M\boldsymbol{y} = \boldsymbol{b}$. The division by $1 - \boldsymbol{v}^{\mathrm{T}}\boldsymbol{u}$ above will become a division by $c = 1 - \boldsymbol{v}^{\mathrm{T}}A^{-1}\boldsymbol{u} = 1 - \boldsymbol{v}^{\mathrm{T}}\boldsymbol{z}$.

Step 1 Solve $A\boldsymbol{z} = \boldsymbol{u}$ and compute $c = 1 - \boldsymbol{v}^{\mathrm{T}}\boldsymbol{z}$.

Step 2 If $c \neq 0$ then $M^{-1}\boldsymbol{b}$ is $\boldsymbol{y} = \boldsymbol{x} + \dfrac{\boldsymbol{v}^{\mathrm{T}}\boldsymbol{x}}{c}\,\boldsymbol{z}$.

Suppose A is easy to work with. A might already be factored into LU by elimination. Then this Sherman-Woodbury-Morrison formula is the fast way to solve $M\boldsymbol{y} = \boldsymbol{b}$. Here are three problems to end the book!

9 Take Steps 1–2 to find $\boldsymbol{y}$ when $A = I$ and $\boldsymbol{u}^{\mathrm{T}} = \boldsymbol{v}^{\mathrm{T}} = [1\;\;2\;\;3]$ and $\boldsymbol{b}^{\mathrm{T}} = [2\;\;1\;\;4]$.

10 Step 2 in this "update formula" claims that $M\boldsymbol{y} = \left(A - \boldsymbol{u}\boldsymbol{v}^{\mathrm{T}}\right)\left(\boldsymbol{x} + \dfrac{\boldsymbol{v}^{\mathrm{T}}\boldsymbol{x}}{c}\,\boldsymbol{z}\right) = \boldsymbol{b}$.

Simplify this to $\dfrac{\boldsymbol{u}\boldsymbol{v}^{\mathrm{T}}\boldsymbol{x}}{c}\,[1 - c - \boldsymbol{v}^{\mathrm{T}}\boldsymbol{z}] = \boldsymbol{0}$. This is true since $c = 1 - \boldsymbol{v}^{\mathrm{T}}\boldsymbol{z}$.

11 When A has a new row $\boldsymbol{v}^{\mathrm{T}}$, $A^{\mathrm{T}}A$ in the least squares equation changes to M:

$$M = \begin{bmatrix} A^{\mathrm{T}} & \boldsymbol{v} \end{bmatrix} \begin{bmatrix} A \\ \boldsymbol{v}^{\mathrm{T}} \end{bmatrix} = A^{\mathrm{T}}A + \boldsymbol{v}\boldsymbol{v}^{\mathrm{T}} = \text{ rank one change in } A^{\mathrm{T}}A.$$

Why is that multiplication correct? The updated $\widehat{\boldsymbol{x}}_{\text{new}}$ comes from Steps 1 and 2.

For reference here are four formulas for M^{-1}. The first two were given above, when the change was $\boldsymbol{u}\boldsymbol{v}^{\mathrm{T}}$. Formulas 3 and 4 go beyond rank one to allow matrices U, V, W.

1 $M = I - \boldsymbol{u}\boldsymbol{v}^{\mathrm{T}}$ and $M^{-1} = I + \boldsymbol{u}\boldsymbol{v}^{\mathrm{T}}/(1 - \boldsymbol{v}^{\mathrm{T}}\boldsymbol{u})$ *(rank 1 change)*

2 $M = A - \boldsymbol{u}\boldsymbol{v}^{\mathrm{T}}$ and $M^{-1} = A^{-1} + A^{-1}\boldsymbol{u}\boldsymbol{v}^{\mathrm{T}}A^{-1}/(1 - \boldsymbol{v}^{\mathrm{T}}A^{-1}\boldsymbol{u})$

3 $M = I - \boldsymbol{U}\boldsymbol{V}$ and $M^{-1} = I_n + U(\boldsymbol{I_m} - \boldsymbol{V}\boldsymbol{U})^{-1}V$

4 $M = A - \boldsymbol{U}\boldsymbol{W}^{-1}\boldsymbol{V}$ and $M^{-1} = A^{-1} + A^{-1}U(\boldsymbol{W} - \boldsymbol{V}\boldsymbol{A}^{-1}\boldsymbol{U})^{-1}VA^{-1}$

Formula **4** is the "matrix inversion lemma" in engineering. Not seen until now! The Kalman filter for solving block tridiagonal systems uses formula **4** at each step.

MATRIX FACTORIZATIONS

1. $\boldsymbol{A = LU} = \begin{pmatrix} \text{lower triangular } L \\ \text{1's on the diagonal} \end{pmatrix} \begin{pmatrix} \text{upper triangular } U \\ \text{pivots on the diagonal} \end{pmatrix}$

   Requirements: No row exchanges as Gaussian elimination reduces square A to U.

2. $\boldsymbol{A = LDU} = \begin{pmatrix} \text{lower triangular } L \\ \text{1's on the diagonal} \end{pmatrix} \begin{pmatrix} \text{pivot matrix} \\ D \text{ is diagonal} \end{pmatrix} \begin{pmatrix} \text{upper triangular } U \\ \text{1's on the diagonal} \end{pmatrix}$

   Requirements: No row exchanges. The pivots in D are divided out to leave 1's on the diagonal of U. If A is symmetric then U is L^{T} and $\boldsymbol{A = LDL^{\mathrm{T}}}$.

3. $\boldsymbol{PA = LU}$ (permutation matrix P to avoid zeros in the pivot positions).

   Requirements: A is invertible. Then P, L, U are invertible. P does all of the row exchanges on A in advance, to allow normal LU. Alternative: $A = L_1 P_1 U_1$.

4. $\boldsymbol{EA = R}$ (m by m invertible E) (any m by n matrix A) = rref(A).

   Requirements: None ! *The reduced row echelon form* R has r pivot rows and pivot columns, containing the identity matrix. The last $m - r$ rows of E are a basis for the left nullspace of A; they multiply A to give $m - r$ zero rows in R. The first r columns of E^{-1} are a basis for the column space of A.

5. $\boldsymbol{S = C^{\mathrm{T}}C}$ = (lower triangular) (upper triangular) with $\sqrt{D}$ on both diagonals

   Requirements: S is symmetric and positive definite (all n pivots in D are positive). This *Cholesky factorization* $C = \mathsf{chol}(S)$ has $C^{\mathrm{T}} = L\sqrt{D}$, so $S = C^{\mathrm{T}}C = LDL^{\mathrm{T}}$.

6. $\boldsymbol{A = QR}$ = (orthonormal columns in Q) (upper triangular R).

   Requirements: A has independent columns. Those are *orthogonalized* in Q by the Gram-Schmidt or Householder process. If A is square then $Q^{-1} = Q^{\mathrm{T}}$.

7. $\boldsymbol{A = X\Lambda X^{-1}}$ = (eigenvectors in X) (eigenvalues in Λ) (left eigenvectors in X^{-1}).

   Requirements: A must have n linearly independent eigenvectors.

8. $\boldsymbol{S = Q\Lambda Q^{\mathrm{T}}}$ = (orthogonal matrix Q) (real eigenvalue matrix Λ) (Q^{T} is Q^{-1}).

   Requirements: S is *real and symmetric*: $S^{\mathrm{T}} = S$. This is the Spectral Theorem.

9. $\boldsymbol{A = BJB^{-1}} =$ (generalized eigenvectors in B) (Jordan blocks in J) (B^{-1}).

Requirements: A is any square matrix. This *Jordan form* J has a block for each independent eigenvector of A. Every block has only one eigenvalue.

10. $$\boldsymbol{A = U\Sigma V^{\mathrm{T}}} = \begin{pmatrix} \text{orthogonal} \\ U \text{ is } m \times m \end{pmatrix} \begin{pmatrix} m \times n \text{ singular value matrix} \\ \sigma_1, \ldots, \sigma_r \text{ on its diagonal} \end{pmatrix} \begin{pmatrix} \text{orthogonal} \\ V \text{ is } n \times n \end{pmatrix}.$$

Requirements: None. This ***Singular Value Decomposition*** (SVD) has the eigenvectors of AA^{T} in U and eigenvectors of $A^{\mathrm{T}}A$ in V; $\sigma_i = \sqrt{\lambda_i(A^{\mathrm{T}}A)} = \sqrt{\lambda_i(AA^{\mathrm{T}})}$.

Those singular values are $\sigma_1 \geq \sigma_2 \geq \cdots \geq \sigma_r > 0$. By column-row multiplication

$$\boldsymbol{A = U\Sigma V^{\mathrm{T}}} = \sigma_1 \boldsymbol{u}_1 \boldsymbol{v}_1^{\mathrm{T}} + \cdots + \sigma_r \boldsymbol{u}_r \boldsymbol{v}_r^{\mathrm{T}}.$$

If S is symmetric positive definite then $U = V = Q$ and $\Sigma = \Lambda$ and $S = Q\Lambda Q^{\mathrm{T}}$.

11. $$\boldsymbol{A^{+} = V\Sigma^{+}U^{\mathrm{T}}} = \begin{pmatrix} \text{orthogonal} \\ n \times n \end{pmatrix} \begin{pmatrix} n \times m \text{ pseudoinverse of } \Sigma \\ 1/\sigma_1, \ldots, 1/\sigma_r \text{ on diagonal} \end{pmatrix} \begin{pmatrix} \text{orthogonal} \\ m \times m \end{pmatrix}.$$

Requirements: None. The *pseudoinverse* A^{+} has $A^{+}A =$ projection onto row space of A and $AA^{+} =$ projection onto column space. $A^{+} = A^{-1}$ if A is invertible. The shortest least-squares solution to $A\boldsymbol{x} = \boldsymbol{b}$ is $\boldsymbol{x}^{+} = A^{+}\boldsymbol{b}$. This solves $A^{\mathrm{T}}A\boldsymbol{x}^{+} = A^{\mathrm{T}}\boldsymbol{b}$.

12. $\boldsymbol{A = QS} =$ (orthogonal matrix Q) (symmetric positive definite matrix S).

Requirements: A is invertible. This *polar decomposition* has $S^2 = A^{\mathrm{T}}A$. The factor S is semidefinite if A is singular. The reverse polar decomposition $A = KQ$ has $K^2 = AA^{\mathrm{T}}$. Both have $Q = UV^{\mathrm{T}}$ from the SVD.

13. $\boldsymbol{A = U\Lambda U^{-1}} =$ (unitary U) (eigenvalue matrix Λ) (U^{-1} which is $U^{\mathrm{H}} = \overline{U}^{\mathrm{T}}$).

Requirements: A is *normal*: $A^{\mathrm{H}}A = AA^{\mathrm{H}}$. Its orthonormal (and possibly complex) eigenvectors are the columns of U. Complex λ's unless $S = S^{\mathrm{H}}$: Hermitian case.

14. $\boldsymbol{A = QTQ^{-1}} =$ (unitary Q) (triangular T with λ's on diagonal) $(Q^{-1} = Q^{\mathrm{H}})$.

Requirements: *Schur triangularization* of any square A. There is a matrix Q with orthonormal columns that makes $Q^{-1}AQ$ triangular: Section 6.4.

15. $$\boldsymbol{F_n} = \begin{bmatrix} I & D \\ I & -D \end{bmatrix} \begin{bmatrix} \boldsymbol{F_{n/2}} & \\ & \boldsymbol{F_{n/2}} \end{bmatrix} \begin{bmatrix} \text{even-odd} \\ \text{permutation} \end{bmatrix} = \text{one step of the recursive } \textbf{FFT}.$$

Requirements: $F_n =$ Fourier matrix with entries w^{jk} where $w^n = 1$: $F_n \overline{F}_n = nI$. D has $1, w, \ldots, w^{n/2-1}$ on its diagonal. For $n = 2^{\ell}$ the *Fast Fourier Transform* will compute $F_n\boldsymbol{x}$ with only $\frac{1}{2}n\ell = \frac{1}{2}n\log_2 n$ multiplications from ℓ stages of D's.

Index

D

Index of Symbols and Computer Codes

Computer Packages

Code Names

Linear Algebra Websites and Email Address

math.mit.edu/linearalgebra Dedicated to readers and teachers working with this book

ocw.mit.edu MIT's OpenCourseWare site including video lectures in 18.06 and 18.085-6

web.mit.edu/18.06 Current and past exams and homeworks with extra materials

wellesleycambridge.com Ordering information for books by Gilbert Strang

linearalgebrabook@gmail.com Direct email contact about this book

Six Great Theorems of Linear Algebra

Dimension Theorem All bases for a vector space have the same number of vectors.

Counting Theorem Dimension of column space + dimension of nullspace = number of columns.

Rank Theorem Dimension of column space = dimension of row space. This is the rank.

Fundamental Theorem The row space and nullspace of A are orthogonal complements in $\mathbf{R}^n$.

SVD There are orthonormal bases ($\boldsymbol{v}$'s and $\boldsymbol{u}$'s for the row and column spaces) so that $A\boldsymbol{v}_i = \boldsymbol{\sigma_i u_i}$.

Spectral Theorem If $A^{\mathrm{T}} = A$ there are orthonormal $\boldsymbol{q}$'s so that $\boldsymbol{Aq_i = \lambda_i q_i}$ and $\boldsymbol{A = Q\Lambda Q^{\mathrm{T}}}$.

LINEAR ALGEBRA IN A NUTSHELL

((*The matrix* A *is* n *by* n))

Nonsingular	Singular
A is invertible	A is not invertible
The columns are independent	The columns are dependent
The rows are independent	The rows are dependent
The determinant is not zero	The determinant is zero
$A\boldsymbol{x} = \boldsymbol{0}$ has one solution $\boldsymbol{x} = \boldsymbol{0}$	$A\boldsymbol{x} = \boldsymbol{0}$ has infinitely many solutions
$A\boldsymbol{x} = \boldsymbol{b}$ has one solution $\boldsymbol{x} = A^{-1}\boldsymbol{b}$	$A\boldsymbol{x} = \boldsymbol{b}$ has no solution or infinitely many
A has n (nonzero) pivots	A has $r < n$ pivots
A has full rank $r = n$	A has rank $r < n$
The reduced row echelon form is $R = I$	R has at least one zero row
The column space is all of $\mathbf{R}^n$	The column space has dimension $r < n$
The row space is all of $\mathbf{R}^n$	The row space has dimension $r < n$
All eigenvalues are nonzero	Zero is an eigenvalue of A
$A^{\mathrm{T}}A$ is symmetric positive definite	$A^{\mathrm{T}}A$ is only semidefinite
A has n (positive) singular values	A has $r < n$ singular values